SYMMETRY IN GRAPHS

This is the first full-length book on the major theme of symmetry in graphs. Forming part of algebraic graph theory, this fast-growing field is concerned with the study of highly symmetric graphs, particularly vertex-transitive graphs, and other combinatorial structures, primarily by group-theoretic techniques. In practice, the street goes both ways and these investigations shed new light on permutation groups and related algebraic structures.

The book assumes a first course in graph theory and group theory but no specialized knowledge of the theory of permutation groups or vertex-transitive graphs. It begins with the basic material before introducing the field's major problems and most active research themes in order to motivate the detailed discussion of individual topics that follows. Featuring many examples and with over 450 exercises, it is an essential introduction to the field for graduate students and a valuable addition to any algebraic graph theorist's bookshelf.

Ted Dobson is Professor at the University of Primorska, Slovenia.

Aleksander Malnič is Professor at the University of Ljubljana, Slovenia.

Dragan Marušič is Professor at the University of Primorska, Slovenia, and the founder of the Slovenian school in Algebraic Graph Theory.

CAMBRIDGE STUDIES IN ADVANCED MATHEMATICS

All the titles listed below can be obtained from good booksellers or from Cambridge University Press. For a complete series listing, visit www.cambridge.org/mathematics.

Already Published
158 D. Huybrechts *Lectures on K3 Surfaces*
159 H. Matsumoto & S. Taniguchi *Stochastic Analysis*
160 A. Borodin & G. Olshanski *Representations of the Infinite Symmetric Group*
161 P. Webb *Finite Group Representations for the Pure Mathematician*
162 C. J. Bishop & Y. Peres *Fractals in Probability and Analysis*
163 A. Bovier *Gaussian Processes on Trees*
164 P. Schneider *Galois Representations and* (φ, Γ)*-Modules*
165 P. Gille & T. Szamuely *Central Simple Algebras and Galois Cohomology* (2nd Edition)
166 D. Li & H. Queffelec *Introduction to Banach Spaces, I*
167 D. Li & H. Queffelec *Introduction to Banach Spaces, II*
168 J. Carlson, S. Müller-Stach & C. Peters *Period Mappings and Period Domains* (2nd Edition)
169 J. M. Landsberg *Geometry and Complexity Theory*
170 J. S. Milne *Algebraic Groups*
171 J. Gough & J. Kupsch *Quantum Fields and Processes*
172 T. Ceccherini-Silberstein, F. Scarabotti & F. Tolli *Discrete Harmonic Analysis*
173 P. Garrett *Modern Analysis of Automorphic Forms by Example, I*
174 P. Garrett *Modern Analysis of Automorphic Forms by Example, II*
175 G. Navarro *Character Theory and the McKay Conjecture*
176 P. Fleig, H. P. A. Gustafsson, A. Kleinschmidt & D. Persson *Eisenstein Series and Automorphic Representations*
177 E. Peterson *Formal Geometry and Bordism Operators*
178 A. Ogus *Lectures on Logarithmic Algebraic Geometry*
179 N. Nikolski *Hardy Spaces*
180 D.-C. Cisinski *Higher Categories and Homotopical Algebra*
181 A. Agrachev, D. Barilari & U. Boscain *A Comprehensive Introduction to Sub-Riemannian Geometry*
182 N. Nikolski *Toeplitz Matrices and Operators*
183 A. Yekutieli *Derived Categories*
184 C. Demeter *Fourier Restriction, Decoupling and Applications*
185 D. Barnes & C. Roitzheim *Foundations of Stable Homotopy Theory*
186 V. Vasyunin & A. Volberg *The Bellman Function Technique in Harmonic Analysis*
187 M. Geck & G. Malle *The Character Theory of Finite Groups of Lie Type*
188 B. Richter *Category Theory for Homotopy Theory*
189 R. Willett & G. Yu *Higher Index Theory*
190 A. Bobrowski *Generators of Markov Chains*
191 D. Cao, S. Peng & S. Yan *Singularly Perturbed Methods for Nonlinear Elliptic Problems*
192 E. Kowalski *An Introduction to Probabilistic Number Theory*
193 V. Gorin *Lectures on Random Lozenge Tilings*
194 E. Riehl & D. Verity *Elements of* ∞*-Category Theory*
195 H. Krause *Homological Theory of Representations*
196 F. Durand & D. Perrin *Dimension Groups and Dynamical Systems*
197 A. Sheffer *Polynomial Methods and Incidence Theory*

Symmetry in Graphs

TED DOBSON
University of Primorska, Slovenia

ALEKSANDER MALNIČ
University of Ljubljana, Slovenia

DRAGAN MARUŠIČ
University of Primorska, Slovenia

CAMBRIDGE
UNIVERSITY PRESS

University Printing House, Cambridge CB2 8BS, United Kingdom

One Liberty Plaza, 20th Floor, New York, NY 10006, USA

477 Williamstown Road, Port Melbourne, VIC 3207, Australia

314–321, 3rd Floor, Plot 3, Splendor Forum, Jasola District Centre,
New Delhi – 110025, India

103 Penang Road, #05–06/07, Visioncrest Commercial, Singapore 238467

Cambridge University Press is part of the University of Cambridge.

It furthers the University's mission by disseminating knowledge in the pursuit of education, learning, and research at the highest international levels of excellence.

www.cambridge.org
Information on this title: www.cambridge.org/9781108429061
DOI: 10.1017/9781108553995

First published 2022

A catalogue record for this publication is available from the British Library.

ISBN 978-1-108-42906-1 Hardback

Contents

Preface

Symmetry and its absence, **asymmetry**, is one of the core concepts in the arts, science, or daily life, for that matter. In everyday language, symmetry – from the Greek "$\sigma\upsilon\mu\mu\epsilon\tau\rho\iota\alpha$ = symmetria," agreement in due proportion, arrangement – usually refers to a sense of beauty, of proportional harmony, and of balance. In trying to present the concept of symmetry in as general a setting as possible, one can think of it as a measure of the inner structural stability of the object in question when confronted with possible external intrusions. When the latter exceed the acceptable robustness threshold and thus break the innate symmetry, a transformation of the object occurs. Given a new life of enriched complexity, the object finds itself in an environmental absence of symmetry and starts a slow but steady ascent to a balanced state of a brand-new stability.

In mathematics, the notion of symmetry, although present in many different areas, is mostly studied by means of **groups** and is therefore arguably an inherently algebraic concept – namely, the set of all transformations/symmetries of a given mathematical object that preserves its inner structure forms a group. Knowing the full set of symmetries of an object is important because it provides the most complete description of its structure.

Historically, many of the most fruitful techniques for studying groups have been developed within the framework of permutation groups and, more generally, **group actions**, preferably on sets enjoying additional inner structure. Among these, transitive actions are perhaps the most natural ones for they represent the building blocks for actions in general. This interplay is nicely recaptured in the combinatorial setting of relational structures, most notably in **graphs**, as the simplest manifestation of such structures. Being one of the core concepts essential both to understanding natural phenomena and the dynamics of social systems and to providing a theoretical framework for efficient mass communication, relational structures (and graphs in particular) with high levels of symmetry are sought for their optimal behavior and high performance.

A mathematical model capturing this situation, and a common thread of this book, is the concept of a **vertex-transitive graph** (together with its counterparts, such as edge-transitive graphs and arc-transitive graphs), and the underlying mathematical discipline is referred to as **algebraic graph theory**. The benefits are twofold. As much as the group-theoretic properties are reflected in the associated vertex-transitive graphs, the various structural properties of the latter are best studied using group-theoretic tools: This is precisely the viewpoint that this book is meant to uphold. Its content is a virtual two-way traffic on a one-way street: While pursuing symmetry properties of combinatorial objects is the main objective of this book, we will, as a byproduct, obtain additional information and understanding of the accompanying permutation groups and other algebraic objects involved in the process.

This philosophy is best reflected in a detailed motivational analysis of the various actions of the alternating group A_5 on the Petersen graph – each corresponding to a specific subgroup of A_5, with the natural action on its vertex set being just one of them. The analysis, which identifies with each subgroup of A_5 a corresponding combinatorial substructure of the Petersen graph, is supposed to serve as an introductory argument. A precursor to the central theme of this book is that mutual, algebraic, and combinatorial benefits are obtained by addressing the intrinsically algebraic concept of symmetry in a combinatorial setting. This analysis is the content of Sections 2.5 and 2.6, and, as it needs only very basic group-theoretic and graph-theoretic concepts, the reader familiar with basic terminology is welcome to jump right into it and start there.

This book has been written assuming that the reader has had a first course in group theory including the Sylow Theorems, as well as a first course in graph theory. A first course in permutation group theory, the main algebraic tools in this text, is **not** assumed. We will develop the permutation group theory tools that we need as we go, with proofs of some (usually longer) results omitted (for example, the O'Nan–Scott Theorem). Basic knowledge of fields and finite fields is assumed (for example, the definition of a field, and the knowledge that finite fields exist and that the group of units in a finite field is cyclic) in some parts as well as some basic linear algebra. Some knowledge of basic number theory would be useful (but mostly what is used in number theory is simply modular arithmetic as seen in group theory).

The book is organized as follows. The first five chapters are on material that we believe everyone who works in this area of algebraic graph theory should have at least some familiarity with. So our intention is that these chapters should more or less be covered in order. That is, each successive chapter assumes you are familiar with the preceding chapters.

Chapter 6 covers additional families of graphs that are fairly common but perhaps not as common as the families in the first five chapters. It is designed to be independent of the material that follows it, and each section of Chapter 6 is independent of the other sections in Chapter 6, so the reader could skip this chapter and only come back to it if needed. Some chapters after Chapter 6 do make use of some material in some sections of Chapter 6.

Chapters 7–13 are each on a different problem in the area of algebraic graph theory. For practical reasons, Chapter 7 should probably be read before Chapter 8. Chapter 7 is on the isomorphism problem for Cayley digraphs and Chapter 8 is on automorphism groups of (mainly) Cayley digraphs. In practice, the isomorphism problem for a fixed group G has been solved before the corresponding automorphism group problem, as the isomorphism problem can be solved using properties of the automorphism group.

Chapter 9 discusses some classification results for vertex-transitive graphs, focusing on classifications of certain orders. Section 9.3 uses results from Section 7.6. The remaining chapters can be read more or less independently. Chapters 10, 11, and 12 focus on topics already encountered in Chapter 3, namely, symmetric graphs, the Hamiltonian problem, and the semiregular elements in automorphism groups of vertex-transitive digraphs. In Chapter 13 we consider graphs with other kinds of symmetry not yet encountered. Finally, in Chapter 14 we discuss some problems we believe deserve a closer look in the future.

As permutation group theory is not a prerequisite for this book, it contains enough permutation group theory to form a short course in that subject. However, as our focus is on graphs with symmetry, the permutation group theory is usually given as needed for a particular graph-theoretic topic. So the permutation group theory is not provided in a self-contained part of the book.

Thanks are owed to many people who have actively contributed to the development of this book. Manuscript forms were used as the basis of several courses taught at the University of Primorska over the last several years. Additionally, quite a few mathematicians have read various parts of the manuscript and offered suggestions on improvements. For this, our thanks go to Rachel Barber, Ademir Hujdurović, István Kovács, Klavdija Kutnar, Štefko Miklavić, Luke Morgan, Rok Požar, Primož Šparl, and Ágnes Szalai.

1

Introduction and Constructions

Let us begin with an informal example. By informal, we mean that we will use terms intuitively, as opposed to formally defining every term we use. Let us consider the symmetries of the cube. This is the usual cube or, if you prefer to have one in your hand, a six-sided die or a Rubik's cube. A common theme that we will see throughout this book, is that when considering symmetries of a graph, it is really helpful to have a clever labeling of the vertex set. As this is a book about symmetries of graphs, to us a clever labeling will be one that shows us, without too much work, many of the symmetries of the graph. We choose to label the vertices of the cube with elements of $\mathbb{Z}_2 \times \mathbb{Z}_2 \times \mathbb{Z}_2$ as in Figure 1.1.

Notice that two vertices of the cube, as labeled in Figure 1.1, are adjacent if and only if their labels are different in exactly one coordinate. Also, for any face of the cube, there is a four-step rotation of the face in either the clockwise or counter-clockwise direction that is a symmetry of the cube (and of course the opposite face is rotated in the same fashion). As interchanging two opposite

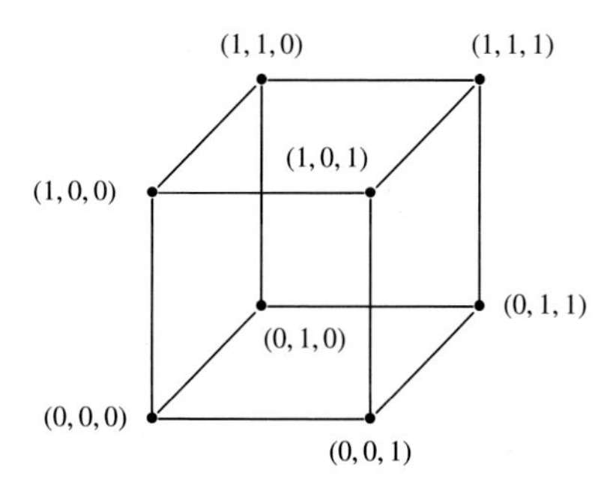

Figure 1.1 The 3-cube.

faces is also a symmetry of the cube, we may map any vertex of the cube to any other vertex of the cube. In the language we will introduce later, this is said as "the cube is vertex-transitive."

In general, once one knows a particular graph is vertex-transitive, a next step in considering all of its symmetries is to think about distinguished *sets* of vertices. By "distinguished," we mean that we can tell the sets are different somehow, under symmetry. We use the word "sets" here as vertex-transitive graphs have no distinguished vertices. For the cube, this means that we cannot distinguish (except by labeling) any corner of the cube from any other corner of the cube. But notice that for every vertex v of the cube, there is a unique vertex of distance three from v. Also, under any symmetry of a graph, vertices at some fixed distance in the graph are at the same distance in the graph under the symmetry. So these four pairs of vertices at distance three are permuted amongst themselves (these are the pairs $(0, 0, 0)$ and $(1, 1, 1)$, $(0, 0, 1)$ and $(1, 1, 0)$, $(0, 1, 0)$ and $(1, 0, 1)$, and $(1, 0, 0)$ and $(0, 1, 1)$. They are easy to remember as the vertex at distance three to (i, j, k) is simply $(i + 1, j + 1, k + 1)$ with arithmetic in each coordinate performed modulo 2). These pairs of vertices are also called *antipodal* vertices. This term is borrowed from geography, where an antipodal point on the earth is the point on the earth opposite the given point. In the language of group theory, this says that the symmetries of the cube *act* on the pairs of antipodal points. We will later also say that a pair of antipodal vertices is a *block* of the symmetries of the cube, and the symmetries of the cube also permute the blocks.

A next obvious step might be to determine the symmetries of the antipodal vertices. That is, to determine how pairs of antipodal vertices are mapped to each other, and also if there are any symmetries, other than the identity, that fix each pair of antipodal points. For symmetries that fix each pair of antipodal points, consider the function that adds 1 modulo 2 in each coordinate of a vertex; that is $f\colon \mathbb{Z}_2^3 \to \mathbb{Z}_2^3$ given by $f(i, j, k) = (i + 1, j + 1, k + 1)$. It is hopefully clear that if two vertices u and v are different in one coordinate, then $f(u)$ and $f(v)$ differ only in exactly the same coordinate. So f is a symmetry of the cube different from the identity that fixes each set of antipodal points. It turns out (we will prove this later), that there are no other such symmetries.

For symmetries of the antipodal vertices, there are four such pairs, so the largest number of such symmetries we can encounter is all of them! That is what we will see happens here. First, observe that no pair of antipodal points is contained in a face of the cube. So if we take a symmetry of a face, or square, and extend that to a symmetry of antipodal vertices (which can always be done as there is a unique antipodal vertex not on a face for each vertex on a face), we will have a symmetry of the cube. The symmetries of a face are just symmetries of the regular 4-gon, so four rotations and four reflections. This gives eight symmetries of the cube. To obtain elements of order 3, it is easiest to imagine that you are holding a cube between two fingers, with a finger on each antipodal vertex. Now spin the cube 120°! The cube does not change, so this gives a symmetry, and this can be repeated three times. We will see later that this is enough to obtain all symmetries of the cube, up to products of symmetries. That is, there are other symmetries, but these other symmetries can be obtained by successively applying symmetries we have already described. Cheating more than a little bit, the symmetries of the pairs of antipodal vertices is all symmetries of a set of 4 elements, so of order 24, while there are two symmetries that fix each set of antipodal vertices. In group-theoretic language, we have the group $S_4 \times \mathbb{Z}_2$ of order 48.

We now turn to the main topics of this chapter. Section 1.1 is mainly concerned with the basic notation and terminology that we will use throughout the book. As the mathematical area that concerns symmetries is group theory, and more specifically permutation group theory, this will include terminology regarding both group theory and graph theory, as well as some examples and basic results.

The central goal of the rest of the chapter is to get examples of vertex-transitive graphs. An old result of Erdős and Rényi (1963) is that almost all graphs have a trivial automorphism group. Thus having any symmetry in a graph is rare, and so instead of looking for graphs with symmetries, we usually start with the symmetries, and then construct graphs that have those symmetries. The most common such construction is a Cayley graph, the topic of Section 1.2. We then give two ways of constructing every vertex-transitive digraph, namely double coset digraphs and orbital digraphs in Sections 1.3 and 1.4, respectively. While these techniques are formally different, as they both construct every vertex-transitive digraph, they should be thought of as being in some sense "the same." We then give the construction for the second most common family of vertex-transitive digraphs, the metacirculant digraphs, in Section 1.5.

One last editorial comment before we get going. You may have noticed we are a little confused in our use of the term "graph" and "digraph." Our main motivation in writing this book is to introduce you to symmetries of graphs. So why do we have digraphs? There are two simple answers. First, there are numerous results in the literature that deal with graphs only, where the proofs hold equally well for digraphs with such dramatic changes as replacing "graph" with "digraph." Hence, in many cases (but not all!), there really is no difference between graphs and digraphs, so why not work with the more general object? Second, there are also many times where a result is only proven for graphs, but in practice the result is needed for digraphs. The general philosophy is then that we are mainly interested in graphs, but when it comes time to prove a theorem, it is best to prove the theorem for digraphs if that is possible.

1.1 Basic Definitions

As this book is about groups acting on graphs, we begin with some of the basic definitions and results from permutation group theory and graph theory.

We emphasize that our permutation multiplication is on the left. That is, $fg(x) = f(g(x))$. Readers should be aware that sometimes in the literature permutation multiplication is written on the right.

A **permutation group** is a subgroup of the **symmetric group** on n letters, denoted $\mathcal{S}_n$. Unless otherwise stated, we will take the n letters that $\mathcal{S}_n$ permutes to be the elements of the set $\mathbb{Z}_n$, the integers modulo n. We denote the group of units in $\mathbb{Z}_n$ under multiplication by $\mathbb{Z}_n^*$, and note that $\text{Aut}(\mathbb{Z}_n) = \{x \mapsto ax : a \in \mathbb{Z}_n^*\}$, the automorphism group of $\mathbb{Z}_n$. More generally, an **action** of a group G on a set X is a function $f\colon G \times X \to G$, with $f(g, x)$ written gx, such that $g(hx) = (gh)x$ and $1x = x$ for every $g, h \in G$ and $x \in X$ (of course 1 is the identity element in G). We will say that G **acts** on X on the left. In this text, unless otherwise stated, all groups and sets are finite, in which case the **degree** of an action is $|X|$, the number of elements in X. A related notion is that of a **permutation representation** of a group G, which is a homomorphism $\phi\colon G \to \mathcal{S}_n$ for some n. A standard result in group theory is that any action of G on X induces a homomorphism $\phi\colon G \to \mathcal{S}_X$ (Dummit and Foote, 2004, Proposition 4.1.1). So any action of G on X induces a corresponding permutation representation of G. We will occasionally abuse terminology and refer to $\phi(G)$ as a permutation representation if the action is clear. An action of G on X is called **faithful** if $\text{Ker}(\phi) = 1$, where the **kernel of the action** ϕ of G on X is denoted $\text{Ker}(\phi)$.

Example 1.1.1 Let n be a positive integer, and define $f\colon \mathbb{Z}_n \times \mathbb{Z}_n \to \mathbb{Z}_n$ by $f(g, x) = g+x$. For $g, h \in \mathbb{Z}_n$ and $x \in \mathbb{Z}_n$, $g+(h+x) = (g+h)+x$ and $0+x = x$ so that f is an action of $\mathbb{Z}_n$ on itself. The degree of this action is $|\mathbb{Z}_n| = n$, and the action is faithful as if $f(g, x) = x$ for all $x \in \mathbb{Z}_n$, then $g = 0$. The corresponding permutation representation of $\mathbb{Z}_n$ is $\phi\colon \mathbb{Z}_n \to \mathcal{S}_n$ given by $\phi(g)$ is the function defined by $x \mapsto g + x \pmod n$. Note that $\phi(\mathbb{Z}_n)$ is the subgroup of $\mathcal{S}_n$ generated by the n-cycle $(0, 1, 2, \ldots, n-1)$, and is usually denoted $(\mathbb{Z}_n)_L$ (see Definition 1.2.12). Similarly, if $H \leq \mathbb{Z}_n$ is a subgroup of order m, then similar arguments show that $k\colon \mathbb{Z}_n \times (\mathbb{Z}_n/H) \to \mathbb{Z}_n \times (\mathbb{Z}_n/H)$ given by $k(g, x + H) = g + (x + H)$

is also an action. If $m > 1$ this action is not faithful, as $k(h, x + H) = x + H$ for every $h \in H$. The degree of this action is then n/m. The corresponding permutation representation of $\mathbb{Z}_n$ is $\delta\colon \mathbb{Z}_n \to S_{n/m}$ given by $\delta(g)$ is the function defined by $x \mapsto x + g \pmod{n/m}$, and is also isomorphic to $(\mathbb{Z}_{n/m})_L$.

Definition 1.1.2 A permutation group $G \leq S_n$ is **transitive** if whenever $x, y \in \mathbb{Z}_n$, then there exists $g \in G$ such that $g(x) = y$.

Typically when discussing permutation groups, we will either start with a subgroup of S_n, or will specify a group and an action of that group on a set X that then induces a natural subgroup of S_X. Similarly, a concept about a permutation group translates into a concept about actions and vice versa, and we will usually refrain from defining analogous concepts in each context. So an action of G on X is **transitive** if for every $x, y \in X$ there is a $g \in G$ such that $gx = y$, and the **degree** of a transitive permutation group $G \leq S_n$ is n.

Example 1.1.3 Both of the actions in Example 1.1.1 are transitive as (using the notation from that example) if $x, y \in \mathbb{Z}_n$ then $f(y-x, x) = y$, and $k(y-x, x+H) = y + H$.

Definition 1.1.4 Let $G \leq S_n$ be transitive, and let the point $x \in \mathbb{Z}_n$. The **stabilizer of** x **in** G, denoted $\mathrm{Stab}_G(x)$, is defined by $\mathrm{Stab}_G(x) = \{g \in G : g(x) = x\}$. That is, $\mathrm{Stab}_G(x)$ is the set of all permutations in G that map x to x.

The stabilizer of x in G is often denoted G_x, and is a subgroup of G (Exercise 1.1.5).

Recall that every permutation in S_n can be written as a product of transpositions, and that this number is always even or always odd.

Definition 1.1.5 A permutation $\rho \in S_n$ is **even** if it can be written as a product of an even number of transpositions, and **odd** if it can be written as a product of an odd number of transpositions. The set of all even permutations in S_n is a subgroup, called the **alternating group on** n **letters**, and is denoted A_n.

Also recall that a cycle of odd length is an even permutation, while a cycle of even length is an odd permutation. For proofs of the above mentioned facts and other information regarding the alternating group (see Dummit and Foote (2004, Section 3.5)).

Example 1.1.6 $\mathrm{Stab}_{S_n}(n-1) = S_{n-1}$, $\mathrm{Stab}_{A_n}(n-1) = A_{n-1}$, and $\mathrm{Stab}_{(\mathbb{Z}_n)_L}(x) = 1$, $x \in \mathbb{Z}_n$.

We now turn to some basic properties of the stabilizer.

Theorem 1.1.7 *Let $G \leq S_n$, $x \in \mathbb{Z}_n$, and $h \in G$. Then $h\mathrm{Stab}_G(x)h^{-1} = \mathrm{Stab}_G(h(x))$. Consequently, if G is transitive, then $\mathrm{Stab}_G(x)$ is conjugate in G to $\mathrm{Stab}_G(y)$ for every $y \in \mathbb{Z}_n$.*

Proof Observe that

$$\begin{aligned}\mathrm{Stab}_G(h(x)) &= \{g \in G : g(h(x)) = h(x)\}\\ &= \{g \in G : h^{-1}gh(x) = x\}\\ &= \{g \in G : h^{-1}gh \in \mathrm{Stab}_G(x)\}\\ &= \{g \in G : g \in h\mathrm{Stab}_G(x)h^{-1}\}\\ &= h\,\mathrm{Stab}_G(x)\,h^{-1}.\end{aligned}$$

For the second statement, as G is transitive, there exists $h \in G$ such that $h(x) = y$. Then $\mathrm{Stab}_G(y) = \mathrm{Stab}_G(h(x)) = h\,\mathrm{Stab}_G(x)\,h^{-1}$. □

The following result is quite useful, and is sometimes called the orbit-stabilizer theorem.

Theorem 1.1.8 *Let $G \leq S_n$, and $x \in \mathbb{Z}_n$. The set $G(x) = \{g(x) : g \in G\}$, is the* **orbit of** x **in** G. *Then $|G| = |G(x)| \cdot |\mathrm{Stab}_G(x)|$, or equivalently, $|G(x)| = [G : \mathrm{Stab}_G(x)]$.*

Proof Define $\phi\colon G \to G(x)$ by $\phi(g) = g(x)$. Note that ϕ is onto (or surjective) by definition, and so $|\phi(G)| = |G(x)|$. Also,

$$\begin{aligned}\phi(g) = \phi(h) &\text{ if and only if } g(x) = h(x)\\ &\text{ if and only if } h^{-1}g(x) = x\\ &\text{ if and only if } h^{-1}g \in \mathrm{Stab}_G(x)\\ &\text{ if and only if } h^{-1}g\mathrm{Stab}_G(x) = 1 \cdot \mathrm{Stab}_G(x) \text{ (as left cosets)}\\ &\text{ if and only if } h \text{ and } g \text{ are in the same left coset of } \mathrm{Stab}_G(x).\end{aligned}$$

Thus $|\phi(G)|$ is the number of left cosets of $\mathrm{Stab}_G(x)$ in G, and so $|G(x)| = [G : \mathrm{Stab}_G(x)]$. As all groups here are finite, $[G : \mathrm{Stab}_G(x)] = |G|/|\mathrm{Stab}_G(x)|$. □

The following application of the orbit-stabilizer theorem is originally due to Miller (1903).

Theorem 1.1.9 *Let G be a transitive group of degree n, and p a prime. The highest power p^k of p dividing n also divides the length of every orbit of a Sylow p-subgroup of G.*

Proof Let P be a Sylow p-subgroup of G, and $x \in \mathbb{Z}_n$. By the orbit-stabilizer theorem, $n = [G : \mathrm{Stab}_G(x)]$ and $[P : \mathrm{Stab}_P(x)] = |P(x)|$. Then,

$$\begin{aligned} n \cdot [\mathrm{Stab}_G(x) : \mathrm{Stab}_P(x)] &= [G : \mathrm{Stab}_G(x)] \cdot [\mathrm{Stab}_G(x) : \mathrm{Stab}_P(x)] \\ &= [G : \mathrm{Stab}_P(x)] \\ &= [G : P] \cdot [P : \mathrm{Stab}_P(x)] \\ &= [G : P] \cdot |P(x)|. \end{aligned}$$

As $\gcd([G : P], p) = 1$, where $\gcd(m, n)$ is the greatest common divisor of m and n, and p^k divides n, we see that p^k divides $|P(x)|$. □

The following immediate corollary is the form of the previous result used most often (for our purposes), and is also due to Miller.

Corollary 1.1.10 *Let p be prime and $k \geq 1$. A Sylow p-subgroup of a transitive group of degree p^k is transitive.*

Definition 1.1.11 A permutation group $G \leq S_n$ is **semiregular** if $\mathrm{Stab}_G(x) = 1$ for every $x \in \mathbb{Z}_n$, and G is **regular** if G is both semiregular and transitive.

The next result follows directly from the orbit-stabilizer theorem as if $G \leq S_n$ is transitive, then G only has one orbit.

Corollary 1.1.12 *A transitive group is regular if and only if its order and degree are the same.*

Example 1.1.13 The group S_n is regular if and only if $n \leq 2$, and A_n is regular if and only if $n = 1$ or 3 but is semiregular for $n = 2$. The group $(\mathbb{Z}_n)_L$ is regular for all positive integers n.

Solution The group S_n is regular by Corollary 1.1.12 if and only if $n! = n$, which is true if and only if $n \leq 2$. Also, $A_1 = 1$ is regular, $A_2 = 1$ is not regular, and for $n \geq 3$, the group A_n is transitive and so regular if and only if $n!/2 = n$, which is true if and only if $n = 3$. Finally, $(\mathbb{Z}_n)_L$ is transitive of order n and so regular. □

A **digraph** Γ is an ordered pair (V, A), where V is a nonempty set (V is the **vertex set of** Γ), and $A \subseteq \{(u, v) : u, v \in V\}$ of ordered pairs of V (A is the **arc set of** Γ). An arc (u, v) of a digraph is usually thought of as being directed from u to v. If $(u, v) \in A(\Gamma)$ and $(v, u) \in A(\Gamma)$, we can identify these two arcs and consider it an **edge** (a loop is also considered an edge). A **graph** Γ is a digraph in which $(u, v) \in A(\Gamma)$ if and only if $(v, u) \in A(\Gamma)$, and in this case we think of the edges as being unordered pairs of vertices, denoted uv. The arc set of a graph Γ is usually called the **edge set** of Γ and denoted by $E(\Gamma)$ instead of $A(\Gamma)$. According to this definition, multiple edges (different edges with the same endpoints) are not allowed. This is not done from any sort of dislike of multiple edges, but more from a desire to start with the simplest definition of a graph, and in most, but not all, situations concerning symmetry in digraphs, multiple edges are irrelevant. This definition does allow loops, but similarly in this book loops are irrelevant most of the time. There are many situations, some of which will be encountered later, where the context calls for modifications to the definition of a graph. In sections or chapters where multiple edges are needed, this will be stated explicitly at the beginning of the section or chapter. In this text, all graphs are finite (so have finite vertex sets). Basic digraph and graph terms and operations such as walks, intersection of digraphs, etc., are defined as usual. See Bollobás (1998), for example.

As is customary, for a graph Γ we denote by $N_\Gamma(u) = \{v : uv \in E(\Gamma)\}$ the set of neighbors in Γ of the vertex u. For a digraph Γ, we denote the outneighbors of u by $N^+_\Gamma(u)$, and the inneighbors by $N^-_\Gamma(u)$. That is, $N^+_\Gamma(u) = \{v : (u, v) \in A(\Gamma)\}$ and $N^-_\Gamma(u) = \{v : (v, u) \in A(\Gamma)\}$. If the graph or digraph Γ is clear, we may simply write $N^+(u)$, etc.

Definition 1.1.14 An **isomorphism** between two digraphs Γ_1 and Γ_2, is a bijection $\phi\colon V(\Gamma_1) \to V(\Gamma_2)$ such that $(\phi(u), \phi(v)) \in A(\Gamma_2)$ if and only if $(u, v) \in A(\Gamma_1)$. We write $\Gamma_1 \cong \Gamma_2$.

Thus an isomorphism is a one-to-one mapping of the vertex set onto the vertex set that preserves arcs.

Any isomorphism ϕ between two digraphs Γ_1 and Γ_2 induces a natural bijection between $A(\Gamma_1)$ and $A(\Gamma_2)$ given by $(u, v) \mapsto (\phi(u), \phi(v))$. We will often abuse notation and simply write $\phi(u, v)$ instead of $(\phi(u), \phi(v))$ for the image of the arc (u, v) under ϕ.

Definition 1.1.15 Any bijection ϕ from $V(\Gamma_1)$ to a set X induces a digraph on X isomorphic to Γ_1. Namely, define a digraph Γ_2 by $V(\Gamma_2) = X$ and $A(\Gamma_2) = \{(\phi(u), \phi(v)) : (u, v) \in A(\Gamma_1)\}$. We will adopt the notational convention of setting $\phi(\Gamma_1) = \Gamma_2$.

An **automorphism** of a digraph is an isomorphism of a digraph with itself. The set of all automorphisms of a digraph Γ is a group under function composition (or permutation multiplication if you prefer – see Exercise 1.1.8), and is called the **automorphism group of** Γ, denoted $\mathrm{Aut}(\Gamma)$. Of course, $\mathrm{Aut}(\Gamma)$ also acts on the arcs of a digraph and the edges of a graph.

Definition 1.1.16 A digraph whose automorphism group is transitive on its vertex set is called a **vertex-transitive digraph**, and a digraph whose automorphism group is transitive on its arc set is called an **arc-transitive digraph**. A graph Γ is **edge transitive** if $\mathrm{Aut}(\Gamma)$ is transitive on its edge set $E(\Gamma)$.

Example 1.1.17 A complete graph K_n of order $n \geq 1$ is vertex-transitive, arc-transitive, and $\mathrm{Aut}(\Gamma) = \mathcal{S}_n$.

Solution As K_n has every arc, if $\sigma \in \mathcal{S}_n$ then $\sigma(u, v) \in A(K_n)$ for every pair of distinct vertices u and v. Thus $\mathrm{Aut}(K_n) = \mathcal{S}_n$ (as $\mathcal{S}_n$ is the "largest" permutation group on n vertices). The group $\mathcal{S}_n$ is certainly transitive, and also if $(u_1, v_1), (u_2, v_2) \in A(\mathcal{S}_n)$, then there is $\sigma \in \mathcal{S}_n$ with $\sigma(u_1, v_1) = (u_2, v_2)$. Thus K_n is vertex-transitive and arc-transitive. □

Example 1.1.18 For $n \geq 1$, define a graph Q_n by $V(Q_n) = \mathbb{Z}_2^n$ and two vertices are adjacent if and only if they differ in exactly one coordinate. The graph Q_n is the n**-dimensional hypercube** or the n**-cube** and is vertex-transitive. The 3-cube, whose automorphism group was discussed at the beginning of this chapter, is shown in Figure 1.1.

Solution Define $\tau_i \colon \mathbb{Z}_2^n \to \mathbb{Z}_2^n$ by $\tau_i(x)$ adds 1 modulo 2 in the ith coordinate of x and is the identity on the other coordinates. If $e \in E(Q_n)$, then $\tau_i(e) \in E(Q_n)$, as in any coordinate, adding a constant (either 0 or 1) does not change whether the coordinates are the same or different. Thus $\tau_i \in \mathrm{Aut}(Q_n)$ for every $1 \leq i \leq n$. Additionally, it is easy to see that any element of $\mathbb{Z}_2^n$ can be transformed into any other element of $\mathbb{Z}_2^n$ by changing (or adding 1) coordinates of the first that differ with the second. Thus $\langle \tau_i : 1 \leq i \leq n \rangle$, the subgroup of $\mathrm{Aut}(Q_n)$ generated by the τ_i, is transitive and Q_n is vertex-transitive. □

The $(n + 1)$-cube can be constructed from the n-cube by taking two copies of the n-cube and joining them by a 1-factor. The vertices of one n-cube has an additional coordinate added with a 1 in the additional coordinate, while the vertices of the other n-cube has an additional coordinate added with a 0 in the additional coordinate. The edges of the 1-factor are then between corresponding vertices in the two copies. That is, the edges of the 1-factor only differ in the additional coordinate. This can be seen in Figure 1.1, where the additional coordinate that is added is the first coordinate, and the two copies of the 2-cube are the top and bottom faces of the cube. This construction for the $(n + 1)$-cube from the n cube is often useful for induction arguments.

Example 1.1.19 The Petersen graph given in Figure 1.2 is vertex-transitive.

Solution It is easy to see that a 72° rotation leaves the Petersen graph invariant, and so any vertex of the "outside" 5-cycle $0, 1, 2, 3, 4, 0$ can be mapped to any other vertex on the "outside" 5-cycle and similarly, any vertex of the "inside" 5-cycle $0', 1', 2', 3', 4', 0'$ can be mapped to any other vertex on the "inside" 5-cycle. It thus only remains to show that there is an

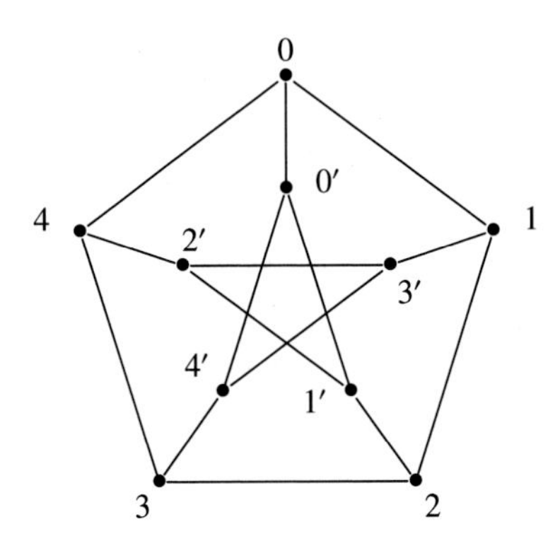

Figure 1.2 The Petersen graph.

automorphism of the Petersen graph that interchanges the outside 5-cycle and the inside 5-cycle. Consider the permutation $(0,0')(1,1')(2,2')(3,3')(4,4')$. Straightforward computations (which you should do!) verify that this map is an automorphism of the Petersen graph and so the Petersen graph is vertex-transitive. □

A word about our notation for cycles. Usually a cycle is denoted by a sequence of vertices, say $v_0v_1,\ldots,v_{n-1}v_0$. You will notice in the previous paragraph we used the slightly nonstandard notation of $v_0,v_1,\ldots,v_{n-1},v_0$, as when using numbers for the labels of vertices, there can often be ambiguity – particularly if the graph has many vertices. For example, in a graph with vertex set $\mathbb{Z}_{25}$ the cycle 01230 could be a cycle of length 4 or two different cycles of length 3! So we will feel free to use commas between vertices in denoting a cycle, or not if it causes no ambiguity. A similar comment holds for cycle notation of permutations.

Before our next discussion, we will need some additional definitions.

Definition 1.1.20 Let Γ be a regular graph. The **valency of** Γ is the number of edges incident with any vertex. We say Γ is **cubic** if it is regular of valency 3. Finally, Γ is **symmetric** if its automorphism group is transitive in its action on both its vertex set and its arc set.

While in this text we are usually concerned with vertex-transitive graphs, from the very beginning of the study of vertex-transitive graphs there has been interest in graphs whose automorphism group also acts transitively on arcs or other subdigraphs in graphs. Indeed, starting in 1932, Foster started compiling what is now known as the **Foster census** (Foster, 1988) of cubic symmetric graphs – these are vertex-transitive and arc-transitive graphs of valency 3. An online and expanded version (though with less information for each graph) can be found in Conder (2006). As we proceed through the text we will see many of the graphs in the Foster census (we have already seen some of them), so we will go ahead and describe how graphs are listed in the Foster census, as well as give some examples. Graphs that we have seen already are the cube, which is known as F8 (the 'F' is for Foster, while 8 is the number of vertices in the cube), the Petersen graph is F10, and K_4 is F4. The dodecahedron, which is shown in Figure 1.14 in the Exercises of this section, is F20A (the addition of the 'A' is because there are two cubic symmetric graphs of order 20, the other being a graph we will see later – the Desargues graph, denoted F20B in the Foster census). We leave a more detailed discussion of the Foster census for Chapter 10.

1.2 Cayley Digraphs

Our main goals for the next four sections is to obtain a library of examples, and to see how to construct vertex-transitive graphs. We begin with Cayley digraphs, as they are by far the most commonly studied and most frequently encountered class of vertex-transitive digraphs. After seeing some of their basic properties, we will give Sabisussi's characterization of Cayley digraphs (Theorem 1.2.20), and show that the Petersen graph is not isomorphic to a Cayley graph of any group (Theorem 1.2.22).

Definition 1.2.1 Let G be a group and $S \subseteq G$. Define a **Cayley digraph of** G, denoted $\mathrm{Cay}(G, S)$, to be the digraph with vertex set $V(\mathrm{Cay}(G, S)) = G$ and arc set $A(\mathrm{Cay}(G, S)) = \{(g, gs) : g \in G, s \in S\}$. We call S the **connection set of** $\mathrm{Cay}(G, S)$.

Note that $(x, y) \in A(\mathrm{Cay}(G, S))$ if and only if $x^{-1}y \in S$ as then $x = g$ and $y = gs$, so $xy^{-1} = s$ (the latter condition is often used in place of the former condition in the definition of a Cayley digraph).

Example 1.2.2 The graph in Figure 1.3 is $\mathrm{Cay}(\mathbb{Z}_{10}, \{1, 3, 7, 9\})$. Note that as the binary operation on $\mathbb{Z}_{10}$ is addition, there is an edge between two vertices if and only if the difference of the labels on the vertices is contained in the set $\{1, 3, 7, 9\}$. Observe that a clockwise rotation of 36° leaves the graph unchanged.

Lemma 1.2.3 *The digraph* $\mathrm{Cay}(G, S)$ *is a graph if and only if* $S = S^{-1}$ *(or, written additively, that* $S = -S$*).*

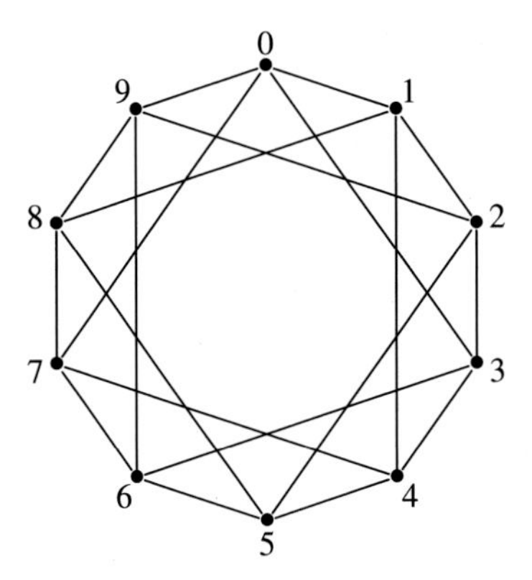

Figure 1.3 The Cayley graph $\mathrm{Cay}(\mathbb{Z}_{10}, \{1, 3, 7, 9\})$.

Proof Suppose $\mathrm{Cay}(G, S)$ is a graph. Then $(g, gs) \in A(\mathrm{Cay}(G, S))$ if and only if $(gs, g) \in A(\mathrm{Cay}(G, S))$ if and only if $(gs)^{-1}g = s^{-1} \in S$. Suppose $S = S^{-1}$. If $(g, gs) \in A(\mathrm{Cay}(G, S))$ then $(gs, gs(s^{-1})) = (gs, s) \in A(\mathrm{Cay}(G, S))$. □

In many situations, whether or not a Cayley digraph has loops does not have any effect. In these cases, the default is usually to exclude loops by also insisting that $1_G \notin S$ (or $0 \notin S$ if G is abelian and the operation is addition).

Many common graphs are isomorphic to Cayley graphs.

Example 1.2.4 The 3-cube, as drawn in Figure 1.1, is $\mathrm{Cay}\left(\mathbb{Z}_2^3, S\right)$, where $S = \{(0, 0, 1), (0, 1, 0), (1, 0, 0)\}$. Notice that for $\mathbb{Z}_2^n$ *every* element is its own inverse, and so Cayley digraphs of $\mathbb{Z}_2^n$ are always graphs.

Example 1.2.5 Let $n \geq 3$ be a positive integer. A cycle of length n is isomorphic to $\mathrm{Cay}(\mathbb{Z}_n, \{1, n - 1\})$.

Example 1.2.6 Let G be a group, $n = |G|$, and $S = G\backslash\{1_G\}$. Then $\mathrm{Cay}(G, S)$ is isomorphic to K_n.

Example 1.2.7 Let $n \geq 1$, and let S be the subset of $\mathbb{Z}_{2n}$ consisting of all n odd numbers. Then $\mathrm{Cay}(\mathbb{Z}_{2n}, S)$ is isomorphic to $K_{n,n}$.

Definition 1.2.8 A digraph Γ is **regular** if the invalence and outvalence of every vertex are the same. We say Γ has **outvalency** d and **invalency** d if every vertex has invalence and outvalence d.

A Cayley digraph without loops whose connection set has d elements is regular with outvalency and invalency d (Exercise 1.2.5).

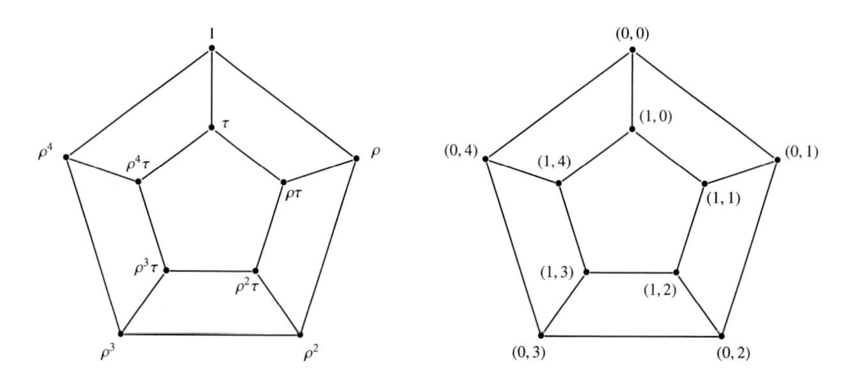

Figure 1.4 (Left) $\mathrm{Cay}(D_{10}, \{\rho, \rho^4, \tau\})$ and (Right) $\mathrm{Cay}(\mathbb{Z}_2 \times \mathbb{Z}_5, \{(0,1), (0,4), (1,0)\})$.

Perhaps the most common Cayley digraphs that one encounters are Cayley digraphs of the cyclic groups $\mathbb{Z}_n$ of order n, as in Example 1.2.2.

Definition 1.2.9 A Cayley (di)graph of $\mathbb{Z}_n$ is called a **circulant** (di)graph of order n.

In our next example we examine a Cayley graph of a dihedral group.

Definition 1.2.10 The **dihedral group** D_{2n} of order $2n$ is the symmetries of a regular n-gon, and has presentation $\langle \rho, \tau : \rho^n = \tau^2 = 1, \tau^{-1}\rho\tau = \rho^{-1} \rangle$.

Example 1.2.11 The Cayley graphs $\mathrm{Cay}(\mathbb{Z}_2 \times \mathbb{Z}_5, \{(0,1), (0,4), (1,0)\})$ and $\mathrm{Cay}(D_{10}, \{\rho, \rho^4, \tau\})$ are isomorphic.

Solution The graph on the left in Figure 1.4 is $\mathrm{Cay}(D_{10}, \{\rho, \rho^4, \tau\})$, while the graph on the right in Figure 1.4 is $\mathrm{Cay}(\mathbb{Z}_2 \times \mathbb{Z}_5, \{(0,1), (0,4), (1,0)\})$. The two graphs are isomorphic by inspection (this phrase simply means "by looking at them"). It is thus possible for a graph to be isomorphic to Cayley graphs of different groups, and being a Cayley graph of G is not invariant under isomorphism. It is easy to see that a rotation of 72° leaves the above graphs the same. Additionally, it is also not hard to see that we can map the "inside" 5-cycle to the "outside" 5-cycle and vice versa. □

Definition 1.2.12 Let G be a group and let $g \in G$. Define $g_L \colon G \to G$ by $g_L(x) = gx$. The map g_L is a **left translation of** G. The **left regular representation of** G, denoted G_L, is $G_L = \{g_L : g \in G\}$. That is, G_L is the group of all left translations of G. It is straightforward to verify that G_L is a group isomorphic to G (Exercise 1.2.4).

Let $x, y \in G$, and let $g = yx^{-1}$. Then $g_L(x) = yx^{-1}x = y$ so that G_L is transitive on G. Also, notice that $g_L(x) = x$ gives $gx = x$ or $g = 1$. So the only element of G_L that fixes a point is the identity. Hence G_L is regular on G.

The permutations mentioned at the end of Example 1.2.2 generate $G_L = (\mathbb{Z}_{10})_L$, while the permutations mentioned at the end of Example 1.2.11 generate $(\mathbb{Z}_{10})_L$ (but not $(D_{10})_L$). To see the generators of $(D_{10})_L$ in Example 1.2.11, observe that there is a reflection that also leaves those graphs the same (the most obvious one is obtained by reflecting about the vertical line through the center of the graph – other reflections can be obtained by rotating that line by 72°). The group $(D_{10})_L$ is then obtained by taking the rotation by 72°, and then mapping the inside 5-cycle to the outside 5-cycle while reflecting.

In general, for an abelian group G, the translations of G will consist of the maps $x \mapsto g + x = x + g$, $g \in G$. That is, $G_L = \{x \mapsto x + g : g \in G\}$. More specifically, for a cyclic group $\mathbb{Z}_n$, we have that $\mathbb{Z}_n$ is generated by the map $x \mapsto x + 1$ (of course instead of 1, one could put any generator of $\mathbb{Z}_n$).

Lemma 1.2.13 *If G is a group and $S \subseteq G$, then $G_L \leq \mathrm{Aut}(\mathrm{Cay}(G, S))$. In particular, a Cayley digraph of a group G is a vertex-transitive digraph.*

Proof Let $g \in G$ and $s \in S$, and so $(g, gs) \in A(\mathrm{Cay}(G, S))$. Let $h \in G$. Then

$$h_L(g, gs) = (hg, hgs) = (g', g's) \in A(\mathrm{Cay}(G, S)),$$

where $g' = hg$. Then $h_L \in \mathrm{Aut}(\mathrm{Cay}(G, S))$ so $G_L \leq \mathrm{Aut}(\mathrm{Cay}(G, S))$. As we have already established that G_L is transitive, $\mathrm{Aut}(\mathrm{Cay}(G, S))$ is transitive, and so $\mathrm{Cay}(G, S)$ is vertex-transitive. □

Some authors will define Cayley digraphs on the right, in such a way that the **right regular representation** G_R is contained in the automorphism group. See Exercise 1.2.13.

We, like most authors, are doing things "backwards" here. Namely, we defined a Cayley digraph of G in terms of the vertices and edges (as is customary in defining a graph!), and then showed that the group G_L is contained in its automorphism group. This is not though, how one should think about how to construct a Cayley digraph. Instead, think of S as being the outneighbors of the identity (which they actually are), and then use the group G_L to "fill out" the rest of the arcs in the digraph by applying the elements of G_L to the arcs between the identity and its neighbors. Note that in this method of constructing a Cayley digraph, G_L is automatically a subgroup of the automorphism group.

Some authors will insist for primarily historical reasons that $\langle S \rangle = G$ when defining a Cayley digraph (although we clearly do not). The following result shows that $\langle S \rangle = G$ is equivalent to insisting that all Cayley digraphs are connected. We note that our notion of a digraph Γ being connected is that there is a directed path between any two vertices. That is, that if $u, v \in V(\Gamma)$ there exists a path $w_1 \ldots w_r$ in Γ with $u = w_1$, $v = w_r$, and $(w_i, w_{i+1}) \in A(\Gamma)$, $1 \le i \le r-1$. Many authors refer to this as Γ being **strongly connected**, with the weaker condition that there is a path between any two vertices in the underlying simple graph of Γ as being a **weakly connected digraph**.

Lemma 1.2.14 *Let G be a group and $S \subseteq G$. A Cayley digraph* $\mathrm{Cay}(G, S)$ *is strongly connected if and only if $\langle S \rangle = G$.*

Proof Suppose $\langle S \rangle = G$. Let $x, y \in G$, and $g = x^{-1}y$. Then there exist $s_1, \ldots, s_r \in S$ such that $s_1 s_2 \cdots s_r = g$. For $1 \le i \le r$, let $v_i = s_1 s_2 \cdots s_i$. Notice that $(v_i, v_i s_{i+1}) = (v_i, v_{i+1})$ so there is an arc in G from v_i to v_{i+1}, $1 \le i \le r-1$. Then $W = 1_G v_1 \ldots v_r$ is a walk in $\mathrm{Cay}(G, S)$ from 1_G to g. Note that $x_L(1_G) = x$ and $x_L(g) = y$, so $x_L(W)$ is a walk in G from x to y. Thus G is strongly connected.

Conversely, suppose $\mathrm{Cay}(G, S)$ is strongly connected. Then for each $g \in G$, there is a path $P_g = v_0 v_1 \ldots v_r$ in $\mathrm{Cay}(G, S)$ from 1_G to g, where $v_0 = 1_G$ and $v_r = g$. As $(v_i, v_{i+1}) \in A(\mathrm{Cay}(G, S))$, $v_{i+1} = v_i s_i$ for some $s_i \in S$, $1 \le i \le r-1$. Then $g = s_1 \cdots s_r$ and $g \in \langle S \rangle$. As g was arbitrary, $\langle S \rangle = G$. □

We now turn to the relationship between Cayley digraphs of G and $\mathrm{Aut}(G)$, the **automorphism group of** G.

Lemma 1.2.15 *Let G and H be isomorphic groups with $\alpha\colon G \to H$ an isomorphism. Let $S \subseteq G$. Then* $\alpha(\mathrm{Cay}(G, S)) = \mathrm{Cay}(H, \alpha(S))$. *In particular, if $H = G$, then* $\alpha(\mathrm{Cay}(G, S))$ *is a Cayley digraph of G with connection set $\alpha(S)$, or equivalently* $\alpha(\mathrm{Cay}(G, S)) = \mathrm{Cay}(G, \alpha(S))$.

Proof Clearly $\alpha(G) = H$ so that $V(\alpha(\mathrm{Cay}(G, S))) = H$. Let $a = (g, gs) \in A(\mathrm{Cay}(G, S))$, where $g \in G$ and $s \in S$. Then

$$\alpha(a) = \alpha(g, gs) = (\alpha(g), \alpha(gs)) = (\alpha(g), \alpha(g)\alpha(s)) = (h, hs'),$$

where $h = \alpha(g)$ and $s' = \alpha(s) \in \alpha(S)$. Then α maps arcs of $\mathrm{Cay}(G, S)$ to arcs of $\mathrm{Cay}(H, \alpha(S))$ bijectively as $|S| = |\alpha(S)|$. So $\alpha(\mathrm{Cay}(G, S)) = \mathrm{Cay}(H, \alpha(S))$, as required. The result follows by setting $H = G$. □

In general, it is quite tedious to check whether or not a permutation is contained in the automorphism group of a graph. For Cayley digraphs of a group G and automorphisms of G this is easy to check!

Corollary 1.2.16 *Let G be a group, $S \subseteq G$, and $\alpha \in \mathrm{Aut}(G)$. Then $\alpha \in$* $\mathrm{Aut}(\mathrm{Cay}(G, S))$ *if and only if $\alpha(S) = S$.*

Example 1.2.17 Which automorphisms of $\mathbb{Z}_{15}$ are contained in the automorphism group of $\Gamma = \mathrm{Cay}(\mathbb{Z}_{15}, \{1, 3, 4, 12\})$?

Solution The automorphisms of $\mathbb{Z}_{15}$ consist of multiplication by units in $\mathbb{Z}_{15}$. That is, the automorphisms of $\mathbb{Z}_{15}$ are the maps $x \mapsto ax$ where $a \in \mathbb{Z}_{15}^*$. As $1 \in S = \{1, 3, 4, 12\}$, we need only check which integers $3, 4, 12$ give rise to automorphisms of Γ. As 4 is the only unit among $\{3, 4, 12\}$, we need only check whether $x \mapsto 4x$ is an automorphisms of Γ as $x \mapsto 1x$ is the identity and is always in $\mathrm{Aut}(\Gamma)$. As $4 \cdot \{1, 3, 4, 12\} = \{4, 12, 1, 3\}$ we see $x \mapsto 4x$ is indeed an automorphism of Γ, and so the automorphisms of $\mathbb{Z}_{15}$ that are contained in $\mathrm{Aut}(\Gamma)$ are $x \mapsto ax$ where $a = 1$ or 4. □

Corollary 1.2.18 *Let G be an abelian group, $G \neq \mathbb{Z}_2^k$ for any $k \geq 1$, and $S \subseteq G$ such that $S = -S$. Then $\mathrm{Aut}(\mathrm{Cay}(G, S)) \neq G_L$, and in particular, the map $\iota\colon G \to G$ given by $\iota(g) = -g$ is in $\mathrm{Aut}(\mathrm{Cay}(G, S))$ and $\iota \neq 1$.*

Proof For a group G, the map $x \mapsto x^{-1}$ is a group automorphism if and only if G is abelian (in which case the map is more appropriately written $x \mapsto -x$ and so is ι). Thus if $\mathrm{Cay}(G, S)$ is a Cayley *graph*, so that $S = -S$, then $\iota \in \mathrm{Aut}(\mathrm{Cay}(G, S))$ by Corollary 1.2.16. We note the only abelian groups such that every element is its own inverse (and for which ι is the identity permutation) are the groups $\mathbb{Z}_2^k$, $k \geq 1$, and for other abelian groups G we have $\iota \notin G_L$ as it fixes the identity in G. Thus if G is abelian, $G \ncong \mathbb{Z}_2^k$, $k \geq 1$, and $\mathrm{Cay}(G, S)$ is a Cayley graph of G, then $\mathrm{Aut}(\mathrm{Cay}(G, S)) \neq G_L$. □

Example 1.2.19 Which automorphisms of $\mathbb{Z}_8$ are contained in the automorphism group of $\Gamma = \mathrm{Cay}(\mathbb{Z}_8, \{2, 6\})$?

Solution Note first that $S = -S$, so Γ is a graph and by Corollary 1.2.18, the map $x \mapsto -x$ is contained in $\mathrm{Aut}(\Gamma)$. The automorphisms of $\mathbb{Z}_8$ consist of multiplication by units in $\mathbb{Z}_8$. Note that if $x \mapsto ax$ is contained in $\mathrm{Aut}(\Gamma)$, then $x \mapsto -ax$ is also contained in $\mathrm{Aut}(\Gamma)$. So we need only check to see which units in $\mathbb{Z}_8$ between 1 and 4 induce automorphisms of Γ as the automorphisms induced by units between 1 and 4 composed with $x \mapsto -x$ give the automorphisms of $\mathbb{Z}_8$ induced by the units between 5 and 7. Computing $3 \cdot \{2, 6\} = \{6, 2\}$ we see that $\mathrm{Aut}(\Gamma)$ contains $x \mapsto ax$ for $a = 1, 3, 5, 7$. □

One of the first major problems considered concerning symmetries in graphs was the question of whether or not there is a Cayley (di)graph of G whose automorphism group is G. The preceding result shows that this is not the case for Cayley graphs of abelian groups that are not elementary abelian 2-groups (a group G is **elementary abelian** if $G \cong \mathbb{Z}_p^k$ for some prime p and $k \geq 1$). This problem is discussed in more detail in Section 5.2.

The following result of Sabidussi (1958) characterizes Cayley digraphs.

Theorem 1.2.20 (Sabidussi's theorem) *A digraph Γ is isomorphic to a Cayley digraph of a group G if and only if* $\mathrm{Aut}(\Gamma)$ *contains a regular subgroup isomorphic to G.*

Proof If $\Gamma \cong \mathrm{Cay}(G, S)$ for some $S \subseteq G$ with $\phi\colon \Gamma \to \mathrm{Cay}(G, S)$ an isomorphism, then by Lemma 1.2.13, $\mathrm{Aut}(\mathrm{Cay}(G, S))$ contains the regular subgroup $G_L \cong G$, namely $\phi^{-1} G_L \phi$ (see Exercise 1.1.11).

For the converse, suppose $\mathrm{Aut}(\Gamma)$ contains a regular subgroup $H \cong G$, with $\delta\colon H \to G$ an isomorphism. Fix $v \in V(\Gamma)$. As H is regular, for each $u \in V(\Gamma)$, by Exercise 1.1.2 there exists a unique $h_u \in H$ such that $h_u(v) = u$. Define $\phi\colon V(\Gamma) \to G$ by $\phi(u) = \delta(h_u)$. The map ϕ is our isomorphism from Γ to a Cayley digraph of G. Essentially, we are picking a vertex v that we will identify (via ϕ) with the identity in G. Then the vertex $u \in V(\Gamma)$ is identified with $\delta(h_u)$. We are left with the technical details to show that this works.

First, note that as each h_u is unique, ϕ is well defined, and is also a bijection as δ is a bijection. Let $U = \{u \in V(\Gamma) : (v, u) \in A(\Gamma)\}$. We claim that $\phi(\Gamma) = \mathrm{Cay}(G, \phi(U))$, where $\phi(\Gamma)$ is the digraph with $V(\phi(\Gamma)) = \{\phi(v) : v \in V(\Gamma)\}$, and $A(\phi(\Gamma)) = \{(\phi(u), \phi(v)) : (u, v) \in A(\Gamma)\}$. As $\phi(V(\Gamma)) = G$, we have $V(\phi(\Gamma)) = G$. Let $a \in A(\phi(\Gamma))$. We must show that $a = (g, gs)$ for some $g \in G$ and $s \in \phi(U)$. As $a \in A(\phi(\Gamma))$, $\phi^{-1}(a) = (u_1, u_2) \in A(\Gamma)$ by Exercise 1.1.11. Let $w \in V(\Gamma)$ such that $h_{u_1}(w) = u_2$. Then $h_{u_1}^{-1}(u_1, u_2) = (v, w) \in A(\Gamma)$ so $w = h_w(v) \in U$. Also, $h_{u_2} = h_{u_1} h_w$ as $h_{u_1} h_w(v) = h_{u_1}(w) = u_2$. Set $g = \delta(h_{u_1})$ and $s = \delta(h_w) \in \phi(U)$. Then

$$
\begin{aligned}
a = \phi(u_1, u_2) &= (\delta(h_{u_1}), \delta(h_{u_2})) = (\delta(h_{u_1}), \delta(h_{u_1} h_w)) \\
&= (\delta(h_{u_1}), \delta(h_{u_1})\delta(h_w)) \\
&= (g, gs),
\end{aligned}
$$

as required. □

A comment is probably in order here about the language that one will see in the literature regarding Cayley digraphs and Sabidussi's theorem. Sabidussi's theorem states that a digraph is *isomorphic to* a Cayley digraph of G if and only if its automorphism group contains a regular subgroup isomorphic to G. It is common for the automorphism group of a digraph containing a regular subgroup isomorphic to G to be used implicitly as the definition of a Cayley digraph, even if the usual definition has been given. That is, it is common to abuse language and say, regardless of the given definition of a Cayley digraph, that a digraph is a Cayley digraph of G if and only if its automorphism group contains a regular subgroup isomorphic to G.

We end this section by giving an example of a vertex-transitive graph that is not isomorphic to a Cayley graph. Our example is the Petersen graph (see Figure 1.2), which together with its complement, are the smallest (in terms of number of vertices) non-Cayley vertex-transitive graphs (although we will not prove the last statement, we will come fairly close with results in Chapter 9). Before proceeding, we prove a group-theoretic result that will simplify the proof (and it or its generalizations are used to simplify many other proofs).

Lemma 1.2.21 *Let $G = \langle \rho, \tau : \rho^m = \tau^n = 1, \tau^{-1}\rho\tau = \rho^a, 0 \le a \le m-1\rangle$. If $\rho_1, \tau_1 \in \langle \rho, \tau\rangle$ such that $\rho_1^m = \tau_1^n = 1$, $\tau_1^{-1}\rho_1\tau_1 = \rho_1^a$, and $\langle \rho_1, \tau_1\rangle = \langle \rho, \tau\rangle$, then there exists $\alpha \in \mathrm{Aut}(G)$ such that $\alpha(\tau) = \tau_1$ and $\alpha(\rho) = \rho_1$.*

Proof As $\rho_1\tau_1 = \tau_1\rho_1^a$ we may write each element of G uniquely as $\tau_1^x\rho_1^y$ for $0 \le x \le n-1$ and $0 \le y \le m-1$. Define $\alpha\colon G \to G$ by $\alpha(\tau^x\rho^y) = \tau_1^x\rho_1^y$. As $\rho\tau = \tau\rho^a$, we may also write each element of G uniquely as $\tau^x\rho^y$, $0 \le x \le n-1$ and $0 \le y \le m-1$. This implies that α is a bijection as $\langle \rho_1, \tau_1\rangle = \langle \rho, \tau\rangle$. Then

$$\tau^x\rho^y\tau^u\rho^v = \tau^x\tau\rho^{ay}\tau^{u-1}\rho^v = \tau^x\tau^2\rho^{a^2y}\tau^{u-2}\rho^v = \ldots = \tau^x\tau^u\rho^{a^uy}\rho^v = \tau^{x+u}\rho^{a^uy+v},$$

and analogously, $\tau_1^{x+u}\rho_1^{a^uy+v} = \tau_1^x\rho_1^y\tau_1^u\rho_1^v$. Then

$$\alpha(\tau^x\rho^y\tau^u\rho^v) = \alpha(\tau^{x+u}\rho^{a^uy+v}) = \tau_1^{x+u}\rho_1^{a^uy+v} = \tau_1^x\rho_1^y\tau_1^u\rho_1^v = \alpha(\tau^x\rho^y)\alpha(\tau^u\rho^v),$$

and $\alpha \in \mathrm{Aut}(G)$. □

An analogous result is true for any group generated by a finite list of elements together with a finite list of defining relations (see Coxeter and Moser [1965, p. 5]).

Theorem 1.2.22 *The Petersen graph is not isomorphic to a Cayley graph of any group.*

Proof We proceed by contradiction and assume the Petersen graph P is a Cayley graph $\mathrm{Cay}(G, S)$, where $|G| = 10$ and $|S| = 3$. As $|G| = 10$, $G = \mathbb{Z}_{10}$ or D_{10} as these are the only groups of order 10. If $G = \mathbb{Z}_{10}$, then by Exercise 1.2.9, the girth of P is at most 4. However, P has girth 5 (by inspection). Thus $P = \mathrm{Cay}(D_{10}, S)$.

Before proceeding it will be useful to review some of the properties of D_{10}. We set $D_{10} = \langle \rho, \tau\rangle$ as in the definition of D_{2n}, and recall that $\langle \rho\rangle \trianglelefteq D_{10}$. Also, $\langle \rho\rangle$ has two left cosets in D_{10}, namely $\langle \rho\rangle$ and $\tau\langle \rho\rangle = \{\tau, \tau\rho, \tau\rho^2, \tau\rho^3, \tau\rho^4\}$. D_{10}, as the symmetries of a regular pentagon, has 5 rotations, each of which is contained in $\langle \rho\rangle$, four of whose elements have order 5 with the other of order 1. As D_{10} has 5 reflections each of order 2, these elements are the elements of the left coset $\tau\langle \rho\rangle$.

We now consider the two possibilities for S, namely that S is three **involutions** (elements of order 2), or is an involution along with an element that is not self-inverse together with its inverse, as P has valency 3. Notice in the latter case the element that is not an involution is contained in $\langle\rho\rangle$. If S is three involutions, then no element of S can be contained in $\langle\rho\rangle$ (which is of order 5), and so every element of S is contained in the left coset $\tau\langle\rho\rangle$. Then each edge of P has one endpoint in $\langle\rho\rangle$ and the other in the left coset $\tau\langle\rho\rangle$. So P is bipartite, which it cannot be as P contains odd cycles. Then S must consist of two elements of order 5 that are inverses of each other, say ρ_1 and ρ_1^{-1}, and an involution, say τ_1, as it is a graph. Then $\rho_1 = \rho^b$ for some $1 \leq b \leq 4$ and $\tau_1 = \tau\rho^c$ for some c. Hence $\rho_1^5 = \tau_1^2 = 1$ and

$$\tau_1^{-1}\rho_1\tau_1 = (\tau\rho^c)^{-1}\rho^b\tau\rho^c = \rho^{-c}\tau^{-1}\rho^b\tau\rho^c = \rho^{-c}\rho^{-b}\rho^c = \rho^{-b} = (\rho_1)^{-1}.$$

By Lemma 1.2.21 there exists $\alpha \in \mathrm{Aut}(D_{10})$ such that $\alpha(\rho_1) = \rho$ and $\alpha(\tau_1) = \tau$. Then $P \cong \mathrm{Cay}\big(D_{10}, \{\rho, \rho^4, \tau\}\big)$ by Lemma 1.2.15. However, by Example 1.2.11, $\mathrm{Cay}\big(D_{10}, \{\rho, \rho^4, \tau\}\big)$ is isomorphic to a Cayley graph of $\mathbb{Z}_{10}$, a contradiction. □

1.3 Double Coset Digraphs

We extend the notion of Cayley digraphs to a more general construction that (theoretically at least) gives all vertex-transitive digraphs. Instead of thinking of the vertices of a digraph as being group elements and then having left multiplication permute the vertices, we think of the vertices as being left cosets of a subgroup of G, and then in the same way left multiplication by elements of G will permute the vertices.

Definition 1.3.1 Let G be a group and let $H \leq G$. Denote the set of left cosets of H in G by G/H. Then G acts on G/H by $g(aH) = (ga)H$, and we will call this action the **left coset action** of G on G/H. The image of the permutation representation of G induced by the left coset action of G on G/H will be denoted by $(G, G/H)$.

The left coset action of G on G/H is an action as $1 \cdot gH = gH$ and $g_1(g_2)gH = (g_1g_2)gH$ for all $g, g_1, g_2 \in G$. Also, $(G, G/H)$ is the permutation group contained in $S_{G/H}$ induced by the left coset action of G on G/H.

Our goal here is extend the notion of a Cayley digraph of G being a vertex-transitive digraph with $G_L \leq \mathrm{Aut}(\Gamma)$ to arbitrary transitive subgroups of S_n. That is, given a transitive group $G \leq S_n$, we seek digraphs Γ of order n such that $G \leq \mathrm{Aut}(\Gamma)$. We will do this with the left coset action of G on G/H where $H = \mathrm{Stab}_G(x)$ with $x \in \mathbb{Z}_n$. This may seem a strange or arbitrary choice but it is not as we shall see later in Theorem 2.4.13 that the image of every permutation representation of G can be obtained in this manner.

So, given an abstract group G and $H \leq G$, we would like to construct digraphs in which the left coset action of G on G/H is faithful and transitive (while the left coset action of G on G/H need not be faithful, a necessary and sufficient condition for this to occur is given in Proposition 1.3.8). In effect, what we would like is to give the faithful action of G on G/H, give the outneighbors of a left coset (H of course), and then use the action to "fill out" the rest of the arcs by letting the left coset action of G on G/H act canonically on the arcs with first coordinate H. This will ensure that the faithful action of G on G/H is contained in the automorphism group of the digraph. Our first naive guess might be the following:

Let G be a group, $H \leq G$ and S a subset of left coset representatives of H in G. We define a digraph $\Gamma(G, H, S)$ by the vertex set that is the set of left cosets of H in G and the arc set is $\{(gH, gsH) : g \in G \text{ and } s \in S\}$.

This guess though does not give a well-defined arc set. For the arc set to be well defined, the arc set must be the same regardless of which left coset representative is chosen for the left cosets gH and gsH.

Lemma 1.3.2 *Let G be a group, $H \leq G$, and S a subset of left coset representatives of H in G. The set $A(\Gamma) = \{(gH, gsH) : g \in G, s \in S\}$ is well defined if and only if $S = \{h_1 s h_2 : h_1, h_2 \in H, s \in S\}$.*

Proof The set $A(\Gamma)$ is well defined if and only if whenever $g \in G$ and $s \in S$

$$\begin{aligned}(gH, gsH) \in A(\Gamma) &\text{ if and only } (gh_1H, gh_1sH) \in A(\Gamma) \text{ for all } h_1 \in H\\ &\text{ if and only } (gh_1H, gh_1sh_2H) \in A(\Gamma) \text{ for all } h_1, h_2 \in H\\ &\text{ if and only } (gH, gh_1sh_2H) \in A(\Gamma) \text{ for all } h_1, h_2 \in H\\ &\text{ if and only } h_1sh_2 \in S \text{ for all } h_1, h_2 \in H.\end{aligned}$$

In the first step, we are simply replacing g with gh_1, and in the second step, we use $gh_1sH = gh_1sh_2H$ for all $h_2 \in H$. Similarly, in the third step, we use $gh_1H = gH$. □

Intuitively, one can also see the naive guess above will not work in the left coset action of G on G/H, as while an element $h \in H$ certainly stabilizes H as $hH = H$, an element of H need not stabilize the other left cosets of H in G, i.e., there is no general reason why if $s \in S$ then $hs \in S$. In contrast, in G_L any element that fixes any vertex must fix every vertex. So, if we just pick our connection set S in any fashion, it may be that there is $s \in S$ such that sH and

tH are in the same orbit of H acting on G/H but that $t \notin S$. The condition that $s \in S$ if and only if $h_1 s h_2 \in S$ also guarantees that we always choose orbits of H acting on the left cosets of H.

Definition 1.3.3 Let G be a group, $H \leq G$, and $s \in G$. The set $HsH = \{h_1 s h_2 : h_1, h_2 \in H\}$ is a **double coset of** H **in** G. Similarly, for $S \subseteq G$, the set $HSH = \{h_1 s h_2 : h_1, h_2 \in H, s \in S\}$ is a union of double cosets of H in G.

The set of double cosets of H in G is a partition of G (Exercise 1.3.13).

Definition 1.3.4 Let G be a group, $H \leq G$ and $S \subseteq G$. Define a digraph $\mathrm{Cos}(G, H, S)$ by $V(\mathrm{Cos}(G, H, S))$ is the set of left cosets of H in G and

$$A(\mathrm{Cos}(G, H, S)) = \{(gH, gaH) : g \in G, a \in HSH\}.$$

The digraph $\mathrm{Cos}(G, H, S)$ is the **double coset digraph of** G **with point stabilizer** H **and connection set** S.

Notice there was a small change in how we defined the arc set as opposed to what might have been expected. As with a Cayley digraph, we still allow the connection set S to be chosen however one wishes. But notice that we require arcs to be of the form (gH, gaH) with $g \in G$ and $a \in$ **HSH** (note the difference between this and $a \in S$, as would be the case in a Cayley digraph).

Many of the facts established for Cayley digraphs have analogues for double coset digraphs, and are listed below:

- for $g \in G$, the map $xH \mapsto gxH$ is an automorphism of $\mathrm{Cos}(G, H, S)$;
- $\mathrm{Cos}(G, H, S)$ is vertex-transitive as if aH, bH are left cosets, then $ba^{-1}aH = bH$. Also, left multiplication of the left cosets of H by ba^{-1} induces an automorphism of $\mathrm{Cos}(G, H, S)$;
- $\mathrm{Cos}(G, H, S)$ has no loops if $S \cap H = \emptyset$, or equivalently, $H \not\subseteq HSH$ (Exercise 1.3.4);
- $(xH, yH) \in A(\mathrm{Cos}(G, H, S))$ if and only if $yH \subseteq xHSH$, or equivalently, $x^{-1}yH \subseteq HSH$ (Exercise 1.3.5);
- $\mathrm{Cos}(G, H, S)$ is a graph if and only if $s^{-1} \in HSH$ for every $s \in S$ (Exercise 1.3.6);
- $N^+(xH) = \{yH : yH \subseteq xHSH\}$ and H has outvalence $|HSH|/|H|$ (Exercise 1.3.8);
- $\mathrm{Cos}(G, H, S) = \mathrm{Cos}(G, H, HSH)$ (Exercise 1.3.15);
- $\mathrm{Cos}(G, H, S)$ is connected if and only if $\langle H, S\rangle = G$ (Exercise 1.3.10);
- if $H = 1$, then $\mathrm{Cos}(G, H, S) \cong \mathrm{Cay}(G, S)$ (Exercise 1.3.11).

Analogues of Lemma 1.2.15 and Corollary 1.2.16 exist, but at the present time we do not have the permutation group-theoretic tools necessary to verify them. See Exercises 4.4.6 and 4.4.7.

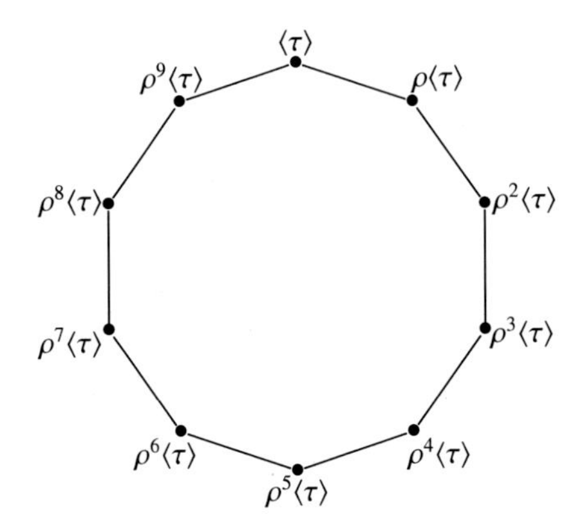

Figure 1.5 The double coset graph $\Gamma = \mathrm{Cos}\left(D_{20}, \langle\tau\rangle, \{\rho,\rho^{-1}\}\right)$.

We remark that some authors refer to these digraphs simply as **coset digraphs**. We call them double coset digraphs to distinguish them from Schrier coset digraphs (which we will not see). Additionally, there is not a single standard definition of a double coset digraph. Indeed, there are two other equivalent definitions, with the least commonly encountered being Sabidussi's original definition (see Exercises 1.3.15 and 1.3.16) when double coset digraphs were first defined (Sabidussi, 1964). Our choice of definition was determined by its being the closest to the definition of a Cayley digraph in the sense that one is allowed to "choose" the connection set from elements of the group.

Example 1.3.5 Let $G = D_{20} = \left\langle \rho, \tau : \rho^{10} = \tau^2 = 1, \tau^{-1}\rho\tau = \rho^{-1}\right\rangle$, $H = \langle\tau\rangle$, and $S = \left\{\rho,\rho^{-1}\right\}$. The graph (it is a graph as $S = S^{-1}$ – see Exercise 1.3.7) $\Gamma = \mathrm{Cos}(G, H, S)$ is drawn in Figure 1.5.

Solution We first calculate HSH, and observe that as every element of D_{20} can be written in the form $\rho^i\tau^j$ with $i \in \mathbb{Z}_{10}$ and $j \in \mathbb{Z}_2$ as $\langle\rho\rangle \trianglelefteq D_{20}$, so can every element of HSH. Note that as $\tau\rho = \rho^{-1}\tau$, $\tau\rho^{-1} = \rho\tau$, $\tau\rho\tau = \rho^{-1}$, and $\tau\rho^{-1}\tau = \rho$, we see $HSH = \langle\tau\rangle\left\{\rho,\rho^{-1}\right\}\langle\tau\rangle = \left\{\rho,\rho^{-1},\rho\tau,\rho^{-1}\tau\right\} = \rho\langle\tau\rangle \cup \rho^{-1}\langle\tau\rangle$. Writing the left cosets of $\langle\tau\rangle$ as $\rho^i\langle\tau\rangle$, $i \in \mathbb{Z}_{10}$, we see $\left(\rho^iH, \rho^jH\right) \in A(\Gamma)$ if and only if $\rho^{i-j} \in \left\{\rho,\rho^{-1}\right\}$. □

Evidently, we have already encountered some double coset digraphs. Our next example, while computationally more involved, shows that the Petersen graph can be constructed as a double coset graph. While the graph in the previous example is isomorphic to a Cayley graph, recall from Theorem 1.2.22 that the Petersen graph is not isomorphic to a Cayley graph of any group. In fact *all* vertex-transitive graphs can be constructed as double coset graphs, as we shall see in Theorem 1.3.9.

Example 1.3.6 Define $\rho, \tau\colon \mathbb{Z}_2 \times \mathbb{Z}_5 \to \mathbb{Z}_2 \times \mathbb{Z}_5$ by $\rho(i, j) = (i, j + 1)$ and $\tau(i, j) = (i + 1, 2j)$. Then $G = \langle \rho, \tau \rangle \le S_{10}$ has order 20, and is transitive with $H = \left\langle \tau^2 \right\rangle$ the stabilizer in G of the point $(0, 0)$. Let $S = \left\{\rho, \rho^4, \tau\right\}$. The double coset graph $\mathrm{Cos}(G, H, S)$ is isomorphic to the Petersen graph.

Solution We first show that $|G| = 20$, and that G is transitive with $\mathrm{Stab}_G(0, 0) = H$. Notice here that arithmetic in the first coordinate is performed modulo 2, while arithmetic in the second coordinate is performed modulo 5. Consequently, $\rho^2(i, j) = (i, j + 2)$, and, in general, $\rho^k(i, j) = (i, j + k)$. Thus $\langle \rho \rangle$ has order 5. Similarly, $\tau^k(i, j) = \left(i + k, 2^k j\right)$. Thus $\tau^k(i, j) = \left(i, 2^k j\right)$ if k is even, and as $2^4 \equiv 1 \pmod 5$ and $2^2 \not\equiv 1 \pmod 5$, we see $\langle \tau \rangle$ has order 4. As τ has order 4, $\tau^{-1} = \tau^3$ and so $\tau^{-1}(i, j) = (i + 1, 3j)$. Then

$$\begin{aligned}\tau^{-1}\rho\tau(i, j) &= \tau^{-1}\rho(i + 1, 2j) = \tau^{-1}(i + 1, 2j + 1)\\ &= (i + 2, 3(2j + 1)) = (i, j + 3) = \rho^3(i, j),\end{aligned}$$

and so $\langle \rho \rangle \trianglelefteq G$. As $\langle \rho \rangle \cap \langle \tau \rangle = 1$, we see $G/\langle \rho \rangle \cong \langle \tau \rangle$ and so $\langle \rho, \tau \rangle$ has order 20.

To see that $\langle \rho, \tau \rangle$ is transitive, let $i_1, i_2 \in \mathbb{Z}_2$ and $j_1, j_2 \in \mathbb{Z}_5$. If $i_1 = i_2$, then $\rho^{j_2 - j_1}(i_1, j_1) = (i_1, j_2)$. If $i_1 \ne i_2$, then let $k \in \mathbb{Z}_5$ such that $k + 2j_1 = j_2$. Then $\rho^k\tau(i_1, j_1) = \rho^k(i_2, 2j_1) = (i_2, j_2)$. Thus $\langle \rho, \tau \rangle$ is transitive. As $\tau^2(0, 0) = (0, 0)$, we have $\mathrm{Stab}_G(0, 0) = \left\langle \tau^2 \right\rangle = H$ has order 2 by the orbit-stabilizer theorem.

Next, we need to determine which left cosets of H are contained in HSH. As $\tau\rho\tau^{-1} = \rho^2$, we have $\tau^2\rho\tau^{-2} = \rho^4$ and so $\tau^2\rho = \rho^4\tau^2$. As $\tau^2\rho\tau^2 = \rho^4$ (note that $\tau^2 = \tau^{-2}$), we have

$$H\rho H = \left\{\rho, \tau^2\rho, \rho\tau^2, \tau^2\rho\tau^2\right\} = \left\{\rho, \rho^4\tau^2, \rho\tau^2, \rho^4\right\} = \rho^4 H \cup \rho H, \text{ and}$$

$$H\rho^4 H = \left\{\rho^4, \tau^2\rho^4, \rho^4\tau^2, \tau^2\rho^4\tau^2\right\} = \left\{\rho^4, \rho\tau^2, \rho^4\tau^2, \rho\right\} = \rho H \cup \rho^4 H,$$

and of course $H\tau H = \tau H$. Thus $HSH = \tau H \cup \rho H \cup \rho^4 H$. The graph is drawn in Figure 1.6. It is the Petersen graph. To make drawing the graph easier, notice that the neighbors of H are $\tau H, \rho H$, and $\rho^4 H$. Applying τ, we see τH is adjacent to H, $\tau\rho H = \rho^2\tau H$, and $\tau\rho^4 H = \rho^3\tau H$. We then multiply on the left repeatedly by ρ to obtain the rest of the edges. □

When constructing a double coset digraph $\mathrm{Cos}(G, H, S)$, one would typically wish that the left coset action of G on G/H is faithful and hence isomorphic

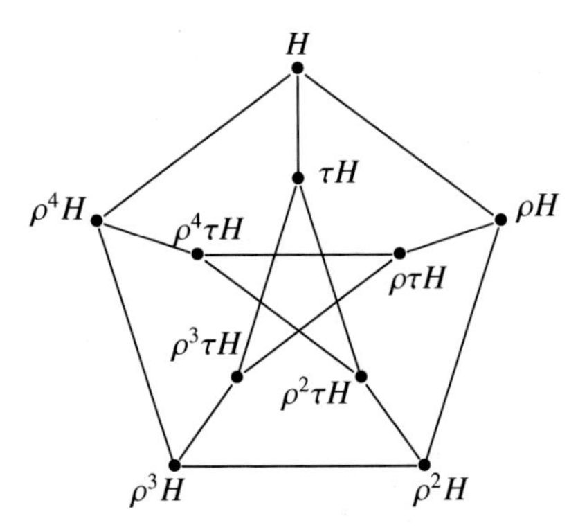

Figure 1.6 The Petersen graph as a double coset graph.

to G. This need not be the case, and the following result gives necessary and sufficient conditions for the left coset action of G on G/H to be faithful. Before proceeding, we need a preliminary definition.

Definition 1.3.7 The largest normal subgroup of G contained in H is called the **core** of H in G, and denoted $\mathrm{core}_G(H)$. That is, $\mathrm{core}_G(H) = \cap_{g\in G} g^{-1}Hg$. If $\mathrm{core}_G(H) = 1$, then H is said to be **core-free** in G.

Proposition 1.3.8 *Let G be a group and $H \leq G$. Then the left coset action of G on G/H is faithful if and only if H is core-free in G.*

Proof We show the contrapositive. That is, the left coset action of G on G/H is not faithful if and only if H contains a nontrivial normal subgroup of G.

Suppose the largest normal subgroup N of G contained in H is nontrivial, with $g \in N$. Let $x \in G$. As $g \in N$, there exists $h_x \in H$ such that $x^{-1}gx = h_x \in N \leq H$ and so $gx = xh_x$. Then $g(xH) = gxH = xh_xH = xH$ and g induces the identity permutation on G/H. Thus the left coset action of G on G/H is not faithful.

For the converse, consider the permutation representation $\phi\colon G \to S_{G/H}$. By definition, the left coset action of G on G/H is not faithful if and only if $\mathrm{Ker}(\phi) = N \neq 1$. Let $n \in N$. Then $ngH = gH$ for every $g \in G$, and in particular, $nH = H$. This occurs if and only if $n \in H$, and so $N = \mathrm{Ker}(\phi) \leq H$, and H is not core-free in G. □

Henceforth, when constructing a double coset digraph $\mathrm{Cos}(G, H, S)$, **we will assume H is core-free in G, or equivalently, that H contains no nontrivial normal subgroups of G.** This also gives that $(G, G/H) \cong G$. Note that Exercise 1.1.9 with this language is that the stabilizer of a point in a transitive subgroup G of $\mathcal{S}_n$ is core-free in G.

We have already seen that a double coset digraph is vertex-transitive. The following result of Sabidussi (1964) shows that the converse is true. Thus a digraph is vertex-transitive if and only if it is the double coset digraph of some group G, some subgroup H of G, and some subset $S \subseteq G$.

Theorem 1.3.9 *Let Γ be a vertex-transitive digraph, and $G \leq \mathrm{Aut}(\Gamma)$ be transitive. Let $u \in V(\Gamma)$ and set $H = \mathrm{Stab}_G(u)$. For $v \in V(\Gamma)$, let $g_v \in G$ such that $g_v(u) = v$. Set $S = \{g_v : (u, v) \in A(\Gamma)\}$. Then $\Gamma \cong \mathrm{Cos}(G, H, S)$.*

Proof Note that, by Exercise 1.1.9, $\mathrm{Stab}_G(v)$ is core-free in G for every $v \in V(G)$, and that g_v exists as G is transitive. First, observe that $gH = g_vH$ if and only if $g(u) = v$. If $g(u) = v$, then $g_v^{-1}g(u) = u$ so $g_v^{-1}g \in H$ and $gH = g_vH$. If $gH = g_vH$ then $g = g_vh$ for some $h \in H$, and, as $h(u) = u$, $g(u) = g_vh(u) = g_v(u) = v$. We now show that $\mathrm{Cos}(G, H, S)$ is well defined. That is, regardless of the choices of g_v, the resulting digraph is the same. By Lemma 1.3.2, this is equivalent to showing that for any $g \in G$ with $g(u) = v$, we have $HgH = Hg_vH$. This follows directly as $gH = g_vH$.

Define $\phi\colon V(\Gamma) \to G/H$ by $\phi(v) = g_vH$. We will show that ϕ is an isomorphism between Γ and $\mathrm{Cos}(G, H, S)$. Note that ϕ is well defined as $gH = g_vH$ if and only if $g(u) = v$. The same reason, together with G being transitive, gives that ϕ is surjective. If $\phi(x) = \phi(y)$ then $g_xH = g_yH$ and $g_y^{-1}g_xH = H$ or, equivalently, $g_y^{-1}g_x \in H$. Then $g_y^{-1}g_x(u) = u$ so $g_y^{-1}(x) = u$ or, equivalently, $g_y(u) = x$. Then $x = y$ and ϕ is injective, and so ϕ is a bijection. Let $(x, y) \in A(\Gamma)$. We need only show that $(\phi(x), \phi(y)) = (g_xH, g_yH) \in A(\mathrm{Cos}(G, H, S))$. Indeed,

$$\begin{aligned}
(x, y) \in A(\Gamma) &\text{ if and only if } \left(g_x^{-1}(x), g_x^{-1}(y)\right) \in A(\Gamma)\\
&\text{ if and only if } \left(u, g_x^{-1}(y)\right) = \left(u, g_x^{-1}g_y(u)\right) \in A(\Gamma)\\
&\text{ if and only if } g_x^{-1}g_yh \in S \text{ for some } h \in H\\
&\text{ if and only if } \left(H, g_x^{-1}g_yH\right) \in A(\mathrm{Cos}(G, H, S))\\
&\text{ if and only if } (g_xH, g_yH) \in A(\mathrm{Cos}(G, H, S)).
\end{aligned}$$

□

In addition to every Cayley digraph being isomorphic to a double coset digraph, there is another relationship between these two classes of digraphs. Sabidussi showed (Sabidussi, 1964, Theorem 4) that a certain "multiple" of a double coset digraph $\mathrm{Cos}(G, H, S)$ is always a Cayley digraph of G; see Corollary 4.3.6.

1.4 Orbital Digraphs

In the previous section we saw how, given a transitive group G (and so the stabilizer H of a point in G), to construct a digraph whose automorphism group contains G. In fact, as every vertex-transitive digraph is isomorphic to a double coset digraph, we can construct all vertex-transitive digraphs that contain G in their automorphism groups simply by constructing double coset digraphs for every possible union of double cosets of H in G. Our goal in this section is to examine another method for constructing every vertex-transitive digraph whose automorphism group contains $G \leq S_n$. So, we begin with a transitive group G. An obvious way to construct a digraph Γ in such a way that $G \leq \mathrm{Aut}(\Gamma)$, is to just pick an arc (or just an edge if one wishes to only construct graphs) and see what other arcs the action of G on that arc yields. That is, we pick an ordered pair $(x, y) \in \mathbb{Z}_n \times \mathbb{Z}_n$, and define a digraph Γ by $V(\Gamma) = \mathbb{Z}_n$ and $A(\Gamma) = \{g(x, y) = (g(x), g(y)) : g \in G\}$. The digraph Γ is an **orbital digraph of** G. You will notice though, that the first part of the title of this section is "Orbital Digraphs," with "Digraphs" plural. Typically, one does not construct *an* orbital digraph of a group $G \leq S_n$, but constructs *all* orbital digraphs of a group $G \leq S_n$, and so intuitively we apply the preceding procedure to every possible arc.

Definition 1.4.1 Let $G \leq S_n$ act on $\mathbb{Z}_n \times \mathbb{Z}_n$ in the canonical way, that is $g(x, y) = (g(x), g(y))$. The orbits of this action are called **orbitals**. One orbital is the diagonal, or $\{(x, x) : x \in \mathbb{Z}_n\}$, and is called the **trivial orbital**. We assume here that $O_1, \ldots, O_r$ are the nontrivial orbitals. Define digraphs $\Gamma_1, \ldots, \Gamma_r$ by $V(\Gamma_i) = \mathbb{Z}_n$ and $A(\Gamma_i) = O_i$. The Γ_i are the **orbital digraphs of** G.

We observe that $r \geq 1$, and that r can be arbitrarily large. For example, if G is regular then $r = |G| - 1$. The number r does not have a name as there is a very related term – the *rank* of a group defined in Definition 1.4.8.

Example 1.4.2 Find the orbital digraphs of the transitive group $G = \{x \mapsto ax + b \pmod{13} : a = 1, 3, 9, b \in \mathbb{Z}_{13}\} \leq \mathrm{AGL}(1, 13)$ of degree 13. Here, as in Exercise 1.3.3, $\mathrm{AGL}(1, 13)$ is a subgroup of S_{13}, and so the "x" in the set brackets is an element of $\mathbb{Z}_{13}$. For example, if $a = 2$, and $b = 1$, the function $x \mapsto ax + b$ is the permutation $(0, 1, 3, 7, 2, 5, 11, 10, 8, 4, 9, 6)(12)$ in S_{13}. The group $\mathrm{AGL}(1, p) = \{x \mapsto ax + b : a \in \mathbb{Z}_p^*, b \in \mathbb{Z}_p\}$ is the **affine general linear group** of dimension 1 over the field of order p.

Solution First observe that G is a group as multiplicatively $\langle 3 \rangle = \{1, 3, 9\}$. As G contains a regular cyclic subgroup $\{x \mapsto x + b \pmod{13} : b \in \mathbb{Z}_{13}\}$ we see by Theorem 1.2.20 that each orbital digraph of G is a circulant digraph of order 13. We thus need only determine the connection sets of each of the orbital digraphs, or equivalently, the neighbors of 0. By the orbit-stabilizer theorem, $|\mathrm{Stab}_G(0)| = 3$ as G has order $3 \cdot 13$. Thus the stabilizer of 0 in G is $\{x \mapsto ax : a = 1, 3, 9\}$. Let α be a generator of $\mathrm{Stab}_G(0)$. The orbits of $\mathrm{Stab}_G(0)$, which we find by choosing an element of $\mathbb{Z}_{13}$ and repeatedly applying α until we obtain the element we began with, are then $\{0\}$, $\{1, 3, 9\}$, $\{2, 6, 5\}$, $\{4, 12, 10\}$, and $\{7, 8, 11\}$. The orbital digraphs of G are then $\mathrm{Cay}(\mathbb{Z}_{13}, \{1, 3, 9\})$, $\mathrm{Cay}(\mathbb{Z}_{13}, \{2, 5, 6\})$, $\mathrm{Cay}(\mathbb{Z}_{13}, \{4, 10, 12\})$ and $\mathrm{Cay}(\mathbb{Z}_{13}, \{7, 8, 11\})$. None of these orbital digraphs is a graph. □

Example 1.4.3 Find the orbital digraphs of the transitive group G generated by $(\mathbb{Z}_5 \times \mathbb{Z}_5)_L$ and $\alpha\colon \mathbb{Z}_5^2 \to \mathbb{Z}_5^2$ given by $\alpha(i, j) = (2i, 4j)$.

Solution Notice that $\alpha^2(i, j) = \alpha(2i, 4j) = (4i, j)$. Similarly, $\alpha^4 = 1$. Also, straightforward computations show that α normalizes $\left(\mathbb{Z}_5^2\right)_L$ (note that $\alpha^{-1}(i, j) = (3i, 4j)$). So $|G| = 4 \cdot 25$. As in the previous example, by Theorem 1.2.20 each orbital digraph of G is a Cayley digraph of $\mathbb{Z}_5^2$, and so we need only find the connection sets of the orbital digraphs, or equivalently, the neighbors of $(0, 0)$. The orbits of $\mathrm{Stab}_G(0, 0)$, which we find by choosing an element of $\mathbb{Z}_5^2$ and repeatedly applying α until we obtain the element we began with, are

$$\{(0, 0)\}, \qquad S_1 = \{(0, 1), (0, 4)\}, \qquad S_2 = \{(0, 2), (0, 3)\},$$

$$S_3 = \{(1, 0), (2, 0), (4, 0), (3, 0)\}, \qquad S_4 = \{(1, 1), (2, 4), (4, 1), (3, 4)\},$$

$$S_5 = \{(1, 2), (2, 3), (4, 2), (3, 3)\}, \qquad S_6 = \{(1, 3), (2, 2), (4, 3), (3, 2)\}, \text{and}$$

$$S_7 = \{(1, 4), (2, 1), (4, 4), (3, 1)\}.$$

So the orbital digraphs of G are the digraphs $\mathrm{Cay}\left(\mathbb{Z}_5^2, S_i\right)$, $1 \leq i \leq 7$. □

Notice that we do not define an orbital digraph corresponding to the trivial orbital. We do this to make our terminology a bit cleaner, but if one wanted to include such a digraph, one could. It would simply be a digraph with vertex set $\mathbb{Z}_n$ with the only arcs being loops at each vertex. Also, we do indeed have that $G \leq \mathrm{Aut}(\Gamma)$ for every orbital digraph Γ of G, as if $g \in G$ and $(x, y) \in A(\Gamma)$, then $g(x, y) \in A(\Gamma)$.

Observe that an orbital digraph is by definition arc-transitive.

Definition 1.4.4 For a set $\Gamma_1, \ldots, \Gamma_r$ of digraphs with the same vertex set, we define their **union**, denoted $\cup_{i=1}^r \Gamma_i$, to be the digraph with vertex set $V(\Gamma_i)$ for any $1 \leq i \leq r$ and arc set $\cup_{i=1}^r A(\Gamma_i)$. A digraph Γ that is a union of orbital digraphs of a group G is called a **generalized orbital digraph of** G.

Example 1.4.5 Let $\Gamma_1 = \text{Cay}(\mathbb{Z}_6, \{3\})$, and $\Gamma_2 = \text{Cay}(\mathbb{Z}_6, \{1,5\})$. Then $\Gamma_1 \cup \Gamma_2 = \text{Cay}(\mathbb{Z}_6, \{1,3,5\})$ as shown in Figure 1.7.

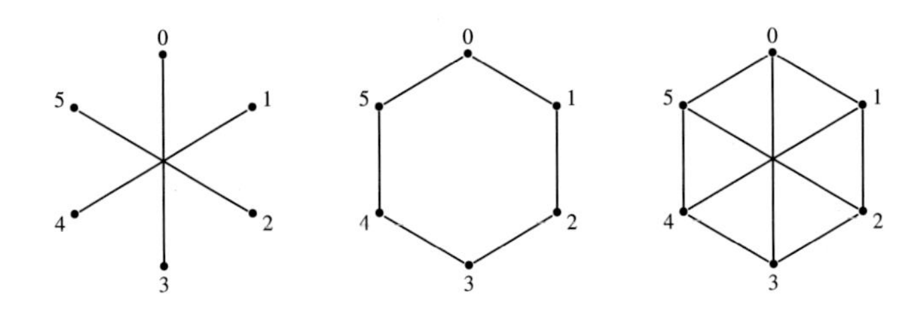

Figure 1.7 (Left) $\text{Cay}(\mathbb{Z}_6, \{3\})$, (Middle) $\text{Cay}(\mathbb{Z}_6, \{1,5\})$, and (Right) their union $\text{Cay}(\mathbb{Z}_6, \{1,3,5\})$.

Also, a digraph Γ with $G \leq \text{Aut}(\Gamma)$ is a union of orbital digraphs of G as the set of orbitals are a partition of the arcs of the complete graph (Exercise 1.4.11). We state this formally to emphasize its importance.

Proposition 1.4.6 *Let $G \leq S_n$ be transitive. Every digraph Γ of order n with $G \leq \text{Aut}(\Gamma)$ is a generalized orbital digraph of G.*

Additionally, this result gives a theoretical method to construct every vertex-transitive digraph Γ whose automorphism group contains G. We simply take all unions of all subsets of the set of orbital digraphs of G. There is a relationship between the arcs in an orbital digraph of G and the orbits of the stabilizer of a vertex.

Lemma 1.4.7 *Let Γ be an orbital digraph of G. For $x \in V(\Gamma)$, $N^+_\Gamma(x)$ is an orbit of* $\text{Stab}_G(x)$*, as is $N^-_\Gamma(x)$.*

Proof As Γ is arc-transitive, for $y_1, y_2 \in N^+_\Gamma(x)$, there exists $g \in G$ (that depends upon y_1 and y_2) such that $g(x, y_1) = (x, y_2)$. Thus g stabilizes x and maps y_1 to y_2. We conclude $N^+_\Gamma(x)$ is an orbit of $\text{Stab}_G(x)$. The proof of the other statement is analogous. □

Definition 1.4.8 An orbit of $\text{Stab}_G(x)$ is called a **suborbit** of G with respect to x. The number of suborbits of G with respect to x is the **rank** of G. The **length** of a suborbit is its order.

As stabilizers are conjugate by Theorem 1.1.7, the length of the suborbits of G with respect to x are independent of the choice of x, and often one will simply see the "suborbits of G." Notice that there is one more (trivial) suborbit of G with respect to x than there are orbital digraphs.

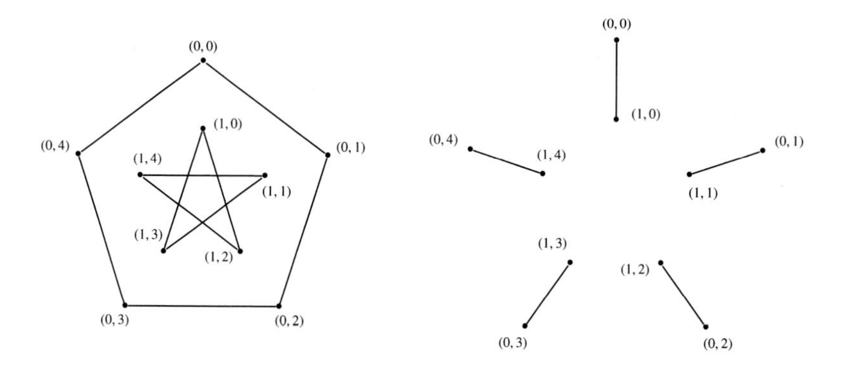

Figure 1.8 The orbital digraphs of $\langle \rho, \tau \rangle$ whose union is the Petersen graph.

In terms of suborbits of G, by Lemma 1.4.7 the neighbors of a vertex in a generalized orbital digraph Γ of G will be a union of suborbits of G. Formally:

Proposition 1.4.9 *Let Γ be a vertex-transitive digraph with $G \leq \text{Aut}(\Gamma)$ transitive. The out- or in-neighbors of a vertex x in Γ are a union of suborbits of G with respect to x.*

While every digraph is a generalized orbital digraph of its automorphism group, it is certainly possible that a digraph is a generalized orbital digraph of a group much smaller than its automorphism group as the next example shows. We show in Example 1.4.15 that the automorphism group of the Petersen graph contains S_5 in its action on 2-subsets of $\mathbb{Z}_5$ (and show in Theorem 2.1.4 that this is the full automorphism group of the Petersen graph).

Example 1.4.10 Define $\rho, \tau \colon \mathbb{Z}_2 \times \mathbb{Z}_5 \to \mathbb{Z}_2 \times \mathbb{Z}_5$ by $\rho(i, j) = (i, j+1)$ and $\tau(i, j) = (i+1, 2j)$. Find the orbital digraphs of the group $H = \langle \rho, \tau \rangle \leq S_{10}$.

Solution We determine the suborbits of H with respect to $(0,0)$. By Lemma 1.4.7 each suborbit O gives the out-neighbors of $(0,0)$ in an orbital digraph of H. The orbital digraphs are then obtained by applying H to the arcs $((0,0), v)$ where $v \in O$. Recall from Example 1.3.6 that $|H| = 20$, $\text{Stab}_H(0,0) = \langle \tau^2 \rangle$, and $\langle \rho \rangle \trianglelefteq H$. Picking elements in $\mathbb{Z}_2 \times \mathbb{Z}_5$ and applying τ^2 to find the orbits that contained the picked elements, we see the suborbits of H with respect to $(0,0)$ are $\{(0,0)\}$, $\{(0,1),(0,4)\}$, $\{(0,2),(0,3)\}$, $\{(1,0)\}$, $\{(1,1),(1,4)\}$, and $\{(1,2),(1,3)\}$. There are then five orbital digraphs of H, one corresponding to each suborbit of H with respect to $(0,0)$ that is not $\{(0,0)\}$. The suborbits of H whose union give the neighbors of $(0,0)$ in the Petersen graph are $\{(0,1),(0,4)\}$ and $\{(1,0)\}$. The orbital digraphs whose union is P corresponding to these suborbits are shown in Figure 1.8. The suborbit $\{(0,2),(0,3)\}$ gives an orbital digraph that is a union of two 5-cycles, while the other suborbits give orbital digraphs that are 10-cycles. □

For our final example, we need some common terminology regarding orbital digraphs, as well as the definition of a transitive constituent.

Definition 1.4.11 Let $G \leq \mathcal{S}_n$ be transitive, and let $(u, v) \in \mathbb{Z}_n \times \mathbb{Z}_n$. The orbitals of G with edges sets $\{g(u, v) : g \in G\}$ and $\{g(v, u) : g \in G\}$ are **paired orbitals**. If these two arc sets are equal, we say the orbital is **self-paired**, and the corresponding orbital digraphs is a graph. An **orbital graph** is either an orbital digraph that is a graph, or the generalized orbital graph that is the union of two paired orbitals.

Definition 1.4.12 Let $G \leq \mathcal{S}_n$ with orbit O, and $g \in G$. Then g induces a permutation on O by restricting the domain of g to O. We denote the resulting permutation in $\mathcal{S}_O$ by g^O. The group $G^O = \{g^O : g \in G\}$ is the **transitive constituent** of G on O.

Lemma 1.4.13 *Let $G \leq \mathcal{S}_n$ and O be an orbit of G. Then G^O is transitive.*

Proof Let $x, y \in O$. As O is an orbit of G, there exists $g \in G$ such that $g(x) = y$. Then $g^O(x) = y$ so G^O is transitive. □

While we have defined orbital digraphs of permutation groups, it is often more convenient to specify an action instead of a permutation group. In this case, we define the **orbital digraph of an action of a group on a set** to be the orbital digraphs of the image of its corresponding permutation representation. Of course, all of the terminology in this section about orbital digraphs of permutation groups will also be used for orbital digraphs of an action, and it will usually be the case that the action is the left coset action of a group G on G/H for some subgroup $H \leq G$.

Definition 1.4.14 For a positive integer n, we define $[1, n] = \{1, 2, \ldots, n\}$ the first n positive integers.

While defining $\mathcal{S}_n$ to permute the set $\mathbb{Z}_n$ is often convenient (for example, when working with circulant digraphs), there are other situations in which no advantage is gained with this convention. For our next example, it is traditional to have $\mathcal{S}_n$ permuting the set $[1, n]$, and we will follow that tradition.

Example 1.4.15 Let $\mathcal{S}_5$ act canonically on the 2-subsets of $[1, 5]$. A **2-subset** is a subset of size 2, and the canonical action is, for a 2-subset $\{x, y\}$ and $g \in \mathcal{S}_5$, $g(\{x, y\}) = \{g(x), g(y)\}$. Show that there are three suborbits with respect to $\{1, 2\}$, all of which are self-paired, and that the nontrivial orbital digraphs are the Petersen graph and its complement.

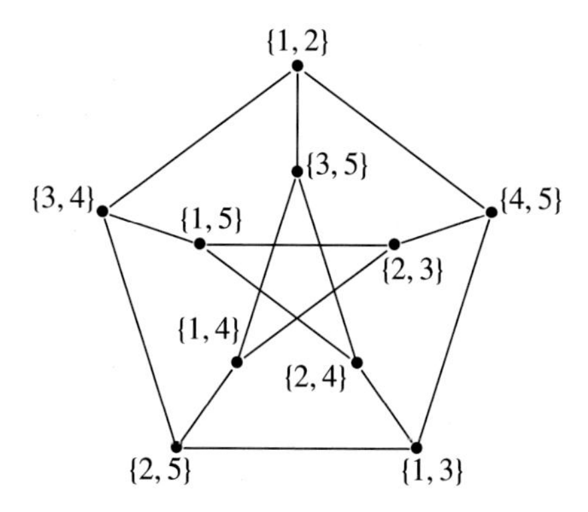

Figure 1.9 The Petersen graph as an orbital digraph of $\mathcal{S}_5$.

Solution Notice that $\mathrm{Stab}_{\mathcal{S}_5}(\{1,2\}) \cong \mathcal{S}_2 \times \mathcal{S}_3$, where $\mathrm{Stab}_{\mathcal{S}_5}(\{1,2\})^{\{1,2\}} = \langle(1,2)\rangle$ and $\mathrm{Stab}_{\mathcal{S}_5}(\{3,4,5\})^{\{3,4,5\}}$ is isomorphic to $\mathcal{S}_3$ permuting $\{3,4,5\}$. There are $\binom{5}{2} = 10$ 2-subsets of $[1,5]$, and three suborbits. The suborbits are $\{\{1,2\}\}$, $\{\{3,4\},\{3,5\},\{4,5\}\}$ and $\{\{2,3\},\{2,4\},\{2,5\},\{1,3\},\{1,4\},\{1,5\}\}$. That the latter two suborbits are self-paired follows from Exercise 1.4.10. The graph of the Petersen graph with this labeling is given in Figure 1.9 and is obtained by applying $\mathcal{S}_5$ to the arcs $(\{1,2\}, v)$ with $v \in \{\{3,4\},\{3,5\},\{4,5\}\}$. □

We will see many more examples of orbital digraphs in Section 2.6.

1.5 Metacirculant Digraphs

Metacirculant graphs were introduced by Alspach and Parsons (1982a), and are probably the most common vertex-transitive digraphs encountered other than Cayley digraphs. As with Cayley digraphs, the easiest way to think about metacirculant digraphs is to think of the neighbors of a fixed vertex, and then to generate the rest of the arcs of the digraph using a natural transitive subgroup of the automorphism group. In the language of orbital digraphs, we simply insist that a metacirculant digraph is a generalized orbital digraph of a particular group that need not be regular. In order to emphasize this, we will follow Alspach and Parsons' approach and introduce the natural transitive subgroup of the automorphism group first, and then derive the appropriate conditions to describe the rest of the arcs of the digraph.

First, the vertex set of a metacirculant digraph will be $\mathbb{Z}_m \times \mathbb{Z}_n$, where m and n are positive integers. Let $\alpha \in \mathbb{Z}_n^*$, and define $\rho, \tau \colon \mathbb{Z}_m \times \mathbb{Z}_n \to \mathbb{Z}_m \times \mathbb{Z}_n$

by $\rho(i,j) = (i, j+1)$ and $\tau(i,j) = (i+1, \alpha j)$. Notice that if $\alpha = 1$, then $\langle \rho, \tau \rangle = (\mathbb{Z}_m \times \mathbb{Z}_n)_L$. The following lemma contains basic properties of the group $\langle \rho, \tau \rangle$.

Lemma 1.5.1 *The group $\langle \rho, \tau \rangle \leq S_{\mathbb{Z}_m \times \mathbb{Z}_n}$ is transitive of degree mn, $\tau\rho\tau^{-1} = \rho^\alpha$, $\langle \rho \rangle \trianglelefteq \langle \rho, \tau \rangle$, $|\tau| = \mathrm{lcm}(a,m)$, $\langle \rho \rangle \cap \langle \tau \rangle = 1$, and $|\langle \rho, \tau \rangle| = n \cdot \mathrm{lcm}(a,m)$, where a is the multiplicative order of α in $\mathbb{Z}_n^*$. (We remind the reader that for positive integers m and n,* $\mathrm{lcm}(m,n)$ *is the* **least common multiple** *of m and n.)*

Proof Think of the sets $V_i = \{(i,j) : j \in \mathbb{Z}_n\}$ as being "blocks," where $i \in \mathbb{Z}_m$. The permutation ρ fixes each V_i set-wise, and permutes the elements of V_i as the natural n-cycle that just adds 1 (in the second coordinate), modulo n. The permutation τ can then be thought of as the permutation that cyclically permutes the V_i, mapping V_i to V_{i+1} (with arithmetic in the subscript performed modulo m). However, as τ maps V_i to V_{i+1}, it does so with a "twist," the specific twist being given by α. We conclude that $\langle \rho, \tau \rangle$ is transitive. Note that $\tau^{-1}(i,j) = \left(i-1, \alpha^{-1} j\right)$ (check this!), and

$$\begin{aligned}\tau\rho\tau^{-1}(i,j) &= \tau\rho\left(i-1, \alpha^{-1}j\right) = \tau\left(i-1, \alpha^{-1}j+1\right)\\ &= \left(i-1+1, \alpha\left(\alpha^{-1}j+1\right)\right) = (i, j+\alpha).\end{aligned}$$

Thus $\tau\rho\tau^{-1} = \rho^\alpha$ and $\langle \rho \rangle \trianglelefteq \langle \rho, \tau \rangle$. We now compute the order of $\langle \rho, \tau \rangle$.

Let $b = \mathrm{lcm}(a,m)$. Note that $\tau^b(i,j) = \left(i, \alpha^b j\right) = (i,j)$ as $a|b$. Hence the order of τ divides b. Conversely, if $\tau^c = 1$, then $\tau^c(0,1) = (c, \alpha^c) = (0,1)$. From the first coordinate, we have $c \equiv 0 \pmod m$, or equivalently c is a multiple of m, and the second coordinate gives us that c is a multiple of a. We conclude c is a multiple of b, and hence $|\tau| = b$. It is easy to see $\langle \rho \rangle \cap \langle \tau \rangle = 1$. Then $|\langle \rho, \tau \rangle| = bn$ by the first isomorphism theorem as $\langle \rho \rangle \trianglelefteq \langle \rho, \tau \rangle$ and $\langle \rho, \tau \rangle / \langle \rho \rangle \cong \langle \tau \rangle$. Note that we now have $|\mathrm{Stab}_{\langle \rho, \tau \rangle}(0,0)| = b/m$ by Theorem 1.1.8. □

The group $\langle \rho, \tau \rangle$ is isomorphic to the **semidirect product** of two cyclic groups (see Example 1.5.6), and we now define the semidirect product.

Definition 1.5.2 Let G be a group. We say G is the **internal semidirect product of** H **and** N, denoted $N \rtimes H$, if there exists $N \trianglelefteq G$ and $H \leq G$ such that $H \cap N = 1$, and $HN = G$.

There is also an external semidirect product defined in Exercise 1.5.11, and the internal and external semidirect products of H and N are isomorphic (Exercise 1.5.12), so we abuse terminology and will typically not mention whether the semidirect product is internal or external. Usually in this text though, the semidirect products we encounter will be internal direct products. For an excellent derivation of the formula given in Exercise 1.5.11 (see Alperin and Bell (1995, Section 1.2).

Before our next example, we remind the reader of the definition of the direct product of two groups, which is a special case of the semidirect product (hence the name of the semidirect product as the prefix "semi" means "half" or "somewhat").

Definition 1.5.3 Let G and H be groups. We define the **direct** or **Cartesian product** of G and H, denoted $G \times H$, to be the group on the set $G \times H$ with binary operation $(g_1, h_1) \cdot (g_2, h_2) = (g_1 g_2, h_1 h_2)$ for every $g_1, g_2 \in G$ and $h_1, h_2 \in H$.

Example 1.5.4 If $H \trianglelefteq G$, $N \trianglelefteq G$, $H \cap N = 1$, and $G = HN$ then the group $N \rtimes H$ is isomorphic to $H \times N$.

Solution The hypothesis implies G is an internal direct product i.e., $HN \cong H \times N$ in the canonical way, of H and N (see Dummit and Foote (2004, Theorem 5.4.9), and the hypothesis is a strengthening of the definition of the internal semidirect product of H and N. □

Example 1.5.5 The dihedral group D_{2n} is isomorphic to $\mathbb{Z}_n \rtimes \mathbb{Z}_2$.

Solution Recall $D_{2n} = \langle \rho, \tau : \rho^n = \tau^2 = 1, \tau^{-1}\rho\tau = \rho^{-1} \rangle$. Then $\langle \rho \rangle \trianglelefteq D_{2n}$, and $\langle \rho \rangle \cdot \langle \tau \rangle = \left\{ \rho^i \tau^j : i \in \mathbb{Z}_n, j \in \mathbb{Z}_2 \right\} = D_{2n}$. Clearly $\langle \rho \rangle \cap \langle \tau \rangle = 1$ so $D_{2n} = \langle \rho \rangle \rtimes \langle \tau \rangle$. As $\langle \tau \rangle \cong \mathbb{Z}_2$ and $\langle \rho \rangle \cong \mathbb{Z}_n$, $D_{2n} \cong \mathbb{Z}_n \rtimes \mathbb{Z}_2$. □

Example 1.5.6 Let m and n be integers with $\alpha \in \mathbb{Z}_n^*$. Define $\rho, \tau \colon \mathbb{Z}_m \times \mathbb{Z}_n \to \mathbb{Z}_m \times \mathbb{Z}_n$ by $\rho(i, j) = (i, j + 1)$ and $\tau(i, j) = (i + 1, \alpha j)$. Let $b = \mathrm{lcm}(a, m)$ (here $a = |\alpha|$, the multiplicative order of α in $\mathbb{Z}_n^*$). Then $\langle \rho, \tau \rangle \cong \mathbb{Z}_n \rtimes \mathbb{Z}_b$.

Solution By Lemma 1.5.1 we have $\langle \rho \rangle \trianglelefteq \langle \rho, \tau \rangle$, $|\tau| = b$, $|\langle \rho, \tau \rangle| = bn$, and $\langle \rho \rangle \cap \langle \tau \rangle = 1$. Also, as $\tau\rho\tau^{-1} = \rho^\alpha$ we additionally have $\langle \tau \rangle \cdot \langle \rho \rangle = \langle \rho, \tau \rangle$, so $\langle \rho, \tau \rangle = \langle \rho \rangle \rtimes \langle \tau \rangle$. Finally, as $\langle \tau \rangle \cong \mathbb{Z}_b$ and $\langle \rho \rangle \cong \mathbb{Z}_n$, $\langle \rho, \tau \rangle$ is isomorphic to $\mathbb{Z}_n \rtimes \mathbb{Z}_b$. □

Returning to digraphs whose automorphism group contains $\langle \rho, \tau \rangle$, as $\langle \rho, \tau \rangle$ is transitive we need only give the outneighbors of one vertex to determine the digraph, and $(0, 0)$ is the natural vertex to select. We let $S_i \subseteq V_i$ for each $0 \le i \le m - 1$ and think of S_i as giving the vertices in V_i that are outneighbors of $(0, 0)$. We now find conditions on the arc set of a digraph Γ with vertex set $V = \mathbb{Z}_m \times \mathbb{Z}_n$ that will ensure that $\langle \rho, \tau \rangle \le \mathrm{Aut}(\Gamma)$.

Lemma 1.5.7 *Let Γ be a digraph with $V(\Gamma) = \mathbb{Z}_m \times \mathbb{Z}_n$, where $m, n \ge 1$ are integers. Define ρ, τ, and α as above, and let $S_i = \{h : ((0, 0), (i, h)) \in A(\Gamma)\}$, $i \in \mathbb{Z}_m$. Then $\langle \rho, \tau \rangle \le \mathrm{Aut}(\Gamma)$ if and only if*

(i) *$\alpha^m S_i = S_i$ for every $i \in \mathbb{Z}_m$, and*
(ii) *$((\ell, j), (\ell + i, k))$ is an arc of Γ if and only if $k - j \in \alpha^\ell S_i$ for every $i \in \mathbb{Z}_m$.*

Proof We are more or less following Alspach and Parsons, and while this is the clearest statement of the lemma, it is not the shortest. Namely, we first show that (i) holds if and only if (ii) holds for $i \in \mathbb{Z}$ (with arithmetic in the subscript of S_i done modulo m). Indeed, suppose that (ii) holds for all $i \in \mathbb{Z}$. Then

$$\begin{aligned} k - j \in \alpha^\ell S_i &\text{ if and only if } ((\ell, j), (\ell + i, k)) \text{ is an arc of } \Gamma \\ &\text{ if and only if } ((\ell + m, j), (\ell + m + i, k)) \text{ is an arc of } \Gamma \\ &\text{ if and only if } k - j \in \alpha^{\ell+m} S_i. \end{aligned}$$

Note that the second "if and only if" above is true as arithmetic in the first coordinate is done modulo m. So

$$\begin{aligned} k - j \in \alpha^\ell S_i &\text{ if and only if } \alpha^{\ell+m} S_i = \alpha^\ell S_i \\ &\text{ if and only if } \alpha^m S_i = S_i, \end{aligned}$$

and condition (i) holds. Reversing the direction of the if and only if statements above, we also see that (i) implies (ii), and so (i) holds if and only if (ii) holds. We thus need only show that $\langle \rho, \tau \rangle \leq \mathrm{Aut}(\Gamma)$ if and only if (ii) holds for all $i \in \mathbb{Z}$.

Suppose $\langle \rho, \tau \rangle \leq \mathrm{Aut}(\Gamma)$. Consider the arc $((\ell, j), (\ell + i, k)) \in A(\Gamma)$, where $i \in \mathbb{Z}$. Note that

$$\rho^{-\alpha^{-\ell} j} \tau^{-\ell}(\ell, j) = \rho^{-\alpha^{-\ell} j}\left(0, \alpha^{-\ell} j\right) = (0, 0),$$

while

$$\rho^{-\alpha^{-\ell} j} \tau^{-\ell}(\ell + i, k) = \rho^{-\alpha^{-\ell} j}\left(i, \alpha^{-\ell} k\right) = \left(i, \alpha^{-\ell}(k - j)\right).$$

By the definition of S_i, we have $((\ell, j), (\ell + i, k)) \in A(\Gamma)$ if and only if $\alpha^{-\ell}(k - j) \in S_i$, or equivalently, $k - j \in \alpha^\ell S_i$.

Conversely, $k - j \in \alpha^\ell S_i$ if and only if $(k + 1) - (j + 1) \in \alpha^\ell S_i$ and so $((\ell, j), (\ell + i, k)) \in A(\Gamma)$ if and only if $((\ell, j + 1), (\ell + i, k + 1)) \in A(\Gamma)$. As $\rho((\ell, j), (\ell + i, k)) = ((\ell, j + 1), (\ell + i, k + 1))$, we have $\rho \in \mathrm{Aut}(\Gamma)$. Similarly, $k - j \in \alpha^\ell S_i$ if and only if $\alpha k - \alpha j \in \alpha^{\ell+1} S_i$, and consequently $((\ell, j), (\ell + i, k)) \in A(\Gamma)$ if and only if $((\ell + 1, \alpha j), (\ell + i + 1, \alpha k)) \in A(\Gamma)$. As $\tau((\ell, j), (\ell + i, k)) = ((\ell + 1, \alpha j), (\ell + i + 1, \alpha k))$, we have $\tau \in \mathrm{Aut}(\Gamma)$. □

We now have the following definition.

Definition 1.5.8 Let $V = \mathbb{Z}_m \times \mathbb{Z}_n$, $\alpha \in \mathbb{Z}_n^*$, and $S_0, \ldots, S_{m-1} \subseteq \mathbb{Z}_n$ such that $\alpha^m S_i = S_i$, $0 \leq i \leq m - 1$. Define the $(m, n, \alpha, S_0, \ldots, S_{m-1})$**-metacirculant digraph** Γ by $V(\Gamma) = \mathbb{Z}_m \times \mathbb{Z}_n$ and $A(\Gamma) = \{((\ell, j), (\ell + i, k)) : k - j \in \alpha^\ell S_i\}$.

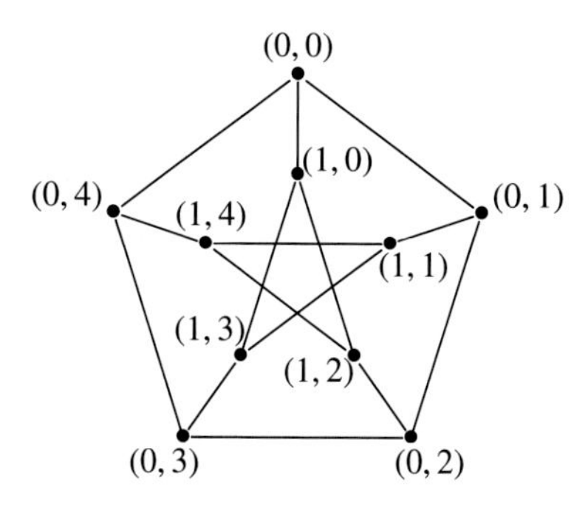

Figure 1.10 The Petersen graph with its metacirculant labeling.

The digraph Γ is usually referred to as an (m, n)-**metacirculant digraph** i.e., it is quite common to delete the reference to α and $S_0, \ldots, S_{m-1}$ if these are not necessary, and it is also common to just refer to a **metacirculant digraph** if the order is also unimportant.

Example 1.5.9 The $(2, 5, 2, \{1, 4\}, \{0\})$-metacirculant graph is drawn in Figure 1.10 and is the Petersen graph. Note that $\alpha^m = 2^2 = 4$, and $4 \cdot \{1, 4\} = \{1, 4\}$ while $4 \cdot \{0\} = \{0\}$.

Example 1.5.10 The $(2, 7, 6, \emptyset, \{1, 2, 4\})$-metacirculant graph is drawn in Figure 1.11. It is known as the **Heawood graph**. Note that $\alpha^m = 6^2 = 1$ (in $\mathbb{Z}_7$), and of course $1 \cdot \emptyset = \emptyset$ and $1 \cdot \{1, 2, 4\} = \{1, 2, 4\}$. As we shall see later, in Theorem 4.5.7, the Heawood graph is a symmetric cubic graph and is F14 in the Foster census.

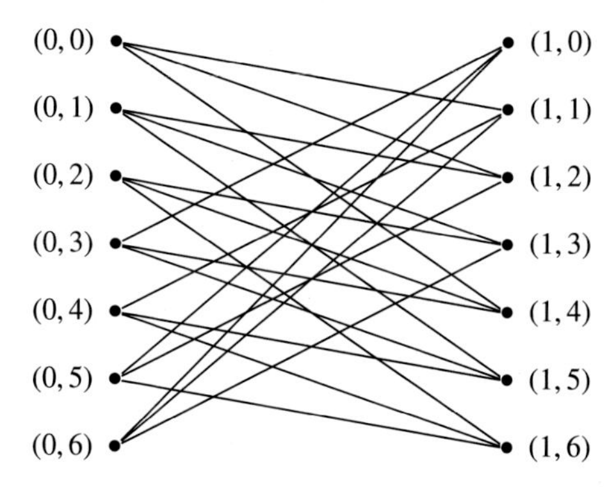

Figure 1.11 The metacirculant labeling of the Heawood graph.

We now restate Lemma 1.5.7 in the language of metacirculant digraphs and obtain the following analogue of Theorem 1.2.20, which we call a theorem due to its importance.

Theorem 1.5.11 *A digraph Γ is an $(m, n, \alpha, S_0, \ldots, S_{m-1})$-metacirculant digraph if and only if $\langle\rho, \tau\rangle \leq \mathrm{Aut}(\Gamma)$.*

It is worthwhile to point out a common trick that can, when working with metacirculant digraphs, simplify some proofs, as we shall see in Theorem 1.5.13. We package this trick as a lemma.

Lemma 1.5.12 *Let Γ be an $(m, n, \alpha, S_0, \ldots, S_{m-1})$-metacirculant digraph, let a be the multiplicative order of $\alpha \in \mathbb{Z}_n^*$, and write $a = xy$, where $x|a$ is maximum such that $\gcd(x, m) = 1$. Then there exists a positive integer z such that $\Gamma = \Gamma(m, n, \alpha^{xz}, S_0, \ldots, S_{m-1})$ and every prime divisor of the multiplicative order of α^{xz} in $\mathbb{Z}_n^*$ is also a prime divisor of m.*

Proof Note that $\tau^x(i, j) = (i + x, \alpha^x j)$, and as $\gcd(x, m) = 1$, there exists a positive integer z such that $xz \equiv 1 \pmod{m}$. So $\gcd(z, m) = 1$. Hence $(\tau^x)^z(i, j) = \tau^{xz}(i, j) = (i + 1, \alpha^{xz} j)$. It now follows by Theorem 1.5.11 that Γ is a $(m, n, \alpha^{xz}, S_0, \ldots, S_{m-1})$-metacirculant digraph. By choice of x, every prime divisor of $|\alpha^x|$ is a divisor of m. If p is a prime divisor of the multiplicative order of α^{xz}, then p is a prime divisor of the multiplicative order of α^x, and so p is a prime divisor of m. □

It is clear (for example by choosing $\alpha = 1$) that some metacirculant digraphs are also Cayley digraphs. An obvious question is whether the class of metacirculant digraphs is contained within the class of Cayley digraphs. We saw in Example 1.5.9 that the Petersen graph is metacirculant, while the Petersen graph is not isomorphic to a Cayley graph by Theorem 1.2.22. We shall also eventually see that the Petersen graph and its complement are the smallest vertex-transitive non-Cayley graphs, where by smallest we mean with fewest vertices. So we shall show that every vertex-transitive digraph of order at most 9 is a Cayley digraph, while the Petersen graph and its complement are the only two non-Cayley graphs of order 10. That all graphs of order 6 are Cayley is Exercise 1.5.4, while Theorem 4.4.1, and Theorems 9.1.1 and 9.1.3, respectively, show all vertex-transitive digraphs of p, p^2, and p^3 are Cayley digraphs. In general, it is very difficult to tell if a particular metacirculant digraph is isomorphic to a Cayley digraph, and so we content ourselves with the following result that gives a necessary condition for a metacirculant digraph to be isomorphic to a Cayley digraph.

Theorem 1.5.13 *Let Γ be an $(m, n, \alpha, S_0, \ldots, S_{m-1})$-metacirculant digraph, a the multiplicative order of $\alpha \in \mathbb{Z}_n^*$, and $c = a/\gcd(a, m)$. If $\gcd(c, m) = 1$, then Γ is isomorphic to a Cayley digraph of the group $\langle \rho, \tau^c \rangle$. Furthermore, this group is abelian if $\gcd(a, m) = 1$ and it is cyclic if $\gcd(a, m) = 1 = \gcd(m, n)$.*

Proof In view of Sabidussi's characterization of Cayley digraphs given in Theorem 1.2.20, it suffices to show that if $\gcd(c, m) = 1$, then $\langle \rho, \tau^c \rangle$ is a regular subgroup, while in addition, if $\gcd(a, m) = 1$ then $\langle \rho, \tau^c \rangle$ is abelian and cyclic if $\gcd(a, m) = 1 = \gcd(m, n)$. By Lemma 1.5.12, we may assume without loss of generality that if $p|a$ is prime, then $p|m$. This implies that if $\gcd(c, m) = 1$, then $c = 1$. We may thus assume without loss of generality that $c = 1$. We conclude $\gcd(a, m) = a$ and so $a|m$. Let $b = \operatorname{lcm}(a, m) = m$. By Lemma 1.5.1 $|\langle \rho, \tau \rangle| = b \cdot n = m \cdot n$, and so $\langle \rho, \tau \rangle$ is regular by Corollary 1.1.12.

If, in addition, $\gcd(a, m) = 1$, then $c = a$ and $a = 1$. This implies $\alpha = 1$, and so τ commutes with ρ. Thus $\langle \tau, \rho \rangle$ is abelian. Additionally, $\langle \tau \rangle \trianglelefteq \langle \rho, \tau \rangle$, and so by Dummit and Foote (2004, Theorem 5.4.9), we have $\langle \rho, \tau \rangle \cong \langle \rho \rangle \times \langle \tau \rangle$. Thus if $\gcd(m, n) = 1$, then $\langle \rho, \tau \rangle$ is cyclic. □

Note that in view of Example 1.5.5, the previous result shows that the Heawood graph is isomorphic to a Cayley graph of D_{14}.

To conclude this section, we would like to point out that the definition of metacirculant digraphs above differs from that given by Alspach and Parsons (1982a), as they only considered graphs Γ without loops. The following result gives conditions on when a metacirulant digraph is a graph without loops, and is essentially Alspach and Parsons' original definition.

Theorem 1.5.14 *Let Γ be an $(m, n, \alpha, S_0, \ldots, S_{m-1})$-metacirculant digraph. The following are equivalent:*

(i) *Γ is a graph without loops, and*
(ii) *the following conditions hold:*
  (a) $0 \notin S_0 = -S_0$,
  (b) $S_{-i} = -\alpha^{-i} S_i$ *for all* $0 \le i \le m - 1$.

Proof To eliminate loops, we add the condition that $0 \notin S_0$. Then Γ is a graph if and only if

$$\begin{aligned}
((\ell, j), (\ell + i, k)) \in A(\Gamma) \ &\text{if and only if} \ ((\ell + i, k), (\ell, j)) \in A(\Gamma)\\
&\text{if and only if} \ \rho^{-\alpha^{-i-\ell}k} \tau^{-i-\ell}((\ell + i, k), (\ell, j)) \in A(\Gamma)\\
&\text{if and only if} \ \left((0,0), \left(-i, -\alpha^{-i-\ell}(k - j)\right)\right) \in A(\Gamma)\\
&\text{if and only if} \ -\alpha^{-i-\ell}(k - j) \in S_{-i}\\
&\text{if and only if} \ k - j \in -\alpha^{i+\ell} S_{-i}.
\end{aligned}$$

As we also have $((\ell, j), (\ell + i, k)) \in A(\Gamma)$ if and only if $k - j \in \alpha^{\ell} S_i$, Γ is a graph if and only if $\alpha^{\ell} S_i = -\alpha^{i+\ell} S_{-i}$, or if and only if $S_{-i} = -\alpha^{-i} S_i$. □

In the literature (and indeed in the original definition of a metacirculant graph), when defining a metacirculant graph, it is not necessary to give all of the sets $S_0, \ldots, S_{m-1}$, only the sets $S_0, \ldots, S_{\mu}$, where $\mu = \lfloor m/2 \rfloor$. Here, for a number x, $\lfloor x \rfloor$ is the greatest integer less than or equal to x. In Exercise 1.5.10 you will show that the sets $S_{\mu+1}, \ldots, S_{m-1}$ can be obtained from the sets $S_1, \ldots, S_{\mu}$.

Finally, it should be pointed out that in the literature there are now other definitions of metacirculant digraphs, some of which are equivalent to the definition presented here, and some of which are not. Some authors (see, for example, Marušič and Šparl (2008)) use the following definition of metacirculant digraphs:

Definition 1.5.15 Let m and n be positive integers. A digraph Γ is an (m, n)-metacirculant digraph if Γ is of order mn, and there exist two automorphisms ρ and τ of Γ such that

(i) $\langle \rho \rangle$ is semiregular and has m orbits,
(ii) τ cyclically permutes the orbits of $\langle \rho \rangle$ and normalizes $\langle \rho \rangle$, and
(iii) τ^m fixes at least one vertex.

It is left as an exercise (Exercise 6.5.12) to show that this an equivalent definition of metacirculant digraphs once the appropriate algebraic tools have been developed. Marušič and Šparl (2008) asked if condition (iii) was indeed necessary, and called graphs satisfying only the first two conditions **weak metacirculants**. Li, Song, and Wang (2013) showed that it was indeed necessary by constructing graphs satisfying the first two conditions but not the third, thus establishing that there are weak metacirculant graphs that are not isomorphic to any metacirculant graph. This result, however, contained an error corrected by Conder, Zhou, Feng, and Zhang (2020a).

Definition 1.5.16 A group G is **metacyclic** if there exists $N \trianglelefteq G$ such that both N and G/N are cyclic.

Note that the first two conditions in Definition 1.5.15 are equivalent to $\langle \rho, \tau \rangle$ being isomorphic to a metacyclic group (Exercise 1.5.14).

1.6 Exercises

1.1.1 Give a regular permutation representation of D_8.

1.1.2 Show that in a regular permutation group $G \leq S_n$, for every $x, y \in \mathbb{Z}_n$, there is a *unique* element of G that maps x to y.

1.1.3 Show that a complete bipartite graph $K_{s,t}$ is vertex-transitive if and only if $s = t$.

1.1.4 Show that Q_n has $n \cdot 2^{n-1}$ edges.

1.1.5 Let $G \leq S_n$ be transitive. Show that $\mathrm{Stab}_G(x) \leq G$ for every $x \in \mathbb{Z}_n$.

1.1.6 Prove that the orbit-stabilizer theorem holds for group actions.

1.1.7 For digraphs, the **outvalence** of a vertex v is the number of arcs with first coordinate v, and **invalence** is defined analogously. For a graph, the **valence** of a vertex $v \in V(\Gamma)$ in a graph Γ is the number of edges incident with v, with loops counted twice. Show that a vertex-transitive graph is always regular, and that in a vertex-transitive digraph the outvalence and invalence at a vertex are always equal.

1.1.8 Show that the set of all automorphisms of a digraph is a group under function composition.

1.1.9 Let $G \leq S_n$ be transitive. Show that $\mathrm{Stab}_G(v)$ contains no nontrivial normal subgroups of G.

1.1.10 Show that an automorphism α of $\mathbb{Z}_n$ is uniquely determined by the value of $\alpha(1)$. Using the fact that the generators of $\mathbb{Z}_n$ are the elements of $\mathbb{Z}_n^*$, show that $\mathrm{Aut}(\mathbb{Z}_n) \cong \mathbb{Z}_n^*$.

1.1.11 Show that if $\phi\colon \Gamma \to \Gamma'$ is a digraph isomorphism, then $\phi^{-1}\colon \Gamma' \to \Gamma$ is also a digraph isomorphism. Also show that if $H \leq \mathrm{Aut}(\Gamma')$, then $\phi^{-1}H\phi \leq \mathrm{Aut}(\Gamma)$.

1.1.12 Show that the permutation $(0, 0')(1, 1')(2, 2')(3, 3')(4, 4')$ is an automorphism of the Petersen graph.

The purpose of the next three exercises is to show that the skeletons of the five Platonic solids (usually simply referred to by the name of the Platonic solid), the tetrahedron (isomorphic to K_4), cube, octahedron, icosahedron, and dodecahedron, are all vertex-transitive. We have already seen that K_4 and the cube are vertex-transitive, and the next three exercises consider the remaining Platonic solids.

1.1.13 Show that the octahedron in Figure 1.12 is vertex-transitive.

1.1.14 Show that the icosahedron graph in Figure 1.13 is vertex-transitive. (Hint: There is an obvious 3-step rotation with four orbits of three vertices each. Choose a three-cycle C whose vertices do not form one of the just obtained orbits in the icosahedron and show that the graph can be redrawn with C in place of the central triangle. In other words, there is a 3-step symmetry about any 3-cycle, not just the obvious ones. The 3-step rotation around C should "fuse" the four previously obtained orbits into one orbit.)

1.1.15 Show that the dodecahedron graph shown in Figure 1.14 is vertex-transitive.

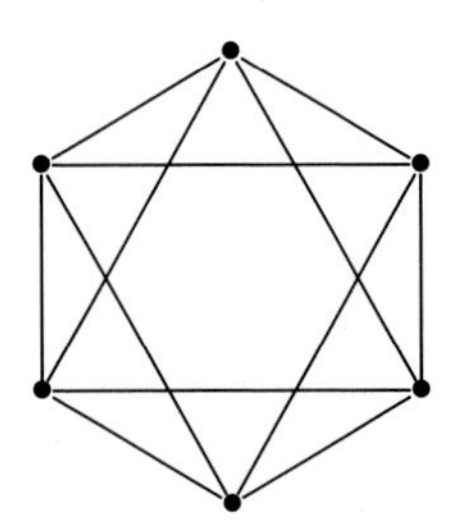

Figure 1.12 Octahedron

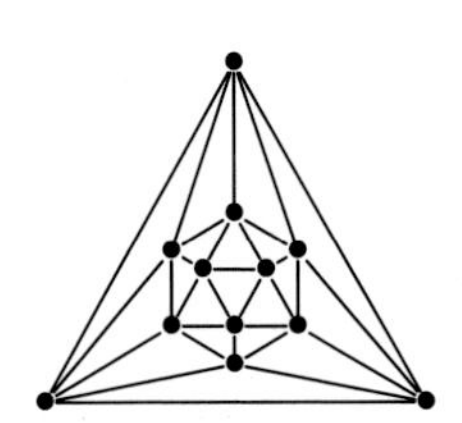

Figure 1.13 Icosahedron

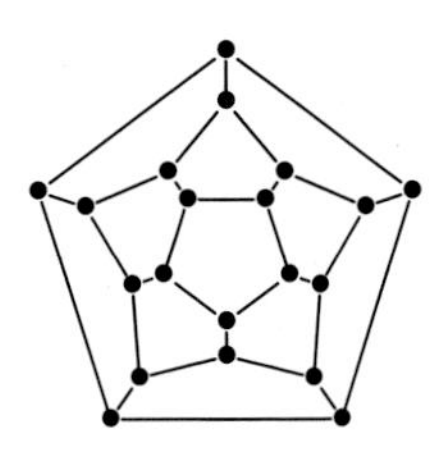

Figure 1.14 Dodecahedron

1.1.16 Show that the dual graph of the dodecahedron is the icosahedron. Use the vertex-transitivity of the dodecahedron to show that the automorphism group of the dodecahedron is transitive on its set of faces. Deduce from this that the icosahedron is vertex-transitive.

1.2.1 Draw the Cayley graph $\mathrm{Cay}(\mathbb{Z}_9, \{1,3,6,8\})$. Which group automorphisms of $\mathbb{Z}_9$ are contained in $\mathrm{Aut}(\mathrm{Cay}(\mathbb{Z}_9, \{1,3,6,8\}))$?

1.2.2 Draw the Cayley graph $\mathrm{Cay}(Q_8, S)$. Here Q_8 is the **quaternion group** with presentation $\langle x, y : x^4 = y^4 = 1, x^2 = y^2, y^{-1}xy = x^{-1}\rangle$, and $S = \{x, y, x^{-1}, y^{-1}\}$.

1.2.3 Let $G = \mathbb{Z}_{12}$. Which of the following connection sets give rise to Cayley graphs as opposed to digraphs: $S = \{1,2,10,11\}$, $\{1,3,6,11\}$, $\{1,3,6,7,11\}$?

1.2.4 Show that G_L is a group isomorphic to G.

1.2.5 Let G be a group and $S \subset G$ that does not contain 1_G. Show that $\mathrm{Cay}(G, S)$ is a regular digraph of outvalency and invalency $|S|$. Additionally, if $S = S^{-1}$ and $\mathrm{Cay}(G, S)$ is a graph, show that it also has valency $|S|$.

1.2.6 Let G be a group and $g \in G$. Show that $\mathrm{Cay}(G, \{g\})$ is a disjoint union of $|G|/|g|$ directed cycles if $|g| \geq 3$, a 1-factor if $|g| = 2$, and a union of $|G|$ loops if $|g| = 1$.

1.2.7 Verify that $(x, y) \in A(\mathrm{Cay}(G, S))$ if and only if $x^{-1}y \in S$.

1.2.8 Draw $\mathrm{Cay}(\mathbb{Z}_{10}, \{1,5\})$ and $\mathrm{Cay}(D_{10}, \{\rho, \tau\})$. Are they isomorphic as the graphs are in Example 1.2.11?

1.2.9 Show that a Cayley graph of an abelian group of valency at least 3 has girth 3 or 4.

1.2.10 How many loopless Cayley digraphs are there of a group G? How many loopless Cayley graphs are there of a group G assuming that $|G| = n$ and G contains m self-inverse elements?

1.2.11 Show that the n-cube Q_n is a Cayley graph of $\mathbb{Z}_2^n$.

1.2.12 Show that the octahedron is a Cayley graph of $\mathbb{Z}_6$.

1.2.13 For G a group, and $S \subseteq G$, the digraph $\mathrm{Cay}_R(G, S)$ has vertex set G and arc set $\{(g, sg) : s \in S, g \in G\}$. Define $g_R \colon G \to G$ by $g_R(x) = xg$. Show $G_R = \{g_R : g \in G\}$ is a subgroup of $\mathrm{Aut}(\mathrm{Cay}_R(G, S))$, and that the map $\iota \colon G \to G$ given by $\iota(g) = g^{-1}$ is an isomorphism between $\mathrm{Cay}_R(G, S)$ and $\mathrm{Cay}(G, S^{-1})$.

1.2.14 Let $n \geq 2$ be an integer, $H = \langle \tau_i : 1 \leq i \leq n \rangle$, where τ_i is defined as in Example 1.1.18, and K the permutation group on $\mathbb{Z}_2^n$ induced by letting $\mathcal{S}_n$ permute the coordinates. Show that $K \leq \mathrm{Aut}(Q_n)$ and that $H \trianglelefteq \langle H, K \rangle$. Deduce that $|\mathrm{Aut}(Q_n)| \geq n! \cdot 2^n$. You will be asked to show in Exercise 5.4.4 that (using different language) $\mathrm{Aut}(Q_n) = \langle H, K \rangle$.

1.3.1 Let $H \leq G$ be core-free in G. Show that the stabilizer of the left coset $1 \cdot H$ in G is the subgroup H. Conclude that the stabilizer of the left coset gH in G is the subgroup gHg^{-1}.

1.3.2 Find the double coset D of $\mathcal{S}_3$ in $\mathcal{S}_4$ that contains the permutation $(0, 1)$. Then draw $\mathrm{Cos}(\mathcal{S}_4, \mathcal{S}_3, \{(0, 1)\})$. Here, $\mathcal{S}_3$ permutes $\{0, 1, 2\}$, and $\mathcal{S}_4$ permutes $\{0, 1, 2, 3\}$.

1.3.3 Let $\mathrm{AGL}(1, 7) = \{x \mapsto ax + b : a \in \mathbb{Z}_7^*, b \in \mathbb{Z}_7\}$. Here, $\mathrm{AGL}(1, 7)$ is a subgroup of $\mathcal{S}_7$, and so the "x" in the set brackets is an element of $\mathbb{Z}_7$. For example, if $p = 7$, $a = 2$, and $b = 1$, the function $x \mapsto ax+b$ is the permutation $(0, 1, 3)(2, 5, 4)(6)$ in $\mathcal{S}_7$. Let $H = \{x \mapsto ax : a = 1, 2, 4\}$, and $D = HSH$ where $S = \{x \mapsto ax : a = 3, 5\}$. Find D explicitly, and draw $\mathrm{Cos}(\mathrm{AGL}(1, 7), H, S)$.

1.3.4 Show that $\mathrm{Cos}(G, H, S)$ has no loops if and only if $S \cap H = \emptyset$, or equivalently, $H \not\subseteq HSH$.

1.3.5 Show that $(xH, yH) \in A(\mathrm{Cos}(G, H, S))$ if and only if $yH \subseteq xHSH$.

1.3.6 Show that $\mathrm{Cos}(G, H, S)$ is a graph if and only if $s^{-1} \in HSH$ for every $s \in S$.

1.3.7 Show that $\mathrm{Cos}(G, H, S)$ is a graph if $S = S^{-1}$.

1.3.8 Show that the set of outneighbors of the vertex xH in $\mathrm{Cos}(G, H, S)$ is the set $N^+(xH) = \{yH : yH \subseteq xHSH\}$, and H has outvalence $|HSH|/|H|$.

1.3.9 Show that $\mathrm{Cos}(G, H, S^{-1})$ is obtained from $\mathrm{Cos}(G, H, S)$ by reversing the direction of every arc.

1.3.10 Show that $\mathrm{Cos}(G, H, S)$ is connected if and only if $\langle H, S \rangle = G$.

1.3.11 Show that $\mathrm{Cos}(G, \{1\}, S)$ is isomorphic to $\mathrm{Cay}(G, S)$.

1.3.12 Let $\alpha \in \mathrm{Aut}(G)$ such that $\alpha(H) = H$. Show that

$$\alpha(\mathrm{Cos}(G, H, S)) = \mathrm{Cos}(G, H, \alpha(S)).$$

1.3.13 Let G be a group and $H, K \leq G$. Define a relation $\sim$ on G by $x \sim y$ if and only if there exists $h \in H$ and $k \in K$ such that $hxk = y$. Show that $\sim$ is an equivalence relation. We call the equivalence classes of $\sim$ **(H, K)-double cosets** of G. Deduce that the set of (H, K)-double cosets of G is a partition of G.

1.3.14 Let G be a group and let $H \leq G$ be the stabilizer of a point x in a faithful and transitive action of G on X. How many double coset digraphs are there for the action of G on X? How many (H, H)-double cosets of G are there?

1.3.15 Show that $\mathrm{Cos}(G, H, S) = \mathrm{Cos}(G, H, HSH)$. Deduce that the following alternative definition of a double coset digraph is equivalent to that given in Definition 1.3.4: Let G be a group, $H \leq G$ and S a subset of G/H whose union is also a union of double cosets of H in G. Define a digraph $\mathrm{Cos}(G, H, S)$ by $V(\mathrm{Cos}(G, H, S)) = G/H$ and $A(\mathrm{Cos}(G, H, S)) = \{(gH, gsH) : sH \in S \text{ and } g \in G\}$.

1.3.16 Let G be a group and $H \leq G$. Define a digraph Γ by $V(\Gamma) = G/H$ and $A(\Gamma) = \{(aH, bH) : aH \neq bH \text{ and } aH \cap bHS = \emptyset\}$. Show $\Gamma \cong \mathrm{Cos}(G, H, S)$, so there is another alternative definition of a double coset digraph.

1.4.1 Find the orbital digraphs of the group $\{x \mapsto ax + b \pmod{15} : a = 1, 4, 7, 13,\ b \in \mathbb{Z}_{15}\}$. Are any orbital digraphs graphs?

1.4.2 Find the orbital digraphs of the group $\langle (\mathbb{Z}_5 \times \mathbb{Z}_5)_L, \delta \rangle$, where $\delta \colon \mathbb{Z}_5^2 \to \mathbb{Z}_5^2$ is given by $\delta(i, j) = (i, j+i)$. Also, show that δ normalizes $(\mathbb{Z}_5 \times \mathbb{Z}_5)_L$ and has order 5. Are any orbital digraphs graphs?

1.4.3 Find the orbital digraphs of $\langle (\mathbb{Z}_{25})_L, \delta \rangle$, where $\delta \colon \mathbb{Z}_{25} \to \mathbb{Z}_{25}$ is given by $\delta(x) = 6x$. Show that δ normalizes $(\mathbb{Z}_{25})_L$ and has order 5. Are any orbital digraph graphs? Compare the orbital digraphs found in this exercise with those found in the previous exercise.

1.4.4 Let G be a permutation group that contains a regular subgroup isomorphic to $\mathbb{Z}_2^k$, $k \geq 1$. Show that every orbital digraph of G is a graph.

1.4.5 Let G be a transitive group that contains a regular abelian group H. Determine a necessary condition for all orbital digraphs of G to be graphs.

1.4.6 Let G be a group. Show that $\mathrm{Cay}(G, S) \cup \mathrm{Cay}(G, T)$ has connection set $S \cup T$.

1.4.7 Let S_n act canonically on the set of all subsets of size 2 of $\mathbb{Z}_n$. Show that S_n has exactly three suborbits in this action.

1.4.8 Determine the suborbits of S_7 acting canonically on 3-subsets of $\mathbb{Z}_7$.

1.4.9 Show that an orbital digraph Γ of G is a graph if and only if the corresponding orbital is self-paired.

1.4.10 Let S be a suborbit of G of size m. Show that if G has no other suborbit of order m then S is self-paired, and hence the corresponding orbital digraph is a graph.

1.4.11 Let $G \leq S_X$ be transitive. Show that the set of arcs of the orbital digraphs of G are a partition of $A(K_{|X|})$. Deduce that an orbital digraph of G is uniquely determined by giving one arc in its arc set.

1.4.12 Prove Proposition 1.4.6.

1.4.13 Prove Proposition 1.4.9.

1.4.14 Show that every vertex-transitive graph is a union of orbital digraphs corresponding to paired orbitals.

1.4.15 Find the orbital digraphs of $G = (\mathbb{Z}_2^3)_L$ whose union is the cube. Find the orbital digraphs of $\langle G, s\rangle$ whose union is the cube where $s\colon \mathbb{Z}_2^3 \to \mathbb{Z}_2^3$ interchanges the second and third coordinates. Then find the orbital digraphs of $\langle G, t\rangle$ whose union is the cube where $t\colon \mathbb{Z}_2^3 \to \mathbb{Z}_2^3$ is given by $t(i, j, k) = (k, i, j)$.

1.4.16 Let $G \leq S_n$ be transitive, and $h \in N_{S_n}(G)$. Here, $N_{S_n}(G) = \{h \in S_n : h^{-1}Gh = G\}$ is the **normalizer** of G in S_n. Show that h permutes the orbital digraphs of G. That is, show that if Γ is an orbital digraph of G, then $h(\Gamma)$ is also an orbital digraph of G. Conclude that any element in S_n centralizing G also permutes the orbital digraphs of G. Recall that h centralizes G if h commutes with every element of G.

1.4.17 Let $G \leq S_n$ be transitive, and let $h \in S_n$ centralize G. Show that if Γ is a generalized orbital digraph of G, then $h \in \mathrm{Aut}(\Gamma)$ if and only if Γ is a union of orbits of h on all orbital digraphs of G.

1.5.1 Draw the (2, 6)-metacirculant digraph $(2, 6, 5, \{1, 4\}, \{0\})$.

1.5.2 Draw the (3, 7)-metacirculant graph $(3, 7, 2, \{1, 6\}, \{0, 1\})$. Notice that we are using the characterization of a metacirculant graph in Theorem 1.5.14 in this exercise, and so only explicitly give $S_0, \ldots, S_\mu$.

1.5.3 Let $m = 8$, $n = 17$, $\alpha = 2$, $S_0 = \{1, 4, 13, 16\}$, $S_4 = \{2, 8\}$, and $S_1 = S_2 = S_3 = S_5 = S_6 = S_7 = \emptyset$. Do these choices yield a metacirculant graph?

1.5.4 Show that up to isomorphism there are exactly eight vertex-transitive graphs of order 6, and all of them are isomorphic to circulant graphs.

1.5.5 Let m and k be positive relatively prime integers. Show that any circulant digraph of order mk is isomorphic to an (m, k)-metacirculant digraph.

1.5.6 Let Γ be a (q, p)-metacirculant digraph where $q > p$ are distinct primes. Show that Γ is isomorphic to a circulant digraph.

1.5.7 Let Γ be an $(m, n, \alpha, S_0, \ldots, S_{m-1})$-metacirculant digraph, $a = |\alpha|$, and $a = xy$, where x is maximal such that $\gcd(x, m) = 1$. Show that the map $(i, j) \mapsto (i, \alpha^y j)$ is an automorphism of Γ.

1.5.8 Show that $\mathrm{AGL}(1, p) \cong \mathbb{Z}_p \rtimes \mathbb{Z}_{p-1}$.

1.5.9 Let $q < p$ be prime with $q|(p-1)$. Show that the nonabelian group of order qp is isomorphic to $\mathbb{Z}_p \rtimes \mathbb{Z}_q$. Conclude that every Cayley digraph of a group G of order qp is isomorphic to a metacirculant digraph.

1.5.10 Let Γ be a $(m, n, \alpha, S_0, \ldots, S_{m-1})$-metacirculant graph. Show that the sets $S_{\mu+1}, \ldots, S_{m-1}$ can be obtained from the sets $S_1, \ldots, S_\mu$.

1.5.11 Let H and N be groups with $\phi\colon H \to \mathrm{Aut}(N)$ a homomorphism. Show that the set $H \times N$ with binary operation $(h_1, n_1)(h_2, n_2) = (h_1h_2, \phi(h_2)(n_1)n_2)$ is a group G by showing that the binary operation is associative, $(1, 1)$ is the identity, and $\left(h^{-1}, (\phi(h))^{-1}\left(n^{-1}\right)\right)$ is the inverse of (h, n). The group G is the **external semidirect product of** H **and** N, and is denoted $N \rtimes H$ (or $N \rtimes_\phi H$ to emphasize the role of ϕ).

1.5.12 For a group G and subgroups $H \leq G$ and $N \trianglelefteq G$, the following are equivalent:

(a) G is isomorphic to the external semidirect product of H and N, where ϕ is the automorphism from H into $\mathrm{Aut}(N)$ induced by conjugation,

(b) $G = N \rtimes_\phi H$,

(c) every element of G can be written uniquely in the form hn, $h \in H, n \in N$,

(d) every element of G can be written uniquely in the form nh, $h \in H, n \in N$.

1.5.13 Show that $N \rtimes_\phi H \simeq N \times H$ provided that $\phi = 1$. That is, the direct product is a special case of the semidirect product.

1.5.14 Show that a digraph Γ is isomorphic to a weak metacirculant digraph if and only if it contains a transitive metacyclic subgroup.

2

The Petersen Graph, Blocks, and Actions of A_5

This chapter has two main themes: the Petersen graph and blocks. Our first goal is to show that the automorphism group of the Petersen graph is isomorphic to a particular action of $\mathcal{S}_5$. We achieve this in Theorem 2.1.4. We then turn to a natural way of dividing transitive permutation groups into two disjoint classes, namely those that have nontrivial blocks (imprimitive groups) and those that do not (primitive groups). This division is, from a practical point of view, quite real, as typically very different techniques are needed to handle many problems in permutation group theory within these two classes. We begin this in Section 2.2 by defining blocks and studying their basic properties. We continue in Section 2.3, where we focus on how to find blocks in a permutation group, including showing, in Theorem 2.3.7, that there is a bijective correspondence between the set of all subgroups of a transitive permutation group G that contain the stabilizer of a point and the set of blocks that contain the point. In Section 2.4 we begin to study transitive groups that do not have nontrivial blocks (we return to these groups again in Section 5.4), as well as a closely related class of groups, quasiprimitive groups. In practice, one can think of quasiprimitive groups as being those imprimitive groups where the techniques used for primitive groups are more appropriate than the usual techniques for imprimitive groups. We also show in Theorem 2.4.13 that any transitive group G can be viewed as the left coset action of G on G/H, where H is the stabilizer of a point in G. We then try to solidify a great deal of the material considered so far by considering the transitive representations of A_5, as promised in the Preface (we remark that we do not consider the transitive representations of $\mathcal{S}_5$ simply because this is a much longer problem, as there are many more conjugacy classes of subgroups of $\mathcal{S}_5$ than in A_5). We show, in Section 2.5, that every such action can be thought of as A_5 permuting natural "combinatorial substructures" of the Petersen graph, one for each action. As we go along, we also discuss all of the blocks of each of these representations. Once we have all of these actions, we calculate the orbital digraphs of most of the permutation groups obtained from actions of A_5 (Section 2.6).

2.1 The Automorphism Group of the Petersen Graph

The Petersen graph is perhaps the most important vertex-transitive graph. Indeed, the Petersen graph has so many interesting properties (some of which are not of interest here) that an entire book (Holton and Sheehan, 1993) has been written that explores some of the graph properties for which the Petersen graph is the most natural or smallest example. Because of this importance, we will spend considerable time on the Petersen graph. We will determine its full automorphism group, and explore the derivation of this group and the group itself from several natural points of view. This investigation will begin in this section and continue in following sections. For the rest of this section, for ease of notation, we will denote the Petersen graph by P.

Perhaps the quickest and from our point of view, least informative, derivation of the automorphism group of the Petersen graph is to invoke two general results from graph theory. The first is a classical result of Whitney (see Ore (1962, Theorem 15.4.1)).

Theorem 2.1.1 (Whitney's theorem) *If Γ is a graph and $|V(\Gamma)| \geq 5$, then* $\mathrm{Aut}(\Gamma) \cong \mathrm{Aut}(L(\Gamma))$, *where $L(\Gamma)$ is the line graph of Γ.*

The second result is that a graph and its complement have the same automorphism group (see Exercise 2.1.1). One then only needs to show that the complement of the line graph of K_5 is the Petersen graph (see Exercise 2.1.2) and, of course, observe that $\mathrm{Aut}(K_5) = \mathcal{S}_5$, and we have $\mathrm{Aut}(P) \cong \mathcal{S}_5$. Note that this argument does not give $\mathrm{Aut}(P)$ as a permutation group but only as an abstract group.

The above argument does though lead to a good method for us – namely, letting the vertices of K_5 be $[1, 5]$, we can think of an edge in K_5 as being an unordered pair of vertices of K_5, or a 2-subsets, chosen from $[1, 5]$. Two edges in the line graph are adjacent if they contain a common endpoint or only if their corresponding 2-subsets have nonempty intersection. In the complement, this would mean that the intersection of the 2-subsets corresponding to the edges is empty. Formally,

Definition 2.1.2 Define a graph Γ by setting $V(\Gamma)$ to be the set of all subsets of size 2 (2-subsets) of $[1, 5]$, and there is an edge between two 2-subsets if and only if they are disjoint.

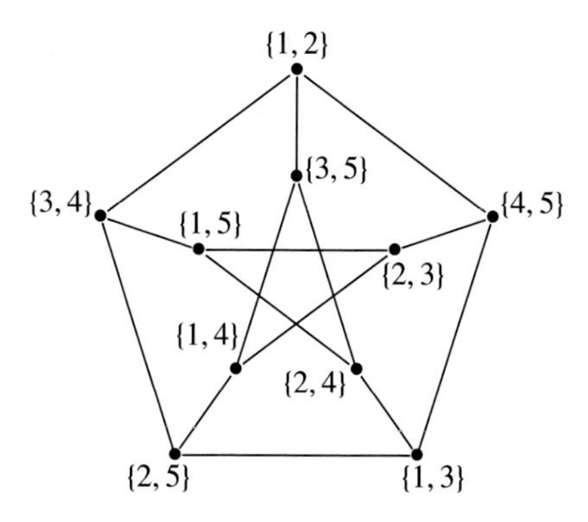

Figure 2.1 The 2-subset labeling of the Petersen graph.

We obtain the graph in Figure 2.1, which is obviously isomorphic to the Petersen graph. This definition of the Petersen graph was first shown in Kowalewski (1917). We call it the **2-subset labeling of** P. Notice that this is exactly the same labeling of the Petersen graph we obtained by considering P as an orbital digraph of $\mathcal{S}_5$ in Example 1.4.15.

Lemma 2.1.3 *$\mathcal{S}_5$ in its action on 2-subsets of $[1,5]$ is a subgroup of the automorphism group of the Petersen graph.*

Proof With the 2-subset labeling of P it is easy to see how $\mathcal{S}_5$ permutes the vertices of P. Let $\delta \in \mathcal{S}_5$ and $\{a,b\}$ be a 2-subset of $[1,5]$. Then $\delta(\{a,b\}) = \{\delta(a),\delta(b)\}$ is a 2-subset of $[1,5]$. So each element δ of $\mathcal{S}_5$ induces a permutation of $V(P)$, which we will denote $\bar{\delta}$. For example, if $\delta = (1,2,3)$, then

$$\bar{\delta} = (\{1,2\},\{2,3\},\{1,3\})(\{1,4\},\{2,4\},\{3,4\})(\{1,5\},\{2,5\},\{3,5\}).$$

Furthermore, $\bar{\delta}$ maps edges of P to edges of P, as if $\{a,b\} \cap \{c,d\} = \emptyset$, then $\{\delta(a),\delta(b)\} \cap \{\delta(c),\delta(d)\} = \emptyset$ as well. If some element of $\mathcal{S}_5$ fixes every 2-subset of $[1,5]$, then it must fix every element of $[1,5]$ as, for example, it fixes $\{1,2\}$ and $\{1,3\}$, and so fixes their intersection $\{1\}$. □

While the 2-subset labeling of P does indeed show how $\mathcal{S}_5$ permutes the vertices of P, it does not establish that $\mathrm{Aut}(P) \cong \mathcal{S}_5$, only that $\mathrm{Aut}(P)$ contains a transitive subgroup isomorphic to $\mathcal{S}_5$. We must show that there is no other automorphism of P, and an obvious way to do this is to give an isomorphism $\phi \colon \mathrm{Aut}(P) \to \mathcal{S}_5$. That is, we need to find five "somethings" in the Petersen graph that are permuted by $\mathrm{Aut}(P)$, and additionally, the only permutation in $\mathrm{Aut}(P)$ that fixes each of these "somethings" is the identity. This will then induce an isomorphism ϕ: Aut(P) $\to$ $\mathcal{S}_5$. We will give two sets of "somethings," one in the proof below, and the other in Theorem 2.4.11.

Theorem 2.1.4 *The automorphism group of the Petersen graph is $\mathcal{S}_5$ in its action on 2-subsets of* $[1,5]$.

Proof We use the 2-subset labeling of P in this proof. Note that in any graph Γ and for any independent set I of Γ, $\alpha(I)$ is an independent set of Γ of the same size for any $\alpha \in \text{Aut}(\Gamma)$. Thus the automorphism group of any graph acts on the independent sets of a fixed size. Call the 5-cycle in P induced by $\{\{1,2\},\{4,5\},\{1,3\},\{2,5\},\{3,4\}\}$ the outside 5-cycle, and the 5-cycle in P induced by the other vertices of P the inside 5-cycle. A 5-cycle has a maximal independent set of size 2, so in a maximal independent set I of P, at most two vertices of I can be on the outer 5-cycle, and at most two on the inner 5-cycle. Thus $|I| \leq 4$. It is easy to find independent sets in P of size 4 – for example, the set I_j is the set of all 2-subsets of $[1,5]$ that contain the fixed element $j \in [1,5]$. There are obviously 5 such independent sets. Now suppose I is an independent set of size 4. There must be two vertices of I on the "outside" 5-cycle and two on the "inside" 5-cycle of P. Let $\rho \in \mathcal{S}_5$ be the 5-cycle $(1,5,3,2,4)$. In its action on 2-subsets of $[1,5]$, $\bar{\rho}$ is the same permutation of the vertices as in the metacirculant labeling of P, i.e., a rotation. After applications of ρ to I we may assume without loss of generality that the vertices $\{3,4\}$ and $\{4,5\}$ are the two vertices of I on the outside 5-cycle. As an independent set of size 2 on the inside 5-cycle is the vertices $\{3,5\}$ and $\{2,3\}$ and their rotations under ρ, we see by inspection that the two vertices of I on the inside 5-cycle are $\{1,4\}$ and $\{2,4\}$. Then $I = I_4$. Thus, the only independent sets in P of size 4 are I_j, $j \in [1,5]$, and $\mathcal{S}_5$ is clearly transitive on $\{I_j : j \in [1,5]\}$ as if $\sigma(j) = k$, then $\sigma(I_j) = I_k$. Let $\delta \in \text{Aut}(P)$ such that $\delta(I_j) = I_j$ for each $j \in [1,5]$. Then for $j,k \in [1,5]$ with $j \neq k$, $\delta(I_j \cap I_k) = I_j \cap I_k = \{\{j,k\}\}$. So δ fixes every 2-subset of $[1,5]$ and so is the identity. Thus the only element of Aut(P) that fixes each maximum independent set is the identity. The action of Aut(P) on $\{I_j : 1 \leq j \leq 5\}$ thus induces an isomorphism ϕ: Aut(P) $\to \mathcal{S}_5$, and the result follows. □

2.2 Blocks and Imprimitive Groups

This section is our first entirely permutation group theoretic section. We will study one of the most basic ideas in permutation group theory, that of a block. We saw an intuitive notion of a "block" in previous sections. Here, we make this notion formal and discuss the basic terms and results concerning blocks. We then turn to results on perhaps the most commonly encountered blocks, namely the orbits of a normal subgroup, and the consequences of these ideas.

Definition 2.2.1 Let X be a set, and let $G \leq \mathcal{S}_X$ be transitive. A subset $B \subseteq X$ is a **block** of G if whenever $g \in G$, then $g(B) \cap B = \emptyset$ or B. If $B = \{x\}$ for some $x \in X$ or $B = X$, then B is a **trivial block**. Any other block is nontrivial. If G has a nontrivial block, then G is **imprimitive**. If G is not imprimitive, we say G is **primitive**.

So a block B of a permutation group $G \leq \mathcal{S}_X$ is a subset of X that is either fixed set-wise by $g \in G$ or moved "away" from B. Note that if B is a block of G, then $g(B)$ is also a block of B for every $g \in G$, and is called a **conjugate block of** B (for an indication of why $g(B)$ is called a **conjugate block** of B, see Exercise 2.3.10). Trivial blocks are called "trivial" as they are blocks of any transitive permutation group.

Definition 2.2.2 The set of all blocks conjugate to B is a partition of X, called a **block system of** G.

There does not seem to be a standard term for what we have defined as a block system of G. Other authors use a **system of imprimitivity** or a G**-invariant partition of** X, or even a **complete block system** for this term.

Example 2.2.3 Let G be a group with $H \leq G$. Then the set of left cosets of H in G is a block system of G_L.

Solution Let $g, k \in G$. Then $g_L(kH) = gkH$ is another left coset of H in G, and, as the left cosets of H in G are a partition of G, $g_L(kH) = kH$ or $g_L(kH) \cap kH = \emptyset$. □

Example 2.2.4 Let Γ be a disconnected vertex-transitive digraph. Then the set of components of Γ is a block system of $\mathrm{Aut}(\Gamma)$.

Solution We know the set of components of Γ are a partition of $V(\Gamma)$ as this is true for any digraph. As any automorphism of Γ maps a component to some component and $\mathrm{Aut}(\Gamma)$ is transitive, we see the set of components is a block system of $\mathrm{Aut}(\Gamma)$. □

Example 2.2.5 The automorphism group of the graph $\Gamma = \mathrm{Cay}(\mathbb{Z}_6, \{2,3,4\})$ in Figure 2.2 has block systems with 2 blocks of size 3 and 3 blocks of size 2.

Solution Observe that $\mathrm{Cay}(\mathbb{Z}_6, \{2,4\}) \leq \Gamma$ is two edge disjoint triangles with vertex sets $\{0,2,4\}$ and $\{1,3,5\}$, and these two triangles are the only triangles

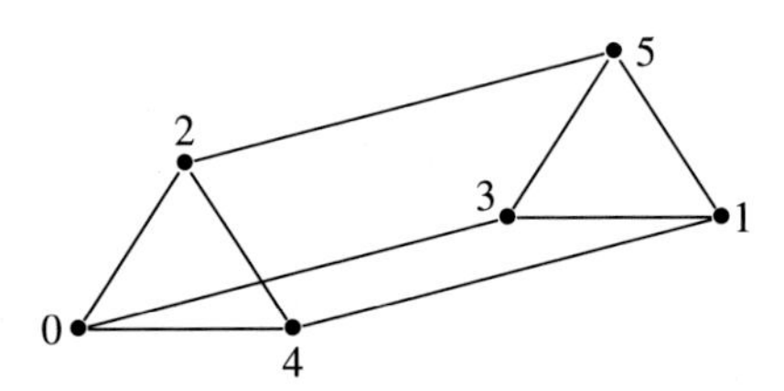

Figure 2.2 Cay($\mathbb{Z}_6$, {2, 3, 4}).

in Γ. As any automorphism of any graph maps triangles to triangles (indeed, an automorphism maps an m-cycle to an m-cycle) and Aut(Γ) is transitive, $\{\{0,2,4\},\{1,3,5\}\}$ is a block system of Aut(Γ) with 2 blocks of size 3. To obtain a block system with 3 blocks of size 2, note that an automorphism of any graph maps edges that are not in a triangle to edges that are not in a triangle (indeed, an automorphism maps edges not in an m-cycle to edges not in an m-cycle). There are three edges in Γ not in triangles, namely the edges 03, 14, and 25. As Aut(Γ) is transitive, the set $\{\{0,3\},\{1,4\},\{2,5\}\}$ is a block system of Aut(Γ) with 3 blocks of size 2. □

We will see examples of transitive groups that only have trivial blocks in Section 2.4.

Any subdigraph Δ of a vertex-transitive digraph Γ such that the vertex sets of all subdigraphs of Γ isomorphic to Δ form a partition of $V(\Gamma)$ will be a block system of Aut(Γ) (Exercise 2.2.3). It is no coincidence that the number of blocks in a block system times the number of elements in a block is the size of the set.

Theorem 2.2.6 *Let $\mathcal{B}$ be a block system of a transitive group $G \leq S_X$. Then every block in $\mathcal{B}$ has the same cardinality, say k. Further, if m is the number of blocks in $\mathcal{B}$ then mk is the degree of G.*

Proof As every block of $\mathcal{B}$ is the image of some block B under an element of G (which is a bijection), every block has the same number of elements. As $\mathcal{B}$ is a partition of X, $mk = |X|$. □

Intuitively, having a block system $\mathcal{B}$ of a transitive group $G \leq S_X$ allows one to think of the elements of G in terms of permutations of smaller sets. Namely, given $g \in G$, g permutes the elements of X, but it also permutes the blocks of $\mathcal{B}$, and if $g(B) = B'$, then g induces a bijection from B to B'. (In Exercise 2.2.6 you will be asked to show that G acts on $\mathcal{B}$ by $g \cdot B = g(B)$. The induced action is then the image of the canonical homomorphism from

G into $S_{\mathcal{B}}$.) In effect, we can coordinatize X, and think of X as $\mathcal{B} \times B$ (note that $|\mathcal{B} \times B| = |X|$ by Theorem 2.2.6). With this labeling, an element of $g \in G$ has the form $g(B,x) = (f(B), h_B(x))$ where f is the permutation on the blocks induced by g, and each h_B is the permutation of B that gives how the block B is mapped to the block $f(B)$. Note that for different blocks B and B', it need not be the case that $h_B = h_{B'}$. After development of the appropriate theory, we will show in the embedding theorem (Theorem 4.3.1) that h_B can be chosen from the "induced action" of the set-wise stabilizer of B in G.

The following result gives perhaps the most common method for finding a block system.

Theorem 2.2.7 *Let $G \le S_X$ be transitive. If $N \trianglelefteq G$, then the set of orbits of N is a block system of G.*

Proof Let $x \in X$ and B be the orbit of N that contains x, so $B = \{n(x) : n \in N\}$. Let $g \in G$, and for $n \in N$, let $n' \in N$ such that $gn = n'g$. Note that n' always exists as $N \trianglelefteq G$, and that $\{n' : n \in N\} = N$ as conjugation by g induces an automorphism of N. Then

$$g(B) = \{gn(x) : n \in N\} = \{n'g(x) : n \in N\} = \{n(g(x)) : n \in N\}.$$

Hence $g(B)$ is the orbit of N that contains $g(x)$, and as the set of orbits of N are a partition of X (indeed, the set of orbits of any group acting on X is a partition of X), $g(B) \cap B = \emptyset$ or B. Thus B is a block, and as every conjugate block $g(B)$ of B is an orbit of N, the set of orbits of N is a block system of G. □

Example 2.2.8 Let A be an abelian group and $N \le A$. Then the set of orbits of $\bar{N}_L$ is a block system of A_L with blocks being the cosets of N in A. Here $\bar{N}_L = \{n_L : n \in N\}$.

Solution By definition, A_L permutes A. As A is abelian, $\bar{N}_L \trianglelefteq A_L$. The set of orbits of $\bar{N}_L$ is a block system of A_L by Theorem 2.2.7. For $a \in A$, the orbit of $\bar{N}_L$ that contains a, written additively, is $\{a + n : n \in N\}$, which is the coset of N that contains a. □

Example 2.2.9 We use the notation from Section 1.5 here. The set of orbits of $\langle\rho\rangle \le \langle\rho, \tau\rangle$ is a block system of $\langle\rho, \tau\rangle$.

Solution By Lemma 1.5.1, we have $\langle\rho\rangle \trianglelefteq \langle\rho, \tau\rangle$, so by Theorem 2.2.7 the set of orbits of $\langle\rho\rangle$ is a block system of $\langle\rho, \tau\rangle$. If Γ is an (m,n)-metacirculant digraph, then the sets $V_i = \{(i,j) : j \in \mathbb{Z}_n\}$ are blocks of $\langle\rho, \tau\rangle$, and our use of the term "block" in this section and in Section 1.5 is consistent. It need not be the case though that V_i is a block of $\mathrm{Aut}(\Gamma)$, as we shall see in Theorem 2.4.9! □

Definition 2.2.10 A block system of a transitive group $G \leq S_X$ that is the set of orbits of a normal subgroup of G is called a **normal block system of** G.

It is not the case that every block system is a normal block system, as we shall see in Example 2.3.9 and Theorem 2.4.11.

Definition 2.2.11 Suppose $G \leq S_X$ is a transitive group with a block system $\mathcal{B}$. Then G has an **induced action on** $\mathcal{B}$. Namely, for $g \in G$, define $g/\mathcal{B}\colon \mathcal{B} \to \mathcal{B}$ by $g/\mathcal{B}(B) = B'$ if and only if $g(B) = B'$, and set $G/\mathcal{B} = \{g/\mathcal{B} : g \in G\}$. We also define the **fixer of** $\mathcal{B}$ **in** G, denoted $\mathrm{fix}_G(\mathcal{B})$, to be $\{g \in G : g/\mathcal{B} = 1\}$. That is, $\mathrm{fix}_G(\mathcal{B})$ is the subgroup of G that fixes each block of $\mathcal{B}$ set-wise.

Observe that $\mathrm{fix}_G(\mathcal{B})$ is the kernel of the induced homomorphism $G \to S_{\mathcal{B}}$, and as such is normal in G. Additionally, by the first isomorphism theorem, $|G| = |G/\mathcal{B}| \cdot |\mathrm{fix}_G(\mathcal{B})|$. If $\mathcal{B}$ is a normal block system of G that is the set of orbits of $N \trianglelefteq G$, then $N \leq \mathrm{fix}_G(\mathcal{B})$ although equality need not hold.

Example 2.2.12 Let G be a group and let $N \trianglelefteq G$. The set of orbits of $\bar{N}_L$ is a block system $\mathcal{B}$ of G_L, $\mathrm{fix}_{G_L}(\mathcal{B}) = \bar{N}_L$, and $G_L/\mathcal{B} \cong G/N$. As above, $\bar{N}_L = \{n_L : n \in N\}$.

Solution By Theorem 2.2.7, the set of orbits of $\bar{N}_L \trianglelefteq G_L$ is a block system $\mathcal{B}$ of G_L. We observed above that $\bar{N}_L \leq \mathrm{fix}_{G_L}(\mathcal{B})$ as $\mathcal{B}$ is the set of orbits of $\bar{N}_L$. As G_L is regular, $\mathrm{fix}_{G_L}(\mathcal{B})$ is semiregular. If $1 \neq h \in \mathrm{fix}_{G_L}(\mathcal{B})$, then $h = g_L$ for some $g \in G$. As $g_L \in \mathrm{fix}_{G_L}(\mathcal{B})$, the orbits of g_L are contained in the orbits of $\bar{N}_L$. So if $g_L(x) = y$, there is $n \in N$ such that $n_L(y) = x$. Then $g_L n_L = 1$ and $g_L = (n_L)^{-1} \in \bar{N}_L$. Thus $\mathrm{fix}_{G_L}(\mathcal{B}) = \bar{N}_L$. To finish, by the first isomorphism theorem,

$$G_L/\mathcal{B} \cong G_L/\mathrm{fix}_{G_L}(\mathcal{B}) = G_L/\bar{N}_L \simeq G/N. \qquad \square$$

Example 2.2.13 If Γ is an (m, n)-metacirculant digraph, then $\langle\rho\rangle \trianglelefteq \langle\rho, \tau\rangle = G \leq \mathrm{Aut}(\Gamma)$, and $\mathcal{B} = \{V_i : i \in \mathbb{Z}_m\}$ is a block system of G (recall $V_i = \{(i, j) : j \in \mathbb{Z}_n\}$). Then $\mathrm{fix}_G(\mathcal{B}) = \langle\rho, \tau^m\rangle$ and $G/\mathcal{B} \cong \mathbb{Z}_m$.

Solution That $\mathcal{B}$ is a block system of G follows from Example 2.2.9. As $\tau^m(i, j) = (i, \alpha^m i)$, $\tau^m \in \mathrm{fix}_G(\mathcal{B})$. Let a be the multiplicative order of α in $\mathbb{Z}_n^*$. Then $\langle\rho, \tau^m\rangle$ has order $n \cdot \mathrm{lcm}(a, m)/m$ as τ has order $\mathrm{lcm}(a, m)$ by Lemma 1.5.1. As G is transitive, $G/\mathcal{B}$ is transitive and so has order at least m. By Lemma 1.5.1, $|G| = \mathrm{lcm}(a, m) \cdot n$. This forces $\mathrm{fix}_G(\mathcal{B}) = \langle\rho, \tau^m\rangle$ and $G/\mathcal{B} \cong \mathbb{Z}_m$ by order arguments. □

Our next goal is Theorem 2.2.19, which gives a sufficient condition to ensure that a block system of G is normal, and exploits the following obvious fact that is stated as a proposition to emphasize its importance. We will also need some additional terminology and a few results before proving Theorem 2.2.19.

Proposition 2.2.14 *If $G \leq S_X$ is transitive with block system $\mathcal{B}$, and $H \leq G$ is transitive, then $\mathcal{B}$ is also a block system of H.*

Definition 2.2.15 Let $G \leq S_X$ be transitive with block systems $\mathcal{B}$ and $\mathcal{C}$. We write $\mathcal{B} \leq \mathcal{C}$ if each block of $\mathcal{B}$ is contained in a block of $\mathcal{C}$, and we say $\mathcal{B}$ is a **refinement** of $\mathcal{C}$ (recall that a **refinement** of a partition P of a set X is simply a partition P' of X such that every set in P is a union of sets in P'). As the trivial block system $\mathcal{D}$ of G consisting of singleton sets is a refinement of every block system of G, $\mathcal{D}$ is called a **trivial refinement** of $\mathcal{B}$. Similarly, $\mathcal{B}$ is a trivial refinement of itself.

Recall that the **center of a group** G, denoted $Z(G)$, is the set of all elements of G that commute with every element of G. That is, $Z(G) = \{g \in G : gh = hg \text{ for all } h \in G\}$. It is easy to verify that $Z(G)$ is a normal, in fact characteristic, subgroup of G.

Lemma 2.2.16 *The center $Z(G)$ of a transitive group $G \leq S_n$ is semiregular.*

Proof Suppose $\alpha \in Z(G)$ and $\alpha(x) = x$ for some $x \in \mathbb{Z}_n$. As G is transitive, for each $y \in \mathbb{Z}_n$ there exists $g_y \in G$ such that $g_y(x) = y$. Then $\alpha g_y = g_y\alpha$ and so

$$y = g_y(x) = g_y\alpha(x) = \alpha g_y(x) = \alpha(y),$$

and so $\alpha(y) = y$ for all $y \in \mathbb{Z}_n$. □

The following result is an immediate consequence of the preceding result, and we call it a theorem to emphasize its importance.

Theorem 2.2.17 *A transitive abelian permutation group is regular.*

We may also immediately deduce the following result.

Corollary 2.2.18 *A transitive abelian permutation group of degree n is self-centralizing in S_n.*

Theorem 2.2.19 *Let $G \leq S_n$ be transitive with an abelian regular subgroup A. Then any block system of G is normal, and is the set of orbits of a subgroup of A.*

Proof Let $\mathcal{B}$ be a block system of G. By Proposition 2.2.14, $\mathcal{B}$ is a block system of A. We only need show that $\mathrm{fix}_A(\mathcal{B})$ has orbits of size $|B|$, $B \in \mathcal{B}$. Now, $A/\mathcal{B}$ is transitive and abelian, and by Theorem 2.2.17, $A/\mathcal{B}$ is regular. Thus $A/\mathcal{B}$ has degree $|\mathcal{B}|$, and so there exists nontrivial $K \leq \mathrm{fix}_A(\mathcal{B})$ of order

$|B|$. Then the set of orbits of K is a block system $\mathcal{C}$ of A by Theorem 2.2.7, and $\mathcal{C} \leq \mathcal{B}$. As K has order $|B|$, we conclude $\mathcal{C} = \mathcal{B}$. □

2.3 More Ways of Finding Blocks

Our main focus in this section is on how to find a block system of a permutation group G without showing that it is the set of orbits of a normal subgroup of G. In Theorem 2.3.7, we will give a theoretical method to find *all* block systems of a permutation group G. Our first result should not be too much of a surprise. We know that for every partition of a set, there is a corresponding equivalence relation and for every equivalence relation there is a corresponding partition. As a block system is a partition that is invariant under a group, there must be an equivalence relation corresponding to the block system, and that equivalence relation must also be invariant under the group. Also, recall that it is often much easier to find the equivalence relation corresponding to a partition than to find the partition directly.

Definition 2.3.1 Let $G \leq S_X$ be transitive. An equivalence relation $\equiv$ on X such that $x \equiv y$ if and only if $g(x) \equiv g(y)$ for all $g \in G$ is called a **G-congruence**.

Theorem 2.3.2 *Let $G \leq S_X$ be transitive. If $\equiv$ is a G-congruence then the set of equivalence classes of $\equiv$ is a block system of G.*

Proof For $x \in X$, let B_x be the equivalence class of $\equiv$ that contains x. Let $x \in X$ and $g \in G$. Then,

$$\begin{aligned} g(B_x) &= \{g(y) : y \in X \text{ and } x \equiv y\} \\ &= \{g(y) : y \in X \text{ and } g(y) \equiv g(x)\} \\ &= B_{g(x)}. \end{aligned}$$

As the set of equivalence classes of $\equiv$ is a partition of X, it follows that $g(B_x) \cap B_x = B_{g(x)} \cap B_x = \emptyset$ or B_x, and so B_x is a block of G. Also, as $g(B_x) = B_{g(x)}$, the set of all blocks conjugate to B_x is just the set of equivalence classes of $\equiv$. □

Theorem 2.2.7 is a special case of the Theorem 2.3.2 (Exercise 2.3.1). Also, the converse of Theorem 2.3.2 is true, but is not typically used (Exercise 2.3.2). Our next main goal is to prove Theorem 2.3.7, which gives a theoretical method for determining all of the block systems of a transitive group G. One of the central ideas is given in the following result.

Lemma 2.3.3 *Let $G \leq S_n$ be transitive and $x \in \mathbb{Z}_n$. Let $H \leq G$ be such that* $\mathrm{Stab}_G(x) \leq H$. *Then the orbit of H that contains x is a block of G.*

Proof Set $B = \{h(x) : h \in H\}$ (so that B is the orbit of H that contains x), and let $g \in G$. We must show that B is a block of G, that is $g(B) = B$ or $g(B) \cap B = \emptyset$. Clearly if $g \in H$, then $g(B) = B$ as B is the orbit of H that contains x and $x \in B$. If $g \notin H$, then, looking for a contradiction, suppose $g(B) \cap B \neq \emptyset$, with $z \in g(B) \cap B$. Then there exists $y \in B$ such that $g(y) = z$ and $h, k \in H$ such that $h(x) = y$ and $k(x) = z$. Then $k(x) = z = g(y) = gh(x)$, and so $gh(x) = k(x)$. Thus $k^{-1}gh \in \mathrm{Stab}_G(x)$. This implies that $g \in k \cdot \mathrm{Stab}_G(x) \cdot h^{-1} \leq H$, a contradiction. Thus if $g \notin H$, then $g(B) \cap B = \emptyset$, and B is a block of G. □

Definition 2.3.4 Let $G \leq S_n$ be a transitive group and B a block of G. The **stabilizer of the block** B, denoted $\mathrm{Stab}_G(B)$, is the set of all $g \in G$ that fix B set-wise. That is, $\mathrm{Stab}_G(B) = \{g \in G : g(B) = B\}$.

It is straightforward to verify that $\mathrm{Stab}_G(B)$ is a subgroup of G. Like stabilizers of points, stabilizers of blocks are conjugate – see Exercise 2.3.10. Also, observe that $\mathrm{Stab}_G(B)/\mathcal{B} = \mathrm{Stab}_{G/\mathcal{B}}(B)$. That is, the action of the stabilizer of B in G on $\mathcal{B}$ is the stabilizer of the point B in $G/\mathcal{B}$. Additionally, $\mathrm{fix}_G(\mathcal{B}) = \cap_{B\in\mathcal{B}}\mathrm{Stab}_G(B)$.

Example 2.3.5 Let G be a group with $H \leq G$. We saw in Example 2.2.3 that the left cosets of H in G are a block system of G_L. Let $\bar{H}_L = \{h_L : h \in H\}$. Then $\mathrm{Stab}_{G_L}(H) = \bar{H}_L$, and if $g \in G$, then $\mathrm{Stab}_{G_L}(gH) = g_L\bar{H}_L g_L^{-1}$.

Proof As $gH = H$ if and only if $g \in H$, we see that $\mathrm{Stab}_{G_L}(H) \leq \bar{H}_L$. As $h_L(H) = H$ for every $h \in H$, $\mathrm{Stab}_{G_L}(H) = \bar{H}_L$. Let $g \in G$. Then $g_L h_L g_L^{-1}(gH) = gH$ so $\mathrm{Stab}_{G_L}(gH) \leq g_L\bar{H}_L g_L^{-1}$. For the other containment, if $k \in G$ is such that $k_L(gH) = gH$, then $kgH = gH$ and so $g^{-1}kgH = H$ and $g^{-1}kg = h \in H$. Then $k = ghg^{-1}$, and, as $(ghg^{-1})_L = g_L h_L g_L^{-1}$, we have $\mathrm{Stab}_{G_L}(gH) = g_L\bar{H}_L g_L^{-1}$. □

Example 2.3.6 Recall from Example 2.2.5 that for $\Gamma = \mathrm{Cay}(\mathbb{Z}_6, \{2,3,4\})$, $\mathrm{Aut}(\Gamma)$ has block systems $\mathcal{B}$ and C with two blocks of size 3 and three blocks of size 2, respectively. Then $\mathrm{Stab}_{\mathrm{Aut}(\Gamma)}(B) \cong S_3$ while $\mathrm{Stab}_{\mathrm{Aut}(\Gamma)}(C) \cong \mathbb{Z}_2 \times \mathbb{Z}_2$, where $B = \{0,2,4\} \in \mathcal{B}$ and $C = \{0,3\} \in C$. Also, $\mathrm{fix}_{\mathrm{Aut}(\Gamma)}(C) \cong \mathbb{Z}_2$ and $\mathrm{Stab}_{\mathrm{Aut}(\Gamma)}(C)^C \cong \mathbb{Z}_2$.

Solution By Exercise 2.2.5, $\mathrm{Aut}(\Gamma) = D_{12} \cong S_3 \times S_2$ (if you have not worked this exercise, feel free to replace $\mathrm{Aut}(\Gamma)$ with D_{12} in this example). So $\mathrm{Aut}(\Gamma) = \{x \mapsto ax + b : a = \pm 1 \text{ and } b \in \mathbb{Z}_6\}$. Note that $-\{0,2,4\} = \{0,4,2\}$ and $2 + \{0,2,4\} = \{2,4,0\}$ so $H = \{x \mapsto ax + b : a = \pm 1 \text{ and } b = 0,2,4\}$ stabilizes $\{0,2,4\}$ and is isomorphic to S_3. Also, $3+\{0,2,4\} \neq \{0,2,4\}$. As $[D_{12} : H] = 2$,

$\mathrm{Stab}_{\mathrm{Aut}(\Gamma)}(B) = H \cong S_3$. Also, $-\{0,3\} = \{0,3\}$ and $3 + \{0,3\} = \{3,0\}$ so $K = \{x \mapsto ax+b : a = \pm 1, b = 0, 3\} \leq \mathrm{Stab}_{\mathrm{Aut}(\Gamma)}(C)$. As $2 + \{0,3\} \neq \{0,3\}$ and $[D_{12} : K] = 3$ is prime, $\mathrm{Stab}_{\mathrm{Aut}(\Gamma)}(C) = K$. As $|K| = 4$ and is clearly not cyclic, $\mathrm{Stab}_{\mathrm{Aut}(\Gamma)}(C) \cong \mathbb{Z}_2 \times \mathbb{Z}_2$. Also, $-\{1,4\} = \{2,5\}$ so $x \mapsto -x$ is not contained in $\mathrm{fix}_{\mathrm{Aut}(\Gamma)}(C)$ while $x \mapsto x+3$ is. So $\mathrm{fix}_{\mathrm{Aut}(\Gamma)}(C) \cong \mathbb{Z}_2$. Finally, $x \mapsto -x$ fixes 0 and 3 so $\mathrm{Stab}_{\mathrm{Aut}(\Gamma)}(C)^C \cong \mathbb{Z}_2$. □

Theorem 2.3.7 *Let $G \leq S_X$ be transitive, and let $x \in X$. Let Ω be the set of all blocks of G that contain x, and S be the set of all subgroups of G that contain* $\mathrm{Stab}_G(x)$. *Define $\phi\colon \Omega \to S$ by $\phi(B) = \mathrm{Stab}_G(B)$. Then ϕ is a bijection, B is the orbit of* $\mathrm{Stab}_G(B)$ *that contains x, and if $B, C \in \Omega$, then $B \subseteq C$ if and only if* $\mathrm{Stab}_G(B) \leq \mathrm{Stab}_G(C)$.

Proof Observe that $\mathrm{Stab}_G(x) \leq \mathrm{Stab}_G(B)$ for every block B with $x \in B$, so ϕ is indeed a map from Ω to S. We first show that ϕ is onto. Let $H \in S$ so that $\mathrm{Stab}_G(x) \leq H$. By Lemma 2.3.3, $B = \{h(x) : h \in H\}$ is a block of G. Then $H \leq \phi(B)$. Looking for a contradiction, suppose there exists $g \in \phi(B)$ such that $g \notin H$. Then $g(B) = B$, and H is transitive in its action on B as B is an orbit of H. Hence there exists $h \in H$ such that $h(x) = g(x)$, and so $h^{-1}g(x) = x \in \mathrm{Stab}_G(x) \leq H$. Thus $h^{-1}g \in H$ so $g \in H$, a contradiction. Thus $\phi(B) = H$ and ϕ is onto.

We now show that ϕ is one-to-one. Suppose $B, C \in \Omega$ and $\phi(B) = \phi(C)$. Then $\mathrm{Stab}_G(B) = \mathrm{Stab}_G(C)$. Assume, by way of contradiction, that $y \in B$ but $y \notin C$. As $\mathrm{Stab}_G(B)$ is transitive on B by Exercise 2.2.7, there exists $h \in \mathrm{Stab}_G(B)$ such that $h(x) = y$. But then $h \in \mathrm{Stab}_G(B) = \mathrm{Stab}_G(C)$ and so y is in the orbit of $\mathrm{Stab}_G(C)$ that contains x, which is C, a contradiction. Thus ϕ is one-to-one and onto, and so a bijection.

Finally, it remains to show that if $B, C \in \Omega$, then $B \subseteq C$ if and only if $\mathrm{Stab}_G(B) \leq \mathrm{Stab}_G(C)$. First suppose $\mathrm{Stab}_G(B) \leq \mathrm{Stab}_G(C)$. Then the orbit of $\mathrm{Stab}_G(C)$ that contains x certainly contains the orbit of $\mathrm{Stab}_G(B)$ that contains x, and so $B \subseteq C$. Conversely, suppose $B \subseteq C$. Let $g \in \mathrm{Stab}_G(B)$. Then $g(x) \in B \subseteq C$, and so $x \in C \cap g(C)$. As C is a block of G, we have $g(C) = C$ so $g \in \mathrm{Stab}_G(C)$. Thus $\mathrm{Stab}_G(B) \leq \mathrm{Stab}_G(C)$. □

Theorem 2.3.7 gives us a theoretical method for determining all block systems of a transitive groups $G \leq S_X$. First, we calculate $\mathrm{Stab}_G(x)$ for some $x \in X$. We then determine the subgroups of G that contain $\mathrm{Stab}_G(x)$. The orbits of these subgroups that contain x will be the set of all blocks of G that contain x. To determine the block systems, we then consider the images of each of the blocks that contain x under elements of G. Notice that $\mathrm{Stab}_G(x)$ corresponds to the trivial block system of G consisting of singletons, while G

itself corresponds to the trivial block system consisting of X. We illustrate this in the following examples.

Example 2.3.8 Let G be a group and $\mathcal{P}$ a partition of G. Then $\mathcal{P}$ is a block system of G_L if and only if $\mathcal{P}$ is the set of left cosets of some subgroup $H \leq G$.

Solution We saw in Example 2.2.3 that the set of left cosets of a subgroup $H \leq G$ is a block system of G_L. Conversely, suppose $\mathcal{P}$ is a partition of G that is a block system of G_L. Let $P \in \mathcal{P}$ with $1 \in P$. By Theorem 2.3.7 there is a subgroup $\bar{H}_L \leq G_L$ such that the orbit of $\bar{H}_L$ that contains 1 is P. Let $H = \{h : h_L \in \bar{H}_L\}$. It is straightforward to verify that as $\bar{H}_L$ is a subgroup of G_L, H is a subgroup of G. As the orbit of $\bar{H}_L$ that contains 1 is H, H is a block of G_L. The set of blocks conjugate to H is then $\mathcal{P}$, and is also the set of left cosets of H in G. □

Example 2.3.9 Define $\rho, \tau \colon \mathbb{Z}_2 \times \mathbb{Z}_5 \to \mathbb{Z}_2 \times \mathbb{Z}_5$ by $\rho(i, j) = (i, j+1)$ and $\tau(i, j) = (i+1, 2j)$. For $i \in \mathbb{Z}_2$, set $V_i = \{(i, j) : j \in \mathbb{Z}_5\}$ and for $j \in \mathbb{Z}_5$, set $V^j = \{(i, j) : i \in \mathbb{Z}_2\}$. Then the only nontrivial block systems of $\langle \rho, \tau \rangle$ are the sets $\{V_i : i \in \mathbb{Z}_2\}$ and $\{V^j : j \in \mathbb{Z}_5\}$. Additionally, the block system $\{V^j : j \in \mathbb{Z}_5\}$ is not normal.

Solution Set $G = \langle \rho, \tau \rangle$. Then $|G| = 20$ by Lemma 1.5.1 and by the orbit-stabilizer theorem we have $|\mathrm{Stab}_G(0,0)| = 2$. Thus $\mathrm{Stab}_G(0,0) = \left\langle \tau^2 \right\rangle$. To apply Theorem 2.3.7, we determine all subgroups of G that contain $\left\langle \tau^2 \right\rangle$.

As $\langle \rho \rangle \trianglelefteq G$ and $\langle \rho \rangle$ has prime order 5, $\langle \rho \rangle$ is the unique Sylow 5-subgroup of G. As $G/\langle \rho \rangle \cong \mathbb{Z}_4$ and $\mathbb{Z}_4$ contains a unique subgroup of order 2 as it is cyclic, we see G contains a unique subgroup of order 10, which is necessarily $\left\langle \rho, \tau^2 \right\rangle$. Additionally, $\langle \tau \rangle$ is a Sylow 2-subgroup of G, and as G is nonabelian (recall $\tau\rho\tau^{-1} = \rho^2$), G cannot have a unique Sylow 2-subgroup. Thus by a Sylow Theorem, G must contain 5 Sylow 2-subgroups. These are $\rho^{-k}\langle \tau \rangle \rho^k$, $0 \leq k \leq 4$. As $\rho^{-k}\tau^2\rho^k(i, j) = (i, 4j + 3k)$, the only subgroup of order 4 of G that contains $\left\langle \tau^2 \right\rangle$ is $\langle \tau \rangle$. As the only possible orders of proper subgroups of G that contain τ^2 are 4 and 10 by the Lagrange theorem, $\left\langle \rho, \tau^2 \right\rangle$ and $\langle \tau \rangle$ are the only proper subgroups of G that properly contain $\left\langle \tau^2 \right\rangle$. Hence the subgroups of G that contain $\left\langle \tau^2 \right\rangle$ are $\left\langle \tau^2 \right\rangle$, $\langle \tau \rangle$, $\left\langle \rho, \tau^2 \right\rangle$, and G.

By Theorem 2.3.7 there are exactly four block systems G, two of which are trivial and, under the correspondence given in Theorem 2.3.7, these are $\mathrm{Stab}_G(0,0) = \langle \tau^2 \rangle$ and G. As the orbit of $\langle \tau \rangle$ that contains $(0,0)$ is V^0 and the orbit $\left\langle \rho, \tau^2 \right\rangle$ that contains $(0,0)$ is V_0, the only nontrivial blocks of G that contain $(0,0)$ are V^0 and V_0. The sets $\{V_i : i \in \mathbb{Z}_2\}$ and $\{V^j : j \in \mathbb{Z}_5\}$ are the block systems of G that contain V_0 and V^0 as ρ fixes V_0 and $\tau(V_i) = V_{i+1}$,

$i \in \mathbb{Z}_2$, while τ fixes V^0 and $\rho(V^j) = V^{j+1}$, $j \in \mathbb{Z}_5$ (addition in the superscript is done modulo 5). As $\langle\tau\rangle$ is not normal in G, $\{V^j : j \in \mathbb{Z}_5\}$ is not a normal block system of G. □

Our next result is concerned with finding refinements of a known block system, or, equivalently, finding blocks that are subsets of known blocks. While one could prove it using Theorem 2.3.7 (Exercise 2.3.3), it is easier to prove by simply using the definition.

Theorem 2.3.10 *Let $G \leq S_n$ be transitive with block system $\mathcal{B}$, and let $B \in \mathcal{B}$. If C is a block of* $\mathrm{Stab}_G(B)^B$*, then C is a block of G.*

Proof Let $g \in G$ and suppose $g(C) \cap C \neq \emptyset$. Then there exists $c \in C$ such that $g(c) \in C$. As $C \subseteq B$, $c \in B$, and so $g(B) \cap B \neq \emptyset$. As B is a block of G, $g(B) = B$, so $g \in \mathrm{Stab}_G(B)$. But then $g(C) = C$ as C is a block of $\mathrm{Stab}_G(B)^B$. □

A consequence of this result is frequently useful, and its proof is left as an exercise (Exercise 2.3.9).

Corollary 2.3.11 *Let $G \leq S_n$ be transitive with block system $\mathcal{B}$. Then $\mathcal{B}$ has no nontrivial refinements if and only if* $\mathrm{Stab}_G(B)$ *is primitive in its action on B, $B \in \mathcal{B}$.*

Recall that a **lattice** is a partially ordered set in which each pair of elements has a greatest lower bound and a least upper bound. The following two results essentially give that the set of all block systems of G are a lattice under $\preceq$.

Theorem 2.3.12 *Let $G \leq S_n$ be transitive with block systems $\mathcal{B}_1, \ldots, \mathcal{B}_k$. Then there is a smallest block system $\mathcal{B}$ of G for which $\mathcal{B}_i \preceq \mathcal{B}$ for every $1 \leq i \leq k$.*

Proof Let $x \in \mathbb{Z}_n$, and $H = \langle \mathrm{Stab}_G(B_i) : B_i \in \mathcal{B}_i \text{ with } x \in B_i\rangle$. Then $H \leq G$ and contains $\mathrm{Stab}_G(x)$ as $\mathrm{Stab}_G(x) \leq \mathrm{Stab}_G(B_i)$ for every $B_i \in \mathcal{B}_i$ with $x \in B_i$. By Theorem 2.3.7 there is a block system $\mathcal{B}$ of G with $\mathcal{B}_i \preceq \mathcal{B}$ for every $1 \leq i \leq k$, and if $B \in \mathcal{B}$ with $x \in B$, then B is the orbit of $H = \mathrm{Stab}_G(B)$ that contains x. Now suppose $\mathcal{C}$ is a block system of G with $\mathcal{B}_i \preceq \mathcal{C}$ with $C \in \mathcal{C}$ such that $x \in C$. Then $\mathrm{Stab}_G(B_i) \leq \mathrm{Stab}_G(C)$ for every $1 \leq i \leq k$ and so $H \leq \mathrm{Stab}_G(C)$. We conclude that $B \subseteq C$ and so $\mathcal{B} \preceq \mathcal{C}$. □

Definition 2.3.13 The block system $\mathcal{B}$ of G in the previous result is the block system **generated** by $\mathcal{B}_1, \ldots, \mathcal{B}_k$.

Lemma 2.3.14 *Let $G \leq S_n$ be transitive, $\mathcal{B}_1, \ldots, \mathcal{B}_k$ be block systems of G, $x \in \mathbb{Z}_n$, and $B_i \in \mathcal{B}_i$ such that $x \in B_i$. Then $\cap_{i=1}^k B_i$ is a block of G.*

Proof Let $D = \cap_{i=1}^k B_i$, and $g \in G$ such that $g(D) \cap D \neq \emptyset$. Then $g(B_i) \cap B_i \neq \emptyset$ for any $1 \leq i \leq k$, so $g(B_i) = B_i$. Hence $g(D) = D$. □

2.4 Primitive and Quasiprimitive Groups

This section is in many ways a continuation of the previous section, but instead of focusing on imprimitive permutation groups, we focus on groups that have no nontrivial block systems or have only block systems that are not the set of orbits of a normal subgroup. As we have not yet seen any examples of primitive groups, we remedy this deficiency first.

Example 2.4.1 S_n is primitive for all $n \geq 1$.

Solution Let $n \geq 1$, and $B \subset \mathbb{Z}_n$ that is not a singleton set. There is no such set if $n \leq 2$ so there are no imprimitive groups of degree 1 or 2. Assume $n \geq 3$. Then $2 \leq |B| < n$, and there is $x, y, z \in \mathbb{Z}_n$ with $x, y \in B$ and $z \notin B$. Then there exists $\delta \in S_n$ with $\delta(x) = x$ and $\delta(y) = z$. Then $\delta(B) \cap B \neq \emptyset$ but $\delta(B) \neq B$, a contradiction. Thus the only blocks of S_n are $\mathbb{Z}_n$ and singleton sets, so S_n is primitive. □

Definition 2.4.2 A permutation group $G \leq S_n$ is **2-transitive** if whenever $(x_1, y_1), (x_2, y_2) \in \mathbb{Z}_n \times \mathbb{Z}_n$ such that $x_1 \neq y_1$ and $x_2 \neq y_2$, then there exists $g \in G$ such that $g(x_1, y_1) = (x_2, y_2)$. That is, a group is 2-transitive provided that for any two nondiagonal ordered pairs, there is a group element that maps one to the other.

Readers should be aware that some authors use **doubly-transitive** for 2-transitive, and some authors use 2-transitive for 2-arc-transitive (see Definition 3.1.1).

Example 2.4.3 The group S_n is 2-transitive for all $n \geq 2$ and the alternating group A_n is 2-transitive for $n \geq 4$.

Solution Let $(x_1, y_1), (x_2, y_2) \in \mathbb{Z}_n \times \mathbb{Z}_n$ with $x_1 \neq y_1$ and $x_2 \neq y_2$. For S_n it is clear that there is $f \in S_n$ with $f(x_1) = x_2$ and $f(y_1) = y_2$ as $x_1 \neq y_1$ and $x_2 \neq y_2$. For A_n, clearly the identity maps (x_1, y_1) to (x_1, y_1). If $x_1 = x_2$ and $y_1 \neq y_2$ then the 3-cycle (y_1, y_2, z), where $z \in \mathbb{Z}_n$ is different from x_1, y_1, and y_2, maps (x_1, y_1) to (x_2, y_2) and is even. A similar map takes (x_1, y_1) to (x_2, y_2) when $x_1 \neq x_2$ and $y_1 = y_2$. Finally, if $x_1 \neq x_2$ and $y_1 \neq y_2$, then the permutation $(x_1, x_2)(y_1, y_2)$ maps (x_1, y_1) to (x_2, y_2) and is even. □

This example easily generalizes to show that any 2-transitive group is primitive (Exercise 2.4.7). Digraphs whose automorphism groups are 2-transitive are not terribly interesting, as a 2-transitive automorphism group of a digraph is necessarily a symmetric group.

Lemma 2.4.4 *Let Γ be a digraph of order n with a 2-transitive automorphism group. Then $\Gamma \cong K_n$ or its complement, and* $\mathrm{Aut}(\Gamma) = S_n$.

Proof Suppose $\mathrm{Aut}(\Gamma)$ is 2-transitive and $A(\Gamma) \neq \emptyset$. Let $(x, y) \in A(\Gamma)$. As $\mathrm{Aut}(\Gamma)$ is 2-transitive, for any ordered pair (w, z) with $w, z \in V(\Gamma)$ and $w \neq z$, there is a $g \in \mathrm{Aut}(\Gamma)$ with $g(x, y) = (w, z)$, in which case $A(\Gamma)$ contains every possible arc. Thus $\Gamma \cong K_n$ and $\mathrm{Aut}(\Gamma) = S_n$. □

Definition 2.4.5 A group $G \leq S_n$ that is primitive but not 2-transitive is **simply primitive** or **uniprimitive**.

The following example is somewhat more interesting.

Example 2.4.6 A transitive group G of prime degree p is primitive.

Solution The only divisors of p are 1 and p, and so the only possible blocks are singleton sets, the entire set that G permutes. □

Our main goal for the rest of this section is modest. We will develop some basic results concerning primitive permutation groups, and give a more interesting example by showing the automorphism group of the Petersen graph, $\mathrm{Aut}(P)$, is uniprimitive. We will then give an example of a related class of groups, quasiprimitive groups. We finish with group-theoretic characterizations of these ideas. We begin with some consequences of results we have already seen for primitive groups.

Theorem 2.4.7 *A nontrivial normal subgroup of a primitive group is transitive.*

Proof Let $G \leq S_n$ be primitive with $N \trianglelefteq G$. Suppose N is not transitive. By Theorem 2.2.7, the set of orbits of N are a block system $\mathcal{B}$ of G and, as N is nontrivial, $\mathcal{B}$ is nontrivial. Thus G is imprimitive, a contradiction. □

Theorem 2.4.8 *A transitive group $G \leq S_n$ is primitive if and only if* $\mathrm{Stab}_G(x)$ *is a maximal subgroup of G for every $x \in \mathbb{Z}_n$.*

Proof By definition, a group G is primitive if and only if its only block systems are trivial, which by Theorem 2.3.7 occurs if and only if the only subgroups of G that contain $\mathrm{Stab}_G(x)$ are $\mathrm{Stab}_G(x)$ and G. So G is primitive if and only if $\mathrm{Stab}(x)$ is maximal in G. □

Theorem 2.4.9 *The automorphism group of the Petersen graph P is primitive on $V(P)$.*

Proof By Example 1.5.9 and Theorem 1.5.11, using the metacirculant labeling of $V(P)$, we see $\langle \rho, \tau \rangle \leq \mathrm{Aut}(P)$ with $\alpha = 2$. In view of Example 2.3.9

and Proposition 2.2.14, we need only show that neither V_0 nor V^0 are blocks of Aut(P). This is easiest to do using the 2-subset labeling of $V(P)$ as in Figure 2.1, in which case $V_0 = \{\{1,2\},\{4,5\},\{1,3\},\{2,5\},\{3,4\}\}$ and $V^0 = \{\{1,2\},\{3,5\}\}$. By Exercise 2.1.3 P is arc-transitive and so edge-transitive. Specifically, we may map the edge $\{\{1,2\},\{3,5\}\}$ to the edge $\{\{1,2\},\{3,4\}\}$. So V^0 is not a block of P. Many permutations will show that V_0 is not a block of Aut(P). We need only find a permutation that maps an element of V_0 to some element of V_0 while simultaneously mapping a vertex in V_0 to a vertex not in V_0. We choose the permutation $\delta = (3,4)$. Then $\delta(\{1,2\}) = \{1,2\}$ and $\delta(\{4,5\}) = \{3,5\}$. □

In light of the previous proof it should be clear why we put "outside" and "inside" in quotes when referring to the "outside" and "inside" 5-cycles of the Petersen graph. Having an outside and inside 5-cycle suggests that the vertices in these 5-cycles are blocks of Aut(P), which they are not (although they are blocks of the subgroup $\langle\rho,\tau\rangle$ of Aut(P)).

We have seen in Theorem 2.4.7 that a normal subgroup of a primitive group is transitive. Is the converse true? That is, is it true that every transitive group all of whose nontrivial normal subgroups are transitive is primitive? The answer as we shall see in our next result is *no*. Before turning to this result, we will need a term.

Definition 2.4.10 A transitive group G is **quasiprimitive** if every nontrivial normal subgroup of G is transitive.

So a primitive group is quasiprimitive by Corollary 2.4.7, and a transitive group G is quasiprimitive but not primitive if it is imprimitive but no nontrivial block system of G is normal.

Recall that the line graph of a graph Γ is denoted $L(\Gamma)$. The line graph of the Petersen graph is given in Figure 2.3 (the reason for the coloring of the vertices will be given in the proof of the next result).

Theorem 2.4.11 *Aut($L(P)$) $\cong S_5$ is a quasiprimitive group that is not primitive, with a nonnormal block system consisting of* 5 *blocks of size* 3.

Proof As P is arc-transitive by Exercise 2.1.3, P is edge-transitive. So Aut(P) acts transitively on $E(P)$ with kernel of index at least $15 = |E(P)|$. As, by Exercise 2.4.13, Aut(P) $= S_5$ has only itself and A_5 as nontrivial normal subgroups, we see the permutation representation $G \le S_{15}$ corresponding to this action is faithful and isomorphic to S_5. It was shown in the proof of Theorem 2.1.4 that there are five maximal independent sets $I_1,\ldots,I_5$ of size 4 in P, and Aut(P) acts transitively on these five independent sets. Hence Aut(P)

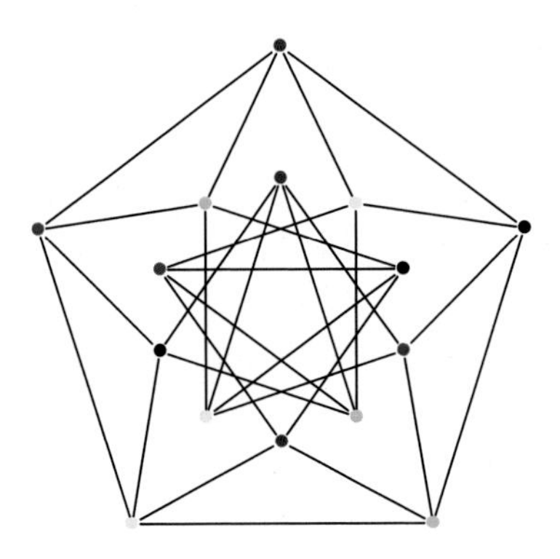

Figure 2.3 The line graph of the Petersen graph.

also acts transitively on the subgraphs E_k of P induced by $V(P)\backslash I_k$, $1 \le k \le 5$. These subgraphs consist of three independent edges, and partition $E(P)$. Each one is colored with a different color in Figure 2.4. Consequently, the set of these five sets of three edges each are a block system $\mathcal{E}$. The vertices of $L(P)$ are colored in Figure 2.3 the same color as they are in Figure 2.4. Also, the "outside" 5-cycle C in $L(P)$ corresponds to the edges of the "outside" 5-cycle in P, the "inside" 5-cycle of P corresponds to the vertices in $L(P)\backslash V(C)$ of valency 4, and the 5 vertices of $L(P)\backslash V(C)$ of valence 2 correspond to the "spoke" edges of P.

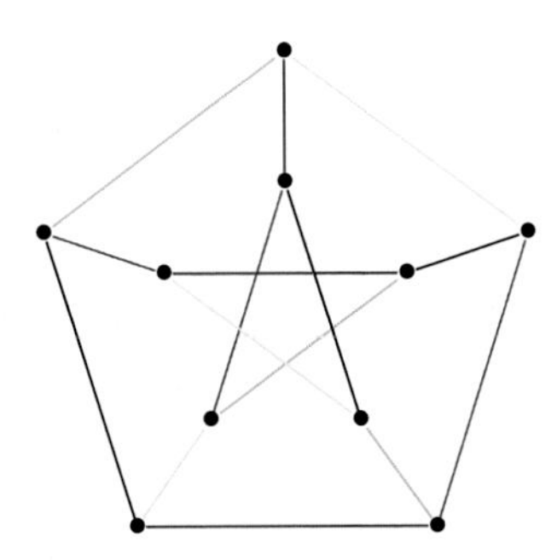

Figure 2.4 The Petersen graph with each edge of E_j a different color.

If we are willing to invoke Whitney's theorem (Theorem 2.1.1) we see $\mathrm{Aut}(L(P)) = G$ and $\mathrm{Aut}(L(P))$ is imprimitive with $\mathcal{E}$ a block system, which is not normal as S_5 only has A_5 as a proper nontrivial normal subgroup.

To see this without Whitney's theorem, we will first show $\mathcal{E}$ is also a block system of $\mathrm{Aut}(L(P))$. Fix $1 \le k \le 5$. Let $\delta \in \mathrm{Aut}(L(P))$ such that $\delta(E_k) \cap E_k \ne \emptyset$

with $e_k \in E_k$ such that $\delta(e_k) = e'_k \in E_k$. Then there exists $g \in G$ such that $g(e'_k) = e_k$ so $g\delta(e_k) = e_k$. Also, observe that $\delta(E_k) = E_k$ if and only if $g\delta(E_k) = E_k$ as $\mathcal{E}$ is a block system of G. Let $B = \{e \in E(P) : \text{dist}_{L(P)}(e, e_k) \le 2\}$. Observe that no two edges in E_k are at distance less than 3 in $L(P)$, so $B \cap E_k = \{e_k\}$. As $g\delta$ fixes e_k, $g\delta(B) = B$ and $g\delta(E(P)\backslash B) = E(P)\backslash B$. It is easy to check that $E(P)\backslash B = E_k\backslash\{e_k\}$ and as $g\delta(e_k) = e_k$, $g\delta(E_k) = E_k$. Hence E_k is a block of $\text{Aut}(L(P))$ and $\mathcal{E}$ is a block system of $\text{Aut}(L(P))$.

Suppose $g \in \text{fix}_{\text{Aut}(L(P))}(\mathcal{E})$, so g can only permute vertices of the same color in Figure 2.3. Observe that each black vertex of $L(P)$ lies in two triangles. However, the colors of the triangles covering a given vertex are *different* for each black vertex, e.g., if e, e' and e'' are the black vertices, such that the two triangles covering e have colors {red, green} and {blue, yellow}, the two triangles covering e' have colors {yellow, green} and {blue, red} and the two triangles covering e'' have colors {red, green} and {blue, yellow}. Since g preserves the colors, g cannot move either of e, e' or e'', hence g fixes all black vertices. The same argument shows that g fixes all of the vertices, so $g = 1$. Then $\mathcal{E}$ is not normal. Hence the action of $\text{Aut}(L(P))$ on the set of blocks of $\mathcal{E}$ given by $g \cdot E = g(E)$ is faithful. Thus there is a injective homomorphism from $\text{Aut}(L(P))$ into $\mathcal{S}_5$. We conclude $\text{Aut}(L(P)) \cong \mathcal{S}_5$ and as its only proper nontrivial normal subgroup is A_5, $\text{Aut}(L(P))$ is quasiprimitive. □

Our next result is a standard result and gives the relationship between a primitive group and its orbital digraphs. It is sometimes called **Higman's theorem** after its discoverer (see Higman (1967, 1.12)).

Theorem 2.4.12 *A transitive permutation group G is primitive if and only if every orbital digraph of G is connected.*

Proof This is the traditional way that this result is stated, but it is much easier to establish the contrapositive in both directions. That is, to prove that a transitive permutation group G is imprimitive if and only if some orbital digraph of G is disconnected.

If some orbital digraph Γ of G is disconnected, then the components of Γ are isomorphic as G is transitive and the components are blocks of $G \le \text{Aut}(\Gamma)$ by Example 2.2.4. Conversely, if $\mathcal{B}$ is a nontrivial block system of G, then let $B \in \mathcal{B}$, and $x, y \in B$ such that $x \ne y$. Such x and y exist as $\mathcal{B}$ is nontrivial. Let Γ be the orbital digraph of G that contains the arc (x, y). If there is some arc in $A(\Gamma)$ between two vertices in different blocks of $\mathcal{B}$, say $a \in B_1$ and $b \in B_2$, then there would exist some $g \in G$ such that $g(x, y) = (a, b)$, in which case B would not be a block of G. Thus there are no arcs between any two vertices in different blocks of $\mathcal{B}$, and Γ is disconnected. □

It turns out that *every* transitive permutation representation of a group G is "equivalent to" the left coset action of G on the set of left cosets of a subgroup of G. We have "equivalent to" in quotes as this term will be defined formally in Definition 4.1.5. In Exercise 4.5.4 you will be asked to verify that the following result shows this. We should also point out that the following result can be obtained from Theorem 4.1.8 and the observation that if $h \in H$, then $hH = H$ (and observing that the left coset action is transitive); see Exercise 4.5.5.

Theorem 2.4.13 *Let G be a group and $H \leq G$. Then $(G, G/H)$ is transitive. Conversely, let $G \leq S_n$ be a transitive group with $H = \mathrm{Stab}_G(0)$. For each $x \in \mathbb{Z}_n$ choose $g_x \in G$ with $g_x(0) = x$. Then for every $g \in G$ and $x \in \mathbb{Z}_n$ we have*

$$g(x) = y \text{ if and only if } gg_xH = g_yH.$$

Proof The first part of this result is clear, as if aH and bH are left cosets of H in G, then $(ba^{-1})aH = bH$ and the left coset action of G on G/H does induce a transitive permutation group on the set of left cosets of H in G. For the converse, for all $g \in G$ and $x \in \mathbb{Z}_n$ we have

$$\begin{aligned} g(x) = y &\text{ if and only if } gg_x(0) = g_y(0) \\ &\text{ if and only if } g_y^{-1}gg_x(0) = 0 \\ &\text{ if and only if } g_y^{-1}gg_x \in H \\ &\text{ if and only if } gg_xH = g_yH. \end{aligned}$$ □

The converse part of the above result can be interpreted to say that if we replace each $x \in \mathbb{Z}_n$ with $g_xH \in G/H$, then g permutes the elements of $\mathbb{Z}_n$ in exactly the same was as multiplication by g permutes the elements of G/H.

It should now be clear in hindsight that Theorem 1.3.9 showing that every vertex-transitive digraph Γ is isomorphic to a double coset digraph is in at least one sense trivial. Namely, once one knows that every transitive group G can be identified with the action of G by left multiplication on the left cosets of $H = \mathrm{Stab}_G(x)$, one can simply relabel the vertices of Γ with left cosets of H in G in such a way that the action of G on the relabeled vertices is exactly the action of G by left multiplication on the left cosets of H. That is essentially how the proof of Theorem 1.3.9 goes, although of course there are details to be checked.

We can now characterize primitive and quasiprimitive groups G in terms of the group-theoretic structure of G.

Theorem 2.4.14 *Let G be a group and $H \leq G$. Then $(G, G/H)$ is primitive and faithful if and only if H is maximal and core-free in G.*

Proof The group $(G, G/H)$ is faithful if and only if H is core-free by Proposition 1.3.8. The stabilizer of H in $(G, G/H)$ is H, and $(G, G/H)$ is primitive if and only if H is maximal by Theorem 2.4.8. □

A frequent application of the previous result is to construct primitive permutation groups using finite simple groups. Of course a subgroup H of a finite simple group G is always core-free in G.

Example 2.4.15 Let G be a finite simple group. If $H \leq G$ is maximal, then $(G, G/H)$ is primitive and faithful.

A fair amount is known about maximal subgroups of simple groups. For maximal subgroups of alternating or symmetric groups, see Liebeck et al. (1987). For classical groups, see Aschbacher (1984) and Kleidman and Liebeck (1990). For exceptional groups of Lie type, see Liebeck and Seitz (1990, 1999, 2004, 2005). For sporadic groups, see Conway et al. (1985) or Wilson et al. (n.d.).

Theorem 2.4.16 *Let G be a group and $H \leq G$. The group $(G, G/H)$ is a faithful representation of G and quasiprimitive but not primitive if and only if H is not maximal in G and every subgroup of G that contains H is core-free in G.*

Proof That $(G, G/H)$ is a faithful representation of G if and only if H is core-free in G is Proposition 1.3.8.

Suppose $(G, G/H)$ is quasiprimitive but not primitive. Then H is not maximal in G by Theorem 2.4.14. As G is quasiprimitive, no nontrivial block system $\mathcal{B}$ of G is normal, and so $\mathrm{Stab}_G(B) \geq H$ is core-free in G where $B \in \mathcal{B}$. As by Theorem 2.3.7 there is a one-to-one correspondence between the block systems of G and the subgroups of G that contain H, all subgroups of G that contain H are core-free in G.

Conversely, suppose H is not maximal in G and every subgroup of G that contains H is core-free in G. As H is not maximal, $(G, G/H)$ is not primitive by Theorem 2.4.14. Let $N \trianglelefteq G$. If every such N is transitive, then $(G, G/H)$ is quasiprimitive and we are finished. Towards a contradiction, assume N is not transitive, so the set of its orbits is a normal block system $\mathcal{B}$ of $(G, G/H)$ by Theorem 2.2.7. Let $B \in \mathcal{B}$ with $H \in B$. Then $N \leq \mathrm{Stab}_{(G,G/H)}(B)$, and so $H \leq \mathrm{Stab}_{(G,G/H)}(B)$ is not core-free in G, a contradiction. □

As in a finite simple group G, every subgroup is core-free in G; we have the following example.

Example 2.4.17 Let G be a finite simple group, and $H \leq G$ that is not maximal. Then $(G, G/H)$ is quasiprimitive but not primitive.

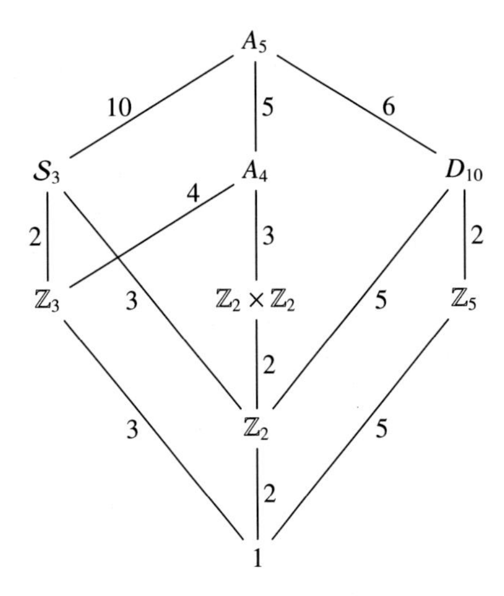

Figure 2.5 The subgroup pattern of A_5.

We will see many applications of these results in Section 2.5 (see the Remarks therein), and so postpone additional examples until then.

2.5 Actions of A_5

As emphasized in the Preface, transitive group actions are the most natural group actions as they represent the building blocks for group actions in general. By Theorem 2.4.13 every action (not necessarily faithful) of a group G can be thought of as a pair (G, H), where $H \leq G$, via the left multiplication action of G on G/H. This leads to a rich interplay between abstract groups on the one hand, and various mathematical objects associated with the left cosets of H in G on the other, such as, for example, digraphs (more precisely, double coset digraphs). Note that we will use the pair (G, H) in this section to emphasize we are choosing a group G and then a subgroup H of G. The notation $(G, G/H)$ is the image of the permutation representation induced by the left coset action of G on G/H.

To illustrate the relationship between groups and graphs, we choose as a starting point the smallest simple group A_5 together with what is often called the **subgroup pattern** of A_5, as shown in Figure 2.5. In the subgroup pattern, all subgroups are listed up to isomorphism. There is a line between two abstract

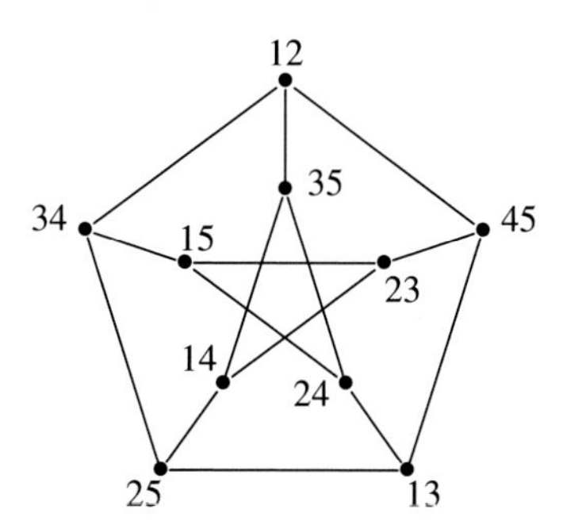

Figure 2.6 The Petersen graph.

groups if the larger group (we have them arranged with the larger groups higher in the pattern) contains a conjugate of the smaller group as a maximal subgroup. It also turns out that in A_5 there is only one subgroup in a given isomorphism class up to conjugation – this is not actually hard to show by taking advantage of the cycle decomposition of elements of A_5. For example, A_5 contains many subgroups of order 2, but all of them are generated by the product of two transpositions, so they all form a conjugacy class in A_5. The edges are labeled with the index of the subgroup below in the subgroup above.

Let $H < A_5$, so H is isomorphic to one $A_4, D_{10}, S_3, \mathbb{Z}_5, \mathbb{Z}_2^2, \mathbb{Z}_3, \mathbb{Z}_2$, or 1. So we have eight different pairs of groups (A_5, H), and so eight different associated group actions. In what follows, a corresponding combinatorial structure will be given for each of these eight pairs. As A_5 is simple, by Example 2.4.17, the left coset action of A_5 on G/H is faithful and quasiprimitive for every proper subgroup $H < A_5$.

Perhaps the most natural environment for seeing A_5 in each of its actions permuting a combinatorial structure is the Petersen graph. The best labeling of the Petersen graph for our purposes here, as with computing its automorphism group, is the 2-subset labeling of the Petersen graph we saw in Section 2.1. **For simplicity of notation we identify the 2-subset $\{i, j\}$ with ij, $i, j \in [1, 5]$, in this section and the next section only, as shown in Figure 2.6.**

As before, let P denote the Petersen graph with its 2-subset labeling, and let $G < \text{Aut}(P)$ be isomorphic to A_5 (so we will use A_5 when referring to a subgroup of S_5, and G when referring to A_5 with its 2-subset action). Our goal is to associate with each pair (G, H), $H < G$, a set S of isomorphic subdigraphs of P, which we will informally call a "**combinatorial substructure**," such that G is also transitive on the subgraphs in S, and H is the stabilizer in G of a subgraph in S. By Theorem 2.4.13, this will show that the left coset action of

A_5 on G/H is "the same" as the action of G on the combinatorial substructure S. We will abuse notation, and also refer to the subgroup of A_5 whose action on 2-subsets is H by H as well. This will cause no confusion as we will make clear whether $H \leq G$ or $H \leq A_5$. Often, to show G is transitive on the combinatorial substructure, we will use the fact that A_5 is transitive on a set X provided that S_5 is transitive on X and $\mathrm{Stab}_{S_5}(x)$ contains an odd permutation. This follows as $|X| = [S_5 : \mathrm{Stab}_{S_5}(x)]$, if $\mathrm{Stab}_{A_5}(x) < \mathrm{Stab}_{S_5}(x)$, then $[A_5 : \mathrm{Stab}_{A_5}(x)] \geq |X|$, which implies A_5 is transitive on X by the orbit-stabilizer theorem. Throughout, we let $n = |G|/|H|$, the degree of the left coset action of G on G/H.

Lemma 2.5.1 *Let $(G, H) = (A_5, A_4)$, so $n = 5$. Then S is the set of five independent sets of vertices $I_j = \{jk : j \neq k\}$, $j \in [1, 5]$. Alternatively, we may take S to be the set of subgraphs $E_j = P[V(P) \backslash I_j]$, $j \in [1, 5]$, each of which is an independent set of three edges.*

Proof We saw in the proof of Theorem 2.1.4 that each I_j is a maximal independent set of vertices in P, that G is transitive on the set $\{I_j : j \in [1, 5]\}$, and that I_1 is stabilized by $\mathrm{Stab}_G(H)$. Thus, as we saw in the proof of Theorem 2.4.11, H also fixes the subgraph induced by the complement of V_1 in $V(P)$; that is, the set of three independent edges $E_1 = \{\{25, 34\}, \{24, 35\}, \{23, 45\}\}$, colored in black in Figure 2.4. (As P has a 1-factor, this set, however, is not a maximal independent set of edges in P.) So the combinatorial substructure of P corresponding to the left coset action of A_5 on A_5/A_4 is the five maximal independent sets of vertices $\{I_1, I_2, I_3, I_4, I_5\}$, or alternatively, of five independent sets of three edges $\{E_1, E_2, E_3, E_4, E_5\}$. Again, note that this set is a partition of the edge set $E(P)$ of P, and is the unique nontrivial block system of G (and in fact the full automorphism group of P as well). □

Lemma 2.5.2 *Let $(G, H) = (A_5, D_{10})$, so $n = 6$. Then S is the set of 1-factors of P, or equivalently, the set of 2-factors of P.*

Proof Let $\rho = (1, 5, 3, 2, 4)$, $\tau = (1, 2)(3, 5)(4)$, and $H = \langle \rho, \tau \rangle$. Straightforward computations show $H \cong D_{10}$. By Sylow's theorem (Dummit and Foote, 2004, Theorem 4.5.18), there are six subgroups of A_5 conjugate to $\langle \rho \rangle$ in A_5, and so by Dummit and Foote (2004, Theorem 4.3.6), the normalizer in A_5 of $\langle \rho \rangle$ has order 10. Hence $H = N_{A_5}(\langle \rho \rangle)$. The 1-factor

$$F = \{\{12, 35\}, \{23, 45\}, \{13, 24\}, \{14, 25\}, \{15, 34\}\},$$

is fixed by $H \leq G$; see Figure 2.7. Note that F can be obtained from $\langle \rho, \tau \rangle$ as an involution $(i, j)(k, \ell) \in H$ can be identified with the edge $\{ij, k\ell\} \in F$. Additionally, H also fixes the complement of F in P. This complement is

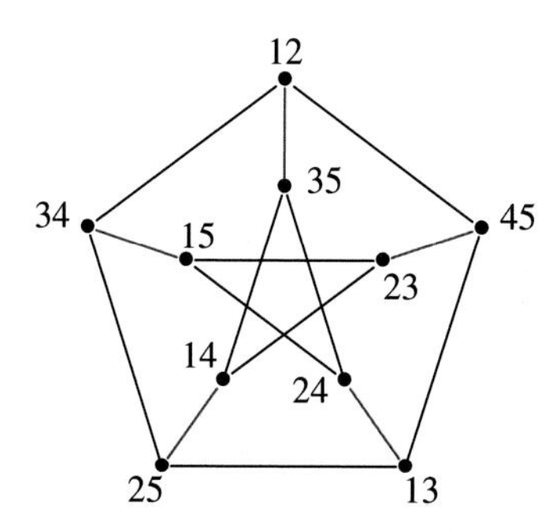

Figure 2.7 The 1-factor $F = \{\{12, 35\}, \{23, 45\}, \{13, 24\}, \{14, 25\}, \{15, 34\}\}$.

two 5-cycles, and their union is a 2-factor of P. Let ρ' generate a Sylow 5-subgroup of G with $\delta^{-1}\langle\rho\rangle\delta = \langle\rho'\rangle$. If $(i, j)(k, \ell) \in N_G(\langle\rho\rangle)$ is an involution, then $\left(\delta^{-1}(i), \delta^{-1}(j)\right)\left(\delta^{-1}(k), \delta^{-1}(\ell)\right)$ is an involution in $N_G(\langle\rho'\rangle)$. We conclude that $\delta^{-1}(F)$ is a 1-factor fixed by $\delta^{-1}N_G(\langle\rho\rangle)\delta$. Hence, as there are six Sylow 5-subgroups in G, there are six corresponding 1-factors (and six complementary 2-factors) of P. By Exercise 2.5.3, there are no other 1-factors in P. Additionally, as A_5 acting by conjugation on the set of Sylow 5-subgroups of A_5 is transitive as Sylow p-subgroups are conjugate, G is transitive on the 1-factors in P (and so on the six complementary 2-factors as well). Hence the combinatorial substructure of P corresponding to the left coset action of A_5 on A_5/D_{10} is the six 1-factors of P, or alternatively, the six complementary 2-factors. □

Lemma 2.5.3 *Let $(G, H) = (A_5, S_3)$, so $n = 10$. Then S is the set of vertices of P. Alternatively, S is the set of claws in P or 6-cycles in P.*

Proof The obvious choice for S is the vertex set $V(P)$. But there are two other choices. Let $\gamma = (3, 4, 5)$, $\tau = (1, 2)(3, 4)$, and $H = \langle\gamma, \tau\rangle$. Straightforward computations show that $H \cong S_3$. The vertex 12, together with its neighbors 34, 35, and 45, are fixed set-wise by H. The subgraph of P induced by these four vertices, isomorphic to $K_{1,3}$, is a **claw**. Further, H also fixes the subgraph of P induced by the vertices at distance 2 from 12, that is, the 6-cycle $14, 23, 15, 24, 13, 25, 14$; see Figure 2.8. Alternatively, removing the vertices of a claw from P gives a 6-cycle. Now, as G is transitive on the vertex set, it is transitive on the set of claws, as well as on the set of 6-cycles in P, as there are exactly ten 6-cycles in P by Exercise 2.5.1. Any of these three sets can be taken as the combinatorial substructure of P corresponding to the left coset action of A_5 on A_5/S_3. In particular, the 6-cycles will prove useful later when we consider the left coset action of A_5 on $A_5/\mathbb{Z}_3$. □

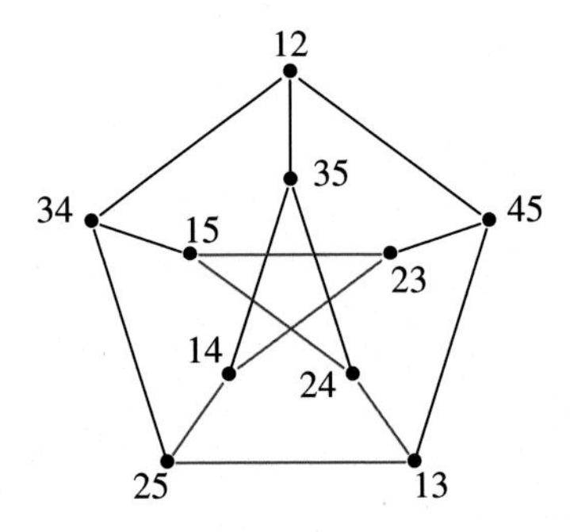

Figure 2.8 The 6-cycle $14, 23, 15, 24, 13, 25, 14$ in the Petersen graph fixed by $H = \langle (3,4,5), (1,2)(3,4) \rangle \cong S_3$.

Remark 2.5.4 Observe that the transitive actions of A_5 of degree 5, 6, and 10 considered in the previous three lemmas are all primitive by Theorem 2.4.14 as S_3, D_{10}, and A_4 are all maximal subgroups of A_5.

For our next pair (G, H), we will need an additional term.

Definition 2.5.5 Let Γ be a graph. An **orientation** of Γ is the digraph obtained by replacing each edge uv of Γ with exactly one of the two directed arcs (u, v) or (v, u).

It should be clear that a graph can have many different orientations. An orientation of the Petersen graph is given in Figure 2.9.

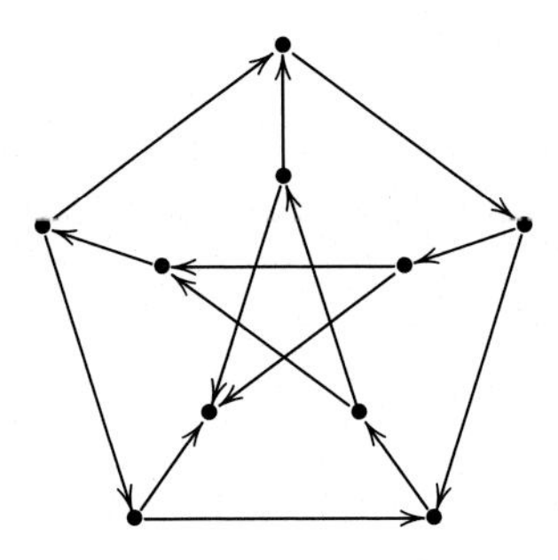

Figure 2.9 An orientation of the Petersen graph.

Lemma 2.5.6 *Let $(G, H) = (A_5, \mathbb{Z}_5)$, so $n = 12$. Then S is the set of oppositely oriented 5-cycles whose union is a 2-factor of P.*

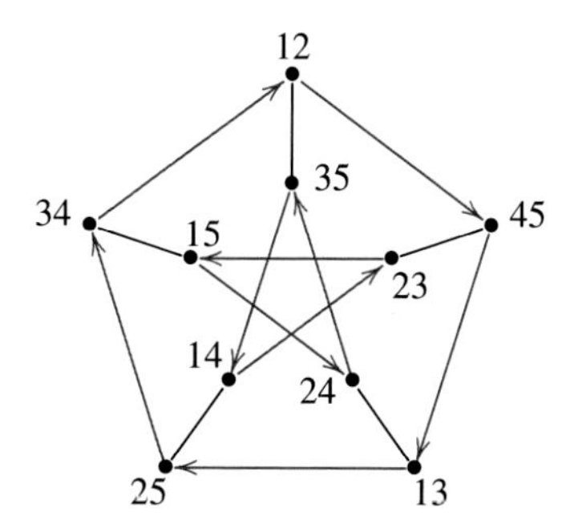

Figure 2.10 A pair of oriented 5-cycles $(C_+, \hat{C}_-)$ in the Petersen graph.

Proof It is tempting to say that this action of G should be on the set of 5-cycles, as there are twelve 5-cycles in P (two for each of the six 2-factors). This is not the case, however, as G is not transitive on the set of 5-cycles. This follows as no odd element of S_5 fixes a 5-cycle. Indeed, the stabilizer of a 5-cycle in S_5 has order 10, and it is easy to see that the canonical rotation and reflection of P fix both the "outside" and "inside" 5-cycles. Both of these permutations though, viewed as elements of S_5, are even. As $|A_5| = 60$, by the orbit-stabilizer theorem G is not transitive on the set of twelve 5-cycles.

To find S, we consider pairs of oriented 5-cycles. Let $H \cong \mathbb{Z}_5$ be the Sylow 5-subgroup of G generated by the permutation ρ with cycle decomposition $(1,5,3,2,4)$ in A_5. We have seen, H fixes the 2-factor consisting of the "outer" 5-cycle $C = 12, 45, 13, 25, 34, 12$ and the "inner" 5-cycle $\hat{C} = 35, 14, 23, 15, 24, 35$. We now orient C in the clockwise direction and $\hat{C}$ in the counter-clockwise direction. This gives a pair of oriented 5-cycles $\left(C_+, \hat{C}_-\right)$ (see Figure 2.10), and by reversing the orientation, the pair $\left(\hat{C}_+, C_-\right)$. We will call such a pair of oppositely oriented 5-cycles a **duad** (the word "duad" is a synonym for "pair"). Straightforward computations show that H stabilizes the duad $\left(C_+, \hat{C}_-\right)$.

Each of the remaining five 2-factors in P can be decomposed into a pair of duads for a total of 12 duads; namely the pairs $\left(D_+, \hat{D}_-\right)$, where D is a 5-cycle in P, and $\hat{D}$ is its complement in a 2-factor of P. To see that G is transitive on duads, notice that the natural reflection through the edge $\{12, 35\}$ (induced by the element $(1,2)(3,5)$ of A_5) maps C_+ to $\hat{C}_-$ and vice versa. As G is transitive on the set of 2-factors in the Petersen graph, the orbit of G that contains C_+ has all 12 duads. They constitute the combinatorial substructure S of P that corresponds to the pair $(A_5, \mathbb{Z}_5)$. □

Remark 2.5.7 The action of A_5 on the set of all twelve duads is imprimitive as $\mathbb{Z}_5$ is not maximal in A_5, with a block system $\mathcal{B}$ with 6 blocks of size 2, where a block is a pair $\left\{\left(D_+, \hat{D}_-\right), \left(\hat{D}_+, D_-\right)\right\}$ of oppositely oriented duads in P. The group $G/\mathcal{B}$ is then the action of G on the two factors of P.

Lemma 2.5.8 *Let $(G, H) = \left(A_5, \mathbb{Z}_2^2\right)$, so $n = 15$. Then $S = E(P)$.*

Proof There are 15 edges in P, and $\mathcal{S}_5$ is transitive on edges as the Petersen graph is arc-transitive by Exercise 2.1.3. The odd permutation $(1, 2)$ in $\mathcal{S}_5$ stabilizes the edge $\{12, 35\}$, so A_5 is transitive on the edges of P. The edge $\{12, 35\}$ is fixed in A_5 by the commuting involutions $(1, 2)(3, 5)$ and $(1, 3)(2, 5)$, and the subgroup they generate is isomorphic to $\mathbb{Z}_2^2$. Thus $E(P)$ is the substructure of P corresponding to this action of A_5. □

Remark 2.5.9 The action of A_5 on edges is imprimitive as $\mathbb{Z}_2^2$ is not maximal in A_5. Additionally, as $\left|A_4 : \mathbb{Z}_2^2\right| = 3$, there are 5 blocks of size 3, with the blocks consisting of the 5 subgraphs of the Petersen graph obtained by removing a maximal independent set.

Lemma 2.5.10 *Let $(G, H) = (A_5, \mathbb{Z}_3)$, so $n = 20$. Then S is the set of oriented 6-cycles in P.*

Proof By Lemma 2.5.3 and its proof, there are ten 6-cycles in P, and hence there are 20 clockwise and counterclockwise oriented 6 cycles. Let $H = \mathbb{Z}_3$ be the Sylow 3-subgroup of G generated by the permutation with cycle decomposition $(3, 4, 5)$ in A_5. Then both orientations of the 6-cycle $C = 13, 24, 15, 23, 14, 25, 13$ have H as their stabilizer in G. Additionally, $(1, 2)(3, 5)$ fixes C as a cycle and interchanges its two orientations, so G is transitive on the 20 oriented 6-cycles. So the substructure of P that corresponds to the left coset action of A_5 on $A_5/\mathbb{Z}_3$ is the set of oriented 6-cycles in P. □

Remark 2.5.11 The action of A_5 on the oriented 6-cycles in P has two block systems as a point stabilizer $\mathbb{Z}_3$ is contained in the maximal subgroups $\mathcal{S}_3$ and A_4 of A_5. One block system has ten blocks of size 2 as $[\mathcal{S}_3 : \mathbb{Z}_3] = 2$ and the other five blocks of size 4 as $[A_4 : \mathbb{Z}_3] = 4$. The blocks of size 2 consist of the ten 6-cycles in the Petersen graph. The blocks of size 4 are more complicated: Fix $i \in \{1, 2, 3, 4, 5\}$, and in four copies of the Petersen graph delete from each one of the four vertices that contains i as well as that vertex's neighbors. A block will consist of the remaining 6-cycles oriented in one direction (see Exercise 2.5.7). Call this block system $\mathcal{B}$. As $\mathcal{S}_3$ is not contained in A_4, the block system with blocks of size 2 is **not** a refinement of the block system with blocks of size 4. This implies that $\mathrm{Stab}_G(\mathcal{B})^B$ has no blocks of size 2, and so is 2-transitive by Theorem 2.3.10 for every $B \in \mathcal{B}$.

Lemma 2.5.12 *Let* $(G, H) = (A_5, \mathbb{Z}_2)$, *so* $n = 30$. *Then* $S = A(P)$.

Proof There are 30 arcs in P and the arc $(12, 35)$ is fixed by the element of A_5 with cycle decomposition $(1, 2)(3, 5)$. By Exercise 2.1.3, S_5 is transitive on the set of arcs of P. As the transposition $(1, 2)$ also fixes $(12, 35)$, G is transitive on the arcs of P. So the set of arcs of P is the substructure of P corresponding to $(A_5, \mathbb{Z}_2)$. □

Remark 2.5.13 The action of A_5 on the set of arcs of P is imprimitive with blocks of size 2 as an arc stabilizer $\mathbb{Z}_2$ is a subgroup of index 2 in $\mathbb{Z}_2^2$. This block system corresponds to the 15 edges of P, viewed as two oppositely oriented arcs, as $\mathbb{Z}_2^2$ stabilizes an edge. There are exactly four choices of m and k for which there are block systems of m blocks of size k, as $\mathbb{Z}_2$ is a proper subgroup of four proper subgroups of A_5. Determining these values of m and k is the aim of Exercise 2.5.4.

To find *all* of the block systems, the subgroup pattern is not enough – we need the subgroup lattice. This is because, by Theorem 2.3.7, there is a one-to-one correspondence between the subgroups that contain the stabilizer of a point and blocks that contains the point (and so block systems). For example, there is a block system with six blocks of size 5 with stabilizer of the arc $(12, 35)$ being $\langle(1, 2)(3, 5)\rangle$ as $\langle(1, 2)(3, 5)\rangle$ is a subgroup of index 5 in D_{10}, which is a maximal subgroup of index 6 in G. The subgroup D_{10} is the stabilizer of a 1-factor in P. But there are *two* 1-factors that contain the edge $\{12, 35\}$, and $\langle(1, 2)(3, 5)\rangle$ stabilizes both of them (one of these 1-factors is the spoke edges of P, while the other is $\{\{12, 35\}, \{25, 34\}, \{13, 45\}, \{15, 24\}, \{14, 23\}\}$). This means that, by Theorem 2.3.7, there are *two different* subgroups of G isomorphic to D_{10} that contain $\langle(1, 2)(3, 5)\rangle$, and so there are *two different* block systems with six blocks of size 5.

Lemma 2.5.14 *Let* $(G, H) = (A_5, 1)$, *so* $n = 60$. *Then* S *is the set of 2-arcs in* P.

Proof The number of 2-arcs in P is 60 and each 2-arc is fixed only by the identity permutation, making the set of 2-arcs of P the substructure of P corresponding to this action of G. □

Combining the lemmas in this section, we have that

Every transitive action of A_5 can be interpreted as the action of A_5 on some combinatorial substructure in the Petersen graph.

2.6 Orbital Digraphs of A_5

Now that we have seen and analyzed all possible left coset actions of A_5 on the left cosets of its proper subgroups, we will find the orbital digraphs of the permutation representations of these actions. As this section can easily be considered a continuation of the previous section, the notational conventions of the previous section will be used here. We will need one more term before we begin.

Definition 2.6.1 Let X be a set and S a collection of subsets of X. The **intersection graph of** S is the graph with vertex set S and edge set $\{S_iS_j : S_i, S_j \in S, i \neq j, \text{ and } S_i \cap S_j \neq \emptyset\}$.

Of course, the complement of the Petersen graph is an example of an intersection graph – it is the intersection graph of the 2-subsets of a 5-element set. Also, given any graph Γ, considering its edges to be 2-subsets of $V(\Gamma)$, the line graph of Γ is the intersection graph of the edges of Γ.

We already know from the previous section that each of the actions of A_5 that we will consider is the action of A_5 on some combinatorial substructure S of the Petersen graph. When constructing orbital digraphs for the permutation representation of the action, we will often use the action of A_5 on S. If so, we will fix one digraph Γ in S, and examine how the stabilizer in A_5 of Γ permutes the other digraphs in S. In this case, the vertex set of the orbital digraph will be the combinatorial substructure S. Typically, we will be able to distinguish between elements of $S\backslash\{\Gamma\}$ via their "interaction" with Γ – usually this is in terms of their intersections. For example, if $\Sigma, \Delta \in S\backslash\{\Gamma\}$ and the subdigraph induced by $A(\Gamma) \cap A(\Sigma)$ is not isomorphic to $A(\Gamma) \cap A(\Delta)$, then Σ and Δ cannot be in the same orbit of the stabilizer in A_5 of Γ. We will commonly have a "small" number of suborbits, and will distinguish the suborbits by fixing an element Γ of S and showing how this fixed element intersects with each orbit of the stabilizer of Γ in A_5. This usually allows us to define a graph, very similar to an intersection graph as defined above, but instead of having edges between S_i and S_j if $S_i \cap S_j \neq \emptyset$, we will have $S_i \cap S_j$ to be some fixed subdigraph.

Lemma 2.6.2 *K_5 is the only orbital digraph of the left coset action of A_5 on A_5/A_4.*

Proof K_5 is the unique orbital digraph of the left coset action of A_5 on A_5/A_4 as this action is 2-transitive. To see this, we give both a group-theoretic argument and a graph-theoretic argument.

The possible suborbits of G have lengths 1 and 4, two of length 1 and one of length 3, one of length 1 and two of length 2, three of length 1 and one of length 2, or all are of length 1. By Exercise 2.6.1, we have $|A_4| \leq 4!$, $3!$, $2! \cdot 2!$, $2!$, or 1, respectively. As $|A_4| = 12$, the only possibility is that G has suborbits of lengths 1 and 4, in which case A_4 fixes a point and is transitive on the other points. So this action of A_5 is 2-transitive by Exercise 2.4.1.

For the graph-theoretic argument, the intersection of any two of the maximum independent sets of vertices I_1, I_2, I_3, I_4, I_5 defined in Lemma 2.5.1 is a unique vertex. So the intersection graph of the set $\{I_1, \ldots, I_5\}$ is K_5. Now apply Exercise 2.6.2. □

Lemma 2.6.3 *K_6 is the only orbital digraph of the left coset action of A_5 on A_5/D_{10}.*

Proof K_6 is the unique orbital digraph of the left coset action of A_5 on A_5/D_{10} as this action is 2-transitive. As in Lemma 2.6.2, we give a group-theoretic and a graph-theoretic argument.

First, a Sylow 5-subgroup Q of A_5 must stabilize a point xH by the orbit-stabilizer theorem. As $|Q| = 5$, Q is transitive on the left cosets of H in G different from xH and this action is 2-transitive by Exercise 2.4.1.

For the graph-theoretic argument, by Lemma 2.5.2 G is transitive on the 1-factors of P. The intersection of any two of the six 1-factors of Γ is a single edge by Exercise 2.5.2, and so the intersection graph of the six 1-factors in P is K_6. As a maximum independent set in the line graph of a graph Γ is a maximum matching in Γ, the intersection graph of the maximum independent sets in $L(K_6)$ is also K_6. The hypothesis of Exercise 2.6.2 holds for the intersection graph of the maximum independent sets in $L(P)$ as G is transitive on the maximum independent sets of $L(P)$ (1-factors of P), as well as their unique (by Exercise 2.6.3) intersections (as P is edge-transitive). By Exercise 2.6.2, the intersection graph of the maximum independent sets in $L(K_6)$ is an orbital digraph of G, and as the maximum independent sets in $L(P)$ are the 1-factors in P, K_6 is an orbital digraph of G. □

Lemma 2.6.4 *The orbital digraphs of the left coset action of A_5 on A_5/S_3 are the Petersen graph and its complement.*

Proof Similar to Exercise 1.4.7, this action is of rank 3 and gives two orbital digraphs, the Petersen graph and its complement. □

The three permutation representations of A_5 considered so far are all primitive. By Theorem 2.4.12 imprimitive permutation representations have a disconnected orbital digraph. Of course, orbital digraphs are vertex-transitive, and so the components of a disconnected orbital digraph are isomorphic.

Definition 2.6.5 Let Γ be a digraph and m a positive integer. By $m\Gamma$ we denote a digraph that is a union of m vertex-disjoint digraphs isomorphic to Γ.

The previous definition does not define $m\Gamma$ uniquely, only up to isomorphism (which is all that is usually needed). Our discussion above shows that the following observation is true.

Lemma 2.6.6 *Let $G \leq S_n$ be transitive and Γ a disconnected orbital digraph of G. Then $\Gamma \cong m\Delta$ for some connected digraph Δ.*

Lemma 2.6.7 *The left coset action of A_5 on $A_5/\mathbb{Z}_5$ is of rank 4 with one orbital digraph isomorphic to $6K_2$ and two isomorphic to the icosahedron.*

Proof By Remark 2.5.7, this action has a block system with blocks of size 2, and each block is a pair of alternately oriented 5-cycles. The orbital digraph whose arc set contains an arc inside a block of size 2 is a disjoint union $6K_2$ of six edges. Further, there are two additional orbital graphs of valency 5 as the stabilizer of a duad is generated by a permutation that is a product of two disjoint 5-cycles, and both are isomorphic to the icosahedron (see Figure 1.13). These orbital digraphs may be constructed by examining intersections of duads. The vertices will be the duads. The intersection of the underlying simple graphs of two duads that are not oppositely oriented has two paths of length two, and two duads are adjacent if the orientations of these two paths in the duads are the same (for one icosahedron) or different (for the other icosohedron). The tedious checking that these two orbital digraphs are icosahedrons is left as Exercise 2.6.4. □

While it is worthwhile to construct the two icosahedrons in the previous result, this is a time-consuming process, and so shortcuts are nice to have. One shortcut is to use known classifications of arc-transitive graphs. For example, in the previous lemma we want to find all arc-transitive graphs of order 12 and valency 5 – we know that we are looking for graphs as the suborbits of size 5 are all self-paired as the adjacency relations given for them in Lemma 2.6.7 are symmetric. In general, for valency 3, we could use the Foster census (but there is no arc-transitive cubic graph of order 12). For valency 5, we can use a recent result of Hua, Feng, and Lee (Hua et al., 2011, Theorem 4.1), which gives that the only 5-valent arc-transitive graphs of order 12 are the icosahedron and the complete bipartite graph with a 1-factor removed. We are then left with the simpler task of determining which of these two graphs is isomorphic to our orbital digraphs.

We take a slight detour. Note that in the previous paragraph we said something obvious that bears repeating. Recall that a **relation** on a set X is a subset of $X \times X$. So the arc set of a digraph is a relation, and we can use the common terms of relations with them. For digraphs, a reflexive arc set means that there are loops at each vertex (so for us not very interesting), a symmetric arc set means the digraph is a graph, and a transitive arc set means that the subdigraphs induced by directed paths of length 2 are always directed triangles.

Lemma 2.6.8 *The orbital digraphs of the left coset action of A_5 on $A_5/\mathbb{Z}_2^2$ are the following: Two disconnected graphs, each isomorphic to the disjoint union of five oriented 3-cycles, and three graphs each isomorphic to the line graph of the Petersen graph.*

Proof By Lemma 2.5.8, the combinatorial substructure on which A_5 acts is $E(P)$. Recall from the proof of Lemma 2.5.8 that $H = \langle (1,2)(3,5),(1,3)(2,5)\rangle$ stabilizes the edge $e = \{12,35\}$. Of course, edges of a fixed distance (with distance computed in the line graph $L(P)$ of the Petersen graph) from e are mapped to edges of the same distance by elements of H. In this action we obtain five orbital graphs. There are two edges $\{15,23\}$ and $\{13,25\}$ at distance three in $L(P)$, and both are fixed by H. These two fixed points give two orbital digraphs that are each a disjoint union of five oriented 3-cycles. These orbital digraphs are unsurprising as there is a block system with 5 blocks of size 3 by Remark 2.5.9 (or Lemma 2.5.1). The other three orbital graphs are all isomorphic to the line graph $L(P)$ of the Petersen graph, with the remaining suborbits of length 4. One orbital graph is the intersection graph of the edges at distance one in $L(P)$, so this orbital graph is so $L(P)$. The edges of P at distance two in $L(P)$ form two suborbits. As the stabilizer of e is H, we see any remaining suborbits of H have order at least 4, but cannot be 8 as $|H| = 4$. That each of these is isomorphic to $L(P)$ is harder to see. We will show this explicitly later in Theorem 6.6.5 in a more general context. □

The connected graphs that we have constructed so far have been well-known graphs. In our next action of A_5, we will find a graph that is not well known, and is not so easy to describe. We will show that it is a union of two non-self-paired orbital digraphs, and we give the neighbors of a vertex (then one uses A_5 to obtain the rest of the edges). Constructing the graph is a task best suited to a computer algebra system such as Magma (Bosma et al., 1997) or GAP (GAP, 2019). These are sophisticated computer programs that can do most routine computations (but that are often very time consuming or even practically impossible to do by hand), and are now everyday tools for most researchers in this field. We will see many more situations where these systems have been used later.

Definition 2.6.9 Let C be a cycle of even length $2n$. The vertices $x, y \in V(C)$ are **antipodal** if $\mathrm{dist}_C(x, y) = n$.

Lemma 2.6.10 *The orbital digraphs of the left coset action A_5 on $A_5/\mathbb{Z}_3$ are the following: There are three disconnected orbital graphs, one isomorphic to $10K_2$ as well as two isomorphic to $5K_4$. Additionally, two orbital graphs are isomorphic to the dodecahedron, and two non-self-paired orbital digraphs whose union is a generalized orbital graph of valency* 6.

Proof Throughout the proof, fix $v = 12$. By Lemma 2.5.10, the combinatorial substructure S of Γ corresponding to this action is the set of 20 oriented 6-cycles in P. Let $\rho = (3, 4, 5)$ and $\tau = (1, 2)(3, 5)$. We saw in Lemmas 2.5.3 and 2.5.10 that $H = \langle\rho\rangle \cong \mathbb{Z}_3$ is the stabilizer in A_5 of the oriented 6-cycle $C_v = (13, 24, 15, 23, 14, 25, 13)$ while τ reverses its orientation (we will often use the notation $(v_1, \ldots, v_r, v_1)$ for a directed cycle with $(v_i, v_{i+1}), (v_r, v_1) \in A(\Gamma)$, $1 \le i \le r - 1$, to remind the reader the cycle is directed). For a directed 6-cycle $C \in S$ let $\hat{C}$ be its oppositely oriented 6-cycle in C. To calculate the orbital digraphs, we shall consider G as acting on S with H stabilizing C_v. Apart from the two orbits $\{C_v\}$ and $\{\hat{C}_v\}$ of length 1, H has six orbits of length 3 as $H \cong \mathbb{Z}_3$ and ρ fixes exactly one 6-cycle. We will show four of these are self-paired and two are non-self-paired.

By Remark 2.5.11, G has blocks of size 2 (the 10 underlying 6-cycles are the blocks of size 2) and 4. As G has blocks of size 2 the graph $10K_2$ is an orbital digraph of G. As G has blocks of size 4, and, as by Remark 2.5.11 the block stabilizer is 2-transitive on a block, $5K_4$ is an orbital graph. The orbital digraph that is $10K_2$ is also easily obtained as the graph with vertex set C and two vertices are adjacent if they are oppositely oriented 6-cycles (the orbital graph that is $5K_4$ is not so easily obtained as an intersection graph). For the other orbital graphs, a more careful analysis is in order.

Choose a vertex $w \in V(\Gamma)$, and let $\mathrm{DC}(w) \in C$ be one of the two directed 6-cycles obtained by deleting w and its neighbors from P, and we choose $\mathrm{DC}(v) = C_v$. Let $x, y \in V(P)$, $x \neq y$, and let C and D be the underlying simple graphs of $\mathrm{DC}(x)$ and $\mathrm{DC}(y)$, respectively. Consider the intersection $C \cap D$. One of three types of intersections occurs.

First, if x is a neighbor of y, then $C \cap D$ is the set of two edges, which are antipodal edges on both cycles C and D, with one such edge having the same orientation on both directed 6-cycles and the other edge the opposite orientation on both directed 6-cycles. See the left-hand graph in Figure 2.11 where $x = 12$, $y = 35$, $C \cap D$ is colored green, the rest of C is colored red, and the rest of D is colored blue. Clearly, once x has been chosen, there are six choices for $\mathrm{DC}(y)$.

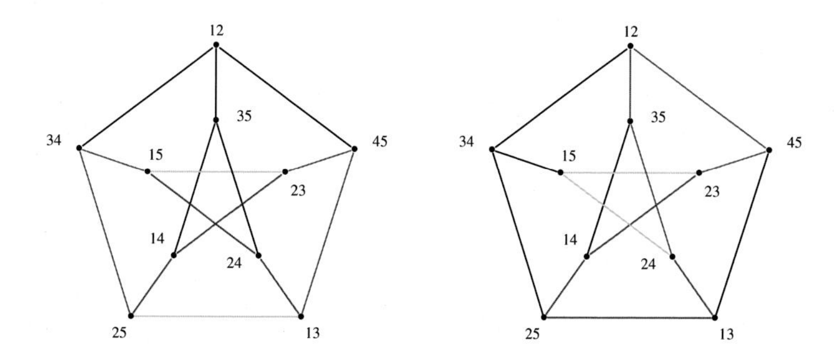

Figure 2.11 Possible intersections of some 6-cycles in P.

Next, if y is at distance 2 from x then $C \cap D$ has two consecutive edges on C and D in common, with two additional types: either the two edges both have the same orientation or opposite orientations in DC(x) and DC(y). Again, there are six directed 6-cycles DC(x) for each of the two types. See the right-hand graph in Figure 2.11 where $x = 12$, $y = 25$, $C \cap D$ is colored green, the rest of C is colored red, and the rest of D is colored blue (the same color scheme as the left-hand side).

To summarize, apart from oppositely oriented pairs of 6-cycles, there are three different types of intersection for pairs of distinct oriented 6-cycles, and once DC(x) has been chosen, each occurs six times among the remaining 18 oriented 6-cycles.

Now consider the 18 pairs (C_v, C_w) where C_w is an oriented 6-cycle different from C_v and $\hat{C}_v$, and denote them by T. By the previous paragraph, there is a partition $\{T_1, T_2, T_3\}$ of T into the three types of intersections of C_v with some C_w. As H maps C_v to C_v and $\hat{C}_v$ to $\hat{C}_v$, H maps each T_i to itself. As H has six orbits of length 3, each T_i is a union of two orbits of H of length 3. This also implies that in an orbital digraph of G corresponding to a suborbit of length 3, the intersection type of each arc is the same. To determine the two orbits of H of length 3 whose union is a T_i, we use computations in A_5.

We begin with the first type, where the arcs are from pairs of oriented 6-cycles (C_x, C_y) with $xy \in A(P)$. The three neighbors of v are $x = 45$, $y = 35$ and $z = 34$. Choose

$$C_x = (15, 34, 25, 14, 35, 24, 15),$$
$$C_y = (13, 45, 23, 15, 34, 25, 13), \text{ and}$$
$$C_z = (14, 35, 24, 13, 45, 23, 14).$$

Straightforward computations show that $\{C_x, C_y, C_z\}$ and $\left\{\hat{C}_x, \hat{C}_y, \hat{C}_z\right\}$ are two orbits of ρ (recall C_v is fixed by ρ). To show that both of these suborbits are self-paired, it suffices to find involutions in G interchanging C_v with C_x and C_v with $\hat{C}_x$. For example, we know that (C_v, C_x) is an arc of one of these orbital digraphs, so an involution interchanging C_v with C_x gives that (C_x, C_v) is also an arc of this orbital digraph. The permutation $(1,5)(2,4)$ interchanges C_v with C_x while the permutation $(1,4)(2,5)$ interchanges C_v with $\hat{C}_x$.

Finally, to determine the two orbital graphs associated with the above two orbits of length 3, first observe that there are three possible candidates. By the Foster census, there are only three nonbipartite arc-transitive graphs of valency 3 on 20 vertices (the Foster census only lists the dodecahedron as it only lists connected graphs), with the other two being $5K_4$ and $2P$. The disconnected graph $2P$ can be immediately eliminated as a possibility as it has two blocks of size 10, but there is no subgroup of A_5 of index 2. To eliminate $5K_4$ simply note the three neighbors of v are at distance 2 from each other, and so the intersection type of the edges C_xC_y, C_xC_z, and C_yC_z is a different type than of C_vC_x. Hence the subgraph induced on the 6-cycles C_v, C_x, C_y, C_z is a claw $K_{1,3}$, not a K_4. It follows that the orbital digraph corresponding to the orbits $\{C_x, C_y, C_z\}$ and $\left\{\hat{C}_x, \hat{C}_y, \hat{C}_z\right\}$ are both dodecahedrons.

We now consider pairs of oriented 6-cycles associated with nonadjacent vertices. The six nonneighbors of v are $a = 13$, $b = 14$, $c = 15$, $d = 25$, $e = 23$, and $f = 24$. Let

$$C_a = (12, 34, 15, 23, 14, 35, 12), \quad C_d = (12, 35, 24, 15, 23, 45, 12),$$
$$C_b = (12, 45, 13, 24, 15, 34, 12), \quad C_e = (12, 34, 25, 13, 24, 35, 12),$$
$$C_c = (12, 35, 14, 25, 13, 45, 12), \quad C_f = (12, 45, 23, 14, 25, 34, 12).$$

Applying ρ, we have that the four orbits of length 3 on the twelve 6-cycles not yet considered in S are as follows: $\{C_a, C_b, C_c\}$, $\left\{\hat{C}_a, \hat{C}_b, \hat{C}_c\right\}$, $\{C_d, C_e, C_f\}$, and $\left\{\hat{C}_d, \hat{C}_e, \hat{C}_f\right\}$.

Exploiting the fact that $(3,4,5)$ in its 2-subset action fixes C_v and is transitive on each of the orbits listed at the end of the previous paragraph, we see the intersection of C_v with any of the cycles from the first and third orbits above has two consecutive edges oriented the same, while the intersection of C_v with the opposite orientation in the second and fourth orbits has two consecutive edges with opposite orientations. We will consider these two situations separately, but need one more preliminary observation first.

We first consider when the intersection of C_v with the opposite orientation in the second and fourth orbits has two consecutive edges with opposite orientations. Observe that the intersections of any two cycles chosen from $\left\{\hat{C}_a, \hat{C}_b, \hat{C}_c\right\}$ all have two consecutive edges oppositely oriented, and the same is true for two cycles chosen from $\left\{\hat{C}_d, \hat{C}_e, \hat{C}_f\right\}$. Then any two cycles chosen from the set $\{C_v, \hat{C}_a, \hat{C}_b, \hat{C}_c\}$ and any two cycles chosen from the set $\{C_v, \hat{C}_d, \hat{C}_e, \hat{C}_f\}$ have intersection with two consecutive edges oppositely oriented.

Now check that $(2,3)(4,5) \in A_5$ maps $\{C_v, \hat{C}_a, \hat{C}_b, \hat{C}_c\}$ to itself and the arc $(C_v, \hat{C}_a) \mapsto (\hat{C}_a, C_v)$. Hence the suborbit $\{\hat{C}_a, \hat{C}_b, \hat{C}_c\}$ is self-paired and is a component of valency 3 in the orbital graph corresponding to the suborbit $\{\hat{C}_a, \hat{C}_b, \hat{C}_c\}$. So the orbital graph is $5K_4$. An analogous argument gives that the orbital graph corresponding to the suborbit $\{\hat{C}_d, \hat{C}_e, \hat{C}_f\}$ is also $5K_4$ (Exercise 2.6.6).

We are now left with suborbits $\{C_a, C_b, C_c\}$ and $\{C_d, C_e, C_f\}$. We first show that there is no permutation in A_5, interchanging C_v and C_a, which will give that $\{C_a, C_b, C_c\}$ is non-self-paired. Indeed, if there is a permutation in A_5 interchanging C_v and C_a, then in its 2-subset action it has even order. So we may assume without loss of generality that its order is a power of 2. That means in A_5 it has order a power of 2, and consequently must be a product of two disjoint 2-cycles. There are 15 of these, and checking each we see none interchange C_v and C_a, so $\{C_a, C_b, C_c\}$ is non-self-paired. Hence $\{C_d, C_e, C_f\}$ is also non-self-paired. Using GAP or Magma, the union of the two non-self-paired orbital digraphs is an arc-transitive Cayley graph of the metacyclic group $\mathbb{Z}_5 \rtimes \mathbb{Z}_4$ whose automorphism group is isomorphic to $S_5 \times \mathbb{Z}_2$. □

Unfortunately, determining the orbital digraphs of the action of A_5 on $\mathbb{Z}_2$ is too involved to do by hand here. Given its complexity, it is probably best in any case to use GAP or Magma. Using Magma, one can calculate that there are 16 suborbits, two of length 1, and 14 of length 2. Of the 14 of length 2, six are self-paired and eight are non-self paired. The self-paired orbitals give orbital graphs, and are as follows. There is one orbital graph isomorphic to $15K_2$, which comes from the block system with blocks of size 2 (or a suborbit of size 1). There are four isomorphic to $6C_5$, one for each block system with blocks of size 5 (we know they are $6C_5$ as the stabilizer of a block is D_{10}). Finally, there are two isomorphic to $10C_3$, as there are two block systems with blocks of size 3 as there are two subgroups of A_5 isomorphic to S_3 that contain $\langle(1,2)(3,4)\rangle$, which is the stabilizer of a point in the left coset action of A_5 on $A_5/\mathbb{Z}_2$. We will not explicitly determine the remaining orbital digraphs, but content ourselves with pointing out they are all digraphs, each vertex of which has in- and outvalence 2.

The remaining action of A_5 is the regular action. With regular actions, the stabilizer of a point stabilizes every point, so all suborbits are single elements of A_5. So all orbital digraphs will be directed cycles of length the order of the element of A_5 chosen, unless the element is self-inverse, in which case the orbital digraph is a graph that is a 1-factor.

2.7 Exercises

2.1.1 Show that for a graph Γ, $\mathrm{Aut}(\Gamma) = \mathrm{Aut}\left(\bar{\Gamma}\right)$.

2.1.2 Show that the Petersen graph is the complement of the line graph of K_5.

2.1.3 Show that the Petersen graph is arc-transitive.

2.2.1 Show that if B is a block of G, then $g(B)$ is a block of G for every $g \in G$.

2.2.2 Show that if $\mathcal{B}$ is a block system of G, then $\mathcal{B}$ is a partition of X.

2.2.3 Let Γ be a vertex-transitive digraph with $\Delta \le \Gamma$. Let C be the collection of all subdigraphs of Γ isomorphic to Δ. If $\mathcal{B} = \{V(C) : C \in C\}$ is a partition of $V(\Gamma)$, then $\mathcal{B}$ is a block system of $\mathrm{Aut}(\Gamma)$.

2.2.4 Suppose Γ is a connected vertex-transitive graph and for each $v \in V(\Gamma)$ there exists a unique vertex u_v whose distance from v is a fixed distance k. Show that $\{\{v, u_v\} : v \in V(\Gamma)\}$ is a block system of $\mathrm{Aut}(\Gamma)$. Deduce that $\mathrm{Aut}(Q_n)$ has a block system with blocks of size 2.

2.2.5 Show that $\mathrm{Aut}(\mathrm{Cay}(\mathbb{Z}_6, \{2, 3, 4\})) \cong S_3 \times S_2 \cong D_{12}$.

2.2.6 Let G be transitive with block system $\mathcal{B}$. Show that G acts on $\mathcal{B}$ by $g \cdot B = g(B)$.

2.2.7 Let $G \le S_X$ be transitive, and suppose B is a block of G. Show that $\mathrm{Stab}_G(B)$ is transitive on B.

2.2.8 Let $G \le S_{np}$ have a block system $\mathcal{B}$ with blocks of prime size p where $p > n$. Show that $\mathcal{B}$ is a normal block system of G.

2.2.9 Let $G \le S_n$ contain $(\mathbb{Z}_n)_L$ and have a block system $\mathcal{B}$ with m blocks of size k. Show that $\mathcal{B}$ is the set of cosets of the unique subgroup of $\mathbb{Z}_n$ of order k.

2.2.10 Let $G \le S_n$ be transitive with blocks B and B' such that $B \cup B'$ has order n. Show that $\{B, B'\}$ is a block system with two blocks of size $n/2$.

2.2.11 Let p be prime and $k \ge 1$ an integer. Show that the number of block systems of $\left(\mathbb{Z}_p^k\right)_L$ with blocks of size p is $\left(p^k - 1\right)/(p-1) = \sum_{i=0}^{k-1} p^i$.

2.2.12 Let X be an orbit of $H \le S_n$ that contains x, and $\delta \in S_n$. Show that $\delta^{-1}(X)$ is the orbit of $\delta^{-1}H\delta$ that contains $\delta^{-1}(x)$. Deduce that if G is transitive with normal block system $\mathcal{B}$, then $\delta^{-1}(\mathcal{B})$ is a normal block system of $\delta^{-1}G\delta$.

2.3.1 Define an appropriate G-congruence and use Theorem 2.3.2 to prove Theorem 2.2.7.

2.3.2 Show that if G is a transitive group with block system $\mathcal{B}$, then there is a G-congruence whose equivalence classes are $\mathcal{B}$.

2.3.3 Prove Theorem 2.3.10 using Theorem 2.3.7.

2.3.4 Let $G \leq \mathcal{S}_n$ with $\mathcal{B}$ a block system. Show that if $\phi \in \mathcal{S}_n$, then $\phi(\mathcal{B})$ is a block system of $\phi G \phi^{-1}$. Deduce that if $\phi \in N_{\mathcal{S}_n}(G)$, then $\phi(\mathcal{B})$ is also a block system of G. Finally, show that if $\mathcal{B}$ is the only block system of G with blocks of a given size, then $\mathcal{B}$ is a block system of $N_{\mathcal{S}_n}(G)$.

2.3.5 Let p and q be distinct primes such that q divides $p - 1$, and G the nonabelian group of order pq. Determine the number of block systems of G_L with blocks of size q and also of size p.

2.3.6 Let $G \leq \mathcal{S}_n$ be transitive, $H \leq G$ contain $\mathrm{Stab}_G(x)$ for some $x \in \mathbb{Z}_n$, and B a block of G. Show that $|B| = [\mathrm{Stab}_G(B) : \mathrm{Stab}_G(x)]$.

2.3.7 Let G be a transitive group of square-free degree (an integer that is **square-free** is not divisible by the square of any prime). Show that G has at most one normal block system with blocks of prime size p. (Hint: Suppose there are at least two such block systems $\mathcal{B}_1$ and $\mathcal{B}_2$. Consider what happens to $\mathrm{fix}_G(\mathcal{B}_2)$ in $G/\mathcal{B}_1$.)

2.3.8 Let $G, H \leq \mathcal{S}_n$ be transitive and let both have $\mathcal{B}$ as a block system. Show that $\mathcal{B}$ is a block system of $\langle G, H\rangle$.

2.3.9 Prove Corollary 2.3.11.

2.3.10 Let $G \leq \mathcal{S}_n$ be transitive with B a block of G. Show that

$$g\mathrm{Stab}_G(B)g^{-1} = \mathrm{Stab}_G(g(B)).$$

2.3.11 Let $R \leq \mathcal{S}_n$ be regular. Show $\mathrm{Stab}_R(B)^B$ is regular for every block B of R.

2.3.12 Let $G \leq \mathcal{S}_n$ be transitive with block system $\mathcal{B}$. Show that $\mathrm{fix}_G(\mathcal{B}) = \cap_{B\in\mathcal{B}}\mathrm{Stab}_G(B)$.

2.3.13 Let Γ be a connected arc-transitive digraph with $G \leq \mathrm{Aut}(\Gamma)$ arc-transitive. Show that if $\mathcal{B}$ is a nontrivial block system of G, then $\Gamma[B]$ has no arcs for every $B \in \mathcal{B}$.

2.3.14 The goal of this exercise is to generalize Theorem 2.2.19. A group G is called a **Dedekind group** if every subgroup of G is normal. Of course, abelian groups are Dedekind groups, and the nonabelian Dedekind groups are called the **Hamiltonian groups**. Show that if H is a transitive group that contains a regular Hamiltonian subgroup, then every block system of H is a normal block system. This shows that Theorem 2.2.19 holds for Dedekind groups.

2.3.15 Let $G \leq \mathcal{S}_n$ be transitive. Define a relation $\equiv$ on $\mathbb{Z}_n$ by $x \equiv y$ if and only if $\mathrm{Stab}_G(x) = \mathrm{Stab}_G(y)$. Show that $\equiv$ is a G-congruence. Then show that if Γ is a connected vertex-transitive graph and $v \in V(\Gamma)$ with $\mathrm{Stab}_{\mathrm{Aut}(\Gamma)}(v) \neq 1$, then there is $x \in V(\Gamma)$ and $\gamma \in \mathrm{Stab}_{\mathrm{Aut}(\Gamma)}(v)$ such that $vx \in E(\Gamma)$ and $\gamma(x) \neq x$.

2.3.16 Let G be a group and $H \leq G$. Let $\mathcal{P}$ be a partition of G. Show that $\mathcal{P}$ is a block system of $(G, G/H)$ if and only if $\mathcal{P}$ is the set of left cosets of some $H \leq K \leq G$.

2.3.17 Let Γ be a connected bipartite vertex-transitive graph with bipartition $\mathcal{B}$. Show that $\mathcal{B}$ is a block system of $\mathrm{Aut}(\Gamma)$.

2.3.18 Let $G \leq S_n$ be transitive with B a block of G that contains a point x. Show that B is a union of orbits of $\mathrm{Stab}_G(x)$.

2.4.1 Show that a group G acting on a set X of order at least two is 2-transitive if and only if $\mathrm{Stab}_G(x)$ is transitive on $X\backslash\{x\}$ for every $x \in X$. Deduce that a 2-transitive group of degree n has order divisible by $n(n-1)$.

2.4.2 Write out the definition of an action of G on X being 2-transitive.

2.4.3 Let $k \geq 1$ be an integer and G a group acting on a set X of order at least $k+1$. We say that G is ***k*-transitive** if for every two ordered k-tuples of distinct element, say $(x_1, \ldots, x_k)$ and $(y_1, \ldots, y_k)$ (so $x_i \neq x_j$ and $y_i \neq y_j$ if $1 \leq i \neq j \leq k$), there is $g \in G$ with $g(x_1, \ldots, x_k) = (y_1, \ldots, y_k)$. Generalize Exercise 2.4.1 by showing that G is k-transitive if and only if the point-wise stabilizer of any set Y of $k-1$ distinct elements of X is transitive on $X\backslash Y$.

2.4.4 Use Exercise 2.4.3 to show that if $G \leq S_n$ is k-transitive then $n!/(n-k)!$ divides $|G|$.

2.4.5 Show that the alternating group A_n is $(n-2)$-transitive, and the symmetric group S_n is n-transitive.

2.4.6 Let p be prime and $\mathrm{AGL}(1,p) = \{x \mapsto ax+b : a \in \mathbb{Z}_p^*, b \in \mathbb{Z}_p\}$. Recalling that the multiplicative subgroup of a finite field is cyclic (Dummit and Foote, 2004, Proposition 9.5.18), use Exercise 2.4.1 to show that $\mathrm{AGL}(1,p)$ is 2-transitive of order $p(p-1)$.

2.4.7 Show that a 2-transitive group is primitive.

2.4.8 Let $q = p^f$ with p a prime and $f \geq 1$, and $d \geq 2$. Denote the field of order q by $\mathbb{F}_q$. Define the general linear group of dimension d over the field $\mathbb{F}_q$, denoted $\mathrm{GL}(d,q)$, to be the group of all invertible $d \times d$ matrices with entries in $\mathbb{F}_q$, with binary operation matrix multiplication. Let $\mathrm{GL}(d,q)$ act on the set of one-dimensional subspaces of the d-dimensional vector space $\mathbb{F}_q^d$ by $A \cdot S = \{A \cdot s : s \in S\}$. Show that there are $n = (q^d-1)/(q-1)$ one-dimensional subspaces of $\mathbb{F}_q^d$ and if $\phi\colon \mathrm{GL}(d,q) \to S_n$ is the permutation representation induced by this action, then $\phi(\mathrm{GL}(d,q))$ is a 2-transitive group and $\mathrm{Ker}(\phi) \cong \mathbb{Z}_q^*$. The group $\phi(\mathrm{GL}(d,q))$ is called the **projective general linear group** of dimension d over the field $\mathbb{F}_q$ and is denoted $\mathrm{PGL}(d, q)$. It is discussed in more detail in Section 4.5.

2.4.9 Let p be a prime and $d \geq 1$. Define a group $\mathrm{AGL}(d, p)$, the **affine general linear group** of dimension d over $\mathbb{F}_p$, to be $\left\langle \left(\mathbb{Z}_p^d\right)_L, \mathrm{GL}(d, p)\right\rangle$. Show that $\left(\mathbb{Z}_p^d\right)_L \trianglelefteq \mathrm{AGL}(d, p)$ and that $\mathrm{AGL}(d, p)$ is 2-transitive.

2.4.10 Show that a transitive group $G \leq S_n$ is quasiprimitive if and only if it is primitive or $\mathrm{Stab}_G(B)$ is core-free in G for every nontrivial block B of G.

2.4.11 Let $G \leq S_n$ be quasiprimitive with block system $\mathcal{B}$. Show that $G/\mathcal{B}$ is quasiprimitive, and $G/\mathcal{B}$ is quasiprimitive and imprimitive if and only if there exists a block system C with $\mathcal{B} < C < \mathbb{Z}_n$.

2.4.12 Let G be quasiprimitive. Show that there is a block system C with G/C primitive and $G/C \cong G$.

2.4.13 Let $H \leq S_n$. Show that either $H \leq A_n$ or half the elements of H are in A_n. Using that A_n is simple, $n \geq 5$, show that A_n is the only proper nontrivial normal subgroup of S_n.

2.5.1 Show that there are exactly ten 6-cycles in the Petersen graph.

2.5.2 Show that the edge $\{12, 35\}$ is contained in exactly two 1-factors of the Petersen graph. (Hint: Remove the edge and its incident vertices from the Petersen graph. Consider the remaining "outside" path of length three, which is the path $34, 25, 13, 45$. Show that there is a unique 1-factor that contains no edges of the this path, no 1-factors that contain one edge of this path, and a unique 1-factor that contains two edges of this path.)

2.5.3 Use Exercise 2.5.2, Exercise 2.1.3, and a counting argument to show that the Petersen graph has exactly six 1-factors. Then show that the automorphism group of the Petersen graph is transitive on its set of 1-factors (and complementary 2-factors).

2.5.4 Determine all possible values of m and k for which there is a block system with m blocks of size k for the action of A_5 on the arcs of the Petersen graph.

2.5.5 Explain, in terms of the combinatorial substructures of P as well as the subgroup structure of A_5, why there are exactly two different block systems of the left coset action of A_5 on $A_5/\mathbb{Z}_2$ with blocks of size 5.

2.5.6 How many different block systems are there of 5 blocks of size 6 in the left coset action of A_5 on $A_5/\mathbb{Z}_2$? Explain your answer in terms of the combinatorial substructures of P as well as the subgroup structure of A_5.

2.5.7 For $1 \le i \le 4$, denote by C_i the 6-cycle remaining when the vertex $\{i,5\}$ and its neighbors are deleted from the Petersen graph (using its 2-subset labeling). Orient C_1 in one direction to obtain $\vec{C}_1$, and then orient the other cycles C_2, C_3, C_4 by giving them the orientation determined by $\gamma\left(\vec{C}_1\right)$, where $\gamma \in A_4$ (considered as the subgroup of $\mathrm{Stab}_{A_5}(5) \le S_5$) to obtain oriented cycles $\vec{C}_2, \vec{C}_3$, and $\vec{C}_4$. Show that $\left\{\vec{C}_1, \vec{C}_2, \vec{C}_3, \vec{C}_4\right\}$ is a block of the left coset action of A_5 on $A_5/\mathbb{Z}_3$.

2.5.8 Determine all block systems of the action of A_5 on 2-arcs of the Petersen graph.

2.6.1 Let $G \le S_n$ with $H = \mathrm{Stab}_G(x)$, $x \in \mathbb{Z}_n$. Let r be the rank of G, and $\ell_1, \ldots, \ell_r$ the lengths of the suborbits of G. Show that H is isomorphic to a subgroup of $\Pi_{i=1}^r S_{\ell_i}$, and, in particular, that $|H| \le \Pi_{i=1}^r \ell_i!$.

2.6.2 Let Γ be a vertex-transitive digraph with $G \le \mathrm{Aut}(\Gamma)$ that is transitive on the set of maximum independent sets of Γ as well as their intersections. Show that if the intersection of two maximum independent sets is unique (that is, no two intersections are the same) and there exists $g \in G$ that interchanges two intersecting maximum independent sets of Γ, then G is transitive on the vertex-set and arc-set of the intersection graph of Γ. Conclude that the intersection graph of Γ is an orbital digraph of G in its action on maximum independent sets of Γ.

2.6.3 Find a pair of 1-factors in the Petersen graph and $g \in A_5$ that interchanges them.

2.6.4 Check that the two orbital digraphs of the left coset action of A_5 on $A_5/\mathbb{Z}_5$ of valence 5 are both isomorphic to the icosahedron.

2.6.5 Let Γ be a connected digraph whose arc set, viewed as a relation, is an equivalence relation. Show that Γ is a complete graph with a loop at every vertex.

2.6.6 Show that the orbital graph corresponding to the suborbit $\left\{\hat{C}_d, \hat{C}_e, \hat{C}_f\right\}$ of the 2-subset action of $(3,4,5) \in A_5$ is $5K_4$.

2.6.7 Let G be a group and $H \le G$. Show that in the left coset action of G on G/H the element $g \in G$ fixes xH if and only if $x^{-1}gx$ normalizes H.

2.6.8 Show that the left coset action of $S_4 \times \mathbb{Z}_2$ on the left cosets of one of its Sylow 3-subgroups Q is faithful on 16 points. Is this action primitive? Find a combinatorial substructure of Q_3 = F8 with 16 subdigraphs that is stabilized by Q and on which $S_4 \times \mathbb{Z}_2$ is transitive. Then find all orbital digraphs of this action.

3

Some Motivating Problems

In this chapter we will pause from more or less providing examples of vertex-transitive graphs, as well as necessary permutation group theory, to have a look at some of the problems that motivated the development of this area of mathematics. In Section 3.1, we start near the beginning with some extremely influential work of Tutte from 1947 on cubic s-arc-transitive graphs. We continue on a similar theme in Section 3.2 by considering graphs that have additional symmetry and are also transitive on edges or arcs. Again, much of the work in that section is related to work of Tutte, but later in his career. We then turn to perhaps the most famous problem in this area of graph theory: Lovász's conjecture that every connected vertex-transitive graph contains a Hamilton path. We will spend four sections on this problem. Section 3.3 is an expository introduction to the problem, while Section 3.4 considers the four known connected vertex-transitive graphs of order at least three that do not have a Hamilton cycle (we will wait until Section 6.1 to show that the Coxeter graph does not have a Hamilton cycle). Section 3.5 is about positive results, and we show that every connected Cayley graph of abelian group of order at least three is Hamiltonian in Theorem 3.5.4. We then pause from our main focus and consider quotient digraphs in Section 3.6, as these are needed for our final section in this chapter on Hamiltonicity. This section on Hamiltonicity has one of the earliest techniques for finding Hamilton cycles in vertex-transitive graphs (and non-vertex-transitive graphs that have some appropriate symmetry). Also in this section we discuss a common way of representing graphs with "nice" automorphisms, and Frucht notation. Finally, Section 3.7 is on a well-known and studied problem of Marušič that was motivated by Lovász's conjecture, the semiregularity problem. Namely, does the automorphism group of every vertex-transitive graph contain a semiregular element? The problems we see here will be revisited at the end of the book, with Chapter 11 on the Lovász conjecture, Chapter 12 on the semiregularity problem, and Chapters 10 and 13 on graphs that combine vertex-transitive with some other form of transitivity.

3.1 Cubic s-Arc-transitive Graphs

In this section, we introduce s-arc-transitive graphs, and following Tutte (1947) (where most of the work in this section was first proved), we mainly consider cubic s-arc-transitive graphs. One aim of this section is to give a third proof that the automorphism group of the Petersen graph P is S_5 in Corollary 3.1.8, and the technique we will develop is commonly used to determine automorphism groups of cubic s-arc-transtive graphs.

Definition 3.1.1 Let $s \geq 1$, and Γ be a digraph. An **s-arc** of Γ is a sequence $x_0, x_1, \ldots, x_s$ of vertices of Γ such that $(x_i, x_{i+1}) \in A(\Gamma)$, $0 \leq i \leq s-1$, and $x_i \neq x_{i+2}$, $0 \leq i \leq s-2$. A digraph Γ is **s-arc-transitive** if $\mathrm{Aut}(\Gamma)$ is transitive on the set of s-arcs of Γ. Also, Γ is **s-arc-regular** if $\mathrm{Aut}(\Gamma)$ is regular on the s-arcs of Γ.

So an s-arc is a directed walk in Γ where no two successive arcs start and end at the same vertex. Note that if $s = 1$, then an s-arc-transitive digraph is simply an arc-transitive digraph. Also, if Γ is s-arc-regular, then $|\mathrm{Aut}(\Gamma)|$ is the number of s-arcs in Γ. Observe that an s-arc-transitive digraph is also an r-arc-transitive digraph for every $r \leq s$ provided Γ has no vertices of invalency or outvalency 1 (this condition guarantees that every arc of Γ is contained in an r-arc of Γ). It is left to the exercises to show that a cycle is s-arc-transitive for every $s \geq 1$ (Exercise 3.1.1), and that K_n is 2-arc-transitive but not 3-arc-transitive for every $n \geq 4$ (Exercise 3.1.3).

Theorem 3.1.2 *The Petersen graph is 3-arc-transitive but it is not 4-arc-transitive.*

Proof In this proof, we will use the 2-subset labeling of the Petersen graph given in Figure 2.1. Suppose x_0, x_1, x_2, x_3 is a 3-arc in P. Set $x_0 = \{a, b\}$, and $x_1 = \{c, d\}$, so $\{a, b\} \cap \{c, d\} = \emptyset$. As $x_0x_2 \notin E(P)$ as the Petersen graph has no triangles, $x_0 \cap x_2 \neq \emptyset$. As $x_0 \neq x_2$, x_2 must contain exactly one element of x_0, say a, and the other element, say f, of $\{1, 2, 3, 4, 5\}$ not contained in $x_0 \cup x_1$. Thus $x_2 = \{a, f\}$. Then $x_2 \cap x_3 = \emptyset$, so $x_3 \subset \{b, c, d\}$ and $x_3 \neq x_1 = \{c, d\}$. Thus x_3 consists of b and one element of x_1, say c, so that $x_3 = \{b, c\}$.

Now suppose y_0, y_1, y_2, y_3 is another 3-arc in P. By arguments in the preceding paragraph, there exists $a', b', c', d', f' \in \{1, 2, 3, 4, 5\}$ such that $y_0 = \{a', b'\}$, $y_1 = \{c', d'\}$, $y_2 = \{a', f'\}$, and $y_3 = \{b', f'\}$. Then the map $\phi \in S_5$ given by

$\phi(a) = a', \phi(b) = b', \ldots, \phi(f) = f'$ maps x_0, x_1, x_2, x_3 to y_0, y_1, y_2, y_3 and P is 3-arc-transitive.

To see that P is not 4-arc-transitive, first observe that there are many 4-arcs that are contained in 5-cycles (simply pick a directed path of length 4 contained in a 5-cycle). Also, the 4-arc $\{1,2\}, \{3,4\}, \{1,5\}, \{2,4\}, \{1,3\}$ is not contained in any 5-cycle in P. As an automorphism of a graph maps a 5-cycle to a 5-cycle, there is no automorphism of P that maps any 4-arc in a 5-cycle to a 4-arc not in a 5-cycle, and so P is not 4-arc-transitive. □

Of course, the Petersen graph is cubic. We proceed for the moment in more generality than just the Petersen graph, as the proofs for the Petersen graph and more general graphs are identical. We will see results concerning s-arc-regularity due to Tutte (1947, 1959) that will be useful, in combination with Lemma 3.1.3, in determining the automorphism groups of some other cubic graphs we will be interested in later as well as a new proof that gives the automorphism group of the Petersen graph.

Lemma 3.1.3 *Let Γ be a cubic s-arc-regular graph. Then*

$$|\mathrm{Aut}(\Gamma)| = 2^s \cdot |E(\Gamma)| = 3 \cdot 2^{s-1} \cdot |V(\Gamma)|.$$

Proof As $\mathrm{Aut}(\Gamma)$ is regular on the s-arcs in Γ, the number of s-arcs is $|\mathrm{Aut}(\Gamma)|$ by Corollary 1.1.12. We thus need to show that the number of s-arcs in Γ is $2^s \cdot |E(\Gamma)|$, because as Γ is cubic, $|E(\Gamma)| = \frac{3}{2} \cdot |V(\Gamma)|$.

Now, choose an edge $e = x_0x_1 \in E(\Gamma)$. We show by induction on $1 \leq r \leq s$ that there are 2^r r-arcs with first arc corresponding to e. There are two possible directions that can be placed on this edge – without loss of generality assume our choice is x_0, x_1. We have established that there are two 1-arcs whose first edge is e, namely x_0, x_1 and x_1, x_0. If there are 2^r r-arcs $y_0, y_1, \ldots, y_r$ with first edge e, then as Γ is cubic there are exactly two neighbors of y_r, say x_{r+1} and y_{r+1}, that are not y_{r-1}. Thus there are exactly two $(r+1)$-arcs $y_0, y_1, \ldots, y_r, x_{r+1}$ and $y_0, y_1, \ldots, y_r, y_{r+1}$ whose first r arcs are $y_0, y_1, \ldots, y_r$. We conclude there are 2^{r+1} $(r+1)$-arcs with first edge e, and by induction there are 2^r r-arcs whose initial edge is e. This implies there are 2^s s-arcs whose inital edge is e, and there are $2^s \cdot |E(\Gamma)|$ s-arcs in Γ. □

We remark that $3 \cdot 2^{s-1}$ is the size of the stabilizer of a vertex in the previous result, and that sometimes the form $3 \cdot 2^{s-1} \cdot |V(\Gamma)|$ is more convenient to use than $2^s \cdot |E(\Gamma)|$.

An obvious question is whether the techniques used to study cubic s-arc-transitive graphs generalize to all valencies. The general answer to this question is "No" (but there are some exceptions – see Section 10.5), and it is in our next result where the crucial property that cubic graphs possess that graphs of other valencies do not, is used. The crucial property is that in a cubic graph Γ, if a subgroup H of $\mathrm{Aut}(\Gamma)$ fixes x and y with $xy \in E(\Gamma)$, and u and v are the other neighbors of y in Γ, then H either fixes both u and v or is transitive on $\{u, v\}$. For higher valencies such a subgroup H need not be transitive on the other neighbors of y in Γ if it is not trivial on the neighbors. See Exercises 3.1.9 and 3.1.11.

Lemma 3.1.4 *Let Γ be a cubic s-arc-transitive graph that is not $(s + 1)$-arc-transitive. If $x_0, \ldots, x_s$ is an s-arc in Γ, and x_{s+1}, y_{s+1} are the neighbors of x_s in Γ distinct from x_{s-1}, then any automorphism of Γ that fixes the s-arc $x_0, \ldots, x_s$ (and so fixes each vertex x_i, $0 \le i \le s$), also fixes x_{s+1} and y_{s+1}.*

Proof We suppose otherwise, and show Γ is $(s + 1)$-arc-transitive. Suppose $x_0, \ldots, x_s$ is an s-arc, with $x_{s-1}, x_{s+1}, y_{s+1}$ the neighbors of x_s, and there exists $\delta \in \mathrm{Aut}(\Gamma)$ that fixes $x_0, \ldots, x_s$ but, say $\delta(x_{s+1}) \ne x_{s+1}$. As δ fixes x_s it must map the neighbors of x_s to the neighbors of x_s, and as $\delta(x_{s-1}) = x_{s-1}$ and $x_{s-1}, x_{s+1}, y_{s+1}$ are the neighbors of x_s, we have that δ interchanges x_{s+1} and y_{s+1}.

Now let $a_0, \ldots, a_{s+1}$, and $b_0, \ldots, b_{s+1}$ be any two $(s+1)$-arcs in Γ. As Γ is s-arc-transitive, there exists $\gamma_a, \gamma_b \in \mathrm{Aut}(\Gamma)$ such that $\gamma_a(a_0, \ldots, a_s) = x_0, \ldots, x_s$ and $\gamma_b(b_0, \ldots, b_s) = x_0, \ldots, x_s$. Then either $\gamma_b^{-1}\gamma_a(a_0, \ldots, a_{s+1}) = b_0, \ldots, b_{s+1}$ or $\gamma_a(a_{s+1}) \ne \gamma_b(b_{s+1})$. In the latter case, both $\gamma_a(a_{s+1})$ and $\gamma_b(b_{s+1})$ must be neighbors of x_s that are not x_{s-1}. So without loss of generality, we may assume $\gamma_a(a_{s+1}) = x_{s+1}$ while $\gamma_b(b_{s+1}) = y_{s+1}$. But then $\gamma_b^{-1}\delta\gamma_a(a_0, \ldots, a_{s+1}) = b_0, \ldots, b_{s+1}$, and Γ is $(s + 1)$-arc-transitive, a contradiction. □

Theorem 3.1.5 *A connected cubic s-arc-transitive graph Γ that is not $(s+1)$-arc-transitive is s-arc-regular.*

Proof Let $x_0, \ldots, x_s$ be an s-arc in Γ. Let N_i, $i \ge 0$, be the set of all vertices of Γ at distance i from x_s. As Γ is connected, there exists some positive integer d such that $N_d \ne \emptyset$ and $\cup_{i=0}^{d} N_i = V(\Gamma)$. Let $H \le \mathrm{Aut}(\Gamma)$ stabilize $x_0, \ldots, x_s$. We claim that for $0 \le i \le d$ and $y_i \in N_i$, there exists an s-arc A in Γ such that all the vertices of A are fixed by H, and the last vertex of A is y_i. This will imply the result as, by Lemma 3.1.4, every neighbor of a vertex in N_i (so all of N_{i+1}) is fixed by H.

To establish the claim, we proceed by induction on $0 \le i \le d$. The claim is true for $i = 0$ as $x_0, \ldots, x_s$ is the required s-arc. Now assume the claim is true for all vertices of Γ at distance $i - 1 \ge 0$ from x_s, and let $y_i \in N_i$. Let $P = x_s y_1 \ldots y_{i-1} y_i$ be a path of length i from x_s to y_i in Γ. Then $y_{i-1} \in N_{i-1}$. By the induction hypothesis, there is an s-arc $z_0, z_1, \ldots, z_s$ in Γ such that $z_0, \ldots, z_s$ are all fixed by H and $z_s = y_{i-1}$. Of course, by Lemma 3.1.4, y_i is fixed by H. If $y_i \neq z_{s-1}$, then $z_1, \ldots, z_s, y_i$ is the required s-arc. Otherwise, $z_s, \ldots, z_0$ is an s-arc in Γ, and z_0 has some neighbor w that is not z_1. By Lemma 3.1.4, w is fixed by H, and $z_{s-1}, \ldots, z_0, w$ is an s-arc in Γ all of whose vertices are fixed by H. Then $w, z_0, \ldots, z_{s-1}$ is an s-arc in Γ all of whose vertices are fixed by H, and $z_{s-1} = y_i$. Our claim, and the result, then follows by induction.

□

Corollary 3.1.6 *Let Γ be a connected cubic arc-transitive graph. Then Γ is s-arc-regular if and only if* $|\mathrm{Aut}(\Gamma)| = 2^s \cdot |E(\Gamma)| = 3 \cdot 2^{s-1} \cdot |V(\Gamma)|$.

Proof Suppose Γ is a connected cubic arc-transitive graph. If Γ is s-arc-regular, then the result follows from Lemma 3.1.3. For the converse, suppose $|\mathrm{Aut}(\Gamma)| = 2^s \cdot |E(\Gamma)|$. As Γ is arc-transitive, Γ is t-arc-transitive but not $(t+1)$-arc-transitive for some $t \geq 1$. By Theorem 3.1.5 we have that Γ is t-arc-regular, and so by Lemma 3.1.3 we have $|\mathrm{Aut}(\Gamma)| = 2^t \cdot |E(\Gamma)|$. We conclude that $t = s$. □

The following result is a combination of Theorem 3.1.2 and Theorem 3.1.5.

Corollary 3.1.7 *The Petersen graph is 3-arc regular.*

We are now ready for our third proof that the automorphism group of the Petersen graph is $\mathcal{S}_5$.

Corollary 3.1.8 Aut(P) *is $\mathcal{S}_5$ in its action on 2-subsets of* $\{1, 2, 3, 4, 5\}$.

Proof By Lemma 2.1.3, the action of $\mathcal{S}_5$ on 2-subsets is a subgroup of Aut(P). We have already noted that the only permutation in $\mathcal{S}_5$ that fixes each 2-subset of $\{1, 2, 3, 4, 5\}$ is the identity. As the Petersen graph is 3-arc-regular by Corollary 3.1.7, by Corollary 3.1.6 $|\mathrm{Aut}(P)| = 2^3 \cdot |E(P)| = 120 = |\mathcal{S}_5|$. □

As mentioned earlier, the study of cubic s-arc-transitive graphs was initiated by Tutte in 1947 in a widely acclaimed and important paper (Tutte, 1947). We state, without proof, two other important results that are contained in this paper.

Theorem 3.1.9 *A cubic s-arc-transitive graph has girth at least* $2s - 2$.

Theorem 3.1.10 *If Γ is a cubic s-arc-transitive graph, then* $s \leq 5$.

Similar to the term cubic graphs for the 3-regular graphs, we have the following for 4-regular graphs.

Definition 3.1.11 A graph is **quartic** if it is regular of valency 4.

Much work has followed from the work of Tutte, and continues today. For example, Djoković and Miller (1980) determined all pairs $(\mathrm{Stab}_G(v), \mathrm{Stab}_G(e))$, in a cubic symmetric graph – there are seven such pairs – see Table 10.1. Conder and Nedela (2009) refined this classification and showed that there are 17 combinations of arc-stabilizers in cubic symmetric graphs and determined the "type" of every cubic symmetric graph on at most 768 vertices; see the first column of Table 10.2. Conder and Dobcsányi (2002) completed and extended the Foster census. Conder then extended this census to all cubic symmetric graphs of order up to 2,048 (Conder, 2006) and then to order 10,000 (Conder, 2011). We mention that there is also a census of all cubic vertex-transitive graphs of order up to 1,280 vertices due to Potočnik, Spiga, and Verret (2013). Additionally, Goldschmidt (1980) determined that in an edge-transitive cubic graph Γ there are 15 pairs of groups $(\mathrm{Stab}_G(u), \mathrm{Stab}_G(v))$ where uv is an edge of Γ. He also showed that $|\mathrm{Stab}_G(v)| \leq 384$. Recently, a census of edge-transitive quartic graphs on up to 512 vertices was obtained by Wilson and Potočnik (2016).

3.2 Arc-, Edge-, and Half-arc-transitive Graphs

In this section we present basic results and constructions for vertex-transitive graphs that are arc-transitive, edge-transitive, and half-arc-transitive in Theorems 3.2.8, 3.2.9, and 3.2.11, respectively. In particular, we give characterizations of each of these kinds of "extra" symmetry in terms of orbital digraphs, double coset digraphs, and orbits of the stabilizer of a point on neighbors. Note that as an s-arc-transitive graph is arc-transitive (provided it has no vertices of valence 1) and an arc-transitive graph is edge-transitive, the Petersen graph is arc-transitive and edge-transitive. The existence of a half-arc-transitive graph is significantly harder to show, as one must not only show the graph is edge-transitive, but one has to show that an automorphism reversing the endpoints of an edge *does not* exist. We show the existence of the smallest half-arc-transitive graph, the Holt graph, in Example 3.2.14.

Definition 3.2.1 A vertex-transitive graph is **half-arc-transitive** if it is edge-transitive but not arc-transitive (recall that we view each edge uv as the identification of the arcs (u, v) and (v, u)).

Clearly an arc-transitive graph is edge-transitive. The graph in Figure 3.1 is vertex-transitive but not edge-transitive as some edges are in 3-cycles and some are not. Also, there are vertex- and edge-transitive graphs that are not arc-transitive; see Example 3.2.14 and Figure 3.3. Also, in older papers the term 1/2-**transitive** is used for half-arc-transitive.

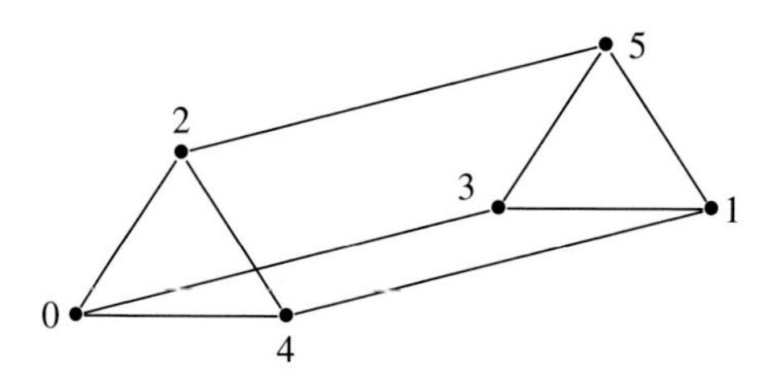

Figure 3.1 $\mathrm{Cay}(\mathbb{Z}_6, \{2,3,4\})$ – a vertex- but not edge-transitive graph.

The following result gives the easiest and perhaps most common method of constructing arc-transitive Cayley graphs. The idea here is the same as for obtaining examples of vertex-transitive digraphs. They may be hard to *find*, but can be easy to *construct*.

Lemma 3.2.2 *Let G be a group and $H \leq \mathrm{Aut}(G)$. Let $\mathcal{O}$ be an orbit of H. Then $\mathrm{Cay}(G, \mathcal{O})$ is an arc-transitive digraph.*

Proof Let G be a group and $H \leq \mathrm{Aut}(G)$. Let $\mathcal{O}$ be an orbit of H. By Corollary 1.2.16, $\alpha \in \mathrm{Aut}(\mathrm{Cay}(G, \mathcal{O}))$ for every $\alpha \in H$. Let $(g_1, h_1), (g_2, h_2) \in A(\mathrm{Cay}(G, \mathcal{O}))$, so $h_i = g_i s_i$ for some $s_i \in \mathcal{O}$, $i = 1, 2$. Then $(g_1)_L^{-1}(g_1, h_1) = (1_G, s_1)$ and $(g_2)_L^{-1}(g_2, h_2) = (1_G, s_2)$. As $s_1, s_2 \in \mathcal{O}$ is an orbit of H, there exists $\alpha \in H$ such that $\alpha(1, s_1) = (1, s_2)$. Then $(g_2)_L \alpha (g_1)_L^{-1}(g_1, h_1) = (g_2, h_2)$ and so $\mathrm{Cay}(G, \mathcal{O})$ is arc-transitive. □

The previous result is often applied when $H \leq \mathrm{Aut}(G)$ is cyclic. So it will be notationally convenient to be able to say an element (here that would be a generator of H) has an orbit. It is defined in the obvious way.

Definition 3.2.3 Let $\alpha \in S_n$. By the **orbit of α that contains** x, we mean the orbit of $\langle \alpha \rangle$ that contains x.

Example 3.2.4 $\mathrm{Cay}(\mathbb{Z}_{12}, \{7, 11\})$, as in Figure 3.2, and $\mathrm{Cay}(\mathbb{Z}_{16}\{1, 3, 9, 11\})$ are arc-transitive digraphs.

Solution Let $\alpha \in \mathrm{Aut}(\mathbb{Z}_{12})$ be given by $\alpha(x) = 5x$. The orbit of α that contains 7 is $\{7, 11\}$, and so the digraph $\mathrm{Cay}(\mathbb{Z}_{12}, \{7, 11\})$ in Figure 3.2 is an arc-transitive digraph. Similarly, let $\alpha \in \mathrm{Aut}(\mathbb{Z}_{12})$ be given by $\alpha(x) = 3x$, then $\mathrm{Cay}(\mathbb{Z}_{16}, \{1, 3, 9, 11\})$ is an arc-transitive digraph as $\{1, 3, 9, 11\}$ is the orbit of α that contains 1. □

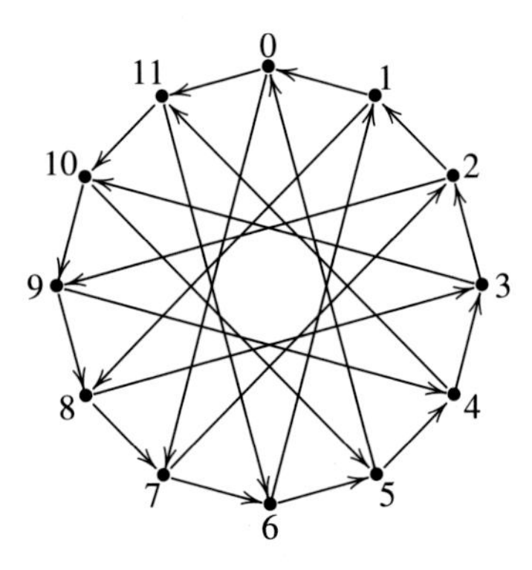

Figure 3.2 The arc-transitive digraph $\mathrm{Cay}(\mathbb{Z}_{12}, \{7, 11\})$.

Example 3.2.5 Let $S = \{(1, 1), (1, 2), (2, 2), (2, 4), (4, 4), (4, 3), (3, 3), (3, 1)\}$. Then $\mathrm{Cay}(\mathbb{Z}_5 \times \mathbb{Z}_5, S)$ is an arc-transitive digraph.

Solution As in the previous example, we need only find an automorphism $\alpha\colon \mathbb{Z}_5 \times \mathbb{Z}_5 \to \mathbb{Z}_5 \times \mathbb{Z}_5$ such that S is an orbit of α. Define α by $\alpha(i, j) = (j, 2i)$. Straightforward computations show that $\alpha \in \mathrm{Aut}(\mathbb{Z}_5^2)$ and S is the orbit of α that contains $(1, 1)$. □

A similar construction holds for double coset digraphs, but we do not yet have the necessary permutation group-theoretic tool – see Exercise 4.4.8.

Next, we investigate the relationship between arc- or edge-transitive digraphs with the orbital digraph construction, as well as double coset digraphs. Often, in practice, when considering arc-transitive or edge-transitive digraphs or graphs, it is common to work with a transitive subgroup $G \leq \mathrm{Aut}(\Gamma)$ that is transitive on arcs or edges instead of the full automorphism group. We thus make the following definitions:

Definition 3.2.6 A digraph Γ is ***G*-arc-transitive** if $G \leq \mathrm{Aut}(\Gamma)$ is vertex-transitive and transitive on $A(\Gamma)$. Similarly, a graph Γ is ***G*-edge-transitive** if $G \leq \mathrm{Aut}(\Gamma)$ is vertex-transitive and also transitive on the edges of Γ. Finally, a graph Γ is G-half-arc-transitive if $G \leq \mathrm{Aut}(\Gamma)$ is G-edge-transitive but not G-arc-transitive.

If Γ is G-arc-transitive or G-edge-transitive, then of course it is arc-transitive or edge-transitive, respectively. However, a G-half-arc-transitive graph need not be a half-arc-transitive graph if G is a proper subgroup of $\mathrm{Aut}(\Gamma)$. We require some notation before proceeding.

Definition 3.2.7 Let G be a group and $H \leq G$ be core-free in G. For $g \in G$, define $\hat{g}_L : G/H \to G/H$ by $\hat{g}_L(aH) = gaH$ for all $a \in G$.

Of course, if $H = 1$, then $\hat{g}_L = g_L$ (and we will use g_L in that case). So we only use that "hat" notation when $H > 1$. Also, $(G, G/H) = \{\hat{g}_L : g \in G\}$.

Theorem 3.2.8 *For a vertex-transitive digraph* Γ *with* $G \leq \mathrm{Aut}(\Gamma)$ *a transitive subgroup, the following are equivalent:*

(i) Γ *is* G*-arc-transitive,*
(ii) Γ *is an orbital digraph of* G*,*
(iii) *for some* $u \in V(\Gamma)$*,* $N^+_\Gamma(u)$ *is an orbit of* $\mathrm{Stab}_G(u)$*, and*
(iv) $\Gamma \cong \mathrm{Cos}(G, H, S)$ *where* $H = \mathrm{Stab}_G(u)$ *for some* $u \in V(\Gamma)$*, and* $S = HsH$ *is a single double coset of* H *for some* $s \in G$*.*

Proof (i) $\Longrightarrow$ (ii): Suppose Γ is a G-arc-transitive digraph. By Proposition 1.4.6, Γ is a generalized orbital digraph of G. If Γ is a union of two (or more) orbital digraphs Γ_1 and Γ_2 of G with $(v, u) \in A(\Gamma_1)$ and $(w, x) \in A(\Gamma_2)$, then there is no $g \in G$ such that $g(v, u) = (w, x)$, a contradiction.

(ii) $\Longrightarrow$ (iii): This is Lemma 1.4.7.

(iii) $\Longrightarrow$ (iv): Suppose $N^+_\Gamma(u)$ is an orbit of $H = \mathrm{Stab}_G(u)$. Let $w \in N^+_\Gamma(u)$ and $g_w \in G$ such that $g_w(u) = w$. We first claim that $S = Hg_wH$ is the set $T = \{g \in G : g(u) = v \text{ with } (u, v) \in A(\Gamma)\}$. To see $S \subseteq T$, for $h_1, h_2 \in H$ note that $h_2 g_w h_1(u) = h_2(w)$ is an outneighbor of u as w is an outneighbor of u and $h_2(u) = u$. For the reverse containment, if $k \in T$, then there exists $v \in V(\Gamma)$ with $(u, v) \in A(\Gamma)$ and $k(u) = v$. As $N^+_\Gamma(u)$ is an orbit of $\mathrm{Stab}_G(u)$, there exists $h_1 \in H$ such that $h_1(w) = k(u)$. Then $h_1 g_w(u) = k(u)$, and so $k^{-1} h_1 g_w = h_2^{-1} \in H$, or $k = h_1 g_w h_2$. Thus $T \subseteq S$ and $T = S$ as claimed. The result now follows by Theorem 1.3.9.

(iv) $\Longrightarrow$ (i) By Proposition 1.3.8, $(G, G/H) \cong G$ as H is core-free in G by Exercise 1.1.9. Let $g_1, g_2 \in G$ and $s_1, s_2 \in HsH$, and so (g_1H, g_1s_1H) and (g_2H, g_2s_2H) are arcs of $\mathrm{Cos}(G, H, S)$. Let $h_1, h_2, h_3, h_4 \in H$ such that $s_1 = h_1 s h_2$ and $s_2 = h_3 s h_4$. Then

$$\hat{g}_2 \hat{h}_3 \hat{h}_1^{-1} \hat{g}_1^{-1}(g_1H, g_1s_1H) = (g_2H, g_2s_2H),$$

and $\hat{g}_2 \hat{h}_3 \hat{h}_1^{-1} \hat{g}_1^{-1} \in \hat{G}$. Hence $\mathrm{Cos}(G, H, S)$ is $\hat{G}$-arc-transitive, so Γ is G-arc-transitive. □

The choice of out-neighbors of v in (iii) of the above theorem is arbitrary, and we could have just as easily used the in-neighbors of v.

Theorem 3.2.9 *For a vertex-transitive graph Γ with $G \leq \mathrm{Aut}(\Gamma)$ a transitive subgroup, the following are equivalent:*

(i) *Γ is G-edge-transitive,*
(ii) *Γ is either a union of one self-paired orbital digraph of G or the union of two paired but not self-paired orbital digraphs of G,*
(iii) *for $u \in V(\Gamma)$, $\{v : uv \in E(\Gamma)\}$ is an orbit of $\mathrm{Stab}_G(u)$ or is the union of two orbits of $\mathrm{Stab}_G(u)$, with those orbits consisting of the in-neighbors of u and the out-neighbors of u in an orbital digraph of G and also in its pair,*
(iv) *if $H = \mathrm{Stab}_G(u)$ for some $u \in V(\Gamma)$ and $\Gamma \cong \mathrm{Cos}(G, H, S)$, then $HSH = H\left\{s, s^{-1}\right\}H$ for some $s \in G$.*

Proof (i) $\Longrightarrow$ (ii): If Γ is G-arc-transitive, then by Theorem 3.2.8 Γ is an orbital digraph of G. As Γ is a graph, Γ is self-paired.

If Γ is not arc-transitive, then set $\Gamma = \cup_{i=1}^{t}\Gamma_i$, where each Γ_i is an orbital digraph of G, $t \geq 2$. As Γ is a graph, each Γ_i is either self-paired, or there exists $1 \leq j \leq t$, $j \neq i$, such that Γ_j is the pair of Γ_i. If some Γ_i is self-paired, then as Γ is edge-transitive and each Γ_i is an orbital digraph of G, $i = 1 = t$. Then $\Gamma = \Gamma_1$ is arc-transitive, a contradiction. Hence each Γ_i is a digraph. Assume without loss of generality that Γ_2 is the pair of Γ_1. If $t \geq 3$, then there exists $(u, v) \in A(\Gamma_1)$ and $(w, x) \in A(\Gamma_3)$ but no $g \in G$ with $g(\{u, v\}) = \{w, x\}$, in which case Γ is not edge-transitive, a contradiction.

(ii) $\Longrightarrow$ (iii): If Γ is a self-paired orbital digraph, then Γ is arc-transitive and, by Theorem 3.2.8, $\{v : (u, v) \in A(\Gamma)\} = \{v : uv \in E(\Gamma)\}$ is an orbit of $\mathrm{Stab}_G(u)$. If $\Gamma = \Gamma_1 \cup \Gamma_2$ where Γ_1 and Γ_2 are different but paired orbital digraphs of G, by Lemma 1.4.7 we see $\{v : (u, v) \in A(\Gamma_i)\}$ is an orbit $\mathcal{O}_i$, of $\mathrm{Stab}_G(u)$, $i = 1, 2$. As Γ_1 and Γ_2 are paired but not equal, the in-neighbors of u in Γ_1 are the out-neighbors of u in Γ_2, and so $\mathcal{O}_2$ is the set of in-neighbors of u in Γ_1, and vice versa.

(iii) $\Longrightarrow$ (iv): Let $u \in V(\Gamma)$, and $H = \mathrm{Stab}_G(u)$. If $\{v : uv \in E(\Gamma)\}$ is an orbit of H, then by Theorem 3.2.8 Γ is arc-transitive and $\Gamma \cong \mathrm{Cos}(G, H, HsH)$, for some $s \in G$. As Γ is a graph, by Exercise 1.3.6 $s^{-1} \in HsH$, and so $HsH = Hs^{-1}H$. Then $H\left\{s, s^{-1}\right\}H = HsH$.

If $\{v : uv \in E(\Gamma)\}$ is the union of two orbits of $\mathrm{Stab}_G(u)$, with those orbits consisting of the in-neighbors of u and the out-neighbors of u in an orbital digraph Γ_1 of G and also in its pair, then $\Gamma_1 \cong \mathrm{Cos}(G, H, HsH)$ for some $s \in G$ by Theorem 3.2.8. The digraph Γ_2 obtained from Γ_1 by reversing the direction of every arc is then a subgraph of Γ and is the pair of Γ_1, and so an orbital digraph. By Exercise 1.3.9, $\Gamma_2 \cong \mathrm{Cos}\left(G, H, Hs^{-1}H\right)$ and so $\Gamma \cong \mathrm{Cos}\left(G, H, H\{s, s^{-1}\}H\right)$.

(iv) $\Longrightarrow$ (i): By Exercise 1.3.6, $\Gamma \cong \mathrm{Cos}(G, H, S)$ is a graph. We have already seen that $(G, G/H) \cong G$. We show that $\Delta = \mathrm{Cos}(G, H, S)$ is $(G, G/H)$-edge transitive. Let $\{g_1H, g_1s_1H\}, \{g_2H, g_2s_2H\} \in E(\Delta)$ for some $g_1, g_2 \in G$ and $s_1, s_2 \in S$. As $\hat{g}_1^{-1}, \hat{g}_2^{-1} \in (G, G/H)$, it suffices to show that there exists $g \in G$ such that $\hat{g}(\{H, s_1H\}) = \{H, s_2H\}$. Write $s_1 = h_1s^ih_2$ and $s_2 = h_3s^jh_4$, where $h_1, h_2, h_3, h_4 \in H$ and $i, j = \pm 1$. If $i = j$, then setting $g = h_3h_1^{-1}$, we see $gH = H$ and $gs_1H = s_2H$. If $i \neq j$ (in which case $-i = j$), then by setting $g = h_3s^{-i}h_1^{-1}$, we see $gH = s_2H$ and $gs_1H = H$. So $\hat{g}(\{H, s_1H\}) = \{H, s_2H\}$ and Δ is edge-transitive. □

Theorem 3.2.10 *For a vertex-transitive graph Γ and $G \leq \text{Aut}(\Gamma)$ that is vertex- and edge-transitive, the following are equivalent:*

(i) *Γ is G-arc-transitive,*
(ii) *Γ is an orbital digraph of G and the corresponding orbital is self-paired,*
(iii) *for $u \in V(\Gamma)$, $\{v : uv \in E(\Gamma)\}$ is an orbit of* $\text{Stab}_G(u)$,
(iv) *G contains an element that interchanges a pair of adjacent vertices, and*
(v) *if $H = \text{Stab}_G(u)$ for some $u \in V(\Gamma)$ and $\Gamma \cong \text{Cos}(G, H, S)$, then $HSH = HsH$ for some $s \in G$ with $s^2 \in H$.*

Proof (i) $\Longrightarrow$ (ii), (ii) $\Longrightarrow$ (iii), and (v) $\Longrightarrow$ (i) follow from Theorems 3.2.8 and 3.2.9 and their proofs.

(iii) $\Longrightarrow$ (iv) Let $uv \in E(\Gamma)$, and Γ_1 the orbital digraph of G that contains (u, v). As $\{v : uv \in E(\Gamma)\}$ is an orbit of $\text{Stab}_G(u)$ and Γ is G-edge-transitive, the only neighbors of u in Γ are the out-neighbors of u in Γ_1. This implies that the in-neighbors of u are also the out-neighbors of u, and so $\Gamma = \Gamma_1$ is a self-paired orbital digraph of G. By Theorem 3.2.8, Γ is G-arc-transitive, and so there exists $g \in G$ with $g(u, v) = (v, u)$.

(iv) $\Longrightarrow$ (v) As Γ is G-edge-transitive and there exists $g \in G$ with $g(u, v) = (v, u)$ for some $uv \in E(\Gamma)$, Γ is G-arc-transitive. Set $H = \text{Stab}_G(u)$. By Theorem 3.2.8 we have $\Gamma \cong \text{Cos}(G, H, S)$ where $S = HsH$ for some $s \in G$, with ϕ, as in Theorem 1.3.9, an isomorphism. Then $(H, sH) \in E(\text{Cos}(G, H, S))$, where $s \in G$ such that $s(u) = v$. We may thus choose $s = g$. But then $g^2(u) = u$ and so $g^2 \in H$. □

Note that in order to show the arc-transitivity of this graph, we did not need to calculate its full automorphism group. The following result more or less follows from Theorem 3.2.9, and its proof is left as an exercise (Exercise 3.2.3).

Theorem 3.2.11 *For a G-edge-transitive graph Γ, the following are equivalent:*

(i) *Γ is G-half-arc-transitive,*
(ii) *Γ is the union of a pair of two non-self-paired orbital digraphs of G,*
(iii) *for $u \in V(\Gamma)$, $\{v : uv \in E(\Gamma)\}$ is a union of two orbits of* $\text{Stab}_G(v)$,
(iv) *no $g \in G$ interchanges the vertices incident with an edge of Γ.*

Some care must be exercised in applying the above results. It is obviously enough to find a transitive subgroup of $\mathrm{Aut}(\Gamma)$ that is arc- or edge-transitive to establish that Γ is arc- or edge-transitive. However, in order to show that a graph is not arc- or edge-transitive, one *must* consider the full automorphism group. For example, Example 1.3.6 shows that the Petersen graph is a double coset graph $\mathrm{Cos}(G, H, D)$ of a group G of order 20, and it is easy to see that D is the union of two double cosets (using the notation of Example 1.3.6 these are $H\langle\tau\rangle H$ and $H\left\{\rho, \rho^4\right\}H$). One may be tempted to conclude by Theorem 3.2.8 that the Petersen graph is not arc-transitive. But, as we saw in Theorem 3.1.2, the Petersen graph is in fact arc-transitive, and if we wanted to show this using Theorem 3.2.8, we chose the wrong transitive subgroup of the automorphism group of the Petersen graph. In order to be sure of choosing the "correct" transitive subgroup, one must consider the full automorphism group.

Regarding graphs, the hypothesis that Γ is vertex-transitive in some of the above theorems is superfluous, as the next result shows.

Proposition 3.2.12 *Let Γ be an arc-transitive digraph without isolated vertices, and $G \leq \mathrm{Aut}(\Gamma)$ such that Γ is G-arc-transitive. Then G is not transitive on $V(\Gamma)$ if and only if Γ is bipartite with all arcs from one bipartition set to the other.*

Proof Suppose first that Γ has a vertex v with positive invalence and outvalence. Let $u, w \in V(\Gamma)$. As Γ has no isolated vertices, there exists an arc with one endpoint at u. If $(u, a_1) \in A(\Gamma)$, then let $(v, a_2) \in A(\Gamma)$. Otherwise, let $(a_1, u), (a_2, v) \in A(\Gamma)$. In any case there exists $g \in G$ such that $g(a_1) = a_2$. This implies $g(u) = v$. Similarly, there exists $h \in G$ with $h(w) = v$, and so $h^{-1}g(u) = w$. Thus G is transitive on $V(\Gamma)$.

Suppose G is not vertex-transitive. As Γ has no isolated vertices, there is an arc at every vertex. By the immediately preceding argument, no vertex of Γ can have positive invalence and outvalence. Let $B_1 \subset V(\Gamma)$ be the set of all vertices of Γ with positive outvalence and B_2 be the set of all vertices of Γ with positive invalence. Then Γ is bipartite with bipartition sets B_1 and B_2, and every arc of Γ is from B_1 to B_2.

Suppose Γ is bipartite with all arcs from one bipartition class to the other. Then there is no $g \in \mathrm{Aut}(\Gamma)$ that maps a vertex of one bipartition class to the other. Thus any $G \leq \mathrm{Aut}(\Gamma)$ that is arc-transitive is not transitive on $V(\Gamma)$. □

We now turn to half-arc-transitive graphs. We begin with a result proven by Tutte (1966, 7.53), which gives that every such graph has even valency.

Theorem 3.2.13 *A half-arc-transitive graph has even valency.*

Proof Let Γ be a half-arc-transitive graph. We first give Γ a natural orientation. That is, we will choose an edge $uv \in E(\Gamma)$ and replace this edge with the arc (u, v). As $\mathrm{Aut}(\Gamma)$ is edge-transitive, for any edge $xy \in E(\Gamma)$, there exists $g \in \mathrm{Aut}(\Gamma)$ such that $g(uv) = xy$. Denote by $\vec{\Gamma}$ the digraph with vertex set $V(\Gamma)$ and arc set $\{g(u, v) : g \in \mathrm{Aut}(\Gamma)\}$. Note that as Γ is not arc-transitive, by Theorem 3.2.11 (iv) it cannot be the case both $(u, v), (v, u) \in A\left(\vec{\Gamma}\right)$. Consequently, for each $xy \in E(\Gamma)$, exactly one of (x, y) and (y, x) is contained in $A\left(\vec{\Gamma}\right)$. Now, each vertex in $V\left(\vec{\Gamma}\right)$ has the same outvalency o and the same invalency i. This implies that $on = in$ where $n = \left|V(\vec{\Gamma})\right|$. Hence $o = i$. Also, observe that $o + i = d$, where d is the valency of a vertex in Γ. Thus d is even, as required. □

In Tutte (1966), Tutte pointed out that it was not known if there were half-arc-transitive graphs. The first examples (an infinite family) were found by Bouwer (1970), with the smallest example having 54 vertices. In 1981, Holt found a 4-valent example on 27 vertices (Holt, 1981). Apparently, the same graph was independently discovered by Doyle (1976). Alspach, Marušič, and Nowitz then showed (Alspach et al., 1994) there was not a half-arc-transitive graph on fewer than 27 vertices. Note that there cannot be a half-arc-transitive graph of valency 1, 2, or 3 (graphs of valency 1 and 2 are arc-transitive as they are disjoint unions of edges or cycles, while graphs of valency 3 are not half-arc-transitive by Theorem 3.2.13). Finally, Xu (1992) showed that up to isomorphism, Holt's graph is the only half-arc-transitive graph of valency 4 with 27 vertices. The heart of our proof follows Holt's proof, the main difference being that we exploit that the Holt graph is a Cayley graph, which Holt did not.

Example 3.2.14 Let $G = \left\langle \rho, \tau : \rho^9 = \tau^3 = 1 \text{ and } \tau^{-1}\rho\tau = \rho^4 \right\rangle$, and $S = \left\{\tau\rho, \tau\rho^8, \tau^2\rho^2, \tau^2\rho^7\right\}$. Let $\alpha \in \mathrm{Aut}(G)$ such that $\alpha(\rho) = \rho^{-1}$ and $\alpha(\tau) = \tau$. The graph $\Gamma = \mathrm{Cay}(G, S)$ is a half-arc-transitive graph of order 27 and valency 4, and $\mathrm{Aut}(\Gamma) = \langle \rho, \tau, \alpha \rangle$ is of order 54.

Solution As $\tau^{-1}\rho\tau = \rho^4$, $\rho\tau = \tau\rho^4$. Then $(\tau\rho)^{-1} = \rho^8\tau^2 = \tau^2\rho^2$, and similarly $\left(\tau\rho^8\right)^{-1} = \tau^2\rho^7$. Then Γ is indeed a 4-valent graph on 27 vertices. This graph, the **Holt graph** or **Doyle–Holt graph**, is shown in Figure 3.3. As $\left(\rho^{-1}\right)^9 = \tau^3 = 1$ and $\tau^{-1}\left(\rho^{-1}\right)\tau = \rho^{-4}$, by Lemma 1.2.21 there exists $\alpha \in \mathrm{Aut}(G)$ such that $\alpha(\rho) = \rho^{-1}$ and $\alpha(\tau) = \tau$. Straightforward computations show that $\alpha(S) = S$, and so $\alpha \in \mathrm{Aut}(\Gamma)$ by Corollary 1.2.16. Checking that $\alpha^{-1}h_L\alpha(g) = \left(\alpha^{-1}(h)\right)_L(g)$ for every $g, h \in G$, we see $\langle \rho_L, \tau_L \rangle \trianglelefteq \langle \rho_L, \tau_L, \alpha \rangle = H$. As $|\alpha| = 2$, we have H has order 54. Additionally, α maps the set $\left\{\tau\rho, \tau^2\rho^2\right\}$ to the set $\left\{\tau\rho^8, \tau^2\rho^7\right\}$, and vice versa. We now show that H is transitive on $E(\Gamma)$.

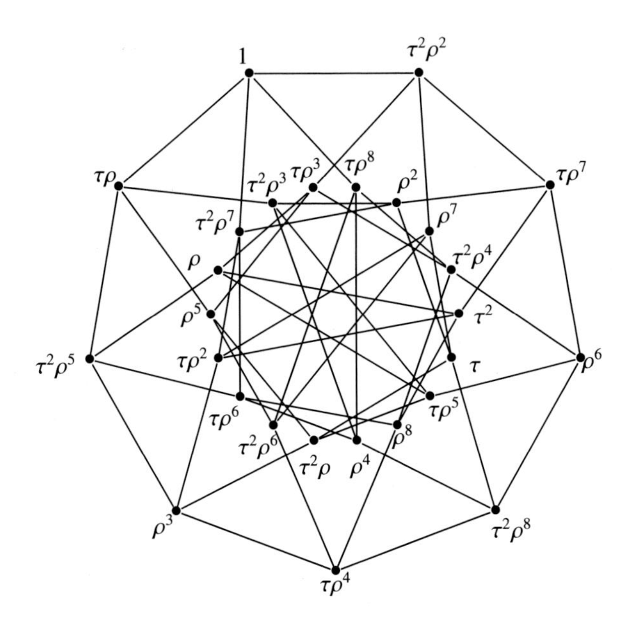

Figure 3.3 The Holt graph.

Let $uv, xy \in E(\Gamma)$, so $v = us_1$ and $y = xs_2$ for some $s_1, s_2 \in S$. Then $\left(u^{-1}\right)_L(uv) = 1s_1$ and $\left(x^{-1}\right)_L(xy) = 1s_2$. If $s_1 = s_2$ or s_2^{-1} set $i = 0$, while if $s_1 \neq s_2$ or s_2^{-1}, set $i = 1$. Then $\alpha^i(s_1) = s_2$ or s_2^{-1} as α maps $\left\{\tau\rho, \tau^2\rho^2\right\}$ to $\left\{\tau\rho^8, \tau^2\rho^7\right\}$ and vice versa. If $\alpha^i(s_1) = s_2$ set $j = 0$ and if $\alpha^i(s_1) = s_2^{-1}$ set $j = 1$. Then $x_L(s_2)_L^j\alpha^i\left(u^{-1}\right)_L(uv) = xy$ and Γ is edge-transitive. It now only remains to show that Γ is not arc-transitive.

Suppose $g \in \mathrm{Aut}(\Gamma)$ and fixes 1. Let Y be the set of all vertices in Γ at distance 2 from 1. These vertices are $\tau^2\rho^5$, $\tau^2\rho^3$, ρ^5, $\tau^2\rho^6$, $\tau^2\rho^4$, ρ^4, ρ^7, $\tau\rho^7$, $\tau\rho^3$, ρ^2, $\tau\rho^6$, and $\tau\rho^2$. As g fixes 1, g maps all vertices at distance 2 from 1 to vertices at distance 2 from 1, and so maps Y to Y. Then g^Y is an automorphism of $\Gamma[Y]$, the subgraph of Γ induced by Y. Note that $\Gamma[Y]$ is the disjoint union of two paths of length 5, colored in red and blue in Figure 3.4. Observe that each vertex at distance 1 in Γ from 1 (colored green in Figure 3.4) is adjacent to exactly one endpoint of a path in $\Gamma[Y]$. Then g^Y maps endpoints of paths in $\Gamma[Y]$ to endpoints of paths in $\Gamma[Y]$.

Now suppose g maps the arc $(\tau\rho, 1)$ to the arc $\left(\tau^2\rho^2, 1\right)$ (and of course g still fixes 1). The vertex $\tau\rho$ is adjacent to the endpoint $\tau^2\rho^5$ of a path in $\Gamma[Y]$ and $\tau^2\rho^2$ is adjacent to the endpoint $\tau\rho^7$ of a path in $\Gamma[Y]$. As $g(\tau\rho) = \tau^2\rho^2$, $g\left(\tau^2\rho^5\right) = \tau\rho^7$. Consequently, g interchanges $\tau\rho^6$ and $\tau^2\rho^3$. These two vertices

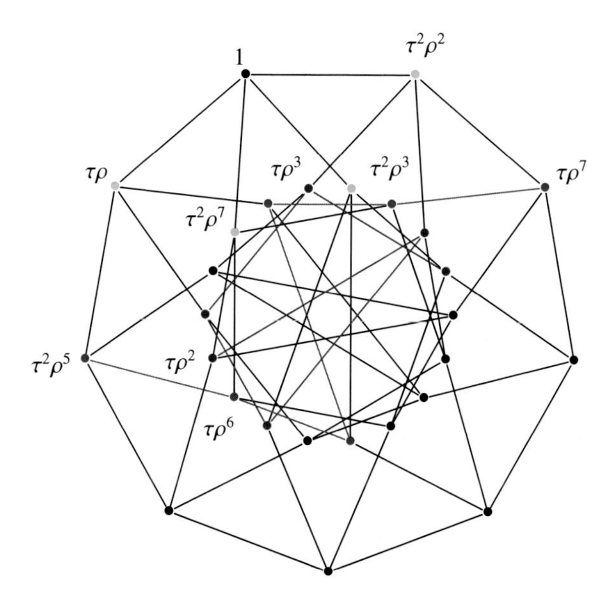

Figure 3.4 The induced subgraph of the Holt graph at distance 2 from the identity.

are both neighbors of $\tau^2\rho^7$, and $\tau^2\rho^7$ is their unique common neighbor. Thus, g fixes $\tau^2\rho^7$. But then $\tau\rho^2$ is also fixed by g as $\tau\rho^2$ is the unique endpoint of a path in $\Gamma[Y]$ adjacent to $\tau^2\rho^7$, and so each vertex on the red path is fixed by g. In particular, $\tau\rho^3$ is fixed by g. This implies that $\tau^2\rho^2$ is fixed by g, a contradiction. This gives that the Holt graph is not arc-transitive, and so is half-arc-transitive. Additionally, it also shows that if $g(1) = 1$ then $g(\tau\rho) \neq \tau^2\rho^2$.

To finish, we must show that $\mathrm{Stab}_{\mathrm{Aut}(\Gamma)}(1) = \langle\alpha\rangle$. Suppose otherwise, and let $g \in \mathrm{Stab}_{\mathrm{Aut}(\Gamma)}(1)\backslash\langle\alpha\rangle$. Then $g(S) = S$. As $g(\tau\rho) \neq \tau^2\rho^2$ and $\alpha(\tau\rho) = \tau\rho^8$, we may assume that $g(\tau\rho) = \tau\rho$ or $g(\tau\rho) = \tau^2\rho^7$. Note that it is not possible for $g(\tau\rho) = \tau^2\rho^7$ as otherwise g^Y is an automorphism of $\Gamma[Y]$, but $\tau\rho$ is adjacent to a leaf vertex (a vertex of degree 1) of the blue path (Figure 3.4) in $\Gamma[Y]$, a nonleaf vertex of the blue path and a non-leaf vertex of the red path, while $\tau^2\rho^7$ is adjacent to two blue nonleaf vertices and one red leaf.

If $g(\tau\rho) = \tau\rho$, then observing that $\tau\rho$ is adjacent to exactly one endpoint of a red or blue path (as in the previous paragraph), we see that g also fixes $\tau^2\rho^5$. Consequently, g must also fix every vertex on the blue path. As $\tau^2\rho^7$ is the unique common neighbor of $\tau\rho^6$ and $\tau^2\rho^3$, which are both on the blue path and hence fixed, we see that g fixes $\tau^2\rho^7$. We now know that g fixes $\tau^2\rho^7$ and three of its four neighbors, and so it must fix the fourth neighbor $\tau\rho^2$. But $\tau\rho^2$ is an endpoint of the red path, and so every vertex on the red path is fixed as well. Noting that a common neighbor of two fixed points of g is a fixed point of g as well as if g fixes three neighbors of a fixed point it fixes all the neighbors, it is straightforward to show $g = 1$, a contradiction. □

We mention that Potočnik, Spiga, and Verret have given a census of quartic half-arc-transitive graphs on up to 1,000 vertices (Potočnik et al., 2015).

3.3 Introduction to the Hamiltonicity Problem

The problem of determining whether or not a given connected vertex-transitive graph has a Hamilton cycle or Hamilton path is one of the oldest and most well-known problems about vertex-transitive graphs, and we will see many more problems that have been posed via attempts to solve this problem. The problem began with what is usually called a conjecture, posed by Lovász (Guy, 1970, Problem Session, Problem 11). We state it below exactly as it was posed by Lovász in 1969.

Problem 3.3.1 "Let us construct a finite, connected, undirected graph that is symmetric and has no simple path containing all elements. A graph is called symmetric, if for any two vertices x, y it has an automorphism mapping x into y."

This problem is usually stated as every connected vertex-transitive graph has a Hamilton path. There are many related conjectures, some of which are contradictory. There are only four known examples of connected vertex-transitive graphs of order at least 3 that are not Hamiltonian. These are the Petersen graph, the Coxeter graph, and the graphs obtained from them by "truncation." We will give three different proofs that the Petersen graph is not Hamiltonian in Theorems 3.4.1, 11.1.4, and Example 11.1.7, and two proofs that the Coxeter graph is not Hamiltonian in Theorem 6.1.5 and Example 11.1.8. In Theorem 3.4.5, we will see that the truncation of a non-Hamiltonian graph is not Hamiltonian. Finally, we have already seen that the Petersen graph is not isomorphic to a Cayley graph in Theorem 1.2.22, that the Coxeter graph is not Cayley is Exercise 6.1.5, and that their truncations are not isomorphic to Cayley graphs are Exercises 3.4.7 and 6.1.8, and so none of these four graphs are isomorphic to Cayley graphs. This has led to the following folklore conjecture.

Conjecture 3.3.2 *Every connected Cayley graph of order at least* 3 *is Hamiltonian.*

Carsten Thomassen (Bermond, 1978b, Conjecture, Section 11) has made the following conjecture.

Conjecture 3.3.3 *There are finitely many connected vertex-transitive graphs that are not Hamiltonian.*

There is not a consensus on whether the Lovász conjecture is correct or not. An example of this is the following conjecture of Babai (1995, p. 1472).

Conjecture 3.3.4 *There is a positive constant c such that there exist infinitely many connected vertex-transitive graphs (or Cayley graphs) Γ without cycles of length at least* $(1 - c)|V(\Gamma)|$.

3.4 Vertex-transitive Graphs That Are Not Hamiltonian

In this section, we will give the first proof that the Petersen graph is not Hamiltonian in Theorem 3.4.1. The rest of the section is devoted to studying graph truncation, as the truncations of both the Petersen graph and the Coxeter graph are vertex-transitive but not Hamiltonian. After showing that the truncation of a non-Hamiltonian graph is not Hamiltonian in Theorem 3.4.5, we then show that the automorphism group of the truncation of a graph is isomorphic as an abstract group to the automorphism group of the graph in Theorem 3.4.9. This explains why the truncation of the truncation of the Petersen graph and Coxeter graph are not also examples of vertex-transitive graphs without Hamilton cycles. They are examples of cubic graphs without Hamilton cycles, but they are not vertex-transitive. We end with a characterization of when the truncation is vertex-transitive in Corollary 3.4.11.

We have already encountered a connected vertex-transitive graph that is not Hamiltonian, namely the Petersen graph.

Theorem 3.4.1 *The Petersen graph is not Hamiltonian.*

Proof We will use the metacirculant labeling of the Petersen graph in the proof as in Figure 3.5. Suppose to the contrary that P has a Hamilton cycle $C = v_0, v_1, \ldots, v_9, v_0$, $v_i \in V(P)$. As P is 3-arc regular, we may assume without loss of generality that $v_0 = (0,0)$, $v_1 = (0,1)$, $v_2 = (0,2)$, and $v_3 = (0,3)$ as $(0,0), (0,1), (0,2), (0,3)$ is a 3-arc in P. Observe that each time a spoke edge is traversed in C going from the outside 5-cycle to the inside 5-cycle, a spoke edge must appear later going from the inside 5-cycle back to the outside 5-cycle. This implies that the spoke edges in C come in pairs, so that there are either 2 or 4 spoke edges in C. As C contains the 3-arc $(0,0), (0,1), (0,2), (0,3)$,

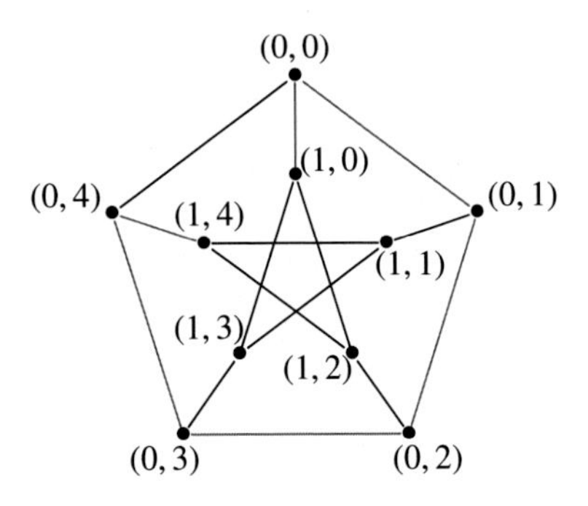

Figure 3.5 The Petersen graph with part of a cycle.

C can only contain 2-spoke edges, and these must be chosen from the spoke edges $(0,0)(1,0)$, $(0,3)(1,3)$, and $(0,4)(1,4)$. Also, C must contain either $(0,3)(0,4)$ and spoke edges $(0,4)(1,4)$ and $(0,0)(1,0)$ or $(0,4)(0,0)$ and spoke edges $(0,3)(1,3)$ and $(0,4)(1,4)$. In the former case (illustrated in Figure 3.5 with the edges used thus far colored in blue), there must then be a path of length 4 on the inner 5-cycle from $(1,0)$ to $(1,4)$, which there is not, a contradiction, while in the latter case there must be path of length 4 on the inner 5-cycle from $(1,3)$ to $(1,4)$, which also does not occur. □

The proof of the previous result is the "standard" proof that the Petersen graph is not Hamiltonian. Usually though, the 3-arc-transitivity of the Petersen graph is not used, in which case there can be four spoke edges in the Hamilton cycle that is assumed to exist. That argument is left as Exercise 3.4.6. We give two less standard proofs in Theorem 11.1.4 and Example 11.1.7.

It is shown in Section 6.1 that the Coxeter graph is not Hamiltonian. As mentioned earlier, there are two more known connected vertex-transitive graphs of order at least 3 that are not Hamiltonian (and a fifth example, K_2). Both are obtained from the Petersen graph and Coxeter graph via truncation.

Definition 3.4.2 For a cubic graph Γ, we define a **truncation** of Γ (or the **vertex inflation** of Γ), denoted $T(\Gamma)$, to be the graph obtained from Γ by replacing each vertex v with a triangle T_v, and if $uv \in E(\Gamma)$, then one vertex of T_v is adjacent to one vertex of T_u and no vertex of T_v is adjacent to more than one vertex outside of T_v.

The truncation of the Petersen graph is shown in Figure 3.6. The name "truncation" is borrowed from geometry, where a solid is truncated if it has some vertex removed by a cut. In our definition it is not completely clear that

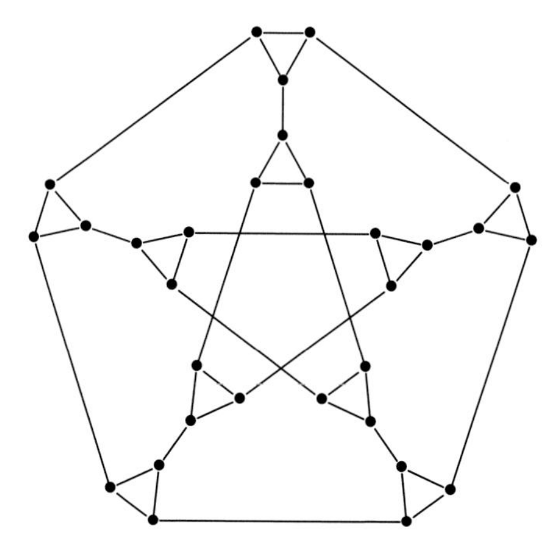

Figure 3.6 The truncation of the Petersen graph.

there is always such a graph as in the definition for any cubic graph. Our first result shows that such a graph always exists.

Lemma 3.4.3 *There is a graph satisfying the definition of the truncation of a cubic graph* Γ.

Proof Surely we may form a graph T whose vertex set is $V(\Gamma) \times \mathbb{Z}_3$ and has edges $\{(v,i)(v,j) : v \in V(\Gamma) \text{ and } i,j \in \mathbb{Z}_3, i \neq j\}$, which is the graph obtained from the empty graph on $V(\Gamma)$ by replacing each vertex $v \in V(\Gamma)$ with a triangle T_v. Suppose now that we successively add edges to T in such a way that one vertex of T_v is adjacent to at most one vertex of T_u whenever $uv \in E(\Gamma)$ until no more such edges can be added, obtaining the graph T'. If every vertex of T' has valency 3, then as every vertex of T has valency 2, whenever $uv \in E(\Gamma)$, each vertex of T_v is adjacent to one vertex of T_u and, by construction, no vertex of T_v is adjacent to more than one vertex outside of T_v. So if T' is regular of valency 3, then T' is a truncation of Γ. Otherwise, T' has a vertex (v,i) of valency 2, $v \in V(\Gamma)$, $i \in \mathbb{Z}_3$. To see that there is no vertex of valency 2 in T', the key observation is that some vertex of T_v is adjacent to some vertex of T_u only if $uv \in E(\Gamma)$, and, there can be at most one edge between T_v and T_u. That is, if (v,i) is of valency 2, then for some neighbor u in $V(\Gamma)$, no vertex of T_v is adjacent to any vertex of T_u and so some vertex (u,j) of T_u also has valency 2, $j \in \mathbb{Z}_3$. Then the edge $(v,i)(u,j)$ can be added to T', contradicting our choice of T'. □

Next, we typically like our definitions to result in *unique* graphs. The situation here is slightly more complicated as while it is true that there are different truncations of a given cubic graph, our next result shows that any two such graphs are isomorphic. It will be convenient to adopt a standard vertex set of each truncation of a cubic graph Γ. We will assume $T_v = \{v\} \times \mathbb{Z}_3$ so that $V(T(\Gamma)) = V(\Gamma) \times \mathbb{Z}_3$.

Lemma 3.4.4 *Let T_1 and T_2 be truncations of a cubic graph Γ. Then $T_1 \cong T_2$ by an element of $S_{V(T(\Gamma))}$ that fixes each T_v set-wise.*

Proof We will give an explicit isomorphism in $S_{V(T(\Gamma))}$ from T_1 to T_2. First, note that to preserve the edges inside a triangle T_v, $v \in V(\Gamma)$, we need only map $\{v\} \times \mathbb{Z}_3$ to $\{v\} \times \mathbb{Z}_3$ for all $v \in V$. Denote by $\hat{T}_1$ and $\hat{T}_2$ the subgraphs of T_1 and T_2 obtained by removing all edges in a T_v for all $v \in V(\Gamma)$. So $\hat{T}_1$ and $\hat{T}_2$ are regular of valency 1 and so 1-factors. Define $\phi\colon V(\Gamma) \times \mathbb{Z}_3 \to V(\Gamma) \times \mathbb{Z}_3$ as follows: for $(v, i)(u, j) \in E\left(\hat{T}_1\right)$, let $\phi(v, i) = (v, i')$ and $\phi(u, j) = (u, j')$ where $(v, i')(u, j') \in E\left(\hat{T}_2\right)$. As $\hat{T}_1$ and $\hat{T}_2$ are 1-factors, ϕ is a well-defined bijection that preserves the sets T_v, $v \in V(\Gamma)$, and so is an isomorphism. □

So from here we will simply say "the truncation of Γ" as these graphs are all isomorphic. We begin our investigation of these graphs with the following general result.

Theorem 3.4.5 *If a cubic graph Γ is not Hamiltonian, then the truncation of Γ is also not Hamiltonian.*

Proof Suppose otherwise, and suppose the truncation $T(\Gamma)$ of Γ is Hamiltonian. We will show that Γ is Hamiltonian. Indeed, suppose $C = x_0 x_1 \dots x_n x_0$ is a Hamilton cycle in $T(\Gamma)$. The key observation in this proof is that if $v \in V(\Gamma)$, then the subgraph of C induced by the vertices of T_v is a path of length 2. In other words, a Hamilton cycle in $T(\Gamma)$ traverses every vertex of T_v the first time it traverses any vertex of T_v (with the possible exception of the very first vertex x_0). This is easy to see, as each time a Hamilton cycle in $T(\Gamma)$ enters a T_v, it must leave T_v, so the number of edges of C with a vertex of a T_v as an endpoint must be even. As there are at most three such edges and there are at least two, there must be exactly two such edges. With this observation in hand, we define a sequence of vertices $y_0 \dots y_n y_0$ in Γ by setting $y_i = v$, where $x_i \in T_v$. Deleting repetitions of symbols in $y_0 \dots y_n y_0$ then yields a cycle (as the only way a vertex can then appear twice is at the beginning and end), which certainly spans Γ as C spanned $T(\Gamma)$. □

While we have not even defined the Coxeter graph yet, we have mentioned that it is a cubic vertex-transitive graph with no Hamilton cycle, so we have the following result.

Corollary 3.4.6 *The truncations of the Petersen graph and the Coxeter graph are not Hamiltonian.*

While the truncation of *any* vertex-transitive cubic graph that is not Hamiltonian will give a cubic graph that is not Hamiltonian, the truncation of such a graph *need not be vertex-transitive*. Indeed, the truncations of the truncations of the Petersen and Coxeter graphs are not Hamiltonian, but it turns out that they are not vertex-transitive. We now investigate when the truncation of a cubic vertex-transitive graph is vertex-transitive.

Theorem 3.4.7 *Let Γ be a cubic graph. Then* $\mathrm{Aut}(T(\Gamma))$ *is isomorphic to a subgroup of* $\mathrm{Aut}(\Gamma)$.

Proof The key here is to observe that the only triangles in $T(\Gamma)$ are the subgraphs T_v, where $v \in V(\Gamma)$. Indeed, let $xyzx$ be any triangle in $T(\Gamma)$, and $v \in V(\Gamma)$ such that $x \in V(T_v)$. Then x is adjacent to only one vertex in $T(\Gamma)$ not in T_v, and so y or z must also be in T_v. If one of y or z is then not in $V(T_v)$ but in $V(T_w)$, then an identical argument gives that two of $\{x, y, z\}$ are in $V(T_w)$, which is not possible. Hence the only triangles in $T(\Gamma)$ are the subgraphs T_v. This implies that the $\mathrm{Aut}(T(\Gamma))$ permutes the set $\mathcal{T} = \{V(T_v) : v \in V(\Gamma)\}$ (and if $\mathrm{Aut}(T(\Gamma))$ is transitive, $\mathcal{T}$ is a block system of $\mathrm{Aut}(T(\Gamma))$).

Let K be the kernel of the action of $\mathrm{Aut}(T(\Gamma))$ on $\mathcal{T}$ (if $\mathrm{Aut}(T(\Gamma))$ is transitive, then $K = \mathrm{fix}_{\mathrm{Aut}(T(\Gamma))}(\mathcal{T})$). We claim that $K = 1$. Indeed, if $K \neq 1$ with $1 \neq \gamma \in K$, then let $T_v \in \mathcal{T}$ such that $\gamma^{T_v} \neq 1$. Then there exists distinct $x, y \in T_v$ such that $\gamma(x) = y$. Now, x is adjacent to some vertex $z \in T_u$, $u \neq v$, and so $\gamma(x)\gamma(z)$ is also an edge from T_v to T_u. However, there is only one edge from a vertex of T_v to a vertex of T_u in $T(\Gamma)$, a contradiction. Thus $K = 1$.

To finish the result, we need only show that if $\gamma \in \mathrm{Aut}(T(\Gamma))$, then $\bar{\gamma} \in \mathrm{Aut}(\Gamma)$, where $\bar{\gamma}$ is the permutation of $\mathcal{T}$ induced by the action of γ on $\mathcal{T}$ (so if $T(\Gamma)$ is vertex-transitive $\bar{\gamma} = \gamma/\mathcal{T}$). So suppose $uv \in E(\Gamma)$. Then some vertex of T_u is adjacent to some vertex of T_v, and as $\gamma \in \mathrm{Aut}(T(\Gamma))$, some vertex of $\gamma(T_u) = T_{\bar{\gamma}(u)}$ is adjacent to some vertex of $\gamma(T_v) = T_{\bar{\gamma}(v)}$. But this occurs if and only if $\bar{\gamma}(u)\bar{\gamma}(v) \in E(\Gamma)$. □

Corollary 3.4.8 *The truncations of the truncations of the Petersen and Coexeter graphs are not vertex-transitive.*

Proof Let Γ be the Petersen or Coxeter graph. If $T(T(\Gamma))$ is vertex-transitive, then 9 divides $|\mathrm{Aut}(T(T(\Gamma)))|$, as $V(T(T(\Gamma))) = V(\Gamma) \times \mathbb{Z}_3 \times \mathbb{Z}_3$. By Theorem 3.4.7 applied twice, we see that 9 divides $|\mathrm{Aut}(\Gamma)|$. However, the automorphism group of the Petersen graphs has order 120 while the automorphism group of the Coxeter graphs has order 336, neither of which are divisible by 9. □

Theorem 3.4.9 *For a cubic graph* Γ, $\mathrm{Aut}(T(\Gamma)) \cong \mathrm{Aut}(\Gamma)$.

Proof In view of Theorem 3.4.7, it suffices to show that each element of $\mathrm{Aut}(\Gamma)$ induces an element of $\mathrm{Aut}(T(\Gamma))$, and that different elements of $\mathrm{Aut}(\Gamma)$ induce different elements of $\mathrm{Aut}(T(\Gamma))$.

For $\gamma \in \mathrm{Aut}(\Gamma)$, define $\hat{\gamma}\colon V(\Gamma) \times \mathbb{Z}_3 \to V(\Gamma) \times \mathbb{Z}_3$ by $\hat{\gamma}(v, i) = (\gamma(v), i)$. Note that $\hat{\gamma}$ is not necessarily an automorphism of $T(\Gamma)$, but $\hat{\gamma}$ certainly maps a triangle T_v to the triangle $T_{\gamma(v)}$. However, as $\gamma \in \mathrm{Aut}(\Gamma)$, $\gamma(T(\Gamma))$ is a truncation of Γ by Definition 3.4.2, so by Lemma 3.4.4 there is $\delta \in \mathcal{S}_{V(\Gamma)\times\mathbb{Z}_3}$ that fixes each T_v set-wise, $v \in V(\Gamma)$, such that $\delta\hat{\gamma}(T(\Gamma)) = T(\Gamma)$. Hence $\delta\hat{\gamma} \in \mathrm{Aut}(T(\Gamma))$, and $\delta\hat{\gamma}(T_v) = T_{\gamma(v)}$. It is easy to see that different elements of $\mathrm{Aut}(\Gamma)$ induce different elements of $\mathrm{Aut}(T(\Gamma))$ as if $\gamma \in \mathrm{Aut}(\Gamma)$ induces γ_1 and γ_2, $\gamma_1 \neq \gamma_2$, then $\gamma_1^{-1}\gamma_2 \neq 1$ fixes each triangle in $T(\Gamma)$, contradicting Theorem 3.4.7. □

For the next result, we will have need of an additional term.

Definition 3.4.10 A permutation $g \in \mathcal{S}_n$ is said to be **semiregular** or is a **semiregular element** if $\langle g\rangle$ is a semiregular subgroup. That is, g is semiregular if, when written as a product of disjoint cycles, every cycle has the same length. If the cycle decomposition of g has m cycles of length n, we will also say g is an (m, n)**-semiregular element**.

Semiregular elements have been studied extensively, and in fact are the subject of Section 3.8 and Chapter 12.

Corollary 3.4.11 *Let* Γ *be a connected vertex-transitive cubic graph. Then the following are equivalent:*

(i) $T(\Gamma)$ *is vertex-transitive,*
(ii) $\mathrm{Aut}(\Gamma)$ *contains an element of order* 3 *with a fixed point, and*
(iii) Γ *is arc-transitive.*

Additionally, $T(T(\Gamma))$ *is not vertex-transitive.*

Proof (i) $\Rightarrow$ (ii) Suppose $T(\Gamma)$ is vertex-transitive. Then $\mathcal{T}$ is a block system of $\mathrm{Aut}(\Gamma)$, and so $\mathrm{Stab}_{\mathrm{Aut}(T(\Gamma))}(T_v)^{T_v}$ is transitive on T_v. As T_v has order 3, $\left(\mathrm{Stab}_{\mathrm{Aut}(T(\Gamma))}(T_v)\right)^{T_v}$ contains an element of order 3. Let $\gamma \in \mathrm{Stab}_{\mathrm{Aut}(T(\Gamma))}(T_v)$ such that γ^{T_v} has order 3. Without loss of generality, we assume γ has order 3 (although we will no longer necessarily have that γ^{T_v} has order 3 – for our purposes we only need an element of order 3 that fixes some T_v). It was shown in the proof of Theorem 3.4.7 that $\mathrm{fix}_{\mathrm{Aut}(T(\Gamma))}(\mathcal{T}) = 1$, so $\gamma/\mathcal{T} \neq 1$. Then $\gamma/\mathcal{T}$ has order 3 and fixes the point v.

(ii) $\Rightarrow$ (iii) Suppose $\mathrm{Aut}(\Gamma)$ contains an element γ of order 3 with a fixed point v. As Γ is connected, some fixed point v of γ is adjacent in Γ to some point that is not fixed by γ, say u. Let (u, w, z) be the 3-cycle in the cycle decomposition of γ that contains u. Then $vu, vw, vz \in E(\Gamma)$. Now let $ab, cd \in E(\Gamma)$. As $\mathrm{Aut}(\Gamma)$ is transitive, there exists $\delta_1 \in \mathrm{Aut}(\Gamma)$ such that $\delta_1(a) = v$, and as $\delta_1(ab) \in E(\Gamma)$, $\delta_1(b) = u, w$, or z. Similarly, there exists $\delta_2 \in \mathrm{Aut}(\Gamma)$ such that $\delta_2^{-1}(c) = v$, and as before it must be the case that $\delta_2^{-1}(d) = u, w$, or z. Then there exists $0 \leq i \leq 2$ such that $\gamma^i(\delta_1(b)) = \delta_2^{-1}(c)$. We then have $\delta_2\gamma^i\delta_1(ab) = cd$, and Γ is edge-transitive. By Theorem 3.2.13, Γ is arc-transitive.

(iii) $\Rightarrow$ (ii) If Γ is a connected cubic edge-transitive graph with $v \in V(\Gamma)$ and $S = \{u \in V(\Gamma) : vu \in E(\Gamma)\}$, then $\mathrm{Stab}_{\mathrm{Aut}(\Gamma)}(v)$ maps S to S and is transitive on S. As 3 is prime, there exists an element $\gamma \in \mathrm{Stab}_{\mathrm{Aut}(\Gamma)}(v)$ of order 3.

(ii) $\Rightarrow$ (i) We will first show that any element of order 3 in $\mathrm{Aut}(T(\Gamma))$ is semiregular. As before, label the vertices of $V(T(\Gamma))$ with elements of $V(\Gamma)\times\mathbb{Z}_3$ in such a way that $T_v = \{(v, i) : i \in \mathbb{Z}_3\}$, $v \in V(\Gamma)$. Suppose $\gamma \in \mathrm{Aut}(T(\Gamma))$ is of order 3 and fixes a point, say (v, i). As γ has order 3, there exists $(w, j) \in V(T(\Gamma))$ such that $\gamma(w, j) \neq (w, j)$. As Γ and consequently $T(\Gamma)$ are connected, there exists a path P in $T(\Gamma)$ from (v, i) to (w, j). We may thus assume without loss of generality that some neighbor of (v, i) is not fixed by γ (as there exists adjacent vertices on P where the first is fixed by γ but the second is not). As γ is of order 3, fixes (v, i), and $\gamma^{T_v} \in \mathcal{S}_3$, γ fixes each vertex in T_v. Then (v, i) is adjacent to at least two vertices that are not in T_v, a contradiction. Hence any element of order 3 in $\mathrm{Aut}(T(\Gamma))$ is semiregular.

Suppose now that Γ has an automorphism γ of order 3 that fixes a point. As Γ is vertex-transitive, $\mathrm{Aut}(T(\Gamma))$ transitively permutes the elements of $\mathcal{T}$, so it suffices to show that there exists $\hat{\gamma} \in \mathrm{Aut}(T(\Gamma))$ such that $\hat{\gamma}$ fixes T_v for some $v \in V(\Gamma)$ and $\hat{\gamma}^{T_v}$ has order 3. Let $\hat{\gamma} \in \mathrm{Aut}(T(\Gamma))$ such that the induced action of $\hat{\gamma}$ on $\mathcal{T}$ is γ, and let v be fixed by γ (so that $\hat{\gamma}(T_v) = T_v$). Then $\hat{\gamma}$ is semiregular, and so $\hat{\gamma}$ permutes the element of T_v as a 3-cycle (and so is transitive). Thus $\mathrm{Aut}(T(\Gamma))$ is transitive.

Finally, as $\mathrm{Aut}(T(\Gamma))$ does not contain an element of order 3 with a fixed point, $T(T(\Gamma))$ is not vertex-transitive. □

The results in this section on graph truncations also hold for vertex-transitive graphs of valencies $d \geq 3$ with a vertex replaced by a complete graph on d vertices in an analogous way (Alspach and Dobson, 2015). Also, a more general notion of truncation was recently introduced by Eiben, Jajcay, and Šparl, where each vertex can be replaced by any graph (Eiben et al., 2019), and a few problems are posed.

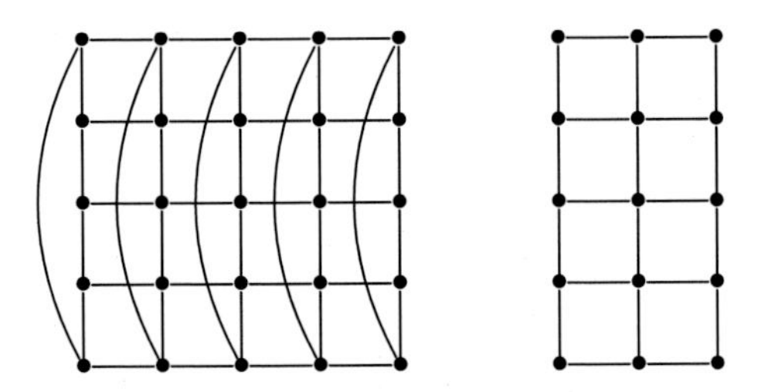

Figure 3.7 The Cartesian products $P_4\Box C_5$ and $P_2\Box P_4$.

3.5 Connected Cayley Graphs of Abelian Groups Are Hamiltonian

In this section we will show that all connected Cayley graphs of abelian groups of order at least 3 are Hamiltonian. This was first done in the early 1970s by Pelikán, and is included in Lovász's book "Combinatorial Problems and Exercises" (Lovász, 1979, Exercise 12.17). Much of the proof is based on the Cartesian product of paths and cycles, which we now define.

Definition 3.5.1 The **Cartesian product** $\Gamma\Box\Delta$ of two graphs Γ and Δ is the graph with vertex set $V(\Gamma)\times V(\Delta)$ and edge set $\{(u,v)(u',v') : u = u' \text{ and } u'v' \in E(\Delta) \text{ or } v = v' \text{ and } uu' \in E(\Gamma)\}$.

We have already encountered some Cartesian products, e.g., the n-cubes Q_n. An example of Cartesian products of a path with a path and a cycle with a path is given in Figure 3.7. We denote a cycle of length n by C_n, and a path of length n by P_n. We shall have need of several basic lemmas regarding which Cartesian products of paths and cycles contains Hamilton cycles.

Lemma 3.5.2 *If n or m is odd, then $P_n\Box P_m$ contains a Hamilton cycle.*

Proof We let $V(P_r) = \mathbb{Z}_{r+1}$ where vertex i is adjacent to vertex $i+1$, $0 \le i \le r-1$. Clearly $P_n\Box P_m \cong P_m\Box P_n$ (via the map that permutes the coordinates), so we may assume without loss of generality that m is odd. Let $Q_i = (1,i)(2,i)\ldots(n-1,i)$, and $P = (0,m)(0,m-1)\ldots(0,0)$. Then

$$(0,0)Q_0Q_1^{-1}\ldots Q_{m-2}^{-1}Q_{m-1}Q_m^{-1}P,$$

is a Hamilton cycle in $P_n\Box P_m$ (see Figure 3.8 where the Hamilton cycle is in red). □

Lemma 3.5.3 *If n is odd and m is even, then $C_n\Box P_m$ contains a Hamilton cycle.*

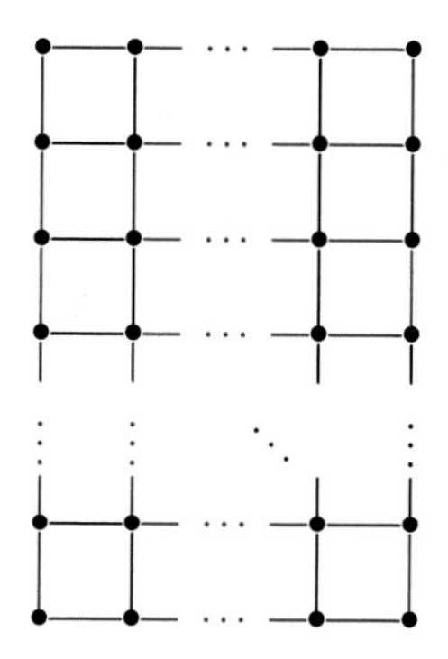

Figure 3.8 A Hamilton cycle in $P_n \Box P_m$.

Proof Let $V(P_m) = \mathbb{Z}_{m+1}$ where vertex i is adjacent to vertex $i+1$, $0 \le i \le m$, and $V(C_n) = \mathbb{Z}_n$, where vertex i is adjacent to vertex $i+1$, $0 \le i \le n-1$. Let $Q_i = (1,i)(2,i)\ldots(n-1,i)$, and $P = (0,m)\ldots(0,0)$. Then

$$(0,0)Q_0^{-1}Q_1Q_2^{-1}\ldots Q_{m-1}Q_m^{-1}P,$$

is a Hamilton cycle in $C_n \Box P_m$. □

With these results in hand, we are ready to prove the main result of this section.

Theorem 3.5.4 *A connected Cayley graph of an abelian group of order at least* 3 *is Hamiltonian.*

Proof Let $\Gamma = \mathrm{Cay}(G,S)$ be connected, where G is abelian of order n and $S = S^{-1} \subset G$. By Lemma 1.2.14, we have $\langle S \rangle = G$. We proceed by induction on $|S|$. If $|S| = 1$, then $S = \{a\}$, $a \in G$, a is a generator of G, and a is self-inverse. We conclude that $G = \mathbb{Z}_2$, a contradiction. If $|S| = 2$, then Γ has valency 2. As it is connected, it is isomorphic to a cycle. Assume the result holds for every generating set of size at most $m-1 \ge 2$, and let S be a self-inverse generating set of G of size m.

In order to apply the induction hypothesis, we now find an appropriate proper subset T of S. If $m = 3$, then S contains a self-inverse element a, and $T = S \backslash \{a\}$ has order 2. If $m > 3$, then let $a \in S$ and set $T = S \backslash \{a, a^{-1}\}$. In any case, let $M = \langle T \rangle$. As G is abelian, $M \trianglelefteq G$. Let k be the smallest positive integer such that $a^k \in M$. Then $G/M \cong \mathbb{Z}_k$. By the induction hypothesis, we see that $\Gamma[M]$ is Hamiltonian as $|T| \ge 2$. Let $c_0c_1\ldots c_rc_0$ be a Hamilton cycle in M. Then $W_i = a^ic_0a^ic_1\ldots a^ic_ra^ic_0$ is a Hamilton cycle in a^iM, $0 \le i \le k-1$. Also, $a^ic_j = c_ja^i$ is adjacent to $a^{i+1}c_j = c_ja^{i+1}$, $0 \le j \le r$ and $0 \le i \le k-1$, and so $Q_j = c_j(ac_j)\dots(a^{k-1}c_j)$ is a path of length $k-1$ in Γ. The subgraph $\Delta = \left(\cup_{i=0}^{k-1} W_i\right) \cup \left(\cup_{j=0}^{r} Q_j\right)$ is then naturally isomorphic to $C_{r+1}\Box P_{k-1}$. As $G/M \cong \mathbb{Z}_k$ we see that Δ is a spanning subgraph of Γ. Finally, Δ contains a Hamilton cycle by Lemma 3.5.2 if $k-1$ or r is odd (as C_{r+1} contains P_r), while if $k-1$ and r are odd, $r+1$ is even and Δ contains a Hamilton cycle by Lemma 3.5.3. □

Given that the preceding theorem was not overly difficult to prove, it is not very surprising that much more is known about Hamilton cycles in Cayley graphs of abelian groups. Before stating the main such result, we will need some additional terminology.

Definition 3.5.5 A graph Γ is **Hamilton connected** if for every $x, y \in V(\Gamma)$, there is a Hamilton path from x to y. A bipartite graph Γ with bipartition (X, Y) is **Hamilton laceable** if for every $x \in X$ and $y \in Y$ there is a Hamilton path in Γ from x to y.

The proof of the following result of Chen and Quimpo (1981) is somewhat long, has many cases, and so has been omitted here. This result is often quite useful, as we shall see in the next section.

Theorem 3.5.6 (Chen–Quimpo theorem) *A connected Cayley graph Γ of a finite abelian group of order at least 3 is Hamilton connected if and only if it is neither a cycle nor bipartite. In the case where Γ is bipartite but not a cycle, Γ is Hamilton laceable.*

There is a nice and useful corollary of the Chen–Quimpo theorem. A generalization of its almost immediate proof is Exercise 3.5.3.

Corollary 3.5.7 *Every edge of every connected Cayley graph of a finite abelian group of order at least 3 is contained in a Hamilton cycle.*

3.6 Block Quotient Digraphs

The main idea behind quotient digraphs is the same as other kinds of quotients – to try to understand the structure and properties (symmetry properties in particular) of a larger digraph in terms of usually much smaller digraphs. Quotient digraphs are most useful when the quotient procedure can be reversed, that is, when the original digraph can be reconstructed from its quotient. As with other kinds of quotients, in general this is a rather hopeless task as objects are not well determined by their quotients. But there are cases when such a reconstruction in the context of digraphs is indeed possible.

We remark that all basic definitions will be formulated for digraphs (as digraphs are occasionally needed as a technical tool), however, further development of the theory as well as applications will be concerned with graphs.

While there are many kinds of quotients (some we will see later and others, including Schrier coset digraphs, we will not see), in this section we will mainly consider block quotients, which are based on the idea of block systems discussed earlier in this chapter. As we will have occasional use of a more general quotient digraph, we start with that definition.

Definition 3.6.1 Let Γ be digraph and $\mathcal{P}$ a partition of $V(\Gamma)$. Define the **quotient digraph of Γ with respect to $\mathcal{P}$**, denoted $\Gamma/\mathcal{P}$, to be the digraph with vertex set $\mathcal{P}$ and arc set $\{(P, P') : P, P' \in \mathcal{P}, P \neq P'$, and $(u, v) \in A(\Gamma)$ for some $u \in P$ and $v \in P'\}$.

Intuitively, the quotient digraph of Γ with respect to $\mathcal{P}$ is obtained by identifying all the vertices in each cell of $\mathcal{P}$, then eliminating loops and multiple arcs.

Example 3.6.2 Let $\Gamma = \mathrm{Cay}(\mathbb{Z}_{10}, \{1, 2, 5, 8, 9\})$ and $\mathcal{P}$ the partition,

$$\{\{0, 1\}, \{2, 3\}, \{4\}, \{5\}, \{6, 7\}, \{8, 9\}\},$$

of $\mathbb{Z}_{10}$. Find the quotient graph $\Gamma/\mathcal{P}$.

Solution The graph Γ is drawn on the left-hand side of Figure 3.9, and the vertices in each cell of $\mathcal{P}$ are colored with the same color, while different cells have different colors. The quotient $\Gamma/\mathcal{P}$ is drawn on the right-hand side of Figure 3.9. Each cell of $\mathcal{P}$ is in this drawing represented by a single vertex with color corresponding to the color of the cell's vertices in Γ. Two vertices in $\Gamma/\mathcal{P}$ are adjacent if and only if in the corresponding cells of $\mathcal{P}$, some vertex of one cell is adjacent to some vertex of the other cell. Note that $\Gamma/\mathcal{P}$ is not vertex-transitive. □

We will mainly be interested in quotient digraphs where the partition is a block system.

Definition 3.6.3 Let Γ be a vertex-transitive digraph whose automorphism group contains a transitive subgroup G with a block system $\mathcal{B}$. Define the **block quotient digraph of Γ with respect to $\mathcal{B}$**, denoted $\Gamma/\mathcal{B}$, to be the digraph with vertex set $\mathcal{B}$ and arc set $\{(B, B') : B, B' \in \mathcal{B}, B \neq B'$, and $(u, v) \in A(\Gamma)$ for some $u \in B$ and $v \in B'\}$.

So the block quotient digraph of Γ with respect to the block system $\mathcal{B}$ of G, $G \leq \mathrm{Aut}(\Gamma)$ is transitive, and is simply the quotient digraph with respect to the partition $\mathcal{B}$.

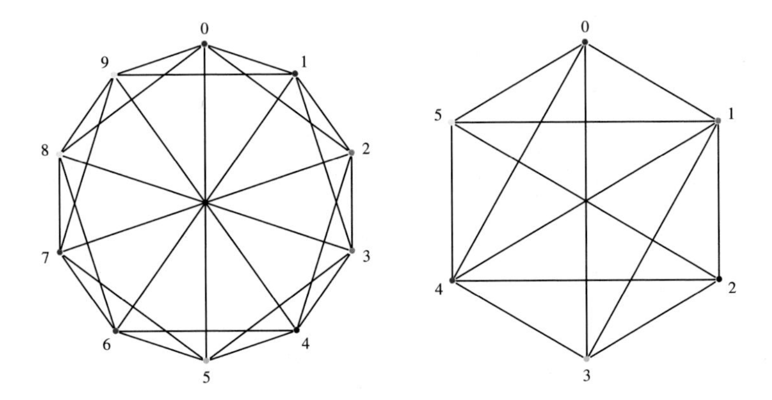

Figure 3.9 (Left) The Cayley graph $\mathrm{Cay}(\mathbb{Z}_{10}, \{1, 2, 5, 8, 9\})$ and (Right) its quotient graph.

Example 3.6.4 There is a transitive subgroup of $\mathrm{Aut}(P)$, where P is the Petersen graph, of order 20 that has a block system $\mathcal{B}$ with blocks of size 2, and $P/\mathcal{B} \cong K_5$.

Solution We consider the metacirculant labeling of P in which case $G = \langle \rho, \tau \rangle \leq \mathrm{Aut}(P)$, where $\rho, \tau\colon \mathbb{Z}_2 \times \mathbb{Z}_5 \to \mathbb{Z}_2 \times \mathbb{Z}_5$ are given by $\rho(i, j) = (i, j + 1)$ and $\tau(i, j) = (i + 1, 2j)$. Then G has a block system $\mathcal{B} = \{V^j : j \in \mathbb{Z}_5\}$ where $V^j = \{(i, j) : i \in \mathbb{Z}_2\}$ by Example 2.3.9. The sets V^j are the sets of the endpoints of the spoke edges. By inspection, for any pair of spoke edges there is an endpoint of one adjacent to an endpoint of the other. Thus $P/\mathcal{B} = K_5$. □

Example 3.6.5 Let $\Gamma = \mathrm{Cay}(\mathbb{Z}_{15}, \{1, 6, 9, 14\})$, as drawn in Figure 3.10. Then $(\mathbb{Z}_{15})_L \leq \mathrm{Aut}(\Gamma)$ has a normal block system $\mathcal{B}$ with blocks of size 3 and $\Gamma/\mathcal{B}$ is a 5-cycle.

Solution Of course $(\mathbb{Z}_{15})_L \leq \mathrm{Aut}(\Gamma)$ as Γ is a Cayley graph of $\mathbb{Z}_{15}$ and $(\mathbb{Z}_{15})_L$ contains a normal subgroup H with orbits of size 3. Hence there is a block system $\mathcal{B}$ of $(\mathbb{Z}_{15})_L$ with blocks of size 3. As $\mathbb{Z}_{15}$ contains a unique subgroup of order 3, namely $\{0, 5, 10\}$, the five blocks of $\mathcal{B}$ are the sets $i + \{0, 5, 10\}$ with $0 \leq i \leq 4$. The vertices of Γ in Figure 3.10 have been colored so vertices have the same color if and only if they are in same block of $\mathcal{B}$. As $\mathbb{Z}_{15}/H \cong \mathbb{Z}_5$, we see that $(\mathbb{Z}_5)_L \leq \mathrm{Aut}(\Gamma/\mathcal{B})$, and so $\Gamma/\mathcal{B}$ is a Cayley digraph of $\mathbb{Z}_5$. As all edges in Γ are between cosets $i + \{0, 5, 10\}$ and $i \pm 1 + \{0, 5, 10\}$, we have $\Gamma/\mathcal{B} = \mathrm{Cay}(\mathbb{Z}_5, \{1, 4\})$ as in Figure 3.10. □

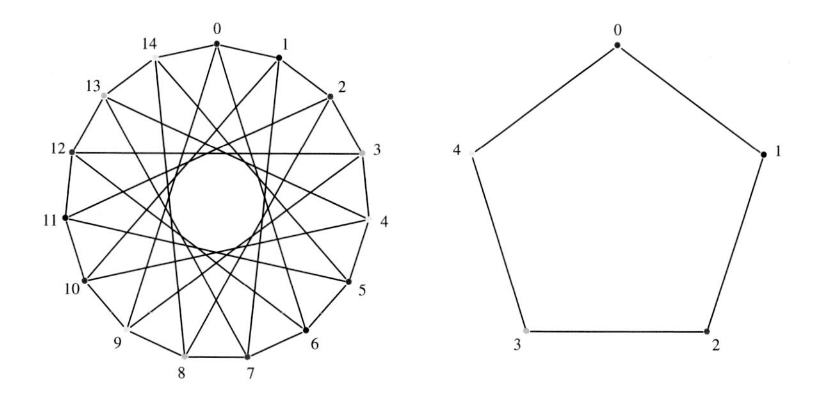

Figure 3.10 (Left) The Cayley graph $\mathrm{Cay}(\mathbb{Z}_{15}, \{1, 6, 9, 14\})$ and (Right) its block quotient graph $\mathrm{Cay}(\mathbb{Z}_5, \{1, 4\})$.

In Definition 3.6.3, the condition that $B \neq B'$ is simply so that a block quotient digraph has no loops. Usually, there is no harm in allowing loops, but in many applications loops do not convey any useful information.

Block quotients inherit transitivity from the original digraph (while some other types of quotients in the literature, which we will not see, do not), and this is one of the main reasons they are useful.

Proposition 3.6.6 *If Γ is a vertex-transitive digraph with $\mathcal{B}$ a block system of G of some transitive group $G \leq \mathrm{Aut}(\Gamma)$, then $\Gamma/\mathcal{B}$ is vertex-transitive. In particular, $G/\mathcal{B} \leq \mathrm{Aut}(\Gamma/\mathcal{B})$.*

Proof Let $(B, B') \in A(\Gamma/\mathcal{B})$ and $g \in G$. Then there exists $(u, v) \in A(\Gamma)$ with $u \in B$, $v \in B'$. Then $(g(u)), g(v)) \in A(\Gamma)$ and $(g(B), g(B')) \in A(\Gamma/\mathcal{B})$. We conclude that $G/\mathcal{B} \leq \mathrm{Aut}(\Gamma/\mathcal{B})$, which implies that $\Gamma/\mathcal{B}$ is vertex-transitive. □

Proposition 3.6.7 *Let Γ be a digraph with $G \leq \mathrm{Aut}(\Gamma)$ transitive with block system $\mathcal{B}$. If G is also transitive on $A(\Gamma)$, then*

(i) *$\Gamma/\mathcal{B}$ is arc-transitive, and*
(ii) *if Γ is connected and $\mathcal{B}$ is nontrivial, then Γ has no arcs with both endpoints contained in a single block of $\mathcal{B}$.*

Proof (i) Let $(B_1, B_2), (B_3, B_4) \in A(\Gamma/\mathcal{B})$, where $B_i \in \mathcal{B}$, $1 \leq i \leq 4$. Then there exists $b_i \in B_i$ with $(b_1, b_2), (b_3, b_4) \in A(\Gamma)$, $1 \leq i \leq 4$. As G is transitive on $A(\Gamma)$, there exists $g \in G$ with $g(b_1, b_2) = (b_3, b_4)$. Then $g/\mathcal{B} \in \mathrm{Aut}(\Gamma/\mathcal{B})$ and $g/\mathcal{B}(B_1, B_2) = (B_3, B_4)$, so $\Gamma/\mathcal{B}$ is arc-transitive.

(ii) Let $B \in \mathcal{B}$. Then there exists an $x \in B$ and $y \notin B$ with $(x, y) \in A(\Gamma)$ as Γ is connected and $\mathcal{B}$ is nontrivial. If $(w, z) \in A(\Gamma)$ for some $w, z \in B$, then there is $g \in \mathrm{Stab}_G(B)$ such that $g(w) = x$, so we may assume without loss of generality that $w = x$. If there is $h \in G$ such that $h(x, z) = (x, y)$ then B is not a block of G, a contradiction. Hence $\Gamma[B]$ has no arcs for every $B \in \mathcal{B}$. □

We leave as Exercise 3.6.5 the analogue for edge-transitive graphs of Proposition 3.6.7.

Somewhat unsurprisingly, stronger results can be proven about block quotient digraphs $\Gamma/\mathcal{B}$ when $\mathcal{B}$ is a normal block system as opposed to a block system of $G \leq \mathrm{Aut}(\Gamma)$. Our next two results will illustrate this.

Definition 3.6.8 If $\mathcal{B}$ is a normal block system of G, then $\Gamma/\mathcal{B}$ is a **normal block quotient digraph**.

The following lemma regarding normal block quotients is due to Praeger (1985), and is a fundamental result in the study of s-arc-transitive graphs. In order to state it in its full generality, we will need some additional notation.

Definition 3.6.9 Let Γ be a digraph and $s \geq 1$. We say Γ is **(G, s)-arc-transitive** if $G \leq \mathrm{Aut}(\Gamma)$ is transitive on $V(\Gamma)$ and the set of s-arcs of Γ.

Obviously if Γ is a (G, s)-arc-transitive digraph then Γ is s-arc-transitive. Additionally, if Γ is (G, s)-arc-transitive, then Γ is (G, t)-arc-transitive for every $1 \leq t \leq s$, provided every vertex of Γ has invalence and outvalence at least 1. We have also seen a variant of this notation earlier as a $(G, 1)$-arc-transitive digraph is a G-arc-transitive digraph. We emphasize to the reader that the following result only applies to graphs, not digraphs that are not graphs.

Lemma 3.6.10 *Let Γ be a connected (G, s)-arc-transitive graph, where $s \geq 2$. If G has a normal block system $\mathcal{B}$ with $|\mathcal{B}| \geq 3$, then*

(i) *$\Gamma/\mathcal{B}$ is $(G/\mathcal{B}, s)$-arc-transitive, and*
(ii) *$\mathrm{fix}_G(\mathcal{B})$ is semiregular, and hence regular on each block of $\mathcal{B}$.*

Proof Let $B \in \mathcal{B}$. Then $\Gamma[B]$ has no edges for every $B \in \mathcal{B}$ by Proposition 3.6.7 (i). As $\Gamma/\mathcal{B}$ is a connected graph by Exercise 3.6.3 and is of order at least 3, $\Gamma/\mathcal{B}$ contains a path $B_0B_1B_2$ of length 2. Thus there exist vertices $u \in B_0$, $v, w \in B_1$, and $z \in B_2$ such that $uv, wz \in E(\Gamma)$. As $\mathcal{B}$ is normal, there exists $g \in \mathrm{fix}_G(\mathcal{B})$ such that $g(w) = v$, so $vx \in E(\Gamma)$ for some $x \in B_2$. Then u, v, x is a 2-arc in Γ.

Now let $B' \in \mathcal{B}$ be distinct from B, and $a \in B$. Suppose there exists a 2-arc b, a, c in Γ with $b, c \in B'$, $b \neq c$. Then there is no $g \in G$ with $g(u, v, x) = b, a, c$ as the edges of b, a, c have endpoints in two blocks of $\mathcal{B}$ while the edges of

u, v, x have endpoints in three. We conclude that each vertex of a block $B \in \mathcal{B}$ is adjacent to at most one vertex in $B' \in \mathcal{B}$. That is, between any two blocks of $\mathcal{B}$ there is either exactly a 1-factor or no edges at all.

If $ab \in E(\Gamma)$, then $\mathrm{Stab}_{\mathrm{fix}_G(\mathcal{B})}(a) = \mathrm{Stab}_{\mathrm{fix}_G(\mathcal{B})}(b)$ as otherwise a or b is adjacent to at least two vertices in one block of $\mathcal{B}$. This implies that if $P = a_1 \dots a_r$ is a path in Γ, then $\mathrm{Stab}_{\mathrm{fix}_G(\mathcal{B})}(a_i) = \mathrm{Stab}_{\mathrm{fix}_G(\mathcal{B})}(a_j)$, $1 \le i, j \le r$. In particular, the endpoints of any path in Γ have the same vertex stabilizer in $\mathrm{fix}_G(\mathcal{B})$. As there is a path between any pair of vertices of Γ, every vertex has the same stabilizer in $\mathrm{fix}_G(\mathcal{B})$, which is necessarily trivial. So $\mathrm{fix}_G(\mathcal{B})$ is semiregular and (ii) follows.

We now show that every s-arc in Γ induces an s-arc in $\Gamma/\mathcal{B}$, and vice versa. Let $x_0, \dots, x_s$ be an s-arc in Γ. Then there exists $B_i \in \mathcal{B}$ with $x_i \in B_i$, and $B_0, \dots, B_s$ is a walk in $\Gamma/\mathcal{B}$. Suppose $B_0, \dots, B_s$ is not an s-arc. Then $B_i = B_{i+2}$ for some $0 \le i \le s-2$. As $x_0, \dots, x_s$ is an s-arc, $x_i \ne x_{i+2}$, and so x_{i+1} is adjacent in Γ to two vertices of $B_i = B_{i+2}$, a contradiction. Hence $B_0, \dots, B_s$ is an s-arc in $\Gamma/\mathcal{B}$. Now let $B_0, \dots, B_s$ be an s-arc in $\Gamma/\mathcal{B}$. Then there exist vertices $x_i \in B_i$ and $y_{i+1} \in B_{i+1}$ such that $(x_i, y_{i+1}) \in A(\Gamma)$ for every $0 \le i \le s-1$. As $\mathrm{fix}_G(\mathcal{B})$ is transitive on $B \in \mathcal{B}$ we may assume without loss of generality (arguing as above) that $y_{i+1} = x_{i+1}$ for every $0 \le i \le s-2$. Then $x_0, \dots, x_s$ is an s-arc in Γ.

Now let $B_0, \dots, B_s$ and $B'_0, \dots, B'_s$ be s-arcs in $\Gamma/\mathcal{B}$. Then there exists corresponding s-arcs $x_0, \dots, x_s$ and $y_0, \dots, y_s$ with $x_i \in B_i$ and $y_i \in B'_i$. As Γ is (G, s)-arc-transitive, there exists $g \in G$ with $g(x_0, \dots, x_s) = y_0, \dots, y_s$. Then $(g/\mathcal{B})(B_0, \dots, B_s) = B'_0, \dots, B'_s$ and $\Gamma/\mathcal{B}$ is $(G/\mathcal{B}, s)$-arc-transitive and (i) holds. □

We remark that Lemma 3.6.10 does not hold for s-arc-transitive digraphs that are not graphs (see Exercise 5.1.7). Additionally, the following example shows that it also does not hold even for graphs when $s = 1$.

Example 3.6.11 Let $S = \{\pm 1, \pm 16, \pm 22\} \subset \mathbb{Z}_{39}$. The graph $\Gamma = \mathrm{Cay}(\mathbb{Z}_{39}, S)$ is a connected G-arc-transitive graph, where $G = \{x \mapsto ax+b : a \in S, b \in \mathbb{Z}_{39}\}$. Additionally, there is a block system $\mathcal{B}$ of G with three blocks of size 13, and $\mathrm{fix}_G(\mathcal{B})$ is not semiregular.

Solution Straightforward computations will show that $aS = S$ for every $a \in S$, and so by Corollary 1.2.16, the maps $x \mapsto ax$ are in $\mathrm{Aut}(\Gamma)$ for every $a \in S$. Then $G \le \mathrm{Aut}(\Gamma)$. Also, S is an orbit of the map $x \mapsto -16x$ (check this!), and so by Lemma 3.2.2, Γ is $(G, 1)$-arc-transitive. As $G \le \left\{x \mapsto ax + b : a \in \mathbb{Z}_{39}^*, b \in \mathbb{Z}_{39}\right\}$, which normalizes $(\mathbb{Z}_{39})_L$ and as $\langle 3 \rangle$ is a characteristic subgroup of $(\mathbb{Z}_{39})_L$ of order 13, $\langle 3 \rangle \trianglelefteq G$. The set of orbits of $\langle 3 \rangle$ is a block system $\mathcal{B}$ of G by Theorem 2.2.7. As $16 \equiv 1 \pmod 3$, the map

$x \mapsto 16x$ fixes each block of $\mathcal{B}$ and so is contained in $\mathrm{fix}_G(\mathcal{B})$. As $x \mapsto 16x$ fixes 0, $\mathrm{fix}_G(\mathcal{B})$ is not semiregular. □

While Lemma 3.6.10 does not hold for $(G, 1)$-arc-transitive graphs, it does hold in the special case where Γ is $(G, 1)$-arc-transitive graph of prime valency, as shown by Lorimer (1984). This result has been especially applied when Γ is a cubic graph.

Lemma 3.6.12 *Let Γ be a connected G-arc-transitive graph of prime valency. If G has a normal block system $\mathcal{B}$ with $|\mathcal{B}| \geq 3$, then*

(i) *$\Gamma/\mathcal{B}$ is arc-transitive, and*
(ii) *$\mathrm{fix}_G(\mathcal{B})$ is semiregular, and hence regular on each block of $\mathcal{B}$.*

Proof By Proposition 3.6.7 (ii) the subgraph $\Gamma[B]$ has no arcs for every $B \in \mathcal{B}$, and $\Gamma/\mathcal{B}$ is arc-transitive. Suppose $\mathrm{Stab}_{\mathrm{fix}_G(\mathcal{B})}(u) \neq 1$ for some $u \in V(\Gamma)$, so $\mathrm{fix}_G(\mathcal{B})$ is not semiregular. Let $v \in V(\Gamma)$ and $h \in \mathrm{Stab}_{\mathrm{fix}_G(\mathcal{B})}(u)$ such that $h(v) \neq v$, and P be a uv-path in Γ. Traversing P from u to v, let $y \neq u$ be the first vertex on P such that $\mathrm{Stab}_{\mathrm{fix}_G(\mathcal{B})}(y) \neq \mathrm{Stab}_{\mathrm{fix}_G(\mathcal{B})}(u)$. Such a y exists as v has this property. Let x be the predecessor of y on P. Then $\mathrm{Stab}_{\mathrm{fix}_G(\mathcal{B})}(x) = \mathrm{Stab}_{\mathrm{fix}_G(\mathcal{B})}(u)$. By Theorem 3.2.8, $\mathrm{Stab}_G(x)$ is transitive on $N^+_\Gamma(x)$, and as Γ is of prime valency p, $|N^+_\Gamma(x)| = p$. Then $\mathrm{Stab}_G(x)$ contains an element g that fixes x and cyclically permutes the elements of $N^+_\Gamma(x)$ as a p-cycle. Let $B \in \mathcal{B}$ contain y. As $y \in N^+_\Gamma(x)$ and $\mathrm{Stab}_{\mathrm{fix}_G(\mathcal{B})}(x)$ does not fix y, x has another outneighbor z contained in $B \in \mathcal{B}$. After raising g to an appropriate power, we may assume $g(z) = y$, in which case $g(B) = B$. Consequently, $N^+_\Gamma(x) \subseteq B$. As Γ is a graph, every neighbor of x is contained in B. Let $B' \in \mathcal{B}$ with $x \in B'$. As $\mathrm{fix}_G(\mathcal{B})$ is transitive on B', every neighbor of every vertex in B' is contained in B. Order arguments then give that every neighbor of every element of B is contained in B'. As Γ is connected, we have $\mathcal{B} = \{B, B'\}$, contradicting $|\mathcal{B}| \geq 3$. □

One weakness of the block quotients we consider here is that in going from a digraph to its normal quotient, much information is lost. In fact, so much information can be lost that we have no hope of recovering the original digraph from its normal quotient. For example, if Γ is a bipartite vertex-transitive graph with bipartition $\mathcal{B}$ that is a normal block system of some subgroup of $\mathrm{Aut}(\Gamma)$, then $\mathrm{Aut}(\Gamma)/\mathcal{B}$ is either an edge or just two vertices! So the normal quotient has many important uses, but if one wishes to have some hope of recovering the original graph from its quotient, the normal quotient is most likely not the quotient one would like to use. Thus we will have need of other, more general quotients.

3.7 The Spiral Path Argument and Frucht Notation

In this section, we will first introduce the spiral path argument, which is one of the most common ways of "lifting" Hamilton cycles from a quotient of a graph back to the graph. This latter idea is a major topic in its own right, and is the subject of Section 11.2. Usually, but not always, "lifting" a Hamilton cycle from a quotient to the original graph involves a Hamilton cycle in the quotient, as well as a semiregular element in the automorphism group of the graph. Finding semiregular elements in the automorphism groups of vertex-transitive graphs is a major problem in algebraic graph theory, and is introduced in Section 3.8 and further studied in Chapter 12. We end this section with a common way of representing a graph via its quotient by a semiregular element, namely Frucht notation.

Definition 3.7.1 Let Γ be a graph with (m,n)-semiregular automorphism

$$\rho = \left(v_0^0, v_1^0, \ldots, v_{n-1}^0\right)\left(v_0^1, v_1^1, \ldots, v_{n-1}^1\right)\cdots\left(v_0^{m-1}, v_1^{m-1}, \ldots, v_{n-1}^{m-1}\right).$$

A **spiral path in Γ with respect to** ρ is a path in Γ of the form $v_{i_0}^0 v_{i_1}^1 \ldots v_{i_{m-1}}^{m-1} v_{i_m}^0$, where $i_j \in \mathbb{Z}_n$ and $i_0 \neq i_m$.

The following lemma is frequently used to construct a Hamilton cycle in a graph Γ.

Lemma 3.7.2 *Let p be prime and Γ a graph with an (m,p)-semiregular automorphism ρ. If Γ contains a spiral path P with respect to ρ, then Γ is Hamiltonian.*

Proof Let $P = v_{i_0}^0 v_{i_1}^1 \ldots v_{i_{m-1}}^{m-1} v_{i_m}^0$ be a spiral path with respect to ρ in Γ. Replacing P by $\rho^{p-i_0}(P)$, we may assume without loss of generality that $i_0 = 0$ and $i_m = r \in \mathbb{Z}_p^*$. Consider the subgraph $W = \rho^0(P)\rho^r(P)\rho^{2r}(P)\ldots\rho^{(p-1)r}(P)$. The graph W is illustrated in Figure 3.11 where each of the paths $\rho^{\ell r}(P)$ is represented with a different color. The black ovals represent orbits of ρ. Note $\rho^{\ell r}(P)$ has first vertex ℓr and last vertex $(\ell+1)r$, for all $0 \le \ell \le p-1$. As p is prime, $\langle r\rangle = \mathbb{Z}_p$. So W is connected. As ρ is a semiregular element and each vertex v_j^i, $i \in \mathbb{Z}_m$, $i \neq 0$, $j \in \mathbb{Z}_p$, is an interior vertex of some $\rho^k(P)$, $0 \le k \le p-1$, we see that v_j^i has valency 2 in W, $i \neq 0$. Also, each vertex $v_{\ell r}^0$ is the first vertex of $\rho^{\ell r}(P)$ and last vertex of $\rho^{(\ell-1)r}(P)$ and is on no other path $\rho^k(P)$, $k \neq (\ell-1)r$ or ℓr, we see that every vertex of W has valency 2. Thus W is a cycle in Γ. As P contains at least one element of each orbit of $\langle\rho\rangle$ and ρ is semiregular of order p, $\cup_{i=0}^{p-1}\rho^{ri}(P) = V(\Gamma)$, and W is a Hamiltonian cycle in Γ. □

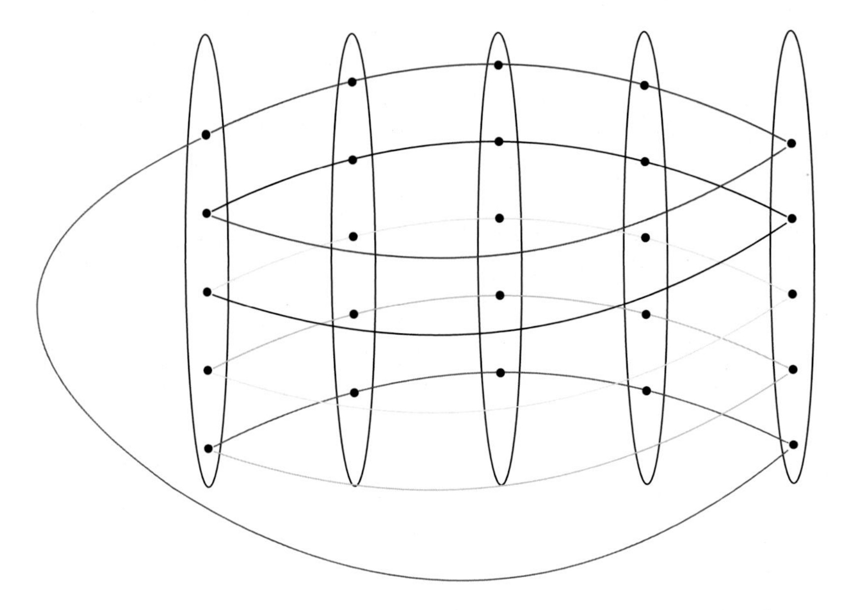

Figure 3.11 A Hamilton cycle from a spiral path with respect to a semiregular element.

Note that if Aut(Γ) contains an (m, k)-semiregular element ρ, then for any prime $p|k$, we have $\rho^{k/p}$ is an (m, p)-semiregular element. So the restriction to (m, p)-semiregular automorphisms in the preceding result is not really a restriction at all. There is one situation in which spiral paths are easy to find.

Lemma 3.7.3 *Let p be prime, and Γ be a graph with an (m, p)-semiregular element ρ. Let $\mathcal{B}$ be the set of orbits of ρ. Suppose*

- *the quotient graph $\Gamma/\mathcal{B}$ contains a Hamilton cycle $B_0 \dots B_{m-1}B_0$, and*
- *there exists $0 \le i \le m-1$ such that a vertex of B_i is adjacent to at least two vertices of B_{i+1}.*

Then Γ contains a spiral path and Γ is Hamiltonian.

Proof We are not assuming Γ is vertex-transitive here, so $\mathcal{B}$ need not be a block system of any subgroup of Aut(Γ). As a Hamilton cycle can begin at any vertex of a graph, we may assume $B_i = B_{m-1}$ and $B_{i+1} = B_0$. Let $v_0 \in B_0$, and for $1 \le i \le m-1$, successively choose $v_i \in B_i$ such that $v_{i-1}v_i \in E(\Gamma)$. Then $v_0v_1 \dots v_{m-1}$ is a path in Γ. As some vertex of B_{m-1} is adjacent to (at least) two vertices of B_0 and $\rho \in \mathrm{Aut}(\Gamma)$, every vertex of B_{m-1} is adjacent in Γ to two vertices of B_0. This means that v_{m-1} is adjacent to some $v_0' \neq v_0$. Then $v_0 \ldots v_{m-1}v_0'$ is a spiral path in Γ. By Lemma 3.7.2, Γ is Hamiltonian. □

Corollary 3.7.4 *Let Γ be a graph, $\rho \in \mathrm{Aut}(\Gamma)$ a semiregular element of order p, and $\mathcal{B}$ the set of orbits of $\langle\rho\rangle$. If every edge of $\Gamma/\mathcal{B}$ is contained in a Hamilton cycle, then Γ is Hamiltonian or the induced subgraph between any two orbits of $\langle\rho\rangle$ is either edgeless or a 1-factor.*

Proof If there are two orbits B_1 and B_2 of $\langle\rho\rangle$ such that the induced subgraph between these two orbits of $\langle\rho\rangle$ is not edgeless and is not a 1-factor, then the edge B_1B_2 in $\Gamma/\mathcal{B}$ is contained in a Hamilton cycle and a vertex of B_1 is adjacent to at least two vertices of B_2. The hypotheses of Lemma 3.7.3 are then satisfied and the result follows. □

We end our discussion on spiral paths with the following general result, originally due to Alspach and Parsons (1982b, Lemma 12).

Corollary 3.7.5 *Let $\Gamma = \Gamma(m, p, \alpha, S_0, \ldots, S_{m-1})$ be a connected metacirculant graph with p a prime, and $|S_i| \geq 2$ for some $1 \leq i \leq m-1$. Then Γ is Hamiltonian.*

Proof Let $\mathcal{B}$ be the block system of $\langle\rho, \tau\rangle$ that is the set of orbits of $\langle\rho\rangle$. Then $\langle\tau\rangle/\mathcal{B}$ is a regular cyclic subgroup, and so $\Gamma/\mathcal{B}$ is isomorphic to a Cayley graph of $\mathbb{Z}_m$ by Theorem 1.2.20. By Corollary 3.5.7, every edge of $\Gamma/\mathcal{B}$ is contained in a Hamilton cycle in $\Gamma/\mathcal{B}$. Then the induced subgraph between the block of $\mathcal{B}$ that contains $(0,0)$ and the block of $\mathcal{B}$ that contains $(i,0)$ is neither empty nor a 1-factor. The result follows by Corollary 3.7.4. □

Alspach (1989) has formalized the spiral path idea in the following way.

Definition 3.7.6 Let Γ be a graph with an (m, n)-semiregular automorphism

$$\rho = \left(v_0^0, v_1^0, \ldots, v_{n-1}^0\right)\left(v_0^1, v_1^1, \ldots, v_{n-1}^1\right)\cdots\left(v_0^{m-1}, v_1^{m-1}, \ldots, v_{n-1}^{m-1}\right).$$

Let $O_i = \left\{v_j^i : j \in \mathbb{Z}_n\right\}$ and $O = \{O_i : i \in \mathbb{Z}_m\}$. Let C be a cycle in Γ/O with initial vertex O_0. The set of terminal vertices of all paths in Γ with initial vertex v_0^0 whose quotient in Γ/O is C is called a **coil** of C.

A Hamilton cycle in Γ can then be constructed from a Hamilton cycle in the quotient graph Γ/O provided its coil contains a vertex v_j^0 where j is relatively prime to n. In Alspach (1989) some sufficient conditions for the Hamiltonicity of Γ were given using the terminology of coils. We now consider Frucht notation.

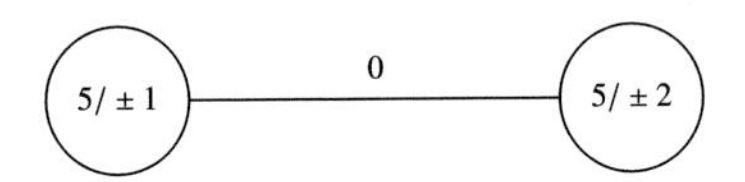

Figure 3.12 The Petersen graph in Frucht notation.

Frucht notation was unsurprisingly introduced by Frucht (1970), and is frequently used to simplify the presentation of graphs, especially those whose automorphism groups contain a semiregular element. We should remark that while we will give a "formal" description of how to represent a graph with respect to a semiregular automorphism using Frucht's notation, the rules for writing a graph in Frucht's notation in the literature is somewhat fluid. Frucht himself did not write down a formal description of his notation, but rather illustrated his ideas using numerous examples, and when introducing this notation wrote "Since the proposed notation is not offered as definitive, any suggestions for improvement will be welcome."

So, suppose Γ is a graph with (m, n)-semiregular automorphism

$$\rho = \left(v_0^0, v_1^0, \ldots, v_{n-1}^0\right)\left(v_0^1, v_1^1, \ldots, v_{n-1}^1\right)\cdots\left(v_0^{m-1}, v_1^{m-1}, \ldots, v_{n-1}^{m-1}\right).$$

We are *not* assuming that $\langle\rho\rangle$ is normal in $\mathrm{Aut}(\Gamma)$ or in any transitive subgroup of $\mathrm{Aut}(\Gamma)$ as with an (m, n)-metacirculant graph. For $i \in \mathbb{Z}_m$, set $O_i = \{v_j^i : j \in \mathbb{Z}_n\}$, so that $\{O_i : i \in \mathbb{Z}_m\}$ is the set of orbits of ρ. Each orbit of ρ is represented by a circle, and inside the circle corresponding to O_i will be the symbol n/T, where $T \subset \mathbb{Z}_n \backslash \{0\}$, and v_j^i is adjacent to v_{j+t}^i for every $j \in \mathbb{Z}_n$ and $t \in T$. In other words, T is the connection set of the natural circulant graph of order n induced by O_i. If $|T|$ is small, it is often the case that the elements of T will be listed instead of T. For example, if $T = \{1, 4\}$, we will feel free to write $n/1, 4$ in place of n/T. If $T = \emptyset$, i.e., the subgraph of Γ induced by O_i has no edges, we will just write n instead of $n/\emptyset$. An arc from orbit O_i to orbit O_k indicates that there are edges between the vertices of O_i and O_k. If this arc is labeled with an integer s, then $\left(v_j^i, v_{j+s}^k\right)$ is an arc of Γ. Usually, the label is omitted if $s = 0$ and the arc replaced with an edge. Finally, several arcs (each with a label) may be replaced by one arc with several labels.

An (m, n)-metacirculant digraph is an ideal graph to represent in Frucht's notation, where the semiregular element is ρ (here $\rho\colon \mathbb{Z}_m \times \mathbb{Z}_n \to \mathbb{Z}_m \times \mathbb{Z}_n$ is defined by $\rho(i, j) = (i, j + 1)$). Hence $O_i = \{(i, j) : j \in \mathbb{Z}_n\} = V_i$. In Figure 3.12, the Petersen graph is represented using Frucht's notation with semiregular element ρ, while in Figure 3.13, the $(4, 13, 8, \{1, 12\}, \{0\}, \{2, 11\}, \{0\})$-metacirculant graph is drawn in Frucht notation.

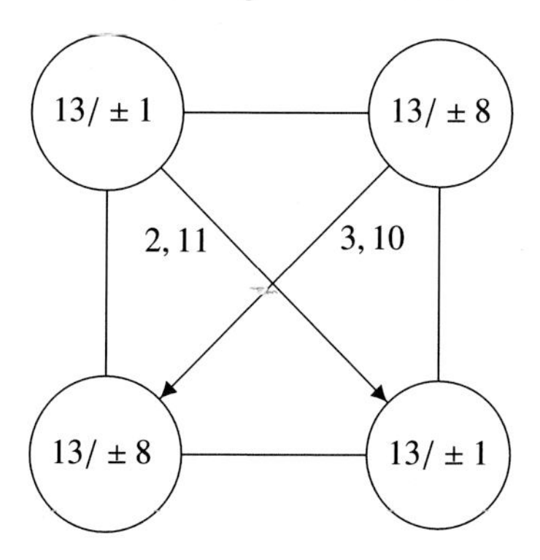

Figure 3.13 The $(4, 13, 8, \{1, 12\}, \{0\}, \{2, 11\}, \{0\})$-metacirculant in Frucht notation.

3.8 Semiregular Elements

In Section 3.7, we saw how a graph having a semiregular automorphism can be useful for constructing Hamilton cycles. With this in mind, Marušič posed what is now known as the semiregularity problem (Marušič, 1981a, Problem 2.4).

Problem 3.8.1 (The semiregularity problem) Is there a vertex-transitive digraph without a nontrivial semiregular automorphism?

One can ask a similar question about a transitive permutation group, and this question has been asked. The point of view, though, is a little different. Rather than asking which transitive permutation groups contain a semiregular element, this problem is usually phrased as which transitive permutation groups do *not* have a semiregular element.

Definition 3.8.2 A transitive group $G \leq \mathcal{S}_n$ is **elusive** if it contains no semiregular elements of prime order.

The name is chosen to suggest that such groups are rare, but at the moment, this is not known for certain (although it is certainly true that such groups can be difficult to construct at the moment). Note that a group contains a semiregular element if and only if it contains a semiregular element of prime order, so an elusive group contains no semiregular elements other than the identity. We begin our study with examples of groups that are not elusive, as well as a few detours to establish important facts about some permutation groups.

Theorem 3.8.3 *Every transitive group G of degree p^k, p a prime and $k \geq 1$, contains a nontrivial semiregular element.*

Proof By Corollary 1.1.10, G has a transitive Sylow p-subgroup P. As P is a nontrivial p-group, P has a nontrivial center $Z(P)$, which is semiregular by Lemma 2.2.16. Then $Z(P)$ contains a nontrivial semiregular element. □

Our next goal is a result that was originally proven by Marušič stating that a transitive group of degree mp, $m \leq p$, necessarily contains a semiregular element of order p (Marušič, 1981a, Theorem 3.4). The method of proof we will use here is due to McKay and Royle (1990, Lemma 4.1), and we choose this proof here to introduce the reader to the orbit-counting lemma, which we prove next. The orbit-counting lemma has many names: it is called the Cauchy–Frobenius lemma after its discoverers, Burnside's lemma after it being proven in Burnside's 1897 text "Theory of Groups of Finite Order," and, amusingly. The Lemma that is not Burnside's. We will also give Marušič's more combinatorial proof in Theorem 12.2.1 when we revisit the semiregularity problem in more detail in Chapter 12.

Definition 3.8.4 For a group G acting on a set X, let X/G denote the set of orbits of G, and for $g \in G$, let $\mathrm{Fix}(g)$ be the set of all elements of X fixed by g. That is, $\mathrm{Fix}(g) = \{x \in X : g(x) = x\}$.

Lemma 3.8.5 (orbit-counting lemma) *For a group G acting on a set X,*

$$|X/G| = \frac{1}{|G|}\sum_{g \in G} |\mathrm{Fix}(g)|.$$

Proof First rewrite $\sum_{g\in G} |\mathrm{Fix}(g)|$ as $|\{(g, x) : g \in G, x \in X, \text{ and } g(x) = x\}|$, and then observe that this is equal to $\sum_{x\in X} |\mathrm{Stab}_G(x)|$. Applying the orbit-stabilizer theorem, we have

$$\begin{aligned}\sum_{g\in G} |\mathrm{Fix}(g)| &= |\{(g, x) : g \in G, x \in X, \text{ and } g(x) = x\}| \\ &= \sum_{x\in X} |\mathrm{Stab}_G(x)| = \sum_{x\in X} \frac{|G|}{|G(x)|} = |G|\sum_{x\in X} \frac{1}{|G(x)|}.\end{aligned}$$

But notice that

$$\sum_{x\in X} \frac{1}{|G(x)|} = \sum_{O\in X/G}\sum_{y\in O} \frac{1}{|G(y)|},$$

and $\sum_{y\in O} \frac{1}{|G(y)|} = \frac{|O|}{|G(y)|} = 1$. Then $\sum_{x\in X} \frac{1}{|G(x)|} = |X/G|$ and the result follows. □

So the orbit-counting lemma says that the number of orbits of the action of G on X is the average number of elements of X fixed by the elements of G.

Theorem 3.8.6 *Let G be a transitive group of degree mp, with $m \leq p$. Then G contains an (m, p)-semiregular element.*

Proof By Theorem 3.8.3, we need only consider when $m < p$. We will apply the orbit-counting lemma to a Sylow p-subgroup P of G. By Theorem 1.1.9, the length of every orbit of P is divisible by p. As $m < p$, every orbit of P has length p, and so P has m orbits of length p. Suppose P does not contain a semiregular element of order p. Then each $\rho \in P$ has at least p fixed points, while the identity has mp fixed points. Then $\sum_{g \in P} |\mathrm{Fix}(g)| \geq mp + (|P| - 1)p$ and so by the orbit-counting lemma the number m of orbits of P is

$$\frac{1}{|P|}\sum_{g \in P} |\mathrm{Fix}(g)| \geq \frac{1}{|P|}(mp + (|P| - 1)p) = \frac{p}{|P|}(m + |P| - 1) \geq p > m,$$

a contradiction. □

Our final goal for this introduction to the semiregularity problem is to show that elusive groups exist. For this we will need some new ideas.

Definition 3.8.7 The **affine geometry** $\mathrm{AG}(d, \mathbb{F})$ of dimension d over the field $\mathbb{F}$ is the vector space $\mathbb{F}^d$, the elements of which are called **points**, together with **affine subspaces**, which are translates of the usual vector subspaces of $\mathbb{F}^d$. That is, if S is a subspace of $\mathbb{F}^d$ and v is a point in the affine geometry (that is, v is a vector in $\mathbb{F}^d$), then $v + S = \{v + s : s \in S\}$ is an affine subspace. We typically write $\mathrm{AG}(d, n)$ for $\mathrm{AG}(d, \mathbb{F})$ when $\mathbb{F}$ is a finite field of order n. A one-dimensional affine subspace is typically called an **affine line**.

We will require a few technical facts concerning $\mathrm{AG}(2, p)$, which we record in the following result.

Lemma 3.8.8 *For any prime p there are $p(p + 1)$ affine lines in* $\mathrm{AG}(2, p)$*, and any two points of* $\mathrm{AG}(2, p)$ *are on a line.*

Proof First, if (u, v) and (x, y) are distinct points in $\mathrm{AG}(2, p)$, then the line ℓ containing all points of the form $a(u - x, v - y) + (x, y)$ for $a \in \mathbb{F}_p$ contains both the points (u, v) (with $a = 1$) and (x, y) (with $a = 0$), and the second part of the result holds.

For the first part, as there are $\binom{p^2}{2}$ distinct pairs of points and each line contains $\binom{p}{2}$ pairs, the number of lines in $\mathrm{AG}(2, p)$ is at least $\binom{p^2}{2}/\binom{p}{2} = p(p+1)$. As there are $p^2 - 1$ nonzero points in $\mathrm{AG}(2, p)$ and each one-dimensional subspace of $\mathbb{F}_p^2$ (as a vector space) contains $p - 1$ nonzero points, there are $\left(p^2 - 1\right)/(p - 1) = p + 1$ one-dimensional subspaces of $\mathbb{F}_p^2$. Denote them by $\ell_1, \ldots, \ell_{p+1}$. These are the affine lines that contain the origin. As there are p^2 translates of each of the lines (one for each element of $\mathbb{F}_p^2$), there are $p^2(p + 1)$ (not necessarily distinct) affine lines of the form $\ell_i + v$ with $1 \leq i \leq p + 1$ and $v \in \mathbb{F}_p^2$. To complete the first part of this result, we will show that each affine line can be written in exactly p ways in this form. Let ℓ be an affine line. Then there exists $(u, v) \in \mathbb{F}_p^2$ and $(x, y) \in \mathbb{F}_p^2$ such that $\ell = \left\{a(u, v) + (x, y) : a \in \mathbb{F}_p\right\}$. Fix $a \in \mathbb{F}_p$, and let $c \in \mathbb{F}_p$. Then

$$(a - c)(u, v) + [(x, y) + c(u, v)] = a(u, v) + (x, y),$$

and as there are p choices for c, each line ℓ can be written as $a(u, v) + (x, y)$ in at least p different ways. Thus there are at most $p(p + 1)$ affine lines. As we have already established that there are at least $p(p + 1)$ affine lines, there are exactly $p(p + 1)$ affine lines. □

Definition 3.8.9 Let q be a prime power and $d \geq 2$. Define the **general linear group of dimension d over the field** $\mathbb{F}$, denoted $\mathrm{GL}(d, \mathbb{F})$, to be the group of all invertible $d \times d$ matrices with entries in $\mathbb{F}$, with binary operation matrix multiplication.

It is straightforward to show using linear algebra that $\mathrm{GL}(d, \mathbb{F})$ is a group.

Lemma 3.8.10 *Let $\mathbb{F}_q$ be the field of order q, and d a positive integer. Then*

$$|\mathrm{GL}(d, \mathbb{F}_q)| = \Pi_{i=0}^{d-1} \left(q^d - q^i\right) = q^{d(d-1)/2}\Pi_{i=0}^{d-1} \left(q^i - 1\right).$$

Proof Let $e_1, \ldots, e_d$ be a basis for $\mathbb{F}_q^d$. As any element A of $\mathrm{GL}(d, \mathbb{F}_q)$ is uniquely determined by $A(e_1), \ldots, A(e_d)$; we need only count the number of choices for $A(e_1), \ldots, A(e_d)$. As $A(e_1)$ can be any element of $\mathbb{F}_q^d$ except for the zero vector, there are $q^d - 1$ choices for $A(e_1)$. As $A(e_2)$ can be any vector in $\mathbb{F}_q^d \backslash \langle A(e_1) \rangle$, there are $q^d - q$ choices for $A(e_2)$. Similarly, $A(e_3)$ can be any vector in $\mathbb{F}_q^d \backslash \langle A(e_1), A(e_2) \rangle$. Also, $\langle A(e_1), A(e_2) \rangle$ has dimension 2 as $A(e_1)$ and $A(e_2)$ are linearly independent as e_1 and e_2 are linearly independent. Thus there are $q^d - q^2$ choices for $A(e_3)$. The result now follows with a straightforward induction argument. □

Definition 3.8.11 Define the **affine general linear group** of dimension d over the field $\mathbb{F}$, denoted $\mathrm{AGL}(d, \mathbb{F})$, to be the group $\left\{x \mapsto Ax + b : A \in \mathrm{GL}(d, \mathbb{F}), \text{ and } b \in \mathbb{F}^d\right\}$ (here $x \in \mathbb{F}^d$).

This group was defined in Exercise 2.4.9 for the special case when $\mathbb{F} = \mathbb{F}_p$. An element of $\mathrm{AGL}(d, \mathbb{F})$ of the form $x \mapsto x + b$ is called a **translation**. The translations T in $\mathrm{AGL}(d, \mathbb{F})$ form a normal subgroup of $\mathrm{AGL}(d, \mathbb{F})$, and $\mathrm{AGL}(d, \mathbb{F})/T \cong \mathrm{GL}(d, \mathbb{F})$. As $T \cap \mathrm{GL}(d, \mathbb{F}) = 1$, we have $\mathrm{AGL}(d, \mathbb{F}) \cong T \rtimes \mathrm{GL}(d, \mathbb{F})$. Note that $\mathrm{AGL}(d, \mathbb{F})$ is transitive on the set S of all affine subspaces of $\mathrm{AG}(d, \mathbb{F})$ of dimension s. This follows as $\mathrm{GL}(d, \mathbb{F})$ will certainly map any s-dimensional subspace of $\mathbb{F}^d$ to any s-dimensional subspace of $\mathbb{F}^d$ as any set of s linearly independent vectors can be mapped to any set of s linearly independent vectors by a linear transformation. But then $\mathrm{AGL}(d, \mathbb{F})$ contains all of the translations in $\mathbb{F}^d$, and so is transitive on S.

Definition 3.8.12 An **automorphism of an affine geometry** $\mathrm{AG}(d, \mathbb{F})$ is a permutation of the points of $\mathrm{AG}(d, \mathbb{F})$ that maps affine subspaces of dimension s to affine subspaces of dimension s.

Thus $\mathrm{AGL}(d, \mathbb{F})$ is a subgroup of the automorphism group of the affine geometry $\mathrm{AG}(d, \mathbb{F})$. We remark that the full automorphism group of the affine geometry $\mathrm{AG}(d, \mathbb{F})$, denoted $\mathrm{A\Gamma L}(d, \mathbb{F})$ is the group generated by $\mathrm{AGL}(d, \mathbb{F})$ and the automorphisms of $\mathrm{AG}(d, \mathbb{F})$ induced by the field automorphisms of $\mathbb{F}$ (see Dixon and Mortimer (1996, §2.8)).

It is customary in the literature to denote $\mathrm{GL}(d, \mathbb{F})$ and $\mathrm{AGL}(\mathbb{F})$ (and other groups we will see later that are derived from $\mathrm{GL}(d, \mathbb{F})$), by $\mathrm{GL}(d, q)$ and $\mathrm{AGL}(d, q)$ when $\mathbb{F} = \mathbb{F}_q$ for some prime power q. We will follow this custom henceforth.

The first example of elusive groups were found by Fein, Kantor, and Schacher (1981).

Theorem 3.8.13 *The group* $\mathrm{AGL}\left(1, p^2\right)$ *has a transitive elusive action on* $p(p+1)$ *points when* $p = 2^t - 1$ *is a Mersenne prime.*

Proof First note that $\mathrm{AGL}\left(1, p^2\right) \leq \mathrm{AGL}(2, p)$ as the underlying set on which both these groups act is the same and every element of $\mathrm{AGL}\left(1, p^2\right)$ is a composition of a translation $(x \mapsto x + b)$ and a linear transformation. So $\mathrm{AGL}\left(1, p^2\right)$ also permutes the one-dimensional affine subspaces (or lines) of $\mathrm{AG}(2, p)$. By Lemma 3.8.8, there are $p(p+1)$ one-dimensional affine subspaces of $\mathrm{AG}(2, p)$. As the translations in $\mathrm{AGL}\left(1, p^2\right)$ and $\mathrm{AGL}(2, p)$ are the same, in order to show that $\mathrm{AGL}\left(1, p^2\right)$ is transitive on the lines of $\mathrm{AG}(2, p)$, it suffices to show that $\mathrm{AGL}\left(1, p^2\right)$ is transitive on the set of lines through the origin in $\mathrm{AG}(2, p)$. But this is clear as each line through the origin is the set of all multiples of some $(u, v) \in \mathbb{Z}_p^2$, and $\mathbb{F}_{p^2}^*$ is cyclic of order $p^2 - 1$ (Dummit and Foote, 2004, Proposition 9.5.18) (and so is transitive on $\mathbb{Z}_p^2 \backslash \{(0, 0)\}$).

Now, as $p = 2^t - 1$ is a Mersenne prime and there are $p(p+1) = p \cdot 2^t$ lines, any semiregular element of $\mathrm{AGL}\left(1, p^2\right)$ in its action on the lines of $\mathrm{AG}(2, p)$ must have order p or order 2. As an element ρ of order p is a translation, say $\rho(u, v) = (u, v) + (x, y)$, for $(x, y) \in \mathbb{F}_p^2$, we see ρ fixes the line $\{a(x, y) : a \in \mathbb{Z}_p\}$. Now let $\tau \in \mathrm{AGL}\left(1, p^2\right)$ be an element of order 2. Write $\tau(x) = ax + b$ with $a \in \mathrm{GL}\left(1, p^2\right) \cong \mathbb{F}_{p^2}^*$ and $b \in \mathbb{F}_{p^2}$. Then $\tau^2(x) = a^2 x + b + ab = x$, and so $\left(a^2 - 1\right)x = -b(a + 1)$ for all $x \in \mathbb{F}_{p^2}$. The only way this equation can be true for all x is if both sides are 0, in which case $a = -1$ or $b = 0$. If $a = -1$, then $\tau(x) = -x + b$. This function has fixed point $2^{-1}b$ (note 2^{-1} exists as p is odd) as $1 - 2^{-1} = 2^{-1}$ as $2\left(1 - 2^{-1}\right) = 1$. If $b = 0$, then $a = \pm 1$, and $f(x) = \pm x$, in which case $x = 0$ is a fixed point. □

3.9 Exercises

3.1.1 Show that a cycle is s-arc-transitive for any $s \geq 1$.

3.1.2 Show that an s-arc-transitive graph Γ is t-arc-transitive for all $1 \leq t \leq s$ provided that Γ has no vertices of valency 1.

3.1.3 Show that K_n is 2-arc-transitive but not 3-arc-transitive for any $n \geq 4$.

3.1.4 Show that any graph that contains an s-cycle and a path of length s (which of course cannot repeat vertices) is not s-arc-transitive. Using this fact show that neither the octahedron nor the icosahedron are 3-arc-transitive.

3.1.5 Show that the octahedron is arc-transitive but that it is not 2-arc-transitive.

3.1.6 Show that the icosahedron is arc-transitive but that it is not 2-arc-transitive.

3.1.7 Show that any permutation of the coordinates of $\mathbb{Z}_2^n$ is an automorphism of Q_n. Show that for $n \geq 3$, Q_n is 2-arc-transitive but that it is not 3-arc-transitive.

3.1.8 Show that the dodecahedron is 2-arc-regular.

3.1.9 Let Γ be an s-arc-transitive graph that is regular of valency k. Let $A = x_0, \ldots, x_s$ be an s-arc in Γ, and H the stabilizer of the s-arc A. Let $y_1, \ldots, y_{k-1}$ be the neighbors of x_s different from x_{s-1}. Show that if H is transitive on $y_1, \ldots, y_{k-1}$, then Γ is $(s+1)$-arc-transitive.

3.1.10 Show that the stabilizer of a vertex v of a 2-arc-transitive graph is 2-transitive on the neighbors of v.

3.1.11 Use Exercise 3.1.9 to prove Lemma 3.1.4.

3.2.1 For a vertex-transitive *graph* Γ, show that the following are equivalent:

(i) Γ is arc-transitive,
(ii) for $u \in V(\Gamma)$, $\{v : (u, v) \in A(\Gamma)\}$ is an orbit of $\mathrm{Stab}_{\mathrm{Aut}(\Gamma)}(u)$, and
(iii) for $u \in V(\Gamma)$, $\{v : (v, u) \in A(\Gamma)\}$ is an orbit of $\mathrm{Stab}_{\mathrm{Aut}(\Gamma)}(u)$.

3.2.2 Let G be a group with $H \leq G$. Show that if $S = HsH$ for some $s \in G$ and that $S = H\left\{t, t^{-1}\right\}H$ for some $t \in G$, then $S = H\left\{s, s^{-1}\right\}H$.

3.2.3 Prove Theorem 3.2.11.

3.2.4 Let Γ be a vertex-transitive graph, $G \leq \mathrm{Aut}(\Gamma)$ be transitive, and $v \in V(\Gamma)$. The permutation group induced by the action of $\mathrm{Stab}_G(v)$ on the neighbors of v is called the **local action of** G. Show that if Γ is connected and G has semiregular local action, then the action of G on the arcs of Γ is also semiregular.

3.2.5 Show that a vertex-transitive graph Γ is half-arc-transitive if and only if there is an orientation $\vec{\Gamma}$ of Γ such that $\mathrm{Aut}(\Gamma) = \mathrm{Aut}\left(\vec{\Gamma}\right)$.

3.2.6 Show that a group has at least one nontrivial self-paired orbital if and only if it has even order.

3.4.1 Show that $\mathrm{Cay}(A_4, \{(0,1,2), (0,2,1), (0,1)(2,3)\})$ is the truncation of K_4. Calculate its automorphism group.

3.4.2 Show that $\mathrm{Cay}(\mathcal{S}_4, \{(0,1,2), (0,2,1), (2,3)\})$ is the truncation of a cube.

3.4.3 Show that a connected vertex-transitive and edge-transitive graph Γ whose automorphism group is imprimitive with B a nontrivial block must satisfy $\Gamma[B] = \overline{K_B}$. Deduce that $T(T(\Gamma))$ is not vertex-transitive for any cubic connected vertex-transitive graph.

3.4.4 Show that the Petersen graph contains a Hamilton path.

3.4.5 Show that the Petersen graph is not 3-edge colorable.

3.4.6 Show that there is no Hamilton cycle in the Petersen graph that contains four spoke edges. Do not use the fact that the Petersen graph is not Hamiltonian to do this.

3.4.7 Use the facts that a group of order 15 is cyclic and A_5 is simple to show that $\mathcal{S}_5$ has no subgroup of order 30. Conclude that the truncation of the Petersen graph is not isomorphic to a Cayley graph.

3.5.1 Show that the Cartesian product of two Hamiltonian graphs is Hamiltonian.

3.5.2 Show that if Γ_1 and Γ_2 are graphs, then $\mathrm{Aut}(\Gamma_1 \square \Gamma_2) \geq \mathrm{Aut}(\Gamma_1) \times \mathrm{Aut}(\Gamma_2)$, but that equality need not be true.

3.5.3 Show that if Γ is Hamilton connected (or Hamilton laceable if Γ is bipartite) and of order at least 3, then every edge of Γ is contained in a Hamilton cycle (or in a Hamilton path if Γ is bipartite).

3.6.1 Find the block quotient digraph of $\mathrm{Cay}(\mathbb{Z}_{21}, \{1,3,5,16,20\})$ with respect to the block system of $(\mathbb{Z}_{21})_L$ with blocks of size 3.

3.6.2 Find the block quotient digraph of $\mathrm{Cay}(\mathbb{Z}_{26}, \{1,5,7,19\})$ with respect to the block system of $(\mathbb{Z}_{26})_L$ with blocks of size 2.

3.6.3 Show that a block quotient digraph of a connected digraph is connected.

3.6.4 The **unitary circulant graph** U_n of order n is Cay($\mathbb{Z}_n, U$), where U is the set of all units in $\mathbb{Z}_n$. Let p be prime, and $\mathcal{B}_i$ be the block system of $(\mathbb{Z}_{p^k})_L$ with blocks of size p^i, $0 \leq i \leq k$. Show that $U_{p^k}/\mathcal{B}_i = U_{p^{k-i}}$ for all $0 \leq i \leq k-1$ so that $U_{p^k}/\mathcal{B}_{p^{k-1}} = K_p$.

3.6.5 Show that a block quotient graph of a vertex- and edge-transitive graph has no edge with both endpoints inside any nontrivial block of any transitive subgroup of its automorphism group, and is vertex- and edge-transitive.

3.6.6 Give an example to show that the block quotient graph of a 2-arc-transitive graph need not be 2-arc-transitive.

3.6.7 Show that Lemma 3.6.10 implies the following result: Let Γ be a connected s-arc-transitive graph, $s \geq 2$. If Γ is a (G, s)-arc-transitive graph with $N \trianglelefteq G$, then one of the following is true:

(a) Γ is bipartite,
(b) N is transitive, or
(c) $\Gamma/\mathcal{B}$ is s-arc-transitive and $\mathrm{fix}_G(\mathcal{B})$ is semiregular.

3.7.1 Show that if Γ is Hamilton connected (or Hamilton laceable if Γ is bipartite), its automorphism group contains a semiregular element ρ such that between two orbits of ρ there are neither no edges nor a 1-factor, then Γ is Hamiltonian.

3.7.2 Let Γ be the $(q, p, \alpha, S_0, \ldots, S_{q-1})$-metacirculant graph, where q and p are distinct primes. Assume that $|\alpha| = q$, $1, p-1 \in S_0$ and $0 \in S_1$. Show that Γ is Hamiltonian.

3.7.3 Show that if Γ is a graph and Aut(Γ) contains a semiregular element ρ such that Γ/ρ is Hamiltonian and the subgraphs of Γ induced by the orbits of $\langle\rho\rangle$ are all Hamiltonian connected, then Γ is Hamiltonian. Observe that the subgraphs of Γ induced by the orbits of $\langle\rho\rangle$ are all isomorphic (and so all are Hamiltonian connected if one is) if there exists a transitive subgroup $G \leq \mathrm{Aut}(\Gamma)$ with $\langle\rho\rangle \trianglelefteq G$.

3.7.4 Draw the Heawood graph in Frucht notation.

3.7.5 Draw the $(5, 31, 2, \{1\}, \{2, 4\}, \{0\}, \{2, 6\}, \{5\})$-metacirculant digraph in Frucht notation.

3.7.6 Draw the $(5, 25, 6, \{1\}, \{2\}, \{0\}, \emptyset, \{2, 3, 22, 23\})$-metacirculant digraph in Frucht notation.

3.7.7 Draw the line graph of the Petersen graph $L(P)$ in Frucht notation, with ρ the $(3, 5)$-semiregular automorphism of $L(P)$ induced by the cycle $(1, 5, 3, 2, 4) \in S_5$.

3.7.8 Let Γ be the $(q, p, \alpha, S_0, \ldots, S_{q-1})$-metacirculant graph, where q and p are distinct primes. Assume that $|\alpha| = q$, $S_0 = \{1, p-1\}, S_1 = \{0\}$, and otherwise $S_i = \emptyset$. Draw Γ in Frucht notation.

3.8.1 Let p be prime and $k \geq 1$. Show that $\mathrm{AGL}(k, p)$ is 2-transitive.

3.8.2 Show that there is a unique line between any two points of $\mathrm{AG}(2, \mathbb{F})$.

3.8.3 Let Γ be a graph with semiregular automorphism ρ. Let B and B' be orbits of $\langle \rho \rangle$. Show that $\Gamma[B, B']$ is regular. Here, $\Gamma[B, B']$ is the subgraph of Γ with vertex set $B \cup B'$ and edge set those edges of Γ with one endpoint in B and the other in B'.

4

Graphs with Imprimitive Automorphism Group

In this chapter, we discuss basic techniques for analyzing groups that are imprimitive. Such a group has a natural coordinate system that lets one think of these groups in terms of two or more permutation groups of smaller degree. This can, in some circumstances, be a very effective method for solving problems about vertex-transitive digraphs by induction. However, the technique does not always work, as we will show.

We begin in Section 4.1 with a seemingly unrelated topic, the notion of "sameness" for permutation groups. Usually, notions of sameness, or isomorphism, are introduced in mathematics as soon as one has defined the relevant object. For us though, this is more complicated as we need notions of sameness for permutation groups and for actions. We then turn in Section 4.2 to what is arguably the most important operation in this area of graph theory, namely wreath products of graphs and digraphs. This operation was designed to produce graphs whose automorphism group is the group-theoretic wreath product of the automorphism groups of its factor graphs, and so we also introduce the group-theoretic wreath product. In Section 4.3, we prove the embedding theorem (Theorem 4.3.1), which is the result that gives the natural coordinate system for imprimitive groups mentioned above. The embedding theorem also gives that every imprimitive permutation group is a subgroup of the wreath product of two specific subgroups of smaller degree, and so from the point of view of using the embedding theorem, the "smallest" transitive groups are those of prime degree. These groups, along with vertex-transitive digraphs of prime order, are the subject of Section 4.4. We will more or less have encyclopedic knowledge of these groups and digraphs. Additionally, in this section we go ahead and calculate the normalizers of permutation groups in the symmetric group in Lemma 4.4.9, as well as considering some related topics. Our final goal in this chapter is to show that this "reduction to smaller groups" approach for imprimitive groups does not work all of the time. We will show in Section 4.5 that the automorphism group of the Heawood graph has automorphism group for which one of the "smaller" groups is not the automorphism group of a digraph – and is 2-transitive but not a symmetric group! So while the natural inductive approach to imprimitive groups given by the embedding theorem can work with imprimitive groups, this is not always the case, as the smaller groups need not be automorphism groups of graphs or digraphs.

4.1 Permutation Isomorphism and Equivalent Representations

While it is customary to discuss the mathematical issue of isomorphism immediately after defining the object under study, we have postponed our discussion of isomorphism of permutation groups as from one point of view it is more subtle than, say, the notion of graph isomorphism. This is because when working with permutation groups, we also work with actions, for example, the induced action of an imprimitive group on its blocks. Where confusion can arise is that there is also a notion of isomorphism for actions that, when one studies permutation groups, is usually couched in the language of permutation representations, and these two notions are not the same. Also, the basic examples of the primary differences between these two notions require some effort, and so are not appropriate for, say, Chapter 1 of this book. We also warn the reader that terminology is not completely standard in this area – this will be discussed in more detail once we have defined all of the terms.

We begin with isomorphisms of permutation groups. As a permutation group is a subgroup G of $\mathcal{S}_X$ permuting a set X, the main notion of "sameness" for permutation groups involves a "relabeling" of the group together with a "relabeling" of the points, that is, the elements of X.

Definition 4.1.1 Let $G \leq \mathcal{S}_X$ and $H \leq \mathcal{S}_Y$. Then G and H are **permutation isomorphic** if there exists a bijection $\lambda\colon X \to Y$ and a group isomorphism $\phi\colon G \to H$ such that $\lambda(g(x)) = \phi(g)(\lambda(x))$ for all $x \in X$ and $g \in G$.

So this definition says that relabeling after applying g to x is the same as applying $\phi(g)$ to x after it has been relabeled.

Clearly, if one permutation group $H \leq \mathcal{S}_X$ is obtained from another $G \leq \mathcal{S}_X$ by simply relabeling X with other elements of X, then G and H are permutation isomorphic as the relabeling gives rise to a bijection $\lambda\colon X \to X$ with $\phi\colon \mathcal{S}_X \to \mathcal{S}_X$ being conjugation by λ, as then $\lambda(g(x)) = \lambda g \lambda^{-1}(\lambda(x))$. Our first result gives a group-theoretic characterization of permutation isomorphic groups in the same symmetric group.

Theorem 4.1.2 *Let $G, H \leq \mathcal{S}_X$. Then G and H are permutation isomorphic if and only if G and H are conjugate in $\mathcal{S}_X$.*

Proof Suppose G and H are conjugate in $\mathcal{S}_X$, with $\lambda \in \mathcal{S}_X$ such that $\lambda G \lambda^{-1} = H$. Clearly $\lambda\colon X \to X$ is a bijection, and conjugation by λ induces an isomorphism $\phi\colon G \to H$ given by $\phi(g) = \lambda g \lambda^{-1}$. Then

$$\lambda(g(x)) = \lambda\left(g\lambda^{-1}\lambda(x)\right) = \lambda g \lambda^{-1}(\lambda(x)) = \phi(g)(\lambda(x)),$$

and G and H are permutation isomorphic.

Conversely, suppose G and H are permutation isomorphic. Then there exists a bijection $\lambda\colon X \to X$ and a group automorphism $\phi\colon G \to H$ such that $\lambda(g(x)) = \phi(g)(\lambda(x))$ for every $g \in G$ and $x \in X$. Then $\lambda \in \mathcal{S}_X$. As $\lambda, g, \phi(g) \in \mathcal{S}_X$, $\lambda g = \phi(g)\lambda$ or $\lambda g \lambda^{-1} = \phi(g)$. Thus $\lambda G \lambda^{-1} = \phi(G) = H$. □

Because of this result, one will often see statements to the effect of "permutation isomorphic groups are the same up to a relabeling of the points" as, of course, conjugation in the same symmetric group *is* simply a relabeling of the points. The statement that "permutation isomorphic groups are the same up to a relabeling of the points" is true, but one must be careful how that is interpreted. Let us choose a specific permutation group to make our discussion more concrete. We will choose $\mathcal{S}_6$ for reasons that will be clear at the end of this section. Before proceeding we will need some common terms.

Definition 4.1.3 Let G be a group. An automorphism $\alpha \in \mathrm{Aut}(G)$ is an **inner automorphism of** G if there exists $h \in G$ such that $\alpha(g) = hgh^{-1}$ for all $g \in G$. The set of inner automorphisms is a normal subgroup of $\mathrm{Aut}(G)$, denoted $\mathrm{Inn}(G)$ – see Exercise 4.5.8. An automorphism $\alpha \in \mathrm{Aut}(G)$ that is not an inner automorphism is an **outer automorphism of** G.

Let us choose a permutation isomorphism from $\mathcal{S}_6$ to $\mathcal{S}_6$. So we have an isomorphism $\phi\colon \mathcal{S}_6 \to \mathcal{S}_6$ and a bijection $\lambda\colon \mathbb{Z}_6 \to \mathbb{Z}_6$ that satisfies the defining equation. The image of $\mathcal{S}_6$ under our permutation isomorphism is, of course, $\mathcal{S}_6$ as a group. We could do nothing in a complicated way by choosing λ in any fashion, and then choosing ϕ to be the inner automorphism of $\mathcal{S}_6$ induced by conjugation by λ. That is, $\phi(g) = \lambda g \lambda^{-1}$. The defining equation is then $\lambda(g) = \lambda(g)$.

There could be an even more complicated situation. What if $\mathcal{S}_6$ has an outer automorphism ϕ that, say, maps the element $g = (0, 1, 2)(3, 4, 5)$ to $(1, 3, 5)$? If this were possible (and it is, as we shall see!), the image of $\mathcal{S}_6$ under such a permutation isomorphism (with any choice of λ) will be $\mathcal{S}_6$ and so we could relabel the points as in Theorem 4.1.2 to obtain $\mathcal{S}_6$. However, we cannot conjugate $\phi(g)$ to obtain g as permutations are conjugate in $\mathcal{S}_n$ if and only if their cycle types are the same (Dummit and Foote, 2004, Proposition 4.3.11). So, we have control with permutation isomorphism of permutation groups as sets using conjugation, but we do not have control over which elements are mapped to which elements using conjugation if ϕ happens to be an outer automorphism of $\mathcal{S}_6$.

Our next result shows that we have already encountered many permutation groups that are permutation isomorphic.

Theorem 4.1.4 *Let $G \leq \mathcal{S}_X$ be a transitive group with block system $\mathcal{B}$, and $B, B' \in \mathcal{B}$. Then $\mathrm{Stab}_G(B)^B$ is permutation isomorphic to $\mathrm{Stab}_G(B')^{B'}$. Additionally, $\mathrm{fix}_G(\mathcal{B})^B$ is permutation isomorphic to $\mathrm{fix}_G(\mathcal{B})^{B'}$.*

Proof Let $B, B' \in \mathcal{B}$. Let $\ell \in G$ such that $\ell(B) = B'$. Define $\lambda\colon B \to B'$ by $\lambda(x) = \ell(x)$. As ℓ maps B bijectively to B', λ is a bijection. Define $\phi\colon \mathrm{Stab}_G(B) \to \mathrm{Stab}_G(B')$ by $\phi(g) = \ell g \ell^{-1}$. As ϕ is obtained by conjugation, ϕ is a group isomorphism. Let $g \in \mathrm{Stab}_G(B)$, and $x \in B$. Then

$$\lambda(g(x)) = \ell g(x) = \ell g \ell^{-1} \ell(x) = \phi(g)\lambda(x),$$

and so $\mathrm{Stab}_G(B)^B$ is permutation isomorphic to $\mathrm{Stab}_G(B')^{B'}$. Analogous arguments will show that the action of $\mathrm{fix}_G(\mathcal{B})^B$ on B is permutation isomorphic to $\mathrm{fix}_G(\mathcal{B})^{B'}$. □

We now consider the more general question of when two actions on the same group G are "the same." As an action induces a permutation representation from the group in a symmetric group and vice versa, a proper context to consider this idea is using the language of permutation representations.

Definition 4.1.5 Let G be a group, and X and Y be sets. Let $\beta\colon G \to \mathcal{S}_X$ and $\delta\colon G \to \mathcal{S}_Y$ be permutation representations of G. We say β and δ are **equivalent permutation representations** of G if there exists a bijection $\lambda\colon X \to Y$ such that $\lambda(\beta(g)(x)) = \delta(g)(\lambda(x))$ for all $x \in X$ and $g \in G$. In this case, we will say that $\beta(G)$ and $\delta(G)$ are **permutation equivalent**.

As the notion of an action is a generalization of the notion of a permutation group, we would expect that if the actions are "the same," then the corresponding permutation groups should be the "the same." Our next result verifies this, showing that equivalent permutation representations of a group G are permutation isomorphic.

Proposition 4.1.6 *Let G be a group, and β and δ be equivalent permutation representations of G. Then $\beta(G)$ is permutation isomorphic to $\delta(G)$.*

Proof Let X and Y be sets, with $\beta\colon G \to S_X$ and $\delta\colon G \to S_Y$ permutation representations of G. Let $\lambda\colon X \to Y$ be such that $\lambda(\beta(g)(x)) = \delta(g)(\lambda(x))$ for all $g \in G$ and $x \in X$. Set $\phi = \delta\beta^{-1}$. Then

$$\lambda(\beta(g)(x)) = \delta(g)(\lambda(x)) = \delta\beta^{-1}\beta(g)(\lambda(x)) = \phi(\beta(g))(\lambda(x)),$$

and $\beta(G)$ is permutation isomorphic to $\delta(G)$. □

To compare permutation isomorphism and permutation equivalence, write the defining equation for permutation isomorphism using the permutation representations β and δ as above. The defining equation becomes $\lambda(\beta(g)(x)) = \phi(\delta(g))(\lambda(x))$. It is apparent that equivalent permutation groups is the special case of permutation isomorphism when $\phi = 1$. With this comparison in mind, we have the following result, which is not surprising in hindsight (and should be compared to Theorem 4.1.2).

Theorem 4.1.7 *Let β and δ be two permutation representations of the group G into the same symmetric group S_X. If β and δ are equivalent permutation representations then there exists $\gamma \in S_X$ such that $\gamma^{-1}\beta(g)\gamma = \delta(g)$ for all $g \in G$.*

Proof This follows easily from the definition by observing that $\lambda(\beta(g)(x)) = \delta(g)(\lambda(x))$ for all $x \in X$ and $g \in G$ gives $\lambda\beta(g) = \delta(g)\lambda$ and the result follows with $\gamma = \lambda^{-1}$. □

The next theorem gives necessary and sufficient conditions for two isomorphic transitive permutation groups of the same degree to be equivalent permutation representations of a group G. In practice, it gives the first method one should try to determine if two such permutation groups are permutation equivalent. Note that if two permutation representations of a group G have different degrees, then they cannot be permutation equivalent as there is no bijection $\lambda\colon X \to Y$.

Theorem 4.1.8 *Let G be a group and X and Y be sets of the same order. Let β and δ be faithful and transitive permutation representations of G into S_X, and S_Y, respectively. Then β and δ are equivalent permutation representations of G if and only if the preimage of the stabilizer of a point in $\beta(G)$ is the preimage of the stabilizer of a point in $\delta(G)$.*

Proof Let $K \leq G$ be such that $\beta(K) = \mathrm{Stab}_{\beta(G)}(z)$, where $z \in X$.

Suppose $\beta(G)$ is permutation equivalent to $\delta(G)$, with $\lambda\colon X \to Y$ a bijection such that $\lambda(\beta(g)(x)) = \delta(g)(\lambda(x))$ for all $x \in X$ and $g \in G$. Let $k \in K$. Then $z = \beta(k)(z)$ implies $\lambda(z) = \lambda(\beta(k)(z)) = \delta(k)(\lambda(z))$ and so $\delta(k)$ stabilizes $\lambda(z)$. Hence $\delta(K) \leq \mathrm{Stab}_{\delta(G)}(\lambda(z))$. As $|\beta(K)| = |\delta(K)| = |\mathrm{Stab}_{\delta(G)}(y)|$ for $y \in Y$, we have $\delta(K) = \mathrm{Stab}_{\delta(K)}(\lambda(z))$.

Conversely, suppose $\delta(K) = \mathrm{Stab}_{\delta(G)}(y)$ for some $y \in Y$. Define $\lambda\colon X \to Y$ by $\lambda(\beta(g)(z)) = \delta(g)(y)$, where $g \in G$. We claim that λ is well defined. Indeed, if $\beta(f)(z) = \beta(g)(z)$, then $\beta\left(f^{-1}g\right)(z) = z$, so $f^{-1}g \in K$ and $\delta\left(f^{-1}g\right)$ stabilizes y. This gives $\delta(f)(y) = \delta(g)(y)$ and λ is well defined. Note λ has domain X and is onto as both $\beta(G)$ and $\delta(G)$ are transitive on X and Y, respectively. If $\lambda(\beta(f)(z)) = \lambda(\beta(g)(z))$ then $\delta(f)(y) = \delta(g)(y)$ so $\delta\left(f^{-1}g\right)$ stabilizes y. Then $\beta\left(f^{-1}g\right)$ stabilizes z and so $\beta(f)(z) = \beta(g)(z)$. Thus λ is a bijection.

Finally, for each $x \in X$ there exists $g_x \in G$ with $\beta(g_x)(z) = x$. Then

$$\lambda(\beta(g)(x)) = \lambda(\beta(g)\beta(g_x)(z)) = \lambda(\beta(gg_x)(z)) = \delta(gg_x)(y) = \delta(g)\delta(g_x)(y).$$

As $\lambda(x) = \lambda(\beta(g_x)(z)) = \delta(g_x)(y)$, we see that $\lambda(\beta(g)(x)) = \delta(g)(\lambda(x))$. □

Now that we have seen permutation isomorphism and equivalent permutation representations, we note that the notation for these ideas is not completely uniform, and for a good reason. The reason is quite simply that both of these ideas are not needed in many applications. For example, if one is only concerned with transitive permutation groups, then there is no need for the less general notion of permutation equivalence. We, however, will need the notion of permutation equivalence to analyze $\mathrm{fix}_G(\mathcal{B})$ for an imprimitive permutation group G with block system $\mathcal{B}$. If $\mathrm{fix}_G(\mathcal{B}) \neq 1$ and acts faithfully on each $B \in \mathcal{B}$, then $\mathrm{fix}_G(\mathcal{B})^B \cong \mathrm{fix}_G(\mathcal{B})$ is permutation isomorphic to $\mathrm{fix}_G(\mathcal{B})^{B'} \cong \mathrm{fix}_G(\mathcal{B})$ for every $B, B' \in \mathcal{B}$, but it need not be the case that the permutation representations of $\mathrm{fix}_G(\mathcal{B})$ on different blocks of $\mathcal{B}$ are equivalent permutation representations. Our goal in the next section is to show that this is exactly what happens when G is the automorphism group of the Heawood graph.

In the next two theorems, we give necessary and sufficient conditions for the left coset action of G on the left cosets of two of its subgroups to be permutation isomorphic and equivalent, respectively. We need a preliminary result.

Lemma 4.1.9 *Let G be a group, $H \leq G$, and $\alpha \in \mathrm{Aut}(G)$. Set $K = \alpha(H)$. Then $\bar{\alpha}\colon G/H \to G/K$ by $\bar{\alpha}(gH) = \alpha(g)K$ is a well-defined bijection.*

Proof To show that $\bar{\alpha}$ is well defined, suppose $g_1, g_2 \in G$ and $g_1H = g_2H$. As $\alpha(H) = K$ and $\alpha \in \mathrm{Aut}(G)$, $\alpha(g_iH) = \alpha(g_i)K$ for $i = 1, 2$, and $\bar{\alpha}$ maps G/H to G/K. Then $\alpha(g_1)K = \alpha(g_1H) = \alpha(g_2H) = \alpha(g_2)K$, and $\bar{\alpha}$ is well defined. That $\bar{\alpha}$ is onto follows as α is onto. Finally, if $\bar{\alpha}(g_1H) = \bar{\alpha}(g_2H)$ then $\alpha(g_1)K = \alpha(g_2)K$, so $\alpha\left(g_2^{-1}g_1\right)K = K$ and $\alpha\left(g_2^{-1}g_1\right) \in K$. As $\alpha(H) = K$, $g_2^{-1}g_1 \in H$ and $g_1H = g_2H$. Thus $\bar{\alpha}$ is one-to-one, and so a bijection. □

Theorem 4.1.10 *Let G be a group with core-free subgroups H and K. Then $(G, G/H)$ is permutation isomorphic to $(G, G/K)$ if and only if there is $\alpha \in \mathrm{Aut}(G)$ with $\alpha(H) = K$.*

Proof If $(G, G/H)$ is permutation isomorphic to $(G, G/K)$, then there exists a bijection $\lambda\colon G/H \to G/K$ and an isomorphism $\phi\colon (G, G/H) \to (G, G/K)$ such that $\lambda(\hat{g}_L(aH)) = \widehat{\phi(g)}_L(\lambda(aH))$ for all $a, g \in G$. As $H = \mathrm{Stab}_{(G,G/H)}(H)$ and $K = \mathrm{Stab}_{(G,G/K)}(K)$ are core-free in G, $(G, G/H)$ and $(G, G/K)$ are both faithful permutation representations of G by Proposition 1.3.8. This gives that $|\mathrm{Stab}_{(G,G/H)}(H)| = |H|$ and $|\mathrm{Stab}_{(G,G/K)}(K)| = |K|$. As λ is a bijection, $|G/H| = |G/K|$, and so $|H| = |K|$. If $\hat{g}_L(aH) = aH$, then

$$\lambda(aH) = \lambda(\hat{g}_L(aH)) = \widehat{\phi(g)}_L(\lambda(aH)) = \phi(g)\lambda(aH),$$

and so $\phi(g) \in \lambda(aH)$. Thus $\phi(H) = \lambda(H)$, and as $1 \in \phi(H)$, $\lambda(H) = K$. Hence $\phi(H) = K$. As $(G, G/H)$ and $(G, G/K)$ are canonically isomorphic to G by the maps π_H and π_K with $\pi_H(\hat{g}_L) = g$ and $\pi_K(\hat{g}_L) = g$ (note that the two $\hat{g}_L$'s are different as their domains are different), there is an automorphism $\alpha\colon G \to G$ given by $\alpha(g) = \pi_K^{-1}\phi\pi_H^{-1}$ and $\alpha(H) = K$.

Conversely, if there exists $\alpha \in \mathrm{Aut}(G)$ with $\alpha(H) = K$ then, by Lemma 4.1.9, α induces a bijection $\lambda\colon G/H \to G/K$ given by $\lambda(aH) = \alpha(a)K$. Define $\phi\colon (G, G/H) \to (G, G/K)$ by $\phi(\hat{g}_L) = \widehat{\alpha(g)}_L$. It is straightforward to verify that ϕ is an isomorphism as α is an isomorphism. Then, for all $a, g \in G$, we have

$$\begin{aligned}\lambda(\hat{g}_L(aH)) &= \lambda(gaH) = \alpha(ga)K = \alpha(g)\alpha(a)K\\ &= \alpha(g)(\lambda(aH)) = \widehat{\alpha(g)}_L(\lambda(aH)) = \phi(\hat{g}_L)(\lambda(aH)),\end{aligned}$$

and $(G, G/H)$ is permutation isomorphic to $(G, G/K)$. □

Theorem 4.1.11 *Let G be a group with core-free subgroups H and K. The permutation representations of G whose images are $(G, G/H)$ and $(G, G/K)$ are equivalent permutation representations of G if and only if H and K are conjugate in G, or equivalently, if and only if there exists $\alpha \in \mathrm{Inn}(G)$ such that $\alpha(H) = K$.*

Proof Suppose $x \in G$ with $x^{-1}Hx = K$. Define $\lambda\colon G/H \to G/K$ by $\lambda(gH) = gHx = gxK$. If $\lambda(gH) = \lambda(g'H)$, then $gHx = g'Hx$ and $gH = g'H$. Thus λ is well defined and injective. Also, λ is onto as if $gK \in G/K$, then $gx^{-1} \in G$ and $\lambda\left(gx^{-1}H\right) = gx^{-1}xK = gK$, so λ is a bijection. Finally, $g(\lambda(aH)) = gaHx = \lambda(g(aH))$ for all $g \in G$ and $aH \in G/H$, and the result follows.

Conversely, suppose the permutation representations of G whose images are $(G, G/H)$ and $(G, G/K)$ are equivalent. As both $(G, G/H)$ and $(G, G/K)$ are transitive, by Theorem 4.1.8 K is the stabilizer of a point in the permutation representation of the left coset action of G on G/H, and so by Theorem 1.1.7 we see that H and K are conjugate in G. □

The previous two results give necessary and sufficient conditions for the construction of faithful and transitive inequivalent permutation representations that are permutation isomorphic.

Corollary 4.1.12 *Let G be a group with a core-free subgroup H, and $\alpha \in \mathrm{Aut}(G)$. Then $(G, G/H)$ and $(G, G/\alpha(H))$ are faithful and transitive permutation isomorphic permutation groups whose permutation representations as images of G are inequivalent if and only if α is an outer automorphism of G such that $\alpha(H)$ is not conjugate to H in G.*

The following result is a version of Theorem 2.4.13 using the language of permutation isomorphism.

Theorem 4.1.13 *Let G be a group and $H \le G$. The left coset action of G on G/H induces a transitive permutation group $(G, G/H)$ on the set of left cosets of H in G. Conversely, let $G \le S_n$ be a transitive group with $H = \mathrm{Stab}_G(0)$. Then G is permutation isomorphic to $(G, G/H)$.*

Proof By Theorem 2.4.13, we need only show the converse. For each $x \in \mathbb{Z}_n$ choose $g_x \in G$ with $g_x(0) = x$. We also know that for every $a \in G$ and $x \in \mathbb{Z}_n$, we have

$$a(x) = y \text{ if and only if } ag_xH = g_yH.$$

Define $\lambda\colon \mathbb{Z}_n \to (G, G/H)$ by $\lambda(x) = g_xH$ and let $\phi\colon G \to G$ be the identity. Then for $a \in G$ and $x \in \mathbb{Z}_n$, we have

$$\lambda(a(x)) = g_{a(x)}H = ag_xH = \phi(a)(\lambda(x)),$$

and the result follows. □

Combining the two previous results, we can now count the number of inequivalent (that is, not equivalent) permutation representations of G that are permutation isomorphic to $(G, G/H)$.

Corollary 4.1.14 *Let G be a group with a core-free subgroup H. The number of inequivalent permutation representations of G that are permutation isomorphic to $(G, G/H)$ is equal to the number of subgroups of G isomorphic to H by an outer automorphism of G but not by an inner automorphism of G.*

Proof By Theorem 4.1.13, up to permutation isomorphism any representation of G that is inequivalent to the representation of G whose image is $(G, G/H)$ is of the form $(G, G/K)$ for some subgroup $K \leq G$. The result follows now by Corollary 4.1.12. □

We now give one of the standard examples of a permutation group with two inequivalent representations, namely the two inequivalent permutation representations of $\mathcal{S}_6$. In the next section we will give another standard example in the context of the automorphism group of the Heawood graph.

Example 4.1.15 The group $\mathcal{S}_6$ has exactly two inequivalent permutation representations of degree 6, and $\mathrm{Aut}(\mathcal{S}_6)/\mathrm{Inn}(\mathcal{S}_6) \cong \mathbb{Z}_2$.

Solution The left coset action of $\mathcal{S}_6$ on the left cosets of any subgroup $H \leq \mathcal{S}_6$ isomorphic to $\mathcal{S}_5$ is core-free in $\mathcal{S}_6$ as A_6 is the only proper nontrivial normal subgroup of $\mathcal{S}_6$ by Exercise 2.4.13. Thus the left coset action of $\mathcal{S}_6$ on $\mathcal{S}_6/H$ is faithful by Proposition 1.3.8. We will now give a subgroup of $\mathcal{S}_6$ isomorphic to $\mathcal{S}_5$ that does not fix a point. By a Sylow theorem, $\mathcal{S}_5$ has exactly six Sylow 5-subgroups (this can also be easily obtained by counting; see Exercise 4.5.2). As for a group G with subgroup H, the number of subgroups of G conjugate in G to H is $[N_G(H) : H]$ (Dummit and Foote, 2004, Proposition 4.3.6), $N_{\mathcal{S}_5}((\mathbb{Z}_5)_L) = \mathrm{AGL}(1,5)$ as it has order 20 and normalizes $(\mathbb{Z}_5)_L$. As $\mathcal{S}_5$ has exactly six Sylow 5-subgroups, and as any two Sylow 5-subgroups of $\mathcal{S}_5$ are conjugate in $\mathcal{S}_5$, $\mathcal{S}_5$ acts transitively by conjugation on its six Sylow 5-subgroups. Also, a subgroup of $\mathcal{S}_5$ that normalizes a Sylow 5-subgroup of $\mathcal{S}_5$ is isomorphic to a subgroup of $\mathrm{AGL}(1,5)$. The only proper nontrivial normal subgroup of $\mathcal{S}_5$ is A_5 by Exercise 2.4.13, and so $\mathrm{AGL}(1,5)$ is core-free in $\mathcal{S}_5$. Let $\beta\colon \mathcal{S}_5 \to \mathcal{S}_6$ be the permutation representation of $\mathcal{S}_5$ in $\mathcal{S}_6$ corresponding to the conjugation action of $\mathcal{S}_5$ on its Sylow 5-subgroups, so the stabilizer of a Sylow p-subgroup in $\beta(\mathcal{S}_5)$ is isomorphic to $\mathrm{AGL}(1,5)$. As $\mathrm{AGL}(1,5)$ is core-free in $\mathcal{S}_5$, $\beta(\mathcal{S}_5) \cong \mathcal{S}_5$ and is transitive. Then $\mathcal{S}_5$ viewed as the stabilizer of a point in $\mathcal{S}_6$ is not conjugate to $\beta(\mathcal{S}_5)$ by any element of $\mathcal{S}_6$, and so the left coset action of $\mathcal{S}_6$ on $\mathcal{S}_6/\mathcal{S}_5$ (stabilizing a point) is permutation inequivalent to the left coset action of $\mathcal{S}_6$ on $\mathcal{S}_6/\beta(\mathcal{S}_5)$ by Corollary 4.1.12.

It remains to show that the only inequivalent actions of $\mathcal{S}_6$ are the two in the previous paragraph, and $\mathrm{Aut}(\mathcal{S}_6)/\mathrm{Inn}(\mathcal{S}_6) \cong \mathbb{Z}_2$. By Corollary 4.1.14, it suffices to show that $\mathrm{Aut}(\mathcal{S}_6)/\mathrm{Inn}(\mathcal{S}_6) \cong \mathbb{Z}_2$. So suppose $\alpha \in \mathrm{Aut}(\mathcal{S}_6)$ and $\alpha \notin \mathrm{Inn}(\mathcal{S}_6)$. As two permutations in $\mathcal{S}_n$ have the same cycle type if and only if they are conjugate in $\mathcal{S}_n$ (Dummit and Foote, 2004, Proposition 4.3.11), α must map some permutation ω in $\mathcal{S}_6$ to an element δ of different cycle type. Let m be

Cycle Decomposition	Order	Number of Elements in $\mathcal{S}_6$
(1)	1	1
$(1,2,3,4,5,6)$	6	120
$(1,2,3)(4,5)$	6	120
$(1,2,3,4,5)$	5	144
$(1,2,3,4)(5,6)$	4	90
$(1,2,3,4)$	4	90
$(1,2,3)(4,5,6)$	3	40
$(1,2,3)$	3	40
$(1,2)(3,4)(5,6)$	2	15
$(1,2)(3,4)$	2	45
$(1,2)$	2	15

Table 4.1 *The conjugacy classes of elements of $\mathcal{S}_6$.*

the number of elements in a conjugacy class of $\mathcal{S}_6$. Note that there are at most two different cycle types of elements of $\mathcal{S}_6$ with m elements of that cycle type by Table 4.1. As an automorphism of $\mathcal{S}_6$ must map a conjugacy class with m elements to a conjugacy class with m elements, we see that α^2 preserves the cycle type of each element of $\mathcal{S}_6$. Then $\text{Aut}(\mathcal{S}_6)/\text{Inn}(\mathcal{S}_6) \cong \mathbb{Z}_2$. □

The method used in the above example was taken mainly from Janusz and Rotman (1982), where one can also find other arguments establishing that $\mathcal{S}_6$ has an outer automorphism of order 2 as well as explicit constructions of an outer automorphism of $\mathcal{S}_6$. The interested reader should also see Howard et al. (2008). In general, it is known that $\mathcal{S}_6$ is the only symmetric group $\mathcal{S}_n$ with $n \geq 4$ with a nontrivial outer automorphism – see, for example, Passman (1968, Theorem 5.7).

4.2 Wreath Products of Digraphs and Groups

Wreath products of graphs were originally introduced by Harary (1959), where the focus was on calculating the automorphism group of the wreath product of two graphs. The wreath product of two digraphs is defined as follows.

Definition 4.2.1 Let Γ_1 and Γ_2 be digraphs. The **wreath product of Γ_1 and Γ_2**, denoted $\Gamma_1 \wr \Gamma_2$, is the digraph with vertex set $V(\Gamma_1) \times V(\Gamma_2)$ and arcs $((u,v),(u,v'))$ for $u \in V(\Gamma_1)$ and $(v,v') \in A(\Gamma_2)$ or $((u,v),(u',v'))$ where $(u,u') \in A(\Gamma_1)$ and $v, v' \in V(\Gamma_2)$.

Some authors refer to the wreath product of digraphs as the **lexicographic product** (presumably because the order in which the digraphs are listed matters up to isomorphism), and denote it by $\Gamma_1[\Gamma_2]$. In older references, it is sometimes called **composition**.

Example 4.2.2 The graph $K_2 \wr \bar{K}_2$ given in Figure 4.1 is a 4-cycle with automorphism group isomorphic to D_8.

Solution Notice that the vertex set of $K_2 \wr \bar{K}_2$ is $\mathbb{Z}_2 \times \mathbb{Z}_2$ as both K_2 and its complement have vertex set $\mathbb{Z}_2$. There is no edge between $(0,0)$ and $(0,1)$ or $(1,0)$ and $(1,1)$ as $\bar{K}_2$ has no edges while every vertex in $\{(0,0),(0,1)\}$ is adjacent to every vertex in $\{(1,0),(1,1)\}$ as there is an edge between 0 and 1 in K_2. The rest of this example is now apparent. □

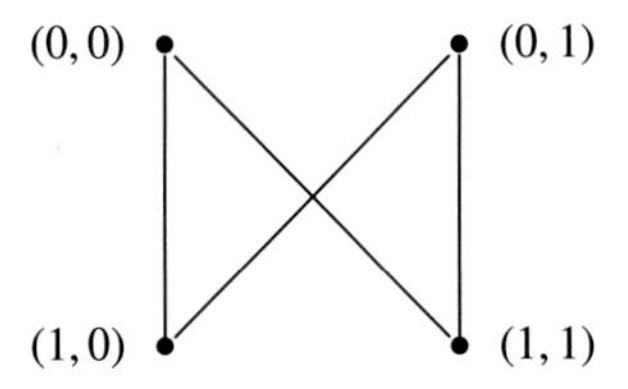

Figure 4.1 The graph $K_2 \wr \bar{K}_2$.

Intuitively, $\Gamma_1 \wr \Gamma_2$ is constructed as follows. First, we have $|V(\Gamma_1)|$ copies of the digraph Γ_2, with these $|V(\Gamma_1)|$ copies indexed by elements of $V(\Gamma_1)$. Next, between corresponding copies of Γ_2 we place every possible directed arc from one copy to another if in Γ_1 there is an arc from one vertex to the other, and no arcs otherwise.

Example 4.2.3 The graph $K_3 \wr \bar{K}_2$ is drawn in Figure 4.2.

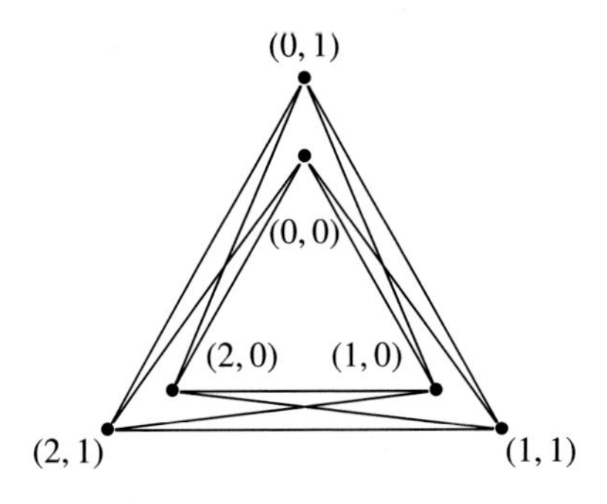

Figure 4.2 The graph $K_3 \wr \bar{K}_2$.

Solution As $V(K_3) = \mathbb{Z}_3$ and $V(\bar{K}_2) = \mathbb{Z}_2$, the vertex set of $K_3 \wr \bar{K}_2$ is $\mathbb{Z}_3 \times \mathbb{Z}_2$. The sets $B_0 = \{(0,0),(0,1)\}$, $B_1 = \{(1,0),(1,1)\}$, and $B_2 = \{(2,0),(2,1)\}$ are independent sets as $\bar{K}_2$ has no edges. Every vertex of B_i is adjacent to every vertex of B_j, $i \neq j$, as K_3 is a complete graph. □

Example 4.2.4 The graph $C_4 \wr C_4$ is drawn in Figure 4.3, where C_4 is a 4-cycle.

Solution As $V(C_4) = \mathbb{Z}_4$, the vertex set of $C_4 \wr C_4$ is $\mathbb{Z}_4 \times \mathbb{Z}_4$. The sets $B_i = \{(i, j) : j \in \mathbb{Z}_4\}$ are induced 4-cycles, while every vertex of B_i is adjacent to every vertex of B_{i+1}, $i \in \mathbb{Z}_4$, as $C_4 = \mathrm{Cay}(\mathbb{Z}_4, \{\pm 1\})$. □

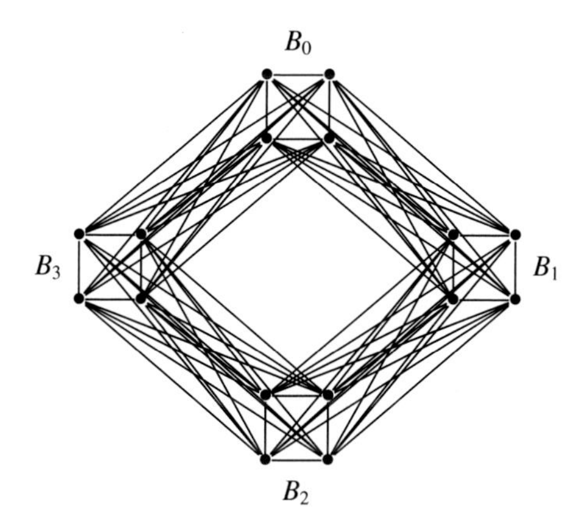

Figure 4.3 The graph $C_4 \wr C_4$.

Note that if Γ_1 or Γ_2 has one vertex, then $\Gamma_1 \wr \Gamma_2 \cong \Gamma_1$ or Γ_2. Such a wreath product is called **trivial**.

There is a corresponding "wreath product" for permutation groups, which as we shall see is intimately related to the automorphism group of the wreath product of two digraphs.

Definition 4.2.5 Let $G \leq S_X$ and $H \leq S_Y$. Define the **wreath product of** G **and** H, denoted $G \wr H$, to be the set of all permutations of $X \times Y$ of the form $(x, y) \mapsto (g(x), h_x(y))$, where $g \in G$ and each $h_x \in H$.

It is straightforward to verify that $G \wr H$ is a group. Additionally, the definition of the group wreath product given above is not the most general definition of the wreath product of two (abstract) groups, and some authors

a	b_0	b_1	Permutation	Element of D_8
0	0	0	1	1
0	0	1	$((1,0),(1,1))$	μ
0	1	0	$((0,0),(0,1))$	$\mu\rho^2$
0	1	1	$((0,0),(0,1))((1,0),(1,1))$	ρ^2
1	0	0	$((0,0),(1,0))((0,1),(1,1))$	$\mu\rho^3$
1	0	1	$((0,0),(1,0),(0,1),(1,1))$	ρ
1	1	0	$((0,0),(1,1),(0,1),(1,0))$	ρ^3
1	1	1	$((0,0),(1,1))((0,1),(1,0))$	$\mu\rho$

Table 4.2 *The elements of* $S_2 \wr S_2$.

call this wreath product the **permutational wreath product**. Finally, some authors reverse the order of G and H when denoting the wreath product.

Example 4.2.6 Consider $S_2 \wr S_2 = (\mathbb{Z}_2)_L \wr (\mathbb{Z}_2)_L$. This group permutes the set $\mathbb{Z}_2 \times \mathbb{Z}_2$ and so is isomorphic to a subgroup of S_4. By definition, this is a group of all permutations of $\mathbb{Z}_2 \times \mathbb{Z}_2$ of the form $(x, y) \mapsto (x + a, j + b_i)$, where $a, b_0, b_1 \in \mathbb{Z}_2$. As there are two choices for each of a, b_0, and b_1, we see that $S_2 \wr S_2$ has order 8. As $|S_4| = 24$, $S_2 \wr S_2$ is a Sylow 2-subgroup of S_4, as is D_8. So $S_2 \wr S_2 \cong D_8$.

Consider the element ρ of $S_2 \wr S_2$ obtained by choosing $a = 1$, $b_0 = 0$, and $b_1 = 1$. It has cycle decomposition $((0,0),(1,0),(0,1),(1,1))$ as in the image of $(0,0)$ under ρ, we add 1 in the first coordinate as $a = 1$, and add 0 in the second coordinate as $b_0 = 0$. The image of $(1,0)$ under ρ is $(0,1)$ because we add 1 in the first coordinate as $a = 1$, and add 1 in the second coordinate as $b_1 = 1$. The image of $(0,1)$ under ρ is $(1,1)$ because we add 1 in the first coordinate as $a = 1$, and add 0 in the second coordinate as $b_0 = 0$. Similarly, the image of $(1,1)$ under ρ is $(0,0)$ because we add 1 in the first coordinate as $a = 1$, and 1 in the second coordinate as $b_1 = 1$. The cycle decompositions of the other elements can be calculated similarly and are given in Table 4.2.

We know that the graph $K_2 \wr \bar{K}_2$ given in Figure 4.1 has automorphism group D_8 as it is a 4-cycle. As ρ, defined in the preceding paragraph, is a 4-cycle, we may think of ρ as being the rotation of a square by 90°. Similarly, we may think of $\mu = ((1,0),(1,1))$ as the reflection about the line through opposite vertices in the square. Then $D_8 = \langle \rho, \mu \rangle$, and the other elements of $S_2 \wr S_2$ in terms of these natural generators of D_8 are given in Table 4.2.

Intuitively, the wreath product $G \wr H$ has elements of G permuting $|X|$ copies of Y, and as an element of G permutes these copies, the copies of Y are mapped to each other via elements of H. Crucially, the elements of H chosen to map copies of Y to each other are chosen independently. Another way to think of $G \wr H$ is that sets $\{(x, y) : y \in Y\}$, $x \in X$, are blocks of $G \wr H$ (because they are!). The elements of G then tell us how to permute these blocks and the choice of element of H tells us how we map a given block to another one.

Example 4.2.7 Consider $G = S_3 \wr (\mathbb{Z}_2)_L$, which is a permutation group on $\mathbb{Z}_3 \times \mathbb{Z}_2$. We will see in Example 5.1.9 that G is the automorphism group of the graph $\Gamma = K_3 \wr \bar{K}_2$. By definition, G is the set of all permutations of the form $(x, y) \mapsto (g(x), h_x(y))$, where $g \in S_3$, and each $h_x \in (\mathbb{Z}_2)_L$. In Γ, the sets $\{i\} \times \mathbb{Z}_2$ are the maximal independent sets of size 2, $i \in \mathbb{Z}_3$. An element of $S_3 \wr (\mathbb{Z}_2)_L$ permutes these three maximal independent sets. As it maps one independent set to another, it can either do this as the identity, or as any other element of $(\mathbb{Z}_2)_L$ (of course there is only one other such element, namely it can add 1 in the second coordinate – this has the effect of interchanging the vertices in the maximal independent set). Whether or not one adds 1 (or interchanges the vertices) is independent of which of these sets is chosen. For example, suppose we map each independent set to itself. The permutation that interchanges the vertices in the independent sets $\{0\} \times \mathbb{Z}_2$ and $\{2\} \times \mathbb{Z}_2$ and fixes the vertices in $\{1\} \times \mathbb{Z}_2$ is in G, and can be written as $(x, y) \mapsto (x, y + b_x)$, where $b_0 = b_2 = 1$ and $b_1 = 0$. There are thus a total of eight elements that map each independent set to itself as there are two choices of maps for each of the three maximal independent sets (interchange or not interchange vertices of the independent set, or add 1 or 0 in the second coordinate). As there are six ways of permuting the maximal independent sets (the order of S_3), there are $48 = |S_3| \cdot |(\mathbb{Z}_2)_L|^3$ elements in G.

Example 4.2.8 The group $(\mathbb{Z}_p)_L \wr (\mathbb{Z}_p)_L \leq S_{p^2}$ has order p^{p+1} and consequently is a Sylow p-subgroup of S_{p^2} when p is prime.

Solution As $(\mathbb{Z}_p)_L \wr (\mathbb{Z}_p)_L = \{(i, j) \mapsto (i + a, j + b_i) : a, b_i \in \mathbb{Z}_p\}$, we see that $|(\mathbb{Z}_p)_L \wr (\mathbb{Z}_p)_L| = p^{p+1}$ as there are p choices for a, as well as p choices for each b_i, and there are exactly p different elements b_i. When p is prime, then, as the multiples of p dividing $p^2!$ are $p, 2p, \ldots, (p-1)p, p^2$, we see the largest power of p dividing $p^2!$ is p^{p+1}. □

The above example is not a coincidence, as Kalužnin (Kaloujnine, 1948) has shown that for any $n \geq 2$ and any prime $p \leq n$, a Sylow p-subgroup of S_n can be written as a wreath product. The next exercise hints at how a Sylow p-subgroup of S_n is constructed. See Exercise 4.2.6 or Passman (1968, p. 10) for the full construction.

Example 4.2.9 Let $n = 22$ and $p = 3$. Let P_3 be a Sylow 3-subgroup of $\mathcal{S}_9$ as constructed in Example 4.2.8. A Sylow 3-subgroup of $\mathcal{S}_{22}$ is isomorphic to $P_3 \times P_3 \times \mathbb{Z}_3$.

Solution The basic idea is as follows. We find the highest power of 3 that is at most 22, so 9, and then choose 9 points and then choose a Sylow 3-subgroup of $\mathcal{S}_9$ that permutes those 9 points. We then find the highest power of 3 on the remaining $22 - 9 = 13$ points that is at most 13, so 9 again, choose 9 points and choose a Sylow 3-subgroup of $\mathcal{S}_9$ that permutes those 9 points, and repeat until the only power of 3 that is at most the number of remaining points is $3^0 = 1$. We then check to make sure the order of the product of the groups chosen is the order of a Sylow 3-subgroup of $\mathcal{S}_{22}$. Formally, we decompose 22 as $2 \cdot 3^2 + 3 + 1$, the 3-adic form of 22 (this simply means writing 22 as $\sum_{i=0}^{2} x_i 3^i$, $x_i \in \mathbb{Z}_3$). We partition $\mathbb{Z}_{22}$ into four sets, two of size 9, say T_1 and T_2, one of size 3, say T_3, and one of size 1, say T_4. We consider the symmetric group $\mathcal{S}_{T_i}$ on T_i, and observe that $\mathcal{S}_{T_1} \times \mathcal{S}_{T_2} \times \mathcal{S}_{T_3} \times \mathcal{S}_{T_4} \leq \mathcal{S}_{22}$. Let $Q_1, \ldots, Q_4$ be Sylow 3-subgroups of $\mathcal{S}_{T_i}$, so $Q_1 \cong Q_2 \cong \mathbb{Z}_3 \wr \mathbb{Z}_3$, $Q_3 = \mathbb{Z}_3$, and $Q_4 = 1$. So $Q_1 \times Q_2 \times Q_3 \times Q_4 \cong P_3 \times P_3 \times \mathbb{Z}_3$. This group has order $3^4 \cdot 3^4 \cdot 3 = 3^9$, and the multiples of 3 that divide 22! are $3, 6, 9, 12, 15, 18, 21$. We conclude that a Sylow 3-subgroup of $\mathcal{S}_{22}$ has order 3^9, and so is isomorphic to $P_3 \times P_3 \times \mathbb{Z}_3$. □

Example 4.2.10 Let $n = 35$ and $p = 5$. Let P_5 be a Sylow 5-subgroup of $\mathcal{S}_{25}$ as constructed in Example 4.2.8. A Sylow 5-subgroup of $\mathcal{S}_{35}$ is isomorphic to $\mathbb{Z}_5 \times \mathbb{Z}_5 \times P_5$.

Solution Decompose 35 as $1 \cdot 5^2 + 2 \cdot 5 + 0$, the 5-adic form of 35. We partition $\mathbb{Z}_{35}$ into three sets, one of size 25, say T_1, and two of size 5, say T_2 and T_3. We consider the symmetric group $\mathcal{S}_{T_i}$ on T_i, and observe that $\mathcal{S}_{T_1} \times \mathcal{S}_{T_2} \times \mathcal{S}_{T_3} \leq \mathcal{S}_{35}$. Let Q_1, Q_2, Q_3 be Sylow 3-subgroups of $\mathcal{S}_{T_i}$, so $Q_1 \cong \mathbb{Z}_5 \wr \mathbb{Z}_5$, and $Q_2 \cong Q_3 \cong \mathbb{Z}_5$. So $Q_1 \times Q_2 \times Q_3 \cong P_5 \times \mathbb{Z}_5 \times \mathbb{Z}_5$. This group has order $5^6 \cdot 5 \cdot 5 = 5^8$, and the multiples of 5 that divide 35! are $5, 10, 15, 20, 25, 30, 35$. We conclude that a Sylow 5-subgroup of $\mathcal{S}_{35}$ has order 5^8, and so is isomorphic to $P_5 \times \mathbb{Z}_5 \times \mathbb{Z}_5$. □

The next example shows one of the reasons that the wreath product is important, as it shows that when computing the automorphism group of a disconnected vertex-transitive digraph, one only needs compute the automorphism group of one of the (isomorphic) components. This reduces the disconnected case to the connected case, which is the most common way of answering questions concerning disconnected digraphs.

Example 4.2.11 A disconnected vertex-transitive digraph Γ is isomorphic to a wreath product $\bar{K}_m \wr C$, where m is the number of components and C is a component, and $\mathrm{Aut}(\Gamma)$ is permutation isomorphic to $\mathcal{S}_m \wr \mathrm{Aut}(C)$.

Solution As Γ is vertex-transitive, each component of Γ is isomorphic to C. We see that $\mathrm{Aut}(\Gamma)$ is permutation isomorphic to $\mathcal{S}_m \wr \mathrm{Aut}(C)$ as we may permute the components in any way we wish, and as we map one component to another, we must preserve the arcs of a component; that is, we map a component to a component via an isomorphism between the components. Clearly, we can label the components so that this map is an automorphism of C. Note that the components of Γ are blocks of $\mathrm{Aut}(\Gamma)$. Thus $\mathrm{Aut}(\Gamma)$ will be imprimitive provided that Γ has arcs. □

The wreath product of two permutation groups can be thought of as a generalization of the direct or Cartesian product of two permutation groups, as we shall see shortly that $G \times H \le G \wr H$.

Example 4.2.12 Let G and H be groups. Then $(G \times H)_L = G_L \times H_L$.

Solution We compare $(G \times H)_L$ and $G_L \times H_L$. The group $(G \times H)_L = \{(x, y) \mapsto (gx, hy) : g \in G, h \in H\}$, while $G_L \times H_L$ is all maps from $G \times H$ to $G \times H$ of the form $(x, y) \mapsto (gx, hy)$ where $g \in G$ and $h \in H$. □

Lemma 4.2.13 *Let $G \le \mathcal{S}_X$ and $H \le \mathcal{S}_Y$. Then $G \times H \le G \wr H$.*

Proof In $G \wr H$, we may choose $h_x = h_{x'}$ for every $x, x' \in X$. Then all functions of the form $(x, y) \mapsto (g(x), h(y))$ are contained in $G \wr H$. So $G \times H \le G \wr H$. □

One theme we have tried to emphasize is that symmetry is rare, and so if one wants a digraph to have a particular kind of symmetry, the best way to do this is to construct the digraph with the symmetry. From that point of view, the next result says we should "expect" the automorphism group of the wreath product of digraphs to be the wreath product of the automorphism groups of the digraphs. There are circumstances though, where other automorphisms can be found. We will determine those circumstances in Section 5.1. The next result is the reason Harary defined the wreath product of graphs the way he did. His definition of the graph wreath product "builds" the wreath product of the automorphism groups into the automorphism group of the wreath product.

Lemma 4.2.14 *Let Γ and Δ be digraphs. Then* $\mathrm{Aut}(\Gamma) \wr \mathrm{Aut}(\Delta) \le \mathrm{Aut}(\Gamma \wr \Delta)$.

Proof As there are two kinds of arcs in $\Gamma \wr \Delta$, we will show each kind is preserved by $\mathrm{Aut}(\Gamma) \wr \mathrm{Aut}(\Delta)$. Let $\gamma \in \mathrm{Aut}(\Gamma) \wr \mathrm{Aut}(\Delta)$. Then there exists $g \in$

$\mathrm{Aut}(\Gamma)$, and $h_x \in \mathrm{Aut}(\Delta)$ for each $x \in V(\Gamma)$, such that $\gamma(x,y) = (g(x), h_x(y))$. Suppose $((x_1,y_1),(x_2,y_2)) \in A(\Gamma \wr \Delta)$. If $x_1 = x_2$, then by definition, $(y_1,y_2) \in A(\Delta)$. Set $x = x_1 = x_2$. Then,

$$\gamma((x_1,y_1)(x_2,y_2)) = ((g(x),h_x(y_1)),(g(x),h_x(y_2))) \in A(\Gamma \wr \Delta),$$

as $g(x) = g(x)$ and $h_x(y_1,y_2) \in A(\Delta)$ as $h_x \in \mathrm{Aut}(\Delta)$. If $x_1 \neq x_2$ then $g(x_1,x_2) \in A(\Gamma)$ as $g \in \mathrm{Aut}(\Gamma)$, and so $((g(x_1),z_1),(g(x_2),z_2)) \in A(\Gamma \wr \Delta)$ for every $z_1,z_2 \in V(\Delta)$. Then,

$$\begin{aligned}\gamma((x_1,y_1),(x_2,y_2)) &= ((g(x_1),h_{x_1}(y_1)),(g(x_2),h_{x_2}(y_2)))\\ &= ((g(x_1),z_1),(g(x_2),z_2)) \in A(\Gamma \wr \Delta),\end{aligned}$$

where $z_1 = h_{x_1}(y_1)$ and $z_2 = h_{x_2}(y_2)$. □

The following result is a special case of Joseph (1995, Lemma 3.11) and shows how to recognize that a vertex-transitive digraph is isomorphic to the wreath product of two digraphs of smaller order in terms of a block system of a transitive subgroup of its automorphism group. This result is an important way that digraphs that are wreath products are recognized in practice.

Theorem 4.2.15 *Let Γ be a vertex-transitive digraph whose automorphism group contains a transitive subgroup G that has a block system $\mathcal{B}$. Then $\Gamma \cong \Gamma/\mathcal{B} \wr \Gamma[B_0]$, $B_0 \in \mathcal{B}$, if and only if whenever $B, B' \in \mathcal{B}$ are distinct then there is an arc (x,x') from a vertex $x \in B$ to a vertex $x' \in B'$ if and only if every pair (x,x') with $x \in B$ and $x' \in B'$ is contained in $A(\Gamma)$.*

Proof Suppose that $\Gamma \cong \Gamma/\mathcal{B} \wr \Gamma[B_0]$, with $B, B' \in \mathcal{B}$ such that $B \neq B'$, and there is an arc (x,x') from a vertex $x \in B$ to the vertex $x' \in B'$. As $\Gamma \cong \Gamma/\mathcal{B} \wr \Gamma[B_0]$, it must be the case that $(B,B') \in A(\Gamma/\mathcal{B})$, and so by the definition of the wreath product, $(x,x') \in A(\Gamma/\mathcal{B} \wr \Gamma[B_0])$ for every $x \in B$ and $x' \in B'$.

Conversely, suppose that whenever $B, B' \in \mathcal{B}$ are distinct then there is an arc (x,x') from a vertex $x \in B$ to a vertex $x' \in B'$ if and only if every arc of the form (x,x') with $x \in B$ and $x' \in B'$ is contained in $A(\Gamma)$. As Γ is vertex-transitive, each of the digraphs $\Gamma[B]$ for $B \in \mathcal{B}$, are isomorphic. Fix $B_0 \in \mathcal{B}$, and for each $B \in \mathcal{B}$ let $\pi_B\colon B \to B_0$ be an isomorphism from $\Gamma[B]$ to $\Gamma[B_0]$. Now define $\phi\colon V(\Gamma) \to \mathcal{B} \times B_0$ by $\phi(v) = (B, \pi_B(v))$, where $v \in B$. Let $w \in B_0$ and $B \in \mathcal{B}$. As π_B is an isomorphism from $\Gamma[B]$ to $\Gamma[B_0]$, it is a bijection, so there exists $v \in B$ such that $\pi_B(v) = w$. Then $\phi(v) = (B,w)$ and so ϕ is onto. As, by Theorem 2.2.6, $|\mathcal{B}| \cdot |B_0| = |V(\Gamma)|$, ϕ is a bijection.

We will show that $\phi(\Gamma) = \Gamma/\mathcal{B} \wr \Gamma[B_0]$. Clearly, $V(\phi(\Gamma)) = \mathcal{B} \times B_0$. Let $(u,v) \in A(\Gamma)$. We consider arcs with both endpoints in a block of $\mathcal{B}$ separately from arcs with endpoints in different blocks of $\mathcal{B}$. First, suppose $u, v \in B \in \mathcal{B}$. Then

$$\begin{aligned}\phi(u,v) \in A(\phi(\Gamma)) &\text{ if and only if } ((B,\pi_B(u)),(B,\pi_B(v))) \in A(\phi(\Gamma))\\ &\text{ if and only if } (\pi_B(u),\pi_B(v)) \in A(\Gamma[B_0]).\end{aligned}$$

Now suppose $u \in B$ and $v \in B'$ with $B \neq B'$ blocks in $\mathcal{B}$. Then $(u, v) \in A(\Gamma)$ for every vertex $u \in B$, and $v \in B'$. Applying ϕ, we see that $((B, x_0), (B', y_0)) \in A(\phi(\Gamma))$ for every $x_0, y_0 \in B_0$. Comparing these arcs to the arcs in $\Gamma/\mathcal{B} \wr \Gamma[B_0]$, we see that the arc sets are identical and $\phi(\Gamma) = \Gamma/\mathcal{B} \wr \Gamma[B_0]$. □

4.3 The Embedding Theorem

The wreath product also plays a central role in the study of imprimitive permutation groups, as evidenced by the next result, known as the **embedding theorem**, due to Krasner and Kalužnin (Krasner and Kaloujnine, 1951).

Theorem 4.3.1 (The embedding theorem) *Let $G \le S_X$ be transitive with block system $\mathcal{B}$. Choose $B_0 \in \mathcal{B}$. Then G is permutation isomorphic to a subgroup of $(G/\mathcal{B}) \wr \left(\mathrm{Stab}_G(B_0)^{B_0}\right)$.*

Proof For each $B \in \mathcal{B}$, there exists $h_B \in G$ such that $h_B(B_0) = B$. Define $\lambda\colon X \to \mathcal{B} \times B_0$ by $\lambda(x) = (B, x_0)$, where $x \in B$ and $x_0 = h_B^{-1}(x)$. Define $\phi\colon G \to (G/\mathcal{B}) \wr \left(\mathrm{Stab}_G(B_0)^{B_0}\right)$ by $\phi(g)(B, x_0) = \left(g(B), h_{g(B)}^{-1} g h_B(x_0)\right)$. Note that $h_{g(B)}^{-1} g h_B(B_0) = h_{g(B)}^{-1} g(B) = B_0$, so ϕ has domain and image as stated. We must show that λ is a bijection, that ϕ is an injective homomorphism (as it is onto its image), and that $\lambda(g(x)) = \phi(g)(\lambda(x))$ for all $x \in X$ and $g \in G$.

To show that λ is a bijection, it suffices to show that λ is injective as $|X| = |\mathcal{B} \times B_0|$ by Theorem 2.2.6. Let $x, x' \in X$ and assume $(B, x_0) = \lambda(x) = \lambda(x')$. Clearly both x and x' are contained in B, and $x_0 = h_B^{-1}(x) = h_B^{-1}(x')$. As h_B is a permutation, it follows that $x = x'$ and λ is injective, and so a bijection.

To show that ϕ is injective, suppose $\phi(g) = \phi(g')$. Applying the definition of ϕ, we see

$$\left(g(B), h_{g(B)}^{-1} g h_B(x_0)\right) = \left(g'(B), h_{g'(B)}^{-1} g' h_B(x_0)\right),$$

for all $B \in \mathcal{B}$ and $x_0 \in B_0$. It immediately follows that $g/\mathcal{B} = g'/\mathcal{B}$ and $h_{g(B)}^{-1} g h_B = h_{g'(B)}^{-1} g' h_B$. Using $g/\mathcal{B} = g'/\mathcal{B}$, we see that $g(B) = g'(B)$, and so $h_{g(B)}^{-1} = h_{g'(B)}^{-1}$. Canceling, we see that $g = g'$ and so ϕ is injective.

Let $g_1, g_2 \in G$. Then

$$\begin{aligned}\phi(g_1)\phi(g_2)(B, x_0) &= \phi(g_1)\left(g_2(B), h^{-1}_{g_2(B)} g_2 h_B(x_0)\right)\\ &= \left(g_1g_2(B), h^{-1}_{g_1(g_2(B))} g_1 h_{g_2(B)}\left(h^{-1}_{g_2(B)} g_2 h_B(x_0)\right)\right.\\ &= \left(g_1g_2(B), h^{-1}_{g_1g_2(B)} g_1 g_2 h_B(x_0)\right)\\ &= \phi(g_1g_2)(B, x_0),\end{aligned}$$

and so ϕ is a homomorphism.

Finally, observe that $\phi(g)(\lambda(x)) = \phi(g)(B, x_0) = \left(g(B), h^{-1}_{g(B)} g h_B(x_0)\right)$ while

$$\lambda(g(x)) = \left(g(B), h^{-1}_{g(B)} g(x)\right) = \left(g(B), h^{-1}_{g(B)} g h_B(x_0)\right),$$

and so $\lambda(g(x)) = \phi(g)(\lambda(x))$ for all $x \in X$ and $g \in G$. □

Intuitively, the embedding theorem says that if G has a block system with m blocks of size k, then we may coordinatize G. That is, we may relabel the set X with elements of $\mathbb{Z}_m \times \mathbb{Z}_k$ in such a way that if $g \in G$, then $g(a, b) = (u(a), v_a(b))$, where $u \in G/\mathcal{B} \le \mathcal{S}_m$ and each $v_a \in \mathrm{Stab}_G(B)^B \le \mathcal{S}_k$, where $B \in \mathcal{B}$ is fixed. The following corollary is immediate, and is, barring any other information about G, the default form of the embedding theorem.

Corollary 4.3.2 *Let $G \le S_n$ be a transitive permutation group with block system $\mathcal{B}$ with m blocks of size k. Then G is permutation isomorphic to a subgroup of $\mathcal{S}_m \wr \mathcal{S}_k$.*

It is not difficult to show that $\Gamma_1 \wr (\Gamma_2 \wr \Gamma_3) = (\Gamma_1 \wr \Gamma_2) \wr \Gamma_3$. That is, that the digraph wreath product is associative (Exercise 4.3.1), and similarly, the permutation group wreath product is also associative (Exercise 4.3.2). The following result shows that primitive groups can be thought of as the "building blocks" of imprimitive groups. Note that in the wreath product of the groups H_i in the following result, parentheses are not necessary as the wreath product is associative.

Corollary 4.3.3 *Let $G \le \mathcal{S}_n$ be transitive. Then there exist integers $k_1, \ldots, k_r$ such that $\prod_{i=1}^r k_i = n$ and primitive groups $H_i \le \mathcal{S}_{k_i}$, $1 \le i \le r$, such that G is permutation isomorphic to a subgroup of $H_1 \wr H_2 \wr \cdots \wr H_r$.*

Proof We proceed by induction on n. If $n = 2$ the result is trivial as $G = \mathcal{S}_2$ is primitive. Let $n \ge 2$ and assume the result holds for all $m \le n$ and let $G \le \mathcal{S}_{n+1}$ be transitive. If G is primitive, then the result is trivial with $k_1 = n + 1$ and $H_1 = G$. So we assume that G is imprimitive with nontrivial block system $\mathcal{B}$ with blocks of minimum size k_r. Then $\mathrm{Stab}_G(B)^B$ is primitive, as otherwise, by Theorem 2.3.10, there is a block of G with blocks of order less than k_r. Then $G/\mathcal{B} \le \mathcal{S}_{(n+1)/k_r}$ and so by the induction hypothesis there exist integers

$k_1,\ldots,k_{r-1}$ such that $\prod_{k=1}^{r-1} k_i = (n+1)/k_r$ and primitive groups $H_i \le S_{k_i}$, $1 \le i \le r-1$, such that $G/\mathcal{B}$ is permutation isomorphic to a subgroup of

$$H_1 \wr H_2 \wr \cdots \wr H_{r-1}.$$

The result now follows by the embedding theorem and induction. □

We saw in Sabidussi's theorem (Theorem 1.2.20) that a digraph Γ is isomorphic to a Cayley digraph if and only if $\mathrm{Aut}(\Gamma)$ contains a regular subgroup. It is thus of interest to us whether or not a permutation group contains a regular subgroup. Our next goal is to use the embedding theorem to show that we may use the wreath product to "make" a non-Cayley digraph into a Cayley digraph. We first clear up one neglected fact about double coset digraphs and the groups $(G, G/H)$ with $H \le G$. Namely, what happens if H is not core-free in G?

Lemma 4.3.4 *Let G be a group, $H \le G$, and $N = \mathrm{core}_G(H)$. Let $\beta\colon G \to (G, G/H)$ and $\delta\colon G/N \to (G/N, (G/N)/(H/N))$ be permutation representations of G and G/N given by $\beta(g) = \hat{g}_L$ and $\delta(gN) = \widehat{gN}_L$. Then β is permutation equivalent to δ.*

Proof Define $\lambda\colon G/H \to (G/N)/(H/N)$ by $\lambda(gH) = g(H/N)$. Note that $|G/H| = |(G/N)/(H/N)|$ and λ is clearly onto. As $|G|$ is finite, λ is a bijection. For $kH \in G/H$ (so $k \in G$), we have

$$\begin{aligned}\lambda(\beta(g)(kH)) &= \lambda(\hat{g}_L(kH)) = \lambda(gkH) = gkH/N \\ &= gN(kH/N) = \widehat{gN}_L(\lambda(kH)) = \delta(g)(\lambda(kH)).\end{aligned}$$

□

In Exercise 4.3.5, you will be asked to show that if $N = \mathrm{core}_G(H)$, then $\mathrm{Cos}(G, H, S) \cong \mathrm{Cos}(G/N, H/N, \{sN : s \in S\})$.

Theorem 4.3.5 *Let G be a group and $H \le G$ be of index m and order k. Then $S_m \wr H_L$ contains a regular subgroup isomorphic to G. Also, $(G, G/H) \wr S_k$ contains a regular subgroup isomorphic to G.*

Proof Let $\bar{H}_L = \{h_L : h \in H\} \le G_L$. Of course, G_L is regular and so $\mathrm{Stab}_{G_L}(1_G) = 1 \le \bar{H}_L$. By Theorem 2.3.7, there is a block system $\mathcal{B}$ of G_L such that $\mathrm{Stab}_{G_L}(B) = \bar{H}_L$, where $B \in \mathcal{B}$ such that $1_G \in B$. As G_L is regular, $\mathrm{Stab}_{G_L}(B)^B \cong H$. Hence $\mathcal{B}$ is the set of left cosets of H in G. By the embedding theorem, we have that G_L is permutation isomorphic to a subgroup of $(G_L/\mathcal{B}) \wr \left(\bar{H}_L^B\right) \cong (G_L/\mathcal{B}) \wr H_L$. As $[G : H] = m$, we see $G_L/\mathcal{B}$ is of degree m. As $(G_L/\mathcal{B}) \wr H_L \le S_m \wr H_L$ the first part of the result follows. As $\mathcal{B}$ is the set of left cosets of H in G, $G_L/\mathcal{B} = (G, G/H)$. So $(G_L/\mathcal{B}) \wr H_L \le (G, G/H) \wr S_H \cong (G, G/H) \wr S_k$ and the second part of the result follows. □

The following result is due to Sabidussi (1964, Theorem 4) and shows that some "multiple" of a double coset digraph is a Cayley digraph. We do not assume in the next result that H is core-free in G.

Corollary 4.3.6 *Let G be a group and $H \leq G$ of order k. The digraph $\mathrm{Cos}(G,H,S) \wr \overline{K_k}$ is isomorphic to a Cayley digraph of G for every $S \subseteq G$ a union of double cosets of H in G.*

Proof By Theorem 4.3.5, $(G, G/H) \wr S_k$ contains a regular subgroup isomorphic to G. As

$$(G, G/H) \wr S_k \leq \mathrm{Aut}(\mathrm{Cos}(G,H,S) \wr S_k),$$

$\mathrm{Aut}\left(\mathrm{Cos}(G,H,S) \wr \overline{K_k}\right)$ is isomorphic to a Cayley digraph of G by Theorem 1.2.20. □

The block system of $G \wr H$ that is the set of orbits of $1_G \wr H$ has properties that not every block system has. These properties can often be exploited.

Theorem 4.3.7 *Let X and Y be sets, $G \leq S_X$ and $H \leq S_Y$ be transitive groups. Then $\mathcal{B} = \{\{(x,y) : y \in Y\} : x \in X\}$ is a normal block system of $G \wr H$. If C is a block system of $G \wr H$, then either $\mathcal{B} \leq C$ or $C \leq \mathcal{B}$. Consequently, $\mathcal{B}$ is the only block system of $G \wr H$ with blocks of size $|Y|$.*

Proof The group $1_{S_X} \wr H$ is a normal subgroup of $G \wr H$, and so, by Theorem 2.2.7, the set $\mathcal{B}$ of its orbits are a normal block system of $G \wr H$. Now let C be a block system of $G \wr H$, and $B \in \mathcal{B}$. Suppose $C \nleq \mathcal{B}$. Then there exists $C \in C$, $B \in \mathcal{B}$ and $u, v \in X \times Y$ such that $u \in B \cap C$ but $v \in C \backslash B$. Let K be the point-wise stabilizer of every point *not* in B. Then K is transitive on B. As $k(v) = v$ for every $k \in K$, we have $k(C) = C$ for every $k \in K$. As K^B is transitive and $u \in B$, $B \subset C$. Hence $\mathcal{B} < C$ and the result follows. □

Definition 4.3.8 Let $B_x = \{x\} \times Y$, $x \in X$, and $\mathcal{B} = \{B_x : x \in X\}$. Then $\mathcal{B}$ is the block system of $G \wr H$ given in the preceding result. It is called the **lexi-partition of $G \wr H$ corresponding to Y**.

4.4 Groups of Prime Degree and Digraphs of Prime Order

As we saw with the embedding theorem, to understand transitive permutation groups of degree n, it is necessary to understand transitive permutation groups of degree k, where $k|m$. As a consequence, one must understand transitive permutation groups of prime degree and vertex-transitive graphs of prime order. Fortunately, such groups and digraphs are very well understood. The following result was first observed by Turner (1967).

Theorem 4.4.1 *A transitive group of prime degree p contains a regular cyclic subgroup. Consequently, every vertex-transitive digraph of prime order p is isomorphic to a Cayley digraph of $\mathbb{Z}_p$.*

Proof Let G be a transitive group of prime degree p. By the orbit-stabilizer theorem, the size of an orbit of a group divides the order of a group. As G is transitive, it has one orbit of size p, and so p divides $|G|$. Then G contains an element of order p by Cauchy's theorem. As $G \leq S_p$, any element of order p must be a p-cycle ρ. Then $\langle\rho\rangle$ is a regular cyclic subgroup, and the result follows by Theorem 1.2.20. □

The main tool for determining the full automorphism group of circulant digraphs of prime order is the following classical result of Burnside (1901). It is extremely useful and is found in most books on permutation group theory; see, for example, Dixon and Mortimer (1996, Theorem 3.5B).

Theorem 4.4.2 (Burnside's theorem) *Let $G \leq S_p$, p a prime, be transitive such that the map $x \mapsto x + 1$ is in G. Then either G is 2-transitive or $G <$ $\mathrm{AGL}(1, p) = \{x \mapsto ax + b : a \in \mathbb{Z}_p^*, b \in \mathbb{Z}_p\} \cong \mathbb{Z}_p \rtimes \mathbb{Z}_p^*$.*

Note that the condition that the map $x \mapsto x + 1$ is contained in G is always true after a relabeling of the set on which G acts as, by Theorem 4.4.1, G contains a p-cycle, say $\rho = (a_0, a_1, \ldots, a_{p-1})$, and so labeling a_i with i will give the p-cycle $(0, 1, \ldots, p - 1)$. Also, we have seen $\mathrm{AGL}(1, p)$ already in Example 1.4.2 and Exercise 2.4.6 (whose solution we give in Corollary 4.4.4). Also, the next result is the first part of Exercise 2.4.1.

Lemma 4.4.3 *A transitive permutation group $G \leq S_n$, $n \geq 2$, is 2-transitive if and only if* $\mathrm{Stab}_G(x)$ *is transitive on $\mathbb{Z}_n\backslash\{x\}$ for some $x \in \mathbb{Z}_n$ (equivalently, every $x \in \mathbb{Z}_n$).*

Proof Suppose G is 2-transitive, and let $x \in \mathbb{Z}_n$. Then for each $y, z \in \mathbb{Z}_n\backslash\{x\}$, there exists $g \in G$ such that $g(x) = x$ and $g(y) = z$. Thus $\mathrm{Stab}_G(x)$ is transitive on $\mathbb{Z}_n\backslash\{x\}$. Conversely, suppose G is transitive on $\mathbb{Z}_n\backslash\{x\}$. Let $x, y, w, z \in \mathbb{Z}_n$ with $x \neq y$ and $w \neq z$. As G is transitive, there exists $g_1 \in G$ such that $g_1(x) = w$. Let $v \in \mathbb{Z}_n$ such that $g_1(v) = z$. As $\mathrm{Stab}_G(x)$ is transitive on $\mathbb{Z}_n\backslash\{x\}$, there exists $g_2 \in \mathrm{Stab}_G(x)$ such that $g_2(y) = v$. Setting $g = g_1g_2$, we see that $g(x) = g_1(x) = w$ and $g(y) = g_1(v) = z$. Thus G is 2-transitive. □

Corollary 4.4.4 $\mathrm{AGL}(1, p)$ *is 2-transitive.*

Proof Let $a \in \mathbb{Z}_p^*$ such that $\langle a \rangle = \mathbb{Z}_p^*$. Then $x \mapsto ax$ is cyclic of order $p-1$, and so must be a $p-1$ cycle with 0 a fixed point. This implies that $\langle x \mapsto ax \rangle$ is transitive on $\mathbb{Z}_p \backslash \{0\}$. The result follows by Lemma 4.4.3. □

In the case of circulant digraphs of prime order, combining the previous result with Lemma 2.4.4 we have the following result, due to Alspach (1973) (this paper also has a result enumerating all circulant digraphs of prime order with a given automorphism group).

Theorem 4.4.5 *Let Γ be a circulant digraph of prime order p. Then either $\Gamma = K_p$ or $\bar{K}_p$ and $\mathrm{Aut}(\Gamma) = \mathcal{S}_p$, or $\mathrm{Aut}(\Gamma) < \mathrm{AGL}(1,p)$.*

This result gives an easy algorithm to determine the automorphism group of a circulant digraph of prime order.

Algorithm 4.4.6 *Let p be prime and $S \subseteq \mathbb{Z}_p$. Let $\Gamma = \mathrm{Cay}(\mathbb{Z}_p, S)$, and $A = \{a \in \mathbb{Z}_p^* : aS = S\}$. Then,*

- *if $|A| = p-1$ then $\mathrm{Aut}(\Gamma) = \mathcal{S}_p$,*
- *if $|A| < p-1$, then $\mathrm{Aut}(\Gamma) = \{x \mapsto ax + b : a \in A, b \in \mathbb{Z}_p\}$.*

Proof By Corollary 1.2.16, the map $x \mapsto ax$, $a \in \mathbb{Z}_p^*$, being an automorphism of $\mathbb{Z}_p$, is contained in $\mathrm{Aut}(\Gamma)$ if and only if $aS = \{as : s \in S\} = S$. If $|A| = p-1$ then $A = \mathbb{Z}_p^*$ and $\mathrm{AGL}(1,p) \le \mathrm{Aut}(\Gamma)$ is 2-transitive by Corollary 4.4.4. Then $\Gamma = K_p$ or $\bar{K}_p$ and $\mathrm{Aut}(\Gamma) = \mathcal{S}_p$ by Lemma 2.4.4. Otherwise, $\{x \mapsto ax + b : a \in A, b \in \mathbb{Z}_p\}$ is the largest subgroup of $\mathrm{AGL}(1,p)$ contained in $\mathrm{Aut}(\Gamma)$ and the result follows by Theorem 4.4.5. □

Example 4.4.7 Let $\Gamma = \mathrm{Cay}(\mathbb{Z}_{17}, \{1,4,13,16\})$. Find $\mathrm{Aut}(\Gamma)$.

Solution As $-S = -\{1,4,13,16\} = \{16,13,4,1\} = S$, Γ is a graph. We need to find those $a \in \mathbb{Z}_{17}^*$ such that $aS = S$. As $1 \in S$, to find such a we need only check which $a \in S$ satisfy $aS = S$. Of course, $1S = S$ and $16S = -S = S$ as Γ is a graph. Also, note that $4S = \{4,16,1,13\} = S$. While we could check to see that $13S = S$, we do not really have to as $13 \equiv -4 \pmod{17}$, so $13S = -4S = -S = S$. Thus $\mathrm{Aut}(\Gamma) = \{x \mapsto ax + b : a = 1,4,13,16, b \in \mathbb{Z}_{17}\}$. Note that $|\mathrm{Aut}(\Gamma)| = 4 \cdot 17 = 68$. □

Before leaving this topic, we will show that $\mathrm{AGL}(1,p) = N_{\mathcal{S}_p}((\mathbb{Z}_p)_L)$. We will have need later of $N_{\mathcal{S}_G}(G_L)$ as well, and as it is not too much extra work, we go ahead now and calculate $N_{\mathcal{S}_n}(G)$ for any transitive group $G \le \mathcal{S}_n$. As every permutation group G is permutation isomorphic to $(G, G/H)$ for some $H \le G$ by Theorem 4.1.13, we may state such a result using that language.

Definition 4.4.8 Let G be a group with subgroup $H \leq G$, and $\alpha \in \mathrm{Aut}(G)$ such that $\alpha(H) = H$. Define $\bar{\alpha}\colon G/H \to G/H$ by $\bar{\alpha}(gH) = \alpha(g)H$.

Recall that $\bar{\alpha}$ is a well-defined permutation of G/H by Lemma 4.1.9.

Lemma 4.4.9 *Let G be a group with subgroup $H \leq G$, and $\bar{A} = \{\bar{\alpha} : \alpha \in \mathrm{Aut}(G)$ and $\alpha(H) = H\}$. Then $N_{S_{G/H}}(G, G/H) = \bar{A}(G, G/H)$.*

Proof Let $\delta \in N_{S_{G/H}}(G, G/H)$. As $(G, G/H)$ is transitive on the set of left cosets of H in G, there exists $\hat{\ell}_1 \in (G, G/H)$ such that $\hat{\ell}_1\delta(H) = H$. As $\delta^{-1}\hat{\ell}_1\delta \in (G, G/H)$, there exists $\hat{\ell} \in (G, G/H)$ with $\delta^{-1}\hat{\ell}_1\delta = \hat{\ell}$, and so $\delta\hat{\ell}(H) = H$. Let $\beta = \delta\hat{\ell}$. As $\hat{\ell}$ and δ normalize $(G, G/H)$, so does β. Define $\alpha\colon G \to G$ by $\alpha(g) = k$ if and only if $\beta^{-1}\hat{g}\beta = \hat{k}$. Let $h \in H$. As $\beta(H) = H, \beta^{-1}\hat{h}\beta(H) = H$, so $\beta^{-1}\hat{h}\beta$ is contained in $(G, G/H)$ and fixes H. Hence $\beta^{-1}\hat{h}\beta = \hat{h}_1$ for some $h_1 \in H$. Then $\alpha(H) = H$. As $\beta^{-1}\hat{g}\hat{h}\beta = \beta^{-1}\hat{g}\beta \cdot \beta^{-1}\hat{h}\beta$, we have $\alpha(gh) = \alpha(g)\alpha(h)$, and so α is a homomorphism. Clearly α, being induced by conjugation, is injective, so $\alpha \in \mathrm{Aut}(G)$.

Note that $\beta^{-1}\hat{g}\beta = \widehat{\alpha(g)}$ for all $g \in G$ and that for $x, g \in G$,

$$\bar{\alpha}^{-1}\hat{g}\bar{\alpha}(xH) = \bar{\alpha}^{-1}(g\alpha(x)H) = \alpha^{-1}(g)(xH),$$

so $\bar{\alpha}^{-1}\hat{g}\bar{\alpha} = \widehat{\alpha^{-1}(g)}$. Thus,

$$\beta^{-1}\bar{\alpha}^{-1}\hat{g}\bar{\alpha}\beta = \beta^{-1}\widehat{\alpha^{-1}(g)}\beta = \hat{g},$$

so $\bar{\alpha}\beta$ commutes with $\hat{g}$ for every $g \in G$. Thus $\bar{\alpha}\beta$ centralizes $(G, G/H)$ so, by Lemma 2.2.16, we have that $\bar{\alpha}\beta$ is semiregular. Finally, as $\beta(H) = \delta\hat{\ell}(H) = H$ and $\alpha(H) = H$, we have $\bar{\alpha}\beta(H) = H$. As $\bar{\alpha}\beta$ is semiregular, we have $\bar{\alpha}\beta = 1$ so $\beta = \bar{\alpha}^{-1}$. Hence $\delta = \beta\hat{\ell}^{-1} = \bar{\alpha}^{-1}\hat{\ell}^{-1}$. Thus $N_{S_{G/H}}(G, G/H) \leq \bar{A}(G, G/H)$. □

Lemma 4.4.10 *An automorphism $\bar{\alpha}$ in $\bar{A}$ is contained in $(G, G/H)$ if and only if $\bar{\alpha}$ is an inner automorphism of G induced by $h \in H$.*

Proof Let $g \in G$. First, $\bar{\alpha}(gH) = hgh^{-1}H$ for some $h \in H$ if and only if $\bar{\alpha}(gH) = hgH = \hat{h}_L(gH)$ and so $\bar{\alpha} = \hat{h}_L$. Now suppose $\hat{g}_L \in \bar{A}$. Then $\hat{g}_L(H) = gH = H$, and so $g \in H$. Thus $\bar{A} \cap (G, G/H) = \{\hat{h}_L : h \in H\}$. □

If $H = 1$ then G is regular, and we have the following result.

Corollary 4.4.11 $N_{S_G}(G_L) = G_L \rtimes \mathrm{Aut}(G)$.

Proof By Lemma 4.4.9 $G_L \trianglelefteq \mathrm{Aut}(G)G_L$, and $G_L \cap \mathrm{Aut}(G) = 1$ by Lemma 4.4.10. The result follows. □

Thus $N_{S_p}((\mathbb{Z}_p)_L) = \mathrm{AGL}(1, p)$ by Exercise 1.5.8. The group $N_{S_G}(G_L)$ has a name:

Definition 4.4.12 The group $N_{S_G}(G_L) = G_L \rtimes \mathrm{Aut}(G)$ is the **holomorph of** G, denoted $\mathrm{Hol}(G)$.

Let G be a group, $H \leq G$ be core-free in G, and $S \subseteq G$. In Exercise 4.4.7, you will be asked to show that if $\alpha \in \mathrm{Aut}(G)$ and $\alpha(H) = H$, then $\alpha \in \mathrm{Aut}(\mathrm{Cos}(G, H, S))$ if and only if $\alpha(HSH) = HSH$. This is the double coset digraph analogue of Corollary 1.2.16 promised in Section 1.3. This allows us to take a small diversion from the main topic of this section, as we are in a position to prove an important result due to Godsil (1981b, Lemma 2.1). In fact, the proof of Lemma 4.4.9 can be extracted from the proof of Godsil (1981b, Lemma 2.1). Before stating the result, we need an additional definition.

Definition 4.4.13 Let G be a group, $H \leq G$, and $S \subseteq G$ such that $S = HSH$. Define $\mathrm{Aut}(G, H, S)$ to be the set of all automorphisms of G that fix H and map HSH to HSH. That is, $\mathrm{Aut}(G, H, S) = \{\alpha \in \mathrm{Aut}(G) : \alpha(H) = H \text{ and } \alpha(HSH) = HSH\}$.

Theorem 4.4.14 *$\mathrm{Aut}(G, H, S)$ is the subgroup of $\mathrm{Aut}(\mathrm{Cos}(G, H, S))$ consisting of all automorphisms of G that fix H that are also automorphisms of $\mathrm{Cos}(G, H, S)$. That is,*

$$\mathrm{Aut}(G, H, S) = \bar{A} \cap \mathrm{Aut}(\mathrm{Cos}(G, H, S)) = N_{S_G}(G, G/H) \cap \mathrm{Stab}_{\mathrm{Aut}(\mathrm{Cos}(G,H,S))}(H).$$

Proof By Exercise 4.4.7, we see that if $\alpha \in \mathrm{Aut}(G)$ and $\alpha(H) = H$ then $\alpha \in \mathrm{Aut}(\mathrm{Cos}(G, H, S))$ if and only if $\alpha(HSH) = HSH$. Thus $\mathrm{Aut}(G, H, S)$ consists of those automorphisms of G that map H to H and are contained in $\mathrm{Aut}(\mathrm{Cos}(G, H, S))$, and the first equality holds. By Lemma 4.4.9, we have $N_{S_{G/H}}(G, G/H) = \bar{A}(G, G/H)$. An element of $N_{S_{G/H}}(G, G/H)$ will stabilize H if and only if it is contained in $\bar{A}\left\{\hat{h}_L : h \in H\right\}$, and by Lemma 4.4.10, $\bar{A}\left\{\hat{h}_L : h \in H\right\} = \bar{A}$. Thus,

$$\bar{A} \cap \mathrm{Aut}(\mathrm{Cos}(G, H, S)) = N_{S_{G/H}}(G, G/H) \cap \mathrm{Stab}_{\mathrm{Aut}(\mathrm{Cos}(G,H,S))}(H). \qquad \square$$

In the previous result, the special case of Cayley graphs is most frequently encountered in the literature. Because of this, we state that special case separately.

Definition 4.4.15 Let G be a group and $S \subseteq G$. Set $\mathrm{Aut}(G, S) = \{\alpha \in \mathrm{Aut}(G) : \alpha(S) = S\}$. Hence $\mathrm{Aut}(G, S)$ is the set of all automorphisms of G that fix S.

Corollary 4.4.16 $\mathrm{Aut}(G, S) = N_{\mathrm{Aut}(\mathrm{Cay}(G,S))}(G_L) \cap \mathrm{Stab}_{\mathrm{Aut}(\mathrm{Cay}(G,S))}(1_G)$.

Returning to the main topic of the section, we can now also state an equivalent form of Burnside's theorem (Theorem 4.4.2) that is often encountered. Before stating this result, we need the following definition.

Definition 4.4.17 For a group G, the **socle of** G, denoted $\mathrm{soc}(G)$, is the subgroup of G generated by all minimal normal subgroups of G.

It is not difficult to show that $\mathrm{soc}(G)$ is the direct product of some (but not necessarily all) nontrivial minimal normal subgroups of G. See Exercise 4.4.2.

Theorem 4.4.18 *Let G be a transitive permutation group of prime degree p. Then either G is 2-transitive with nonregular nonabelian simple socle or G has a normal Sylow p-subgroup.*

Proof As in Theorem 4.4.1, relabel the set G permutes so that the map $x \mapsto x + 1$ is contained in G. Then, by Theorem 4.4.2, either G is 2-transitive, or $G < \mathrm{AGL}(1, p)$. As $\mathrm{AGL}(1, p) = N_{S_p}((\mathbb{Z}_p)_L)$ by Corollary 4.4.11, we have that either G is 2-transitive or G has a normal Sylow p-subgroup. In the latter case we are finished. If G is 2-transitive, then another theorem of Burnside (Dixon and Mortimer, 1996, Theorem 4.1A) states that the socle of a 2-transitive group is either a regular elementary abelian p-group P or is a nonregular nonabelian simple group. In the latter case, the result follows. In the former case, as $G \leq S_p$ and $|S_p| = p!$, we have that P is cyclic of order p. Then G normalizes in S_p a cyclic group of order p, and the result also follows. □

As Burnside's theorem (Theorem 4.4.2) was so effective in determining the automorphism groups of circulant digraphs of prime order, one might wonder if there are generalizations of this result to other orders. The first is due to Jones (1979) who proved that if $G \leq S_{p^2}$ with an elementary abelian Sylow p-subgroup, then $\mathrm{soc}(G) = T_1 \times T_2$, where T_1 and T_2 are simple groups of degree p (so either isomorphic to $\mathbb{Z}_p$ or a nonregular nonabelian 2-transitive simple group). Li and Sim (2001) proved that a transitive group G with a regular metacyclic Sylow p-subgroup P is 2-transitive or that $P \trianglelefteq G$. We remark that this result is not stated in Li and Sim (2001) as they were only interested in automorphism groups of graphs, but this result can be extracted from their proofs. Dobson and Witte (2002) showed that up to conjugation there are exactly $2p - 1$ transitive p-subgroups of S_{p^2}, and all but three of them (including $\mathbb{Z}_p \times \mathbb{Z}_p$) have the property that any transitive group G with Sylow p-subgroup P is either 2-transitive with nonregular nonabelian simple socle or contains a normal Sylow p-subgroup. Dobson (2005) has shown that a transitive group G of odd prime power degree p^k whose Sylow p-subgroup P has the property that every minimal transitive subgroup of P is cyclic is

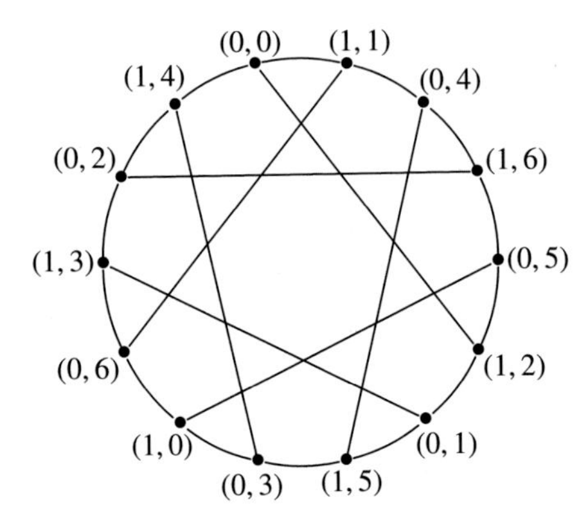

Figure 4.4 The Heawood graph with its metacirculant labeling.

either 2-transitive with nonabelian nonregular simple socle or $P \trianglelefteq G$. Finally, Dobson (2009) has shown that a transitive group with a regular abelian Sylow p-subgroup, p a sufficient large odd prime, has a normal subgroup that is a direct product of cyclic groups whose orders are power of p and nonregular nonabelian 2-transitive simple groups whose degrees are power of p. Dobson, Li, and Spiga then removed all conditions on p in the previous result (Dobson et al., 2012) .

4.5 The Automorphism Group of the Heawood Graph

As mentioned previously, the automorphism group of the Heawood graph will give an example of a graph where the automorphism group can be "reduced" to two smaller groups, but where one is not the automorphism group of a graph. In particular, we will show in Corollary 4.5.8 that the automorphism group of the Heawood graph is $G = \mathrm{PGL}(3,2) \rtimes \mathbb{Z}_2$, and has a block system $\mathcal{B}$ with two blocks of size 7, $\mathrm{fix}_G(\mathcal{B})$ is faithful on $B \in \mathcal{B}$, and $\mathrm{fix}_G(\mathcal{B})^B = \mathrm{PGL}(3,2)$. But $\mathrm{PGL}(3,2)$ is 2-transitive but not a symmetric group and so is not the automorphism group of any graph or digraph. So this "reduction" to groups of smaller degree (made explicit in the next chapter) does not work very well as it reduces the automorphism group to two groups, one of which is not the automorphism group of a graph (and so induction does not work smoothly).

We first would like to point out that the metacirculant drawing of the Heawood graph that we saw in Example 1.5.10 is not the "usual drawing" of the Heawood graph. We give this in Figure 4.4 with its metacirculant labeling of vertices as in Example 1.5.10. Note that with the metacirculant drawing of the Heawood graph in Exercise 1.5.10, it is easy to see that the Heawood graph is bipartite – and it is not so easy to see that the Heawood graph is bipartite in Figure 4.4. However, it is easy to see that the Heawood graph is Hamiltonian in Figure 4.4, while with the metacirculant drawing this is not clear by inspection. So, as with the Petersen graph, viewing this graph from different points of view can make finding different graph properties easier. In order to determine the automorphism group of the Heawood graph, we will need yet another point of view, with this new point of view coming from linear algebra and geometry. Our first few results are in more generality than is needed here, as the proofs of the more general results are exactly the same as the specific results that we require. Throughout this section, let q be a prime power, $\mathbb{F}_q$ the unique field of order q, and $d \geq 2$.

Definition 4.5.1 For a subspace S of $\mathbb{F}_q^d$, we denote by $S^\perp$ the **orthogonal complement of S**. That is, $S^\perp = \left\{w \in \mathbb{F}_q^d : w \cdot v = 0 \text{ for every } v \in S\right\}$.

Recall that $S^\perp$ is a subspace of the vector space $\mathbb{F}_q^d$.

Lemma 4.5.2 *Let $A \in \mathrm{GL}(d,q)$, and S be a subspace of $\mathbb{F}_q^d$. Then $A(S^\perp)^\perp = \left(A^{-1}\right)^T(S)$.*

Proof First, recall that if $w, v \in \mathbb{F}_q^d$ are column vectors, then the dot product of w and v, $w \cdot v$, can also be written as $w^T v$, where for a matrix B, B^T denotes the transpose of B. Let $w_1, \ldots, w_r$ be a basis for $S^\perp$, so that $A(S^\perp)$ has basis $Aw_1, \ldots, Aw_r$. To show that $A(S^\perp)^\perp = \left(A^{-1}\right)^T(S)$, it suffices to show that $\left(A^{-1}\right)^T v$ is orthogonal to Aw_i for any $1 \leq i \leq r$ and $v \in S$, as $\dim(S) + \dim(S^\perp) = d$. Then,

$$(Aw_i) \cdot \left(A^{-1}\right)^T v = (Aw_i)^T \left(A^{-1}\right)^T v = w_i^T A^T \left(A^{-1}\right)^T v = w_i^T v = 0. \qquad \square$$

Definition 4.5.3 A **point** in $\mathbb{F}_q^d$ is a one-dimensional subspace, while a **hyperplane** is the orthogonal complement of a point (so a subspace of $\mathbb{F}_q^d$ of dimension $d-1$). A **scalar matrix** is a matrix of the form αI_d, where $\alpha \in \mathbb{F}_q^*$.

Note that the number of points and hyperplanes of $\mathbb{F}_q^d$ are the same. The set Z of all scalar matrices is a normal subgroup of $\mathrm{GL}(d,q)$, and each element of Z fixes each point and hyperplane of $\mathbb{F}_q^d$. The notation Z is chosen as it turns out that Z is the center of $\mathrm{GL}(d,q)$ – see Alperin and Bell (1995, Proposition 6.4).

Definition 4.5.4 Define the **projective general linear group of dimension** d **over the field** $\mathbb{F}_q$, denoted $\mathrm{PGL}(d,q)$, to be the quotient group $\mathrm{GL}(d,q)/Z$. The set of all one-dimensional subspaces, or points, of $\mathbb{F}_q^d$ whose hyperplanes are the $(d-1)$-dimensional subspaces of $\mathbb{F}_q^d$ with incidence relation given by containment, is a **projective space** or a **projective geometry of dimension** $d-1$, denoted $\mathrm{PG}(d-1,q)$.

The group $\mathrm{PGL}(d,q)$ is transitive on points, as for nonzero $v, w \in \mathbb{F}_q^d$, there is a linear transformation A with $Av = w$, and so A maps the one-dimensional subspace p_1 spanned by v to the one-dimensional subspace p_2 spanned by w. As every element of Z fixes every one-dimensional subspace, any element of the left coset AZ maps p_1 to p_2. As $\mathrm{PGL}(d,q)$ permutes the set P of points of $\mathbb{F}_q^d$ transitively, it also permutes the set H of hyperplanes of $\mathbb{F}_q^d$ transitively. Thus $\mathrm{PGL}(d,q)$ is also a permutation group on the hyperplanes of $\mathbb{F}_q^d$.

Note that the representations of $\mathrm{PGL}(d,q)$ on points and hyperplanes are inequivalent if $d > 2$ by Theorem 4.1.8, as the subgroup of $\mathrm{PGL}(d,q)$ that stabilizes a fixed point $p \in P$ is transitive on the remaining points in P (and so by Lemma 4.4.3, $\mathrm{PGL}(d,q)$ is 2-transitive on points), and so fixes no hyperplane $h \in H$ as the dimension of h is at least 2. See Exercise 4.1.4. It should also be clear that both of these actions are faithful – see Exercise 4.1.5.

By Lemma 3.8.10, $\mathrm{GL}(d,q)$ has order $\Pi_{i=0}^{d-1}(q^d - q^i)$, and so $\mathrm{PGL}(d,q)$ has order $\frac{1}{q-1}\Pi_{i=0}^{d-1}(q^d - q^i)$. Some final words about terminology. It is clear that as we "project" from $\mathbb{F}_q^d$ to $\mathrm{PG}(d-1,q)$, the dimension is reduced by 1; hence the use of $d-1$ in $\mathrm{PG}(d-1,q)$. Because a point is an object of dimension 1 in $\mathbb{F}_q^d$, a point in $\mathrm{PG}(d-1,q)$ has dimension 0, and so is also called a **projective point**.

Definition 4.5.5 Let q be a prime power, and $d \geq 2$. Let P and H be the sets of points and hyperplanes, respectively, of the vector space $\mathbb{F}_q^d$. Define a graph Γ by $V(\Gamma) = P \cup H$ and a point is adjacent to a hyperplane in Γ if and only if the point is contained in the hyperplane. These graphs will be denoted by $B(\mathrm{PG}(d-1,q))$, and are called the **Levi graphs** or **point-hyperplane incidence graphs of** $\mathbb{F}^d$.

As $\mathbb{F}_q^d$ contains $(q^d-1)/(q-1)$ one-dimensional subspaces or points (each such point contains $q-1$ nonzero distinct vectors), so $|P| = (q^d-1)/(q-1)$. As the number of hyperplanes and points are the same, $|H| = (q^d-1)/(q-1)$ as well. Hence $|V(\Gamma)| = 2(q^d-1)/(q-1)$. In Figure 4.5, we draw Γ for the case where $d = 3$ and $q = 2$ (so $\mathbb{F}_2 = \{0, 1\}$). Note that $(q^d-1)/(q-1) = 7$ so that Γ has 14 vertices. Unsurprisingly, it is the Heawood graph.

The group $\mathrm{PGL}(d,q)$ has a natural action on $P \cup H$, where, for $A \in \mathrm{GL}(d,q)$, we have $(AZ)(S) = AS$, where $S \in P \cup H$. This makes sense as the image of a one-dimensional subspace under a linear transformation is a one-dimensional subspace, and the image of a hyperplane under a linear transformation is also

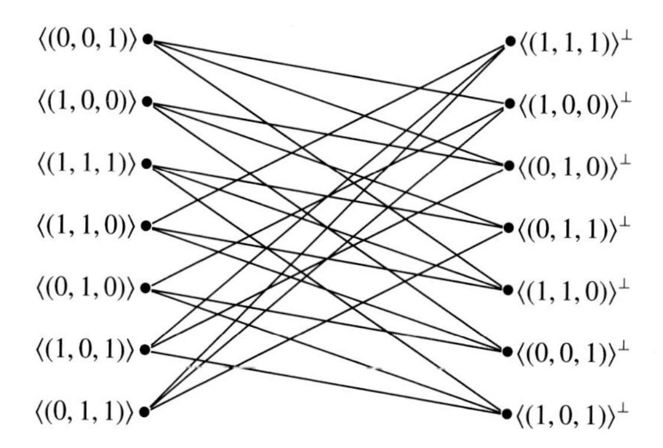

Figure 4.5 The Heawood graph labeled with the points and hyperplanes of $\mathbb{F}_2^3$.

a hyperplane. Henceforth, when we say $\mathrm{PGL}(d,q)$ acts on $P \cup H$ canonically, we will mean this action.

Define $\tau\colon P \cup H \to P \cup H$ by $\tau(S) = S^{\perp}$. Note that τ is well defined, as the subspace orthogonal to a point is a hyperplane, while the subspace orthogonal to a hyperplane is a point.

Theorem 4.5.6 *Let q be a prime power and $d \geq 3$. Then,*

$$\mathrm{PGL}(d,q) \rtimes \mathbb{Z}_2 \cong \langle \mathrm{PGL}(d,q), \tau\rangle = G \leq \mathrm{Aut}(B(\mathrm{PG}(d-1,q)),$$

is transitive. Also, $\mathcal{B} = \{P, H\}$ is a block system of G, $\mathrm{fix}_G(\mathcal{B}) = \mathrm{PGL}(d,q)$, $\mathrm{PGL}(d,q)$ acts faithfully on P and H, and the permutation representations induced by the action of $\mathrm{PGL}(d,q)$ *on P and H are inequivalent.*

Proof Let $g \in \mathrm{PGL}(d,q)$, $p \in P$, and $h \in H$ with $p \subset h$. Clearly the point $g(p)$ is contained in the hyperplane $g(h)$. Thus $\mathrm{PGL}(d,q) \leq \mathrm{Aut}(\Gamma)$. Also, $|\tau| = 2$ as $(S^{\perp})^{\perp} = S$ for every subspace S of $\mathbb{F}_q^d$. As every vector in $h^{\perp}$ is orthogonal to every vector in h, and as $p \subset h$, every vector in $h^{\perp}$ is orthogonal to every vector in p. Thus $h^{\perp} \subset p^{\perp}$ and so if $ph \in E(\Gamma)$, then $\tau(ph) = \tau(p)\tau(h) \in E(\Gamma)$. So $\tau \in \mathrm{Aut}(\Gamma)$, and $G \leq \mathrm{Aut}(B(\mathrm{PG}(d-1,q)))$.

Let $S \in P \cup H$. Then for $g \in \mathrm{PGL}(d,q)$, we have

$$\tau^{-1}g\tau(S) = \tau^{-1}g(S^{\perp}) = \tau^{-1}(g(S^{\perp})) = g(S^{\perp})^{\perp}.$$

Let $A \in \mathrm{GL}(d,q)$ such that $g = AZ$ (recall that the elements of $\mathrm{PGL}(d,q)$ are cosets of Z in $\mathrm{GL}(d,q)$). Then $A(S^{\perp})^{\perp} = \left(A^{-1}\right)^T(S)$ by Lemma 4.5.2. Thus $g(S^{\perp})^{\perp} = \left(g^{-1}\right)^T(S)$, where $\left(g^{-1}\right)^T = \left(A^{-1}\right)^T Z$. This implies that $\tau^{-1}g\tau = \left(g^{-1}\right)^T$ and $\mathrm{PGL}(d,q) \trianglelefteq G$. Then $G \cong \mathrm{PGL}(d,q) \rtimes \mathbb{Z}_2$ by the definition of the internal semidirect product. Finally, $\mathrm{PGL}(d,q)$ is transitive on P and H. As τ interchanges P and H, we have that G is transitive on $P \cup H$. The set of orbits $\{P, H\}$ of $\mathrm{PGL}(d,q) \trianglelefteq G$ is a block system $\mathcal{B}$ of G with two blocks. Then $\mathrm{fix}_G(\mathcal{B}) = \mathrm{PGL}(d,q)$, $\mathrm{PGL}(d,q)$ is faithful on each $B \in \mathcal{B}$, and $\mathrm{PGL}(d,q)^P$ is inequivalent to $\mathrm{PGL}(d,q)^H$ by Theorem 4.1.8. □

To calculate the automorphism groups of the graphs $B(\mathrm{PG}(d-1,q))$, more machinery is needed in general. This machinery will be developed in Section 6.4. We now focus exclusively on the Heawood graph as we already have the machinery necessary to calculate its automorphism group. So henceforth, $d = 3$ and $\mathbb{F}_q = \{0,1\}$. We will use the same machinery, developed by Tutte, that we used to determine the automorphism group of the Petersen graph in Section 2.1. We thus consider the s-arc-regularity of the Heawood graph.

Theorem 4.5.7 *The Heawood graph is 4-arc regular.*

Proof For brevity, we denote the Heawood graph by its designation F14 in the Foster census. By Theorem 3.1.5, it suffices to show that F14 is 4-arc-transitive but not 5-arc-transitive. As $(0,2),(1,4),(0,0),(1,1),(0,4),(1,6)$ is a 5-arc contained in a 6-cycle while $(0,0)$, $(1,1)$, $(0,4)$, $(1,5)$, $(0,3)$, $(1,0)$ is a 5-arc not contained in a 6-cycle (using the metacirculant labeling of F14 as in Figure 4.4), F14 is not 5-arc-transitive. To show that F14 is 4-arc-transitive, we use the labeling of Figure 4.5. Suppose a_i, b_i, c_i, d_i, f_i is a 4-arc in F14, $i = 1,2$. As τ is an automorphism of F14, we may assume without loss of generality that $a_i \in P$, in which case $a_i, c_i, f_i \in P$ and $b_i, d_i \in H$, $i = 1,2$. As each hyperplane in $\mathbb{F}_2^3$ has dimension 2, we have $\langle a_i, c_i\rangle = b_i$ and $\langle c_i, f_i\rangle = d_i$ (we are abusing notation here by thinking of a_i, for example, as not only a point, but as the vector that generates the point). Then a_i and c_i are linearly independent, as are c_i and f_i. The only possible way in which a_i and f_i are not linearly independent is if $a_i = f_i$, but in that case a_i, b_i, c_i, d_i, f_i is a 4-cycle while F14 has girth 6. Thus $\{a_i, c_i, f_i\}$, $i = 1,2$, is a basis for $\mathbb{F}_2^3$, and so there is $T \in \mathrm{GL}(3,2)$ such that $T(a_1) = a_2$, $T(c_1) = c_2$, and $T(f_1) = f_2$ (thinking of a_i, c_i, and f_i as vectors). Then $T(b_1) = b_2$, $T(d_1) = d_2$, and so $T(a_1,b_1,c_1,d_1,f_1) = a_2,b_2,c_2,d_2,f_2$. □

Corollary 4.5.8 *The automorphism group of the Heawood graph is isomorphic to* $\mathrm{PGL}(3,2) \rtimes \mathbb{Z}_2$.

Proof As the Heawood graph is 4-arc-regular, it is arc-transitive by Exercise 3.1.2. Also, $|\mathrm{PGL}(3,2)| = 7\cdot 6\cdot 4$ and so $|\mathrm{PGL}(3,2) \rtimes \mathbb{Z}_2| = 2^4\cdot 3\cdot 7$ (see Exercise 4.1.1). The result then follows by Corollary 3.1.6. □

In the literature, one will also see $\mathrm{Aut}(\mathrm{F14}) \cong \mathrm{PSL}(3,2) \rtimes \mathbb{Z}_2$ (the group $\mathrm{PSL}(3,2)$ is defined in Section 5.4 and $\mathrm{PSL}(3,2) = \mathrm{PGL}(3,2)$). It also turns

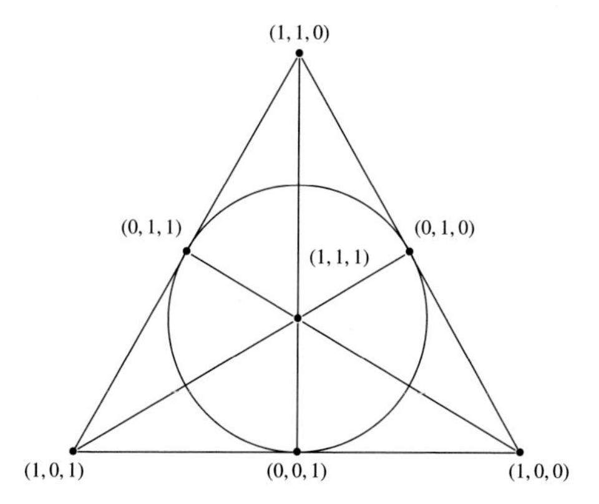

Figure 4.6 The Fano Plane.

out that $\mathrm{PGL}(2,7) \cong \mathrm{PGL}(3,2) \rtimes \mathbb{Z}_2$, so sometimes one sees $\mathrm{Aut}(\mathrm{F14}) = \mathrm{PGL}(2,7)$, but from our point of view there is no "natural" action of $\mathrm{PGL}(2,7)$ on $V(\mathrm{F14})$.

One last thing before we leave the Heawood graph. As we have seen, the Heawood graph is intimately related to the vector space $\mathbb{F}_2^3$. There is a standard way of representing this vector space (and so extracting the necessary information to construct the Heawood graph) geometrically, the **Fano plane**, shown in Figure 4.6. The vertices are labeled with the nonzero elements of $\mathbb{F}_2^3$. Notice that the vertices on a straight line are the nonzero elements of a hyperplane, while the vertices on the circle are also the nonzero elements of a hyperplane. As there are six straight lines and one circle, and $\mathbb{F}_2^3$ contains seven hyperplanes, every nonzero element of $\mathbb{F}_2^3$ and every hyperplane is represented in the Fano plane, as well as which nonzero elements are contained in which hyperplane. Finally, the circle in the Fano plane is usually referred to as a "line" so that we can simply say that the nonzero elements of a hyperplane are the vertices of a line in the Fano plane. We will also use the Fano plane in constructing the Coxeter graph in Section 6.1.

4.6 Exercises

4.1.1 Show that semidirect product $N \rtimes H$ has order $|H| \cdot |N|$.

4.1.2 Show that the Heawood graph is a Cayley graph of the dihedral group of order 14.

4.1.3 Let $G \leq S_X$ have normal block systems $\mathcal{B}$ and C. Let $x \in X$ and

$B_x \in \mathcal{B}$ and $C_x \in C$ such that $x \in B_x \cap C_x$. Show that the block system consisting of $B_x \cap C_x$ and all of its conjugate blocks need not be a normal block system of G.

4.1.4 Show that $\mathrm{PGL}(d,q)$ is 2-transitive on the points and hyperplanes of $\mathbb{F}_q^d$. (Hint: Given points p_1, p_2, p_3, is there a matrix in $\mathrm{GL}(d,q)$ that maps p_1 to p_1 and p_2 to p_3?)

4.1.5 Show that $\mathrm{PGL}(d,q)$ is faithful on points and hyperplanes by first considering its action on the points generated by the canonical basis vectors and the vector consisting of all 1s.

4.2.1 Draw the graph $K_4 \wr \bar{K}_3$.

4.2.2 Let p be prime. Show that the unitary circulant graph of order p^k is isomorphic to $K_p \wr \bar{K}_{p^{k-1}}$.

4.2.3 Show that $G \wr H$ has order $|G| \cdot (|H|)^{|X|}$, where $G \leq S_X$.

4.2.4 Show that a Sylow p-subgroup of S_{p^k}, $k \geq 1$ is $\mathbb{Z}_p^{\wr k} = \mathbb{Z}_p \wr \mathbb{Z}_p \wr \cdots \wr \mathbb{Z}_p$, where the wreath product is taken k times.

4.2.5 Show that a Sylow p-subgroup of S_{p^k}, $k \geq 1$, has order p^ℓ, where $\ell = (p^k - 1)/(p-1)$.

4.2.6 Let n be a positive integer and p a prime. Let $n = a_0 + a_1 p + a_2 p^2 + \cdots + a_k p^k$. Show that a Sylow p-subgroup of S_n is isomorphic to $\Pi_{i=1}^k (\mathbb{Z}_p^{\wr i})^{a_i}$, where $\mathbb{Z}_p^{\wr 0} = 1$. (Hint: The highest power of p that divides $n!$ is $\sum_{i=1}^{\infty} \lfloor n/p^i \rfloor$; see, for example, Burton (2010, Theorem 6.9).)

4.2.7 Let $G \leq S_n$ be transitive with block system $\mathcal{B}$ whose blocks are of prime size p. If $G/\mathcal{B}$ is 2-transitive and there exists $g \in \mathrm{Stab}_G(B) \cap \mathrm{Stab}_G(B')$ for distinct blocks $B, B' \in \mathcal{B}$, and g in its action on B is a p-cycle while g in its action on B' is not a p-cycle, then any vertex-transitive digraph Γ of order n with $G \leq \mathrm{Aut}(\Gamma)$ can be written as a wreath product $\Gamma_1 \wr \Gamma_2$, where $\Gamma_1 = K_{n/p}$ or $\bar{K}_{n/p}$ and Γ_2 is a vertex-transitive digraph of prime order p.

4.3.1 Verify that the graph wreath product is associative.

4.3.2 Verify that the permutation group wreath product is associative.

4.3.3 Give an example of a permutation group G with a normal block system $\mathcal{B}$ such that $\mathrm{Stab}_G(B)^B \neq \mathrm{fix}_G(\mathcal{B})^B$.

4.3.4 Let $G \leq S_n$ be transitive and $\mathcal{B}$ be the unique block system of G with blocks of size k. Suppose $\phi \in S_n$ such that $\mathcal{B}$ is also a block system of $\langle G, \phi G \phi^{-1} \rangle$. Show that $\phi \in S_m \wr S_k$, where $n = mk$.

4.3.5 Let G be a group, $H \leq G$, and $S \subseteq G$ with $S = HSH$. Let $N = \mathrm{core}_G(H)$. Show that $\lambda\colon G/H \to (G/N)/(H/N)$ given by $\lambda(gH) = g(H/N)$ as in Lemma 4.3.4 is an isomorphism from $\mathrm{Cos}(G,H,S)$ to $\mathrm{Cos}(G/N, H/N, \{sN : s \in S\})$.

4.4.1 Let G be a group, M a minimal normal subgroup of G, and $N \trianglelefteq G$. Show that either $M \leq N$ or $\langle M, N\rangle \cong M \times N$.

4.4.2 Let G be a group. Show that $\mathrm{soc}(G)$ is a direct product of some minimal transitive subgroups of G.

4.4.3 Calculate the automorphism group of $\mathrm{Cay}(\mathbb{Z}_{13}, \{2, 5, 6\})$.

4.4.4 Calculate the automorphism group of $\mathrm{Cay}(\mathbb{Z}_{17}, \{1, 3, 7, 12\})$.

The purpose of the next four exercises is to verify previously known facts about Cayley digraphs for double coset digraphs.

4.4.5 Let G_1 and G_2 be isomorphic groups with $\alpha\colon G_1 \to G_2$ an isomorphism. Let $H \leq G_1$, and $S \subseteq G$ such that $HSH = S$. Then $\alpha(\mathrm{Cos}(G_1, H, S)) \cong \mathrm{Cos}(G_2, \alpha(H), \alpha(S))$. In particular, if $G_1 = G_2$ then $\alpha(\mathrm{Cos}(G, H, S))$ is a double coset digraph of G_1 with connection set $\alpha(S)$, or, equivalently, $\alpha(\mathrm{Cos}(G_1, H, S)) = \mathrm{Cay}(G_1, \alpha(H), \alpha(S))$.

4.4.6 Let G be a group and $H < G$. Let $\alpha \in \mathrm{Aut}(G)$ such that $\alpha(H) = H$. Show that $\alpha(\mathrm{Cos}(G, H, S)) = \mathrm{Cos}(G, H, \alpha(S))$ (compare to Lemma 1.2.15).

4.4.7 Let G be a group and $H < G$. Show that $\alpha \in \mathrm{Aut}(G)$ such that $\alpha(H) = H$ is contained in $\mathrm{Aut}(\mathrm{Cos}(G, H, S))$ if and only if $\alpha(HSH) = HSH$ (compare to Corollary 1.2.16).

4.4.8 Let G be a group and $H < G$. Let $\alpha \in \mathrm{Aut}(G)$ such that $\alpha(H) = H$ and O be an orbit of α acting on the left cosets of H in G. Show that $\mathrm{Cos}(G, H, O)$ is arc-transitive (compare to Lemma 3.2.2).

4.5.1 Show that S_5 has two inequivalent transitive actions of degree 60.

4.5.2 Let $n \geq 2$. Show that there are $(n-1)!$ n-cycles in S_n. Deduce that there are $(n-1)!/\varphi(n)$ regular cyclic subgroups of S_n, where φ is **Euler's phi function**. Euler's phi function $\varphi\colon \mathbb{N} \to \mathbb{N}$ is defined by $\varphi(n)$ is the number of positive integers less than n that are relatively prime to n.

4.5.3 Write out a detailed proof that $\mathrm{fix}_G(\mathcal{B})^B$ is permutation isomorphic to $\mathrm{fix}_G(\mathcal{B})^{B'}$, as in Theorem 4.1.4.

4.5.4 Use Theorem 2.4.13 to show that every transitive permutation group G is permutation equivalent to the permutation representation induced by left multiplication of left cosets of the stabilizer of a point in G.

4.5.5 Use Theorem 4.1.8 to prove Theorem 2.4.13.

4.5.6 Show that any transitive subgroup of $\mathrm{AGL}(1, p)$ has order cp for some $c|(p-1)$, and use Hall's Theorem (Dummit and Foote, 2004, Exercise 6.1.33) to show that any two transitive and faithful permutation representations of a subgroup of $\mathrm{AGL}(1, p)$ of order cp and degree p are permutation equivalent representations.

4.5.7 Let $G \leq S_n$ be transitive and $x, y \in \mathbb{Z}_n$. Let Γ be the orbital digraph of G that contains (x, y). Let $H \leq S_n$ be permutation isomorphic to G via the bijection $\lambda \colon \mathbb{Z}_n \to \mathbb{Z}_n$ and isomorphism $\phi \colon G \to H$. Show that $\lambda(\Gamma)$ is the orbital digraph of H containing the arc $(\phi(x), \phi(y))$.

4.5.8 Show that for a group G, $\mathrm{Inn}(G) \trianglelefteq \mathrm{Aut}(G)$.

5

The End of the Beginning

This chapter is, in the view of the authors, the end of the beginning, in that the flavor of subsequent chapters is different. In Chapter 6, we introduce and study more specialized families of graphs. In the rest of the book, each chapter focuses on what the authors consider a major problem in the study of vertex-transitive graphs, with three chapters revisiting and expanding on problems we saw in Chapter 3.

To start with in this chapter, in Section 5.1 we finish the study of wreath products by determining the full automorphism groups of vertex-transitive graphs that can be written as wreath products. As mentioned earlier, while we are primarily concerned with graphs, most of the results on vertex-transitive graphs also work without much change for vertex-transitive digraphs. For some problems, most arguments also work in a larger class, namely vertex-transitive color digraphs. In Section 5.2 we study their automorphism groups, which Wielandt (1969) calls **2-closed groups**. In Section 5.3, we introduce **generalized wreath product** digraphs. These digraphs are closely related to digraphs that are wreath products, and feature prominently in the study of isomorphisms between, and automorphism groups of, circulant digraphs in Sections 7.7 and 8.5, respectively. In Section 5.4, we give a brief introduction to the O'Nan–Scott theorem, which gives the structure of primitive groups. This result, together with the classification of the finite simple groups (CFSG), is a powerful tool that in some cases allows for the complete determination of primitive groups satisfying certain conditions. Many problems on vertex-transitive graphs can, at the moment, only be solved using applications of the O'Nan–Scott theorem and the CFSG. Finally, in Section 5.5 we consider covers. Intuitively, with covers we are interested in the relationship between a graph and quotients of the graph, and, in particular, whether an automorphism of a quotient graph will "lift" to an automorphism of the original graph.

5.1 The Automorphism Group of Wreath Product Graphs

Note that $K_r \wr K_s = K_{rs}$. Also, $\mathcal{S}_r \wr \mathcal{S}_s \neq \mathcal{S}_{rs}$, and so we have an example where the automorphism group of the wreath product of two graphs is not the wreath product of the corresponding graph automorphism groups. A similar statement holds for complements of complete graphs. We shall see that while these are not the only such examples, they do capture the idea behind all such examples.

We are now ready to determine the automorphism group of a wreath product graph. Harary (1959) initially claimed that unless Γ_1 and Γ_2 were complete or the complements of complete graphs, then $\mathrm{Aut}(\Gamma_1 \wr \Gamma_2) = \mathrm{Aut}(\Gamma_1) \wr \mathrm{Aut}(\Gamma_2)$. Sabidussi (1959) observed that this was not quite correct, and gave necessary and sufficient conditions for $\mathrm{Aut}(\Gamma_1 \wr \Gamma_2) = \mathrm{Aut}(\Gamma_1) \wr \mathrm{Aut}(\Gamma_2)$. Dobson and Morris (2009) later determined $\mathrm{Aut}(\Gamma_1 \wr \Gamma_2)$ when Γ_1 and Γ_2 are vertex-transitive (Theorem 5.1.8), although much of the proof that is given here is Sabidussi's. Theorem 5.1.8 also holds for color digraphs, which are defined in Definition 5.2.1. We begin with some preliminary lemmas and definitions.

Lemma 5.1.1 *Let Γ be a vertex-transitive graph. Define an equivalence relation $\equiv$ on $V(\Gamma)$ by $x \equiv y$ if and only if $N_\Gamma(x) = N_\Gamma(y)$. If an equivalence class of $\equiv$ has r elements, then $\Gamma \cong \Gamma_1 \wr \bar{K}_r$ for some vertex-transitive graph Γ_1.*

Proof It is straightforward to verify that $\equiv$ is an equivalence relation. Additionally, if $\gamma \in \mathrm{Aut}(\Gamma)$ and $x \equiv y$, then $N_\Gamma(\gamma(x)) = N_\Gamma(\gamma(y))$, so $\equiv$ is an $\mathrm{Aut}(\Gamma)$-congruence. By Theorem 2.3.2, the set of equivalence classes of $\equiv$ is a block system $\mathcal{B}$ of $\mathrm{Aut}(\Gamma)$. If $r = 1$, then the result is trivial with $\Gamma_1 = \Gamma$. If $r \geq 2$ with say $x, y \in V(\Gamma)$, $x \neq y$, and $x \equiv y$, then $xy \notin E(\Gamma)$ as Γ is a graph and has no loops. Hence $\Gamma[B] = \bar{K}_r$ for every $B \in \mathcal{B}$. By Theorem 4.2.15, it only remains to show that if $B, B' \in \mathcal{B}$ and some vertex of B is adjacent to some vertex of B', then every vertex of B is adjacent to every vertex of B'. Suppose $x \in B$, $y \in B'$ and $xy \in E(\Gamma)$. Then $x'y \in E(\Gamma)$ for every $x' \equiv x$, and so every vertex of B is adjacent to y. Similarly, $xy' \in E(\Gamma)$ for every $y' \equiv y$. Hence every vertex of B' is adjacent to x. Then every vertex of B is adjacent to every vertex of B', as required. □

The equivalence relation given in the statement of the previous lemma has appeared several times in the literature (Kotlov and Lovász (1996); Sabidussi (1964); Wilson (2003)), and graphs for which the equivalence classes of this relation are not simply the diagonal have been studied. Note that $\equiv$ can also be defined for non-vertex-transitive graphs, but in that case the equivalence classes need not all have equal size.

Definition 5.1.2 A graph Γ is **reducible** or **unworthy** if there exist distinct vertices $u, v \in \Gamma$ such that $u \equiv v$. A graph that is not reducible or unworthy is **irreducible** or **worthy**, respectively.

The "reducible" terminology is due to Sabidussi (1964), while the "unworthy" terminology is due to Wilson (2003). Wilson calls them "unworthy" as an unworthy graph contains an automorphism that is a transposition. The same idea was also studied by Kotlov and Lovász (1996) who called vertices $u, v \in V(\Gamma)$ with $u \equiv v$ and $u \neq v$ **twins**.

The proof of the following result is quite similar to the proof of the preceding result, and is left as an exercise. The equivalence relation given in the next lemma should be thought of as a "complementary" equivalence relation to the one in the preceding lemma. This follows, as if $N_\Gamma(x) = N_\Gamma(y)$ then $N_{\bar{\Gamma}}(x) \cup \{x\} = N_{\bar{\Gamma}}(y) \cup \{y\}$.

Lemma 5.1.3 *Let Γ be a vertex-transitive graph. Define an equivalence relation $\equiv$ on $V(\Gamma)$ by $x \equiv y$ if and only if $N_\Gamma(x) \cup \{x\} = N_\Gamma(y) \cup \{y\}$. If an equivalence class of $\equiv$ has r elements, then $\Gamma \cong \Gamma_1 \wr K_r$ for some vertex-transitive graph Γ_1.*

We are now ready to begin the proof itself, which is quite long. As a consequence, the proof is broken down into a series of lemmas. We begin with the final notation we will need in this section (which will only be used in this section).

Definition 5.1.4 Let Γ_1 and Γ_2 be vertex-transitive digraphs, and $\mathcal{B}$ the lexi-partition of $\mathrm{Aut}(\Gamma_1) \wr \mathrm{Aut}(\Gamma_2)$ corresponding to $V(\Gamma_2)$. For each $x \in V(\Gamma_1)$ we denote the block $\{(x, y) : y \in V(\Gamma_2)\}$ of $\mathcal{B}$ by B_x. Similarly, if $\gamma \in \mathrm{Aut}(\Gamma_1 \wr \Gamma_2)$ (which need not be in $\mathrm{Aut}(\Gamma_1) \wr \mathrm{Aut}(\Gamma_2)$), then for $x \in V(\Gamma_1)$ and $y \in V(\Gamma_2)$ we denote the block of $\mathcal{B}$ that contains $\gamma(x, y)$ by $B_{\gamma(x,y)}$.

Lemma 5.1.5 *Let Γ_1 and Γ_2 be vertex-transitive graphs with Γ_1 connected. Let $\Gamma = \Gamma_1 \wr \Gamma_2$, $x \in V(\Gamma_1)$, $\mathcal{B}$ be the lexi-partition of $\mathrm{Aut}(\Gamma_1) \wr \mathrm{Aut}(\Gamma_2)$ corresponding to $V(\Gamma_2)$, and $B_x \in \mathcal{B}$. Assume there exists $\gamma \in \mathrm{Aut}(\Gamma)$ such that $\gamma(B_x) \cap B_x \neq \emptyset$ but $\gamma(B_x) \neq B_x$. Let $C_x = \{c \in V(\Gamma_1) : B_c \cap \gamma(B_x) \neq \emptyset\} \subseteq V(\Gamma_1)$. If $\Gamma_1[C_x]$ is not a complete graph, then there exist integers $r, s \geq 2$ and graphs Γ_1' and Γ_2' with Γ_2' connected such that $\Gamma_1 \cong \Gamma_1' \wr \bar{K}_r$ and $\Gamma_2 \cong \bar{K}_s \wr \Gamma_2'$.*

Proof Suppose $\Gamma_1[C_x]$ is not a complete graph. Then there exists $c_1, c_2 \in C_x$, $c_1 \neq c_2$, such that $c_1c_2 \notin E(\Gamma_1)$. As $c_1, c_2 \in C_x$, there exists $(x, y_1), (x, y_2) \in B_x$ such that $\gamma(x, y_1) = (c_1, w_1) \in B_{c_1}$ and $\gamma(x, y_2) = (c_2, w_2) \in B_{c_2}$, where $B_{c_1}, B_{c_2} \in \mathcal{B}$. As Γ_1 is connected, x is adjacent to some vertex z in Γ_1, and so $(x, y_1)(z, v)(x, y_2)$ is a path of length 2 in $\Gamma_1 \wr \Gamma_2$, where $v \in V(\Gamma_2)$. Hence

$\text{dist}_{\Gamma_1 \wr \Gamma_2}((x, y_1), (x, y_2)) \leq 2$. As $\gamma((x, y_1)(z, v)(x, y_2)) = (c_1, w_1)\gamma(z, v)(c_2, w_2)$, we see that $\text{dist}_{\Gamma_1}(c_1, c_2) \leq 2$. As $c_1c_2 \notin E(\Gamma_1)$, we have $\text{dist}_{\Gamma_1}(c_1, c_2) = 2$. As no vertex of B_{c_1} is adjacent to any vertex of B_{c_2}, $\text{dist}_{\Gamma_1 \wr \Gamma_2}((c_1, w_1), (c_2, w_2)) = 2$. Consequently, applying γ^{-1} to a path of length 2 from (c_1, w_1) to (c_2, w_2), we see that $\text{dist}_{\Gamma_1 \wr \Gamma_2}((x, y_1), (x, y_2)) = 2$. Then $y_1y_2 \notin E(\Gamma_2)$.

Now let W be the set of common neighbors in $\Gamma_1 \wr \Gamma_2$ of (x, y_1) and (x, y_2), so $W = N_{\Gamma_1 \wr \Gamma_2}(x, y_1) \cap N_{\Gamma_1 \wr \Gamma_2}(x, y_2)$. As $c_1c_2 \notin E(\Gamma_1)$, (c_1, w_1) and (c_2, w_2) have no common neighbors in $B_{c_1} \cup B_{c_2}$, and so $\gamma(W) = (N_{\Gamma_1}(c_1) \cap N_{\Gamma_1}(c_2)) \times V(\Gamma_2)$.

Claim 1: (x, y_1) and (x, y_2) have no common neighbors in B_x. We proceed by contradiction so assume y_1 and y_2 have a common neighbor in Γ_2. Let W' be the set of common neighbors of (x, y_1) and (x, y_2) in B_x. Note that $1 \leq |W'| < |V(\Gamma_2)|$ as $(x, y_1), (x, y_2) \notin W'$ and y_1 and y_2 have a common neighbor in Γ_2. Then $W = W' \cup (N_{\Gamma_1}(x) \times V(\Gamma_2))$, and $\gamma(W) = (N_{\Gamma_1}(c_1) \cap N_{\Gamma_1}(c_2)) \times V(\Gamma_2)$. We conclude that $|\gamma(W)|$ is divisible by $|V(\Gamma)|$ but that $|W|$ is not as $1 \leq |W'| < |V(\Gamma_2)|$, a contradiction establishing the claim.

Claim 2: The neighbors of c_1 and c_2 in Γ_1 are identical. By Claim 1, (x, y_1) and (x, y_2) have no common neighbors in B_x. This implies that $|W| = \deg_{\Gamma_1}(x) \cdot |V(\Gamma_2)|$. As $\gamma(W) = (N_{\Gamma_1}(c_1) \cap N_{\Gamma_1}(c_2)) \times V(\Gamma_2)$, $\deg_{\Gamma_1}(x) = |N_{\Gamma_1}(c_1) \cap N_{\Gamma_1}(c_2)|$, and as Γ_1 is regular, the neighbors of c_1 and c_2 in Γ_1 are identical, establishing the claim.

Claim 3: y_1 and y_2 are in different components of Γ_2. We proceed by contradiction so suppose y_1 and y_2 are in the same component of Γ_2. Set $x_0 = y_1$ and $x_t = y_2$, and let $y_1x_1x_2 \ldots x_{t-1}y_2$ be a shortest path from y_1 to y_2 in Γ_2.

Suppose first there are $0 \leq i < j \leq t$ such that $\gamma(x, x_i), \gamma(x, x_j) \in B \in \mathcal{B}$. Choose i so that i is minimal and j so that j is maximal. Note that as $c_1 \neq c_2$, it cannot be that $i = 0$ and $j = t$. We will consider the case where $i \neq 0$, the case where $j \neq t$ being similar. Then $\gamma(x, x_{i-1}) \notin B$. As $B_{\gamma(x,x_i)} = B$ we have that every vertex of $B_{\gamma(x,x_{i-1})}$ is adjacent to every vertex of B. In particular, $\gamma(x, x_{i-1})$ is adjacent to $\gamma(x, x_j)$, and so x_{i-1} is adjacent to x_j in Γ_2. This contradicts $x_0 \ldots x_t$ being a shortest path from $x_0 = y_1$ to $x_t = y_2$ in Γ_2.

The only remaining possibility is that if $0 \leq i < j \leq t$, then $B_{\gamma(x,x_i)}$ is distinct from $B_{\gamma(x,x_j)}$. Setting $\gamma(x, x_i) = (u_i, z_i)$, we have that $u_0u_1 \ldots u_t$ is a path of length t in Γ_1, and as $u_0 = c_1$, $u_t = c_2$, and the neighbors of c_1 and c_2 in Γ_1 are identical, we see that $t = 2$. Thus (x, y_1) and (x, y_2) have a common neighbor in B_x, a contradiction to Claim 1, which establishes Claim 3.

Let $\Gamma_2' = C$ be a component of Γ_2, and suppose Γ_2 has s components. As Γ_2 is vertex-transitive, the components of Γ_2 are isomorphic, and so $\Gamma_2 \cong \bar{K}_s \wr \Gamma_2'$. By Lemma 5.1.1 and Claim 2, we have $\Gamma_1 = \Gamma_1' \wr \bar{K}_r$, where $r > 1$. □

Lemma 5.1.6 *Let Γ_1 and Γ_2 be vertex-transitive graphs with Γ_1 connected. Let $\Gamma = \Gamma_1 \wr \Gamma_2$, $x \in V(\Gamma_1)$, $\mathcal{B}$ be the lexi-partition of* $\text{Aut}(\Gamma_1) \wr \text{Aut}(\Gamma_2)$

corresponding to $V(\Gamma_2)$*, and* $B_x \in \mathcal{B}$*. Assume there exists* $\gamma \in \mathrm{Aut}(\Gamma)$ *such that* $\gamma(B_x) \cap B_x \neq \emptyset$ *but* $\gamma(B_x) \neq B_x$*. Let* $C_x = \{c \in V(\Gamma_1) : B_c \cap \gamma(B_x) \neq \emptyset\} \subseteq V(\Gamma_1)$*. If* $\Gamma_1[C_x]$ *is a complete graph, then there exist integers* $r, s \geq 2$ *and graphs* Γ_1' *and* Γ_2' *with* $\bar{\Gamma}_2'$ *connected such that* $\Gamma_1 \cong \Gamma_1' \wr K_r$ *and* $\Gamma_2 \cong K_s \wr \Gamma_2'$*.*

Proof Suppose $\Gamma_1[C_x]$ is a complete graph. For $c \in C_x$, let $M_c = \{m : \gamma(x,m) \in B_c\}$. So M_c is the set of the second coordinate of all vertices in B_x that are mapped by γ to vertices in B_c. Let $c_1, c_2 \in C_x$, $c_1 \neq c_2$ (which is possible as $|C_x| \geq 2$ as $\gamma(B_x) \cap B_x \neq \emptyset$ but $\gamma(B_x) \neq B_x$), and $(x, y_1), (x, y_2) \in B_x$ such that $(x, y_1) \in M_{c_1}$ and $(x, y_2) \in M_{c_2}$. As $\Gamma_1[C_x]$ is complete, we have $\gamma(x, y_1)\gamma(x, y_2) \in E(\Gamma)$, and so $(x, y_1)(x, y_2) \in E(\Gamma)$. Hence $y_1 y_2 \in E(\Gamma_2)$. Thus every vertex of M_{c_1} is adjacent to every vertex of M_{c_2}. As $\{M_c : c \in B_x\}$ is a partition of $V(\Gamma_2)$, we see that $\bar{\Gamma}_2$ is disconnected, and being a vertex-transitive graph, $\bar{\Gamma}_2 = \bar{K}_s \wr \bar{\Gamma}_2'$ for some connected vertex-transitive graph $\bar{\Gamma}_2'$, where $s = |V(\Gamma_2)|/|V(\Gamma_2')|$. To finish the result, it only remains to show that $\equiv$ defined on $V(\Gamma_1)$ as in Lemma 5.1.3 has equivalence classes with $r > 1$ elements. We will show that C_x is contained in an equivalence class of $\equiv$.

Let $c_1, c_2 \in C_x$ such that $c_1 \neq c_2$, and $a \in V(\Gamma_1)$ such that $c_1 a \in E(\Gamma_1)$. If $a \in C_x$, then as $\Gamma_1[C_x]$ is complete, $c_2 a \in E(\Gamma_1)$ provided $a \neq c_2$. If $a \notin C_x$, then there exists $w_1 \in V(\Gamma_2)$ such that $(c_1, w_1)(a, z) \in E(\Gamma)$ for every $z \in V(\Gamma_2)$. Then $\gamma^{-1}((c_1, w_1)(a, z)) = (x, y_1)\gamma^{-1}(a, z)$ for some $y_1 \in V(\Gamma_2)$, and, as $a \notin C_x$, we have $\gamma^{-1}(a, z) \notin B_x$. Then every vertex of B_x is adjacent to every vertex of $B_{\gamma^{-1}(a,z)}$, and, in particular, $(x, y_2)\gamma^{-1}(a, z) \in E(\Gamma)$, where $\gamma(x, y_2) \in B_{c_2}$. Then $\gamma\big((x, y_2)\gamma^{-1}(a, z)\big) = (c_2, e)(a, z) \in E(\Gamma)$ for some $e \in V(\Gamma_2)$, and so $c_2 a \in E(\Gamma_1)$. As $c_1 c_2 \in E(\Gamma_1)$, we have $c_1 \equiv c_2$, and the result follows. □

Lemma 5.1.7 *Let* Γ_1 *and* Γ_2 *be vertex-transitive graphs. Let* $\Gamma = \Gamma_1 \wr \Gamma_2$ *and* $\mathcal{B}$ *be the lexi-partition of* $\mathrm{Aut}(\Gamma_1) \wr \mathrm{Aut}(\Gamma_2)$ *corresponding to* $V(\Gamma_2)$*. If* $\mathcal{B}$ *is a block system of* $\mathrm{Aut}(\Gamma)$*, then* $\mathrm{Aut}(\Gamma) = \mathrm{Aut}(\Gamma_1) \wr \mathrm{Aut}(\Gamma_2)$*.*

Proof By Lemma 4.2.14, we have $\mathrm{Aut}(\Gamma_1) \wr \mathrm{Aut}(\Gamma_2) \leq \mathrm{Aut}(\Gamma)$. If $\mathcal{B}$ is a block system of $\mathrm{Aut}(\Gamma)$, then $\mathrm{Stab}_{\mathrm{Aut}(\Gamma)}(B)^B$ is a subgroup of $\mathrm{Aut}(\Gamma_2)$, and for any $\gamma \in \mathrm{Aut}(\Gamma)$, $\gamma/\mathcal{B}$ is an automorphism of Γ_1. By the embedding theorem $\mathrm{Aut}(\Gamma) \leq \mathrm{Aut}(\Gamma_1) \wr \mathrm{Aut}(\Gamma_2)$, and so $\mathrm{Aut}(\Gamma) = \mathrm{Aut}(\Gamma_1) \wr \mathrm{Aut}(\Gamma_2)$. □

Suppose Γ is a finite graph (we emphasize that Γ is finite here as if Γ is not finite, then what follows need not be true), and $\Gamma \cong \Gamma' \wr K_u$ for some $u \geq 2$ and graph Γ'. If $\Gamma' \cong \Gamma'' \wr K_v$ for some $v \geq 2$ and graph Γ'', then $\Gamma \cong \Gamma'' \wr K_{uv}$. Repeating this process, if necessary, we see there is a maximum integer r such that $\Gamma \cong \Gamma' \wr K_r$, for some graph Γ'. The same argument in the complement shows that if $\Gamma \cong \Gamma' \wr \bar{K}_u$ for some $u \geq 2$ and graph Γ', there is also a maximum integer r such that $\Gamma \cong \Gamma' \wr \bar{K}_r$ for some graph Γ'.

Similarly, suppose that Γ is a finite graph, and $\Gamma \cong K_u \wr \Gamma'$ for some $u \geq 2$ and graph Γ'. If $\Gamma' \cong K_v \wr \Gamma''$ for some $v \geq 2$ and graph Γ'', then $\Gamma \cong K_{uv} \wr \Gamma''$. Repeating this process, if necessary, we see there is a maximum integer s such that $\Gamma \cong K_s \wr \Gamma'$ for some graph Γ'. The same argument in the complement shows that if $\Gamma \cong \bar{K}_u \wr \Gamma'$ for some $u \geq 2$ and graph Γ', there is a maximum integer s such that $\Gamma \cong \bar{K}_s \wr \Gamma'$ for some graph Γ'.

Theorem 5.1.8 *Let Γ_1 and Γ_2 be vertex-transitive graphs. If $\Gamma = \Gamma_1 \wr \Gamma_2$ and* $\mathrm{Aut}(\Gamma) \neq \mathrm{Aut}(\Gamma_1) \wr \mathrm{Aut}(\Gamma_2)$, *then there exist positive integers $r > 1$ and $s > 1$ and vertex-transitive graphs Γ_1' and Γ_2' for which $\Gamma_1 \cong \Gamma_1' \wr K_r$, $\Gamma_2 \cong K_s \wr \Gamma_2'$ or $\Gamma_1 \cong \Gamma_1' \wr \bar{K}_r$, $\Gamma_2 \cong \bar{K}_s \wr \Gamma_2'$. In either case, if r and s are chosen to be maximum, then* $\mathrm{Aut}(\Gamma) \cong \mathrm{Aut}(\Gamma_1') \wr \left(S_{rs} \wr \mathrm{Aut}(\Gamma_2')\right)$.

Proof If Γ_1 is disconnected, then we consider $\bar{\Gamma}_1 \wr \bar{\Gamma}_2 = \overline{\Gamma_1 \wr \Gamma_2}$ (Exercise 5.1.11). We thus assume that Γ_1 is connected. If the lexi-partition $\mathcal{B}$ of $\mathrm{Aut}(\Gamma_1) \wr \mathrm{Aut}(\Gamma_2)$ corresponding to $V(\Gamma_2)$ is a block system of $\mathrm{Aut}(\Gamma)$, then the result follows by Lemma 5.1.7. We thus assume $\mathcal{B}$ is not a block system of $\mathrm{Aut}(\Gamma)$.

Fix $x \in V(\Gamma_1)$, and $B_x \in \mathcal{B}$. Then there exists $\gamma \in \mathrm{Aut}(\Gamma)$ such that $\gamma(B_x) \cap B_x \neq \emptyset$ but $\gamma(B_x) \neq B_x$. This implies that $|V(\Gamma_1)|, |V(\Gamma_2)| \geq 2$. Let $C_x = \{c \in V(\Gamma_1) : B_c \cap \gamma(B_x) \neq \emptyset\} \subseteq V(\Gamma_1)$. As γ exists, we have $|C_x| \geq 2$.

First suppose $\Gamma_1[C_x]$ is not complete. By Lemma 5.1.5 there exist integers $r, s \geq 2$ and graphs Γ_1' and Γ_2' with both Γ_1' and Γ_2' connected such that $\Gamma_1 \cong \Gamma_1' \wr \bar{K}_r$ and $\Gamma_2 \cong \bar{K}_s \wr \Gamma_2'$. We assume r is chosen to be maximum. Then,

$$\Gamma_1 \wr \Gamma_2 \cong (\Gamma_1' \wr \bar{K}_r) \wr (\bar{K}_s \wr \Gamma_2') = \Gamma_1' \wr (\bar{K}_r \wr \bar{K}_s) \wr \Gamma_2' \cong \Gamma_1' \wr (\bar{K}_{rs} \wr \Gamma_2').$$

Let $\mathcal{D}$ be the lexi-partition of $\Gamma_1' \wr (\bar{K}_{rs} \wr \Gamma_2')$ corresponding to $V(\bar{K}_{rs} \wr \Gamma_2')$. Suppose $\mathcal{D}$ is not a block system of $\mathrm{Aut}(\Gamma)$. Then there exists $x' \in \mathrm{Aut}(\Gamma_1')$, $D_{x'} \in \mathcal{D}$, and $\gamma' \in \mathrm{Aut}(\Gamma)$ such that $\gamma'(D_{x'}) \cap D_{x'} \neq \emptyset$ and $\gamma'(D_{x'}) \neq D_{x'}$. Let $E_{x'} = \{y \in V(\Gamma_1') : D_y \cap \gamma'(D_{x'}) \neq \emptyset\} \subseteq V(\Gamma_1')$. As r was chosen to be maximum, by Lemma 5.1.5 we have $\Gamma_1'[E_{x'}]$ is complete. By Lemma 5.1.6 applied to $\Gamma_1' \wr (\bar{K}_{rs} \wr \Gamma_2')$, we have $\bar{K}_{rs} \wr \Gamma_2'$ is connected, a contradiction. Thus $\mathcal{D}$ is a block system of $\mathrm{Aut}(\Gamma)$, and by Lemma 5.1.7,

$$\mathrm{Aut}(\Gamma) = \mathrm{Aut}\left(\Gamma_1'\right) \wr \mathrm{Aut}\left(\bar{K}_{rs} \wr \Gamma_2'\right) = \mathrm{Aut}\left(\Gamma_1'\right) \wr \left(S_{rs} \wr \mathrm{Aut}\left(\Gamma_2'\right)\right),$$

with the last equality following from Example 4.2.11. The result follows in this case.

Suppose $\Gamma_1[C_x]$ is complete. By Lemma 5.1.6, there exist integers $r, s \geq 2$ and graphs Γ_1' and Γ_2' with $\bar{\Gamma}_2'$ connected such that $\Gamma_1 \cong \Gamma_1' \wr K_r$ and $\Gamma_2 \cong K_s \wr \Gamma_2'$. Assume $r, s \geq 2$ are chosen be maximum. Then,

$$\Gamma \cong \left(\Gamma_1' \wr K_r\right) \wr \left(K_s \wr \Gamma_2'\right) \cong \Gamma_1' \wr \left(K_{rs} \wr \Gamma_2'\right).$$

Let $\mathcal{D}$ be the lexi-partition of $\mathrm{Aut}(\Gamma_1') \wr \mathrm{Aut}\left(K_{rs} \wr \Gamma_2'\right)$ corresponding to $V\left(K_{rs} \wr \Gamma_2'\right)$. Suppose $\mathcal{D}$ is not a block system of $\mathrm{Aut}(\Gamma)$. Then there exists $x' \in V\left(\Gamma_1'\right)$, $D_{x'} \in \mathcal{D}$, and $\gamma' \in \mathrm{Aut}(\Gamma)$ such that $\gamma'(D_{x'}) \cap D_{x'} \neq \emptyset$ and $\gamma'(D_{x'}) \neq D_{x'}$. Let $E_{x'} = \left\{y \in V\left(\Gamma_1'\right) : D_y \cap \gamma(D_{x'}) \neq \emptyset\right\} \subseteq V(\Gamma_1')$. As r was chosen to be maximum, by Lemma 5.1.6 we have that $\Gamma_1'[E_{x'}]$ is not complete. By Lemma 5.1.6, we have that $K_{rs} \wr \Gamma_2'$ is disconnected, a contradiction. Thus $\mathcal{D}$ is a block system of $\mathrm{Aut}(\Gamma)$, and by Lemma 5.1.7,

$$\mathrm{Aut}(\Gamma) \cong \mathrm{Aut}\left(\Gamma_1'\right) \wr \mathrm{Aut}\left(\bar{K}_{rs} \wr \Gamma_2'\right) = \mathrm{Aut}\left(\Gamma_1'\right) \wr \left(\mathcal{S}_{rs} \wr \mathrm{Aut}\left(\Gamma_2'\right)\right). \qquad \square$$

Example 5.1.9 Calculate the automorphism group of $K_3 \wr \bar{K}_2$ as in Figure 4.2.

Solution As both 2 and 3 are primes, the only possible values for r and s, as in Theorem 5.1.8, are $r = 3$ and $s = 2$. As it is not the case that both K_3 and $\bar{K}_2$ are both complete or have no edges, by Theorem 5.1.8 $\mathrm{Aut}(K_3 \wr \bar{K}_2) = \mathrm{Aut}(K_3) \wr \mathrm{Aut}(\bar{K}_2) = \mathcal{S}_3 \wr \mathcal{S}_2 = \mathcal{S}_3 \wr (\mathbb{Z}_2)_L$. □

Example 5.1.10 Calculate the automorphism group of $C_4 \wr C_4$ as in Figure 4.3.

Solution First, observe that $C_4 = K_{2,2}$ can be written as the nontrivial wreath product $K_2 \wr \bar{K}_2$. So $C_4 \wr C_4 = K_2 \wr \bar{K}_2 \wr K_2 \wr \bar{K}_2$. We observe that we do not need to write parentheses in this expression as the wreath product is associative (Exercise 4.3.1). We see that no two successive factors are both complete graphs or their complements, and so by Theorem 5.1.8 $\mathrm{Aut}(C_4 \wr C_4) = \mathrm{Aut}(C_4) \wr \mathrm{Aut}(C_4) = D_8 \wr D_8$. □

Example 5.1.11 Calculate the automorphism group of $K_5 \wr C_4$.

Solution As noted above, $C_4 = K_2 \wr \bar{K}_2$, so

$$K_5 \wr C_4 = K_5 \wr (K_2 \wr \bar{K}_2) = (K_5 \wr K_2) \wr \bar{K}_2 = K_{10} \wr \bar{K}_2.$$

The maximum r for which K_5 can be written as $\Gamma_1' \wr K_r$ is $r = 5$ with $\Gamma_1' = K_1$ (so we will ignore Γ_1') and the maximum s for which C_4 can be written as $K_s \wr \Gamma_2'$ is $s = 2$ with $\Gamma_2' = \bar{K}_2$. Now applying Theorem 5.1.8, we have $\mathrm{Aut}(K_5 \wr C_4) = \mathcal{S}_{10} \wr \mathcal{S}_2 = \mathcal{S}_{10} \wr (\mathbb{Z}_2)_L$. □

5.2 2-Closed Permutation Groups

We now consider a topic very much related to the orbital digraphs considered in Section 1.4. Suppose we have a transitive group G and would like to find the smallest "automorphism group of a graph" that contains G – we will see why "automorphism group of a graph" is in quotes shortly. This is actually a very common situation in practice. That is, one will have some information about the automorphism group of a graph (and usually this information is exactly a transitive subgroup), and would like to know any additional automorphisms that G being a subgroup of the automorphism group guarantees will exist – we will see a way in which this occurs in Lemma 5.3.11 – but already have a trivial example. If G is 2-transitive, then we know that the graph is either complete or has no edges, and every additional permutation is contained in the automorphism group. Simply picking an orbital digraph of G to approximate this smallest automorphism group may give an automorphism group that is *much* larger than G. For example, if G is cyclic of composite order, then G is not primitive as it has nontrivial normal subgroups, and so will have some disconnected orbital digraphs by Theorem 2.4.12. Such automorphism groups are wreath products and much larger than a cyclic group. The solution to this problem is not to choose one orbital digraph, but to choose *all* of them. The price we will pay for this choice is that the resulting object is not necessarily a graph or digraph, but a *color* graph or digraph.

Definition 5.2.1 A **color digraph** is a digraph in which each arc has been assigned a color – typically a color is just an integer.

Example 5.2.2 Any digraph Γ can be thought of as a color digraph by coloring arcs of K_n with color 1 if they are in $A(\Gamma)$, and with color 0 if they are not in $A(\Gamma)$. In a similar way, one can always think of a color digraph of order n as being obtained by coloring the arcs of K_n.

Definition 5.2.3 Let Γ be a color digraph. An **automorphism** of Γ is a bijection $\phi\colon V(\Gamma) \to V(\Gamma)$ such that $(x, y) \in A(\Gamma)$ and is of color i if and only if $\phi(x, y) = (\phi(x), \phi(y)) \in A(\Gamma)$ is of color i. It is straightforward to verify that the set of all automorphisms of a color digraph Γ is a group, the **automorphism group** of Γ, denoted $\mathrm{Aut}(\Gamma)$.

Example 5.2.4 A natural way of constructing color digraphs is with the orbital digraphs of a group G. Given the orbital digraphs $\Gamma_1, \ldots, \Gamma_r$ of a group G of order n, we color $(x, y) \in A(K_n)$ with color i if and only if $(x, y) \in A(\Gamma_i)$ to obtain the color digraph Γ. Then $\mathrm{Aut}(\Gamma) = \cap_{i=1}^{r}\mathrm{Aut}(\Gamma_i)$, and as $G \leq \mathrm{Aut}(\Gamma_i)$ for all $1 \leq i \leq r$, we have $G \leq \mathrm{Aut}(\Gamma)$.

This idea behind Example 5.2.4 is usually discussed using different language, which we now introduce.

Definition 5.2.5 Let X be a set and $G \leq \mathcal{S}_X$. The **2-closure of** G, denoted $G^{(2)}$, is the largest subgroup of $\mathcal{S}_X$ whose orbits on $X \times X$ are the same as the orbits of G. If $G^{(2)} = G$, we say G is **2-closed.**

The 2-closure was introduced by Wielandt (1969) – this hard to find reference is also available in its entirety in the more easily accessible Wielandt (1994) (Wielandt also introduced a similar notion of k-closure, $k \geq 2$, but we will not need k-closure until Section 7.6). It should be clear that with Γ as in Example 5.2.4, $\mathrm{Aut}(\Gamma) = G^{(2)}$ as an orbit of G on $X \times X$ is either the arc set of an orbital digraph of G or the diagonal $\{(x,x) : x \in X\}$. So some authors define a 2-closed group G to be either the automorphism group of a color digraph, or the intersection of the automorphism groups of the orbital digraphs of G. Our next result shows that the automorphism group of any vertex-transitive digraph is 2-closed.

Proposition 5.2.6 *Let Γ be a vertex-transitive digraph. Then* $\mathrm{Aut}(\Gamma)$ *is 2-closed.*

Proof Let $\Gamma_1, \ldots, \Gamma_r$ be the orbital digraphs of $\mathrm{Aut}(\Gamma)$. If $\gamma \in \mathrm{Aut}(\Gamma)^{(2)}$, then $\gamma \in \mathrm{Aut}(\Gamma_i)$, $1 \leq i \leq r$, as $\mathrm{Aut}(\Gamma)^{(2)} = \cap_{i=1}^{r}\mathrm{Aut}(\Gamma_i)$. By Proposition 1.4.6, Γ is a generalized orbital digraph of $\mathrm{Aut}(\Gamma)$, with, say, $\cup_{i=1}^{s}\Gamma_i = \Gamma$ for some $1 \leq s \leq r$. But then $\gamma \in \mathrm{Aut}(\Gamma)$ as $\gamma \in \mathrm{Aut}(\Gamma_i)$, $1 \leq i \leq s$. □

While $G^{(2)}$ is the automorphism group of a color digraph, it need not be the automorphism group of a digraph or graph.

Example 5.2.7 The group $(\mathbb{Z}_2 \times \mathbb{Z}_2)_L$ is 2-closed but is not the automorphism group of a digraph or graph.

Solution As every element of $\mathbb{Z}_2 \times \mathbb{Z}_2$ is self-inverse, every Cayley digraph of $\mathbb{Z}_2 \times \mathbb{Z}_2$ is a graph. By inspection, there are four pairwise nonisomorphic Cayley graphs of $\mathbb{Z}_2 \times \mathbb{Z}_2$, namely K_4, $\bar{K}_4$, a 1-factor, and a cycle of length 4. The former two graphs have automorphism group $\mathcal{S}_4$ while the latter two graphs have automorphism group $\mathbb{Z}_2 \wr \mathbb{Z}_2$.

To see that $(\mathbb{Z}_2 \times \mathbb{Z}_2)_L$ is 2-closed, consider the color digraph Γ given in Figure 5.1. This color digraph Γ is the union of the three orbital digraphs of $(\mathbb{Z}_2 \times \mathbb{Z}_2)_L$, with each orbital digraph of $(\mathbb{Z}_2 \times \mathbb{Z}_2)_L$ colored in a different color. Hence $(\mathbb{Z}_2 \times \mathbb{Z}_2)_L \leq \mathrm{Aut}(\Gamma)$. Consider the orbital digraph Γ_1 of $(\mathbb{Z}_2 \times \mathbb{Z}_2)_L$ shown in green in Figure 5.1. The transposition $\gamma = ((1,0),(1,1)) \in \mathrm{Aut}(\Gamma_1)$ by inspection, but $\gamma \notin \mathrm{Aut}(\Gamma_2)$, where Γ_2 is the orbital digraph of $(\mathbb{Z}_2 \times \mathbb{Z}_2)_L$ shown in red in Figure 5.1. As $\mathrm{Aut}(\Gamma_1) \cong \mathbb{Z}_2 \wr \mathbb{Z}_2 \cong \langle(\mathbb{Z}_2 \times \mathbb{Z}_2)_L, \gamma\rangle$ is of order 8, we see that $\mathrm{Aut}(\Gamma_1) \cap \mathrm{Aut}(\Gamma_2) = (\mathbb{Z}_2 \times \mathbb{Z}_2)_L$. Hence $\mathrm{Aut}(\Gamma) = (\mathbb{Z}_2 \times \mathbb{Z}_2)_L$ is 2-closed. □

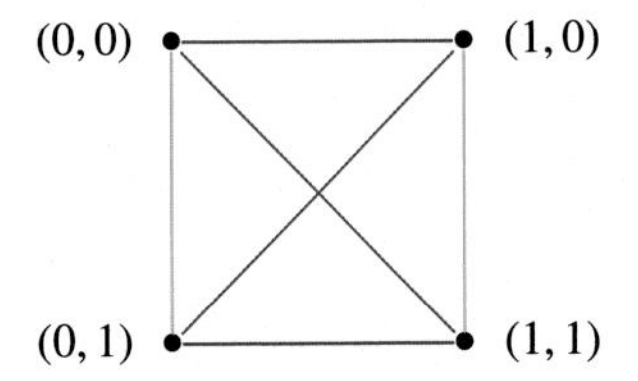

Figure 5.1 A Cayley color digraph of $\mathbb{Z}_2 \times \mathbb{Z}_2$ with automorphism group $\mathbb{Z}_2 \times \mathbb{Z}_2$.

Perhaps the first major problem that was considered and subsequently solved on symmetries in graphs was the question: Is every regular group the automorphism group of a digraph or graph?

Definition 5.2.8 If $\Gamma = \mathrm{Cay}(G, S)$ is a digraph and $\mathrm{Aut}(\Gamma) = G_L$, then Γ is a **digraphical regular representation**, or **DRR** of G, while if Γ is a graph and $\mathrm{Aut}(\Gamma) = G_L$, then Γ is a **graphical regular representation of** G, or **GRR**.

We saw in Corollary 1.2.18 that if G is an abelian group that is not an elementary abelian 2-group, then $\mathrm{Aut}(\Gamma) \neq G_L$ for every Cayley graph Γ of G (there will be an extra "reflection" in the automorphism group at a minimum). Additionally, Nowitz (1968) has shown that the generalized dicyclic groups are also not GRRs – see Exercise 5.2.8.

Definition 5.2.9 A **generalized dicyclic group** is generated by an abelian group A together with an element b not in A such that $b^4 = 1$, $1 \neq b^2 \in A$, and $b^{-1}ab = a^{-1}$ for all $a \in A$.

Godsil (1981a) completed the classification of finite groups that are not GRRs, with the previously mentioned families being the only two infinite families. There are ten additional examples of groups that are not GRRs, usually of small order. Babai (1980) showed that every group is a DRR, with the exceptions of the elementary abelian groups of order 4, 8, 9, and 16, and the quaternion group.

If $G \leq S_n$ is regular, then $G^{(2)} = G$ (Exercise 5.2.1). Hence, a regular group that has no DRR is 2-closed but not the automorphism group of a digraph. Hence, the regular representations of the elementary abelian groups of order 4 (we already have seen this in Example 5.2.7), 8, 9, and 16, and the quaternion group are 2-closed but not the automorphism group of a digraph.

Recently there has been a small revival on problems related to the DRR and GRR problems. See Doyle et al. (2018); Morris and Spiga (2018b); Morris and Tymburski (2018); Spiga (2018).

We now turn to some useful results regarding 2-closed groups, with the first appearing in Wielandt's original lecture notes (Wielandt, 1969).

Theorem 5.2.10 *Let X be a set and $G \leq S_X$ be transitive. A partition $\mathcal{B}$ of X is a block system of $G^{(2)}$ if and only if it is a block system of G.*

Proof Clearly if $\mathcal{B}$ is a block system of $G^{(2)}$, then $\mathcal{B}$ is a block system of G as $G \leq G^{(2)}$. Conversely, suppose $\mathcal{B}$ is a block system of G. Let $\Gamma_1, \ldots, \Gamma_r$ be the orbital digraphs of G, and assume, without loss of generality, that $\Gamma_1, \ldots, \Gamma_k$, $k \leq r$, have the property that $1 \leq j \leq k$ if and only if Γ_j contains an arc with both endpoints contained in some $B \in \mathcal{B}$. Then for $1 \leq j \leq k$, every arc of Γ_j has both endpoints contained in some $B \in \mathcal{B}$ as $\mathcal{B}$ is a block system of G. Then each Γ_j, $1 \leq j \leq k$, is disconnected, and its components are a block system C_j of $\mathrm{Aut}(\Gamma_j)$ with $C_j \leq \mathcal{B}$. As C_j is a block system of $\mathrm{Aut}(\Gamma_j)$ and $G^{(2)} \leq \mathrm{Aut}(\Gamma_j)$, we have that C_j is a block system of $G^{(2)}$ for every $1 \leq j \leq k$. As for every $x, y \in B$, $x \neq y$, there exists $1 \leq j \leq k$ (depending on x and y) such that the arc $(x, y) \in A(\Gamma_j)$, we have $x, y \in C_j \in C_j$. Let C be the block system of $G^{(2)}$ generated by $C_1, \ldots, C_k$. Then the block of C that contains x contains every point of $B \in \mathcal{B}$ with $x \in B$. As $C \leq \mathcal{B}$ as $C_j \leq \mathcal{B}$ for every $1 \leq j \leq k$, $C = \mathcal{B}$ and $\mathcal{B}$ is a block system of $G^{(2)}$ by Theorem 2.3.12. □

The following result is originally due to Kalužnin and Klin (1976) (in Russian), and is also in Cameron et al. (2002).

Theorem 5.2.11 *Let $G \leq S_X$ and $H \leq S_Y$ be transitive. Then,*

(i) *$(G \wr H)^{(2)} = G^{(2)} \wr H^{(2)}$, and*
(ii) *with the canonical action on $X \times Y$, $(G \times H)^{(2)} = G^{(2)} \times H^{(2)}$.*

Proof (i) Let $\Gamma_1, \ldots, \Gamma_r$ be the orbital digraphs of $G \wr H$, and $\mathcal{B}$ the lexi-partition of $G \wr H$ corresponding to Y. Then $\mathcal{B}$ is also a block system of $(G \wr H)^{(2)}$ by Theorem 5.2.10.

If Γ_j, $1 \leq j \leq r$, has an arc whose endpoints are contained in block of $\mathcal{B}$, then Γ_j is disconnected, with each component of Γ_j contained in some block of $\mathcal{B}$. Let $B \in \mathcal{B}$, with $B = \{(x, y) : y \in Y\}$, where $x \in X$. Define the digraph Γ_B by $V(\Gamma_B) = Y$ and $A(\Gamma_B) = \{(y, y') : ((x, y), (x, y')) \in A(\Gamma_j)\}$. We observe that $\Gamma_B = \Gamma_{B'}$ for every $B' \in \mathcal{B}$ as Γ_j is an orbital digraph of $G \wr H$. Also, $\Gamma_j = \bar{K}_m \wr \Gamma_B$, where $m = |X|$. As $\mathcal{B}$ is a block system of $G \wr H$, we have $\mathrm{Stab}_{G \wr H}(B)^B$ is permutation isomorphic to H. As Γ_j is arc-transitive by Theorem 3.2.8, $\Gamma_j[B] \cong \Gamma_B$ is arc-transitive. Then Γ_B is H-arc-transitive, and, by Theorem 3.2.8, Γ_B is an orbital digraph of H. Also, $S_m \wr \mathrm{Aut}(\Gamma_B)$ is the largest subgroup of $\mathrm{Aut}(\Gamma_j)$ that has $\mathcal{B}$ as a block system.

If Γ_j, $1 \le j \le r$, does not contain an arc whose endpoints are contained in a block of $\mathcal{B}$, then as $G \wr H \le \mathrm{Aut}(\Gamma_j)$, Γ_j is the wreath product of a digraph Γ'_j with $G \le \mathrm{Aut}(\Gamma'_j)$, and $\bar{K}_\ell$, where $|B| = \ell$. As Γ_j is arc-transitive, Γ'_j is arc-transitive, and as $G \le \mathrm{Aut}(\Gamma'_j)$, $\mathrm{Aut}(\Gamma'_j)$ is an orbital digraph of G. Also, $\mathrm{Aut}(\Gamma'_j) \wr \mathcal{S}_\ell$ is the largest subgroup of $\mathrm{Aut}(\Gamma_j)$ that has $\mathcal{B}$ as a block system.

For $1 \le j \le r$, let L_j be the largest subgroup of $\mathrm{Aut}(\Gamma_j)$ that has $\mathcal{B}$ as a block system. As $(G \wr H)^{(2)}$ has $\mathcal{B}$ as a block system and is contained in the automorphism group of every orbital digraph of $G \wr H$, we see that $(G \wr H)^{(2)} \le \cap_{j=1}^r L_j$. Let $\Gamma_1^1, \ldots, \Gamma_g^1$ be the orbital digraphs of G and $\Gamma_1^2, \ldots, \Gamma_h^2$ be the orbital digraphs of H, where g and h are positive integers. Then,

$$(G \wr H)^{(2)} = \cap_{j=1}^r L_j = \left(\cap_{j=1}^g \mathrm{Aut}\left(\Gamma_j^1 \wr \bar{K}_\ell\right)\right) \cap \left(\cap_{j=1}^h \mathrm{Aut}\left(\bar{K}_m \wr \Gamma_j^2\right)\right).$$

As $(G \wr H)^{(2)}$ has $\mathcal{B}$ as a block system by Theorem 5.2.10, the largest subgroup of $\cap_{j=1}^g \mathrm{Aut}\left(\Gamma_j^1 \wr \bar{K}_\ell\right)$ that is contained in $(G \wr H)^{(2)}$ is

$$\cap_{j=1}^g \left(\mathrm{Aut}\left(\Gamma_j^1\right) \wr \mathcal{S}_\ell\right) = \left(\cap_{j=1}^g \mathrm{Aut}\left(\Gamma_j^1\right)\right) \wr \mathcal{S}_\ell = G^{(2)} \wr \mathcal{S}_\ell.$$

Similarly, the largest subgroup of $\cap_{j=1}^h \mathrm{Aut}\left(\bar{K}_m \wr \Gamma_j^2\right)$ that is contained in $(G \wr H)^{(2)}$ is the group $\mathcal{S}_m \wr \left(\cap_{j=1}^h \mathrm{Aut}\left(\Gamma_j^2\right)\right) = \mathcal{S}_m \wr H^{(2)}$. As

$$\left(G^{(2)} \wr \mathcal{S}_\ell\right) \cap \left(\mathcal{S}_m \wr H^{(2)}\right) = \left(G^{(2)} \cap \mathcal{S}_m\right) \wr \left(\mathcal{S}_\ell \cap H^{(2)}\right) = G^{(2)} \wr H^{(2)},$$

we have that (i) holds.
(ii) First consider $H \wr G \le \mathcal{S}_{Y \times X}$, which is all elements of the form $(y, x) \mapsto (h(y), g_y(x))$ with $h \in H$ and $g_y \in G$. Denote by $\overline{H \wr G}$ all those permutations of $X \times Y$ of the form $(x, y) \mapsto (g_y(x), h(y))$ with $g_y \in G$ and $h \in H$. The purpose of $\overline{H \wr G}$ is to obtain a group naturally isomorphic to $H \wr G$, but whose vertex set is $X \times Y$, and not $Y \times X$. Clearly $\overline{H \wr G} \cong H \wr G$. Then $(G \wr H) \cap \left(\overline{H \wr G}\right) = G \times H$, and so $(G \times H)^{(2)} \le (G \wr H)^{(2)}$ and $(G \times H)^{(2)} \le \left(\overline{H \wr G}\right)^{(2)}$. We conclude

$$(G \times H)^{(2)} \le (G \wr H)^{(2)} \cap \left(\overline{H \wr G}\right)^{(2)} = \left(G^{(2)} \wr H^{(2)}\right) \cap \overline{H^{(2)} \wr G^{(2)}} = G^{(2)} \times H^{(2)},$$

with the middle equality holding by the first part of this result. Now suppose $\alpha \in G^{(2)}$ and $\beta \in H^{(2)}$. Let $x_1, x_2 \in X$ and $y_1, y_2 \in Y$. Then there exists $g \in G$ and $h \in H$ such that $\alpha(x_1, x_2) = g(x_1, x_2) = (g(x_1), g(x_2))$ and $\beta(y_1, y_2) = h(y_1, y_2) = (h(y_1), h(y_2))$. Then,

$$\begin{aligned}(\alpha, \beta)((x_1, y_1)(x_2, y_2)) &= (\alpha(x_1), \beta(y_1))(\alpha(x_2), \beta(y_2)) \\ &= (g(x_1), h(y_1))(g(x_2), h(y_2)).\end{aligned}$$

Hence the orbits of $G^{(2)} \times H^{(2)}$ on $(X \times Y)^2$ are contained in the orbits of $G \times H$ on $(X \times Y)^2$, in which case $G^{(2)} \times H^{(2)} \le (G \times H)^{(2)}$. Thus $(G \times H)^{(2)} = G^{(2)} \times H^{(2)}$. □

5.3 Generalized Wreath Products

As we shall eventually see in Section 8.5, generalized wreath products are one of the three main families of circulant digraphs, at least in terms of their automorphism groups. Additionally, in Section 7.7 we will see that they are also a central focus of the isomorphism problem for circulant digraphs. In order to define them and see their relationship to wreath products, we will need the following characterization of when a Cayley digraph of an abelian group is isomorphic to a wreath product. We will need one piece of terminology.

Definition 5.3.1 Let Γ be a digraph. For $(u, v) \in A(\Gamma)$ we say that v is **out adjacent** to u and that u is **in adjacent** to v.

Proposition 5.3.2 *Let A be an abelian group and $S \subseteq A$. The digraph $\Gamma = \mathrm{Cay}(A, S)$ can be written as a nontrivial wreath product if and only if there exists $1 < H < A$ such that $S \setminus H$ is a union of cosets of H in A.*

Proof Suppose $\Gamma \cong \Gamma_1 \wr \Gamma_2$, where both Γ_1 and Γ_2 have order at least 2. If Γ is a graph, then by Theorem 5.1.8 we may assume without loss of generality that $\mathrm{Aut}(\Gamma) = \mathrm{Aut}(\Gamma_1) \wr \mathrm{Aut}(\Gamma_2)$. If Γ is a digraph, then the same assumption can be made (Dobson and Morris, 2009, Theorem 5.7). Then the lexi-partition $\mathcal{B}$ corresponding to $V(\Gamma_2)$ is a normal block system of $\mathrm{Aut}(\Gamma)$ by Theorem 4.3.7. By Theorem 2.2.19 and Example 2.2.8, there exists $H \leq A$ such that $\mathcal{B}$ is the set of cosets of H in A. Let $a \in A \setminus H$ and consider the coset $a + H \neq H$. As $\Gamma = \Gamma_1 \wr \Gamma_2$, every vertex of H is either in adjacent or out adjacent to every vertex of $a + H$ or to no vertices of $a + H$, so that $S \setminus H$ is a union of cosets of H.

Conversely, suppose there exists $1 < H < A$ such that $S \setminus H$ is a union of cosets of H in A. Then the cosets of H in A are a block system $\mathcal{B}$ of A_L as $\bar{H}_L \trianglelefteq A_L$, where $\bar{H}_L = \{h_L : h \in H\} \leq A_L$. Let $a + H, b + H \in \mathcal{B}$, $a + H \neq b + H$, and assume $(a + h, b + h') \in A(\Gamma)$ for some $h, h' \in H$. As $S \setminus H$ is a union of cosets of H in A, $(b - a) + H \subseteq S \setminus H$ as $b - a \neq 0$ as $b - a \notin H$. We conclude that $(a + h, b + h') \in A(\Gamma)$ for every $h, h' \in H$. The result follows by Theorem 4.2.15. □

We now wish to generalize wreath products of Cayley digraphs of abelian groups.

Definition 5.3.3 Let A be an abelian group, and $S \subseteq A$. Let $\Gamma = \mathrm{Cay}(A, S)$. We say that Γ is a (K, L)**-generalized wreath product** or, more usually simply a **generalized wreath product**, if there exist $K \leq L \leq A$ such that $S \setminus L$ is a union of cosets of K.

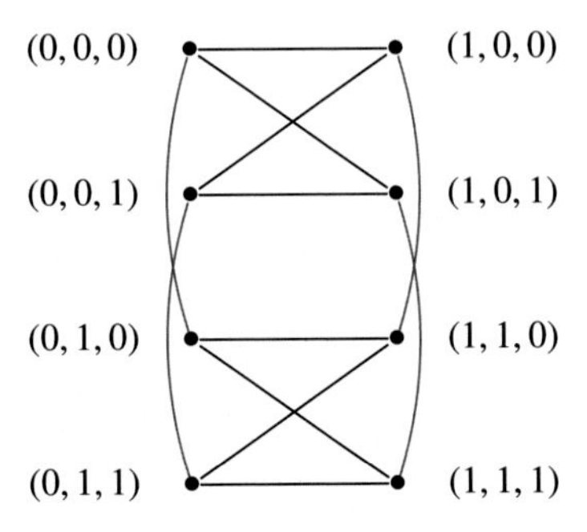

Figure 5.2 $\mathrm{Cay}\left(\mathbb{Z}_2^3, \{(1,0,0),(1,0,1),(0,1,0)\}\right)$, a generalized wreath product.

Example 5.3.4 The graph $\mathrm{Cay}\left(\mathbb{Z}_2^3, \{(1,0,0),(1,0,1),(0,1,0)\}\right)$ shown in Figure 5.2 is a (K,L)-generalized wreath product where $K = \langle(0,0,1)\rangle$ and $L = \langle(0,0,1),(0,1,0)\rangle$.

Solution The graph Γ is drawn in Figure 5.2, and all edges of colors blue and black are in $E(\Gamma)$. The set $S\setminus L = \{(1,0,0),(1,0,1)\} = (1,0,0) + \langle(0,0,1)\rangle$ is a coset of K. So Γ is a (K,L)-generalized wreath product. In Figure 5.2, the edges of the Cayley graph $\mathrm{Cay}\left(\mathbb{Z}_2^3, S\setminus L\right)$ are drawn in black, while the edges in blue are the edges of $\mathrm{Cay}\left(\mathbb{Z}_2^3, S\cap L\right)$. □

If $L = K$, then a (K,L)-generalized wreath product is simply a wreath product by Proposition 5.3.2. Note that there is no actual "product" here, the name being chosen to suggest that "generalized wreath products" share some of the same properties of wreath products. We have defined generalized wreath products only for Cayley digraphs of abelian groups as, until very recently, there was no analogue of Proposition 5.3.2 for nonabelian groups. However, recently Barber and Dobson (2022) showed that a double coset digraph $\mathrm{Cos}(G,H,S)$ is isomorphic to a nontrivial wreath product if and only if there is $H < K < G$ such that S is a union of double cosets of K in G. Consequently, they were able to define generalized wreath products for double coset digraphs.

The different colored edges in Figure 5.2 illustrate how one can think about generalized wreath products. The edges in black are the edges of the graph $\mathrm{Cay}\left(\mathbb{Z}_2^3, \{(1,0,0),(1,0,1)\}\right)$, which is the wreath product of two edges with $\bar{K}_2$, while the edges in blue are contained within the cosets of L. In general, one can always "decompose" a (K,L)-generalized wreath product $\mathrm{Cay}(A,S)$ into two Cayley digraphs of A, one with connection set consisting of $S\cap L$ and the other with connection set $S\setminus L$. Our first result shows the automorphism group of a generalized wreath product contains the intersection of the automorphism groups of such wreath products, and is the analogue to the fact that the automorphism group of a wreath product contains the wreath product of the automorphism groups. Before proceeding, we will need some more notation.

Notation 5.3.5 *Let $G \leq S_n$ be transitive with normal block system $\mathcal{B} \preceq \mathcal{C}$. Let $g \in \mathrm{fix}_G(\mathcal{B})$. By $g|_C$, $C \in \mathcal{C}$, we mean the element of S_n such that $g|_C(x) = g(x)$ if $x \in C$, while $g|_C(x) = x$ if $x \notin C$. By $\mathrm{fix}_G(\mathcal{B})|_C$ we mean $\{g|_C : g \in \mathrm{fix}_G(\mathcal{B})\}$.*

Note that we are abusing notation somewhat here. The notation $g|_C$ usually means the restriction of g to C, i.e., we simply restrict the domain of the function g to C. Here, we are restricting g to C, but then make its domain the same as before by redefining it everywhere else to be the identity.

Theorem 5.3.6 *Let A be an abelian group and $S \subseteq A$. Let $\Gamma = \mathrm{Cay}(A, S)$, and suppose Γ is a (K, L)-generalized wreath product where $K \leq L < A$. Then,*

(i) $\mathrm{Aut}(\Gamma) \geq \mathrm{Aut}(\mathrm{Cay}(A, S\setminus L)) \cap \mathrm{Aut}(\mathrm{Cay}(A, S \cap L))$,
(ii) $\mathrm{Cay}(A, S\setminus L) \cong \mathrm{Cay}(A/K, \{aK : aK \subseteq S \text{ and } aK \not\subseteq L\}) \wr \mathrm{Cay}(K, \emptyset)$,
(iii) $\mathrm{Cay}(A, S \cap L) \cong \mathrm{Cay}(A/L, \emptyset) \wr \mathrm{Cay}(L, S \cap L)$, *and*
(iv) *let $\mathcal{B}$ and $\mathcal{C}$ denote the set of cosets of K and L in A, respectively. If $\mathcal{B}$ is a block system of $G \leq \mathrm{Aut}(\Gamma)$, then $\mathrm{fix}_G(\mathcal{B})|_C \leq \mathrm{Aut}(\Gamma)$ for every $C \in \mathcal{C}$.*

Proof (i) That

$$\mathrm{Aut}(\Gamma) \geq \mathrm{Aut}(\mathrm{Cay}(A, S\setminus L)) \cap \mathrm{Aut}(\mathrm{Cay}(A, S \cap L))$$

is almost trivial. In fact, if $\{U, V\}$ is any partition of S, then

$$\mathrm{Aut}(\mathrm{Cay}(A, U)) \cap \mathrm{Aut}(\mathrm{Cay}(A, V)) \leq \mathrm{Aut}(\mathrm{Cay}(A, S)),$$

as if

$$\gamma \in \mathrm{Aut}(\mathrm{Cay}(A, U)) \cap \mathrm{Aut}(\mathrm{Cay}(A, V)), \tag{5.1}$$

and $(x, y) \in A(\mathrm{Cay}(A, S))$, then $(x, y) \in A(\mathrm{Cay}(A, U))$ or $A(\mathrm{Cay}(A, V))$. As Equation (5.1) holds, $\gamma(x, y) \in A(\mathrm{Cay}(A, U)))$ or $A(\mathrm{Cay}(A, V))$. Hence $\gamma(x, y) \in A(\mathrm{Cay}(A, S))$ and $\gamma \in \mathrm{Aut}(\Gamma)$.
(ii) Notice that $\mathrm{Cay}(A, S \cap L)$ is disconnected as every arc will have both endpoints contained within a coset of L and, of course, the subdigraphs induced by different cosets are isomorphic. Then $\mathrm{Cay}(A, S \cap L) \cong \mathrm{Cay}(A/L, \emptyset) \wr \mathrm{Cay}(L, S \cap L)$.
(iii) By the definition of a generalized wreath product, $S\setminus L$ is a disjoint union of cosets of K and so $\mathrm{Cay}(A, S\setminus L) \cong \mathrm{Cay}(A/K, \{aK : aK \subset S \text{ and } aK \not\subseteq L\}) \wr \mathrm{Cay}(K, \emptyset)$.
(iv) It is straightforward to verify that $\mathrm{fix}_G(\mathcal{B})|_C$ is contained in both $\mathrm{Cay}(A, S \cap L)$ and $\mathrm{Cay}(A, S\setminus L)$. □

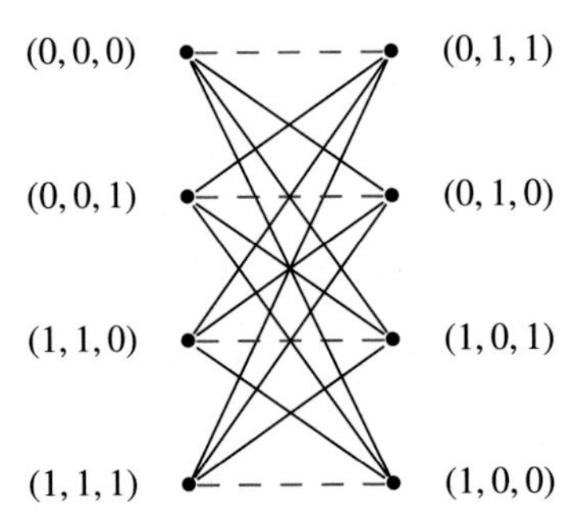

Figure 5.3 $\text{Cay}(\mathbb{Z}_2^3, \{(1,0,0),(1,0,1),(0,1,0)\})$ redrawn.

We will show later, in Example 5.3.18 that there are generalized wreath product digraphs where equality holds in Theorem 5.3.6 (i). Recalling our general notion that one should not expect symmetries that are not "built into" the automorphism group, this suggests that $\text{Aut}(\Gamma_1)\cap\text{Aut}(\Gamma_2)$ is the "expected" automorphism group of a generalized wreath product. Of course, it need not be the case that $\text{Aut}(\Gamma) = \text{Aut}(\text{Cay}(A, S\setminus L))\cap\text{Aut}(\text{Cay}(A, S\cap L))$ in the preceding result. There is the obvious example of a $K_{mn} = K_m \wr K_n$ and $\mathcal{S}_{mn} = \text{Aut}(K_{mn}) \neq \mathcal{S}_m \wr \mathcal{S}_n$ with $m, n > 1$. We next show that there are more such examples.

Example 5.3.7 The graph $\Gamma = \text{Cay}\left(\mathbb{Z}_2^3, \{(1,0,0),(1,0,1),(0,1,0)\}\right)$ in Figure 5.2 has automorphism group $\mathcal{S}_2 \times \mathcal{S}_4$ and is a generalized wreath product.

Solution The graph Γ here is the same graph as in Example 5.3.4. It can be redrawn as in Figure 5.3 (its edges are the black edges), where we see that Γ is $K_2 \wr \bar{K}_4$ with a 1-factor removed (the dashed edges in red, denoted $4K_2$). The sets $B_1 = \{(0,0,0),(0,0,1),(1,1,0),(1,1,1)\}$ and $B_2 = \mathbb{Z}_2^3\setminus B_1$ are the unique maximal independent sets of Γ of size 4, and as they are disjoint, $\mathcal{B} = \{B_1, B_2\}$ is a block system of $\text{Aut}(\Gamma)$ (Exercise 5.3.7). It is easy to see that $\mathcal{S}_2 \times \mathcal{S}_4 \leq \text{Aut}(\Gamma)$ as $\mathcal{S}_2\times\mathcal{S}_4$ is contained in the automorphism group of both $K_{4,4}$ and $4K_2$. As $\mathcal{B}$ is a block system of $\text{Aut}(\Gamma)$, by Corollary 4.3.2 we have $\text{Aut}(\Gamma) \leq \mathcal{S}_2 \wr \mathcal{S}_4$. If $\gamma \in \text{fix}_{\text{Aut}(\Gamma)}(\mathcal{B})$, then there exists $\delta \in \text{fix}_{\mathcal{S}_2\times\mathcal{S}_4}(\mathcal{B})$ such that $(\gamma\delta)^{B_1} = 1$. Then $(\gamma\delta)^{B_2} = 1$ as well, and so $\text{Aut}(\Gamma) = \mathcal{S}_2 \times \mathcal{S}_4$. We shall see more graphs like Γ later, as it is also a **deleted wreath product**, another class of graphs defined in Definition 8.5.4. □

There is another perspective from which one can see that a generalized wreath product "generalizes" the wreath product. Namely, a vertex-transitive digraph can be thought of as a wreath product if its automorphism group has a transitive subgroup with a block system and there is a subgroup that fixes point-wise all of the vertices not in one block and is transitive on the remaining block. An analogous statement (of course with a small variation) is true for vertex-transitive digraphs that are generalized wreath products. Namely, for such a generalized wreath product its automorphism group has a transitive subgroup H with a block system and there is a subgroup that fixes point-wise all of the vertices not in one block and whose orbits on the remaining block are blocks of H.

Our next result is a general tool for constructing additional automorphisms of a digraph under certain circumstances, and will also allow for the construction of some generalized wreath products. It is also useful in other contexts as we shall see. We will need some preliminary definitions.

Definition 5.3.8 Let $G \leq S_n$ such that G has block systems $\mathcal{B} \preceq \mathcal{C}$. Then the set of all blocks of $\mathcal{B}$ whose union is a block $C \in \mathcal{C}$ is a block of $G/\mathcal{C}$, denoted $C/\mathcal{B}$. The set $\{C/\mathcal{B} : C \in \mathcal{C}\}$ is a block system of $G/\mathcal{B}$, and is the block system of $G/\mathcal{B}$ **induced** by $\mathcal{C}$.

It is straightforward to verify that $\mathcal{C}/\mathcal{B}$ is a block system of $G/\mathcal{B}$ (Exercise 5.3.3).

Example 5.3.9 Let n be a positive integer that can be written as $n = mk$ with $m > 1$ and k composite. Let $1 < \ell < k$ divide k. Then $(\mathbb{Z}_n)_L$ has normal block systems $\mathcal{B} \prec \mathcal{C}$, $\mathcal{B}$ has blocks of size ℓ, and $\mathcal{C}$ has blocks of size k. Also, $(\mathbb{Z}_n)_L/\mathcal{B}$ has a normal block system $\mathcal{C}/\mathcal{B}$ with blocks of size k/ℓ.

Solution It is clear that $(\mathbb{Z}_n)_L$ has block systems $\mathcal{B} \prec \mathcal{C}$ with $\mathcal{B}$ having blocks of size ℓ and $\mathcal{C}$ having blocks of size k as $\mathbb{Z}_n$ has unique normal subgroups H and K of orders ℓ and k, and as $1 < \ell < k$ divides k, $H < K$. The blocks of $\mathcal{B}$ are cosets of H and the blocks of $\mathcal{C}$ are cosets of K by Example 2.3.8. Also, each coset $a + K$ of K in G is a union of k/ℓ cosets of H, and so the blocks in $\mathcal{C}/\mathcal{B}$ have blocks of size k/ℓ. □

Definition 5.3.10 Let G be a transitive group that has a normal block system $\mathcal{B}$ with blocks of prime size p. Then $\mathrm{fix}_G(\mathcal{B})^B$, $B \in \mathcal{B}$, is a transitive group of prime degree p, and so contains a p-cycle. Define a relation $\equiv$ on $\mathcal{B}$ by $B \equiv B'$ if and only if whenever $\gamma \in \mathrm{fix}_G(\mathcal{B})$ then γ^B is a p-cycle if and only if $\gamma^{B'}$ is also a p-cycle. It is straightforward to verify that $\equiv$ is an equivalence relation (Exercise 5.3.1). Let C be an equivalence class of $\equiv$ and $E_C = \cup_{B \in C} B$ (remember that the equivalence classes of $\equiv$ consist of *blocks* of $\mathcal{B}$). Let $\mathcal{E} = \{E_C : C \text{ is an equivalence class of } \equiv\}$.

Lemma 5.3.11 *Let Γ be a digraph with $G \leq \mathrm{Aut}(\Gamma)$ that has a normal block system $\mathcal{B}$ with blocks of prime size p. Let $\mathcal{E}$ be defined as in Definition 5.3.10.*

Then $\mathcal{E}$ is a block system of G, and for every $\gamma \in \mathrm{fix}_G(\mathcal{B})$, $\gamma|_E \in \mathrm{Aut}(\Gamma)$ for every $E \in \mathcal{E}$.

Proof We will first show that $\mathcal{E}/\mathcal{B}$ is a block system of $G/\mathcal{B}$ by showing that $\equiv$ is a $G/\mathcal{B}$-congruence and applying Theorem 2.3.2. This will give that $\mathcal{E}$ is a block system of G. We thus need to show that $B \equiv B'$ if and only if $g(B) \equiv g(B')$ for every $g \in G$. Suppose $B \equiv B'$ but $g(B) \not\equiv g(B')$ for some $g \in G$. Then there exists $\gamma \in \mathrm{fix}_G(\mathcal{B})$ such that $\gamma^{g(B)}$ is a p-cycle but $\gamma^{g(B')}$ is not a p-cycle, and so has order relatively prime to p. Raising γ to that power, we may assume without loss of generality that $\gamma^{g(B)}$ is the p-cycle $(x_0, x_1, \ldots, x_{p-1})$, while $\gamma^{g(B')} = 1$. Then,

$$\left(g^{-1}\gamma g\right)^B = \left(g^{-1}(x_0), g^{-1}(x_1), \ldots, g^{-1}(x_{p-1})\right),$$

as $g^{-1}\gamma g\left(g^{-1}(x_i)\right) = g^{-1}(x_{i+1})$. So $\left(g^{-1}\gamma g\right)^B$ is a p-cycle, and analogous arguments show $\left(g^{-1}\gamma g\right)^{B'} = 1$. This contradicts $B \equiv B'$ and so $g(B) \equiv g(B')$. Hence, if $B \equiv B'$ then $g(B) \equiv g(B')$. Reversing the preceding argument, we may also conclude that if $g(B) \equiv g(B')$ then $B \equiv B'$, and so $\mathcal{E}$ is a block system of G.

Now suppose $B \not\equiv B'$. We will first show that if $(x_0, y_0) \in A(\Gamma)$ for some $x_0 \in B$ and $y_0 \in B'$, then $(x, y) \in A(\Gamma)$ for every $x \in B$ and $y \in B'$. As $B \not\equiv B'$, there is $\gamma \in \mathrm{fix}_G(\mathcal{B})$ such that γ^B is a p-cycle while $\gamma^{B'}$ is not a p-cycle. As above, raising γ to the power $|\gamma^{B'}|$, which is relatively prime to p, we may assume without loss of generality that $\gamma^{B'} = 1$. As γ^B is a p-cycle, we may write $\gamma^B = (x_0, x_1, \ldots, x_{p-1})$, i.e., we are writing γ^B as a p-cycle starting at x_0. Applying γ to the arc (x_0, y_0), we obtain the arc (x_1, y_0), and applying γ to the arc (x_0, y_0) r times, we obtain the arc (x_r, y_0). We conclude that $(x, y_0) \in A(\Gamma)$ for every $x \in B$. Now, there exists $\delta \in \mathrm{fix}_G(\mathcal{B})$ such that $\delta^{B'}$ is a p-cycle. Applying δ to each of the arcs (x, y_0) $p-1$ times (similar to above), we have the arcs $(x, y) \in A(\Gamma)$ for every $x \in B$ and $y \in B'$.

Now, let $\gamma \in \mathrm{fix}_G(\mathcal{B})$, and consider the map $\gamma|_E$, $E \in \mathcal{E}$. If $a = (x_0, y_0) \in A(\Gamma)$ and both $x_0, y_0 \in E$, then surely $\gamma|_E(a) = \gamma(a) \in A(\Gamma)$. Similarly, if both $x_0, y_0 \notin E$, then $\gamma|_E(a) = a \in A(\Gamma)$. If $x_0 \in E$ but $y_0 \notin E$, then let $B, B' \in \mathcal{B}$ such that $x_0 \in B$ and $y_0 \in B'$. As $y_0 \notin E$, $B \not\equiv B'$, and so by arguments in the previous paragraph $(x, y) \in A(\Gamma)$ for every $x \in B$ and $y \in B'$. As $\gamma|_E$ fixes each block of $\mathcal{B}$ set-wise, $\gamma|_E(x_0) \in B$ and $\gamma|_E(y_0) = y_0$. So $\gamma|_E(x_0, y_0) \in A(\Gamma)$. An analogous argument will show that $\gamma|_E(a) \in A(\Gamma)$ if $x_0 \notin E$ but $y_0 \in E$. As in every case, $\gamma|_E \in A(\Gamma)$, we have $\gamma|_E \in \mathrm{Aut}(\Gamma)$ establishing the result. □

Definition 5.3.12 We call the block system $\mathcal{E}$ given in Definition 5.3.10 the **$\mathcal{B}$-fixer block system of G.**

The preceding Lemma can be generalized – see Exercises 5.3.5 and 5.3.6. We next give an application of the preceding result to the semiregularity problem.

Corollary 5.3.13 *Let $m \geq 1$ and p be a prime. Let $G \leq S_{mp}$ have a normal block system $\mathcal{B}$ with m blocks of prime size p. Then $\mathrm{fix}_{G^{(2)}}(\mathcal{B})$ contains a semiregular element.*

Proof Let Γ_i be an orbital digraph of G, so $G \leq \mathrm{Aut}(\Gamma_i)$. Let $\mathcal{E}$ be the $\mathcal{B}$-fixer block system of G. By Lemma 5.3.11, $\gamma|_E \in \mathrm{Aut}(\Gamma_i)$ for every $\gamma \in \mathrm{fix}_G(\mathcal{B})$ and $E \subset \mathcal{E}$. Let $M_i \leq \mathrm{Aut}(\Gamma_i)$ be the largest subgroup of $\mathrm{Aut}(\Gamma_i)$ that has $\mathcal{B}$ as a block system. Then $\gamma|_E \in \mathrm{fix}_{M_i}(\mathcal{B})$. For each $E \in \mathcal{E}$, let $B_E \in \mathcal{B}$ such that $B_E \subseteq E$. Choose $\gamma_E \in \mathrm{fix}_G(\mathcal{B})$ such that $(\gamma_E)^B$ is a p-cycle. Then $(\gamma_E)^{B'}$ is a p-cycle for every $B' \in \mathcal{B}$ with $B' \subseteq E$. So $\pi = \Pi_{E \in \mathcal{E}}(\gamma_E)|_E$ is a product of m p-cycles and consequently semiregular, and is contained in $\mathrm{fix}_{M_i}(\mathcal{B}) \leq \mathrm{Aut}(\Gamma_i)$. As Γ_i is an arbitrary orbital digraph of G, π is contained in $\mathrm{fix}_{M_j}(\mathcal{B}) \leq \mathrm{Aut}(\Gamma_j)$ for every orbital digraph Γ_j of G, and so $\pi \in \mathrm{fix}_{G^{(2)}}(\mathcal{B})$. □

We next give a sufficient condition for a Cayley digraph of an abelian group A to be a generalized wreath product. We first prove a more general result that holds for all vertex-transitive digraphs, and the sufficient condition will follow as a consequence.

Lemma 5.3.14 *Let Γ be a digraph such that $H \leq \mathrm{Aut}(\Gamma)$ is transitive and is the largest subgroup of $\mathrm{Aut}(\Gamma)$ that has a normal block system $\mathcal{B}$ with m blocks of prime size p. Suppose $\mathrm{fix}_H(\mathcal{B})$ is not faithful on $B \in \mathcal{B}$. Then there exists a block system $\mathcal{E} \geq \mathcal{B}$ of H and subgraphs $\Gamma_1, \Gamma_2 \leq \Gamma$ such that:*

(i) *$\Gamma_1 \cong \Gamma_1' \wr \bar{K}_p$ and $\Gamma_2 \cong \bar{K}_r \wr \Gamma_2'$, where Γ_1' is the digraph obtained from $\Gamma/\mathcal{B}$ by removing every arc with both endpoints contained within a block of $\mathcal{E}/\mathcal{B}$, and $\Gamma_2' = \Gamma[E]$, $E \in \mathcal{E}$, and $r = |\mathcal{E}|$, and*
(ii) *$H = H^{(2)} \cong \mathrm{Aut}(\Gamma_1) \cap \mathrm{Aut}(\Gamma_2)$.*

Proof As $\mathrm{fix}_H(\mathcal{B})$ does not act faithfully on $B \in \mathcal{B}$ and $\mathrm{fix}_H(\mathcal{B})^{B'}$ is primitive (as it is of prime degree), we see by Theorem 2.4.7 that the kernel K of the action of $\mathrm{fix}_H(\mathcal{B})$ on B is transitive on each block $B' \in \mathcal{B}$ where it is nontrivial. We conclude that a Sylow p-subgroup of $\mathrm{fix}_H(\mathcal{B})$ has order at least p^2. By Lemma 5.3.11, there exists a block system $\mathcal{E} \geq \mathcal{B}$ of H such that $h|_E \in H^{(2)}$ for every $h \in \mathrm{fix}_H(\mathcal{E})$. Without loss of generality, we assume that $\mathcal{E}$ is the block system of H with this property that has blocks of smallest size.

Let Γ_2' and Γ_2 be defined as above, and let $\Gamma_1 = \Gamma \backslash A(\Gamma_2)$. As $h|_E \in H^{(2)}$ for every $h \in \mathrm{fix}_H(\mathcal{B})$ and $E \in \mathcal{E}$, in Γ, between any two blocks $B_1, B_2 \in \mathcal{B}$ contained in different blocks of $\mathcal{E}$, we have either every arc from B_1 to B_2, or every arc from B_2 to B_1, or no arcs between B_1 to B_2. Also, such arcs in Γ are also arcs in Γ_1. In Γ_1, there are no arcs between any two blocks of $\mathcal{B}$ contained in a fixed block of $\mathcal{E}$. Hence in Γ_1, for two distinct blocks $B_1, B_2 \in \mathcal{B}$ there is either every arc from B_1 to B_2, or every arc from B_2 to B_1, or no arcs between B_1 and B_2. Thus $\Gamma_1 = \Gamma'_1 \wr \bar{K}_p$ by Theorem 4.2.15. As every arc between blocks in $\mathcal{B}$ in different blocks of $\mathcal{E}$ is contained in $A(\Gamma_1)$, $\Gamma_1/\mathcal{B} = \Gamma/\mathcal{B}$ (except perhaps in arcs with both endpoints in $E/\mathcal{B}$, $E \in \mathcal{E}$). As Γ'_1 contains no arcs in $E/\mathcal{B}$, Γ'_1 is $\Gamma/\mathcal{B}$ with all arcs with both endpoints in $E/\mathcal{B}$ removed, $E \in \mathcal{E}$, and (i) follows.

To see (ii), first note that, by Theorem 5.2.10, $\mathcal{B}$ is a block system of $H^{(2)}$. As H is maximal in Aut(Γ) with $\mathcal{B}$ as a block system, this implies $H = H^{(2)}$. (ii) now follows from Exercise 5.3.8. □

In the special case where Γ is a Cayley digraph of an abelian group A, we have the following result.

Corollary 5.3.15 *Let A be an abelian group, $S \subseteq A$ such that there is $A_L \leq H \leq$ Aut(Cay(A, S)) with a normal block system $\mathcal{B}$ with blocks of prime size p that are the cosets of $K \leq A$. If $\mathrm{fix}_H(\mathcal{B})$ is not faithful on $B \in \mathcal{B}$, then there exists a block system $\mathcal{E} \geq \mathcal{B}$ of H where $\mathcal{E}$ is the set of cosets of $K \leq L < A$ and Γ is a (K, L)-generalized wreath product.*

Proof Let Γ = Cay(A, S). By Theorem 2.2.19, all block systems of H are normal as $A_L \leq H$. By Lemma 5.3.14 there are subgraphs $\Gamma_1, \Gamma_2 \leq \Gamma$, $\Gamma_1 = \Gamma'_1 \wr \bar{K}_p$, and $\Gamma_2 = \bar{K}_r \wr \Gamma'_2$ for digraphs Γ'_1 and Γ'_2 defined as follows. There is a block system $\mathcal{E} \geq \mathcal{B}$ of H with $\Gamma'_2 \cong \Gamma[E]$ for any $E \in \mathcal{E}$, and Γ'_1 is $\Gamma/\mathcal{B}$ where every arc with both endpoints contained within a block of $\mathcal{E}/\mathcal{B}$ is removed. Note that as $\mathcal{E}$ is normal, there exists $L \leq A$ such that $\mathcal{E}$ is the set of cosets of L in A. By the definition of Γ_1, we see that $S \setminus L$ is a union of cosets of K. □

We now give the promised example of a generalized wreath product digraph Γ = Cay(A, S), A an abelian group, which can be decomposed into digraphs Γ_1 = Cay($A, S \setminus L$) and Γ_2 = Cay($A, S \cap L$) so that Aut(Γ) = Aut(Γ_1) ∩ Aut(Γ_2), and Γ is not a wreath product. We will then have an example of a Cayley digraph that is a generalized wreath product but is not a wreath product whose full automorphism group is the "expected automorphism group." Before doing so, we need an additional permutation group-theoretic idea.

Definition 5.3.16 A group G of order n is called a **Burnside group** if whenever $H \leq S_n$ and contains a regular subgroup isomorphic to G, then H is either imprimitive or 2-transitive.

Unsurprisingly, Burnside groups were introduced by Burnside. The following is a useful result regarding Burnside groups (see Dixon and Mortimer (1996, Theorem 3.5A) for a proof of this result).

Theorem 5.3.17 *An abelian group of composite order with a nontrivial cyclic Sylow p-subgroup is a Burnside group.*

A consequence of the CFSGs is that all 2-transitive groups are known (Cameron, 1981, Theorem 5.3). So if G is a Burnside group, then a transitive permutation group that contains G is either imprimitive or known. Additionally, the primitive groups that contain a regular abelian subgroup have been determined by Li (see Theorem 5.4.14), and, with Song and Zhang, also determined quasiprimitive groups that contain a dihedral subgroup (Song et al., 2014, Theorem 3.3) (all such groups are 2-transitive). Finally, Liebeck, Praeger, and Saxl have determined the regular subgroups of primitive groups (Liebeck et al., 2010).

Example 5.3.18 Let p be prime, $K = \left\langle p^2 \right\rangle < \mathbb{Z}_{p^3}$, and $K < L = \langle p \rangle < \mathbb{Z}_{p^3}$. Define digraphs $\Gamma = \mathrm{Cay}\left(\mathbb{Z}_{p^3}, \{p, 1 + K\}\right), \Gamma_1 = \mathrm{Cay}\left(\mathbb{Z}_{p^3}, \{p\}\right)$, and $\Gamma_2 = \mathrm{Cay}\left(\mathbb{Z}_{p^3}, 1 + K\right)$. Then Γ is a (K, L)-generalized wreath product and $\mathrm{Aut}(\Gamma) = \mathrm{Aut}(\Gamma_1) \cap \mathrm{Aut}(\Gamma_2)$ has order p^{p+2}.

Solution Clearly Γ is a (K, L)-generalized wreath product. As $\mathbb{Z}_{p^3}$ is a Burnside group by Theorem 5.3.17, $\mathrm{Aut}(\Gamma)$ is either imprimitive or 2-transitive. If $\mathrm{Aut}(\Gamma)$ is 2-transitive, then $\Gamma = K_{p^3}$ (as it has some arcs), and so $p+1 = p^3-1$, which is not possible. Thus $\mathrm{Aut}(\Gamma)$ is imprimitive, with nontrivial block system $\mathcal{B}$ with blocks of size p or p^2. As $\mathrm{Aut}(\Gamma)$ contains $(\mathbb{Z}_{p^3})_L$, by Theorem 2.2.19 the only nontrivial block systems of $\mathrm{Aut}(\Gamma)$ are the cosets of K (with blocks of size p) or cosets of L (with blocks of size p^2). If $\mathcal{B}$ has blocks of size p^2, then $\mathrm{Stab}_{\mathrm{Aut}(\Gamma)}(B)^B$, $B \in \mathcal{B}$, is contained in $\mathrm{Aut}(\Gamma[B])$, and $\Gamma[B]$ is a directed cycle of length p^2. We conclude that $\mathrm{Stab}_{\mathrm{Aut}(\Gamma)}(B)^B \cong \mathbb{Z}_{p^2}$, and by Theorem 2.3.10, $\mathrm{Aut}(\Gamma)$ also has blocks of size p. We thus assume $\mathcal{B}$ has blocks of size p.

Let C be the block system of $\left(\mathbb{Z}_{p^3}\right)_L$ with blocks of size p^2, and so is the set of cosets of L. By Theorem 5.3.6 (i) we have $\mathrm{fix}_{\left(\mathbb{Z}_{p^3}\right)_L}(\mathcal{B})|_C \leq \mathrm{Aut}(\Gamma)$ for every $C \in C$, and so a Sylow p-subgroup of $\mathrm{fix}_{\mathrm{Aut}(\Gamma)}(\mathcal{B})$ has order at least p^p. This implies that $\mathrm{fix}_{\mathrm{Aut}(\Gamma)}(\mathcal{B})$ is not faithful in its action on $B \in \mathcal{B}$, and $\mathrm{Aut}(\Gamma)$ has order p^{p+2} if and only $\mathrm{fix}_{\mathrm{Aut}(\Gamma)}(\mathcal{B})$ has order p^p and $\mathrm{Aut}(\Gamma)/\mathcal{B}$ has order p^2. By Lemma 5.3.14 (with $H = \mathrm{Aut}(\Gamma)$), we have $\mathrm{Aut}(\Gamma) = \mathrm{Aut}(\Gamma_1) \cap \mathrm{Aut}(\Gamma_2)$. As $\mathrm{Aut}(\Gamma_1) \cong S_p \wr \mathbb{Z}_{p^2}$ and $\mathrm{Aut}(\Gamma_2) \cong \mathbb{Z}_{p^2} \wr S_p$, by Theorem 5.1.8 it follows that $\mathrm{fix}_{\mathrm{Aut}(\Gamma)}(\mathcal{B})$ has order p^p and $\mathrm{Aut}(\Gamma)/\mathcal{B}$ has order p^2. □

We have now seen examples of (K, L)-generalized wreath products $\Gamma = \text{Cay}(A, S)$ where $\text{Aut}(\Gamma) = \text{Aut}(\Gamma_1) \wr \text{Aut}(\Gamma_2)$ and $\text{Aut}(\Gamma) \neq \text{Aut}(\Gamma_1) \wr \text{Aut}(\Gamma_2)$ with $\Gamma_1 = \text{Cay}(A, S \backslash L))$ and $\Gamma_2 = \text{Cay}(A, S \cap L))$ (although in the latter case we did not mention it when we saw the example). In the latter case, if $\text{Aut}(\Gamma) = S_m \times G$ is of order n where $m = k\ell$, $k, \ell > 1$ (so m is composite), then there is a subgroup K of A of order k and $K < L$ is of order n/ℓ so that Γ is a (K, L)-generalized wreath product but $\text{Aut}(\Gamma) \neq \text{Aut}(\Gamma_1) \wr \text{Aut}(\Gamma_2)$ (Exercise 8.5.6). The graph given in Example 5.3.4 is of this type. Also, if Γ is a (K, L)-generalized wreath product but is also an (H, L)-generalized wreath product or a (K, M)-generalized wreath product for some $1 < H < K$ or $L < M < A$, then Γ will also not have automorphism group $\text{Aut}(\Gamma_1) \wr \text{Aut}(\Gamma_2)$ (Exercise 5.3.9). It is not known if there are other examples. Note that, by Exercise 5.3.8, $\text{Aut}(\Gamma) = \text{Aut}(\Gamma_1) \wr \text{Aut}(\Gamma_2)$ if the cosets of L in A are a block system of $\text{Aut}(\Gamma)$ (and that the converse is true! See Exercise 5.3.10). This condition on when $\text{Aut}(\Gamma) = \text{Aut}(\Gamma_1) \wr \text{Aut}(\Gamma_2)$ seems unsatisfactory as it is a condition on $\text{Aut}(\Gamma)$. Ideally, we would like necessary and sufficient conditions only involving the digraph (analogous to Theorem 5.1.8 for wreath products), or equivalently, its connection set S. This leads to the following natural problem.

Problem 5.3.19 Let G be an abelian group, $\Gamma = \text{Cay}(G, S)$, and suppose Γ is a (K, L)-generalized product where $K \leq L < G$. Determine necessary and sufficient conditions on Γ (or S) so that $\text{Aut}(\Gamma) = \text{Aut}(\text{Cay}(G, S \backslash L)) \wr \text{Aut}(\text{Cay}(G, S \cap L))$.

We will see generalized wreath products again in Chapters 7 and 8, especially in Theorem 8.5.1 where we will see that the automorphism groups of circulant digraphs fall into three broad families, one of which is the automorphism groups of circulant digraphs that are generalized wreath products.

Very little is known about Problem 5.3.19, but Dobson and Morris (2009) have shown that for each circulant digraph $\text{Cay}(\mathbb{Z}_n, S)$ of square-free order n that is a generalized wreath product, there is $L \leq \mathbb{Z}_n$ (that depends upon S) such that $\text{Aut}(\text{Cay}(\mathbb{Z}_n, S) = \text{Aut}(\text{Cay}(\mathbb{Z}_n, S \backslash L)) \wr \text{Aut}(\text{Cay}(G, S \cap L))$.

5.4 Primitive Groups and the O'Nan–Scott Theorem

We saw in Corollary 4.3.3 that any transitive permutation group can be embedded into a wreath product of primitive permutation groups. Thus, in order to understand imprimitive groups, it is necessary to understand primitive groups (although there are more things that need to be known than just primitive groups, such as how to "put them together" to form an imprimitive group). In this section, we introduce the main tool in studying and constructing primitive permutation groups, the O'Nan–Scott theorem. This theorem reduces the study of primitive permutation groups to the study of finite simple groups and their actions. As all finite simple groups are given in the CFSG, together these two results are powerful tools in studying primitive permutation groups. We have already discussed results that give all primitive permutation groups satisfying various properties, such as the degree of the primitive permutation group or whether or not a primitive permutation group contains a regular subgroup. These results, and others like them, are all proven using the O'Nan–Scott theorem and CFSG.

The reader should only consider this section at most a very brief introduction to these topics. We will not provide a proof of the O'Nan–Scott theorem, as this can be found elsewhere, for example in Dixon and Mortimer (1996). Additionally, to fully understand how to use these techniques, much more knowledge about the finite simple groups than we will provide is usually needed, and we have only defined a few of the finite simple groups! So our goal is to provide only a brief glimpse into what these results are and how they are used. We will start with one of the simplest examples of primitive groups, constructed as the automorphism group of graphs, that is not more or less just a finite simple group acting by left multiplication on a maximal subgroup, which we have seen by Theorem 2.4.14 will give a primitive group. We will then state a weak form of the O'Nan–Scott theorem, and develop enough theory to solve the same problem using the O'Nan–Scott theorem and CFSG.

Definition 5.4.1 For $n \geq 2$ and $d \geq 2$, define a graph $H(d,n)$ by $V(H(d,n)) = \mathbb{Z}_n^d$, where two vertices of $H(d,n)$ are adjacent if and only if they differ in exactly one coordinate. The graph $H(d,n)$ is the **Hamming graph** of dimension d over $\mathbb{Z}_n$.

So the n-dimensional hypercubes are examples of Hamming graphs. That is, $Q_n = H(n,2)$. The graph $H(2,3)$ is drawn in Figure 5.4.

Lemma 5.4.2 *Let $\sigma \in S_n$, and define $\sigma_1, \sigma_2, \tau \colon \mathbb{Z}_n \times \mathbb{Z}_n \to \mathbb{Z}_n \times \mathbb{Z}_n$ by $\sigma_1(i,j) = (\sigma(i), j)$, $\sigma_2(i,j) = (i, \sigma(j))$, and $\tau(i,j) = (j,i)$. Then,*

$$\langle S_n \times S_n, \tau\rangle \cong \langle \sigma_1, \sigma_2, \tau : \sigma \in S_n\rangle \leq \mathrm{Aut}(H(2,n)).$$

Proof First observe that for there to be an edge between two vertices in $H(2,n)$, it must be the case that the ordered pairs representing the vertices are the same in one coordinate and different in the other. For $\sigma \in S_n$ and $e \in E(H(2,n))$, σ_1, σ_2, and τ applied to e is a set of two ordered pairs the same in one coordinate and different in the other. Thus $\sigma_1, \sigma_2, \tau \in \mathrm{Aut}(H(2,n))$. It is clear that $\langle \sigma_1, \sigma_2 : \sigma \in S_n\rangle \cong S_n \times S_n$, and the result follows. □

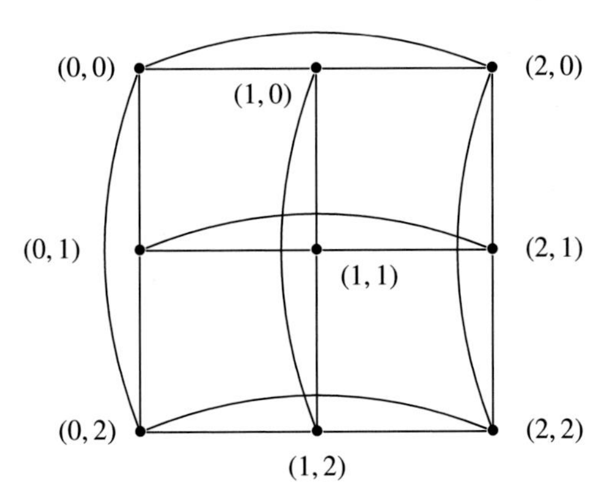

Figure 5.4 The graph $H(2,3)$.

Our first goal of this section is to prove that $\mathrm{Aut}(H(2,n)) = \langle S_n \times S_n, \tau\rangle$ is primitive. We start by showing the latter group is primitive.

Lemma 5.4.3 *For $n \geq 3$, $G = \langle S_n \times S_n, \tau\rangle$ is primitive.*

Proof For $k = 0, 1, 2$, let O_k be the set of all elements of $\mathbb{Z}_n \times \mathbb{Z}_n$ with exactly k coordinates 0. Note that $\mathrm{Stab}_G((0,0)) = \langle S_{n-1} \times S_{n-1}, \tau\rangle$ where, here, S_{n-1} permutes the $n-1$ elements of $[1, n-1]$. Let $(i, j), (i', j') \in O_0$ (so none of i, i', j, j' are zero). Then there exists $\delta, \gamma \in S_n$ that both fix 0, $\delta(i) = i'$, and $\gamma(j) = j'$. Then $(\delta, \gamma) \in \mathrm{Stab}_G((0,0))$ and maps (i, j) to (i', j'), and $\mathrm{Stab}_G((0,0))$ is transitive on O_0. Clearly, $\mathrm{Stab}_G((0,0))$ is also transitive on O_2. As $1 \times S_{n-1} \leq \mathrm{Stab}_G((0,0))$, we see that $\mathrm{Stab}_G((0,0))$ is transitive on $F = \{(0,y) : y \in [1, n-1]\}$, and as τ maps F to $\{(x,0) : x \in [1,n]\}$, we see that $\mathrm{Stab}_G((0,0))$ is transitive on O_1. Also, every block of G that contains $(0,0)$ is the union of some subset of $\{O_k : 0 \leq k \leq 2\}$ by Exercise 2.3.18. Let $\omega \in S_n$ be the transposition $(0,1)$, and define $\omega' : \mathbb{Z}_n \times \mathbb{Z}_n \to \mathbb{Z}_n \times \mathbb{Z}_n$ by $\omega'(i, j) = (\omega(i), j)$. Clearly $\omega' \in S_n \times S_n$. Then ω' maps some elements of O_1 to O_2 and maps some elements of O_2 to O_1. Hence the only blocks of G that contain $(0,0)$ are trivial, and G is primitive. □

Theorem 5.4.4 *For $n \geq 3$, $\mathrm{Aut}(H(2,n)) = \langle S_n \times S_n, \tau\rangle$ is primitive and isomorphic to $S_2 \wr S_n$ as an abstract group.*

Proof Set $\Gamma = H(2,n)$. In view of Lemmas 5.4.2 and 5.4.3, and the fact that $\langle S_n \times S_n, \tau\rangle$ has order $2 \cdot (n!)^2 = |S_2 \wr S_n|$, we need only show that $\mathrm{Aut}(\Gamma) \cong S_2 \wr S_n$. We need to find $2n$ objects in Γ so that the action of $\mathrm{Aut}(\Gamma)$ on these $2n$ objects is $S_2 \wr S_n$, and the only permutation that fixes these $2n$ objects is the identity. The $2n$ objects are induced subgraphs of Γ isomorphic to K_n. Setting $F_i = \{i\} \times \mathbb{Z}_n$ and $S_i = \mathbb{Z}_n \times \{i\}$, it is easy to see that $\Gamma[F_i] \cong \Gamma[S_i] \cong K_n$, $i \in \mathbb{Z}_n$.

Now suppose $S \subset \mathbb{Z}_n \times \mathbb{Z}_n$ and $\Gamma[S] = K_n$. As $\text{Aut}(\Gamma)$ is transitive, we may assume without loss of generality that $(0,0) \in S$. Observe that the neighbors of $(0,0)$ in Γ are $\{(0,j) : j \in \mathbb{Z}_n\} \cup \{(j,0) : j \in \mathbb{Z}_n\}$. If $(0,0) \neq (0,i) \in S$, then as the only common neighbors of $(0,0)$ and $(0,i)$ in Γ are in $F_0 = \{(0,j) : j \in \mathbb{Z}_n\}$, we have $S = F_0$. Similarly, if $(0,0) \neq (i,0) \in S$, then $S = S_0$. Thus Γ contains exactly $2n$ induced subgraphs isomorphic to K_n.

Define a graph $K(\Gamma)$ by $V(K(\Gamma)) = \{F_i, S_i : i \in \mathbb{Z}_n\}$ and $E(K(\Gamma)) = F_iS_j$, $i,j \in \mathbb{Z}_n$ ($K(\Gamma)$ is the **clique graph** of Γ). So $K(\Gamma) \cong K_{n,n} \cong K_2 \wr \bar{K}_n$, and $\text{Aut}(K(\Gamma)) \cong \mathcal{S}_2 \wr \mathcal{S}_n$. Additionally, $F_i \cap S_j = \{(i,j)\}$ for all $i,j \in \mathbb{Z}_n$. So any permutation in $\text{Aut}(\Gamma)$ that fixes each clique on n vertices also fixes each (i,j), $i,j \in \mathbb{Z}_n$, and so is the identity. Clearly, if $\gamma \in \text{Aut}(\Gamma)$, then $\gamma(C) \cong K_n$ for any $C \in V(K(\Gamma))$. Hence γ induces a permutation on $V(K(\Gamma))$. As F_i and S_j have exactly one vertex in common, $\gamma(F_i)$ and $\gamma(S_j)$ have exactly one vertex in common, and so $\gamma(F_iS_j) \in E(K(\Gamma))$, $i,j \in \mathbb{Z}_n$. We conclude that γ induces an automorphism of $K(\Gamma)$. This implies that $\text{Aut}(\Gamma) \cong \text{Aut}(K(\Gamma)) \cong \mathcal{S}_2 \wr \mathcal{S}_n$, establishing the result. □

Some remarks are in order. First, the automorphism groups of the graphs $H(2,n)$ were first studied by Chao (1965), who proved that $\text{Aut}(H(2,n))$ is primitive, but did not calculate $\text{Aut}(H(2,n))$. The graphs $H(d,n)$ are examples of the Cartesian product of d graphs. Recall from Definition 3.5.1 that the **Cartesian product** $\Gamma \square \Delta$ of two graphs Γ and Δ is the graph with vertex set $V(\Gamma) \times V(\Delta)$ and edge set

$$\{(u,v)(u',v') : u = u' \text{ and } vv' \in E(\Gamma) \text{ or } v = v' \text{ and } uu' \in E(\Delta)\}.$$

Then $H(d,n)$ is the Cartesian product of K_n with itself d times. Our result gives a specific example of a more general theorem on automorphism groups of Cartesian products that we state without proof here (see Imrich et al. (2008, Corollary 15.6) for the proof).

Theorem 5.4.5 *Let $\Gamma_1, \ldots, \Gamma_r$ be graphs. If no Γ_i, $1 \le i \le r$ can be written as the Cartesian product of two smaller nontrivial graphs, then the automorphism group of the Cartesian product $\Gamma_1 \square \Gamma_2 \square \cdots \square \Gamma_r$ is isomorphic to the automorphism group of the disjoint union of the graphs $\Gamma_1, \ldots, \Gamma_r$.*

Theorem 5.4.4 points out that some wreath products have a different transitive action that is primitive. Our goal now is to study this new action, which is called the **product action**. The basic idea behind the product action is as follows. Let X and Y be sets, with $H \le \mathcal{S}_X$ and $K \le \mathcal{S}_Y$. In $H \wr K$ in its usual action, there is a subgroup isomorphic to $K^{|X|}$, and each copy of K independently permutes one of the natural $|X|$ copies of Y in $X \times Y$. In the product action of $H \wr K$, elements will act on $Y^{|X|}$. We first apply an element of the $|X|$ copies of K that independently permutes the elements in the coordinates of $Y^{|X|}$, and then permute the $|X|$ coordinates of $Y^{|X|}$ using elements of H. (We remark our choice of applying independently the elements of $K^{|X|}$ and then permuting the coordinates is determined by our choosing in the definition of the wreath product to apply elements of K to the second coordinate of the sets $\{x\} \times Y$ and then apply an element of H to X.)

Let $n = |X|$, and, for convenience, assume $X = [1,n]$. Formally, for an element $\gamma \in H \wr K$, with $\gamma(x,y) = (h(x), k_x(y))$, $h \in H$ and $k_x \in K$, we let γ act on Y^n by,

$$\gamma(y_1, y_2, \ldots, y_n) = \left(k_{h^{-1}(1)}\left(y_{h^{-1}(1)}\right), k_{h^{-1}(2)}\left(y_{h^{-1}(2)}\right), \ldots, k_{h^{-1}(n)}\left(y_{h^{-1}(n)}\right)\right). \quad (5.2)$$

Notice that in the first coordinate, we are applying $k_{h^{-1}(1)}$ to $y_{h^{-1}(1)}$, as the coordinate of $(y_1, \ldots, y_n)$ that is in the first coordinate after h permutes the coordinates, is the element in coordinate $h^{-1}(1)$ before applying h. As we apply the elements of K first and then permute the coordinates with an element of H, the element $k_{h^{-1}(1)}\left(y_{h^{-1}(1)}\right)$ is in the coordinate that h maps to the first coordinate. Similarly for the other coordinates.

Recall that the action of a group G on a set Z must satisfy $(gh)(z) = g(hz)$ and $e(z) = z$ where $g, h \in G$, $z \in Z$ and $e \in G$ is the identity. Clearly our definition of γ acting on Y^n satisfies the latter condition. For the former condition, suppose $\delta \in H \wr K$ with $\delta(x,y) = \left(\hat{h}(x), \hat{k}_x(y)\right)$, where $\hat{h} \in H$ and each $\hat{k}_x \in K$. As $(\gamma\delta)(x,y) = \left(h\hat{h}(x), k_{\hat{h}(x)}\hat{k}_x(y)\right)$, we see that

$$\begin{aligned}\gamma\delta(y_1, \ldots, y_n) &= \left(k_{\hat{h}((h\hat{h})^{-1}(1))}\hat{k}_{(h\hat{h})^{-1}(1)}\left(y_{(h\hat{h})^{-1}(1)}\right), \ldots, k_{\hat{h}((h\hat{h})^{-1}(n))}\hat{k}_{(h\hat{h})^{-1}(n)}\left(y_{(h\hat{h})^{-1}(n)}\right)\right)\\ &= \left(k_{h^{-1}(1)}\hat{k}_{(h\hat{h})^{-1}(1)}\left(y_{(h\hat{h})^{-1}(1)}\right), \ldots, k_{h^{-1}(n)}\hat{k}_{(h\hat{h})^{-1}(n)}\left(y_{(h\hat{h})^{-1}(n)}\right)\right).\end{aligned}$$

The computation for $\gamma(\delta(y_1, \ldots, y_n))$ is a little trickier. We will write it down and then explain it.

$$\begin{aligned}\gamma(\delta(y_1, \ldots, y_n)) &= \gamma\left(\hat{k}_{\hat{h}^{-1}(1)}\left(y_{\hat{h}^{-1}(1)}\right), \ldots, \hat{k}_{\hat{h}^{-1}(n)}\left(y_{\hat{h}^{-1}(n)}\right)\right)\\ &= \left(k_{h^{-1}(1)}\hat{k}_{(h\hat{h})^{-1}(1)}\left(y_{(h\hat{h})^{-1}(1)}\right), \ldots, k_{h^{-1}(n)}\hat{k}_{(h\hat{h})^{-1}(n)}\left(y_{(h\hat{h})^{-1}(n)}\right)\right).\end{aligned}$$

Notice that in the first coordinate of the second line of the displayed equation we have $y_{\hat{h}^{-1}h^{-1}(1)} = y_{\left(h\hat{h}\right)^{-1}(1)}$ as the element of $(y_1, \ldots, y_n)$ that is in the first coordinate of $\gamma(\delta(y_1, \ldots, y_n))$ is y_x, $1 \leq x \leq n$, where $h\hat{h}(x) = 1$, and so $x = \hat{h}^{-1}h^{-1}(1) = \left(h\hat{h}\right)^{-1}(1)$. We have that $k_{h^{-1}(1)}\hat{k}_{\left(h\hat{h}\right)^{-1}(1)}$ is applied to $y_{\left(h\hat{h}\right)^{-1}(1)} = y_{\hat{h}^{-1}h^{-1}(1)}$ as the map $\hat{k}_{\left(h\hat{h}\right)^{-1}(1)}$ is applied to $y_{\left(h\hat{h}\right)^{-1}(1)}$ before being permuted to the coordinate $h^{-1}(1)$ by $\hat{h}$. Then $\hat{k}_{\left(h\hat{h}\right)^{-1}(1)}\left(y_{(h\hat{h})^{-1}(1)}\right)$ has $k_{h^{-1}(1)}$ applied to it before being permuted by h to the first coordinate. The other coordinates are similar. Thus $\gamma(\delta(y_{x_1}, \ldots y_{x_n})) = \gamma\delta(y_{x_1}, \ldots y_{x_n})$ and Equation (5.2) does indeed give an action.

The conditions on G and H under which $G \wr H$ with the product action gives a primitive permutation group are as follows. The interested reader is referred to Dixon and Mortimer (1996, Lemma 2.7A) for a proof of this result.

Theorem 5.4.6 *Let $G \leq S_X$ with $|X| = m$, and $H \leq S_Y$ be nontrivial. Then $G \wr H$ acting on Y^m with the product action is primitive if and only if*

(i) *H is primitive but not regular on Y, and*
(ii) *G is transitive on X, and X is finite.*

The O'Nan–Scott theorem gives the structure of all primitive groups. We do not give a precise statement of the result, as many of the cases do not occur frequently in the study of vertex-transitive graphs (at least not yet), and a detailed explanation of the cases we will not encounter would be lengthy. For additional details and a proof, see either Dixon and Mortimer (1996) or Liebeck et al. (1988). Recall from Definition 4.4.17 that for a group G, the socle of G, denoted $\mathrm{soc}(G)$, is the subgroup of G generated by all minimal normal subgroups of G.

Theorem 5.4.7 (The O'Nan–Scott theorem) *Let G be a primitive group of degree n. Then $\mathrm{soc}(G) = T^m$ for some simple group T and $m \geq 1$. Additionally, one of the following is true:*

(i) *$n = p^k$ for some prime p, $\mathrm{soc}(G)$ is elementary abelian, and $G \leq \mathrm{AGL}(k, p)$, or*
(ii) *$\mathrm{soc}(G) = T$ for some nonabelian simple group T and $G \leq \mathrm{Aut}(T)$, or*
(iii) *$m \geq 2$, $n = |T|^{m-1}$, and G is of "diagonal type," or*
(iv) *$m \geq 2$, and there is a divisor $d \neq m$ of m and a primitive group H with socle T^d and degree ℓ such that $G \leq S_d \wr H$ with the product action. The degree is $n = \ell^{m/d}$, or*
(v) *$m \geq 6$, $n = |T|^m$ and T^m is regular.*

The O'Nan–Scott theorem, together with CFSG, allows for the solution of many longstanding problems in permutation group theory, many of which have applications to the study of vertex-transitive graphs. One such problem, for example, was to determine all primitive groups of degree $2p$, p a prime, and, in particular, are S_5 in its action on 10 points, i.e., are the automorphism group of the Petersen graph and its normal primitive subgroup A_5 the only ones that are uniprimitive? The answer is yes (Liebeck and Saxl, 1985a), and the only known proof is using the O'Nan–Scott theorem and CFSG.

We now illustrate this type of argument by showing that $\mathrm{Aut}(H(2,n)) = S_2 \wr S_n$, $n \geq 4$, with the product action (assuming that $S_2 \wr S_n \leq \mathrm{Aut}(H(2,n))$). First, observe that S_n certainly contains a regular cyclic subgroup, and so $S_n \times S_n \leq S_2 \wr S_n \leq \mathrm{Aut}(H(2,n))$ contains a transitive abelian subgroup isomorphic to $\mathbb{Z}_n \times \mathbb{Z}_n$. The primitive permutation groups that contain a regular abelian subgroup have been determined by Li (2003, Theorem 1.1) (we remark that Jones independently also found the primitive permutation groups that contain a regular cyclic subgroup (Jones, 2002, Theorem 3)). Before turning to the statement of this result, it will be useful to define certain matrix groups (one of which we briefly encountered in Section 4.5).

Definition 5.4.8 The **special linear group of dimension** d **over the field** $\mathbb{F}_q$, denoted $\mathrm{SL}(d,q)$, is the subgroup of $\mathrm{GL}(d,q)$ of all matrices of determinant 1. That is, $\mathrm{SL}(d,q)$ is the group of invertible $d \times d$ matrices with determinant 1, where the entries in the matrices are elements of $\mathbb{F}_q$.

Lemma 5.4.9 *Let* $d \geq 1$ *and* q *be a prime power. Then,*

$$|\mathrm{SL}(d,q)| = \frac{1}{q-1}\Pi_{i=0}^{d-1}\left(q^d - q^i\right).$$

Proof As there are $q-1$ possible determinants for matrices in $\mathrm{GL}(d,q)$, it suffices to show $|\mathrm{GL}(d,q)| = \Pi_{i=0}^{d-1}(q^d - q^i)$, which is Lemma 3.8.10. □

Recalling the order of $\mathrm{PGL}(d,q)$ (see p. 160), we have $|\mathrm{PGL}(d,q)| = |\mathrm{SL}(d,q)|$. We digress briefly from our main aim of this discussion to show that $\mathrm{SL}(d,q)$ and $\mathrm{PGL}(d,q)$ are not necessarily isomorphic, as well as complete our catalog of basic matrix groups.

We begin our digression with an additional matrix group. If $\gcd(d,q-1) \neq 1$, then as $\mathbb{F}_q^*$ is cyclic, there exists $\alpha \in \mathbb{F}_q^*$ whose order is $\gcd(d,q-1)$. Then the matrix αI_d is a scalar matrix with determinant 1. Hence, in this case the diagonal matrices in $\mathrm{SL}(d,q)$ form a nontrivial normal subgroup $Z \cap \mathrm{SL}(d,q)$ of order $\gcd(d,q-1)$ (recall from Section 4.5 that the diagonal matrices in $\mathrm{GL}(d,q)$ are denoted by Z).

Definition 5.4.10 The **projective special linear group of dimension** d **over** $\mathbb{F}_q$ is defined to be the quotient group $\mathrm{SL}(d,q)/(Z \cap \mathrm{SL}(d,q))$, and is denoted $\mathrm{PSL}(d,q)$.

Note that the use of "P" in $\mathrm{PSL}(d,q)$ and $\mathrm{PGL}(d,q)$ is consistent – in each case we are forming the quotient group using the canonical normal subgroup consisting of scalar matrices. The group $\mathrm{PSL}(d,q)$ is simple unless $d = 2$ and $q = 2$ or 3 (Alperin and Bell, 1995, Theorem 2.6.8).

Lemma 5.4.11 *Let* $d \geq 1$ *and* q *be a prime power. Then*

$$|\mathrm{PSL}(d,q)| = \frac{1}{(q-1)\gcd(d,q-1)}\Pi_{i=0}^{d-1}\left(q^d - q^i\right).$$

Proof We saw above that $Z \cap SL(d,q)$ is a normal subgroup of $SL(d,q)$ of order $\gcd(d, q-1)$. As $|SL(d,q)| = \frac{1}{q-1}\Pi_{i=0}^{d-1}\left(q^d - q^i\right)$, the result follows by the first isomorphism theorem. □

It is now easy to see, using Exercise 5.4.1, that $PGL(d,q) \not\cong SL(d,q)$ if $\gcd(d,q) \neq 1$ as in that case $Z(SL(d,q)) \neq 1$ while $Z(PGL(d,q)) = 1$.

There are two final groups that are commonly encountered. First, observe that applying a field automorphism α of $\mathbb{F}_q$ to a vector $(v_1, \ldots, v_d)$ in the canonical fashion of $\alpha(v_1, \ldots, v_d) = (\alpha(v_1), \ldots, \alpha(v_d))$ certainly induces a permutation $\bar{\alpha}$ of the vectors in $\mathbb{F}_q^d$. However, this permutation cannot be represented as a linear transformation as, if $\alpha \neq 1$ and $a \in \mathbb{F}_q$, then

$$\alpha(a(v_1, \ldots, v_d)) = \alpha(a)(\alpha(v_1), \ldots, \alpha(v_d)) \neq a(\alpha(v_1), \ldots, \alpha(v_d)).$$

Definition 5.4.12 The **general semilinear group of dimension** d **over** $\mathbb{F}_q$, denoted $\Gamma L(d,q)$, is defined to the group generated by $GL(d,q)$ together with all permutations of the vectors induced by field automorphisms of $\mathbb{F}_q$.

Lemma 5.4.13 *Let $d \geq 1$ and q be a prime power. Then* $GL(d,q) \trianglelefteq \Gamma L(d,q)$.

Proof Let $A \in GL(d,q)$, with $A = [a_{i,j}]$, $v = (v_1, \ldots, v_d) \in \mathbb{F}_q^d$, $\alpha \in \mathrm{Aut}(\mathbb{F}_q)$, and $\bar{\alpha}$ the permutation of $\mathbb{F}_q^d$ induced by α. Then,

$$\begin{aligned}
\bar{\alpha}^{-1}A\bar{\alpha}(v^T) &= \bar{\alpha}^{-1}A(\alpha(v), \ldots, \alpha(v_d))^T \\
&= \bar{\alpha}^{-1}\Big(\sum_{j=1}^{d} a_{1,j}\alpha(v_j), \sum_{j=1}^{d} a_{2,j}\alpha(v_j), \ldots, \sum_{j=1}^{d} a_{d,j}\alpha(v_j)\Big) \\
&= \Big(\sum_{j=1}^{d} \alpha^{-1}(a_{1,j})v_j, \sum_{j=1}^{d} \alpha^{-1}(a_{2,j})v_j, \ldots, \sum_{j=1}^{d} \alpha^{-1}(a_{d,j})v_j\Big) \\
&= [\alpha^{-1}(a_{i,j})]v^T.
\end{aligned}$$

We conclude that $\bar{\alpha}^{-1}A\bar{\alpha}$ is the matrix obtained from A by applying α^{-1} to each entry. Thus $GL(d,q) \trianglelefteq \Gamma L(d,q)$. □

Of course, α also induces a permutation $\hat{\alpha}$ of the one-dimensional subspaces of $\mathbb{F}_q^d$. The group generated by $\hat{\alpha}$ and $PGL(d,q)$ is the **projective semilinear group of dimension** d **over** $\mathbb{F}_q$ and denoted $P\Gamma L(d,q)$. We leave it to Exercise 5.4.2 to show that $P\Gamma L(d,q)$ normalizes $PGL(d,q)$ – in fact, $P\Gamma L(d,q)$ is the normalizer in $S_{(q^d-1)/(q-1)}$ of $PSL(d,q)$ (see Dieudonné (1980) for the automorphisms of $PSL(d,q)$ and then apply Lemma 4.4.9).

With these groups in hand, we are now ready to state the following result of Li, that gives all primitive groups that contain a regular abelian subgroup

(Li, 2003, Theorem 1.1). We state this result more or less using the notation that Li used, which is often called the "Atlas notation," after its use in the *Atlas of Finite Groups* (Conway et al., 1985) (it is defined there along with all of the notation from the Atlas, which is often used in the literature, on page xx). The only part of this notation that we will need is the notation $G \cdot H$. This means that this is a group with normal subgroup G, whose quotient by G is equal to H. That is, $L = G \cdot H$ means $G \trianglelefteq L$ and $L/G \cong H$. One will often see this also used with the British notation for multiplication, namely $G.H$ (in his paper, Li uses the British notation for multiplication).

Theorem 5.4.14 *Let G be a primitive permutation group of degree n. Then G contains a regular abelian subgroup H if and only if*

(i) $G \leq \mathrm{AGL}(d, p)$, *where p is prime, $d \geq 1$, and $n = p^d$, or*
(ii) $G = (T_1 \times \cdots \times T_\ell) \cdot O \cdot P$, $H = H_1 \times \cdots \times H_\ell$, *and* $n = m^\ell$, *where* $\ell \geq 1$, $H_i < T_i$ *with* $|H_i| = m$, $T_1 \cong \cdots \cong T_\ell$, $O \leq \mathrm{Out}(T_1) \times \cdots \times \mathrm{Out}(T_\ell)$, $P \leq \mathcal{S}_\ell$ *is transitive, and, additionally, one of the following holds:*
   (a) $(T_i, H_i) = (\mathrm{PSL}(2, 11), \mathbb{Z}_{11}), (M_{11}, \mathbb{Z}_{11}), \left(M_{12}, \mathbb{Z}_2^2 \times \mathbb{Z}_3\right)$, *or* $(M_{23}, \mathbb{Z}_{23})$,
   (b) $(T_i, H_i) = \mathrm{PGL}(d, q)$, *and* $H_i = \mathbb{Z}_{(q^d-1)/(q-1)}$,
   (c) $T_i = \mathrm{P\Gamma L}(2, 8)$ *and* $\mathbb{Z}_9 = H_i \nleq \mathrm{PSL}(2, 8)$,
   (d) $T_i = \mathcal{S}_m$ *or* A_m, *and H_i is an abelian group of order m.*

The above result is a simplification of Li's actual result, but the full result is not needed here and requires additional definitions. Also, all of the groups in Theorem 5.4.14 (ii) that do not have simple socle also satisfy Theorem 5.4.7 (iv) and so the action is the product action. We have not yet seen the Mathieu groups M_{11}, M_{12}, or M_{23}, nor the action of PSL(2, 11) of degree 11. We will not have need of these groups or actions here. The interested reader can find additional information concerning these groups and actions in Dixon and Mortimer (1996, Chapter 6 and Example 7.5.2). See Wilson (2009) for additional information concerning the finite simple groups.

We now give a different proof that $\mathrm{Aut}(H(2, n)) \cong \mathcal{S}_2 \wr \mathcal{S}_n$ using Theorem 5.4.14 (and assuming that $\mathcal{S}_2 \wr \mathcal{S}_n \leq \mathrm{Aut}(H(2, n))$):

Proof If $\mathrm{Aut}(H(2, n))$ satisfies Theorem 5.4.14 (i), then $n = p^d$ for some prime p and $d \geq 1$. As $\mathcal{S}_2 \wr \mathcal{S}_{p^d} \leq \mathrm{Aut}(H(2, n))$, by Exercise 4.2.5 we have a Sylow p-subgroup of $\mathrm{Aut}\left(H(2, p^d)\right)$ has order at least $p^{2\ell}$, where $\ell = (p^d - 1)/(p - 1)$. On the other hand, AGL$(2d, p)$ has Sylow p-subgroup of order p^{2d^2+d} by Lemma 3.8.10. Hence $2\left(p^d - 1\right)/(p - 1) \leq 2d^2 + d$. Using the fact that the left-hand side of this equation is exponential in d while the right-hand side is a quadratic in d, we see that $d = 1$, or d and p are small. The cases where d and

p are small can be handled by other means (it is not unusual when applying the O'Nan–Scott theorem that small cases have to be handled separately, often by a computer algebra system). Suppose $d = 1$, and $\mathcal{S}_2 \wr \mathcal{S}_p \leq \mathrm{AGL}(2, p)$. As $|\mathcal{S}_2 \wr \mathcal{S}_p| = 2(p!)^2$ and $|\mathrm{AGL}(2, p)| = p^2 \left(p^2 - 1\right)\left(p^2 - p\right)$, this implies $[(p-2)!]^2$ divides $(p+1)/2$. This again gives p is small, so other techniques could be used.

If Theorem 5.4.14 (ii) occurs, then either $\ell = 1$ and $\mathrm{soc}(\mathrm{Aut}(H(2, n)))$ is simple or $\ell = 2$ and Theorem 5.4.14(ii)(d) holds. In the former case, according to Theorem 5.4.14, the only primitive group of degree n^2 with simple socle and a regular subgroup isomorphic to $\mathbb{Z}_n \times \mathbb{Z}_n$ is A_{n^2} or $\mathcal{S}_{n^2}$. Neither of these outcomes are possible as $H(2, n)$ is not a complete graph or a complement. In the latter case, we have $G \leq \mathcal{S}_2 \wr \mathcal{S}_n$ with the product action, while $\mathcal{S}_2 \wr \mathcal{S}_n \leq G$. Hence $\mathrm{Aut}(H(2, n)) = \mathcal{S}_2 \wr \mathcal{S}_n$ with the product action. □

This simple example of an application of a consequence of the O'Nan–Scott theorem is probably not the best possible application, as we already have a combinatorial argument. In general, as a matter of style and understanding, when a combinatorial argument can be found, it is preferable to use an argument using a consequence of the O'Nan–Scott theorem, as typically such consequences also rely on CFSG, a theorem whose proof is not understood by many mathematicians. However, there are definitely situations where such arguments are not only appropriate, but are the *only* tools available (typically, for example, when one only knows that one has a primitive group). We shall encounter some such situations later.

Many consequences of the O'Nan–Scott theorem are in the literature, and have applications to the study of vertex-transitive graphs. Some are motivated by problems concerning vertex-transitive graphs. Cameron has determined all 2-transitive groups (Cameron, 1981). Liebeck and Saxl have determined all primitive groups of odd degree (1985b), as well as all primitive groups of degree mp (1985a), where $m < p$ and p is prime, while Liebeck, Praeger, and Saxl have determined all primitive groups that contain a regular subgroup (Liebeck et al., 2010). Finally, Li and Seress have determined all primitive groups of square-free degree (Li and Seress, 2003).

As a final remark, we would like to point out that Praeger has proven a version of the O'Nan–Scott theorem for quasiprimitive groups (Praeger, 1993).

5.5 Covers

The basic idea behind covers is to try to obtain symmetry information about a "large" graph from symmetry information of a quotient of the "large" graph. Presumably, symmetry information about the quotient is easier to obtain as it has fewer vertices and edges. A major problem with this approach though, is that this does not always work. That is, there are graphs and quotients of the graph where the symmetries of the quotient seem unrelated to the symmetries of the larger graph. For example, there are Cayley graphs whose automorphism group is as small as possible, but which have a quotient that is a complete graph! We leave the verification of this statement to Chapter 8, where we will develop some tools that allow for this fact to be verified easily (Exercise 8.3.7). While the problem is difficult in general, we shall see that there are, of course, situations where this idea works well.

Our goal for this section on covers is modest. We will consider the "nicest" possible case, which not coincidentally is also the first case considered in the literature. So what is the nicest possible case to consider? When considering a quotient, in order to obtain information about the original graph, it is crucial that no symmetry information about the original graph is lost. One way to think about this is that all information about the edges in the large graph has to be somehow recoverable from its smaller quotient. If we wish to keep our definition of a graph as it is, i.e., no multiple edges, loops, etc., which in the simplest possible case we do, we need there to be either no edges inside a block or every edge – we choose no edges. Additionally, we need for the number of edges between blocks, if there are any at all, to be the same, and, to avoid multiple edges between blocks (which we do not allow for normal quotient graphs), we need that number to be one. So, inside a block there will be no edges and between blocks either no edges or a 1-factor. These conditions by themselves will not guarantee all symmetry information is contained in the quotient, but they are clearly necessary. One final comment, we have actually already seen graphs with this property in Lemma 3.6.10. We leave as an exercise (Exercise 5.5.1) to show that graphs that satisfy the hypothesis of Lemma 3.6.10 do have no edges inside a block and either no edges or a 1-factor between different blocks.

It is common to express this notion in terms of a mapping from the larger graph to its quotient.

Definition 5.5.1 Let $\tilde{\Gamma}$ and Γ be graphs. A **homomorphism** is a function $h\colon V\left(\tilde{\Gamma}\right) \to V(\Gamma)$ such that $h(uv) = h(u)h(v) \in E(\Gamma)$ for every $uv \in E\left(\tilde{\Gamma}\right)$.

So a homomorphism maps vertices to vertices and edges to edges, but it need not be (and in this section usually will not be) a bijection.

Example 5.5.2 Let Γ be a graph with $G \leq \mathrm{Aut}(\Gamma)$ having $\mathcal{B}$ as a block system. If $\Gamma[B]$ has no edges, then the map $h\colon \Gamma \to \Gamma/\mathcal{B}$ given by $h(v) = B_v$, where $v \in B_v$, is an onto homomorphism.

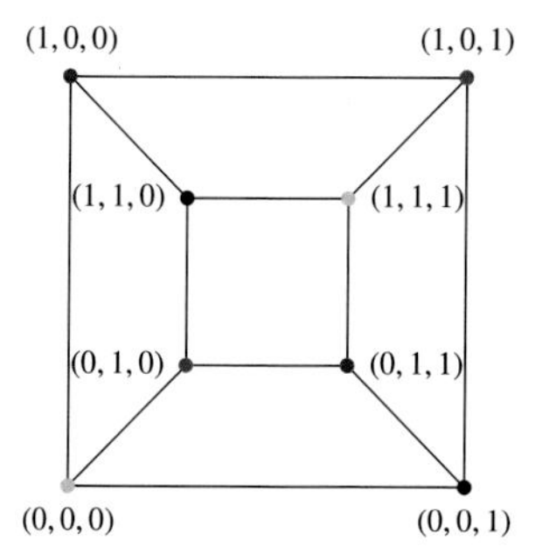

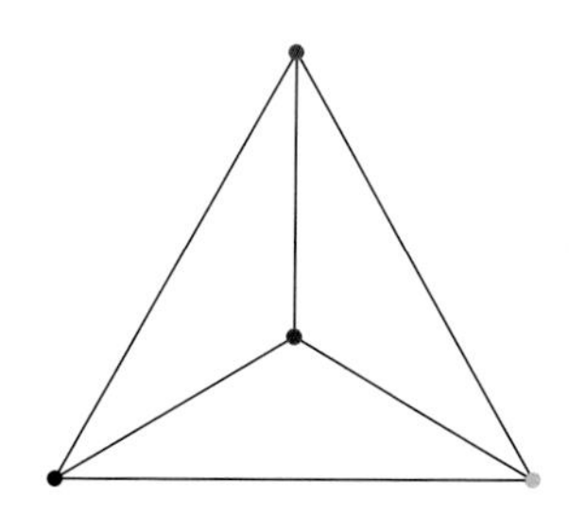

Figure 5.5 The image of a regular Z_2-covering projection of the cube (Left) $Q_3 = F8$ is (Right) K_4.

Solution We have $h(uv) = h(u)h(v) = B_u B_v \in E(\Gamma/\mathcal{B})$ if and only if there is an edge between some vertex of B_u and some vertex of B_v. This is true as the only edges in Γ are between blocks of $\mathcal{B}$, i.e., $uv \in E(\Gamma)$ only if $B_u \neq B_v$. Note that if we added a loop to each vertex of $\Gamma/\mathcal{B}$, then we could drop the hypothesis that $\Gamma[B]$ has no edges. □

Definition 5.5.3 Let $\wp\colon \tilde{\Gamma} \to \Gamma$ be an onto graph homomorphism. By "onto," we mean that $\wp$ is an onto map of vertices and edges of Γ. The sets $\wp^{-1}(v)$, denoted fib_v, are the (**vertex**) **fibers**, $v \in V(\Gamma)$. The **arc fibers** are the sets $\wp^{-1}(u, v)$, $(u, v) \in A(\Gamma)$, and similarly, the **edge fibers** are the sets $\wp^{-1}(uv)$, $uv \in E(\Gamma)$. We say that $\wp$ is a **regular covering projection** if there exists a semiregular subgroup $C \leq \text{Aut}\left(\tilde{\Gamma}\right)$ such that the each vertex and edge fiber of $\wp$ is an orbit of C. The graph $\tilde{\Gamma}$ is called the **covering graph** and the graph Γ is called the **base graph**. If C is isomorphic to an abstract group N, then a regular covering projection may be called an N-**covering** projection, and to emphasize this we sometimes write $\wp_N$ instead of just $\wp$.

Notice that before this section a map between vertex-sets was usually called ϕ while the symbol ρ was usually used for a semiregular element (particularly with a metacirculant graph). We will use $\wp$ for a regular covering projection as that is the symbol most commonly used in the literature for a regular covering projection. The symbol $\wp$ is called the "Weierstrauss p." It is used as, while we would like to use p for "projection," p in mathematics usually means "prime."

Example 5.5.4 The image of a regular Z_2-covering projection of the cube (Left) $Q_3 = F8$ is (Right) K_4.

Solution The cube is the left-hand graph of Figure 5.5, and while a different drawing of the cube, it is the same labeling used in Figure 1.1. We discussed at the beginning of the book and in Exercise 1.2.14, using the notation of semidirect products, that $\mathrm{Aut}(Q_3) \cong \mathbb{Z}_2^3 \rtimes \mathcal{S}_3$. We first give a more formal proof of this.

Define $t\colon \mathbb{Z}_2^3 \to \mathbb{Z}_2^3$ by $t(i,j,k) = (i+1, j+1, k+1)$. Then t is an involution, interchanging antipodal vertices (vertices at distance 3). By Exercise 2.2.4, $\mathrm{Aut}(Q_3)$ has a block system $\mathcal{B}$, where each block of $\mathcal{B}$ is a pair of antipodal vertices in Q_3. So $\mathcal{B}$ contains the sets,

$$\{(0,0,0),(1,1,1)\}, \{(0,0,1),(1,1,0)\}, \{(0,1,0),(1,0,1)\}, \{(1,0,0),(0,1,1)\}.$$

Then $t \in \mathrm{fix}_{\mathrm{Aut}(Q_3)}(\mathcal{B})$, and it is not hard to see that $\mathrm{fix}_{\mathrm{Aut}(Q_3)}(\mathcal{B}) = \langle t\rangle$. Note that $\mathrm{Aut}(Q_3)/\mathcal{B} \le \mathcal{S}_4$ and has order at least 24 as $\mathbb{Z}_2^3 \rtimes \mathcal{S}_3$ has order 48. Thus $\mathrm{Aut}(Q_3)/\mathcal{B} = \mathcal{S}_4$, and it is now not hard to prove that $\mathrm{Aut}(Q_3) = \mathbb{Z}_2^3 \rtimes \mathcal{S}_3 \cong \mathcal{S}_4 \times \mathbb{Z}_2$.

The blocks of $\mathcal{B}$ are the vertex fibers (elements in the same block of $\mathcal{B}$ are colored with same color in the left-hand side of Figure 5.5) of the onto homomorphism $\wp\colon Q_3 \to K_4$ given by $\wp(v) = B_v$, where $v \in B_v \in \mathcal{B}$ (so we are using $V(K_4) = \mathcal{B}$ here). The image of a block of $\mathcal{B}$ under $\wp$ is colored with the same color as the elements of the block in the right-hand side of Figure 5.5. Observe that $\wp^{-1}(B_v) = B_v$ (where the first B_v is considering B_v as a vertex of $\Gamma/\mathcal{B}$, and the second B_v is the block of $\mathcal{B}$ that contains v), so $\wp^{-1}(B_v)$ is an orbit of τ. Finally, for any two of the colors red, blue, black, and green, we see that $\wp^{-1}(e)$ is always an edge orbit of t for any edge $e \in E(K_4)$ (in fact, it is a 1-factor between two blocks of $\mathcal{B}$). So Q_3 is a regular $\mathbb{Z}_2$-cover of K_4. □

Example 5.5.5 Let Γ be a graph with $G \le \mathrm{Aut}(\Gamma)$ transitive and $\mathcal{B}$ a block system of G that is the set of orbits of a semiregular subgroup $C \le \mathrm{fix}_G(\mathcal{B})$. The map $h\colon \Gamma \to \Gamma/\mathcal{B}$ given by $h(v) = B_v$, where $v \in B_v$, is a regular covering projection if and only if $\Gamma[B]$ has no edges, $B \in \mathcal{B}$, and between any two blocks of $\mathcal{B}$ there is either a 1-factor or no edges.

Solution Suppose h is a regular covering projection. As h is an onto homomorphism, if $x, y \in B \in \mathcal{B}$ such that $xy \in E(\Gamma)$, then $h(xy) = h(x)h(y) \in E(\Gamma/\mathcal{B})$. As $h(x)h(y) = B_xB_x$ we see that $\Gamma/\mathcal{B}$ has a loop, a contradiction. Thus $\Gamma[B]$ has no edges for every $B \in \mathcal{B}$. Now suppose $x \in B$, $y \in B'$ with $B, B' \in \mathcal{B}$ and $B \ne B'$. As h is a regular covering projection, $h^{-1}(B_xB_y) = \{c(xy) : c \in C\}$, and as C is semiregular, $\left|h^{-1}(B_xB_y)\right| = |C|$. We conclude that if there is an edge between two different blocks B and B' in $\Gamma/\mathcal{B}$, then there are exactly $|C|$ edges between B and B' in Γ. Hence there is a 1-factor between B and B'.

Conversely, suppose $\Gamma[B]$ has no edges, and between any two blocks of $\mathcal{B}$ there is either a 1-factor or no edges. We saw that h is an onto homomorphism in Example 5.5.2. The vertex fibers $h^{-1}(B)$ are the blocks of $\mathcal{B}$, which are the orbits of C. Similarly, if $B_xB_y \in E(\Gamma/\mathcal{B})$ then the edge fiber $h^{-1}(B_xB_y) = \{c(xy) : c \in C\}$ for any $xy \in E(\Gamma)$ with $x \in B_x$ and $y \in B_y$. □

Our next goal is to give, in Lemma 5.5.16, a method for constructing each regular N-covering projection. We begin with some definitions.

Definition 5.5.6 Let Γ be a connected graph and N a finite group. A function $\zeta\colon A(\Gamma) \to N$, given by

$$\zeta(v,u) = (\zeta(u,v))^{-1},$$

is a **voltage function** and N a **voltage group**.

A voltage function is sometimes called a **voltage assignment**.

Definition 5.5.7 Let Γ be a graph and N a finite group. For $\zeta\colon A(\Gamma) \to N$ a voltage function, define the **derived graph**, denoted $\mathrm{Cov}(\Gamma;\zeta)$, to be the graph with vertex set $V(\Gamma) \times N$, and edge set

$$\{(u,a)(v,a\cdot\zeta(u,v)) : uv \in E(\Gamma) \text{ and } a \in N\}.$$

When we only care that Γ is a derived graph, but do not care what N or the voltage function is, we write $\mathrm{Cov}(\Gamma)$, and call $\mathrm{Cov}(\Gamma)$ a **cover** of Γ. If we care what N is we will call $\mathrm{Cov}(\Gamma)$ an ***N*-cover**.

It is not hard to see why the condition $\zeta(v,u) = (\zeta(u,v))^{-1}$ in a voltage function must hold. As we wish for $\mathrm{Cov}(\Gamma;\zeta)$ to be a graph, if the edge

$$(u,a)(v,a\cdot\zeta(u,v)) \in E(\mathrm{Cov}(\Gamma;\zeta)),$$

the edge

$$(v,a\cdot\zeta(u,v))(u,a) \in E(\mathrm{Cov}(\Gamma;\zeta)),$$

and it must be the case that $(a\cdot\zeta(u,v))\cdot\zeta(v,u) = a$. So $\zeta(v,u)$ must be $(\zeta(u,v))^{-1}$.

Lemma 5.5.8 *Let Γ be a graph, N a finite group, and $\zeta\colon A(\Gamma) \to N$ a voltage function. Define $\wp_\zeta\colon \mathrm{Cov}(\Gamma;\zeta) \to \Gamma$ to be a projection in the first coordinate. Then $\wp_\zeta$ is an onto graph homomorphism and a regular N-covering projection.*

Proof The projection $\wp_\zeta$ is obviously an onto graph homomorphism. For each $n \in N$, define $\hat{n}_L\colon V(\Gamma)\times N \to V(\Gamma)\times N$ by $\hat{n}_L(v,a) = (v,na)$. Each $\hat{n}_L \in \mathrm{Aut}(\mathrm{Cov}(\Gamma;\zeta))$ and $\wp_\zeta^{-1}(v) = \{(v,n) : n \in N\}$ is an orbit of the semiregular subgroup $\{\hat{n}_L : n \in N\}$. So $\wp_\zeta$ is a regular N-covering projection (which can be identified with $\wp_N$). □

Example 5.5.9 Q_3 is a cover of the cube K_4.

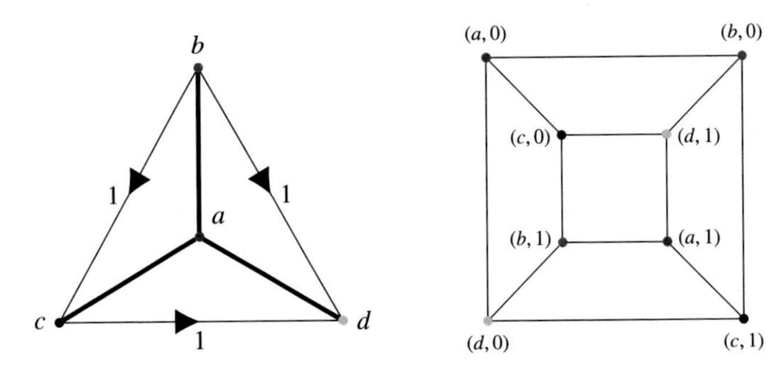

Figure 5.6 (Left) The complete graph K_4 with voltage assignment $\zeta\colon A(K_4) \to \mathbb{Z}_2$, where the bold edges carry the voltage 0, and (Right) the derived graph Q_3.

Solution Let $V(K_4) = \{a, b, c, d\}$ and let T be the spanning tree in K_4 consisting of the edges ab, ac, and ad as shown in the left-hand graph of Figure 5.6. Let $\zeta\colon A(K_4) \to \mathbb{Z}_2$ be the voltage function such that all arcs contained in T carry the voltage $0 \in \mathbb{Z}_2$ while all the arcs not in T carry the voltage $1 \in \mathbb{Z}_2$, as shown in Figure 5.6. Note that in this figure, the edges that have been given nonzero voltage have been directed. For example, the edge bc has been drawn as the arc (b, c) and labeled with 1. We do this so we know that the voltage on the arc (c, b) is the inverse in $\mathbb{Z}_2$ of 1. So the voltage on the arc (c, b) is also 1. Hence, when drawing a voltage graph, the edges that have non-identity voltage are drawn directed and the voltage of the drawn arc is used as a label. The computations for the neighbors of, say, the vertex $(d, 0)$ are

$$\begin{aligned}
&(a, 0 + \zeta(d, a)) = (a, 0 + 0) = (a, 0),\\
&(b, 0 + \zeta(d, b)) = (b, 0 + 1) = (b, 1), \text{ and}\\
&(c, 0 + \zeta(d, c)) = (c, 0 + 1) = (c, 1).
\end{aligned}$$

Then $\mathrm{Cov}(K_4; \zeta)$ is a graph with vertex set $\{(x, i) : x \in V(K_4), i \in \mathbb{Z}_2\}$ and edge set $\{(a, i)(b, i), (a, i)(c, i), (a, i)(d, i), (b, i)(d, i + 1), (b, i)(c, i + 1), (c, i)(d, i + 1) : i \in \mathbb{Z}_2\}$, and is isomorphic to the 3-cube Q_3 (see Figure 5.6). □

As we shall see, much information about whether an automorphism of Γ has a corresponding automorphism in $\mathrm{Cov}(\Gamma; \zeta)$ depends upon properties of closed walks in $\mathrm{Cov}(\Gamma; \zeta)$. So we need to extend voltage assignments to walks, and formalize the correspondence of walks in Γ and $\mathrm{Cov}(\Gamma; \zeta)$.

Definition 5.5.10 Let $\zeta\colon A(\Gamma) \to N$ be a voltage function. Then ζ naturally extends to walks in Γ by defining the **voltage ζ_W of a walk** $W = w_1 w_2 \ldots w_r$ as

$$\zeta_W = \zeta(w_1, w_2)\zeta(w_2, w_3) \cdots \zeta(w_{r-1}, w_r).$$

Definition 5.5.11 Let $\wp_\zeta \colon \mathrm{Cov}(\Gamma;\zeta) \to \Gamma$ be a regular N-covering projection, and $W = w_1, \ldots, w_n$ a walk in Γ. For $a \in N$, the walk

$$\tilde{W}_a = (w_1, a), (w_2, a \cdot \zeta(w_1, w_2)), \ldots, (w_n, a \cdot \Pi_{i=1}^{n-1} \zeta(w_i, w_{i+1}))$$

in $\mathrm{Cov}(\Gamma;\zeta)$ is the **lift** of W at a.

Remark 5.5.12 It is not hard to show, using induction on the length of a walk, that for each walk W from u to v in Γ, there is a unique walk $\tilde{W}$ over W starting at an arbitratry vertex $\tilde{u} \in \mathrm{fib}_u$. This is the single most important characteristic property of covers, and is called the **unique walk lifting property**. Also note that, for a walk W starting at $\tilde{u} \in \mathrm{fib}_u$, the terminal vertex of the walk $\tilde{W}$ is conveniently $(v, \zeta(W))$.

Example 5.5.13 In Example 5.5.9, the walk $W = b, c, d, a, b$ in K_4 has lift

$$\tilde{W}_0 = (b, 0), (c, 1), (d, 0), (a, 0), (b, 0),$$

at 0 and $\zeta_W = 0$. (Note that to find the voltages of subwalks we simply walk in K_4 and sum the voltages as we traverse the walk.) Similarly, the walk $V = b, c, d, b$ in K_4 has lift $\tilde{V}_1 = (b, 1), (c, 0), (d, 1), (b, 0)$ at 1 and $\zeta_V = 1$. Note that both W and V are closed walks in K_4, while $\tilde{W}_0$ is a closed walk but $\tilde{V}_1$ is not.

The following result should be clear, and its proof is left as Exercise 5.5.6.

Lemma 5.5.14 *Let $\wp_\zeta \colon \mathrm{Cov}(\Gamma;\zeta) \to \Gamma$ be a regular N-covering projection of graphs. A closed walk W in Γ lifts to a closed walk ζ_W if and only if $\zeta(W) = 1$.*

Notice that in the above result, if $\zeta_W = 1$, W will lift to $|N|$ closed walks in $\mathrm{Cov}(\Gamma;\zeta)$, one for each vertex in fib_b, where b is any fixed vertex of W.

Definition 5.5.15 Two regular covering projections $\wp \colon \tilde{\Gamma} \to \Gamma$ and $\wp' \colon \tilde{\Gamma}' \to \Gamma$ of a graph Γ are **isomorphic** if there exists an automorphism $\alpha \in \mathrm{Aut}(\Gamma)$ and an isomorphism $\tilde{\alpha} \colon \tilde{\Gamma} \to \tilde{\Gamma}'$ such that $\alpha \circ \wp = \wp' \circ \tilde{\alpha}$. We write this as $(\alpha, \tilde{\alpha}) \colon \wp \to \wp'$ is an isomorphism of $\wp$. In particular, if $\alpha = 1$, then $\wp$ and $\wp'$ are said to be **equivalent**.

Lemma 5.5.16 *Every regular N-covering projection $\wp_N \colon \tilde{\Gamma} \to \Gamma$ can be constructed, up to equivalence, from Γ and a voltage function.*

Proof Label the vertices in each vertex fiber fib_u by the elements of N in the following way. First, choose a vertex v_u in fib_u and label it with $1 \in N$. Then label the other vertices in fib_u with the label by means of the regular action of N^{fib_u} with N viewed as a group of automorphisms of $\tilde{\Gamma}$. That is, for each $v' \in \mathrm{fib}_u$, there is $n \in N$ with $n(v) = v'$. We label v' with n, and call the resulting graph $\tilde{\Gamma}'$. This relabeling defines an isomorphism $\tilde{\alpha} \colon \tilde{\Gamma} \to \tilde{\Gamma}'$. Next, we show that $\tilde{\Gamma}'$ can be constructed from Γ using a voltage function. Define a voltage function ζ as follows: if (u, v) is an arc in Γ, let $(\tilde{u}, \tilde{v})$ be an arc in $\tilde{\Gamma}$ that projects onto (u, v). If the label of $\tilde{u}$ is $m \in N$ and the label of $\tilde{v}$ is $n \in N$, then let $\zeta(u, v) = m^{-1}n$. Suppose $x \in N$ such that $(u, 1)(v, x) \in E(\mathrm{Cov}(\Gamma; \zeta))$ and $(u, m)(v, n) \in E(\mathrm{Cov}(\Gamma; \zeta))$. Applying the definition of a voltage graph, we see that $\tilde{\Gamma}' = \mathrm{Cov}(\Gamma; \zeta))$ and the result follows. □

Exercise 5.5.7 asks you to show that the covering digraph $\mathrm{Cov}(\Gamma; \zeta)$ is connected if and only if Γ is connected and every element of N is the voltage of some closed walk at some fixed vertex $b \in V(\Gamma)$. Additionally, in Exercise 5.5.8 you will show that if $\tilde{\Gamma}$ is connected, then one can change an existing voltage function ζ to ζ', where ζ' is trivial on the arcs of an arbitrary spanning tree T, and the values on the arcs not in T generate the group N; this can be done in such a way that $\wp_{\zeta'} : \mathrm{Cov}(\Gamma; \zeta') \to \Gamma$ is equivalent to $\wp_\zeta$. For an extensive treatment of graph coverings, see Gross and Tucker (1987); Malnič et al. (2000).

Definition 5.5.17 Let $\wp : \tilde{\Gamma} \to \Gamma$ be a regular covering projection. An **automorphism** of $\wp$ is an isomorphism $(\alpha, \tilde{\alpha}) : \wp \to \wp$ of $\wp$ onto itself. A more common way of expressing this is that α **lifts** (and that $\tilde{\alpha}$ **projects**) along $\wp$. We also say that $\wp$ is **α-admissible**. If G is a subgroup of $\mathrm{Aut}(\Gamma)$ such that $\wp$ is α-admissible for all $\alpha \in G$, then $\wp$ is **G-admissible**. The set of all such lifts is a group $\tilde{G} \leq \mathrm{Aut}(\tilde{\Gamma})$, the **lift** of G. The lift of the identity subgroup of $\mathrm{Aut}(\Gamma)$ is the **group of covering transformations** and is denoted $\mathrm{CT}(\wp)$.

Note that if both Γ and $\tilde{\Gamma}$ are connected, then $\mathrm{CT}(\wp_N) \cong N$. Also, if G lifts to a subgroup $\tilde{G}$ of $\mathrm{Aut}(\tilde{\Gamma})$, then $\mathrm{CT}(\wp)$ is a normal subgroup of $\tilde{G}$, and so $\mathrm{CT}(\wp) \to \tilde{G} \to G$ is a short exact sequence.

We leave it to the reader as Exercise 5.5.10 to prove that if $\wp$ is G-admissible, then any covering equivalent to $\wp$ is also G-admissible. So the question of which automorphisms lift can be studied by considering an equivalent derived covering. A necessary and sufficient condition for $\wp_\zeta : \mathrm{Cov}(\Gamma; \zeta) \to \Gamma$ to be G-admissible can be stated combinatorially in terms of voltages. We will need more terminology.

Definition 5.5.18 Let $b \in V(\Gamma)$ be a fixed vertex of a connected graph Γ, called the **base vertex**, and let $\Pi(\Gamma, b)$ be the set of all closed walks in Γ starting at b. Further, if T is an arbitrary spanning tree in Γ, then the unique closed walk at b formed by T and an arc $(u, v) \in A(\Gamma)$ not in T is called the **fundamental closed walk** at b relative to T and the arc (u, v).

Theorem 5.5.19 (The basic lifting lemma) *Let Γ be a connected graph, N a finite group, ζ a voltage function, $\tilde{\Gamma} = \mathrm{Cov}(\Gamma; \zeta)$, and $\wp_\zeta : \mathrm{Cov}(\Gamma; \zeta) \to \Gamma$ be a* *regular N-covering projection. Then $\alpha \in \mathrm{Aut}(\Gamma)$ has a lift if and only if for all closed walks $W \in \Pi(\Gamma, b)$ the following implication holds:*

$$\zeta_W = 1 \Rightarrow \zeta_{\alpha(W)} = 1.$$

Proof Suppose $\alpha \in \mathrm{Aut}(\Gamma)$ lifts to $\tilde{\alpha}$. Let W be a closed walk at b with trivial voltage $\zeta_W = 1$. As $\zeta_W = 1$, by Lemma 5.5.14 this walk lifts as a closed walk at all vertices in the fiber fib_b. Let $\tilde{W}$ be the lift of W at the vertex $\tilde{b} \in \mathrm{fib}_b$. The automorphism $\tilde{\alpha}$ maps $\tilde{W}$ to a closed walk $\tilde{\alpha}\left(\tilde{W}\right)$. As $\tilde{\alpha}$ is the lift of α, the walk $\tilde{\alpha}\left(\tilde{W}\right)$ is the lift of $\alpha(W)$. As $\tilde{\alpha}\left(\tilde{W}\right)$ is closed, we have $\zeta_{\alpha(W)} = 1$.

Conversely, suppose that for any closed walk $W \in \Pi(\Gamma, b)$ with $\zeta_W = 1$ we have $\zeta_{\alpha(W)} = 1$. Define $\phi\colon N \to N$ by

$$\phi(\zeta_W) = \zeta_{\alpha(W)},$$

where $W \in \Pi(\Gamma, b)$. We show that $\phi \in \mathrm{Aut}(N)$. To see that ϕ is well defined, let $W_1, W_2 \in \Pi(\Gamma, b)$ with $\zeta_{W_1} = \zeta_{W_2}$. Then $\zeta_{W_1}\zeta_{W_2}^{-1} = 1$, and so $\zeta_{W_1 W_2^{-1}} = 1$ where W_2^{-1} is the inverse walk of the walk W_2. By hypothesis, $1 = \zeta_{\alpha\left(W_1 W_2^{-1}\right)} = \zeta_{\alpha(W_1)\alpha(W_2)^{-1}}$, and hence $\zeta_{\alpha(W_1)} = \zeta_{\alpha(W_2)}$. Thus ϕ is well defined. As $\tilde{\Gamma}$ is connected, by Exercise 5.5.7 for each $a \in N$ there exists $W_a \in \Pi(\Gamma, b)$ such that $\zeta_{W_a} = a$. So the domain of ϕ is N. To see ϕ is onto, first note that there is a bijective correspondence between $\Pi(\Gamma, b)$ and $\alpha(\Pi(\Gamma, b))$ as $\alpha(\Pi(\Gamma, b)) = \Pi(\Gamma, \alpha(b))$. Then there is $W_a' \in \Pi(\Gamma, b)$ such that $\alpha\left(W_a'\right) = W_a$, and

$$\phi\left(W_a'\right) = \zeta_{\alpha(W_a')} = \zeta_{W_a} = a.$$

Hence ϕ is onto. As N is finite, ϕ is a bijection. Finally, let $a, c \in N$. Then,

$$\begin{aligned}\phi(ac) &= \phi(\zeta_{W_a}\zeta_{W_c}) = \phi(\zeta_{W_a W_c}) = \zeta_{\alpha(W_a W_c)}\\ &= \zeta_{\alpha(W_a)\alpha(W_c)} = \zeta_{\alpha(W_a)}\zeta_{\alpha(W_c)} = \phi(a)\phi(c),\end{aligned}$$

and $\phi \in \mathrm{Aut}(N)$.

We now construct a lift $\tilde{\alpha}$ of α as follows. For an arbitrary vertex (v, a) in $\tilde{\Gamma}$, and W an arbitrary walk from the vertex v to the base vertex b in Γ, set

$$\tilde{\alpha}(v, a) = \left(\alpha(v), \phi(a\,\zeta_W)\zeta_{\alpha(W)}^{-1}\right).$$

See Figure 5.7 for an illustration. Clearly, $\tilde{\alpha}$ maps vertices in fib_v to vertices in $\mathrm{fib}_{\alpha(v)}$, and it satisfies the condition $\wp_\zeta \circ \tilde{\alpha} = \alpha \circ \wp_\zeta$ by construction.

We next show that $\tilde{\alpha}$ is a well-defined bijection. To see that $\tilde{\alpha}$ is well defined (and so does not depend on the choice of walk W from v to b), let W_1 and W_2 be walks from v to b. Then $W_2^{-1}W_1 \in \Pi(\Gamma, b)$ is a closed walk at b, so $\phi\left(\zeta_{W_2^{-1}W_1}\right) = \zeta_{\alpha\left(W_2^{-1}W_1\right)}$. As $\phi \in \mathrm{Aut}(N)$ we have $\phi(\zeta_{W_2})^{-1}\phi(\zeta_{W_1}) = \zeta_{\alpha(W_2)}^{-1}\zeta_{\alpha(W_1)}$, and hence $\phi(\zeta_{W_1})\zeta_{\alpha(W_1)}^{-1} = \phi(\zeta_{W_2})\zeta_{\alpha(W_2)}^{-1}$. Multiplying both sides of this equation

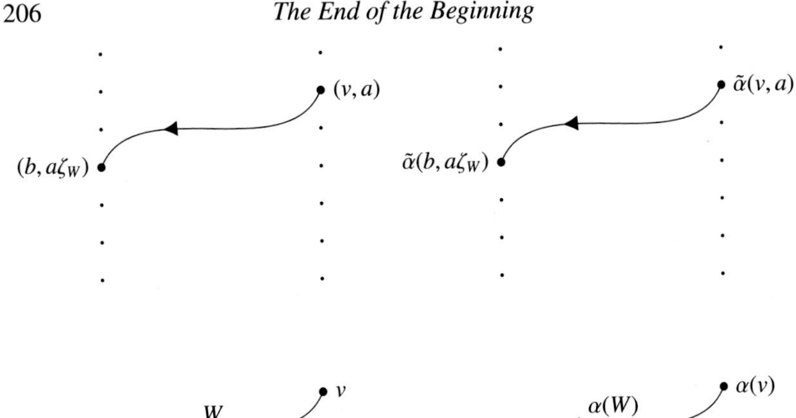

Figure 5.7 Constructing the lifted automorphism.

on the left by $\phi(a)$, we see that $\phi(a\zeta_{W_1})\zeta^{-1}_{\alpha(W_1)} = \phi(a\zeta_{W_2})\zeta^{-1}_{\alpha(W_2)}$. Thus $\tilde{\alpha}$ does not depend on the choice of W, and is well defined.

To show that $\tilde{\alpha}$ is injective, suppose $\phi(a_1\,\zeta_W)\zeta^{-1}_{\alpha(W)} = \phi(a_2\,\zeta_W)\zeta^{-1}_{\alpha(W)}$. Equivalently, $\phi(a_1)\phi(\zeta_W)\zeta^{-1}_{\alpha(W)} = \phi(a_2)\phi(\zeta_W)\zeta^{-1}_{\alpha(W)}$, so $\phi(a_1) = \phi(a_2)$, and $a_1 = a_2$. Thus $\tilde{\alpha}$ is injective, and, as the fibers are finite, it is a bijection.

It remains to show that $\tilde{\alpha}$ preserves adjacency. Choose the walk from u to b to be $(u, v)W$ (recall that W is an arbitrary walk from v to b). Let $\big((u, a), \big(v, a\zeta_{(u,v)}\big)\big) \in A(\tilde{\Gamma})$. Set

$$c = \phi(a\zeta_{(u,v)W})\zeta^{-1}_{\alpha((u,v)W)} = \phi(a\zeta_{(u,v)}\zeta_W)\zeta^{-1}_{\alpha(W)}\zeta^{-1}_{\alpha(u,v)}, \text{ and}$$

$$d = \phi((a\zeta_{(u,v)})\zeta_W)\zeta^{-1}_{\alpha(W)} = \phi(a\zeta_{(u,v)}\zeta_W)\zeta^{-1}_{\alpha(W)}.$$

So $d = c\zeta_{(\alpha(u),\alpha(v))}$, and $\tilde{\alpha}((u, a), (v, a\zeta_{(u,v)})) = ((\alpha(u), c), (\alpha(v), d)) \in A(\tilde{\Gamma})$. Hence $\tilde{a} \in \mathrm{Aut}(\tilde{\Gamma})$. □

Remark 5.5.20 The condition in Theorem 5.5.19 need only be tested on the fundamental closed walks at b relative to an arbitrary spanning tree T.

Remark 5.5.21 The following is an equivalent form of Theorem 5.5.19: $\alpha \in \mathrm{Aut}(\Gamma)$ lifts if and only if there exists an automorphism $\phi \in \mathrm{Aut}(N)$ of the voltage group N such that $\phi(\zeta_W) = \zeta_{\alpha(W)}$ for every closed walk W. Using Remark 5.5.20, this can be restated as: $\alpha \in \mathrm{Aut}(\Gamma)$ lifts if and only if the mapping of voltages $\zeta_W \to \zeta_{\alpha(W)}$ of fundamental closed walks at b extends to an automorphism of the voltage group N.

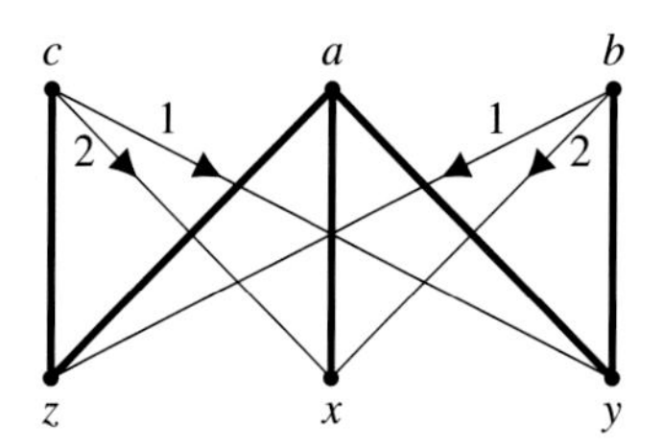

Figure 5.8 The complete bipartite graph $K_{3,3}$ with the voltage function $\zeta\colon A(K_{3,3}) \to \mathbb{Z}_3$, where the bold edges carry the voltage 0.

Example 5.5.22 Let Γ be the complete bipartite graph $K_{3,3}$ with vertex set $V(\Gamma) = \{a, b, c, x, y, z\}$, and let T be the spanning tree consisting of the edges $\{ax, ay, az, by, cz\}$, as shown in Figure 5.8. Let $\zeta\colon A(\Gamma) \to \mathbb{Z}_3$ be a voltage function such that the arcs on T carry the voltage 0 while

$$\zeta(b, x) = 2,\ \zeta(b, z) = 1,\ \zeta(c, y) = 1,\ \zeta(c, x) = 2.$$

Then $\alpha = (a, x)(b, y)(c, z) \in \mathrm{Aut}(\Gamma)$ lifts to an automorphism of $\mathrm{Cov}(\Gamma; \zeta)$.

Solution There are four fundamental cycles in Γ with respect to the spanning tree T and base vertex a: $W_1 = aybza$, $W_2 = azcxa$, $W_3 = aybxa$, and $W_4 = azcya$, whose voltages are $\zeta_{W_1} = 1$, $\zeta_{W_2} = 2$, $\zeta_{W_3} = 2$ and $\zeta_{W_4} = 1$. By Remark 5.5.21, to show that α lifts to an automorphism of $\mathrm{Cov}(\Gamma, \zeta)$ it suffices to check that the mapping of voltages $\zeta_W \to \zeta_{\alpha(W)}$ assigned to the fundamental closed walks extends to an automorphism of $\mathbb{Z}_3$. The voltages of the images of the fundamental cycles W_i, $i \in \{1, 2, 3, 4\}$ when mapped by α are:

i	W_i	ζ_{W_i}	$\alpha(W_i)$	$\zeta_{\alpha(W_i)}$
1	$aybza$	1	$xbycx$	2
2	$azcxa$	2	$xczax$	1
3	$aybxa$	2	$xbyax$	1
4	$azcya$	1	$xczbx$	2

The mapping of voltages $\zeta_W \to \zeta_{\alpha(W)}$, where $W \in \{W_i : i \in \{1, 2, 3, 4\}\}$, clearly extends to the automorphism $k \mapsto 2k$ of the voltage group $\mathbb{Z}_3$, and so α lifts to an automorphism of $\mathrm{Cov}(\Gamma, \zeta)$. Checking the lifting condition for all automorphisms of Γ, we see that *every* automorphism of Γ lifts to an automorphism of $\mathrm{Cov}(\Gamma, \zeta)$. The covering graph is isomorphic to the Pappus graph F18. □

5.6 Exercises

5.1.1 Calculate $\mathrm{Aut}\left(C_4 \wr \bar{K}_6\right)$.

5.1.2 Calculate $\mathrm{Aut}\left(K_5 \wr K_{5,5} \wr \bar{K}_5\right)$.

5.1.3 Show that $(\mathbb{Z}_m)_L \wr (\mathbb{Z}_n)_L$ contains a regular subgroup isomorphic to $\mathbb{Z}_{mn}$.

5.1.4 Show that for any two groups G and H, $G_L \wr H_L$ contains $(G \times H)_L$. Deduce that the wreath product of two Cayley digraphs is a Cayley digraph.

5.1.5 Let Γ_1, Γ_2 be digraphs, and $\Gamma = \Gamma_1 \wr \Gamma_2$ be connected, vertex-transitive, and not a complete graph. Show that Γ is arc-transitive if and only if Γ_1 is arc-transitive and Γ_2 is the complement of a complete graph.

5.1.6 Let Γ_1, Γ_2 be digraphs, and $\Gamma = \Gamma_1 \wr \Gamma_2$ be disconnected and vertex-transitive. Show that Γ is arc-transitive if and only if Γ_1 is the complement of a complete graph and each component of Γ_2 is arc-transitive.

5.1.7 Let $\vec{C}_m$ be the directed cycle of length m. Show that $\vec{C}_m \wr \overline{K}_n$ is s-arc-transitive for every $1 \leq s \leq m-1$ but does not satisfy the conclusion of Lemma 3.6.10.

5.1.8 Let $\vec{C}_m$ be the directed cycle of length m. Show that $\vec{C}_m \wr \overline{K}_n$ is not s-arc-transitive for every $s \geq m$.

5.1.9 Let Γ be an s-arc-transitive digraph, $n \geq 2$ an integer, and $s \geq 1$. Show that $\Gamma \wr \overline{K}_n$ is arc-transitive, but not necessarily s-arc-transitive.

5.1.10 Prove Lemma 5.1.3.

5.1.11 Show that $\mathrm{Aut}\left(\bar{\Gamma}_1 \wr \bar{\Gamma}_2\right) = \mathrm{Aut}\left(\overline{\Gamma_1 \wr \Gamma_2}\right)$.

5.1.12 Let Γ be a vertex-transitive graph and define the equivalence relation $\equiv$ on $V(\Gamma)$ by $x \equiv y$ if and only if $N_\Gamma(x) = N_\Gamma(y)$. Let $\mathcal{R}$ be the block system of $\mathrm{Aut}(\Gamma)$ with blocks the equivalence classes of $\equiv$. Show that $\Gamma/\mathcal{R}$ is irreducible (or in the alternative notation, $\Gamma/\mathcal{R}$ is worthy).

5.1.13 Show that the automorphism group of a disconnected digraph is not elusive.

5.2.1 Show that if G is regular, then $G^{(2)} = G$.

5.2.2 Show that if G is 2-transitive of degree n, then $G^{(2)} = \mathcal{S}_n$. Deduce that $(\mathrm{AGL}(1,p))^{(2)} = \mathcal{S}_p$.

5.2.3 Show that for p and q primes, $(\mathrm{AGL}(1,p) \times \mathrm{AGL}(1,q))^{(2)} = \mathcal{S}_p \times \mathcal{S}_q$.

5.2.4 Let P be a Sylow p-subgroup of $\left\{x \mapsto ax + b \pmod{p^2} : a \in \mathbb{Z}_{p^2}, b \in \mathbb{Z}_{p^2}^*\right\}$, which by Corollary 4.4.11 is $N_{\mathcal{S}_{p^2}}\left(\left(\mathbb{Z}_{p^2}\right)_L\right)$. Show that $|P| = p^3$ and P has a block system with blocks of size p. Then show that every orbital digraph of P is a nontrivial wreath product with $\mathcal{B}$ the lexi-partition. Conclude that $P^{(2)} = \mathbb{Z}_p \wr \mathbb{Z}_p$.

5.2.5 Show that if P is a p-group, p a prime, then $P^{(2)}$ is also a p-group.

5.2.6 Show that $\mathbb{Z}_2^3$ is not a DRR.

5.2.7 Show that Q_8 is not a DRR.

5.2.8 Let G be a generalized dicyclic group, and $\Gamma = \mathrm{Cay}(G, S)$ for some $S \subseteq G$ with $S = S^{-1}$. We use the notation of Definition 5.2.9. Let $n = |A|$ (note that as $1 \neq b^2 \in A$ and $b^4 = 1$, n is even). Show that every element of G can be written uniquely as ab^j where $a \in A$ and $j = 0$ or 1, and $|G| = 2n$. Define $\alpha\colon G \to G$ by $\alpha\left(a^i\right) = a^i$ and $\alpha\left(a^ib\right) = \left(a^ib\right)^{-1}$. Show that $\alpha \in \mathrm{Aut}(G)$ and $\alpha \in \mathrm{Aut}(\Gamma)$. Conclude that no Cayley graph of a generalized dicyclic group is a GRR.

5.2.9 Let $G \leq S_n$ be transitive and 2-closed. Let $H \leq G$ be transitive with block system $\mathcal{B}$ (note $\mathcal{B}$ need not be a block system of G). Let $K \leq G$ be maximal such that $\mathcal{B}$ is a block system of that K. Show that K is 2-closed.

5.3.1 Show that the relation $\equiv$ defined in Definition 5.3.10 is an equivalence relation.

5.3.2 Show that every generalized wreath product graph of order 8 that is isomorphic to a circulant graph of order 8 is isomorphic to a nontrivial wreath product.

5.3.3 Let G be a transitive group with $\mathcal{B} \leq C$ block systems. Show that $C/\mathcal{B}$ is a block system of $G/\mathcal{B}$.

5.3.4 Let A be an abelian group, $S \subset A$, and $1 < K \leq L < A$ such that $\Gamma = \mathrm{Cay}(A, S)$ is a (K, L)-generalized wreath product. Let $\mathcal{B} \leq C$ be the block systems of A_L that are of the set of cosets of K and L, respectively. Let $H \leq \mathrm{Aut}(\Gamma)$ be a maximal that has both $\mathcal{B}$ and C as block systems. Show that if $C \in C$ and $\gamma \in \mathrm{Stab}_H(C)$ such that γ^C fixes each block of $\mathcal{B}$ contained in C, then $\gamma|_C \in H$.

5.3.5 Let G be a transitive group with normal block system $\mathcal{B}$ with blocks of size k. Assume that $\mathrm{fix}_G(\mathcal{B})^B$ is quasiprimitive, where $B \in \mathcal{B}$. Define an equivalence relation $\equiv'$ on $\mathcal{B}$ by $B \equiv' B'$ if and only if whenever $H \trianglelefteq \mathrm{fix}_G(\mathcal{B})$ then H^B is transitive if and only if $H^{B'}$ is transitive. Show that the set of equivalence classes of $\equiv'$ is a block system of G, and if k is prime then $B \equiv' B'$ if and only if $B \equiv B'$, where $\equiv$ is defined as in Definition 5.3.10. In this case we also call the equivalence classes of $\equiv'$ the **$\mathcal{B}$-fixer block system of** G.

5.3.6 Use Exercise 5.3.5 to show that the following generalization of Lemma 5.3.11 is true:

Let G be a transitive group with normal block system $\mathcal{B}$ such that $\mathrm{fix}_G(\mathcal{B})^B$, $B \in \mathcal{B}$, is quasiprimitive. Let $\mathcal{E}$ be the $\mathcal{B}$-fixer block system of G. Then $\mathcal{E}$ is a block system of G and $g|_E \in \mathrm{Aut}(\Gamma)$ for every $g \in G$ and $E \in \mathcal{E}$.

5.3.7 Let Γ be a vertex-transitive digraph. Let $\mathcal{B}$ be the set of all independent sets of Γ of size k. Show that if $\mathcal{B}$ is a partition of $V(\Gamma)$, then $\mathcal{B}$ is a block system of $\mathrm{Aut}(\Gamma)$.

5.3.8 Let Γ be a vertex-transitive digraph and $G \leq \mathrm{Aut}(\Gamma)$ be transitive and maximal with block system $\mathcal{B}$. Let Γ_2 be the spanning subdigraph of Γ consisting of all arcs within any block $B \in \mathcal{B}$, and $\Gamma_1 = \Gamma\backslash A(\Gamma_2)$. Then $G^{(2)} = \mathrm{Aut}(\Gamma_1) \cap \mathrm{Aut}(\Gamma_2)$.

For the next two exercises, let A be an abelian group, and $S \subset A$ such that $\Gamma = \mathrm{Cay}(A, S)$ is a (K, L)-generalized wreath product. Set $\Gamma_1 = \mathrm{Cay}(A, S\backslash L)$ and $\Gamma_2 = \mathrm{Cay}(A, S \cap L)$.

5.3.9 If $1 < H < K$ such that Γ is also an (H, L)-generalized wreath product or $L < M < A$ such that Γ is also a (K, M)-generalized wreath product, then Γ does not have automorphism group $\mathrm{Aut}(\Gamma_1) \cap \mathrm{Aut}(\Gamma_2)$.

5.3.10 Show that if $\mathrm{Aut}(\Gamma) = \mathrm{Aut}(\Gamma_1) \wr \mathrm{Aut}(\Gamma_2)$ then the cosets of L in A are a block system of $\mathrm{Aut}(\Gamma)$.

For the following two exercises, we will use the following more general notion of a generalized wreath product. Let Γ be a vertex-transitive digraph. Suppose there exists a transitive group $G \leq \mathrm{Aut}(\Gamma)$ with normal block systems $\mathcal{B} \preceq C$. If, for each $v \in V(\Gamma)$ with $C_v \in C$ such that $v \in C_v$, the set of all neighbors of v not in C_v is a union of blocks of $\mathcal{B}$, then we say that Γ is a **$(\mathcal{B}, C)$-generalized wreath product.**

5.3.11 Show that the above definition of a generalized wreath product is equivalent to the definition given before the exercises if G is a transitive abelian group.

5.3.12 State and prove a generalization of Theorem 5.3.6 using the definition of a generalized wreath product given above.

5.4.1 Show that the center of $\mathrm{GL}(n, q)$ is the group of all scalar matrices. (Hint: To show that every element of the center of $\mathrm{GL}(n, q)$ is scalar, let $e_{i,j}$ be the noninvertible matrix consisting of 1 in the ijth entry and 0s everywhere else, and $E_{ij} = I_n + e_{i,j}$. Show that if $i \neq j$, then $E_{i,j} \in \mathrm{GL}(n, q)$. If $A \in Z(\mathrm{GL}(n, q))$, then $AE_{i,j} = E_{i,j}A$ so $A(I_n + e_{ij}) = A + Ae_{ij}$ and $(I_n + e_{ij})A = A + e_{i,j}A$. Conclude that A must be diagonal. Now show that the only diagonal matrices in $Z(\mathrm{GL}(n, q))$ are the scalar matrices by considering the $n \times n$ invertible matrix obtained from the identity by switching two rows.)

5.4.2 Show that $\mathrm{PGL}(d, q) \trianglelefteq \mathrm{P\Gamma L}(d, q)$ and $\mathrm{PSL}(d, q) \trianglelefteq \mathrm{P\Gamma L}(d, q)$.

5.4.3 Let $H \leq S_n$ have a block system $\mathcal{B}$ and $K \trianglelefteq H$ such that

(a) $K \leq \mathrm{fix}_H(\mathcal{B})$,
(b) K is a direct product of isomorphic nonabelian simple groups T^ℓ, and
(c) $K^B = T$ is transitive for every $B \in \mathcal{B}$.

Let L be a minimal nontrivial normal subgroup of K. Show that the set $\mathrm{supp}(L) = \{x \in \mathbb{Z}_n : h(x) \neq x$ for some $h \in H\}$ is a block of G.

5.4.4 Show that $\mathrm{Aut}(Q_n)$ is isomorphic to $S_n \wr \mathbb{Z}_2$.

5.4.5 Show that the Hamming graph $H(d, n)$ is regular of valency $d(n-1)$. Deduce that no Hamming graph is self-complementary.

5.4.6 Use the O'Nan–Scott theorem to deduce the following result due to Wielandt (see Jones (1979) for this and more): Let p be prime and $G \leq S_{p^2}$ be primitive with a regular elementary abelian Sylow p-subgroup P. Show that $P \trianglelefteq G$ or G contains a normal subgroup H of index 2 such that H is permutation isomorphic to a subgroup of $S_p \times S_p$.

5.4.7 Use the O'Nan–Scott theorem to show that if n is square-free and not prime, then every primitive group of degree n has simple nonabelian socle.

5.4.8 Show that $\mathbb{Z}_3^2$ is not a DRR.

5.4.9 Show that $\mathrm{GL}(d, \mathbb{F})$ is transitive on the set of nonzero vectors in $\mathbb{F}^d$. Then show that $\mathrm{SL}(d, \mathbb{F})$ is 2-transitive on the set of one-dimensional subspaces of $\mathbb{F}^d$. Conclude that $\mathrm{PSL}(d, \mathbb{F})$ is 2-transitive on the set of one-dimensional subspaces of $\mathbb{F}^d$.

5.5.1 Show that a graph that satisfies the hypothesis of Lemma 3.6.10 has no edges inside a block of $\mathcal{B}$ and either no edges or a 1-factor between different blocks of $\mathcal{B}$.

5.5.2 Verify that the edge set of $\mathrm{Cov}(K_4, \zeta)$ given in Example 5.5.9 is correct.

5.5.3 Draw the $\mathbb{Z}_3$-cover $\mathrm{Cov}(K_4, \zeta)$ for the voltage assignments given in Figure 5.9.

5.5.4 Draw the $\mathbb{Z}_3$-cover $\mathrm{Cov}(\mathrm{Cay}(\mathbb{Z}_6\{1, 3, 5\}), \zeta)$ for the voltage assignments given in Figure 5.10.

5.5.5 Let $\wp\colon \tilde{\Gamma} \to \Gamma$ be a regular covering projection with semiregular subgroup $C \leq \mathrm{Aut}\left(\tilde{\Gamma}\right)$ whose orbits are the sets $\wp^{-1}(v)$, $v \in V(\Gamma)$. Show that the arc fibers and edge fibers are the arc and edge orbits of C.

5.5.6 Prove Lemma 5.5.14.

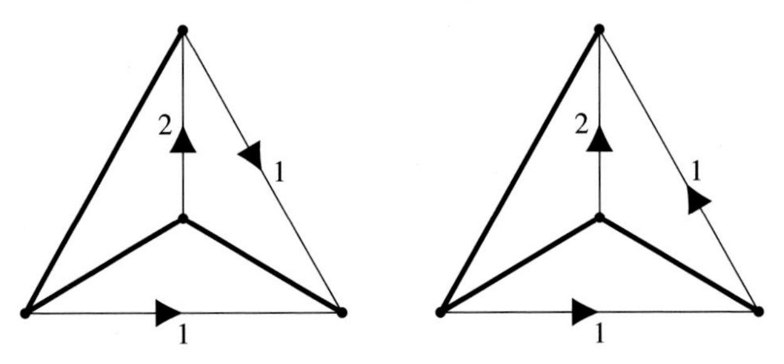

Figure 5.9 Two voltage assignments for K_4.

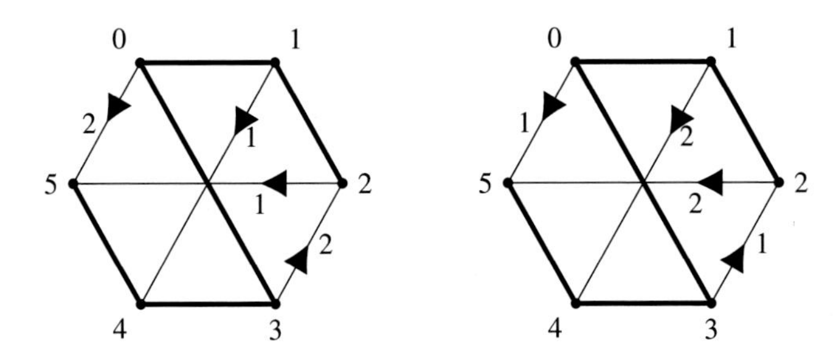

Figure 5.10 Two voltage assignments for $\mathrm{Cay}(\mathbb{Z}_6, \{1, 3, 5\})$.

5.5.7 Show that the covering graph $\mathrm{Cov}(\Gamma; \zeta)$ is connected if and only if Γ is connected and every element of N appears as the voltage of some closed walk at some fixed chosen vertex $b \in V(\Gamma)$.

5.5.8 Show that if $\tilde{\Gamma}$ is connected, then one can change an existing voltage function ζ to ζ', where ζ' is trivial on the arcs of an arbitrary spanning tree T, and the values on the arcs not in T generate the group N. Additionally, show that this can be done in such a way that $\wp_{\zeta'} : \mathrm{Cov}(\Gamma; \zeta') \to \Gamma$ is equivalent to $\wp_\zeta$.

5.5.9 Let $\wp : \tilde{\Gamma} \to \Gamma$ be a regular covering projection. Show that $\mathrm{CT}(\wp)$ is a group.

5.5.10 Let $\wp : \tilde{\Gamma} \to \Gamma$ be a regular covering projection. Prove that if $\wp$ is G-admissible, then any covering equivalent to $\wp$ is also G-admissible.

5.5.11 Let Γ be the graph in Figure 5.11. Let $\zeta : V(\Gamma) \to \mathbb{Z}_5$ be the voltage function on Γ where $\zeta(a, b) = \zeta(a, c) = \zeta(a, d) = \zeta(b, e) = \zeta(b, f) = 0$ while $\zeta(c, d) = 1$ and $\zeta(e, f) = 2$.

(a) Draw the corresponding derived graph $\tilde{\Gamma} = \mathrm{Cov}(\Gamma, \zeta)$.

(b) Which of the automorphisms $(a, b)(c, e)(d, f)$, $(a, b)(c, f)(d, e)$,

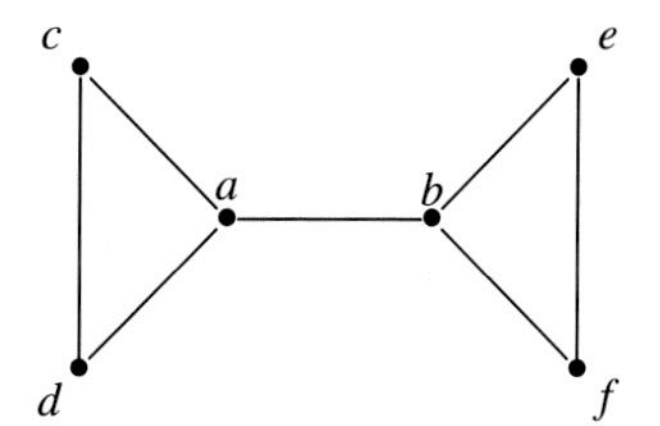

Figure 5.11 The graph Γ in Exercise 5.5.11.

$(a, b)(c, e, d, f)$, $(c, d)(e, f) \in \mathrm{Aut}(\Gamma)$ lift to an automorphism of the derived graph $\tilde{\Gamma}$?

(c) Let $\tilde{\Gamma}'$ be the graph obtained from $\tilde{\Gamma}$ by deleting vertices of valency 2 and replacing the corresponding paths of length 3 by an edge. Convince yourself that $\tilde{\Gamma}'$ is the Petersen graph.

5.5.12 Prove that a regular covering projection preserves valency. Show that the converse is false by giving an example of a valency-preserving onto homomorphism $h\colon Y \to X$ that is not a regular covering and yet $|h^{-1}(v)|$, $v \in V_X$, is constant.

5.5.13 Find all onto homomorphisms $h\colon C_{2n} \to C_n$ that satisfy $|h^{-1}(v)| = 2$ for all $v \in V_{C_n}$.

5.5.14 Find all onto homomorphisms $h\colon C_{kn} \to C_n$ that satisfy $|h^{-1}(v)| = k \geq 3$ for all $v \in V_{C_n}$.

6

Other Classes of Graphs

In Section 6.1 we consider the Coxeter graph. In the rest of this chapter we discuss other common classes of graph. Our goals in each section are to give the automorphism groups of the graphs, determine their isomorphism classes, characterize which of the graphs are isomorphic to Cayley graphs, and to determine which are Hamiltonian – but there are some exceptions. In some sense our goals are arbitrary – there are other properties one could consider for each class – but we are focusing on what some would call the "main" problems. There are, of course, many other classes of graphs we could have included, but we chose the ones we have as they seem to be the main classes that are most often encountered. For some applications and problems, other classes may well be more important. So perhaps the proper way to view this is that in this chapter you are starting to expand your library of classes of graphs that commonly appear in the literature, and as you work more in the area, you will add additional classes, and perhaps ones more suited to the problems you are interested in.

A few words about why this chapter is Chapter 6 (instead of earlier or later). It is not earlier as the first five chapters of the book cover what we consider to be "basic knowledge" that researchers in the area would either know or know about. We could have placed it anywhere else in the book, as this chapter is independent of those that follow (but some of the following chapters use information from this one), so the reader may wish to come back to this chapter later.

We begin not with a class of graphs, but with the Coxeter graph!

6.1 The Coxeter Graph

It was mentioned in Chapter 3 that the Coxeter graph, as well as its truncation, is a connected vertex-transitive graph without a Hamiltonian cycle. In this

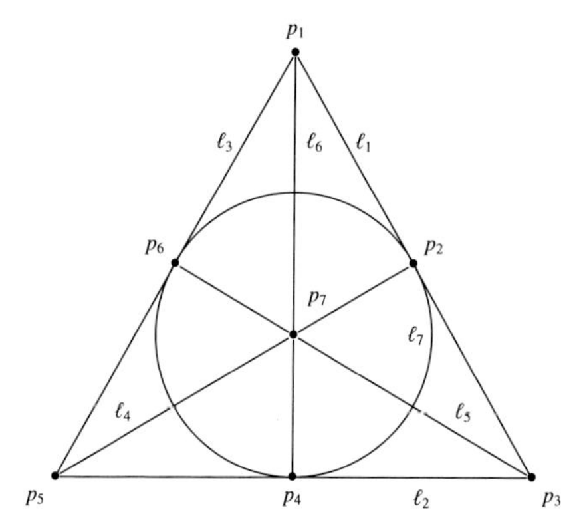

Figure 6.1 The points and lines of the Fano plane labeled.

section we will show this, after defining the Coxeter graph and determining its full automorphism group.

We define the Coxeter graph using the Fano plane in Figure 6.1, basically following the strategy used to find the automorphism group of the Heawood graph in Section 4.5. That is, we define it in a way that will allow us to determine a large subgroup of its automorphism group fairly easily. We will then show that it is 3-arc-regular and apply Corollary 3.1.6 to show this large subgroup is its full automorphism group. Doing this will show that the Coxeter graph is a cubic symmetric graph. It is denoted F28 in the Foster census. We will usually refer to the Coxeter graph as F28 except in statements of results.

Definition 6.1.1 Let P be the set of points of the Fano plane (or the one-dimensional subspaces of $\mathbb{F}_2^3$), and L the set of lines through the origin of the Fano plane (or the hyperplanes of $\mathbb{F}_2^3$). We construct a graph as follows. For the vertices, we take all ordered pairs (p, ℓ), where $p \in P$ and $\ell \in L$ such that the point p does not lie on the line ℓ. For the edge set, we say that (p, ℓ) is adjacent to (p', ℓ') if and only if $\ell \cup \ell' \cup \{p, p'\} = P$. That is, the set of points on ℓ and ℓ' together with the points p and p' give every point on the Fano plane. The resulting graph is known as the Coxeter graph and is shown in Figure 6.2, using the labeling of the points and lines of the Fano plane as shown in Figure 6.1.

Who discovered the Coxeter graph is something of a mystery. Tutte (1960) gives credit to Coxeter. Coxeter, however, disagreed and gives the credit independently to Foster and Conway (Coxeter, 1983). It is unquestionably though named after Coxeter! Tutte (1998) also wrote that Coxeter called him

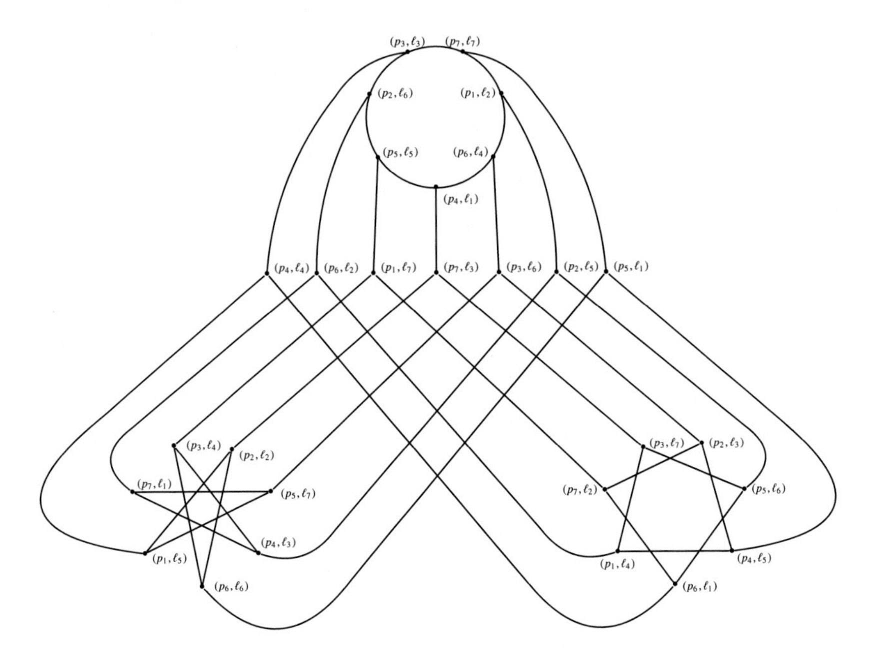

Figure 6.2 The Coxeter graph.

on the telephone to enquire about a certain symmetric graph of order 28 and girth 7. After hearing a description of the construction of this new graph, Tutte replied "Yes, that is what graph-theorists call the Coxeter graph." Tutte then reports that this motivated Coxeter to write a new paper entitled "My Graph" (Coxeter 1983).

Lemma 6.1.2 *The Coxeter graph is a cubic graph with* 28 *vertices.*

Proof As there are seven points and seven lines in the Fano plane, and each line contains exactly three points, there are $7 \cdot 7 - 3 \cdot 7 = 28$ ordered pairs (p, ℓ) where $p \in P$ is a point of the Fano plane that does not lie on the line $\ell \in L$. Thus the Coxeter graph has 28 vertices. Note that any two distinct lines on the Fano plane intersect in exactly one point, so for any lines $\ell, \ell' \in L$, we have $|\ell \cup \ell'| \leq 5$ with equality if and only if ℓ and ℓ' are distinct. Let $(p, \ell), (p', \ell') \in V(\text{F28})$ be adjacent. As $\ell \cup \ell' \cup \{p, p'\} = P$, $|\ell \cup \ell'| \leq 5$, and $|\{p, p'\}| \leq 2$, it must be the case that ℓ and ℓ' are distinct, p' is not on ℓ, p is not on ℓ', and $p \neq p'$. We conclude that F28 is regular of valency 3. That is, for each fixed line ℓ and point p not on ℓ, there are exactly three lines ℓ_1, ℓ_2, ℓ_3

with one point in $\ell \cup \{p\}$, and for each ℓ_i there is exactly one point p_i such that $\ell \cup \ell_i \cup \{p, p_i\} = P$, $1 \le i \le 3$. □

For our next result, we remind the reader that $\mathrm{PGL}(3,2) \cong \mathrm{PSL}(3,2)$, and, by Corollary 4.5.8 and the discussion following it, the automorphism group of the Heawood graph is also $\mathrm{PGL}(2,7) \cong \mathrm{PSL}(3,2) \rtimes \mathbb{Z}_2$.

Theorem 6.1.3 *The automorphism group of the Coxeter graph contains a subgroup isomorphic to* $\mathrm{PGL}(2,7) \cong \mathrm{PSL}(3,2) \rtimes \mathbb{Z}_2$.

Proof Define $\tau\colon P \times L \to P \times L$ by $\tau(p,\ell) = (\ell^\perp, p^\perp)$, and let $k \in K \cong \mathrm{PSL}(3,2)$ act on $P \times L$ by $k(p,\ell) = (k(p), k(\ell))$. Similar to Theorem 4.5.6, it is straightforward to show $\langle \tau, K\rangle \cong \mathrm{PSL}(3,2) \rtimes \mathbb{Z}_2$. We first show that $\langle \tau, K\rangle$ maps $V(\mathrm{F28})$ to itself.

Suppose $(p,\ell) \in V(\mathrm{F28})$. If $\tau(p,\ell) \notin V(\mathrm{F28})$, then $\ell^\perp \in p^\perp$. Setting q to be the nonzero vector in $\ell^\perp$, we see that q is perpendicular to p. Additionally, a vector $x \in \ell$ if and only if x is perpendicular to q as $\ell^\perp = \{0, q\}$, so that $p \in \ell$, a contradiction. Thus $\tau(p,\ell) \in V(\mathrm{F28})$. As $\mathrm{PSL}(3,2)$ preserves incidence, if $(p,\ell) \in P \times L$ is such that $p \notin \ell$, then $k(p) \notin k(\ell)$ for any $k \in \mathrm{PSL}(3,2)$. Thus $\langle \tau, K\rangle$ maps $V(\mathrm{F28})$ to itself.

Now suppose (p,ℓ) and (q,m) are adjacent vertices in F28. Note that this necessarily implies that p and q are different, as $\ell \cup m$ contains five points of P. Also, neither p nor q are contained in $\ell \cup m$. By Exercise 6.1.1, we have $p + q$ is contained in $\ell \cap m$.

Let $g \in \mathrm{PSL}(3,2)$. Clearly $g(p) \ne g(q)$, and $g(p+q) = g(p) + g(q)$. Thus $g(p) + g(q)$ is contained in $g(\ell)$ and $g(m)$, and $g(p), g(q) \notin g(\ell) \cup g(m)$. Thus $g(\ell)$ and $g(m)$ are the two lines (by Exercise 6.1.1) that do not contain $g(p)$ and $g(q)$. Then $g(\ell) \cup g(q) \cup \{g(p), g(q)\} = P$, and $\mathrm{PSL}(3,2) \le \mathrm{Aut}(\mathrm{F28})$.

It remains to show that $\tau \in G$. We already know that $l^\perp \notin p^\perp$ and $m^\perp \notin q^\perp$ as τ maps $V(\mathrm{F28})$ to itself. We must show that $p^\perp \cup q^\perp \cup \ell^\perp \cup m^\perp = P$. First, the union of any two distinct lines contains five points of P. Suppose $\ell^\perp \in p^\perp \cup q^\perp$. As $\ell^\perp \notin p^\perp$, we have $\ell^\perp \in q^\perp$. That is, $\ell^\perp \cdot q = 0$. So any point orthogonal to each point in ℓ is also orthogonal to q. But any point in P is orthogonal to exactly three points in P (its orthogonal complement), and these all lie on a line. Thus $q = \ell^\perp \in \ell$, which contradicts $q \notin \ell$. Hence $\ell^\perp \notin p^\perp \cup q^\perp$. Similarly, $m^\perp \notin p^\perp \cup q^\perp$, so $p^\perp \cup q^\perp \cup \ell^\perp \cup m^\perp = P$, and $\tau \in \mathrm{Aut}(\mathrm{F28})$. □

Theorem 6.1.4 *The Coxeter graph is 3-arc regular and has automorphism group* $\mathrm{PGL}(2,7) \cong \mathrm{PSL}(3,2) \rtimes \mathbb{Z}_2$.

Proof We first show that F28 is arc-transitive. Suppose $((p,\ell),(p',\ell'))$ and $((q,m),(q',m'))$ are arcs in F28, with p, p', q, q' points, and ℓ, ℓ', m, m' lines.

As $((p,\ell),(p',\ell'))$ is an arc of F28, $p, p' \notin \ell \cup \ell'$, and $p \neq p'$. By Exercise 6.1.1, there are exactly two lines that do not contain either p or p', and $p + p'$ is contained in both. We conclude that ℓ and ℓ' are these two lines, and so $\ell \cap \ell' = p + p'$. Similarly, $m \cap m' = q + q'$. Let $r \in \ell$ such that $r \neq p + p'$, and $r' \in m$ such that $r' \neq q + q'$. Let $A \in \mathrm{GL}(3,2)$ such that A maps p to q, p' to q', and r to r'. Then $A(p + p') = q + q'$, and so $A(\ell) = m$. As ℓ' is a line that does not contain p or p', $A(\ell')$ is a line that does not contain $A(p) = q$ or $A(p') = q'$ and is not $m = A(\ell)$. As m and m' are the only lines that do not contain q or q', $A(\ell') = m'$. This implies that F28 is arc-transitive.

Let s be the smallest positive integer such that F28 is not $(s+1)$-arc transitive. By Theorem 3.1.5, we have that F28 is s-arc-regular, and by Corollary 3.1.6, we see Aut(F28) has order $28 \cdot 3 \cdot 2^{s-1}$.

Clearly there are 4-arcs of F28 that are contained in 7-cycles. Also, the 4-arc $(p_4,\ell_4),(p_1,\ell_5),(p_5,\ell_7),(p_3,\ell_6),(p_6,\ell_4)$ is not contained in a 7-cycle. Thus F28 is not 4-arc-transitive, and so $|\mathrm{Aut(F28)}| \leq 28 \cdot 3 \cdot 2^{3-1} = 336$. As $\mathrm{PGL}(2,7) \leq G$ and is of order 336, $\mathrm{Aut(F28)} = \mathrm{PGL}(2,7)$. □

With the automorphism group of the Coxeter graph in hand, it is not difficult to show that the Coxeter graph is not isomorphic to a Cayley graph of any group. This is left as an exercise (Exercise 6.1.8).

We now turn to a well-known and unusual property of the Coxeter graph that we have already seen the Petersen graph possesses. Namely, F28 is not Hamiltonian, although it does contain a Hamilton path. At the present time these two graphs, together with two graphs obtained from these graphs by truncation (Section 3.4), are the only known vertex-transitive graphs of order at least three that are not Hamiltonian. That F28 is not Hamiltonian was first shown by Tutte (1960), and the proof we present here is a modification of a proof due to Biggs (1973). Biggs proved that there are exactly 84 1-factors in F28, and that the automorphism group of F28 is a transitive group acting on these 1-factors. As in a cubic graph, a Hamilton cycle is the complement of a 1-factor, we then need only find a single 1-factor in F28, delete it from F28, and then observe that the resulting 2-factor is a disjoint union of two cycles, and not a Hamilton cycle. We first show that F28 has exactly 84 1-factors, and in doing so will show that there are at most 4 orbits of the automorphism group of F28 on the set of all 1-factors of F28. We then check that the complements of a representative from each of these orbits in F28 is not a Hamiltonian cycle. We will also present an alternative proof in Example 11.1.8.

Before proceeding, there are some specific automorphisms of F28 that we will need. In order to define the automorphisms easily, we will need a different natural labeling of the vertices of F28. First, we let

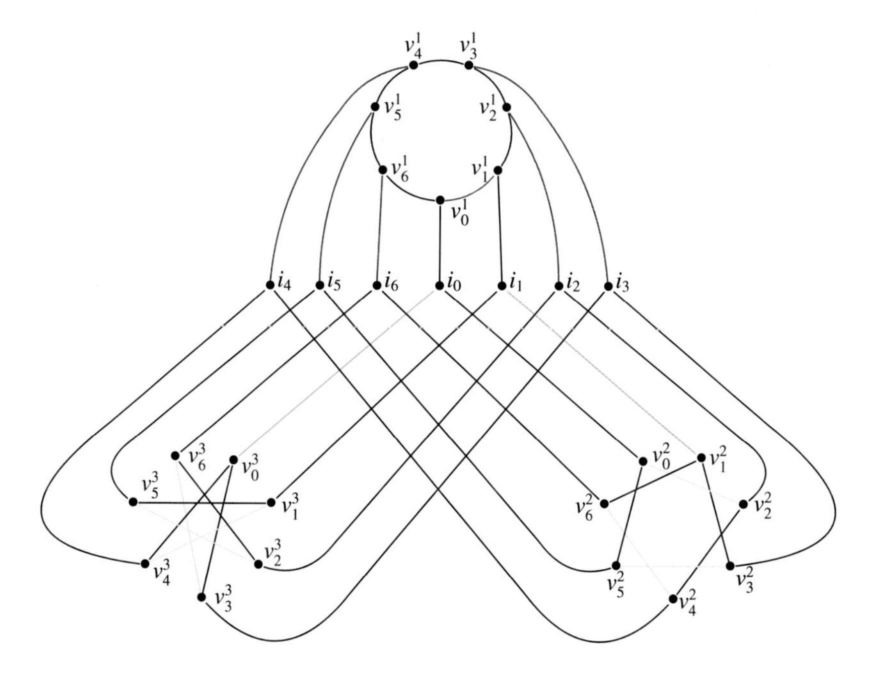

Figure 6.3 Edges of a matching M of F28 with two of the 7-cycles in F28 disjoint from I containing three edges of M.

$$I = \{(p_4, \ell_4), (p_6, \ell_2), (p_1, \ell_7), (p_7, \ell_3), (p_3, \ell_6), (p_2, \ell_5), (p_5, \ell_1)\}.$$

For ease of notation, we set $I = \{i_j : j \in \mathbb{Z}_7\}$ as in Figure 6.3. We denote the three 7-cycles of F28 with the vertices of I removed by C_1, C_2, and C_3. We label the vertices of C_j, $1 \leq j \leq 3$, with elements of v_k^j, $k \in \mathbb{Z}_7$, as in Figure 6.3. The maps $\delta, \rho \colon V(\text{F28}) \to V(\text{F28})$ defined by $\delta\left(v_k^j\right) = v_{2k}^{j+1}$ and $\delta(i_k) = i_{2k}$ (where we adopt the convention that $v_{2k}^{3+1} = v_{2k}^1$) and $\rho\left(v_k^j\right) = v_{k+1}^j$ and $\rho(i_k) = i_{k+1}$, are automorphisms of F28 (Exercise 6.1.3).

Theorem 6.1.5 *The Coxeter graph is not Hamiltonian.*

Proof Let M be a 1-factor in F28. Of course, M contains seven edges incident with the seven vertices in I, and so M must also contain seven edges that are contained in $E(C_1) \cup E(C_2) \cup E(C_3)$. Denote the seven edges incident with the seven vertices in I by E_I. Note that the endpoints of the vertices in E_I cannot be contained in a single C_i, $1 \leq i \leq 3$, as then the remaining seven edges of M is a 1-factor of the disjoint union of two 7-cycles. This implies that each C_i contains at least one edge of M, and of course each C_i contains at most three edges of M. Let n_i be the number of edges of M in $E(C_i)$, $1 \leq i \leq 3$. As there are seven edges in $M \backslash E_I$, we see $n_1 + n_2 + n_3 = 7$. As $1 \leq n_i \leq 3$, $1 \leq i \leq 3$, we see either two of n_1, n_2, and n_3 are 3 and the other is 1 or one of n_1, n_2, n_3 is 3 and the other two are 2. We will consider each of these two cases separately, and show that in each case there are exactly 42 different 1-factors.

Case 1. Two of n_1, n_2, and n_3 are 3 and the other is 1. Then exactly one of C_1, C_2, and C_3 contains exactly one edge e of M, and as there are 21 edges in $E(C_1) \cup E(C_2) \cup E(C_3)$, there are exactly 21 choices for $e = xy$. As δ and ρ preserve the decomposition of V(F28) into C_1, C_2, C_3, and I, applying appropriate powers of δ and then ρ, we may assume without loss of generality that $e = v_0^1 v_1^1$, (shown in blue in Figure 6.3). It must then be the case that the five edges $v_k^1 i_k$ are contained in M for $k = 2, 3, 4, 5, 6$ (shown in red in Figure 6.3). The edges e' and e'', respectively, of M with endpoints i_0 and i_1 must then have their other endpoints in $V(C_2) \cup V(C_3)$. As $n_1 = n_2 = 3$, we have i_0 is adjacent to a vertex of C_2 and i_1 is adjacent to a vertex of C_3 or vice versa (the case where i_0 is adjacent to a vertex of C_3 is shown in green in Figure 6.3). As there are two choices for e' and then e'' is determined, there are exactly 42 choices for the edges of M so far. As the remaining vertices to be matched induce two edge disjoint paths of length 5 that can be matched in exactly one way (shown in yellow in Figure 6.3), we see there are exactly 42 1-factors of F28 with $n_1 = n_2 = 3$ and $n_3 = 1$. Additionally, as F28 is edge-transitive, we see that the action of the automorphism group of F28 on the set of such 1-factors has at most two orbits (depending on the choice of e' and e''). Finally, it is clear that removing this 1-factor from F28 yields a union of two 14-cycles, and this can also be verified for the other choice of e' and e''.

Case 2. One of n_1, n_2, and n_3 is 3 and the other two are 2. If $1 \leq j \leq 3$ such that C_j contains three edges of M, then there is a unique vertex $v \in V(C_j)$ that is not matched by an edge of M with both endpoints in C_j. As there are 21 vertices in $V(C_1) \cup V(C_2) \cup V(C_3)$, there are 21 choices for v. Again, by applying a power of δ and then a power of ρ, we may assume without loss of generality that $v = v_0^1$, and so v_0^1 is matched in M by the edge $v_0^1 i_0$. Then $v_1^1 v_2^1, v_3^1 v_4^1, v_5^1 v_6^1$ are in M. Also, v_0^2 and v_0^3 are matched by edges contained in C_2 and C_3, respectively, and so one of $v_0^2 v_2^2$ and $v_5^2 v_0^2$ and one of $v_0^3 v_3^3$ and $v_4^3 v_0^3$ are contained in M. There are now four cases to consider depending upon which of the edges in the immediately preceding sentence are contained in M. We will consider the cases where $v_0^2 v_2^2$ and $v_0^3 v_3^3$ are in M and $v_0^2 v_2^2$ and $v_4^3 v_0^3$ are in M, with the other two cases being analogous. We will show the former case yields exactly 21 different 1-factors (with the case $v_5^2 v_0^2$ and $v_4^3 v_0^3$ in M being analogous), and that the latter case is impossible (with the case $v_5^2 v_0^2$ and $v_0^3 v_3^3$ in M being analogous). As before, there will be at most two orbits of the

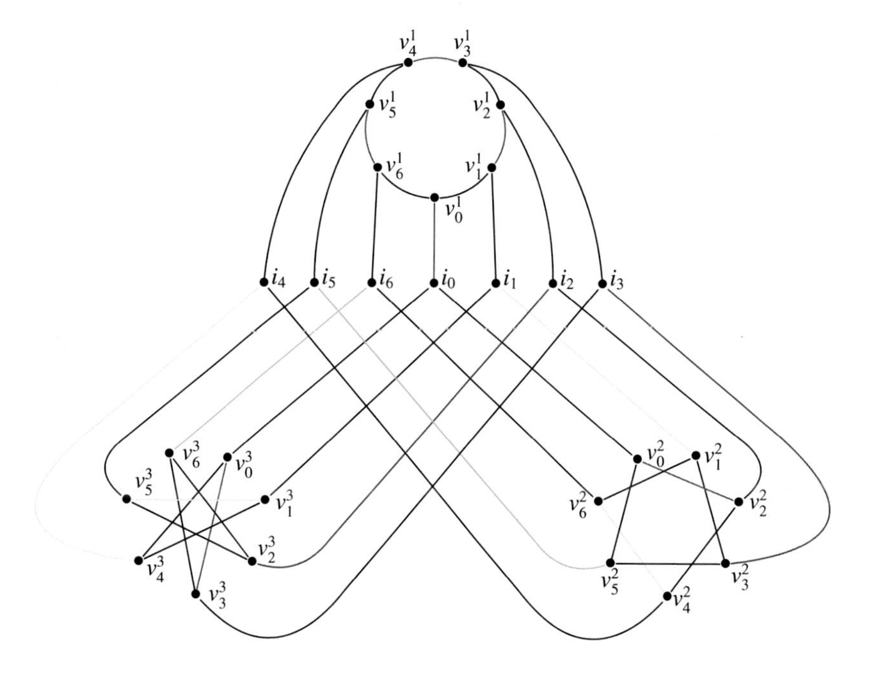

Figure 6.4 Two of n_1, n_2, and n_3 are 2, and $v_0^2v_2^2$ and $v_0^3v_3^3$ are in M.

automorphism group of F28 on the set of 1-factors of F28 as F28 is vertex-transitive, and removing an appropriate 1-factor from F28 will again yield two 14-cycles.

Case 2a. $v_0^2v_2^2$ and $v_0^3v_3^3$ are in M. The edges thus far determined in M are colored blue in Figure 6.4. The edge in M incident with i_2 cannot be $i_2v_2^2$ or $i_2v_2^1$, and so must be $i_2v_2^3$. A similar argument shows that the edge $i_3v_3^2$ is also in M. These edges are colored red in Figure 6.4. We now see that the edge of M incident with v_5^2 is $i_5v_5^2$, and similarly, the edge of M incident with v_6^3 is $i_6v_6^3$. These edges are colored green in Figure 6.4. The edge of M incident with v_5^3 is now contained in C_3 and cannot be v_2^3 as it is already matched in M. So $v_5^3v_1^3 \in M$, which forces $i_4v_4^3 \in M$. These edges are colored yellow in Figure 6.4. Also, v_4^2 must also be matched with in edge from C_2 as its neighbor outside C_2 is already matched in M, and cannot be v_2^2, which is also matched by M. Thus $v_4^2v_6^2 \in M$, and consequently $i_1v_1^2 \in M$, and so M is determined. As before, we see that after removing the edges of M from F28, we are left with two 14-cycles, and that all 21 such 1-factors are contained in an orbit of the automorphism group of F28 acting on the set of all such 1-factors, as F28 is vertex-transitive.

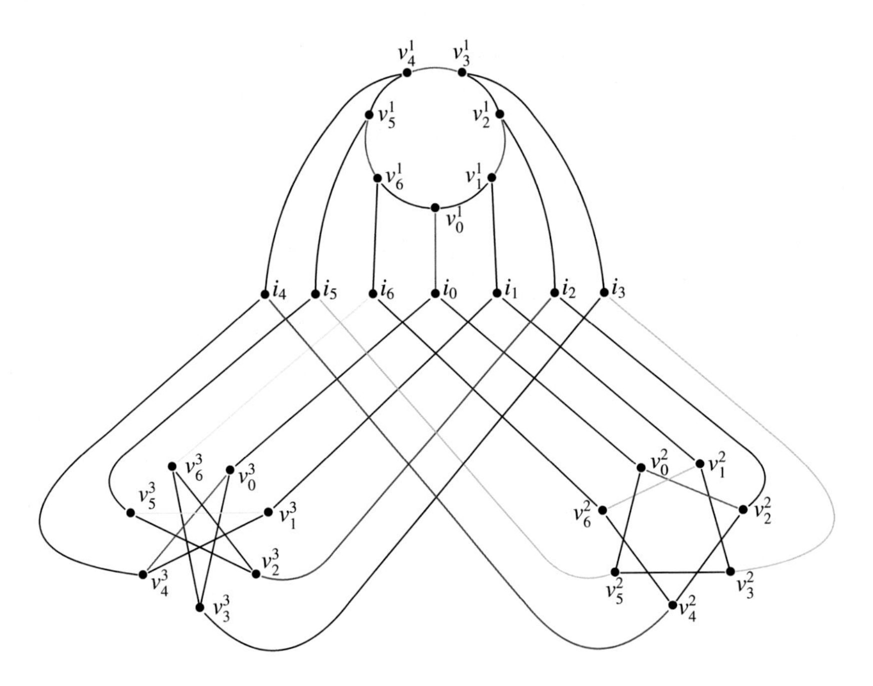

Figure 6.5 Two of n_1, n_2, and n_3 are 2, and $v_0^2v_2^2$ and $v_4^3v_0^3$ are in M.

Case 2b. $v_0^2v_2^2$ and $v_4^3v_0^3$ are in M. The edges thus far determined in M are colored blue in Figure 6.5. Similarly to the previous subcase, the edges $i_2v_2^3$ and $i_4v_4^2$ are contained in M, and are colored red in Figure 6.5. Note $v_6^2v_4^2$ is not an edge in M. So the edge of M incident with v_6^2 is either $v_6^2v_1^2$ or $v_6^2i_6$.

Suppose $v_6^2v_1^2 \in M$. Then no other edge of M is contained in C_2, and so the edges $v_5^2i_5$ and $i_3v_3^2$ are in M. These three edges are colored green in Figure 6.5. The edge incident with v_5^3 we can see now must be $v_5^3v_1^3$, colored in yellow in Figure 6.5. Also, the edge incident with i_6 must be $i_6v_6^3$, also colored yellow in Figure 6.5. As there are then no other edges in C_3 that can be in M, no edge of M is incident to v_3^3, a contradiction.

So $v_6^2i_6 \in M$. M is now determined. The vertex v_6^3 must be incident to a vertex in C_3, and as v_2^3 is already matched, $v_6^3v_3^3 \in M$. This then forces $v_5^3i_5, v_1^3i_1 \in M$, which in turn forces $i_3v_3^2 \in M$. Now only the two nonadjacent vertices v_1^2 and v_5^2 are not matched, a contradiction. □

Applying Theorem 3.4.5 we have the following result.

Corollary 6.1.6 *The truncation of the Coxeter graph is vertex-transitive but not Hamiltonian.*

Note that by Corollary 3.4.8, while the truncation of the truncation of the Coxeter graph is not Hamiltonian, this graph is not vertex-transitive.

6.2 Kneser Graphs and Odd Graphs

The Kneser graphs were introduced and first studied by Kneser (1955).

Definition 6.2.1 Let $k \geq 2$ and $n \geq 2k + 1$ be positive integers. The **Kneser graphs** $K(n, k)$ are the graphs whose vertex set is all k-subsets chosen from $[1, n] = \{1, \ldots, n\}$, and two vertices ($k$-subsets) are adjacent if and only if their intersection is empty.

The Petersen graphs is $K(5, 2)$, and just as the Petersen graph is isomorphic to the complement of the line graph of K_5, the graph $K(n, 2)$ is isomorphic to the complement of the line graph of K_n. Kneser (1955) conjectured that the chromatic number of $K(n, k)$ was exactly $n-2k+2$. This conjecture was verified in a celebrated paper of Lovász (1978).

The Kneser graphs share some properties with the Petersen graph, the most obvious being that $\mathrm{Aut}(K(n, k))$ contains a transitive subgroup isomorphic to $\mathcal{S}_n$ with the action being that of $\mathcal{S}_n$ on k-subsets of $[1, n]$. It also turns out that this is always the full automorphism group of the Kneser graphs, as we will show using the Erdős–Ko-Rado theorem (Erdős et al., 1961). A short proof of this theorem can be found in Katona (1972).

Theorem 6.2.2 *If $n \geq 2k + 1$ and $\mathcal{A}$ is a collection of k-subsets of $[1, n]$ such that every pair of k-subsets in $\mathcal{A}$ has nontrivial intersection, then $|\mathcal{A}| \leq \binom{n-1}{k-1}$. Additionally, the largest sets of k-subsets that have nontrivial intersection are the k-subsets that contain a fixed element of $[1, n]$.*

We now find the automorphism groups of the graphs $K(n, k)$.

Theorem 6.2.3 $\mathrm{Aut}(K(n, k)) = \mathcal{S}_n$.

Proof As two vertices (k-subsets) of $K(n, k)$ are adjacent if and only if their intersection is empty, the Erdős–Ko–Rado theorem gives an upper bound on the size of a maximal independent set in $K(n, k)$. This upper bound is attained by the set I_j of all k-subsets of $[1, n]$ that contain $j \in [1, n]$. Each I_j clearly has $\binom{n-1}{k-1}$ elements, and as each element of I_j contains j, there are no edges between elements of I_j. Furthermore, we also have that the sets I_j, $j \in [1, n]$ are the only maximal independent sets. Note that a graph automorphism maps an independent set to an independent set of the same size, so $\mathrm{Aut}(K(n, k))$ permutes the sets I_j with $j \in [1, n]$ and induces a natural homomorphism into $\mathcal{S}_n$. The only thing remaining to do is to show that the only element of $\text{Aut}(K(n,k))$ that fixes every I_j, $j \in [1,n]$, is the identity. But this is easy, as if $g \in \text{Aut}(K(n,k))$ fixes each I_j, it fixes $\cap_{S \in I_j} S = \{j\}$. □

Note that the main difference between our argument here and the argument given for $\text{Aut}(K(5,2)) = \mathcal{S}_n$ in Section 2.1 (the automorphism group of the Petersen graph), is that here we invoked the Erős–Ko–Rado theorem, while in Section 2.1 we verified the information given by the Erős–Ko–Rado theorem for the case $n = 5$ and $k = 2$ directly.

There has been some work done on determining which of the Kneser graphs are Hamiltonian. At the moment, the best known is by Chen (2003) (this paper also contains a good survey on the state of the problem), and states that $K(n,k)$ is Hamiltonian provided that $n \geq \frac{3k+1+\sqrt{5k^2-2k+1}}{2}$. Note $\frac{3k+1+\sqrt{5k^2-2k+1}}{2} < 2.62k + 1$.

Godsil (1980) has determined which $K(n,k)$ are Cayley graphs.

Theorem 6.2.4 *The graphs $K(n,k)$ are not isomorphic to Cayley graphs except in the following cases:*

(i) $k = 2$, *n is a prime power, and* $n \equiv 3 \pmod 4$, *and*
(ii) $k = 3$, *and* $n = 8$ *or* 32.

Furthermore, if $n = 2k + 1$ or $k \geq 5$, a transitive subgroup G of $\text{Aut}(K(n,k))$ *is isomorphic as an abstract group to A_n or $\mathcal{S}_n$, or one of the following is true:*

(i) $k = 2$ *and* $G \cong \text{AGL}(1,5)$,
(ii) $k = 4$, *and either* $G \cong \text{PGL}(2,8)$ *or* $\text{P}\Gamma\text{L}(2,8)$, *or*
(iii) $n = 24$ *and* $G \cong M_{24}$, *the Mathieu group of degree* 24.

The above statement is slightly stronger than that given by Godsil. In particular, Godsil only asserted that a transitive group G of $\text{Aut}(K(n,k))$ is isomorphic as an abstract group to A_n or $\mathcal{S}_n$ provided that $n = 2k + 1$ and $k \neq 2$ or 4. In the case of $n = 2k + 1$ and $k = 2$ or 4, the groups not isomorphic to A_n or $\mathcal{S}_n$ are given in Beaumont and Peterson (1955) (we shall see the main idea in this work in the next section when we consider which Johnson graphs are Cayley), while the form stated here is given in Livingstone and Wagner (1965). If $k \geq 5$, Livingstone and Wagner (1965, Theorem 2) showed that a group transitive on the k-subsets of $[1,n]$ is also k-transitive. A consequence of the CFSGs is that all 2-transitive groups are known, and that the only 5-transitive group that is not alternating or symmetric is M_{24} of degree 24 – see Cameron (1981, Theorem 5.3).

Definition 6.2.5 Define the **odd graph** O_k, $k \geq 2$, to be the graph $K(2k+1, k)$.

Thus the Petersen graph is O_2. The last part of the previous result states, amongst other things, that not only are the odd graphs not Cayley graphs, but that if $k \neq 2$ or 4, they are not metacirculant graphs (recall that the Petersen graph is a $(2,5)$-metacirculant graph) by Theorem 1.5.11.

As O_2 is the Petersen graph, it is not Hamiltonian. However, recently Mütze, Nummenpalo, and Walczak showed that O_k is Hamiltonian for all $k \geq 3$ (Mütze et al., 2018).

Theorem 6.2.6 *The odd graphs O_k are 3-arc transitive.*

Proof Let x_0, x_1, x_2, x_3 be a 3-arc in O_k (so each x_i is a k-subset of $[1, 2k+1]$). Then $x_0 \cap x_1 = x_1 \cap x_2 = x_2 \cap x_3 = \emptyset$. Thus $x_0 \cup x_1$ contains all but one element, say v, of $[1, 2k+1]$, while $x_2 \neq x_0$. This is only possible if x_2 contains a subset y_2 of x_0 of size $k-1$, together with the vertex v. Similarly, x_3 cannot contain any element of x_2, but cannot be x_1. This can only happen if x_3 is the one element u of x_0 not in x_2, together with a subset y_1 of x_1 of size $k-1$. Let w be the element of $x_1 \backslash y_1$. Then $\{y_1, y_2, \{u, v, w\}\}$ is a partition of $[1, 2k+1]$ and

$$\begin{aligned}
x_0 &= y_2 \cup \{u\}, \\
x_1 &= y_1 \cup \{w\}, \\
x_2 &= y_2 \cup \{v\}, \text{ and} \\
x_3 &= y_1 \cup \{u\}.
\end{aligned}$$

Now let z_0, z_1, z_2, z_3 be another 3-arc in O_k. Arguing as above, there are $y_1' \subset z_1$, $y_2' \subset z_0$, and $\{u', v', w'\}$ such that $z_0 = y_2' \cup \{u'\}$, $z_1 = y_1' \cup \{w'\}$, $z_2 = y_2' \cup \{v'\}$, and $z_3 = y_1' \cup \{u'\}$, and $\{y_1', y_2', \{u', v', w'\}\}$ is a partition of $[1, 2k+1]$. As $\mathcal{S}_n$ is n-transitive by Exercise 2.4.5, there exists $\omega \in \mathcal{S}_n$ such that $\omega(y_1) = y_1'$, $\omega(y_2) = y_2'$, $\omega(u) = u'$, $\omega(v) = v'$, and $\omega(w) = w'$. Then $\omega(x_0, x_1, x_2, x_3) = z_0, z_1, z_2, z_3$ so that O_k is 3-arc transitive. □

The odd graphs in some sense predate the Kneser graphs, as the odd graph O_3 was studied in Kowalewski (1917) – the same paper where the Petersen graph was first drawn as $O_2 = K(5, 2)$. See Biggs (1979) for more information on the odd graphs.

6.3 The Johnson Graphs

The Johnson graphs are defined somewhat similarly to the Kneser graphs – instead of having two k-sets be disjoint to have an edge between them, we

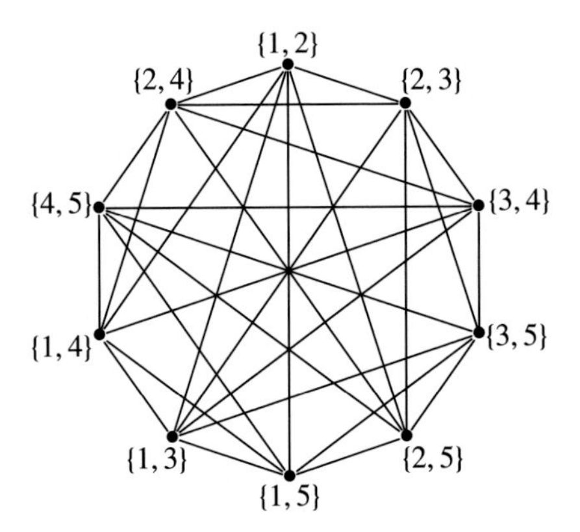

Figure 6.6 The Johnson graph $J(5, 2)$.

require the size of their intersection to be $k - 1$. They are named after Selmer Johnson.

Definition 6.3.1 The **Johnson graphs** $J(n, k)$ are the graphs whose vertex set $V_{n,k}$ consists of all k-subsets chosen from $[1, n]$, and two vertices (k-element subsets) are adjacent if and only if their intersection has $k - 1$ elements.

For $k = 1$, the Johnson graph is simply a complete graph. For $k = 2$ the Johnson graph is the complement of the Kneser graph $K(n, 2)$, and so the Petersen graph is $\overline{J(5, 2)}$. The Johnson graph $J(5, 2)$ is shown in Figure 6.6. The graph $J(n, 2)$ is also the line graph of the complete graph K_n. The graphs $J(n, 2)$ are sometimes called the **triangular graphs** and denoted $T(n)$, while the graphs $J(n, 3)$ are sometimes called the **tetrahedral graphs**.

6.3.1 Automorphism Groups of Johnson Graphs

To begin, we will have need of a natural isomorphism between some Johnson graphs that is also an automorphism of $J(2k, k)$.

Lemma 6.3.2 *The map $\iota \colon V_{n,k} \to V_{n,n-k}$ given by $\iota(S) = [1, n] \backslash S$, is an isomorphism between $J(n, k)$ and $J(n, n - k)$. Also, ι is an automorphism of $J(2k, k)$.*

Proof If $K_1, K_2 \in V_{n,k}$ such that $|K_1 \cap K_2| = k - 1$, then $[1, n] \backslash K_1$ and $[1, n] \backslash K_2$ are sets of size $n - k$, and $([1, n] \backslash K_1) \cap ([1, n] \backslash K_2)$ consists of the $n - k - 1$ points that are not contained in $K_1 \cup K_2$. So ι is an isomorphism between $J(n, k)$ and $J(n, n - k)$, and an automorphism of $J(2k, k)$. □

The induced action of $\mathcal{S}_n$ on k-subsets induces an automorphism of $J(n,k)$ for every k and n.

We will first compute the automorphism groups of the Johnson graphs – these are always the automorphisms we have just seen. This proof is somewhat long, and so it will be useful to discuss the strategy of the proof, as this can often be obscured by the details. The proof is based on an analysis of a graph $C(n,k)$ of $J(n,k)$, obtained from maximal cliques of $J(n,k)$ in the following way. The vertices of $C(n,k)$ are the maximal cliques of $J(n,k)$ of maximum size, and two such cliques are adjacent if and only if their intersection is non-empty. One may ask why we will focus on $C(n,k)$ (other than, of course, the stupid answer of "it works"). The answer is that we will be able to completely describe the maximal cliques of $J(n,k)$, and their relationship to one another. It turns out that the maximal cliques come in two sizes unless $n = 2k$ – we choose the maximum size for no particular reason. Maximal cliques of minimal size would also work in the proof. We shall see that two such cliques are either disjoint or intersect in exactly one vertex of $J(n,k)$ – the one vertex will be crucial, as this will allow us to show that every automorphism of $J(n,k)$ induces an automorphism of $C(n,k)$. Finally, unless $n = 2k$, we will show $C(n,k) \cong J(n,k-1)$, and if $n = 2k$, $C(n,k)$ is a disconnected graph with two components both isomorphic to $J(n,k-1)$. We thus proceed by induction on k. One moral to this story is that if one is looking for some "structure" in a graph to try to determine its automorphism group, a structure has a much better chance of being used to determine automorphisms of the graph if one can describe precisely when the structure occurs. If one cannot do this, the structure one is examining is unlikely to help and should probably be discarded. We now investigate the structure of the maximal cliques of $J(n,k)$. Also, as mentioned earlier, if $k = 1$, then $J(n,1) \cong K_n$ and so $\mathrm{Aut}(J(n,1)) \cong \mathcal{S}_n$. We thus assume $k \geq 2$, and as $\iota(J(n,1)) = J(n,n-1)$, we also assume $k \leq n-2$. So henceforth, we will only consider when $2 \leq k \leq n-2$.

We begin by setting some notation that we will use throughout the proof. For $u, v \in [1,n]$, $K \in V_{n,k}$, $u \in K$, $v \notin K$, let $Y_{u,v}(K) = (K\backslash\{u\}) \cup \{v\}$. Let $\mathcal{Y}_u(K) = \{Y_{u,v}(K) : v \notin K\} \cup \{K\}$, and $\mathcal{Z}_v(K) = \{Y_{u,v}(K) : u \in K\} \cup \{K\}$. Note that $\mathcal{Y}_u(K)$ is the set of all k-subsets of $[1,n]$ that contain the $(k-1)$-subset $K\backslash\{u\}$.

Lemma 6.3.3 *Let $2 \leq k \leq n-2$. Every maximal clique in $J(n,k)$ is either $\mathcal{Y}_u(K)$ or $\mathcal{Z}_v(K)$ for some $K \in V_{n,k}$, $u \in K$, and $v \notin K$. Additionally, $|\mathcal{Y}_u(K)| = n-k+1$ while $|\mathcal{Z}_v(K)| = k+1$.*

Proof Clearly $Y_{u,v}(K) \in V_{n,k}$ and K and $Y_{u,v}(K)$ are adjacent. Also, the only neighbors of K in $J(n,k)$ are of the form $Y_{u,v}(K)$, $u \in K$, $v \notin K$. It is also clear that $Y_{u,v}$ is adjacent to $Y_{u,v'}$ if $v' \neq v$. Thus the subgraph $G_u(K)$ of $J(n,k)$ induced by $\mathcal{Y}_u(K)$ is a clique in $J(n,k)$ that contains K. Observe that $|\mathcal{Y}_u(K)| = n - k + 1$.

Suppose $X \in V_{n,k}$, $X \notin \mathcal{Y}_u(K)$, and X is adjacent to every vertex of $G_u(K)$. Let $X \cap K = S$. As $XK \in E(J(n,k))$, we see $|S| = k - 1$. If $u \in S$, then $|S \setminus \{u\}| = k - 2$, and $(S \setminus \{u\}) \subseteq Y_{u,v}(K)$ for every $v \notin K$. As X is adjacent to $Y_{u,v}(K)$ for every $v \notin K$, $(S \setminus \{u\}) \cup \{v\} \subset X$ for every $v \notin K$. However, X then has order $k - 1 + n - k = n - 1$, a contradiction. So $u \notin S$, and S contains every element of $K \setminus \{u\}$. This gives $X = (K \setminus \{u\}) \cup \{v\}$ for some $v \notin K$, and $X = Y_{u,v}(K)$, a contradiction. Thus $G_u(K)$ is a maximal clique of $J(n,k)$.

Similarly, $\mathcal{Z}_v(K)$ also induces a maximal clique in $J(n,k)$ of order $k + 1$ (Exercise 6.3.2).

We need only show that any clique C of size at least 2 is contained in $\mathcal{Y}_u(K)$ or $\mathcal{Z}_v(K)$. Let K be a vertex of C. Then any other vertex of C, being adjacent to K, is of the form $Y_{u,v}(K)$. If $|C| = 2$, then C is contained in $\mathcal{Y}_u(K)$ and $\mathcal{Z}_v(K)$, and so in $\mathcal{Y}_u(K)$ or $\mathcal{Z}_v(K)$. If $|C| \geq 3$, then let $Y_{a,b}(K), Y_{u,v}(K) \in V(C)$ be distinct. As $Y_{u,v}(K)$ is adjacent to $Y_{a,b}(K)$ we see that either $a = u$ or $b = v$. If $|V(C)| = 3$, then C is in $\mathcal{Y}_u(K)$ or $\mathcal{Z}_v(K)$. If $|V(C)| \geq 4$, then let $Y_{x,y}(K)$ be any other vertex in C not K, $Y_{u,v}(K)$, or $Y_{a,b}(K)$. As $Y_{x,y}(K)$ is adjacent to the preceding three vertices, if $a = u$ then $x = u$ or if $b = v$ then $y = v$. We conclude that C is contained in $\mathcal{Y}_u(K)$ or $\mathcal{Z}_v(K)$. □

Lemma 6.3.4 *Let $K, L \in V_{n,k}$, $u \in K$, $w \in L$ such that $\mathcal{Y}_u(K) \neq \mathcal{Y}_w(L)$ are cliques in $J(n,k)$. Then $\mathcal{Y}_u(K) \cap \mathcal{Y}_w(L)$ contains at most one vertex in $J(n,k)$.*

Proof First, suppose $K = L$. If $u \neq w$, then no element of $\mathcal{Y}_u(K)$ except K contains u while every element of $\mathcal{Y}_w(L)$ does, so $\mathcal{Y}_u(K) \cap \mathcal{Y}_w(L) = \{K\}$. If $u = w$, then $\mathcal{Y}_u(K) = \mathcal{Y}_w(L)$, contrary to the hypothesis.

Suppose $K \neq L$. Let $v, v' \in [1,n]$ such that $v \notin K$ and $v' \notin L$, and suppose $Y_{u,v}(K) = Y_{w,v'}(L)$ for some $u \in K$ and $w \in L$. Note that $L = (Y_{w,v'}(L) \setminus \{v'\}) \cup \{w\}$, and so

$$L = (((K \setminus \{u\}) \cup \{v\}) \setminus \{v'\}) \cup \{w\}.$$

If $v = v'$, then $L = (K \setminus \{u\}) \cup \{w\} = Y_{u,w}(K)$. Similarly, $K = Y_{w,u}(L)$. As $Y_{u,v}(K) = Y_{w,v}(L)$ for some $v \notin K \cup L$, $Y_{u,v}(K) = Y_{w,v}(L)$ for every $v \notin K \cup L$. Hence $\mathcal{Y}_u(K) = \mathcal{Y}_w(L)$, a contradiction. If $v \neq v'$, then $L = (K \setminus \{u, v'\}) \cup \{v, w\}$. Then

$$K \cap L = K \setminus \{u, v'\} = L \setminus \{v, w\}.$$

Now assume $Y_{u,t}(K) = Y_{w,t'}(L)$. Arguing as above, we see that $K \cap L = K\backslash\{u, t'\}$ and $L = (K \cap L) \cup \{w, t\}$. We conclude that $\{u, t'\} = \{u, v'\}$ and $\{w, t\} = \{v, w\}$ as $u, v', t' \in K$ and $w, t \in L$. This implies that $|\mathcal{Y}_u(K) \cap \mathcal{Y}_w(L)| = 1$. □

Lemma 6.3.5 *If $n = 2k$ and $k \geq 2$, then two cliques $\mathcal{Y}_u(K)$ and $\mathcal{Z}_v(L)$ either do not intersect or intersect in at least two sets.*

Proof As $n-k+1 = k+1$, the cliques $\mathcal{Z}_v(K)$ have the same size as the cliques $\mathcal{Y}_u(K)$. Suppose $M \in \mathcal{Z}_v(L) \cap \mathcal{Y}_u(K)$. If $L = K$, then both $\mathcal{Y}_u(K)$ and $\mathcal{Z}_v(L)$ contain $K = L$ and $Y_{u,v}(K)$, and the result follows. So we assume $K \neq L$.

If $K \neq M \neq L$, then $M = (K\backslash\{u\}) \cup \{v'\}$ for some $v' \notin K$ and $M = (L\backslash\{u'\}) \cup \{v\}$ for some $u' \in L$. We consider various cases.

If $v = v'$ then $v \notin K \cup L$ and $K \setminus \{u\} = L \setminus \{u'\}$. Hence there are $v_1, \ldots, v_{k-1} \in K \cap L$ such that $K = \{v_1, \ldots, v_{k-1}, u\}$ and $L = \{v_1, \ldots, v_{k-1}, u'\}$. Thus $L = Y_{u,u'}(K)$ and $M \neq L \in \mathcal{Z}_v(L) \cap \mathcal{Y}_u(K)$. The result follows.

If $u = u'$ and $v \neq v'$, then there exists $v_1, \ldots, v_{k-2} \in K \cap L$ with $K = \{v_1, \ldots, v_{k-2}, u, v\}$ and $L = \{v_1, \ldots, v_{k-2}, u, v'\}$. Then $Y_{v',v}(L) = K$, and $M \neq K \in \mathcal{Z}_v(L) \cap \mathcal{Y}_u(K)$. The result follows.

If $v \neq v'$ and $u \neq u'$, then $|K \cap L| = k - 2$. Set $K \cap L = \{v_1, \ldots, v_{k-2}\}$. Of course, $u \in K$ and $u' \in L$, and as $v, v' \in M$, we see $v \in K$ and $v' \in L$. Then $K = \{v_1, \ldots, v_{k-2}, u, v\}$, $L = \{v_1, \ldots, v_{k-2}, u', v'\}$, and so $Y_{u,u'}(K) = \{v_1, \ldots, v_{k-2}, v, u'\} = Y_{v',v}(L) \neq M$. So $M \neq Y_{u,u'}(K)$ are in $\mathcal{Y}_u(K) \cap \mathcal{Z}_v(L)$.

If $M = Y_{u,v'}(K) = L$, then $L = (K \setminus \{u\}) \cup \{v'\}$. By the definition of $\mathcal{Z}_v(L)$, we have $v \notin L$, and so $v' \neq v$. Also, every element of $L \setminus \{v'\}$ is contained in K and $u \in K$ is not contained in L. This gives that $Y_{v',v}(L) = Y_{u,v}(K)$, and so $|\mathcal{Z}_v(L) \cap \mathcal{Y}_u(K)| \geq 2$.

If $M = Y_{z,v}(L) = K$, then $K = (L\backslash\{z\}) \cup \{v\}$. By the definition of $\mathcal{Y}_v(L)$, we still have $v \notin L$, so $v \neq z$. Hence every element of $K \setminus \{v\}$ is contained in L and $z \in L$ is not contained in K. This gives that $Y_{u,z}(K) = Y_{u,v}(L)$, and so $|\mathcal{Z}_v(L) \cap \mathcal{Y}_u(K)| \geq 2$. □

Definition 6.3.6 Define the **clique graph** $C(n, k)$ of $J(n, k)$ by letting the vertices of $C(n, k)$ be the cliques of $J(n, k)$ of size $n - k + 1$ (so $\mathcal{Y}_u(K)$ if $n \neq 2k$ and all maximal cliques of $J(n, k)$ if $n = 2k$), and two cliques of $J(n, k)$ of size $n - k + 1$ are adjacent if and only if their intersection is a single vertex.

Lemma 6.3.7 *For $n \neq 2k$, $C(n, k) \cong J(n, k - 1)$, while if $n = 2k$, $C(n, k)$ has two components, each isomorphic to $J(n, k - 1)$.*

Proof Let n_c be the number of distinct cliques $\mathcal{Y}_u(M)$ in $J(n, k)$, where $M \in V_{n,k}$, and n_K to be the number of distinct cliques $\mathcal{Y}_u(L)$ that contain a fixed vertex K of $J(n, k)$. We count the number of pairs $(\mathcal{Y}_u(K), L)$, where $\mathcal{Y}_u(K)$ is a maximum clique and $L \in \mathcal{Y}_u(K)$ in two ways. We first choose $\mathcal{Y}_u(K)$ and then L, and obtain $n_c(n - k + 1)$. Choosing L first, and then $\mathcal{Y}_u(K)$, we obtain $\binom{n}{k} n_K$. Hence $n_c \cdot (n - k + 1) = \binom{n}{k} \cdot n_K$. Set $T_u(K) = \cap \mathcal{Y}_u(K) = K \setminus \{u\}$. Then $|T_u(K)| = k - 1$, and $\mathcal{Y}_u(K) = \{T_u(K) \cup \{v\} : v \notin K\}$. Thus $\mathcal{Y}_u(K)$ is uniquely determined by $T_u(K)$. Hence n_K is simply the number of $k - 1$ subsets of a set of size k, and $n_K = k$. Then,

$$n_c = \binom{n}{k-1} = |V_{n,k-1}| = |V(J(n, k-1))|.$$

Let G be the subgraph of $C(n, k)$ induced by all $\mathcal{Y}_u(K)$, $K \in V_{n,k}$ and $u \in K$. Hence if $n \neq 2k$, we have $G = C(n, k)$. We showed in the previous paragraph that $|V(G)| = |V(J(n, k - 1))|$, and now wish to show that $G \cong V(J(n, k - 1))$. Identify each clique $\mathcal{Y}_u(K)$ with the $(k-1)$-subset $T_u(K)$. We need to show that two distinct cliques $\mathcal{Y}_u(K)$ and $\mathcal{Y}_w(L)$ are adjacent in G if and only if $T_u(K)$ and $T_w(L)$ have intersection of size $k - 2$. Now, $\mathcal{Y}_u(K)$ and $\mathcal{Y}_w(L)$ are adjacent in G if and only if $\mathcal{Y}_u(K)$ and $\mathcal{Y}_w(L)$ have a common k-subset M by Lemma 6.3.4. This occurs if and only if $T_u(K)$ and $T_w(L)$ are both subsets of M, which occurs if and only if $T_u(K) \cup T_w(L) = M$ (as $T_u(K) \neq T_w(L)$, each is a $(k - 1)$-subset, and M has size k). This occurs if and only if $T_u(K)$ contains exactly one vertex not in $T_w(L)$ and $T_w(L)$ contains exactly one vertex not in $T_u(K)$. Hence $T_u(K) \cap T_w(L)$ is a $(k - 2)$-subset, and $G \cong J(n, k - 1)$. The result then follows if $n \neq 2k$. If $n = 2k$, then ι (recall $\iota\colon V_{n,k} \to V_{n,n-k}$ is given by $\iota(S) = [1, n] \setminus S$) maps G to the subgraph of $C(2k, k)$ induced by all $\mathcal{Z}_v(L)$, $L \in K_{2k,k}$, $v \notin L$, and there are no edges between any vertex $\mathcal{Y}_u(K)$ and $\mathcal{Z}_v(L)$, $K, L \in K_{2k,k}$, $u \in K$, $v \in L$ by Lemma 6.3.5. □

Theorem 6.3.8 *If n is odd or if n is even and $k \neq n/2$, then* $\mathrm{Aut}(J(n, k)) = \mathcal{S}_n$ *with its natural action on k-subsets. If n is even then* $\mathrm{Aut}(J(n, n/2)) = \mathcal{S}_n \times \langle \iota \rangle$.

Proof We may assume without loss of generality that $k \leq n/2$. We proceed by induction on $1 \leq k \leq n/2$. As $J(n, 1) = K_n$, the result is trivial if $k = 1$. Assume the result is true for $k - 1 \geq 1$. By Lemma 6.3.7, if $n \neq 2k$ the clique graph $C(n, k)$ of $J(n, k)$ is isomorphic to $J(n, k - 1)$, while if $n = 2k$, $C(n, k)$ has two components each isomorphic to $J(n, k - 1)$. If $n \neq 2k$, set $G = C(n, k)$, while if $n = 2k$ let G be the subgraph of $C(n, k)$ induced by all cliques of the form $\mathcal{Y}_u(K)$, $K \in V_{n,k}$, $u \in K$. We claim that $\mathrm{Aut}(J(n, k)) \cong \mathrm{Aut}(G)$, in which case $\mathrm{Aut}(G) \cong \mathrm{Aut}(J(n, k-1))$. If $2k \neq n$, then, by induction, $\mathrm{Aut}(J(n, k - 1)) = \mathcal{S}_n$, and for those values of k the result will follow by induction.

Any automorphism of $J(n, k)$ induces an automorphism of G as such an automorphism maps two cliques that intersect in one k-subset to two cliques that intersect in one k-subset. Also, the only automorphism of $J(n, k)$ that induces the identity on G fixes every clique of the form $\mathcal{Y}_u(K)$, $u \in K$ and $K \in V_{n,k}$, and so fixes the intersection of any two such cliques. As the intersection of two such cliques is either empty or an element of $V_{n,k}$ by Lemma 6.3.4, we see that an automorphism of $J(n, k)$ that induces the identity on G fixes every element of $V_{n,k}$. Thus there is an injective homomorphism from $\mathrm{Aut}(J(n, k))$ to $\mathrm{Aut}(G)$. If $n \neq 2k$, then $G = C(n, k)$ and, as we know that $\mathcal{S}_n$ in its action on k-subsets is contained in $\mathrm{Aut}(J(n, k))$, we see that $\mathrm{Aut}(C(n, k)) = \mathrm{Aut}(J(n, k-1)) = \mathcal{S}_n$ and the result follows. If $n = 2k$, then clearly the action of $\mathcal{S}_n$ as a subgroup of $J(2k, k)$ permutes the cliques of the form $\mathcal{Y}_u(K)$, and so we have $\mathrm{Aut}(G) = \mathcal{S}_n$.

Now, as $C(n, k) = \bar{K}_2 \wr G$, $\mathrm{Aut}(C(n, k)) = \mathbb{Z}_2 \wr \mathrm{Aut}(G) = \mathbb{Z}_2 \wr \mathcal{S}_n$. Let $\gamma \in \mathrm{Aut}(C(n, k))$ that is the induced action of an automorphism of $J(2k, k)$. We may assume without loss of generality that γ fixes each component of $C(n, k)$, as $\iota \in \mathrm{Aut}(J(2k, k))$ permutes these components. As $\mathcal{S}_n \leq \mathrm{Aut}(J(2k, k))$ induces every automorphism of G, we may also assume without loss of generality that γ fixes every vertex of G. But then γ fixes every clique of the form $\mathcal{Y}_u(K)$, $u \in K$, $K \in V_{n,k}$, and so fixes the intersection of any two such cliques. As above, we conclude that γ is the identity on $K_{2k,k} = V(J(2k, k))$. Thus every such γ is contained in the group $\langle \mathcal{S}_n, \iota \rangle$. That this group is isomorphic to $\mathcal{S}_n \times \langle \iota \rangle$ is left as Exercise 6.3.3. □

We remark that Ramras and Donovan (2011) recently gave a different elementary proof showing that $\mathrm{Aut}(J(n, k)) = \mathcal{S}_n$ when $n \neq 2k$.

6.3.2 Johnson Graphs That are Cayley Graphs

Observe that unless n is even and $k = n/2$, then the automorphism group of a Johnson graph $J(n, k) \cong J(n, n-k)$, $k < n$, is the same as the automorphism group of the Kneser graph $K(n, k)$. Hence, for these values of n and k, a Johnson graph $J(n, k)$ is a Cayley graph if and only if $K(n, k)$ is a Cayley graph. Thus Theorem 6.2.4 also tells us which Johnson graphs are Cayley graphs if $k \neq n/2$.

In the case where n is even and $k = n/2$, Theorem 6.2.4 tells us nothing. Our goal now is to determine which Johnson graphs $J(2k, k)$ are isomorphic to Cayley graphs. If $k = 1$, then $J(2, 1)$ is simply an edge and so is a Cayley graph. If $k = 2$, $J(4, 2)$ has order 6, and every vertex-transitive graph of order 6 is isomorphic to a Cayley graph. The arguments for $k \geq 3$ will take considerably more effort.

Our strategy though is clear. As $\langle \iota \rangle \trianglelefteq \mathrm{Aut}(J(2k, k))$, the set of its orbits is a block system $\mathcal{P}$. Any regular subgroup of $\mathrm{Aut}(J(2k, k))$ will either be a regular subgroup or a transitive subgroup with stabilizer of order 2 of $\mathrm{Aut}(J(2k, k))/\mathcal{P}$.

Hence in either case, $\mathrm{Aut}(J(2k,k))/\mathcal{P}$ will have a transitive subgroup of "small" order. We will show that this rarely happens. To do this, we will generalize the result of Beaumont and Peterson (1955) that Godsil used to determine which Kneser graphs are Cayley, and which groups are transitive on the set of all partitions of a particular "shape." Here, the "shape" of a partition is simply the sizes of the sets contained in it. We remark that which groups are transitive on partitions of any fixed shape have been determined in Dobson and Malnič (2015) (from which our proof is extracted), and independently in André et al. (2016). Our first task will be to determine the shape of the partition we will need. We begin with the following definition.

Definition 6.3.9 Let $k \geq 2$. A **folded Johnson graph**, denoted $\bar{J}(2k,k)$, has vertex set the set of partitions of $[1,2k]$ into two disjoint k-subsets, and two such partitions are adjacent if and only if they have a common refinement with sets of size $1, k-1, k-1$, and 1.

Observe that if $k = 2$ or 3, then $\bar{J}(2k,k)$ is a complete graph. Note that $\bar{J}(2k,k)$ has $\frac{1}{2}\binom{2k}{k}$ vertices. The automorphism group of a folded Johnson graph is $\mathrm{Aut}(J(2k,k))/\mathcal{P}$!

Theorem 6.3.10 *Let $k \geq 4$. The set of orbits of $\langle\iota\rangle$ is a block system $\mathcal{P}$ of* $\mathrm{Aut}(J(2k,k))$*, and each block is a set of size k together with its complement. Also, $J(2k,k)/\mathcal{P} = \bar{J}(2k,k)$.*

Proof Clearly, the vertex set of $J(2k,k)/\mathcal{P}$ is the vertex set of $\bar{J}(2k,k)$. By Theorem 6.3.8, we see $\langle\iota\rangle \trianglelefteq \mathrm{Aut}(J(2k,k))$, and so the set of its orbits is a block system $\mathcal{P}$ of $\mathrm{Aut}(J(2k,k))$ by Theorem 2.2.7. As ι maps a set to its complement, each block of $\mathcal{P}$ is a set of size k and its complement, which is also of size k. Hence each block of $\mathcal{P}$ contains a unique set that contains the symbol $2k$. We may thus identify a block $B \in \mathcal{P}$ uniquely by specifying the set of size k that contains $2k$. Additionally, every set of size k that contains $2k$ can be written $\{2k\} \cup A$, where $A \subset [1,2k-1]$. It remains to show that $E(J(2k,k)/\mathcal{P}) = E\left(\bar{J}(2k,k)\right)$.

The block of $\mathcal{P}$ identified with $\{2k\} \cup A$ is adjacent in $\bar{J}(2k,k)$ to the block of $\mathcal{P}$ identified with $\{2k\} \cup A'$, where $A, A' \subset [1,2k-1]$, if and only if either $|(\{2k\} \cup A) \cap (\{2k\} \cup A')| = k-1$, or $\left|(\{2k\} \cup A) \cap \left(\overline{\{2k\} \cup A'}\right)\right| = k-1$, where $\overline{\{2k\} \cup A'}$ is the complement of $\{2k\} \cup A'$.

In the former case, $|(\{2k\} \cup A) \cap (\{2k\} \cup A')| = k-1$ if and only if $|A \cap A'| = k-2$. Set $B = (A \cap A') \cup \{2k\}$, $C = [1,2k-1]\backslash(A \cup A')$, $D = A\backslash B$ and $E = A'\backslash B$. A straightforward argument will show that $\{B,C,D,E\}$ is a partition of $[1,2k]$ (Exercise 6.3.4), and that it refines both partitions of $[1,2k]$ consisting of $\{2k\} \cup A$ and its complement and $\{2k\} \cup A'$ and its complement (Exercise 6.3.5). Then $|A \cap A'| = k - 2$ if and only if $D = \{x\}$ and $E = \{y\}$ for some $x, y \in [1, 2k]$, which occurs if and only if $\{B, C, D, E\}$ is a common refinement of the blocks of $\mathcal{P}$ identified with $\{2k\} \cup A$ and $\{2k\} \cup A'$.

In the latter case, as $2k \notin \overline{\{2k\} \cup A'}$, $A \subset \overline{\{2k\} \cup A'}$, with $\overline{\{2k\} \cup A'} = A \cup \{x\}$, for some $x \in [1, 2k-1]$. Hence $\{\{2k\}, \{x\}, A, A'\}$ is a partition of $[1, 2k]$. Then $x \notin A$ so that $x \notin \{2k\} \cup A$. We conclude that $\overline{\{2k\} \cup A} = [1, 2k] \setminus (A \cup \{2k\}) = A' \cup \{x\}$. Then $\{\{2k\}, \{x\}, A, A'\}$ is a refinement of the two partitions consisting of $\{2k\} \cup A$ and its complement and $\{2k\} \cup A'$ and its complement. Thus $E(J(2k, k)/\mathcal{P}) = E(\bar{J}(2k, k))$. □

Definition 6.3.11 Let $G \leq S_n$, and $1 \leq k \leq n$. We say G is **k-set-transitive**, or **k-homogeneous**, if G is transitive on the set $V_{n,k}$ of all subsets of $[1, n]$ of size k. We say G is **set-transitive**, or **homogeneous**, if G is k-set-transitive for every $1 \leq k \leq n$.

So a 1-set-transitive group is just a transitive group, and a k-transitive group is k-set-transitive. The notion of k-set-transitivity is weaker than that of k-transitivity in general. Usually, only values $1 \leq k \leq \lfloor n/2 \rfloor$, are considered, as it should be clear that a group is k-set-transitive if and only if it is $(n - k)$-set transitive for $k \neq n$. Beaumont and Peterson (1955) determined all set-transitive groups (see also Livingstone and Wagner (1965)). We shall have need of an intermediate result (Beaumont and Peterson, 1955, Theorem 11).

Theorem 6.3.12 *Let $G \leq S_n$. If G is $\lfloor n/2 \rfloor$-set-transitive, then $G \cong A_n, S_n$ or* AGL(1, 5), PGL(2, 5), PGL(2, 8) *or* PΓL(2, 8) *with $n = 5, 6, 9,$ and 9, respectively. These groups are all set-transitive.*

We shall have need of the following number-theoretic result that is proved in the discussion following (Beaumont and Peterson, 1955, Theorem 8).

Lemma 6.3.13 *Unless $2k = 4, 6, 10, 14,$ or 16, there exists a prime p such that $k < p < 4k/3$.*

Let $\mathcal{P}$ be the set of partitions of $[1, 2k]$ into two disjoint k-subsets. We denote an element of $\mathcal{P}$ by $\left(K, \bar{K}\right)$, where $K \in K_{2k,k}$. Then S_{2k} is transitive on $\mathcal{P}$. Over the course of a few results, we will determine all subgroups of S_{2k} transitive on $\mathcal{P}$.

Theorem 6.3.14 *If $G \leq S_{2k}$, $k \geq 3$, is transitive on $\mathcal{P}$ then one of the following is true:*

(i) *G is transitive on $[1, 2k]$,*
(ii) *$G \cong A_{2k-1}$ or S_{2k-1} and fixes a point,*

(iii) $k = 3$, $G \cong \mathrm{AGL}(1,5) \leq \mathcal{S}_6$ *and fixes a point, or*

(iv) $k = 5$, *and either* $G \cong \mathrm{PGL}(2,8) \leq \mathcal{S}_{10}$ *or* $G \cong \mathrm{P\Gamma L}(2,8) \leq \mathcal{S}_{10}$ *and fixes a point.*

Proof Suppose that G is intransitive on $[1,2k]$. By Lemma 6.3.13, there exists a prime divisor p of $\binom{2k}{k}$ such that $k < p < 4k/3$ unless $2k = 4, 6, 10, 14$, or 16. As p is odd, p also divides $\frac{1}{2}\binom{2k}{k}$. Then $5 \leq p < \frac{1}{2}\binom{2k}{k}$. For $2k = 6, 10, 14$, or 16, one can check that there exists a prime divisor $p \geq 5$ of $\frac{1}{2}\binom{2k}{k}$ that is strictly less than $\frac{1}{2}\binom{2k}{k}$. Thus for all values of k as in the hypothesis, there exists a prime divisor $p \geq 5$ of $\frac{1}{2}\binom{2k}{k}$ that is strictly less than $\frac{1}{2}\binom{2k}{k}$. As G is transitive on $\mathcal{P}$, p divides $|G|$, and so G contains an element of order p. We conclude that some orbit O of G in its action on $[1,2k]$ has size $r \geq p$.

If $r < 2k$ is even then there are two elements of $[1,2k]$ that are not in O. Partition O into two subsets L and M such that $|L| = r/2$ and $|M| = r/2$, and let K_1 be any k-subset of $[1,2k]$ that contains L and no element of M is contained in K_1. Choose $x \in L$, $y \in \bar{K}_1$ such that $y \notin M$, and set $K_2 = (K_1 \backslash \{x\}) \cup \{y\}$. Then K_2 contains $r/2 - 1$ elements of O while $\bar{K}_2$ contains $r/2 + 1$ elements of O. As the image of any set under any element of G cannot change the number of elements of O contained in that set, we see there is no element of G that maps $(K_1, \bar{K}_1)$ to $(K_2, \bar{K}_2)$, a contradiction.

If $r < 2k - 1$ is odd, partition O into two subsets L and M, where $|L| = (r+3)/2$ and $|M| = (r-3)/2$. Observe that as $r \geq 5$, $(r-3)/2 \geq 1$ and as $r < 2k - 1$, $(r+3)/2 \leq k$. Let $K_1 \in V_{2k,k}$ that contains L and no elements of M. Then K_1 contains $(r+3)/2$ elements of O, while $\bar{K}_1$ contains $(r-3)/2$ elements of O. Let $x \in L$, and $y \in [1,2k]$ such that $y \notin O$. Set $K_2 = (K_1 \setminus \{x\}) \cup \{y\}$. Then K_2 contains $(r+1)/2$ elements of O while $\bar{K}_2$ contains $(r-1)/2$ elements of O. As applying an element of G to a set cannot change the number of elements of O that it contains, there is no $g \in G$ with $g(K_1, \bar{K}_1) = (K_2, \bar{K}_2)$. Hence G is not transitive on $\mathcal{P}$. So if r is odd, then $r = 2k - 1$.

Finally, if $r = 2k - 1$, then let $x \in [1,2k]$ such that $x \notin O$. Given $(K_1, \bar{K}_1)$ and $(K_2, \bar{K}_2) \in \mathcal{P}$, with $x \in K_1 \cap K_2$, the permutation in $\mathcal{S}_{2k}$ that fixes x and maps $K_1 \setminus \{x\}$ to $K_2 \setminus \{x\}$, also maps K_1 to K_2, and consequently maps $\bar{K}_1$ to $\bar{K}_2$. Hence any subgroup of $\mathcal{S}_{2k-1}$ that is $(k-1)$-set-transitive on $[1,2k]\backslash\{x\}$ will be transitive on $\mathcal{P}$. Conversely, an element of G must map any $k-1$ elements of $[1,2k]\backslash\{x\}$ to any $k-1$ elements of $[1,2k]\backslash\{x\}$, and so if G is transitive on $\mathcal{P}$ but is not transitive on $[1,2k]$, then it is contained in $\mathcal{S}_{2k-1}$ and is $(k-1)$-set-transitive. The result then follows by Theorem 6.3.12. □

We next further restrict G when it is transitive on $\mathcal{P}$ and $[1,2k]$.

Lemma 6.3.15 *If $G \leq \mathcal{S}_{2k}$ is transitive on $\mathcal{P}$ and is also transitive on* $[1, 2k]$, *then G is primitive on* $[1, 2k]$.

Proof Suppose otherwise, and let $\mathcal{B}$ be a nontrivial block system of G with ℓ blocks of size m (so $\ell m = 2k$).

If ℓ is even, then $\ell/2$ is an integer. Let K_1 be the union of $\ell/2$ blocks of $\mathcal{B}$, and so $\bar{K_1}$ is also the union of $\ell/2$ blocks of $\mathcal{B}$. Choose $x \in K_1$ and $y \in \bar{K_1}$, and set $K_2 = (K_1 \backslash \{x\}) \cup \{y\}$. Then K_2 is not a union of blocks of $\mathcal{B}$. There can then be no element of G that maps $\left(K_1, \bar{K_1}\right)$ to $\left(K_2, \bar{K_2}\right)$ as every element of G maps blocks to blocks, a contradiction.

If ℓ is odd, then m is even, and as $\mathcal{B}$ is nontrivial, $\ell \geq 3$. Choose $B_1, B_2, B_3 \in \mathcal{B}$ distinct. Observe that $2k = m(\ell-1)/2 + m/2$. Let K_1 be the k-subset of $[1, 2k]$ that contains half of the elements of B_1, all of B_2, and the remaining elements are chosen as a union of blocks of $\mathcal{B}$ distinct from B_3. Let $x \in B_2 \cap K_1$ and $y \in B_3 \subset \bar{K_1}$. Let $K_2 = (K_1 \backslash \{x\}) \cup \{y\}$. Then K_2 and $\bar{K_2}$ both contain a different number of blocks of $\mathcal{B}$ than K_1 and $\bar{K_1}$, and so no element of G can map $\left(K_1, \bar{K_1}\right)$ to $\left(K_2, \bar{K_2}\right)$, a contradiction. □

The proof of the following result is a modified version of the proof used by Beaumont and Peterson (1955).

Theorem 6.3.16 *If $G \leq \mathcal{S}_{2k}$ is transitive on* $[1, 2k]$, $k \geq 4$, *and is transitive on $\mathcal{P}$, then $G = A_{2k}$ or $\mathcal{S}_{2k}$.*

Proof By Lemma 6.3.15 we see that G is primitive. By Lemma 6.3.13, unless $2k = 10, 14$, or 16, there exists a prime p such that $k < p < 4k/3$. Let $p = 13$, if $2k = 16$, $p = 11$ if $2k = 14$, and $p = 7$ if $2k = 10$. Then p divides $\binom{2k}{k}$ and so p divides $\frac{1}{2}\binom{2k}{k}$. As G is transitive on $\mathcal{P}$, p divides $|G|$. Hence there exists an element $\delta \in G$ of order p. In $\mathcal{S}_{2k}$, δ must be a p-cycle as $p > k$. We conclude that $\langle\delta\rangle$ is primitive on its nontrivial orbit of prime size p, and fixes point-wise the other $2k - p$ points. It then follows by Burnside (1897, Theorem I, p. 199) that G is $(2k - p + 1)$-transitive.

If $k < p < 4k/3$, then G is at least $2k/3 + 1$ transitive. By Burnside (1897, p. 152), any group of degree n that is at least $(n/3 + 1)$-transitive must contain A_n. Thus $G = A_{2k}$ or $\mathcal{S}_{2k}$. We thus need only consider when $2k = 10$, 14, or 16.

Suppose $2k = 10, 14$, or 16. By Miller (1915), if $2k = \ell q + r$, where $q > \ell$ is prime and $r > \ell$, then a subgroup of $\mathcal{S}_{2k}$ not containing A_{2k} can be at most r-transitive unless $\ell = 1$ and $r = 2$. Writing $16 = 1 \cdot 13 + 3$, $14 = 1 \cdot 11 + 3$, and $10 = 1 \cdot 7 + 3$, we see either the result follows or G is at most triply transitive. As G is also $(2k - p + 1)$-transitive, we see that G is 4-transitive, a contradiction. □

Lemma 6.3.17 *Let $G \leq \mathcal{S}_6$ be transitive on* $[1, 6]$. *Then G is transitive on $\mathcal{P}$ if and only if $G \cong A_5$ or $\mathcal{S}_5$ of degree* 6, *or A_6 or $\mathcal{S}_6$.*

Proof Suppose $G \leq \mathcal{S}_6$ is transitive on $\mathcal{P}$. As $\frac{1}{2}\binom{6}{3} = 10$, we see that 5 divides $|G|$. Let $\delta \in G$ be of order 5. If G is a transitive subgroup of $\mathcal{S}_6$, then G is 2-transitive as G contains a 5-cycle (so the stabilizer of a point is transitive on the remaining points), and the only 2-transitive group of degree 6 that is not A_6 or $\mathcal{S}_6$ is A_5 or $\mathcal{S}_5$ of degree 6 by Cameron (1981, Theorem 5.3) (or see Dixon and Mortimer (1996, Appendix B)).

Clearly A_6 and $\mathcal{S}_6$ are transitive on $\mathcal{P}$ as they are 3-transitive. Additionally, it can be verified by hand that A_5 of degree 6 is also transitive on $\mathcal{P}$. We remark that generators of A_5 of degree 6 can be found in Beaumont and Peterson (1955), where this group is called H_2 (one can then list every single element of H_2 using a computer algebra system and manually check that H_2 is transitive on $\mathcal{P}$). Thus $\mathcal{S}_5$ is also transitive on $\mathcal{P}$. □

Combining Theorem 6.3.14, Theorem 6.3.16, and Lemma 6.3.17, we have the following result.

Theorem 6.3.18 *Let $k \geq 3$ and $G \leq \mathcal{S}_{2k}$ be transitive on $\mathcal{P}$. Then one of the following is true:*

(i) $k = 3$, $G \cong A_5$ *or* $\mathcal{S}_5$ *of degree* 6, *or* $G \cong \mathrm{AGL}(1, 5)$ *fixes a point in* $\mathcal{S}_6$,
(ii) $k = 5$ *and* $G \cong \mathrm{PGL}(2, 8)$, *or* $\mathrm{P\Gamma L}(2, 8)$ *and fixes a point in* $\mathcal{S}_{10}$,
(iii) $G = A_{2k-1}$ *or* $\mathcal{S}_{2k-1}$ *and fixes a point in* $\mathcal{S}_{2k}$, *or*
(iv) $G = A_{2k}$ *or* $\mathcal{S}_{2k}$.

Corollary 6.3.19 *$\bar{J}(2k, k)$ is isomorphic to a Cayley graph if and only if $k \leq$ 3. Additionally, if $k \geq 4$ and $k \neq 5$, then the only transitive subgroups of* $\mathrm{Aut}\left(\bar{J}(2k, k)\right)$ *are $\mathcal{S}_{2k}$, A_{2k}, $\mathcal{S}_{2k-1}$, and A_{2k-1}.*

Proof If $k \leq 3$, then $\bar{J}(2k, k)$ is a complete graph and so clearly a Cayley graph. If $k \geq 4$, then by Theorem 6.3.10 and Theorem 6.3.8 we have $\mathrm{Aut}\left(\bar{K}(2k, k)\right) = \mathcal{S}_{2k}$ in its natural action on $\mathcal{P}$. If $k = 5$, then $\left|V\left(\bar{J}(2k, k)\right)\right| =$ 126, and by Theorem 6.3.18, $\mathcal{S}_{2k}$ acting on $\mathcal{P}$ contains no regular subgroups as none of 252, 512, 10!, 10!/2, 9!, or 9!/2 are equal to 126. The result then follows by Theorem 6.3.18. □

Corollary 6.3.20 *$J(2k, k)$ is isomorphic to a Cayley graph if and only if $k \leq 2$. Additionally, if $k \geq 4$, then the only transitive subgroups of* $\mathrm{Aut}(J(2k, k))$ *are of index* 1, 2, *or* 4.

Proof As $J(2, 1) = K_2$, $J(2, 1)$ is Cayley, while $J(4, 2)$ is the complement of a 1-factor, and thus has automorphism group $\mathcal{S}_3 \wr \mathcal{S}_2$, which contains a regular cyclic subgroup. Suppose $k \geq 3$ and $\mathrm{Aut}(J(2k,k)) = \mathcal{S}_{2k} \times \langle \iota \rangle$ contains a regular subgroup R. Then $\mathcal{P}$ is a block system of $\mathrm{Aut}(J(2k,k))$, and $\mathcal{S}_{2k} \cong \mathrm{Aut}(J(2k,k))/\mathcal{P}$. Additionally, R is then a transitive subgroup of $\mathcal{S}_{2k}$ in its action on $\mathcal{P}$ of order r or $2r$, where $r = \frac{1}{2}|R| = \frac{1}{2}\binom{2k}{k}$. If $k = 3$, then $r = 10$ and $2r = 20$, and by Theorem 6.3.18 the only transitive subgroup of $\mathcal{S}_6$ that is transitive on $\mathcal{P}$ of order 10 or 20 is $\mathrm{AGL}(1,5)$, which is not a transitive subgroup of $\mathcal{S}_6$. Thus $J(6,3)$ is not Cayley. We henceforth assume $k \geq 4$.

By Theorem 6.3.18, the only subgroups of $\mathcal{S}_{2k}$ transitive on both $[1,2k]$ and $\mathcal{P}$ are A_{2k} and $\mathcal{S}_{2k}$, and order arguments easily show neither of the groups have order r or $2r$. As $\mathrm{fix}_{\mathrm{Aut}(J(2k,k))}(\mathcal{B}) = \langle \iota \rangle$ is of order 2, we see that for any transitive subgroup $T \leq \mathcal{S}_{2k} \times \langle \iota \rangle = \mathrm{Aut}(J(2k,k))$ has index dividing 4. □

It turns out that Johnson graphs are not just Hamiltonian, but by a recent result of Alspach are Hamilton connected. The interested reader should see Alspach (2013) for the short proof of this fact.

There are a variety of graphs similar to the Johnson graphs that are considered in the literature. First, many authors do not insist that in order for there to be an edge between two k-subsets of $[1,n]$ that they intersect in $k-1$ points, and introduce a third parameter $s \leq k-1$ that gives the size of the intersection of two k-subsets when there is an edge between them. For example, $J(n,k,3)$ would be the graph with vertex set the k-subsets of $[1,n]$ and there is an edge between two k-subsets if and only if their intersection has three elements. Note then that $J(n,k,0) = K(n,k)$, a Kneser graph. The reader interested in Johnson graphs defined in this way is referred to Godsil and Royle (2001) (also see Jones and Jajcay (2016)).

Another class of graphs are the **Grassman graphs**. Instead of vertices being k-subsets of an n-element set as with the Johnson graphs, the Grassman graphs have vertex sets the k-dimensional subspaces of an n-dimensional vector space over a field $\mathbb{F}$. Two k-dimensional subspaces are adjacent if their intersection is a $(k-1)$-dimensional subspace. The interested reader is referred to Brouwer et al. (1989, Section 9.3) for more information on the Grassman graphs.

6.4 Levi Graphs

In this section, we examine a technique for using the automorphism groups of other combinatorial objects to calculate the automorphism groups of vertex-transitive graphs. This technique is widely used for some graphs, especially for the graphs $B(\mathrm{PG}(d-1,q))$ in Section 4.5. We will consider this topic in fairly exhaustive detail as, while the technique is widely used, the details of the technique are not, as far as the authors know, contained in any readily accessible reference in algebraic graph theory. There is, however, an excellent treatment of the relationship between graphs and configurations in Pisanski and Servatius (2013).

A word of warning. Levi graphs arise in the literature in a wide variety of situations, and relationships between these situations are not typically spelled out. Our intention here is not simply to get to the desired results that we wish to present with the shortest presentation, but to help prepare the reader to extract information from the literature as well. As a consequence, this section is somewhat heavy on terminology, and it will take us some time to get to our desired results.

Like most good ideas, Levi graphs appear in a wide range of areas in combinatorics, often with different terminology. Rather than immediately presenting the most general form of Levi graphs (these are called **incidence graphs**), we will begin with the first area where Levi graphs appeared. This was in a seminal paper of Coxeter (1950), building on work of Levi (1942). Coxeter was interested in Levi graphs as they provided a unique way of representing a **configuration**, which we now define.

We have already seen some examples of Levi graphs – the Heawood graph is one for example, as are all of the graphs $B(\mathrm{PG}(d-1,q))$ in Section 4.5.

Definition 6.4.1 A **configuration** is an ordered triple $(P, L, \mathcal{I})$, where P is a set whose elements are called **points**, L is a set of size d whose elements are called **lines**, and $\mathcal{I}$ is an **incidence relation** (or subset of $P \times L$). If $(p, \ell) \in \mathcal{I}$, we say the point p is on, or incident with, the line ℓ or the line ℓ contains, or is incident with, the point p. Additionally, the following conditions must hold:

- each of the points is on exactly q lines,
- each line contains exactly k points, and
- two points are on at most one line and two lines contain at most one point in common.

Example 6.4.2 Any regular simple graph Γ is a configuration with points the vertices of Γ and lines the edges. A point and line are incident if the point is the endpoint of line (edge), and vice versa.

Proof As Γ is simple, two points are on at most one line, and two lines are incident with at most one point. Again, as Γ is simple, any two points are on at most one line, and each line contains exactly two points (or $k = 2$). Finally, as Γ is regular, each point is incident with exactly q lines, where q is the valency of Γ. □

Definition 6.4.3 Let P have size p. A configuration with the parameters p, q, d, k, with q, d, k, as above, will be called a (p_q, d_k)**-configuration** and if p and d are not important, a $[q, k]$-configuration.

It is common for configurations to be drawn as a diagram in the plane. When this is done, the points are a set of points in the plane and the lines are a set of lines in the plane. The incidence structure is the obvious implicit one: A point is incident to a line that contains it and a line is incident with the points it contains. One difference though, is that not all of the points in the plane that are on lines are points of the configuration, and those points are usually clear from the drawing.

Example 6.4.4 The Fano plane in Figure 6.1 is a configuration, where the points are the set $\{p_i : 1 \leq i \leq 7\}$ and lines are $\{\ell_i : 1 \leq i \leq 7\}$ (so our notation here is consistent with our previous notation). The Fano plane is then a $(7_3, 7_3)$-configuration.

Solution Recall a point is a one-dimensional subspace and a line is a hyperplane. Each point is contained in exactly three lines, so $q = 3$, and each line contains three points, so $d = 3$. □

Example 6.4.5 The configuration in Figure 6.7 is a $(10_3, 10_3)$ configuration, known as the Desargues configuration.

Solution In the figure, points are vertices and lines are straight line segments that contain three vertices. We do not extend the lines infinitely (although we could have if we wanted). By inspection, there are 10 points, 10 lines, each point is on 3 lines, and each line contains 3 points. So the Desargues configuration is a $(10_3, 10_3)$-configuration. □

Obviously, we are thinking of configurations as being generalizations of planes (from where our natural conditions on an incidence relation are derived). This definition of a configuration is taken from Grünbaum (2009), where the interested reader is referred to for much more information on configurations. We remark that Grünbaum calls our configurations "combinatorial configurations," and that other types of configurations exist.

Definition 6.4.6 Given a configuration C we construct a (necessarily bipartite) graph Γ_C as follows. The vertices of Γ_C are the set $P \cup L$, and a point is adjacent to a line if and only if the point is on the line. We call Γ_C the **Levi graph** of C, or the **point-line incidence graph** of C.

As points are not on points and lines are not on lines, Γ_C is bipartite with P and L the natural bipartition sets.

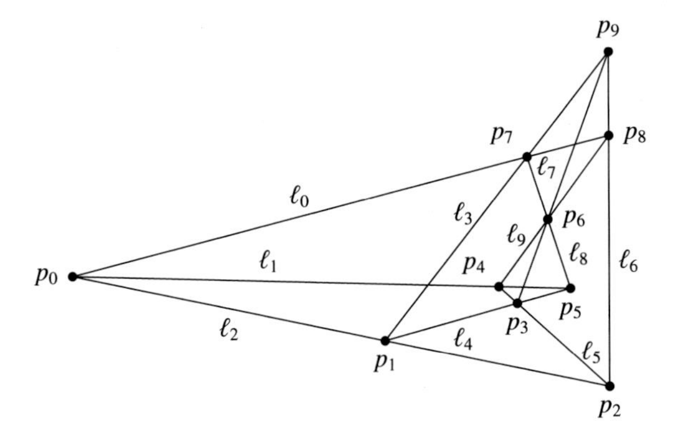

Figure 6.7 The Desargues configuration.

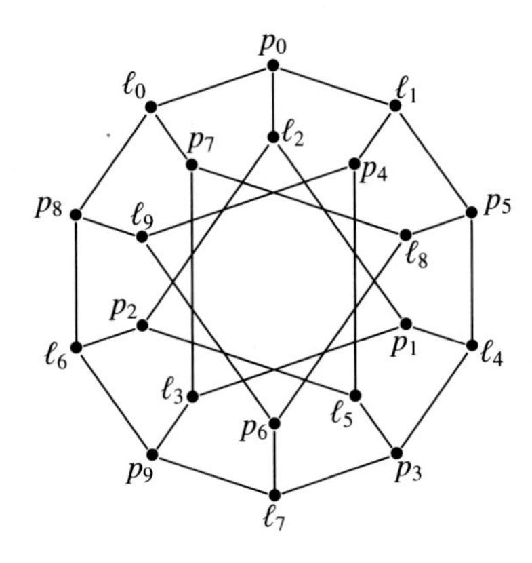

Figure 6.8 The Levi graph of the Desargues configuration.

Example 6.4.7 The Heawood graph is the Levi graph of the Fano plane.

Example 6.4.8 The Levi graph of the Desargues configuration is shown in Figure 6.8. This graph is called the Desargues graph and is F20B in the Foster census.

Typically, we think of the vertices of Γ_C corresponding to points as "white" vertices while the vertices of Γ_C corresponding to lines as "black" vertices. The relationship between bipartite graphs and $[q,k]$-configurations is given in the following result.

Theorem 6.4.9 *A bipartite graph Γ is isomorphic to the Levi graph Γ_C of a $[q,k]$-configuration C if and only if*

- *all black vertices have valency k while all white vertices have valency q,*
- *Γ_C has girth at least* 6.

Proof If Γ is the Levi graph of a $[q,k]$-configuration C, then clearly Γ is bipartite, and so contains no odd cycles. A 4-cycle in Γ would correspond to a point incident with two common lines that are incident to another common point. However, two points can be incident with at most one line, and vice versa, so Γ does not contain a 4-cycle. Thus Γ has girth at least 6. As each point is incident with exactly k lines and each line is incident with exactly q points, we see that all the black vertices have valency k while all the white vertices have valency q.

Conversely, let Γ be a bipartite graph of girth at least 6. Let all vertices in one cell of the bipartition of Γ be "white" while all of the vertices in the other cell are "black." Our hypothesis then implies that all black vertices have valency k while all white vertices have valency q. We define an incidence relation in the natural way, by letting a white vertex w (a "point") be incident with a black vertex b (a "line") if there is an edge between w and b in Γ. To establish that this relation is an incidence relation, we note that this relation is symmetric as Γ is a graph, not a digraph. Also, incidence only involves a point and a line by definition, and two points can be incident with at most one line and two lines are incident with at most one point as Γ has girth at least 6. Thus our notion of incidence does give an incidence relation, and consequently a configuration C. Finally, as each white vertex is adjacent to q black vertices, each point is on exactly k lines, while as each black vertex is adjacent to k white vertices, we see each line contains exactly k points. Thus C is a $[q,k]$-configuration. □

We now turn to "automorphisms" of configurations, and note before we start that the standard terminology for configurations is (naturally) inconsistent with that of vertex-transitive graphs.

Definition 6.4.10 Define a **symmetry** of a configuration to be a bijective mapping $f\colon P \cup L \to P \cup L$ such that p is incident with ℓ if and only if $f(p)$ is incident with $f(\ell)$, for every point p and line ℓ. The set of all symmetries of a configuration C is a group under function composition, the **group of symmetries** of C.

Thus a "symmetry" of a configuration is what we would normally call an "automorphism." While in geometry it is not possible for a symmetry to map a point to a line or vice versa, in a configuration it is indeed possible, and so we distinguish two types of symmetries, those that map points to points and lines to lines, and those map points to lines and vice versa.

Definition 6.4.11 A symmetry of a configuration C that maps points to points and lines to lines is called an **automorphism of** C, while a symmetry that maps points to lines and vice versa is called a **duality**. The set of all automorphisms of C is group under function composition, called the **automorphism group of** C, and is denoted by $\mathrm{Aut}(C)$.

We are now in a position to state and prove one of the main results of this section, which shows that the group of symmetries of a configuration is the same as the automorphism group of its Levi graph.

Theorem 6.4.12 *The group of symmetries of a configuration C is equal to the automorphism group of its Levi graph Γ_C.*

Proof As the vertex set of Γ_C is $P \cup L$ and a symmetry of a configuration is a permutation of $P \cup L$ as well, we see that the group of symmetries G of a configuration and $\mathrm{Aut}(\Gamma_C)$ are contained in the same symmetric group. Now, $g \in G$ if and only if whenever $p \in P$ and $\ell \in L$ such that p is incident to ℓ, then $g(p)$ is incident to $g(\ell)$ (note that we are not assuming that $g(p)$ is a point and that $g(\ell)$ is a line). This occurs if and only if whenever $p \in P$ and $\ell \in L$ such that $p\ell \in E(\Gamma_C)$, we have $g(p)g(\ell) \in E(\Gamma_C)$. Thus $g \in G$ if and only if $g \in \mathrm{Aut}(\Gamma_C)$ and the result follows. □

In many cases we can do a bit better than the previous result. In order to state this improvement, we will need one more natural term.

Definition 6.4.13 A configuration is **connected** if and only if its corresponding Levi graph is connected.

Lemma 6.4.14 *In a connected configuration C, every symmetry of C is either an automorphism or a duality.*

Proof Suppose otherwise, and let C be a connected configuration such that some symmetry f of C is neither an automorphism nor a duality. Then there is a symmetry f of C that maps some point u to some point v while also mapping some point w to some line ℓ. Let P be a (u, ℓ)-path in Γ_C. As P is a path in a bipartite graph from a white vertex to a black vertex, it has odd length. However, $f(P)$ is a $(f(u), f(\ell))$-path in Γ_C from a white vertex to a white vertex, and so has even length, a contradiction. □

Corollary 6.4.15 *In a connected configuration C with a duality of order* 2, $\mathrm{Aut}(\Gamma_C) \cong \mathrm{Aut}(C) \rtimes \mathbb{Z}_2$.

Proof By Lemma 6.4.14, every symmetry of C is either an automorphism or a duality. It is easy to see that $\mathrm{Aut}(C) \trianglelefteq G$, where G is the group of symmetries of C, and $G/\text{Aut}(C)$ has order dividing 2. As C has a duality τ of order 2, we have $G/\text{Aut}(C)$ has order 2, $G = \langle\tau\rangle\text{Aut}(C)$, and $\langle\tau\rangle \cap \text{Aut}(C) = 1$. As $\langle\tau\rangle \cong \mathbb{Z}_2$, we see that $\text{Aut}(\Gamma_C) \cong \text{Aut}(C) \rtimes \mathbb{Z}_2$ by Theorem 6.4.12. □

We remark that not every duality of a configuration is of order 2. The Foster census gives that F40 is a cubic bipartite graph of girth 8, and so by Theorem 6.4.9 it is the Levi graph of a configuration. Conder, Estélyi, and Pisanski have shown that the graph F40 has a duality of order 4 (Conder et al., 2018).

Recalling from Section 4.5 that the automorphism group of the Heawood graph is $\text{PSL}(3,2) \rtimes \mathbb{Z}_2$, it is very suggestive that the preceding result is the result we need to calculate the automorphism group of the Heawood graph by viewing it as a Levi graph (some additional information is required though, which we will see below). This is correct, but the preceding results will, at most, deal with the graphs $B(\text{PG}(2,q))$. In order to consider the case where $d \geq 4$, i.e., $B(\text{PG}(d-1,q))$, we will need a new combinatorial object, which can be viewed as a generalization of configurations. This is because when $d \geq 4$, a hyperplane has dimension three, and so more than one hyperplane contains each pair of different points.

Before turning to this generalization of configurations, let us mention again that there are more generalizations of Levi graphs of configurations. That there is a great deal of similarity between these ideas will be apparent as we continue. Many of the proofs we will see are very similar to proofs we have seen (some, but not all, need some modifications), and some are so similar as to be left as exercises.

Definition 6.4.16 Let V be a set, and k, r, and λ be positive integers. A **block design** D is a set of k-subsets of V, called **blocks**, such that the number r of blocks that contain a fixed $v \in V$ is independent of v, and the number λ of blocks containing distinct $u, v \in V$ is also independent of the choice of u, v.

For the record, in design theory, what we have defined here would formally be called a 2-design or a balanced incomplete block design (BIBD). The notation is usually a (v,k,λ)-design or a (v,b,k,r,λ)-design, where $v = |V|$ and b is the number of blocks. As we are primarily concerned with k, r, and, λ we will say a **design with parameters** k**,** r**, and** λ, or just a design.

As $\lambda \geq 1$, it follows that $k \geq 2$. Note that we could think of the elements of V as being "points" and the blocks as "lines" with the canonical incidence relation of a point v is incident with a block W if v is contained in W. While not particularly standard in design theory, we will in fact do this – refer to the elements of V as points and the blocks as lines in order to make our notation consistent with that of configurations. The insistence on each block of D being a subset of size k of V can then be rephrased as "each of the lines contains k points" while "the number r of blocks that contain a fixed $v \in V$ is independent of v" can be rephrased as "each of the points is on exactly r lines." Setting $\lambda = 1$, the statement "the number of blocks λ containing distinct $u, v \in V$ is also independent of the choice of u, v" simply means that "two points are on *exactly* λ lines." This condition is stronger than the corresponding condition for configurations, as nothing in the definition of a configuration requires that any two points lie on a line, just that any two points lie on at *most* one line. Thus a block design with $\lambda = 1$ is simply a configuration (although the converse is not necessarily true because, as mentioned above, two points of a configuration do not have to be on a line), and designs can be thought of as certain kinds of configurations generalized to "higher dimensions" (higher than dimension 2), although design theorists would not think of designs in this fashion.

Definition 6.4.17 A **symmetric design** is a design in which the size of V and the number of blocks are the same (if $\lambda = 1$, then in the language of configurations the number of points and lines are the same).

Example 6.4.18 Let q be a prime power and $d \geq 2$. The projective space $\mathrm{PG}(d-1, q)$ is a symmetric design.

Solution We think of the projective points of $\mathrm{PG}(d-1, q)$ as being "points" and the hyperplanes of $\mathrm{PG}(d-1, q)$ as being "blocks" or "lines." Obviously, each hyperplane in $\mathrm{PG}(d-1, q)$ contains the same number of projective points, and each projective point is contained in the same number of hyperplanes by simply mapping one projective point to another – this maps the hyperplanes that contain the first projective point to the hyperplanes that contain the second. Also, given two distinct projective points u and v and two other distinct projective points x and y, there is a linear transformation mapping u to x and v to y as we may think of a projective point as a basis vector. Thus the number of hyperplanes that contains u and v is independent of the choice of u and v, and so $\mathrm{PG}(d-1, q)$ is a design. Finally, as a hyperplane is the orthogonal complement of a projective point and vice versa, the number of projective points and hyperplanes are the same. We conclude that $\mathrm{PG}(d-1, q)$ is a symmetric design. □

Definition 6.4.19 Define a **symmetry** of a design as a bijective mapping $f\colon P \cup L \to P \cup L$ such that p is incident with ℓ if and only if $f(p)$ is incident with $f(\ell)$, for every point p and line ℓ.

Again, a "symmetry" of a design is what we would normally call an "automorphism." The terminology is similar to that of configurations.

Definition 6.4.20 Let D be a design. A symmetry that maps points of D to points of D and blocks of D to blocks of D is called an **automorphism**, while a symmetry that maps points to lines and vice versa is called a **duality**. The set of all automorphisms of a design D is a group under function composition, called the **automorphism group of** D, and is denoted Aut(D). The set of all symmetries of D is also a group under function composition, the **group of symmetries** of D.

Considering the symmetric design PG$(d-1,q)$, we see that the action of PGL(d,q) on the lines of PG$(d-1,q)$ contains PGL(d,q), and so is 2-transitive. Additionally, the map τ defined on p. 161 is a duality of the symmetric design PG$(d-1,q)$.

Definition 6.4.21 Let C be a design. Define a (necessarily bipartite) graph Γ_C as follows: The vertices of Γ_C are the set $P \cup L$, and a point is adjacent to a line if and only if the point is on the line. We call Γ_C the **Levi graph** of C.

As points are not on points and lines are not on lines, Γ_C is bipartite with P and L the natural bipartition sets. Typically, we think of the vertices of Γ_C corresponding to points as "white" vertices while the vertices of Γ_C corresponding to lines as "black" vertices. Thus the Levi graph of a design is the exact analogue to the Levi graph of a configuration and if $\lambda = 1$, the two definitions of Levi graphs of configurations and symmetric designs are the same.

Our goal now is to replicate the theory of Levi graphs of configurations to Levi graphs of designs. We begin with an analogue of Theorem 6.4.9.

Theorem 6.4.22 *A bipartite graph Γ is the Levi graph Γ_C of a design with parameters k, r, and λ if and only if*

- *all black vertices have valency k while all white vertices have valency r, and*
- *Γ_C contains no $K_{2,\lambda+1}$ with two vertices white and $\lambda+1$ black, and for any two white vertices u and v, u and v are contained in exactly one $K_{2,\lambda}$ in Γ_C.*

Proof If Γ is the Levi graph of a design with parameters k, r, and λ, then clearly Γ is bipartite. If Γ contains a $K_{2,\lambda+1}$ with two vertices white and $\lambda+1$ black, then two points are contained on $\lambda+1$ lines, while with our chosen parameters, two points must be contained on λ lines. Thus Γ contains no $K_{2,\lambda+1}$. Additionally, as any two points u and v are contained in exactly λ lines, there must be a $K_{2,\lambda}$ in Γ that contains the white vertices u and v (notice that if there is more than one $K_{2,\lambda}$ containing the white vertices u and v, then the white vertices u and v are contained in a $K_{2,\lambda+1}$). As each point is on exactly k lines and each line is incident with exactly r points, we see all of the black vertices have valency k while all of the white vertices have valency r.

Conversely, let Γ be a bipartite graph such that one of the cells of the bipartition, which we call "white", has the property that Γ contains no $K_{2,\lambda+1}$ with two vertices white and $\lambda + 1$ black, and any two white vertices are contained in exactly one $K_{2,\lambda}$ in Γ_C. We let all vertices in the other cell of the bipartition be "black." Our hypothesis then implies that all black vertices have valency k while all white vertices have valency r. We let V be the set of white vertices, and label each black vertex with the its k neighbors. Then each black vertex is labeled as a k-subset of the white vertices. Let D be the set of all labels of black vertices, or "blocks."

We now show that D is a design with parameters k, r, and λ. As each white vertex has valency r, the number of blocks that contains a fixed vertex is r (and so is independent of the choice of point). Finally, as Γ contains no $K_{2,\lambda+1}$ with two vertices white and $\lambda + 1$ black, and for any two white vertices u and v, u and v are contained in exactly one $K_{2,\lambda}$ in Γ_C, the number of blocks containing points u and v is always λ, and so is independent of the choice of u and v. □

As with a configuration, we say that a design is **connected** if its corresponding Levi graph is connected. In contrast to configurations, we have:

Lemma 6.4.23 *Every design with $k, \lambda \geq 2$ is connected.*

Proof Let D be a design defined on the set V with parameters $k, \lambda \geq 2$ and $r \geq 1$. Let Γ be the Levi graph of D. Let $u, v \in V$, $u \neq v$, and L a line of D that contains both u and v. Then uLv is a path in Γ from u to v. Let L_1 and L_2 be different lines of D. Then there is $u \in L_1$ and $v \in L_2$ that are distinct. There is also a line L_3 of D that contains both u and v, and $L_1uL_3vL_2$ is a walk in Γ from L_1 to L_2. Finally, let u be a point and L a line. Then there exists $v \in V$ such that $v \neq u$ as $k \geq 2$. Let L_1 be a line that contains u and v. Then uL_1vL is a walk in Γ from u to L. □

The proof of the following result is almost identical to Lemma 6.4.14, and is left as an exercise (Exercise 6.4.1).

Lemma 6.4.24 *In a design D, every symmetry of D is either an automorphism or a duality.*

Similar to the previous result, the proof of the following result is identical to Corollary 6.4.15, and is also left as an exercise (Exercise 6.4.2).

Corollary 6.4.25 *In a design D with a duality of order 2,* $\mathrm{Aut}(\Gamma_D) \cong \mathrm{Aut}(D) \rtimes \mathbb{Z}_2$.

The following result gives the relationship between the automorphism group G of a symmetric design that is 2-transitive on points, and the action of G on blocks or lines.

Lemma 6.4.26 *Let D be a symmetric design. If* $\mathrm{Aut}(D)$ *is 2-transitive but not $\mathcal{S}_V$ in its action on V, then the image of the permutation representation of the action of* $\mathrm{Aut}(D)$ *on the lines of D is a faithful inequivalent permutation representation of* $\mathrm{Aut}(D)$.

Proof If the action of $\mathrm{Aut}(D)$ on lines is not faithful, then the kernel K of the action is normal in $\mathrm{Aut}(D)$ and so transitive on V by Theorem 2.4.7. Let L be a line with v a point in L. Let $u \in V$. Then there is $k \in K$ such that $k(v) = u$. As K is the kernel of the action on lines, $k(L) = L$ and $u \in L$, so each line of D is V. Then D has a unique line V, and so $\mathrm{Aut}(D) = \mathcal{S}_V$, a contradiction. Thus $\mathrm{Aut}(D)$ acts faithfully on the lines of D.

If the actions of $\mathrm{Aut}(D)$ on points and lines are equivalent permutation representations of $\mathrm{Aut}(D)$, then by Theorem 4.1.8, we have $\mathrm{Stab}_{\mathrm{Aut}(D)}(v) = \mathrm{Stab}_{\mathrm{Aut}(D)}(L)$ for some point v and line L. As $\mathrm{Stab}_G(v)$ is transitive on $V\backslash\{v\}$ and $\mathrm{Stab}_G(v)$ fixes L, either $L = V$, $L = V\backslash\{v\}$, or each line is a singleton. If $L = V$, every line is V, and $\mathrm{Aut}(D) \cong \mathcal{S}_V$ as above, a contradiction. If each line is a singleton, then $\mathrm{Aut}(D) \cong \mathcal{S}_V$, also a contradiction. If $L = V\backslash\{v\}$, every line contains $|V| - 1$ points, and the set of lines is $\{V\backslash\{v\} : v \in V\}$. It is then easy to check that $\mathrm{Aut}(D) = \mathcal{S}_V$, a contradiction. □

The following result of Kantor (1985) uses CFSG to give all symmetric designs whose automorphism group is 2-transitive on points.

Theorem 6.4.27 *Let D be a symmetric design with $r > 2k$ such that* $\mathrm{Aut}(D)$ *is 2-transitive on points. Then D is one of the following:*

(i) *a projective space,*
(ii) *the unique Hadamard design with $r = 11$ and $k = 5$,*
(iii) *a unique design with $r = 176$, $k = 50$ and $\lambda = 14$, or*
(iv) *a design with $r = 2^{2m}$, $k = 2^{m-1}(2^{m-1} - 1)$ and $\lambda = 2^{m-1}(2^{m-1} - 1)$, of which there is exactly one for each $m \geq 2$.*

Recall that a projective space was defined in Section 4.5. The "Hadamard design" mentioned in Theorem 6.4.27 (ii) comes from an unusual action of $\mathrm{PSL}(2, 11)$ of degree 11, not the usual action of $\mathrm{PSL}(2, 11)$ on projective space that is of degree $\left(11^2 - 1\right)/(11 - 1) = 12$ – see Dixon and Mortimer (1996, Example 7.5.2). See Kantor (1985) for information about the designs mentioned in Theorem 6.4.27 (iii) and (iv).

We should point out one consequence of CFSG that will clear up any remaining ambiguity regarding the action of Aut(D) on the lines of a symmetric design D when the action of Aut(D) is 2-transitive. By Lemma 6.4.26, the action of Aut(D) on the lines in such a case induces an inequivalent permutation representation of the action of Aut(D) on points. As a consequence of CFSG, all 2-transitive groups are determined and are listed in Cameron (1981, Table 1), along with the number of inequivalent permutation representations of a 2-transitive group – it is always one or two. Thus in our case, the action of Aut(D) on lines is uniquely determined by the action of Aut(D) on points. In particular, if D is a projective space, then we may always take the action of Aut(D) on points to be the natural action on projective points, in which case the action on lines is always the natural action on hyperplanes.

Theorem 6.4.28 $\mathrm{Aut}(B(\mathrm{PG}(d-1,q))) = \mathrm{P\Gamma L}(d,q) \rtimes \mathbb{Z}_2$.

Proof By the fundamental theorem of projective geometry (Artin, 1988, Theorem 2.26), whose proof is outside the scope of this book, $\mathrm{Aut}(\mathrm{PG}(d-1,q)) = \mathrm{P\Gamma L}(d,q)$. The result follows using the duality τ and Lemma 6.4.25. □

The reader interested in more information about Levi graphs should see the books by Grünbaum (2009) and Pisanski and Servatius (2013).

6.5 The Generalized Petersen Graphs

Generalized Petersen graphs were introduced by Coxeter (1950), and next studied by Watkins (1969) who was interested in cubic graphs without proper 3-edge colorings. A proper 3-edge coloring of a trivalent graph is called a **Tait coloring**. The Petersen graph is not Tait colorable (Exercise 3.4.5), so it is reasonable to ask whether "similar" type graphs are Tait colorable.

Definition 6.5.1 For integers k and n satisfying $1 \leq k \leq n-1$, $2k \neq n$, define the **generalized Petersen graph** $\mathrm{GP}(n,k)$ to be the graph with vertex set $\{u_i, v_i : i \in \mathbb{Z}_n\}$ and edge set $\{u_i u_{i+1}, u_i v_i, v_i v_{i+k} : i \in \mathbb{Z}_n\}$ (arithmetic in the subscripts is performed modulo n).

The Petersen graph is $\mathrm{GP}(5,2)$, while the graph in Figure 6.9 is $\mathrm{GP}(10,4)$.

Watkins (1969), conjectured that every generalized Petersen graph except the Petersen graph itself has a Tait coloring and also verified his conjecture in some special cases. Watkins' conjecture was shown to be true by Castagna and Prins (1972). The generalized Petersen graphs have been extensively studied, and much of interest to us is known.

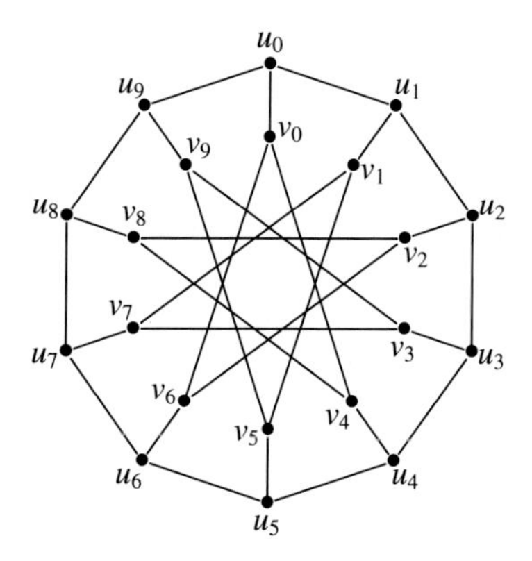

Figure 6.9 The generalized Petersen graph GP(10, 4).

6.5.1 The Automorphism Groups of the Generalized Petersen Graphs

Most of the work in this subsection is due to Frucht, Graver, and Watkins (Frucht et al., 1971), and so is most of our notation. The proof is of independent interest to us, as it nicely illustrates one method of determining automorphism groups of graphs, which is to focus on how particular isomorphic subgraphs are permuted by the automorphism group. Here, the main argument will be based on the number of 8-cycles in a generalized Petersen graph, and of course 8-cycles are preserved under automorphisms. Some special cases will also be handled in a similar way but with different subgraphs. There are natural sets of edges that will be crucial to our determination of the automorphism groups of generalized Petersen graphs, which we now define.

Definition 6.5.2 An edge of the form u_iu_{i+1} is an **outer edge**, an edge of the form v_iv_{i+k} is an **inner edge**, and an edge of the form u_iv_i is a **spoke edge** of a generalized Petersen graph $\mathrm{GP}(n, k)$. For brevity (and to be consistent with Frucht et al. (1971)), we denote $\mathrm{Aut}(\mathrm{GP}(n, k))$ by $A(n, k)$, and let $B(n, k)$ be the subgroup of $A(n, k)$ that preserves spoke edges, i.e., maps spoke edges to spoke edges. For convenience, let $\mathcal{I} = \{v_i : i \in \mathbb{Z}_n\}$ be the **inside vertices** and $\mathcal{O} = \{u_i : i \in \mathbb{Z}_n\}$ the **outside vertices**.

For example, with GP(10, 4) as above, the inside vertices form two 5-cycles, while the outside vertices form a 10-cycle. Straightforward computations will prove the following result.

Lemma 6.5.3 *Let n be an integer and $k < n/2$. The maps $\rho, \delta\colon V(\mathrm{GP}(n,k)) \to V(\mathrm{GP}(n,k))$ given by $\rho(u_i) = u_{i+1}$, $\rho(v_i) = v_{i+1}$, $\delta(u_i) = u_{-i}$ and $\delta(v_i) = v_{-i}$ are automorphisms of* $\mathrm{GP}(n,k)$.

Lemma 6.5.4 *If $\gamma \in A(n,k)$ fixes any of the sets of inner edges, outer edges, or spoke edges, then it either fixes all three or fixes the set of spoke edges and interchanges the sets of inner and outer edges.*

Proof Suppose γ fixes either the set of inner edges or outer edges. Then it fixes the set of inner or outer vertices. As each spoke edge is incident with an inner and outer vertex, γ must fix the spoke edges, and consequently the outer or inner edges. So $A(n,k)$ fixes all three sets of edges.

Suppose γ fixes the set of spoke edges but does not fix the sets of inner or outer edges. Then some inner edge must be mapped to an outer edge or vice versa. As the subgraph of $\mathrm{GP}(n,k)$ with all spoke edges removed induced by the outer edges is a cycle and so connected, γ maps all outer edges to inner edges, and consequently maps all inner edges to outer edges. □

Throughout the rest of this subsection, for $k \le n/2$, define $\tau\colon V(\mathrm{GP}(n,k)) \to V(\mathrm{GP}(n,k))$ by $\tau(u_i) = v_{ki}$ and $\tau(v_i) = u_{ki}$.

Lemma 6.5.5 *If $k^2 \not\equiv \pm 1 \pmod n$, then $B(n,k) = \langle \rho, \delta \rangle$, while if $k^2 \equiv \pm 1 \pmod n$, then $B(n,k) = \langle \rho, \delta, \tau \rangle$.*

Proof First, as $\mathrm{GP}(n,k)[\mathcal{O}]$ is an n-cycle whose automorphism is isomorphic to D_{2n}, the subgroup F of $A(n,k)$ that fixes $\mathcal{O}$ (and consequently $\mathcal{I}$) set-wise must satisfy $F^{\mathcal{O}} = \langle \rho, \delta \rangle^{\mathcal{O}} \cong D_{2n}$. If $\tau \in A(n,k)$, then $\tau^2 \in F$, $\tau^2(u_i) = u_{k^2 i}$, and $(\tau^2)^{\mathcal{O}} \in \langle \rho, \delta \rangle^{\mathcal{O}}$. As τ^2 fixes u_0, and the only elements of $\langle \rho, \delta \rangle$ that fix u_0 are δ and the identity, we see that $k^2 \equiv \pm 1 \pmod n$. Conversely, τ maps $u_i v_i$ to the spoke edge $u_{ki} v_{ki}$, $u_i u_{i+1}$ to the inner edge $v_{ki} v_{ki+k}$, and $v_i v_{i+k}$ to $u_{ki} u_{ki+k^2}$. So $\tau(v_i v_{i+1}) \in E(\mathrm{GP}(n,k))$ provided $k^2 \equiv \pm 1 \pmod n$. Thus $\tau \in A(n,k)$ if and only if $k^2 \equiv \pm 1 \pmod n$. This argument also shows that $\tau \in A(n,k)$ if and only if $\tau \in B(n,k)$.

We next show that $F = \langle \rho, \delta \rangle$. Indeed, we have already seen that $F^{\mathcal{O}} = \langle \rho, \delta \rangle^{\mathcal{O}}$. Also, as the only edges between the inside and outside vertices are spokes that form a 1-factor, F acts faithfully on $\mathcal{O}$, and so $F = \langle \rho, \delta \rangle$. This finishes the result by Lemma 6.5.4 unless some element $\gamma \in A(n,k)$ fixes the set of spoke edges and interchanges the sets of inner and outer edges.

Now suppose some $\gamma \in A(n,k)$ interchanges $\mathcal{O}$ and $\mathcal{I}$. Then $\gamma(u_0 u_1) = v_{rk} v_{rk \pm k}$ for some $r \in \mathbb{Z}_n$. As $\rho(\mathcal{O}) = \mathcal{O}$, replacing γ with $\rho^{-rk}\gamma$, we may assume $\gamma(u_0 u_1) = v_0 v_{\pm k}$. As $\delta \in A(n,k)$ and maps $\mathcal{O}$ to $\mathcal{O}$, we may additionally assume that $\gamma(u_0 u_1) = v_0 v_k$. As γ maps the n-cycle $u_0, u_1, \ldots, u_{n-1}, u_n$ to the n-cycle $v_0, v_k, \ldots, v_{(n-1)k}, v_0$ we see that $\gamma = \tau$. Hence $B(n,k) = \langle \rho, \delta, \tau \rangle$. □

Notice that if $\tau \in B(n,k)$ then $B(n,k)$ is vertex-transitive. Hence if $k^2 \equiv \pm 1$ then $\mathrm{GP}(n,k)$ is vertex-transitive. We now begin to examine what happens if there is an automorphism of $\mathrm{GP}(n,k)$ that does not preserve spoke edges.

Lemma 6.5.6 *The following are equivalent:*

(i) *There exists $\gamma \in A(n,k)$ that maps a spoke edge to a nonspoke edge,*
(ii) *$A(n,k) > B(n,k)$, and*
(iii) *$\mathrm{GP}(n,k)$ is edge-transitive.*

Proof It follows from the definition of $B(n,k)$ that (i) implies (ii). It is also easy to see that (iii) implies (i). It only remains to show that (ii) implies (iii).

Suppose $A(n,k) > B(n,k)$. Then there exists $\gamma \in A(n,k)$ that maps a spoke edge to an outer or inner edge. We assume γ maps a spoke edge to an outer edge – the case where γ maps a spoke edge to an inner edge is similar. Now, by Lemma 6.5.4, $B(n,k)$ has either two or three edge orbits, with two edge orbits if $\tau \in B(n,k)$, one consisting of the outer and inner edges and the other of spoke edges, and three otherwise, with orbits the outer, inner, and spoke edges. In the former case $\gamma \in A(n,k)$ gives one edge orbit, or that $\mathrm{GP}(n,k)$ is edge-transitive. In the latter case, assume $\mathrm{GP}(n,k)$ is not edge-transitive. Then $\gamma \in A(n,k)$ combines the spoke and outer edges into an edge orbit. The inner edges form a separate edge orbit, which contradicts Lemma 6.5.4. □

We will need some additional notation before proceeding.

Definition 6.5.7 Let C be a cycle in $\mathrm{GP}(n,k)$. Let $r(C)$ be the number of outer edges in C, $s(C)$ the number of spoke edges in C, and $t(C)$ the number of inner edges of C. Let C_j be the set of all cycles in $\mathrm{GP}(n,k)$ of length j, $j \geq 3$. We set $R(j) = \sum_{C \in C_j} r(C)$, $S(j) = \sum_{C \in C_j} s(C)$, and $T(j) = \sum_{C \in C_j} t(C)$.

The following lemma will be needed and its straightforward proof is left as an exercise (Exercise 6.5.1).

Lemma 6.5.8 *Let Γ be a graph and $H \leq \mathrm{Aut}(\Gamma)$. Let O be an orbit of the action on H on the edges of Γ. Any two edges of O are contained in exactly the same number of cycles of length j, $j \geq 3$.*

Corollary 6.5.9 *If $\mathrm{GP}(n,k)$ is edge-transitive, then $R(j) = S(j) = T(j)$ for $j \geq 3$.*

Proof We first apply Lemma 6.5.8 with $H = \langle \rho \rangle$ which has edge orbits the outer, spoke, and inner edges. Hence there are constants c, c', and c'' so that each outer, spoke, and inner edge is contained in c, c', and c'' cycles of length j. As there are n outer, spoke, and inner edges, $R(j) = cn$, $S(j) = c'n$, and

Type	Representative Z	Conditions	#	r	s	t
1	$u_0, u_1, u_2, u_3, u_4, u_5, v_5, v_0$	$k = 5$ or $n - k = 5$	n	5	2	1
2	$u_0, u_1, v_1, v_{k+1}, v_{2k+1}, v_{3k+1}, v_{4k+1}, v_0$	n or $2n = 5k + 1$	n	1	2	5
2′	$u_0, v_0, v_k, v_{2k}, v_{3k}, v_{4k}, v_1, u_1$	n or $2n = 5k - 1$	n	1	2	5
3	$u_0, u_1, u_2, u_3, u_4, v_4, v_{\frac{n+4}{2}}, v_0$	$n = 2k + 4$	n	4	2	2
4	$u_0, u_1, u_2, v_2, v_{k+2}, v_{2k+2}, v_{3k+2}, v_0$	n or $2n = 4k + 2$	n	2	2	4
4′	$u_2, u_1, u_0, v_0, v_k, v_{2k}, v_{3k}, v_2$	$n = 4k - 2$	n	2	2	4
5	$u_0, u_1, u_2, u_3, v_3, v_{\frac{n+6}{3}}, v_{\frac{2n+3}{3}}, v_0$	$n = 3k + 3$	n	3	2	3
5′	$u_0, v_0, v_{\frac{n+3}{3}}, v_{\frac{2n+6}{3}}, v_3, u_3, u_2, u_1$	$n = 3k - 3$	n	3	2	3
6	$u_0, u_1, v_1, v_{\frac{n}{2}}, u_{\frac{n}{2}}, u_{\frac{n}{2}+1}, v_{\frac{n}{2}+1}, v_0$	$n = 2k + 2$	$\frac{n}{2}$	2	4	2
7	$v_0, v_k, \ldots, v_{7k}$	$n = 8k$	$\frac{n}{8}$	0	0	8
7′	$v_0, v_k, \ldots, v_{7k}$	$3n = 8k$	$\frac{n}{8}$	0	0	8
8	$u_0, u_1, \ldots, u_7$	$n = 8$	1	8	0	0
9	$u_0, u_1, v_1, v_{1+k}, u_{1+k}, u_k, v_k, v_0$	$n \geq 4$	n	2	4	2

Table 6.1 *Types of* 8*-cycles in a generalized Petersen graph.*

$T(j) = c''n$. Applying Lemma 6.5.8 with $H = A(n, k)$, every edge is contained in the same number of cycles of length j, and so $c = c' = c''$. Hence $R(j) = S(j) = T(j)$. □

We now turn to the main argument. The previous result says that the number of spoke edges and outer (inner) edges contained in all cycles of a fixed length are the same. We will focus first on cycles of fixed length 8. An obvious question arises: Why focus on 8-cycles? The short answer is because this works! The longer answer is that most $\mathrm{GP}(n, k)$ contain 8-cycles, and 8-cycles are neither long nor short. So there are probably not too many of them nor is it overly complicated to construct them. Before turning to our result about 8-cycles in generalized Petersen graphs, we will need some notation.

Definition 6.5.10 We say that two cycles C_1 and C_2 in $\mathrm{GP}(n, k)$ are of the same **type** if there exists $\gamma \in \langle \rho, \delta \rangle$ such that $\gamma(C_1) = C_2$.

Lemma 6.5.11 *Let $n \geq 4$ and $2 < k < n/2$. Then every generalized Petersen graph* $\mathrm{GP}(n, k)$ *contains an* 8*-cycle, with all types as is given in Table 6.1.*

Proof Much of the idea of this proof is similar to the proof of Theorem 3.4.1 showing that the Petersen graph does not contain a Hamilton cycle. Let C be an 8-cycle in $\mathrm{GP}(n, k)$ that contains u_0, or that contains v_0 if C contains no vertex u_i, $i \in \mathbb{Z}_n$. Then C must contain an even number $s(C)$ of spoke edges. The only possibilities for the number of spoke edges are then $s(C) = 0, 2$, and 4.

If $s(C) = 0$, then $r(C) = 8$ and $t(C) = 0$ or $r(C) = 0$ and $t(C) = 8$. In the former case $C = u_0, u_1, \ldots, u_7, u_0$ and is of type 8 (note that in Table 6.1

we only list distinct vertices of the cycle for brevity). In the latter case, $C = v_0, v_k, \ldots, v_{7k}, v_0$. Additionally, it must be the case that $8k \equiv 0 \pmod{n}$, and k generates a subgroup of $\mathbb{Z}_n$ of order 8. Hence $k = n/8, 3n/8, 5n/8$, or $7n/8$, and as $k < n/2$, we see that $k = n/8$ or $k = 3n/8$ and that C is of type 7 or 7′. If C is of type 7 or 7′, then $\rho^i(C) = \rho^j(C)$ if and only $i \equiv j \pmod{k}$. As all cycles of type 7 or 7′ form a partition of the inner edges and as C has eight edges, there are $n/8$ cycles of type 7 or 7′.

If $s(C) = 2$, then we will consider when $r(C) = 2$ and $t(C) = 4$, with the other case being analogous. As $s(C) = 2$, the two outside edges of C are incident and form a path of length 2, and the four inside edges of C form a path of length 4. As $\rho \in \mathrm{Aut}(\mathrm{GP}(n,k))$, we may assume u_0, u_1, u_2 is a path contained in C. Then

$$C = u_0, u_1, u_2, v_2, v_{\pm k+2}, v_{\pm 2k+2}, v_{\pm 3k+2}, v_{\pm 4k+2}, u_0,$$

in which case $v_{\pm 4k+2} = v_0$, and $\pm 4k + 2 \equiv 0 \pmod{n}$. If $4k + 2 \equiv 0 \pmod{n}$, then C is of type 4. As $k/2 < n$, we conclude that $4k + 2 = n$ or $4k + 2 = 2n$. As $\rho^i(C) = \rho^j(C)$ if and only if $i = j$, there are n cycles of type 4. If $-4k + 2 \equiv 0 \pmod{n}$, then as $k/2 < n$, it follows that $n = 4k - 2$. As $k > 2$, $v_{-k+2} \neq v_0$. Also, $-(4 - i)k + 2 \equiv ik \pmod{n}$, and so traversing C backwards starting at u_2, we have $C = u_2, u_1, u_0, v_0, v_k, v_{2k}, v_{3k}, v_{4k}, u_2$. Hence C is of type 4′. Finally, as $\rho^i(C) = \rho^j(C)$ if and only if $i = j$, there are n cycles of type 4′.

If $s(C) = 4$, then $r(C) = 2$, and $t(C) = 2$. After an application of δ, if needed, we may assume u_0, u_1 is a path on C. Then

$$C = u_0, u_1, v_1, v_{1\pm k}, u_{1\pm k}, u_{1\pm k\pm 1}, v_{1\pm k\pm 1}, v_{1\pm k\pm 1\pm k}, u_0,$$

and so $1 \pm k \pm 1 \pm k \equiv 0 \pmod{n}$. There are eight possibilities for choosing $\pm k$, ± 1, and $\pm k$. Two give $2k \equiv 0 \pmod{n}$, two give $2 \equiv 0 \pmod{n}$, all of which are contradictions. Choosing $-k, 1, -k$ gives $2k \equiv 2 \pmod{n}$. This gives $k = 1$ or $2k = n + 2$, both of which are excluded by hypothesis. Choosing $k, -1, -k$, we see that $C = u_0, u_1, v_1, v_{1+k}, u_{1+k}, u_k, v_k, v_0, u_0$ is always an 8-cycle in $\mathrm{GP}(n,k)$ of type 9. Clearly there are n cycles of type 9 in $\mathrm{GP}(n,k)$. Choosing $-k, -1, k$, the cycle C is then $u_0, u_1, v_1, v_{1-k}, u_{1-k}, u_{-k}, v_{-k}, v_0, u_0$. But then $\rho^k(C)$, written starting at $\rho^k(u_{-k})$ and traversed backwards, is $u_0, u_1, v_1, v_{1+k}, u_{1+k}, u_k, v_k, v_0, u_0$, and so this is one of the type 9 cycles already counted. The only remaining choice is $k, 1, k$, which gives $2 + 2k \equiv 0 \pmod{n}$. Then $k < n/2$, $n = 2k + 2$, and $k = \frac{n}{2} - 1$. Hence $C = u_0, u_1, v_1, v_{\frac{n}{2}}, u_{\frac{n}{2}}, u_{\frac{n}{2}+1}, v_{\frac{n}{2}+1}, v_0, u_0$ and C is of type 6. There are $n/2$ such cycles as if $i \equiv j \pmod{n/2}$, then $\rho^i(C) = \rho^j(C)$. □

Lemma 6.5.12 *If* $2 < k < n/2$ *and*

$$(n,k) \neq (8,3), (10,3), (12,5), (13,5), (24,5), \text{ or } (26,5),$$

then $B(n, k) = A(n, k)$.

Proof As pairs $(8, 1), (8, 2)$, and $(8, 3)$ have all been excluded, we have $n \neq 8$. Our strategy is to show, for all allowed pairs (n, k), that the conclusion of Corollary 6.5.9 does not hold for cycle lengths 8, in which case $\mathrm{GP}(n, k)$ is not edge-transitive. Then by Lemma 6.5.6 we will have $A(n, k) = B(n, k)$. So it suffices to show that $R(8) \neq S(8)$, $R(8) \neq T(8)$, or $S(8) \neq T(8)$.

First suppose $n = 8k$. By Lemma 6.5.11, $\mathrm{GP}(8k, k)$ contains cycles of type 8, 9, and if $k = 5$, also of type 1. If $k = 5$, then the number of outer edges of $\mathrm{GP}(40, 5)$ contained in 8-cycles is $R(8) = 7n$ while the number of spoke edges of $\mathrm{GP}(40, 5)$ contained in 8-cycles is $S(8) = 6n$. Then $R(8) \neq S(8)$. If $k \neq 5$, then $R(8) = 2n$ while $S(8) = 4n$.

Now suppose $n \neq 8k$, $R(8) = T(8)$, and $S(8) = T(8)$. Note that $\mathrm{GP}(n, k)$ cannot contain 8-cycles of both types 2 and $2'$, or 4 and $4'$, or 5 and $5'$. For $1 \leq i \leq 7$, let $x_i = 0$ if $\mathrm{GP}(n, k)$ contains no cycles of type i or i', and $x_i = 1$ if $\mathrm{GP}(n, k)$ does contain cycles of type i or i'. By Lemma 6.5.11 we obtain the following three equations in seven unknowns:

$$R(8) = 5nx_1 + nx_2 + 4nx_3 + 2nx_4 + 3nx_5 + nx_6 + 2n,$$

$$S(8) = 2nx_1 + 2nx_2 + 2nx_3 + 2nx_4 + 2nx_5 + 2nx_6 + 4n,$$

$$T(8) = nx_1 + 5nx_2 + 2nx_3 + 4nx_4 + 3nx_5 + nx_6 + nx_7 + 2n.$$

As $R(8) = T(8)$,

$$4x_1 + 2x_3 = 4x_2 + 2x_4 + x_7.$$

As the left-hand side of the previous equation is even, $x_7 = 0$. Similarly, as the left-hand side is either 0, 2, 4, or 6, it must then follow that $x_2 = x_1$ and $x_4 = x_3$. If $x_1 = x_2 = 1$, then by the conditions given in Lemma 6.5.11, we must have $k = 5$, in which case $n = 12, 13, 24$, or 26, and all such pairs (n, k) are excluded. If $x_3 = x_4 = 1$, then $k = 3$ and $n = 10$, again an excluded pair. Thus $x_1 = x_2 = x_3 = x_4 = 0$. Combining this with the assumption that $S(8) = T(8)$, we see $x_5 = x_6 + 2$, which has no solution as x_5 and x_6 are either 0 or 1. □

We now consider a case not covered by the preceding result. Other such cases are similar to what has already been shown and are left as exercises (Exercises 6.5.2, 6.5.3, and 6.5.5). In all of these cases considering 8-cycles to determine $\mathrm{Aut}(\mathrm{GP}(n, k))$ does not work, and so some other graph-theoretic structure must be used. In the following result, we focus on paths of length 4.

Lemma 6.5.13 $\mathrm{Aut}(\mathrm{GP}(26, 5)) = B(26, 5)$.

Proof As $k^2 \equiv -1 \pmod{26}$, $B(26,5)$ is transitive by Lemma 6.5.5. It suffices to show that $\mathrm{Stab}_{\mathrm{Aut(GP(26,5))}}(u_0) \leq B(26,5)$. Let $\gamma \in \mathrm{Stab}_{\mathrm{Aut(GP(26,5))}}(u_0)$. It is straightforward to show that there are 14 vertices of GP(26, 5) connected to u_0 by a path of length 4 (Exercise 6.5.4), and that there are three ways in which this can be done, depending on the number of such paths:

(i) the vertices $u_{10}, u_{16}, v_7, v_9, v_{17}$, and v_{19} are connected to u_0 by exactly one path of length 4,
(ii) the vertices $u_6, u_{20}, v_3, v_{11}, v_{15}$, and v_{23} are connected to u_0 by exactly two paths of length 4, and
(iii) the vertices u_4 and u_{22} are connected to u_0 by exactly three paths of length 4.

Then γ fixes each of these sets of vertices, and in particular, $\gamma(u_4) = u_4$ and $\gamma(u_{22}) = u_{22}$, or $\gamma(u_4) = u_{22}$ and $\gamma(u_{22}) = u_4$. Of course, $\delta \in B(n,k)$ fixes u_0 and interchanges u_4 and u_{22}, so after replacing γ with $\gamma\delta$, if necessary, we may assume without loss of generality that γ fixes u_0, u_4, and u_{22}. We will show that $\gamma = 1$, which will establish the result.

As u_0 is connected to only u_4 and u_{22} by exactly three paths of length 4, u_i is connected to only u_{i+4} and u_{i-4} by exactly three paths of length 4. If γ fixes u_i and u_{i-4}, then γ must fix u_{i+4}. Arguing inductively, this implies that γ fixes u_{4i} for $i \in \mathbb{Z}_{26}$. That is, γ fixes all u_{2i} with $i \in \mathbb{Z}_{26}$. As the only common neighbor of u_i and u_{i+2} is u_{i+1}, γ must also fix all u_{2i+1}, and so γ fixes all u_i, $i \in \mathbb{Z}_{26}$. Then $\gamma = 1$ and the result follows. □

We are now ready to give the full automorphism groups of all generalized Petersen graphs. We will need to define one more map. For $(n,k) = (4,1)$, $(8,3)$, $(12,5)$, and $(24,5)$, define $\sigma\colon V(\mathrm{GP}(n,k)) \to V(\mathrm{GP}(n,k))$ by

$$\sigma(u_{4i}) = u_{4i}, \quad \sigma(v_{4i}) = u_{4i+1}, \quad \sigma(u_{4i+1}) = u_{4i-1}, \quad \sigma(u_{4i-1}) = v_{4i},$$

$$\sigma(u_{4i+2}) = v_{4i-1}, \quad \sigma(v_{4i-1}) = v_{4i+5}, \quad \sigma(v_{4i+1}) = u_{4i-2}, \text{ and } \sigma(v_{4i+2}) = v_{4i-6},$$

for all i.

Theorem 6.5.14 *Let $2 \leq 2k < n$. Then,*

(i) $(n,k) = (4,1)$, GP(4, 1) *is the cube* F8 *with automorphism group* $\mathcal{S}_3 \wr \mathcal{S}_2 \cong \mathbb{Z}_2 \times \mathcal{S}_4$,
(ii) $(n,k) = (5,2)$, GP(5, 2) *is the Petersen graph* F10 *with automorphism group* $\mathcal{S}_5$,
(iii) $(n,k) = (8,3)$, GP(8, 3) *is the Möbius–Kantor graph* F16 *with automorphism group* $\mathrm{Aut(GP(8,3))} = \langle \rho, \tau, \delta, \sigma \rangle$,

(iv) $(n, k) = (10, 2)$, GP(10, 2) *is the dodecahedron* F20A *with automorphism group* $A_5 \times \mathcal{S}_2$,

(v) $(n, k) = (10, 3)$, GP(10, 3) *is the Desargues graph* F20B *with automorphism group* $\mathcal{S}_5 \times \mathcal{S}_2$,

(vi) $(n, k) = (12, 5)$, GP(12, 5) *is the Nauru graph* F24 *with automorphism group* $\langle \rho, \tau, \delta, \sigma \rangle$,

(vii) $(n, k) = (24, 5)$, GP(24, 5) *is* F48 *with automorphism group* $\langle \rho, \tau, \delta, \sigma \rangle$,

(viii) (n, k) *is not one of the pairs above and*

 (a) *if* $k^2 \not\equiv \pm 1 \pmod n$, *then* $\mathrm{Aut}(\mathrm{GP}(n, k)) = \langle \rho, \delta \rangle$,

 (b) *if* $k^2 \equiv \pm 1 \pmod n$, *then* $\mathrm{Aut}(\mathrm{GP}(n, k)) = \langle \rho, \tau, \delta \rangle$.

Proof Combining Lemmas 6.5.12 and 6.5.13 together with Exercises 6.5.2, 6.5.3, and 6.5.5, we have $\mathrm{Aut}(\mathrm{GP}(n, k)) = B(n, k)$ or (n, k) is one of the seven pairs $(4, 1)$, $(5, 2)$, $(8, 3)$, $(10, 2)$, $(10, 3)$, $(12, 5)$, or $(24, 5)$. If $\mathrm{Aut}(\mathrm{GP}(n, k)) = B(n, k)$, then the result follows by Lemma 6.5.5. It only remains to consider the seven exceptional pairs listed earlier.

If $(n, k) = (4, 1)$, then GP(4, 1) is the three-dimensional hypercube, or just the cube, whose full automorphism group is isomorphic to $\mathcal{S}_3 \wr \mathcal{S}_2$ by Exercise 5.4.4, a result first shown in Frucht (1936/37). Thus (i) follows.

If $(n, k) = (5, 2)$, then GP(5, 2) is the Petersen graph whose automorphism group is $\mathcal{S}_5$ by Theorem 2.1.4 and (ii) follows.

If $k^2 \equiv 1 \pmod n$, then tedious, though straightforward, computations will show that σ is an automorphism of GP(n, k), for $(n, k) = (4, 1)$, $(8, 3)$, $(12, 5)$, and $(24, 5)$, and clearly $\sigma \notin B(n, k)$. As $\sigma(u_{4i}v_{4i}) = u_{4i}u_{4i+1}$, GP$(n, k)$ is edge-transitive by Lemma 6.5.6. As $\tau(u_0v_0) = v_0u_0$, GP(n, k) is arc-transitive by Theorem 3.2.10. Then GP(n, k) is s-arc-regular for some $s \geq 1$ by Theorem 3.1.5. By Exercises 6.5.8, 6.5.9, and 6.5.10, the graphs GP(8, 3), GP(12, 5), and GP(24, 5) are 2-arc-regular. The automorphism groups of these graphs then have order $6 \cdot 16$, $6 \cdot 24$, and $6 \cdot 48$ by Corollary 3.1.6. Order arguments then give that $\mathrm{Aut}(\mathrm{GP}(n, k)) = \langle \rho, \tau, \delta, \sigma \rangle$. Finally, each of these graphs is the unique symmetric cubic graph of their order, and so their names and designations are as given by the Foster census.

If $(n, k) = (10, 2)$ then GP(10, 2) is the dodecahedron (see Exercise 6.5.11). The map ρ^2 is transitive on each of the two 5-cycles in GP$(10, 2)[\mathcal{I}]$. By Exercise 1.1.16, Aut(GP(10, 2)) is transitive on its set of faces. As every 2-arc is contained in a 5-cycle and every 5-cycle is a face, we see that GP(10, 2) is 2-arc-transitive. Also, there are paths of length 3 in GP(10, 2) that are contained in 5-cycles, and some that are not. Hence GP(10, 2) is 2-arc regular, and so $|\mathrm{Aut}(\mathrm{GP}(10, 2))| = 6 \cdot 20$. Frucht (1936/37) showed that Aut(GP(10, 2)) has automorphism group isomorphic to $A_5 \times \mathcal{S}_2$.

Finally, if $(n,k) = (10,3)$, then GP(10, 3) is the Desargues graph, which is drawn as a generalized Petersen graph in Figure 6.8 (although it is labeled as a Levi graph). As $\tau \in \text{Aut}(\text{GP}(10,3))$, it is arc-transitive, and so arc-regular for some $s \geq 1$. It is also easy to find 4-arcs in GP(10, 3) contained in 6-cycles and other 4-arcs not contained in 6-cycles. It turns out that GP(10, 3) is 3-arc regular. It is F20B in the Foster census, and Coxeter (1950) showed that its full automorphism group is isomorphic to $\mathcal{S}_5 \times \mathcal{S}_2$. □

As $\delta \in \text{Aut}(\text{GP}(n,k))$, it is not difficult to see that $\text{GP}(n,k) = \text{GP}(n, n-k)$ for every $k \in \mathbb{Z}_n$, $k \neq n/2$. Thus the restriction that $2 \leq 2k < n$ does not mean that some automorphism groups of generalized Petersen graphs are not known. Additionally, this restriction is not given in the statements of results in Frucht et al. (1971), as it is incorporated into the definition of a generalized Petersen graph in that paper (but not in Watkins' original paper (1969)). Also, $\langle\tau\rangle \leq \langle\rho,\sigma,\delta\rangle$; see Frucht et al. (1971).

We can extract which generalized Petersen graphs are vertex-transitive from Theorem 6.5.14, a result first obtained in Frucht et al. (1971).

Theorem 6.5.15 *For $2 \leq 2k < n$, GP(n,k) is vertex-transitive if and only if $k^2 \equiv \pm 1 \pmod n$ or $n = 10$ and $k = 2$.*

We also observe that if (n,k) is not one of the seven exceptional pairs, then no automorphism of GP(n,k) maps a spoke edge u_iv_i to an outer or inner edge of GP(n,k), and so the seven exceptional pairs (n,k) are the only edge-transitive generalized Petersen graphs by Lemma 6.5.6.

6.5.2 The Cayley Generalized Petersen Graphs

With the automorphism groups of the generalized Petersen graphs in hand, it is not difficult to determine which generalized Petersen graphs are Cayley graphs using Theorem 1.2.20. Somewhat surprisingly, which generalized Petersen graphs are Cayley graphs was not determined until 1995 by Nedela and Škoviera (1995a), who used the classification of orientable regular embeddings of the n-dipole, the graph consisting of two vertices and n parallel edges. The proof below is due to Lovrečič Saražin (1997).

Theorem 6.5.16 *A generalized Petersen graph GP(n,k) is a Cayley graph of a group G if and only if $k^2 \equiv 1 \pmod n$. The group G can always be chosen to be D_{2n}, or $\mathbb{Z}_2 \times \mathbb{Z}_n$.*

Proof By Theorem 6.5.14, unless $k^2 \equiv \pm 1 \pmod n$ or (n,k) is one of the seven exceptional pairs given in that result, Aut(GP(n,k)) is not transitive.

As GP(10, 2) was shown to be non-Cayley by Watkins (1990), it suffices to only consider when $k^2 \equiv \pm 1 \pmod n$. Note $\mathrm{GP}(n,k)$ is isomorphic to a $(2,n)$-metacirculant graph with $\alpha = k$.

Suppose $k^2 \equiv 1 \pmod n$. Then $a = |k|$ divides 2, and $c = a/\gcd(a,2) = 1$. By Theorem 1.5.13, $\mathrm{GP}(n,k)$ is a Cayley graph of $G = D_{2n}$ if $k \equiv -1 \pmod n$, or $G = \mathbb{Z}_2 \times \mathbb{Z}_n$ if $k = 1$.

Suppose $k^2 \equiv -1 \pmod n$. First, the Petersen graph GP(5, 2) is not Cayley by Theorem 1.2.22, and GP(10, 3) is also not Cayley – see Watkins (1990) for a proof of this. By Theorem 6.5.14, $\mathrm{Aut}(\mathrm{GP}(n,k)) = B(n,k) = \langle\rho,\tau\rangle$. Suppose $H < B(n,k)$ is regular. Note that $|B(n,k)| = 4n$ as $|\rho| = n$, $|\tau| = 4$, and $\langle\rho\rangle \trianglelefteq B(n,k)$. By the orbit-stabilizer theorem $|\mathrm{Stab}_{B(n,k)}(u_0)| = 2$, and so $\mathrm{Stab}_{B(n,k)}(u_0) = \left\{\tau^2, 1\right\} = \mathrm{Stab}_{B(n,k)}(v_0)$. As $\langle\rho\rangle \trianglelefteq B(n,k)$, any element of $B(n,k)$ can be written as $\tau^i\rho^j$, $i, j \in \mathbb{Z}$. If $\tau^i\rho^j(u_0) = v_0$, then $\tau^{-1}\left(\tau^i\rho^j\right)(u_0) = u_0$ and $\tau^{i-1}\rho^j \in \left\{\tau^2, 1\right\}$. We conclude that $i - 1 \equiv 0 \pmod 2$ and $j \equiv 0 \pmod n$. Thus the only elements of $B(n,k)$ that map u_0 to v_0 are τ and τ^{-1}, neither of which are in H. Thus H is not transitive, a contradiction. □

We remark that in the preceding result some of the exceptional graphs are Cayley graphs of other groups. For example, we saw in Example 1.2.4 that the cube is also a Cayley graph of $\mathbb{Z}_2^3$, while Coxeter has shown that GP(24, 5) is also a Cayley graph of GL(2, 3) (Coxeter, 1986) amongst other groups.

6.5.3 Isomorphisms Between Generalized Petersen Graphs

As mentioned above, as $\delta \in \mathrm{Aut}(\mathrm{GP}(n,k))$, $\mathrm{GP}(n,k) = \mathrm{GP}(n,n-k)$ for every $k \in \mathbb{Z}_n$, $k \neq n/2$. Also, if $k \in \mathbb{Z}_n^*$, then the map $\kappa\colon I \cup O \to I \cup O$ given by $\kappa(u_i) = v_{k^{-1}i}$ and $\kappa(v_i) = v_{k^{-1}i}$ maps edges of the form u_iv_i to $u_{k^{-1}i}v_{k^{-1}i}$, and so as a set fixes such edges, maps edges of the form $v_{ki}v_{ki+k}$ to u_iu_{i+1}, and maps edges of the form u_iu_{i+1} to $u_{k^{-1}i}u_{k^{-1}i+k^{-1}}$. Thus if k is relatively prime to n, we have $\kappa(\mathrm{GP}(n,k)) = \mathrm{GP}\left(n,k^{-1}\right)$. We will show that these types of isomorphisms (and their compositions) between generalized Petersen graphs are the only possible ones.

While our main result (Theorem 6.5.18) is interesting, it also points out a technique we will use later to determine isomorphisms between some Cayley graphs. Namely, we will show that an isomorphism between two generalized Petersen graphs normalizes $\langle\rho\rangle$, and then calculate the normalizer of $\langle\rho\rangle$ in the symmetric group (actually we will prove things in the reverse order). In Chapter 7, we will study when all isomorphisms of Cayley graphs of a group G normalize G_L, and, by Corollary 4.4.11, $N_{S_G}(G_L) = G_L \rtimes \mathrm{Aut}(G)$. As it will not be much additional work, we will also calculate the normalizer of another group whose normalizer we will need later.

Lemma 6.5.17 *Let $m, n \geq 2$ be integers, $\rho\colon \mathbb{Z}_m \times \mathbb{Z}_n \to \mathbb{Z}_m \times \mathbb{Z}_n$ by $\rho(i,j) = (i, j+1)$, and $H = \{(i,j) \mapsto (i, j+a_i) : a_i \in \mathbb{Z}_n\}$. Then $N_{S_{mn}}(\langle\rho\rangle) = \{(i,j) \mapsto (\omega(i), \beta j + b_i) : \omega \in S_m, \beta \in \mathbb{Z}_n^*, b_i \in \mathbb{Z}_n\}$ and $N_{S_{mn}}(H) = S_m \wr N(n)$, where $N(n) = \{x \mapsto ax + b : a \in \mathbb{Z}_n^*, b \in \mathbb{Z}_n\}$.*

Proof Let $a_i \in \mathbb{Z}_n$, $\beta_i \in \mathbb{Z}_n^*$, and $\omega \in S_n$. Define $\gamma, \delta, \phi\colon \mathbb{Z}_m \times \mathbb{Z}_n \to \mathbb{Z}_m \times \mathbb{Z}_n$ by $\gamma(i,j) = (i, j+a_i)$, $\delta(i,j) = (\omega(i), j)$, and $\phi(i,j) = (i, \beta_i j)$. Straightforward computations give $\delta^{-1}\gamma\delta(i,j) = (i, j + a_{\omega(i)})$, and $\phi^{-1}\gamma\phi(i,j) = \left(i, j + \beta_i^{-1} a_i\right)$. Hence δ and ϕ normalize H, as does H itself, and so $S_m \wr N(n) \leq N_{S_{mn}}(H)$. This also shows that δ commutes with ρ, and for $\beta \in \mathbb{Z}_n^*$ the map $(i,j) \mapsto (i, \beta j)$ normalizes $\langle\rho\rangle$. It is easy to verify that H centralizes $\langle\rho\rangle$. Thus $\{(i,j) \mapsto (\omega(i), \beta j + b_i) : \omega \in S_m, \beta \in \mathbb{Z}_n^*, b_i \in \mathbb{Z}_n\} = N \leq N_{S_{mn}}(\langle\rho\rangle)$.

Obviously $\langle\rho\rangle \trianglelefteq N_{S_{mn}}(\langle\rho\rangle)$ and $H \trianglelefteq N_{S_{mn}}(H)$, and as the orbits of H are the same as the orbits of $\langle\rho\rangle$, both of these normalizers have a block system $\mathcal{B}$ that is the set of orbits of $\langle\rho\rangle$. Let $B \in \mathcal{B}$ that contains $(0,0)$. Of course, $H^B = \langle\rho\rangle^B \trianglelefteq \mathrm{Stab}_{N_{S_{mn}(\langle\rho\rangle)}}(B)^B$, and as $\langle\rho\rangle^B \cong (\mathbb{Z}_n)_L$, we have $\mathrm{Stab}_{N_{S_{mn}(\langle\rho\rangle)}}(B)^B \cong (\mathbb{Z}_n)_L \rtimes \mathrm{Aut}(\mathbb{Z}_n)$ by Corollary 4.4.11. As $\mathrm{Aut}(\mathbb{Z}_n) = \{x \mapsto ax : a \in \mathbb{Z}_n^*\}$ and $N \leq N_{S_{mn}}(\langle\rho\rangle)$, we conclude that

$$\mathrm{Stab}_{N_{S_{mn}(\langle\rho\rangle)}}(B)^B \leq \{(0,j) \mapsto (0, \beta j + b_0) : \beta \in \mathbb{Z}_n^*, b_0 \in \mathbb{Z}_n\} \cong N(n).$$

Similarly, $\mathrm{Stab}_{N_{S_{mn}(H)}}(B)^B$ is isomorphic to a subgroup of $N(n)$. By the embedding theorem, we have that both $N_{S_{mn}}(\langle\rho\rangle)$ and $N_{S_{mn}}(H)$ are subgroups of $S_m \wr N(n)$. This completes the proof that $N_{S_{mn}}(H) = S_m \wr N(n)$.

Now let $g \in N_{S_{mn}}(\langle\rho\rangle) \leq S_m \wr N(n)$. Then $g(i,j) = (\omega(i), \beta_i j + b_i)$, $\omega \in S_m$, $\beta_i \in \mathbb{Z}_n^*$, and $b_i \in \mathbb{Z}_n$. As $N \leq N_{S_{mn}}(\langle\rho\rangle)$, for some $\nu \in N$, $\nu g(i,j) = (i, \beta_i j)$. For $i \in \mathbb{Z}_m$, define $z_i\colon \mathbb{Z}_m \times \mathbb{Z}_n$ by $z_i(i,j) = (i, j+1)$ while $z_i(i', j) = (i', j)$, $i' \neq i$. Then $\rho = \Pi_{i=0}^{m-1} z_i$, and $(\nu g)\rho(\nu g)^{-1}(i,j) = (i, j + \beta_i) \in \langle\rho\rangle$. We conclude that $\beta_i = \beta_{i'}$ for every $i, i' \in \mathbb{Z}_m$, and so $\nu g \in N$. Thus $g \in N$ and $N_{S_{mn}}(\langle\rho\rangle) = N$. □

We are now ready to prove the following result that characterizes exactly when two generalized Petersen graphs are isomorphic.

Theorem 6.5.18 *The generalized Petersen graphs* $\mathrm{GP}(n,k)$ *and* $\mathrm{GP}(n,\ell)$ *are isomorphic if and only if either* $k \equiv \pm\ell \pmod n$ *or* $k\ell \equiv \pm 1 \pmod n$.

Proof If (n,k) is one of the seven exceptional pairs given by Theorem 6.5.14, then the result is certainly true as $1 \leq k \leq n/2$ and $\mathrm{GP}(n,k) = \mathrm{GP}(n,-k)$. We thus assume (n,k) is not one of the seven exceptional pairs given by Theorem 6.5.14, and so $\mathrm{Aut}(\mathrm{GP}(n,k)) \leq \langle\rho, \tau, \delta\rangle$.

Let ϕ be an isomorphism from $\text{GP}(n,k)$ to some other generalized Petersen graph. Let $F = \langle \rho, \delta \rangle$. Then either $F = \text{Aut}(\text{GP}(n,k))$ or $\tau \in \text{Aut}(\text{GP}(n,k))$ in which case $\tau^2 = \delta$ or 1. If $\tau^2 = \delta$, then F is the unique subgroup of $\text{Aut}(\text{GP}(n,k))$ of index 2 that contains a semiregular cyclic group of order n. As $\text{Aut}(\phi(\text{GP}(n,k))) = \text{Aut}(\text{GP}(n,k))$ by Theorem 6.5.14 and $\text{Aut}(\phi(\text{GP}(n,k))) = \phi^{-1}\text{Aut}(\text{GP}(n,k))\phi$ by Exercise 1.1.11, if $F = \text{Aut}(\text{GP}(n,k))$ or $\tau^2 = \delta$, then ϕ normalizes F. As $F = \langle \rho, \delta \rangle \cong D_{2n}$, which has a unique cyclic subgroup $\langle \rho \rangle$ of order n, $\langle \rho \rangle$ is characteristic in F. Thus ϕ normalizes $\langle \rho \rangle$. Finally, if $\tau^2 = 1$, then $\langle \rho \rangle \trianglelefteq \text{Aut}(\text{GP}(n,k))$ and $\text{Aut}(\text{GP}(n,k))/\langle \rho \rangle \cong \mathbb{Z}_2^2$. Then every subgroup of index 2 in $\text{Aut}(\text{GP}(n,k))$ has $\langle \rho \rangle$ as its unique normal cyclic subgroup of order n unless n is even, $n/2$ is odd, and $k = \pm 1$. In the latter case, we are finished as the only possibilities for k are $k = \pm 1$. In the former case, as $\text{Aut}(\text{GP}(n,k)) = \text{Aut}(\phi(\text{GP}(n,k)))$, we see ϕ normalizes $\langle \rho \rangle$. So in every remaining case, ϕ normalizes $\langle \rho \rangle$.

It will now be convenient from a computational point of view to switch from the usual labeling of a generalized Petersen graph to a more "metacirculant" labeling. So, relabel the vertices of $\text{GP}(n,k)$ with elements of $\mathbb{Z}_2 \times \mathbb{Z}_n$ in such a way that $\mathcal{O} = \{(1,i) : i \in \mathbb{Z}_n\}$, $\mathcal{I} = \{(0,i) : i \in \mathbb{Z}_n\}$, and the edge set of $\text{GP}(n,k)$ is $\{(1,i)(1,i+1), (0,i)(0,i+k), (0,i)(1,i) : i \in \mathbb{Z}_n\}$. Then $\rho(i,j) = (i, j+1)$, $\delta(i,j) = (i,-j)$, and $\tau(i,j) = (i+1, kj)$, if $\tau \in B(n,k)$.

By Lemma 6.5.17, we have $\phi(i,j) = (i+a, \beta j + b_i)$, where $a \in \mathbb{Z}_2, \beta \in \mathbb{Z}_n^*$, and $b_i \in \mathbb{Z}_n$. As $\rho \in B(n,k)$, we can and do assume without loss of generality that $b_0 = 0$. We now consider the cases $a = 0$ and $a = 1$ separately.

If $a = 0$, then ϕ maps $\text{GP}(n,k)[\mathcal{O}]$ to $\text{GP}(n,\ell)[\mathcal{O}]$, and these graphs are equal (as the outside cycles of a generalized Petersen graph are always equal). As the automorphism group of cycle is a dihedral group, we have that $\beta = \pm 1$. As $\phi(0,0) = (0,0)$ and ϕ maps spoke edges to spoke edges, we see that $\phi(1,0) = (1,0)$ and so $b_1 = 0$ as well. Then $\phi = \delta$, and $\text{GP}(n,k)[\mathcal{I}] = \text{GP}(n,\ell)[\mathcal{I}]$. Thus $\ell = \pm k$.

If $a = 1$, then ϕ maps $\text{GP}(n,k)[\mathcal{O}]$ to $\text{GP}(n,\ell)[\mathcal{I}]$. As $\phi((1,0)(1,1)) = (0,0)(0,\ell)$ or $(0,0)(0,-\ell)$, we have $\beta = \pm\ell$. As ϕ maps spoke edges to spoke edges and $\phi(1,0) = (0,0)$, we have $\phi(0,0) = (1,0)$ and again $b_1 = 0$. Then $\phi((0,0)(0,k)) = (1,0)(1,\pm k\ell) = (1,0)(1,1)$ or $(1,0)(1,-1)$ and so $k\ell = \pm 1$, completing the proof. □

Theorem 6.5.18 was first proven by Boben, Pisanski, and Žitnik (Boben et al., 2005). The above theorem was also proven in the case where k and ℓ are relatively prime to n by Steimle and Staton (2009) using very different techniques, and they additionally assumed that $2 \leq k, \ell \leq n-2$.

6.5.4 Hamiltonian Generalized Petersen Graphs

Determining which generalized Petersen graphs are Hamiltonian is difficult. Several authors contributed to the solution of this problem, finally solved by Alspach (1983). We content ourselves with simply stating the final result, and referring the interested reader to Alspach (1983) for a detailed history of the problem.

Theorem 6.5.19 *The generalized Petersen graph* $\mathrm{GP}(n, k)$ *is not Hamiltonian if and only if it is*

(i) $\mathrm{GP}(n, n/2)$, $n = 0 \pmod 4$ *and* $n > 8$, *or*
(ii) *isomorphic to* $\mathrm{GP}(n, 2)$, *for* $n \equiv 5 \pmod 6$.

Applying Theorem 6.5.18, we see that $\mathrm{GP}(n, 2) \cong \mathrm{GP}(n, n-2) \cong \mathrm{GP}(n, (n-1)/2) \cong \mathrm{GP}(n, (n+1)/2)$, when $n \equiv 5 \pmod 6$. We also remark that in our definition of generalized Petersen graphs, the case $k = n/2$ is not allowed as the resulting graph is not cubic. As the preceding result implies, authors occasionally modify the definition of the generalized Petersen graph in this way if it is convenient (see, for example, Alspach (1983)). It may be interesting to know that when $\mathrm{GP}(n, k)$ is not Hamiltonian, Bondy has shown that it is **hypo-Hamiltonian** (Bondy, 1972, Corollary 3.1). A graph is hypo-Hamiltonian if it is not Hamiltonian but every graph obtained by deleting a vertex is Hamiltonian.

6.5.5 Concluding Remarks

Finally, one can ask are there constructions of families of graphs similar to the generalized Petersen graphs? The answer is, of course, yes, with many appearing in the literature. We will discuss three that are in some senses "closest" to generalized Petersen graphs. First, one could drop the condition that the outside cycle is a circulant graph with connection set ± 1. This was done by Boben, Pisanski, and Žitnik, who called such graphs ***I*-graphs** (Boben et al., 2005). Amongst other things, they determined which I-graphs are vertex-transitive, calculated their automorphism groups, and determined the isomorphism classes of I-graphs. Second, instead of having one 1-factor between the outside and inside vertices, we could have two such 1-factors. This was done by Wilson (2008), who called such graphs **rose window graphs**. See Definition 10.5.10. Rose window graphs can be more complicated than either of the previous two families, and Wilson found four families of edge-transitive rose window graphs. He conjectured that these were the only edge-transitive rose window graphs, which was verified by Kovács, Kutnar, and Marušič

(Kovács et al., 2010). The full automorphism groups of edge-transitive rose window graphs are also computed in Kovács et al. (2010). Finally, one could delete the outside and inside edges, and replace them with two additional 1-factors from the outside vertices to the inside vertices (and obtain a bipartite cubic graph with ρ in its automorphism group). This was done in a specific case by Boreham, Bouwer, and Frucht (Boreham et al., 1974) (who did not give them a name). Amongst other things, they determined the isomorphism classes of such graphs as well as their full automorphism groups when the number of vertices is $2p$, p a prime.

6.6 Fermat Graphs

Fermat graphs (also called **Marušič–Scapellato** graphs after Marušič and Scapellato (1992a) who discovered them) are important as they are the only examples of vertex-transitive graphs of order a product of two distinct primes whose automorphism groups are quasiprimitive and imprimitive. So amongst other things, these graphs are perhaps the simplest family of graphs whose automorphism group has this property. This of course clearly demonstrates that the notion of a quasiprimitive groups is more general than the notion of a primitive group. We mostly follow Marušič and Scapellato (1993). Fermat graphs are generalized orbital digraphs of particular actions of $\mathrm{SL}(2, 2^k)$, $k \geq 2$, one for each divisor m of $2^k - 1$. We first obtain the imprimitive action of $\mathrm{SL}(2, 2^k)$ with blocks of size $2^k - 1$, and from this action will obtain the others.

As $\mathrm{SL}\left(2, 2^k\right) \leq \mathrm{GL}\left(2, 2^k\right)$, $\mathrm{SL}\left(2, 2^k\right)$ also permutes the vectors of $\mathbb{F}_{2^k}^2 \setminus \{(0,0)\}$. For our first imprimitive action, the set being permuted is $\mathbb{F}_{2^k}^2 \setminus \{(0,0)\}$ of order $2^{2k} - 1 = \left(2^k + 1\right)\left(2^k - 1\right)$. Earlier, we defined a projective point to be a one-dimensional subspace. With this definition, it is clear that the zero vector is contained in every projective point. Here, we think of $\mathrm{SL}\left(2, 2^k\right)$ as being transitive not only on the projective points $\mathrm{PG}\left(1, 2^k\right)$, but simultaneously on the nonzero vectors in $\mathbb{F}_{2^k}^2$. So in the discussion that follows, we will consider a projective point to be all of the nonzero vectors in a one-dimensional subspace, and so each projective point will consist of $2^k - 1$ vectors. The projective points will be the blocks, and the vectors the elements of the blocks.

As there are $2^k - 1$ vectors in a projective point, there are $2^k + 1$ projective points. It is traditional to identify the projective points with elements of $\mathbb{F}_{2^k} \cup \{\infty\}$ in the following way. The nonzero vectors in the one-dimensional subspace generated by $(1, 0)$, will be identified with ∞. Any other one-dimensional subspace is generated by a vector of the form $(c, 1)$, where $c \in \mathbb{F}_{2^k}$. The nonzero vectors in the one-dimensional subspace generated by $(c, 1)$ will be identified with c. To distinguish the field element c from the projective point c, the projective point c will be in bold. Before establishing some basic facts about the action of $\mathrm{SL}\left(2, 2^k\right)$ on $\mathbb{F}_{2^k}^2 \backslash \{0\}$, we fix some notation that will be used throughout this section.

For $a \in \mathbb{F}_{2^k}^*$ and $b \in \mathbb{F}_{2^k}$, let

$$h_b = \begin{bmatrix} 1 & b \\ 0 & 1 \end{bmatrix}, \quad \text{and} \quad k_a = \begin{bmatrix} a & 0 \\ 0 & a^{-1} \end{bmatrix}.$$

Let $H = \{h_b : b \in \mathbb{F}_{2^k}\}$ and $K = \left\{k_a : a \in \mathbb{F}_{2^k}^*\right\}$. Of course, each element of H and K has determinant 1, so $H, K \leq \mathrm{SL}\left(2, 2^k\right)$.

Lemma 6.6.1 *For $k \geq 2$, $\mathrm{SL}\left(2, 2^k\right)$ is transitive on vectors in $\mathbb{F}_{2^k}^2 \backslash \{(0,0)\}$ with block system $\mathrm{PG}\left(1, 2^k\right)$ of $2^k + 1$ blocks of size $2^k - 1$. Additionally,*

(i) $\mathrm{Stab}_{\mathrm{SL}(2,2^k)}(1,0) = H \cong \mathbb{Z}_2^k$, *has order* 2^k,
(ii) $\mathrm{Stab}_{\mathrm{SL}(2,2^k)}(\infty) = H \rtimes K \cong \mathbb{Z}_2^k \rtimes \mathbb{Z}_{2^k-1}$, *has order* $2^k \cdot (2^k - 1)$,
(iii) $\mathrm{SL}(2, 2^k)/\mathrm{PG}(1, 2^k) \cong \mathrm{PSL}(2, 2^k)$ *is 2-transitive, and*
(iv) $\mathrm{fix}_{\mathrm{SL}(2,2^k)}(\mathrm{PG}(1, 2^k)) = 1$.

Proof First, we have already discussed that $\mathrm{SL}\left(2, 2^k\right)$ is transitive on vectors in $\mathbb{F}_{2^k}^2 \setminus \{(0,0)\}$. We also know that $\mathrm{PG}\left(1, 2^k\right)$ is a block system, as the elements of $\mathrm{PG}\left(1, 2^k\right)$ are one-dimensional subspaces of $\mathbb{F}_{2^k}$ with the zero vector removed (so any two elements of $\mathrm{PG}\left(1, 2^k\right)$ are either disjoint or the same).

Of course, $\mathrm{PSL}\left(2, 2^k\right) \leq \mathrm{PGL}\left(2, 2^k\right)$, and $\gcd\left(2, 2^k - 1\right) = 1$. Thus,

$$\left|\mathrm{PSL}\left(2, 2^k\right)\right| = \left|\mathrm{PGL}\left(2, 2^k\right)\right| = \left|\mathrm{SL}\left(2, 2^k\right)\right| = 2^k\left(2^k - 1\right)\left(2^k + 1\right).$$

(See p. 160 for $|\mathrm{PGL}(n, q)|$, Lemma 5.4.11 for $|\mathrm{PSL}(n, q)|$, and Lemma 5.4.9 for $\left|\mathrm{SL}\left(2, 2^k\right)\right|$.) Then $\mathrm{PSL}\left(2, 2^k\right) = \mathrm{SL}\left(2, 2^k\right)/\mathrm{PG}\left(1, 2^k\right)$ is 2-transitive on the projective points $\mathrm{PG}\left(1, 2^k\right)$ by Exercise 5.4.9. Thus (iii) and (iv) hold.

As the underlying additive group of $\mathbb{F}_{2^k}$ is an elementary abelian 2-group, straightforward computations show that H is an elementary abelian 2-group of order 2^k, and K is a cyclic group of order $2^k - 1$. Also, $h(1,0) = (1,0)$ and $k_a(1,0) = (a,0)$ for every $h \in H$ and $k_a \in K$, so both H and K stabilize the projective point ∞. Straightforward computations also show that K normalizes H, and clearly $H \cap K = 1$. As K normalizes H, HK is a group. Then $HK \cong H \rtimes K$ and is of order $2^k \cdot \left(2^k - 1\right)$. As $\left|\mathrm{PG}\left(1, 2^k\right)\right| = 2^k + 1$, we see $HK = \mathrm{Stab}_{\mathrm{SL}(2,2^k)}(\infty)$ by the orbit-stabilizer theorem. Thus (ii) follows. Again, by the orbit-stabilizer theorem, $H = \mathrm{Stab}_{\mathrm{SL}(2,2^k)}(1,0)$ and (i) follows. □

Of course there are more subgroups of $H \rtimes K$ that contain H, and by Lemma 2.3.3 each such subgroup will give a block system of $\mathrm{SL}\left(2, 2^k\right)$ that refines

$\mathrm{PG}\left(1,2^k\right)$ (although we will find these other block systems in a way that does not involve Lemma 2.3.3). The other imprimitive permutation representations of $\mathrm{SL}\left(2,2^k\right)$ that we seek are the induced actions on these other block systems of $\mathrm{SL}\left(2,2^k\right)$.

Lemma 6.6.2 *For $k \geq 2$, $\mathrm{Stab}_{\mathrm{SL}(2,2^k)}(\infty)^\infty = K^\infty$ is a regular cyclic subgroup of order 2^k-1. Consequently, for each divisor ℓ of 2^k-1, there is a unique block system $\mathcal{D}_\ell$ of $\mathrm{SL}\left(2,2^k\right)$ with blocks of size ℓ that is a refinement of $\mathrm{PG}\left(1,2^k\right)$. Additionally, the blocks of $\mathcal{D}_\ell$ contained in ∞ are the orbits of the unique subgroup of K^∞ of order ℓ.*

Proof Note that for $c \in \mathbb{F}_{2^k}^*$ we have $h_b(c,0) = (c,0)$. As $\mathrm{Stab}_{\mathrm{SL}(2,2^k)}(\infty) = H \rtimes K$ by Lemma 6.6.1, and every element of $H \rtimes K$ can be written as hk, $h \in H$, $k \in K$, we see $\left(\mathrm{Stab}_{\mathrm{SL}(2,2^k)}(\infty)\right)^\infty = K^\infty$. Let d be a generator of $\mathbb{F}_{2^k}^*$. Then $k_d^i(c,0) = \left(d^i c, 0\right)$ for every $c \in \mathbb{F}_{2^k}^*$ and positive integer i. We conclude that $(k_d)^\infty$ has order $|d| = 2^k-1$ and is transitive, and so K^∞ is cyclic and regular of order $2^k - 1$ as a transitive abelian group is regular by Theorem 2.2.17.

For each divisor ℓ of 2^k-1, $\left(\mathrm{Stab}_{\mathrm{SL}(2,2^k)}(\infty)\right)^\infty$ has a unique normal subgroup L of order ℓ, and so by Theorem 2.2.7 the set of orbits of L are a block system of $\left(\mathrm{Stab}_{\mathrm{SL}(2,2^k)}(\infty)\right)^\infty$. By Theorem 2.3.10, there is a block system of $\mathrm{SL}\left(2,2^k\right)$ with blocks of size ℓ and it is a refinement of $\mathrm{PG}\left(1,2^k\right)$ as each orbit of L is contained in ∞. Finally, suppose $\mathcal{C} \preceq \mathrm{PG}\left(1,2^k\right)$ has blocks of size ℓ. The set of blocks of $\mathcal{C}$ contained in ∞ is a block system $\mathcal{D}$ of $\left(\mathrm{Stab}_{\mathrm{SL}(2,2^k)}(\infty)\right)^\infty$. As $\left(\mathrm{Stab}_{\mathrm{SL}(2,2^k)}(\infty)\right)^\infty$ is a regular cyclic subgroup, it follows by Theorem 2.2.19 that $\mathcal{D}$ is the set of orbits of the unique subgroup of $\left(\mathrm{Stab}_{\mathrm{SL}(2,2^k)}(\infty)\right)^\infty$ of order ℓ. Thus there is only one block system of $\mathrm{SL}\left(2,2^k\right)$ with blocks of size ℓ. □

The following notation will be used for the remainder of this section. Let ℓ divide $2^k - 1$. We denote the unique block system of $\mathrm{SL}\left(2,2^k\right)$ with blocks of size ℓ that is a refinement of $\mathrm{PG}\left(1,2^k\right)$ by $\mathcal{D}_\ell$. Additionally, set $m = \left(2^k - 1\right)/\ell$, and $L \leq K$ be the unique subgroup of order ℓ. The following result summarizes much of our work thus far.

Theorem 6.6.3 *Let $k \geq 2$, ℓ divide $2^k - 1$ and $m = \left(2^k - 1\right)/\ell$. Then*

$$\mathrm{SL}\left(2,2^k\right)/\mathcal{D}_\ell \cong \mathrm{SL}\left(2,2^k\right),$$

and has a block system $\mathrm{PG}\left(1,2^k\right)/\mathcal{D}_\ell$ with 2^k+1 blocks of size m. Additionally, $\mathrm{Stab}_{\mathrm{SL}(2,2^k)/\mathcal{D}_\ell}(D_\ell) = (H \rtimes L)/\mathcal{D}_L \cong H \rtimes L$, where $D_\ell \in \mathcal{D}_\ell$ with $(1,0) \in D_\ell \subseteq \infty$.

Proof By Lemma 6.6.2, $\mathrm{SL}\left(2,2^k\right)$ has a unique block system $\mathcal{D}_\ell$ with blocks of size ℓ that is a refinement of $\mathrm{PG}\left(1,2^k\right)$. As each projective point of $\mathrm{PG}\left(1,2^k\right)$

contains $2^k - 1$ vectors, $\mathrm{SL}\left(2, 2^k\right)/\mathcal{D}_\ell$ has $\mathrm{PG}\left(1, 2^k\right)/\mathcal{D}_\ell$ as a block system with blocks of size $\left(2^k - 1\right)/\ell = m$. We have $\mathrm{fix}_{\mathrm{SL}(2,2^k)}\left(\mathrm{PG}\left(1, 2^k\right)\right) = 1$ by Lemma 6.6.1 (iv). As $\mathcal{D}_\ell \leq \mathrm{PG}\left(1, 2^k\right)$ and $\mathrm{fix}_{\mathrm{SL}(2,2^k)}(\mathcal{D}_\ell) \leq \mathrm{fix}_{\mathrm{SL}(2,2^k)}\left(\mathrm{PG}\left(1, 2^k\right)\right) = 1$, we have $\mathrm{fix}_{\mathrm{SL}(2,2^k)}(\mathcal{D}_\ell) = 1$ and so $\mathrm{SL}\left(2, 2^k\right)/\mathcal{D}_\ell \cong \mathrm{SL}\left(2, 2^k\right)$. Also, $H = \mathrm{Stab}_{\mathrm{SL}(2,2^k)}(1, 0)$ by Lemma 6.6.1 (i), and thus $H/\mathcal{D}_\ell \leq \mathrm{Stab}_{\mathrm{SL}(2,2^k)/\mathcal{D}_\ell}(D_\ell)$. By Lemma 6.6.2, the blocks of $\mathcal{D}_\ell$ contained in ∞ are the orbits of the unique subgroup of K^∞ of order ℓ, which is L^∞. Thus $(L/\mathcal{D}_\ell)^{\infty/\mathcal{D}_\ell} = 1$ and $L/\mathcal{D}_\ell \leq \mathrm{Stab}_{\mathrm{SL}(2,2^k)/\mathcal{D}_\ell}(D_\ell)$. By Lemma 6.6.1 (ii), we have $\mathrm{Stab}_{\mathrm{SL}(2,2^k)}(\infty) \cong H \rtimes K$, and thus $(H \rtimes L)/\mathcal{D}_\ell \leq \mathrm{Stab}_{\mathrm{SL}(2,2^k)/\mathcal{D}_\ell}(D_\ell)$. Finally, by Lemma 6.6.2, if $k_a \in K\backslash L$ then $k_a(D_\ell) \neq D_\ell$, and so $\mathrm{Stab}_{\mathrm{SL}(2,2^k)/\mathcal{D}_\ell}(D_\ell) = (H \rtimes L)/\mathcal{D}_\ell$. □

We may now formally define Fermat digraphs.

Definition 6.6.4 A generalized orbital digraph of $\mathrm{SL}\left(2, 2^k\right)/\mathcal{D}_\ell$ is called a **Fermat digraph**.

As Fermat digraphs are generalized orbital digraphs of $\mathrm{SL}\left(2, 2^k\right)/\mathcal{D}_\ell$ for some $\ell | \left(2^k - 1\right)$, in order to understand their structure we need information about the orbital digraphs of $\mathrm{SL}\left(2, 2^k\right)/\mathcal{D}_\ell$. We first find the suborbits, and then show that all suborbits that do not consist of singletons are isomorphic and are self-paired.

Theorem 6.6.5 $\mathrm{SL}\left(2, 2^k\right)/\mathcal{D}_\ell$ *has* m *suborbits of length* 1 *and* m *suborbits of length* 2^k. *Additionally, if* S *is a suborbit of* $\mathrm{SL}\left(2, 2^k\right)/\mathcal{D}_\ell$ *with respect to* $D_\ell \in \mathcal{D}_\ell$ *with* $D_\ell \subseteq \infty$, *then for every projective point* $\mathbf{c} \in \mathrm{PG}\left(1, 2^k\right)$ *with* $c \in \mathbb{F}_{2^k}$, *we have* $|S \cap (\mathbf{c}/\mathcal{D}_\ell)| = 1$.

Proof By Theorem 6.6.3, $\mathrm{Stab}_{\mathrm{SL}(2,2^k)/\mathcal{D}_\ell}(D_\ell) = (H \rtimes L)/\mathcal{D}_\ell$. As the action of H on ∞ is trivial by Lemma 6.6.2, the action of H on $\infty/\mathcal{D}_\ell$ is also trivial. Additionally, $L/\mathcal{D}_\ell$ is also trivial on the blocks of $\mathcal{D}_\ell$ contained in ∞ as they are the orbits of L by Lemma 6.6.2. Thus $\mathrm{Stab}_{\mathrm{SL}(2,2^k)/\mathcal{D}_\ell}(D_\ell)$ is trivial on $\infty/\mathcal{D}_\ell$, and so the m elements of $\infty/\mathcal{D}_\ell$ are all suborbits of length 1.

Let $D_0 \in \mathcal{D}_\ell$ that is contained in $\mathbf{0}$. Let $a \in \mathbb{F}_{2^k}$ such that a generates $\mathbb{F}_{2^k}^*$. The block D_ℓ of $\mathcal{D}_\ell$ that contains $(1, 0)$ is the orbit of the unique subgroup $L = \langle k_{a^m} \rangle$ of K of order ℓ. So every block of $\mathcal{D}_\ell$ contained in ∞ has the form $k_{a^j}(D_\ell) = \{(a^{mi+j}, 0) : 0 \leq i < \ell\}$, for some $0 \leq j \leq m - 1$. Let

$$\iota = \begin{bmatrix} 0 & -1 \\ 1 & 0 \end{bmatrix},$$

so that $\iota \in \mathrm{SL}\left(2, 2^k\right)$. Then $D_0 = \iota k_{a^j}(D_\ell) = \{(0, a^{mi+j}) : 0 \leq i < \ell\}$ for some $0 \leq j \leq m - 1$, and L also fixes D_0. This then implies that $\mathrm{Stab}_{(H \rtimes K)/\mathcal{D}_\ell}(D_0) = L/\mathcal{D}_\ell$. As $\mathrm{SL}\left(2, 2^k\right)$ is 2-transitive on $\mathrm{PG}\left(1, 2^k\right)$, the suborbit S of $\mathrm{SL}\left(2,2^k\right)/\mathcal{D}_\ell$ with respect to D_ℓ that contains D_0 contains at least one element of $\mathcal{D}_\ell$ that is a subset of each projective point that is not ∞, and the number $t \geq 1$ of blocks of $\mathcal{D}_\ell$ that are a subset of a projective point is the same for each projective point that is not ∞. Thus $|S| = t2^k$. We then have $\mathrm{Stab}_{\mathrm{Stab}_{\mathrm{SL}(2,2^k)/\mathcal{D}_\ell}}(D_0) = L/\mathcal{D}_\ell$ as, by Theorem 6.6.3, $\mathrm{Stab}_{\mathrm{SL}(2,2^k)/\mathcal{D}_\ell}(D_\ell) = (H \rtimes L)/\mathcal{D}_\ell$. Hence by the orbit-stabilizer theorem, $|S| = [\mathrm{Stab}_{\mathrm{SL}(2,2^k)/\mathcal{D}_\ell}(D_0) : L] = |H| = 2^k$. Then $t = 1$. Thus every suborbit S of $\mathrm{SL}\left(2,2^k\right)/\mathcal{D}_\ell$ that contains a block of $\mathcal{D}_\ell$ contained in $\mathbf{0}$ satisfies $|S \cap (\mathbf{c}/\mathcal{D}_\ell)| = 1$ for every projective point $\mathbf{c} \in \mathrm{PG}\left(1,2^k\right)$ with $c \in \mathbb{F}_{2^k}$. As H stabilizes D_ℓ and is transitive on $\mathrm{PG}\left(1,2^k\right)\backslash\{\infty\}$, the result follows. □

Lemma 6.6.6 *Any two orbital digraphs of* $\mathrm{SL}\left(2,2^k\right)/\mathcal{D}_\ell$ *with respect to* $D_\ell \subseteq \infty$ *of valency* 2^k *are isomorphic, and they are all self-paired.*

Proof Let T be the set of all orbital digraphs of $\mathrm{SL}\left(2,2^k\right)/\mathcal{D}_\ell$ with respect to $D_\ell \subseteq \infty$ of valency 2^k, and $\Gamma \in T$. For each $x \in \mathbb{F}_{2^k}^*$, let xI be the diagonal matrix all of whose diagonal entries are x. As we have seen, $Z = \left\{xI : x \in \mathbb{F}_{2^k}^*\right\}$ centralizes $\mathrm{SL}\left(2,2^k\right)$. So $xI/\mathcal{D}_\ell$ centralizes $\mathrm{SL}\left(2,2^k\right)/\mathcal{D}_\ell$.

Suppose $xI/\mathcal{D}_\ell \in \mathrm{Aut}(\Gamma)$. Then $xIk_{x^{-1}}/\mathcal{D}_\ell \in \mathrm{Aut}(\Gamma)$, and

$$xIk_{x^{-1}} = \begin{bmatrix} 1 & 0 \\ 0 & x^2 \end{bmatrix}.$$

If $xIk_{x^{-1}}/\mathcal{D}_\ell \neq 1$, then $(xIk_{x^{-1}})^\infty = 1$ while $(xIk_{x^{-1}})^\mathbf{1} \neq 1$. But then $|S \cap (\mathbf{1}/\mathcal{D}_\ell)| \geq 2$, where S is a suborbit of $\mathrm{SL}\left(2,2^k\right)/\mathcal{D}_\ell$ with respect to D_ℓ. This contradicts Theorem 6.6.5, so $xI/\mathcal{D}_\ell = 1$.

If $xI/\mathcal{D}_\ell \neq 1$, then $xI/\mathcal{D}_\ell \notin \mathrm{Aut}(\Gamma)$. By Exercise 1.4.16, $xI/\mathcal{D}_\ell$ permutes the orbital digraphs of $\mathrm{SL}\left(2,2^k\right)$, and so maps one of valency 2^k to another of valency 2^k. As a projective point is simply all scalar multiples of a nonzero vector, and multiplication by xI multiplies a vector by the scalar x, Z fixes every projective point and is transitive on every projective point. As it is semiregular by Lemma 2.2.16, $Z/\mathcal{D}_\ell$ is semiregular of order $\left(2^k-1\right)/\ell = m$. So the natural action of $Z/\mathcal{D}_\ell$ on T is transitive, and the elements of T are pairwise isomorphic.

Finally, as each pair of digraphs in T are isomorphic, every element of T is either self-paired or a digraph. They cannot all be digraphs, as m is odd, so they are all self-paired and graphs. □

Marušič and Scapellato also gave a more traditional definition of Fermat digraphs, and proved this definition is equivalent to the orbital digraph definition (Marušič and Scapellato, 1993, Proposition 4.1). We are content to simply state the definition and result without proof.

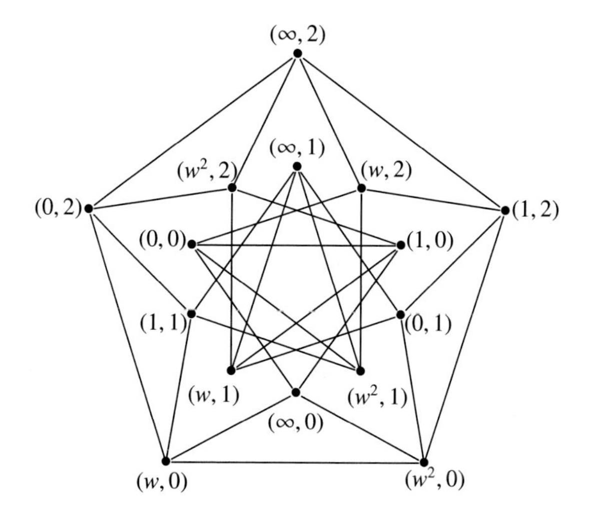

Figure 6.10 The Fermat graph $X(5, 3, \emptyset, \{0\})$.

Definition 6.6.7 Let $k \geq 2$, m be a divisor of $2^k - 1$, w a fixed generator of $\mathbb{F}_{2^k}$, and $S, T \subseteq \mathbb{Z}_m$. Define a digraph $X = X(k, m, S, T)$ by $V(X) = \mathrm{PG}\left(1, 2^k\right) \times \mathbb{Z}_m$ and for each $r \in \mathbb{Z}_m$ and $x \in \mathrm{PG}\left(1, 2^k\right)$, the neighbors of (x, r) are

$$(\infty, r + s) \text{ and } (b, r + t) \qquad \text{if } x = \infty,$$
$$(x, r + s), (\infty, r - t), \text{ and } (x + w^i, -r + t + 2i) \quad \text{if } x \neq \infty,$$

where $s \in S$, $t \in T$, $b \in \mathbb{F}_{2^k}$, and $i \in \mathbb{Z}_{2^k - 1}$.

The Fermat graph $X(5, 3, \emptyset, \{0\})$ is shown in Figure 6.10.

Theorem 6.6.8 *A generalized orbit digraph of* $\mathrm{SL}\left(2, 2^k\right)/\mathcal{D}_\ell$ *is isomorphic to* $X(k, m, S, T)$ *for some* $S, T \subset \mathbb{Z}_m$ *where* $m = \left(2^k - 1\right)/\ell$.

As mentioned earlier, Fermat graphs play a large role in the classification of vertex-transitive graphs of order a product of two distinct primes. Marušič and Scapellato (1992a) have shown that any vertex-transitive graph of order a product of two distinct primes that is not isomorphic to metacirculant graph but has an imprimitive automorphism group is a Fermat graph. This occurs when $p = 2^k + 1$ is a Fermat prime (hence the name chosen for these graphs), and q is any prime divisor of $2^k - 1$.

Information about the automorphism groups of Fermat digraphs is incomplete. Praeger, Wang, and Xu (Praeger et al., 1993, Lemma 4.9) have determined the full automorphism groups of edge-transitive Fermat graphs of order pq, $p = 2^k + 1$ a Fermat prime and $q | \left(2^k - 1\right)$ is also prime, whose automorphism group is contained in $\mathrm{Aut}\left(\mathrm{SL}\left(2, 2^k\right)\right)$. The full automorphism groups of all Fermat digraphs of order pq is given in Dobson et al. (2020a). It is known though that not all Fermat graphs have automorphism group contained in $\mathrm{Aut}\left(\mathrm{SL}\left(2, 2^k\right)\right)$, as Marušič and Scapellato (1994b, p. 192–193) have observed that $\mathrm{SL}\left(2, k^2\right) \leq \mathrm{Sp}(4, k)$ for all prime powers k, and is transitive. The group $\mathrm{Sp}(4, k)$ in this action is primitive. We are then left with an obvious problem.

Problem 6.6.9 Determine the automorphism groups of Fermat digraphs.

The isomorphism problem for Fermat graphs of order pq, where p and q are distinct primes, was first solved in Dobson (2016). A better solution, involving the diagonal matrices, can be found in Dobson et al. (2020a). We have another obvious problem.

Problem 6.6.10 Determine necessary and sufficient conditions for two Fermat digraphs to be isomorphic.

We postpone discussing that most Fermat graphs are Hamiltonian until Theorem 11.3.4, by which time we will have developed the appropriate tools for the proof.

6.7 Exercises

6.1.1 Show that both the points $(0, 0, 1)$ and $(0, 1, 0)$ are not contained in exactly two lines of the Fano plane, and that $(0, 0, 1) + (0, 1, 0) = (0, 1, 1)$ is contained in both of them. Deduce that any two points p and p' in the Fano plane are both not contained in exactly two lines of the Fano plane, and that $p + p'$ is contained in both of them.

6.1.2 Show that the 4-arc $(p_4, \ell_4), (p_1, \ell_5), (p_5, \ell_7), (p_3, \ell_6), (p_6, \ell_4)$ is not contained in a 7-cycle of the Coxeter graph.

6.1.3 Show that δ defined by $\delta\left(v_k^j\right) = v_{2k}^{j+1}$ and $\delta(i_k) = i_{2k}$ is an automorphism of the Coxeter graph, where we adopt the convention that $v_{2k}^{3+1} = v_{2k}^1$. Also, show that the map ρ defined by $\rho\left(v_k^j\right) = v_{k+1}^j$ and $\rho(i_k) = i_{k+1}$ is also an automorphism of the Coxeter graph.

6.1.4 Draw the Frucht diagram of the Coxeter graph with respect to the semiregular element ρ defined in Exercise 6.1.3.

6.1.5 Show that the Coxeter graph is not isomorphic to a Cayley graph. (Hint: Show that a group of order 28 has a unique Sylow 7-subgroup. Then show that the normalizer of a Sylow 7-subgroup in $\mathrm{PGL}(2, 7)$ has order 42.)

6.1.6 Let $n_1 = 3$ and $n_2 = n_3 = 2$ as in the proof (of Theorem 6.1.5) that the Coxeter graph is not Hamiltonian. Show that there is exactly one 1-factor containing $v_5^2 v_0^2$, $v_4^3 v_0^3$, and $v_0^1 i_0$.

6.1.7 Let $n_1 = 3$ and $n_2 = n_3 = 2$ as in the proof that the Coxeter graph is not Hamiltonian. Show that there are no 1-factors containing $v_5^2 v_0^2$, $v_3^3 v_0^3$, and $v_0^1 i_0$.

6.1.8 Show that the truncation of the Coxeter graph is not a Cayley graph. (Hint: Show that PSL(3, 2) has eight Sylow 7-subgroups and a group of order 84 has a unique Sylow 7-subgroup. Then show $\mathrm{PSL}(3,2) \rtimes \mathbb{Z}_2$ has no subgroup of order 84.)

6.2.1 What is the valency of a vertex in $K(n, k)$?

6.2.2 Draw the graph $K(6, 2)$.

6.2.3 Show that if $n \neq 2k + 1$, then $K(n, k)$ is arc-transitive but not 2-arc-transitive.

6.2.4 Draw the graph O_3.

6.2.5 Show that the graph O_k is not 4-arc transitive, $k \geq 2$.

6.3.1 What is the valency of a vertex in $J(n, k)$? In $\bar{J}(2k, k)$?

6.3.2 Show that the subgraph of $J(n, k)$ induced by the set $\mathcal{Z}_v$ of Theorem 6.3.8 is a maximal clique of $J(n, k)$ of order $k + 1$.

6.3.3 Show that $\langle \mathcal{S}_{2k}, \iota \rangle \cong \mathcal{S}_{2k} \times \langle \iota \rangle$. Begin by showing that ι commutes with every permutation of $\mathcal{S}_{2k}$ in its action on k-subsets.

6.3.4 Let $k \geq 2$ be an integer, $A, A' \subset [1, 2k - 1]$, $B = (A \cap A') \cup \{2k\}$, $C = [1, 2k - 1] \backslash (A \cup A')$, $D = A \backslash B$, and $E = A' \backslash B$. Show that if the sets B, C, D, and E are all nonempty, then $\{B, C, D, E\}$ is a partition of $[1, 2k]$.

6.3.5 Let $k \geq 2$, and A, A', B, C, D, E be as in Exercise 6.3.4 with $|A| = |A'| = k - 1$. Show that if $\{B, C, D, E\}$ is a partition, then it refines the partitions of $[1, 2k]$ consisting of $A \cup \{2k\}$ and its complement, and $A' \cup \{2k\}$ and its complement.

6.3.6 Show that $J(n, k)$ is arc-transitive, but not 2-arc-transitive.

6.3.7 Show that $\mathcal{S}_4 \times \mathcal{S}_2 \cong \mathcal{S}_3 \wr \mathcal{S}_2$.

6.4.1 Prove Lemma 6.4.24.

6.4.2 Prove Lemma 6.4.25.

6.4.3 Draw the Levi graph of the Möbius–Kantor configuration, which is given in Figure 6.11. This graph is known as the **Möbius–Kantor graph**, and is F16 in the Foster census.

6.4.4 Draw the Levi graph of the Pappus configuration, which is given in Figure 6.12. This graph is the **Pappus graph**.

6.4.5 Draw the Levi graph of the projective space PG(2, 3).

6.5.1 Prove Lemma 6.5.8.

6.5.2 Show that if $n \neq 4$, then $B(n, 1) = A(n, 1)$. (Hint: The proof is similar to the proof of Lemma 6.5.11 with 4-cycles playing the role of 8-cycles.)

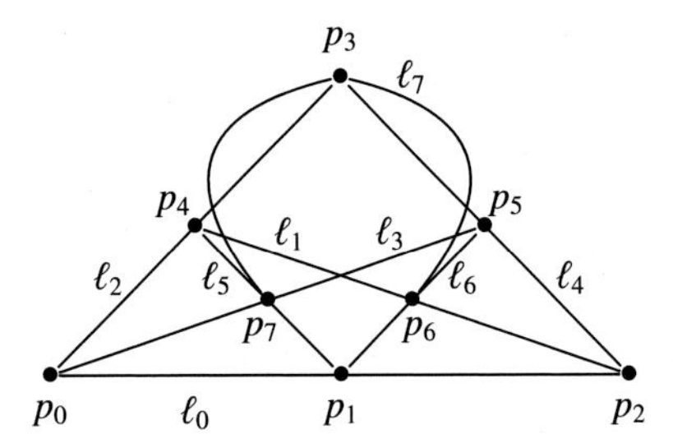

Figure 6.11 The Möbius–Kantor configuration.

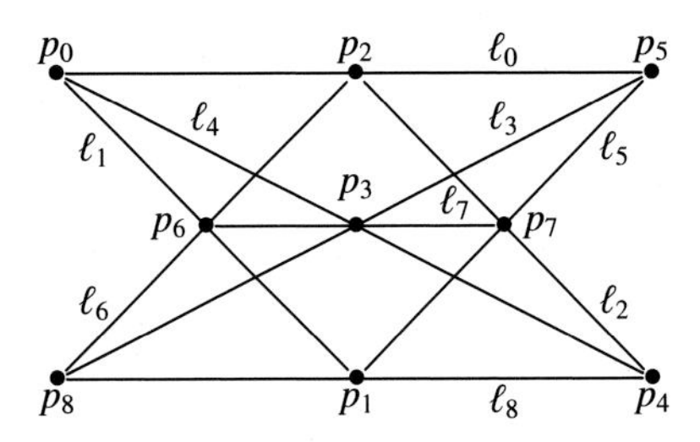

Figure 6.12 The Pappus configuration.

6.5.3 Show that if $n \neq 5$ or 10, then $B(n, 2) = A(n, 2)$. (Hint: The proof is similar to the proof of Lemma 6.5.11 with 5-cycles playing the role of 8-cycles.)

6.5.4 Show that there are 14 vertices of GP(26, 5) connected to u_0 by a path of length 4.

6.5.5 Show that $B(13, 5) = A(13, 5)$. The proof is similar to the proof of Lemma 6.5.11 with 7-cycles playing the role of 8-cycles.

6.5.6 Show that $\langle \rho, \tau, \delta, \sigma \rangle \leq \text{Aut}(\text{GP}(n, k))$ is transitive on the set of 2-arcs of GP(n, k) for $(n, k) = (4, 1), (8, 3), (12, 5)$, and $(24, 5)$.

6.5.7 Let Γ be a graph with semiregular automorphism ρ with two orbits B_0 and B_1 of size n. Suppose $\Gamma[B_0]$ and $\Gamma[B_1]$ are cycles and $\Gamma[B_0, B_1]$ is a 1-factor. Show $\Gamma \cong \text{GP}(n, k)$ for some positive integer k.

6.5.8 Show that GP(8, 3) is 2-arc-regular. (Hint: Show that u_4 is the only vertex x with six paths of length 4 from u_0 to x. Let $\omega \in \text{Aut}(\text{GP}(8, 3))$ such that ω fixes the 2-arc v_0, u_0, u_1. Show that $\omega = 1$.)

6.5.9 Show that GP(12, 5) is 2-arc-regular. (Hint: Count the paths of length 4 from u_0 to the other vertices of GP(12, 5). Let $\omega \in \text{Aut}(\text{GP}(12, 5))$

fix the vertices of a 2-arc in GP(12, 5) that contains u_0. Show that $\omega = 1$.)

6.5.10 Show that GP(24, 5) is 2-arc-regular.

6.5.11 Show that GP(10, 2) is isomorphic to the dodecahedron.

6.5.12 Show that Definition 1.5.15 is equivalent to the definition of metacirculant digraphs given in Definition 1.5.8. For the more difficult implication, use Lemma 6.5.17 to show that after relabeling the set ρ and τ permute with $\mathbb{Z}_m \times \mathbb{Z}_n$, one may assume $\rho(i, j) = (i, j + 1)$ and $\tau(i, j) = (i + 1, \alpha j + b_i)$, $\alpha \in \mathbb{Z}_n^*$, $b_i \in \mathbb{Z}_n$. Using that τ^m has a fixed point and so τ has an orbit of length m, conclude that $\sum_{i=0}^{m-1} \alpha^i b_i \equiv 0 \pmod n$, and that $\tau^m(i, j) = (i, \alpha^m j)$. Finally, show that there is an element γ in the centralizer in S_{mn} of $\langle\rho\rangle$ such that $\gamma^{-1}\tau\gamma(i, j) = (i + 1, \alpha j)$.

6.6.1 Show that H is transitive on the set $\mathrm{PG}\left(1, 2^k\right) \backslash \{\infty\}$.

7

The Cayley Isomorphism Problem

While this chapter is about isomorphisms between Cayley graphs, the problem of finding isomorphisms between simple graphs is a much more widely known problem. Recently, in an already famous result, Babai (2016) showed that the graph isomorphism problem can be solved in quasipolynomial time. In Section 7.1, we very briefly discuss this result. We then turn to the main topic of this chapter, and, in Section 7.2, we introduce the Cayley isomorphism problem. We then develop basic tools on the Cayley isomorphism problem, including Babai's characterization of the CI-property in Lemma 7.3.1. Section 7.4 is devoted to giving examples of non-CI-groups with respect to graphs, and we focus on cyclic non-CI-groups. Next, in Section 7.5, we determine all CI-groups with respect to digraphs of order a product of two primes. In Section 7.6, we show one direction of Pálfy's result characterizing which groups are CI-groups with respect to *every* class of combinatorial objects. We finish the chapter with a mainly expository section describing Muzychuk's solution to the isomorphism problem for circulant digraphs.

Our approach to the Cayley isomorphism problem will be entirely group theoretic. There is, though, another technique that is commonly used, broadly referred to as the **method of Schur**, first introduced by Schur (1933). The first use of the method of Schur to obtain results on the isomorphism problem was by Klin and Pöschel (1981) where they showed, amongst other results, that for distinct primes p and q, $\mathbb{Z}_{qp}$ is a CI-group with respect to digraphs. It is also the technique Muzychuk used for his solution to the isomorphism problem for circulant digraphs that we will see in Section 7.7. It is strongly recommended that if one is interested in the Cayley isomorphism problem, the method of Schur should be studied in addition to the group-theoretic approach used here. An introduction to the method of Schur can be found in Wielandt (1964, Chapter IV) and Scott (1987, Chapter 13), with more recent information and applications to algebraic combinatorics in Muzychuk and Ponomarenko (2009).

7.1 Graph Isomorphims in Quasipolynomial Time

Before discussing the Cayley isomorphism problem, it is natural to discuss the general graph isomorphism problem. Recently, Babai (2016) showed the graph isomorphism problem has quasipolynomial time complexity. We will briefly discuss his result, and to do so we will need to define a few terms. We begin with Big O notation.

Definition 7.1.1 Let $f, g\colon R \mapsto \mathbb{R}$, where $R \subseteq \mathbb{R}$ is unbounded, such that $g(x) > 0$ for all sufficiently large values of x. We write $f(x) = O(g(x))$ (as $x \to \infty$) if for all sufficiently large values of x, $|f(x)| \le cg(x)$ for some constant c. That is, there exists $x_0 \in \mathbb{R}$ such that $|f(x)| \le cg(x)$ for all $x \ge x_0$. The symbol $O(g(x))$ is read "Big O of $g(x)$."

Note that $2^{O(\log n)}$ is at most some polynomial in n, as there exists some constant c such that $O(\log n) \le c \log n$, so $2^{O(\log n)} \le (2^{\log n})^c = n^c$ (note that these statements are true for sufficiently large n, although henceforth we will not write or mention that). A function f is **quasipolynomial** if $f(n) = O\left(2^{(\log n)^{O(1)}}\right)$. So quasipolynomial functions are clearly larger than polynomial functions. The **computational complexity of an algorithm** is the number of steps (in the worst possible case) to execute the algorithm, based on its input. It is usually given asymptotically. For graphs and digraphs, the input is usually the number of vertices or edges – observe that the number of edges is a polynomial in the number of vertices. Babai proved the following result.

Theorem 7.1.2 *There exists a quasipolynomial time algorithm to determine if two graphs of order n are isomorphic.*

As the proof of this result is around 80 pages long, we will content ourselves with a brief overview of the proof. First, Babai considers the **string isomorphism problem**.

Definition 7.1.3 A **string** is a finite sequence of elements chosen from a set S, usually called an **alphabet**. Two strings $s = s_1 s_2 \dots s_r$ and $t = t_1 t_2 \dots t_r$ are **isomorphic** by $g \in S_r$ if $g(s) = g(s_1)g(s_2)\dots g(s_r) = t$. The **string isomorphism problem** asks if given two strings of the same length r and $G \le S_r$, is there $g \in G$ with $g(s) = t$?

A graph Γ of order n can easily be encoded as a string of length $n(n-1)/2$ with alphabet $\mathbb{Z}_2$. Simply label the edges of K_n with integers $1, \dots, n(n-1)/2$

and set $s_i = 1$ if the edge labeled i is in $E(\Gamma)$ and $s_i = 0$ otherwise. If Δ is another graph of order n, then $\Gamma \cong \Delta$ if and only if their corresponding strings are isomorphic by some $g \in \mathcal{S}_n$, where the action of $\mathcal{S}_n$ is the induced action on the edges of K_n.

The basic structure of the proof to solve the string isomorphism problem is due to Luks (1982). We remark that Luks uses the term "color" for "alphabet." Let s and t be two strings over an alphabet of the same length r, and $G \leq \mathcal{S}_r$. If G is intransitive, then we may reduce the string isomorphism problem to the induced action of G on each of its orbits to corresponding substrings of s and t. So we may assume G is transitive. Then choose a block system $\mathcal{B}$ of G with blocks of largest size. This gives that $G/\mathcal{B}$ is a primitive group of order G/K, where $K = \mathrm{fix}_G(\mathcal{B})$.

Luks more or less reduces the string isomorphism problem to checking $[G : K]$ instances of finding which elements of K are string isomorphisms between particular strings. So if $[G : K]$ is "large" in comparison with n, this cannot be done efficiently. It is this part of Luks algorithm that Babai improves. His improvement uses CFSG, and, in particular, a result of Cameron (1981) that gives all primitive groups that are "large." For a more thorough overview of the result and its proof see Babai (2016). For the proof itself, see Babai (2015).

7.2 Introduction and Basic Definitions

The Cayley isomorphism problem for graphs began in 1967, when Ádám (1967) conjectured that two circulant digraphs $\mathrm{Cay}(\mathbb{Z}_n, S)$ and $\mathrm{Cay}(\mathbb{Z}_n, S')$ are isomorphic if and only if there exists $m \in \mathbb{Z}_n^*$ such that $mS = S'$ (where $mS = \{ms : s \in S\}$). If $mS = S'$, it is often said that $\mathrm{Cay}(\mathbb{Z}_n, S)$ and $\mathrm{Cay}(\mathbb{Z}_n, S')$ are isomorphic by a **multiplier**. Recall that, by Lemma 1.2.15, the map $x \mapsto mx$ is a group automorphism of $\mathbb{Z}_n$ for each $m \in \mathbb{Z}_n^*$. Hence, Ádám conjectured that two circulant digraphs of order n are isomorphic if and only if they are isomorphic by a group automorphism of $\mathbb{Z}_n$. Also by Lemma 1.2.15, if one were to write down a shortest possible list $\mathcal{L}$ of permutations such that any two Cayley digraphs of $\mathbb{Z}_n$ are isomorphic if and only if they are isomorphic by a permutation in $\mathcal{L}$, then the group automorphisms of $\mathbb{Z}_n$ would necessarily have to be included in $\mathcal{L}$. So from this point of view, Ádám conjectured that this $\mathcal{L}$ is as short as possible for circulant digraphs.

It was quickly shown by Elspas and Turner (1970) that Ádám's conjecture is not true by giving the following example.

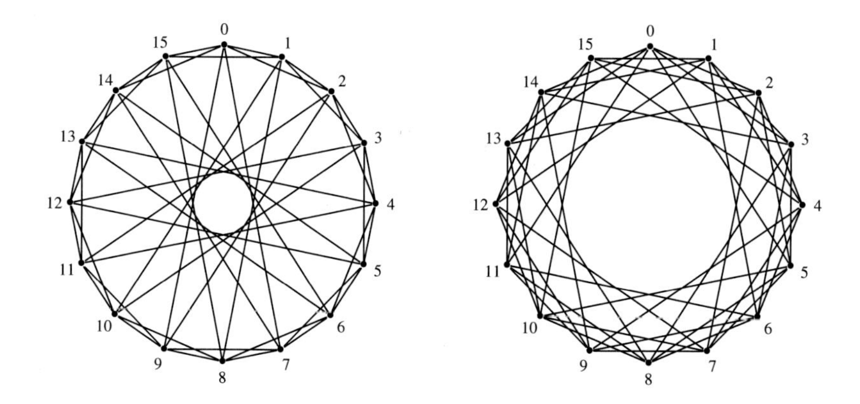

Figure 7.1 Two circulant graphs not isomorphic by a multiplier.

Example 7.2.1 There are two isomorphic circulant graphs of order 16 that are not isomorphic by a multiplier.

Solution The circulant graphs $\Gamma_1 = \text{Cay}(\mathbb{Z}_{16}, \{1, 2, 7, 9, 14, 15\})$ and $\Gamma_2 = \text{Cay}(\mathbb{Z}_{16}, \{2, 3, 5, 11, 13, 14\})$ are shown in Figure 7.1. If Γ_1 and Γ_2 are isomorphic by a multiplier, then there exists $m \in \mathbb{Z}_{16}$ such that $m\{1, 2, 7, 9, 14, 15\} = \{2, 3, 5, 11, 13, 14\}$. As both Γ_1 and Γ_2 are graphs, the map $x \mapsto -x$ is an automorphism of both Γ_1 and Γ_2. Thus for graphs, $mS = S'$ if and only if $(-m)S = S'$. We may thus assume without loss of generality that $m \leq 8$, and as $m \in \mathbb{Z}_{16}^*$, $m = 1, 3, 5, 7$. As 1 is in the connection set of Γ_1, m is in the connection set of Γ_2, so $m = 3, 5$. As $3 \cdot 2 = 6 \notin S'$, and $5 \cdot 2 = 10 \notin S'$, where S' is the connection set of Γ_2, we see that Γ_1 and Γ_2 are not isomorphic by a multiplier. Finally, straightforward though tedious computations will show that the map defined by $x \mapsto x$ if x is even and $x \mapsto x + 4$ if x is odd is an isomorphism from Γ_1 to Γ_2. □

Elspas and Turner (1970) also gave an example of two circulant digraphs of order 8 that are not isomorphic by a multiplier; see Example 7.7.31. After Elspas and Turner's examples showing that Ádám's original conjecture was false, the conjecture quickly turned into a problem and was generalized to Cayley graphs of groups that were noncyclic. We remark that one reason that Ádám's conjecture was not abandoned was that Turner (1967) had already verified Ádám's conjecture for circulant graphs of prime order.

Problem 7.2.2 Which groups G have the property that two Cayley (di)graphs of G are isomorphic if and only if they are isomorphic by a group automorphism of G?

Definition 7.2.3 A group G that has the property that any two Cayley (di)graphs of G are isomorphic if and only if they are isomorphic by a group automorphism of G is called a **CI-group with respect to (di)graphs**.

One may wonder why we do not just say "CI-group" instead of "CI-group with respect to (di)graphs." This is because it is possible to ask the same question about other classes of "combinatorial objects," e.g., combinatorial designs, once one has a notion of a "Cayley object" or "Cayley design." In fact, the question had already been considered for designs in the 1930s by Bays (1930, 1931) and Lambossy (1931).

Definition 7.2.4 A **combinatorial object** is an ordered pair (V, E), where V is the vertex set and $E \subseteq V \cup 2^V \cup 2^{2^V} \cup \ldots$ is the edge set. Here, for a set X, 2^X is the set of all subsets of X. An **isomorphism** between two combinatorial objects is a bijection that maps the vertex and edge set of one to the vertex and edge set of the other, and an **automorphism** is an isomorphism of a combinatorial object with itself.

We realize that we are being more than a little vague in our discussion of combinatorial objects, but a precise understanding of them is not that important for us here. The interested reader is referred to Muzychuk (1999) for more precise information on combinatorial objects. We define a "Cayley object" so that it agrees with Sabidussi's characterization of Cayley digraphs given in Theorem 1.2.20.

Definition 7.2.5 A **Cayley object** X of a group G in a class $\mathcal{K}$ of combinatorial objects is one in which $G_L \leq \text{Aut}(X)$, the automorphism group of X.

One more term is necessary before we proceed.

Definition 7.2.6 Let Γ be a Cayley (di)graph of G such that if Γ' is any Cayley (di)graph of G, then Γ and Γ' are isomorphic if and only if they are isomorphic by a group automorphism of G. Such a Cayley (di)graph of G is called a **CI-(di)graph of** G. Similarly, for a Cayley object X of G in some class of combinatorial objects $\mathcal{K}$, we say that X is a **CI-object of** G if and only if whenever X' is another Cayley object of G in $\mathcal{K}$, then X and X' are isomorphic if and only if $\alpha(X) = X'$ for some $\alpha \in \text{Aut}(G)$.

Evidently, G is a CI-group with respect to (di)graphs if and only if every Cayley (di)graph of G is a CI-(di)graph.

7.3 Basic Tools

We begin with the development of the main tool that is used almost universally in proving that groups are CI-groups with respect to (di)graphs (and other combinatorial objects). The version stated here is due to Babai (1977), although Alspach and Parsons (1979) also had a version of this result for cyclic groups. While working through the proof of the following result, observe that we do not use any combinatorial information – the proof is group theoretic, and this is why we will not require much information about combinatorial objects.

Lemma 7.3.1 *Let X be a Cayley object of G in some class $\mathcal{K}$ of combinatorial objects. Then the following are equivalent:*

(i) *X is a CI-object of G in $\mathcal{K}$,*
(ii) *whenever $\phi \in S_G$ such that $\phi^{-1}G_L\phi \leq \mathrm{Aut}(X)$, G_L and $\phi^{-1}G_L\phi$ are conjugate in* $\mathrm{Aut}(X)$.

Before proceeding to the proof, it might be useful to discuss the second condition in the previous result, as, at first, the condition seems very strange. Additionally, one may wonder why there is only one combinatorial object in the statement, not two. As the kind of combinatorial object is irrelevant, our discussion will only be about graphs.

Suppose Γ and Γ' are isomorphic Cayley graphs of G, with $\phi\colon \Gamma \to \Gamma'$ an isomorphism. Then ϕ is a bijection, so ϕ^{-1} is also a bijection, and in fact maps Γ' back to Γ (Exercise 1.1.11). For any automorphism γ' of Γ', we have that $\gamma'\phi$ is also an isomorphism from Γ to Γ', and as ϕ^{-1} maps Γ' back to Γ, $\phi^{-1}\gamma'\phi$ is an automorphism of Γ. So $\mathrm{Aut}(\Gamma)$ contains a conjugate of $\mathrm{Aut}(\Gamma')$. This is why the lemma does not mention Γ'. Also, as $g_L \in \mathrm{Aut}(\Gamma')$ for every $g \in G$, we see that $\phi^{-1}G_L\phi \leq \mathrm{Aut}(\Gamma)$.

Now, what about conjugating $\phi^{-1}G_L\phi$ by an element of $\mathrm{Aut}(\Gamma)$? Intuitively, one should think that conjugation of $\phi^{-1}G_L\phi$ by an automorphism of Γ replaces the isomorphism ϕ with a (hopefully nicer) isomorphism. For example, if $\gamma \in \mathrm{Aut}(\Gamma)$, then $\phi\gamma\colon \Gamma \to \Gamma'$ is an isomorphism. By our previous argument, we then see that $(\phi\gamma)^{-1}G_L(\phi\gamma) = \gamma^{-1}\phi^{-1}G_L\phi\gamma$, and so the effect of replacing ϕ with our (hopefully nicer) new isomorphism $\phi\gamma$ on $\phi^{-1}G_L\phi$ is to conjugate it by γ. Finally, if we can conjugate $\phi^{-1}G_L\phi$ by an automorphism of Γ and get G_L, our new (and nicer!) isomorphism of Γ and Γ' normalizes G_L – but we know what that normalizer is by Corollary 4.4.11! To the proof.

Proof (i) $\Rightarrow$ (ii): Suppose X is a CI-object of G in $\mathcal{K}$ and $\phi \in S_G$ such that $\phi^{-1}G_L\phi \leq \mathrm{Aut}(X)$. Similar to Exercise 1.1.11, we have $\mathrm{Aut}(\phi(X)) =$

$\phi\text{Aut}(X)\phi^{-1}$, so $\phi\left(\phi^{-1}G_L\phi\right)\phi^{-1} = G_L \leq \text{Aut}(\phi(X))$. Thus $\phi(X)$ is a Cayley object of G. As X is a CI-object of G, there exists $\alpha \in \text{Aut}(G)$ such that $\alpha(X) = \phi(X)$. Then $\phi^{-1}\alpha \in \text{Aut}(X)$ as $\phi^{-1}: X' \to X$ is an isomorphism. Finally,

$$\left(\phi^{-1}\alpha\right)^{-1}\left(\phi^{-1}G_L\phi\right)\left(\phi^{-1}\alpha\right) = \alpha^{-1}G_L\alpha = G_L,$$

by Corollary 4.4.11.
(ii) $\Rightarrow$ (i): Suppose $\phi: X \to X'$ is an isomorphism for two Cayley objects X and X' in $\mathcal{K}$. As $G_L \leq \text{Aut}(X')$, we have $\phi^{-1}G_L\phi \leq \text{Aut}(X)$. By hypothesis, there exists $\delta \in \text{Aut}(X)$ such that $\delta^{-1}\phi^{-1}G_L\phi\delta = G_L$. Note that $\phi\delta: X \to X'$ is an isomorphism, and that $\phi\delta$ normalizes G_L. By Corollary 4.4.11, we have $\phi\delta = g_L\alpha$ for some $\alpha \in \text{Aut}(X)$ and $g \in G$. As $g_L \in \text{Aut}(X')$, we have $g_L^{-1}\phi\delta = \alpha$ and $g_L^{-1}\phi\delta(X) = X'$. □

A couple of observations are in order here. First, in the structure of the above proof, what is so special about G_L (as opposed to something that is not a regular group)? Nothing. So could we replace G_L with a nonregular group and invoke Lemma 4.4.9 instead of Corollary 4.4.11 and get a version of Lemma 7.3.1 for non-Cayley vertex-transitive graphs? Absolutely, and in fact this has been done by Tyshkevich and Tan (1987) (this paper is in Russian, but the proof is no different than as advertised). The second observation is similar to the first, but more general. Suppose we have two graphs Γ_1 and Γ_2 (not necessarily even vertex-transitive) of order n whose automorphism groups both contain a common (hopefully large) subgroup H, with $\phi: \Gamma_1 \to \Gamma_2$ an isomorphism. Then $\phi^{-1}H\phi \leq \text{Aut}(\Gamma_1)$, and if there exists a $\delta \in \text{Aut}(\Gamma_1)$ such that $\delta^{-1}\phi^{-1}H\phi\delta = H$, then $\phi\delta: \Gamma_1 \to \Gamma_2$ is an isomorphism that is also contained in $N_{S_n}(H)$. If one can then calculate $N_{S_n}(H)$, then one has solved the isomorphism problem (although the solution will only be useful if $N_{S_n}(H)$ is not too large). This may seem a far-fetched approach, but it is precisely the technique used to solve the isomorphism problem for generalized Petersen graphs in Theorem 6.5.18.

Before turning to other tools, we now obtain our first CI-group. The proof does not use any combinatorial information – it is entirely group theoretic. This means that it holds *all* combinatorial objects.

Definition 7.3.2 Let G be a group. We say G is a **CI-group** if it is a CI-group with respect to every class of combinatorial objects.

The following result is due to Babai (1977), and was obtained by Turner (1967) in the special case of graphs.

Theorem 7.3.3 *Let p be a prime. Then $\mathbb{Z}_p$ is a CI-group.*

Proof Let X be a Cayley object of $\mathbb{Z}_p$ in some class $\mathcal{K}$ of combinatorial objects, and $\phi \in S_p$ such that $\phi^{-1}(\mathbb{Z}_p)_L\phi \leq \text{Aut}(X)$. Notice that $(\mathbb{Z}_p)_L$ has order p, and that S_p has order $p!$. Also observe that the highest power of p that divides $p!$ is p. We conclude that $(\mathbb{Z}_p)_L$ and $\phi^{-1}(\mathbb{Z}_p)_L\phi$ are Sylow p-subgroups of S_p, and so are Sylow p-subgroups of Aut(X). Consequently, by a Sylow theorem $(\mathbb{Z}_p)_L$ and $\phi^{-1}(\mathbb{Z}_p)_L\phi$ are conjugate in Aut(X). The result then follows by Lemma 7.3.1. □

Example 7.3.4 Is $\text{Cay}(\mathbb{Z}_{11}, \{1, 5, 6, 7\}) \cong \text{Cay}(\mathbb{Z}_{11}, \{1, 2, 4, 10\})$? Explain.

Solution According to Theorem 7.3.3, these two digraphs are isomorphic if and only if there is a multiplier $m \in \mathbb{Z}_{11}^*$ such that

$$m \cdot \{1, 5, 6, 7\} = \{m, 5m, 6m, 7m\} = \{1, 2, 4, 10\}.$$

We need only check $m = 2, 4, 10$ as $1 \in \{1, 5, 6, 7\}$. For $m = 2$, $2 \cdot \{1, 5, 6, 7\} = \{2, 10, 1, 3\}$, for $m = 4$, $4 \cdot \{1, 5, 6, 7\} = \{4, 9, 2, 6\}$, and for $m = 10$, $10 \cdot \{1, 5, 6, 7\} = \{10, 6, 5, 4\}$. We conclude that these two digraphs are not isomorphic. □

Example 7.3.5 Let $S_1 = \{2, 5, 6, 7, 8, 11\}$ and $S_2 = \{1, 3, 4, 9, 10, 12\}$. Is $\text{Cay}(\mathbb{Z}_{13}, S_1) \cong \text{Cay}(\mathbb{Z}_{13}, S_2)$? Explain. Are they graphs or digraphs?

Solution According to Theorem 7.3.3, these two digraphs are isomorphic if and only if there is $m \in \mathbb{Z}_{13}^*$ such that $m \cdot S_1 = S_2$. Noting that

$$2 \cdot \{2, 5, 6, 7, 8, 11\} = \{4, 10, 12, 1, 3, 9\},$$

we see that the digraphs are isomorphic. Also, these are graphs as

$$-S_1 = \{11, 8, 7, 6, 5, 2\} = S_1,$$

and they are complements of one another. Additionally, notice that the function $x \mapsto 2x$ in $\mathbb{Z}_{13}^*$ has cycle decomposition $(2, 4, 8, 3, 6, 12, 11, 9, 5, 10, 7, 1)$, and that the connection sets of these graphs are formed by taking every other element of this cycle. □

Definition 7.3.6 A graph Γ is **self-complementary** if Γ is isomorphic to its complement. That is, $\Gamma \cong \bar{\Gamma}$.

The graph in Example 7.3.5 is a self-complementary circulant graph. There has been quite a bit of interest in self-complementary circulants. See, for example, Alspach et al. (1999); Fronček et al. (1996); Jajcay and Li (2001); Li et al. (2014); Liskovets and Pöschel (2000); Sachs (1962), with the main question being whether or not they can be constructed using a construction introduced by Suprunenko (1985). Let G be a group and $\alpha \in \text{Aut}(G)$ such that α has exactly one fixed point and each cycle in the cycle decomposition of α has length a multiple of four. For each cycle $C = (a_0, a_1, \ldots, a_{4n-1})$ of α, let $C_e = \{a_m : m \text{ is even}\}$ and $C_o = \{a_m : m \text{ is odd}\}$. Choose one of these two sets, and call it C_0. Denote the other one by C_1. Let $S_i = \{C_i : C \text{ is a cycle in the cycle decomposition of } \alpha\}$. Let $\Gamma = \mathrm{Cay}(G, S_0)$. Then Γ is self-complementary with $\alpha(\Gamma) = \bar{\Gamma} = \mathrm{Cay}(G, S_1)$ (Exercise 7.4.6).

The notion of a self-complementary graph has been generalized in two ways. First, a self-complementary graph necessarily has an odd number of vertices. An analogous graph on an even number n of vertices would be a partition of K_n with a 1-factor removed into two edge disjoint isomorphic subgraphs, and is called an **almost self-complementary graph**. This has been investigated in Dobson and Šajna (2004); Potočnik and Šajna (2006, 2007, 2009); Zhou (2017). Second, instead of decomposing K_n into two edge disjoint isomorphic subgraphs, we could ask if this can be done for k edge disjoint isomorphic subgraphs. Such a decomposition is called a **homogeneous factorization of** K_n. Finally, one could also replace the graph K_n with any graph and ask the same question. See Cuaresma et al. (2008); Giudici et al. (2007, 2008); Li and Praeger (2002); Li et al. (2009); Praeger et al. (2009); Xu (2017) for the precise definitions and known results.

The proof of the following result has a similar flavor to the proof of Theorem 7.3.3 (and generalizes it if one only considers digraphs), and shows that graphs of "small" valency always have the CI-property. The result is due to Li (1998, Theorem 3.1), and generalizes an earlier result of Babai (1977, Theorem 3.6) giving the same result for p-groups. We will need some additional terms, and a preliminary lemma that will also be useful later.

Definition 7.3.7 Let π be a set of prime numbers. A group G is called a **π-group** if every element in G has order a product of powers of primes in π.

So a group G is a π-group if a prime p divides the order of an element of G only if $p \in \pi$.

Definition 7.3.8 Let π be a set of primes, and G a group of order n. We say that $H \leq G$ is a **Hall π-subgroup** of G if H is a π-group and $|G|/|H|$ is relatively prime to p for every $p \in \pi$. That is, if $p^a|n$, $a \geq 0$, then p^a divides $|H|$ for every $p \in \pi$.

A Hall π-subgroup is a generalization of a Sylow p-subgroup to more than one prime, but Hall π-subgroups need not exist. Gross has shown that odd order Hall π-subgroups of G are conjugate in G (Gross, 1987).

Lemma 7.3.9 *Let Γ be a weakly connected vertex-transitive digraph with each vertex of outvalence r, $v \in V(\Gamma)$, and $G \leq \mathrm{Aut}(\Gamma)$ be transitive. Then $|\mathrm{Stab}_G(v)|$ is not divisible by any prime larger than r.*

Proof Suppose $|\mathrm{Stab}_G(v)|$ is divisible by a prime $q > r$. Then there is $\gamma \in \mathrm{Stab}_G(v)$ of order q. Let $w \in V(\Gamma)$ with $\gamma(w) \neq 1$. As Γ is weakly connected, there is a vw-path P in the underlying simple graph of Γ, and P has a first vertex u that is not fixed by γ. Let $x \in V(\Gamma)$ precede u on P. Then either (x, u) or (u, x) is an arc of Γ, and x is inadjacent or outadjacent in Γ to every vertex in thc orbit of Γ that contains u. As in a vertex-transitive digraph, the invalence and outvalence of every vertex is the same, every vertex of Γ has outvalence at least q, a contradiction. □

Theorem 7.3.10 *Let G be a group of odd order n, $S \subseteq G$, and p the smallest prime divisor of n. If $\Gamma = \mathrm{Cay}(G, S)$ is weakly connected and $|S| \leq p - 1$, then Γ is a CI-digraph of G.*

Proof If $p = 2$, then as Γ is connected, it is a directed cycle whose automorphism group is regular and isomorphic to $\mathbb{Z}_n$. Then $G \cong \mathbb{Z}_n$, and Γ is a CI-digraph of G by Lemma 7.3.1. If $p \geq 3$, then by Lemma 7.3.9, no prime $q \geq p$ divides the order of $\mathrm{Stab}_{\mathrm{Aut}(\Gamma)}(v)$. Thus $\gcd(|G|, |\mathrm{Stab}_G(x)|) = 1$ and G_L is a Hall π-subgroup of $\mathrm{Aut}(\Gamma)$, where π is the set of primes dividing $|G|$. Now let $\phi \in \mathcal{S}_G$ such that $\phi^{-1}G_L\phi \leq \mathrm{Aut}(\Gamma)$. Then $\phi^{-1}G_L\phi$ is also a Hall π-subgroup of $\mathrm{Aut}(\Gamma)$ and so $\phi^{-1}G_L\phi$ and G_L are conjugate in $\mathrm{Aut}(\Gamma)$ by the previously mentioned result of Gross. The result follows by Lemma 7.3.1. □

The following result is due to Babai and Frankl (1978), and allows one to conclude that if H is not a CI-group with respect to digraphs, then no group that contains H as a subgroup is a CI-group with respect to digraphs.

Theorem 7.3.11 *Let G be a CI-group with respect to digraphs and $H \leq G$. Then H is a CI-group with respect to digraphs.*

Proof Let $\mathrm{Cay}(H, S_1)$ and $\mathrm{Cay}(H, S_2)$ be isomorphic Cayley digraphs of H. As $\mathrm{Cay}(H, S_1)$ is a CI-digraph of H if and only if its complement is a CI-digraph of H (Exercise 7.4.3) we may assume $\mathrm{Cay}(H, S_1)$ and $\mathrm{Cay}(H, S_2)$ are both connected by replacing them with their complements if necessary. By Lemma 1.2.14, we have $\langle S_1\rangle = \langle S_2\rangle = H$. Then $\mathrm{Cay}(G, S_1)$ and $\mathrm{Cay}(G, S_2)$ are isomorphic Cayley digraphs of G, so there exists $\alpha \in \mathrm{Aut}(G)$ such that $\mathrm{Cay}(G, S_2) = \alpha(\mathrm{Cay}(G, S_1)) = \mathrm{Cay}(G, \alpha(S_1))$. Hence $\alpha(S_1) = S_2$, and so $H = \langle S_2\rangle = \alpha(\langle S_1\rangle) = \alpha(H)$. The restriction of α to H is then an automorphism of H that is an isomorphism from $\mathrm{Cay}(H, S_1)$ to $\mathrm{Cay}(H, S_2)$. □

The following result was shown in Dobson and Morris (2015) and was also given in Babai and Frankl (1978) in the special case where H is a characteristic subgroup of G.

Theorem 7.3.12 *Let G be a CI-group with respect to digraphs and $H \trianglelefteq G$. Then G/H is a CI-group with respect to digraphs.*

Proof Let $m = |H|$, and $\mathrm{Cay}(G/H, S_1)$ and $\mathrm{Cay}(G/H, S_2)$ be isomorphic. If $\mathrm{Cay}(G/H, S_1) \not\cong \Gamma_1 \wr K_\ell$ for some digraph Γ_1 and $\ell \geq 2$, then $\mathrm{Cay}(G/H, S_2) \not\cong \Gamma_2 \wr K_\ell$ for any digraph Γ_2 and $\ell \geq 2$. In this case, define $T_i = \{gh : gH \in S_i, h \in H\} \cup (H \backslash \{1_G\})$, $i = 1, 2$. Then $\mathrm{Cay}(G, T_i) \cong \mathrm{Cay}(G/H, S_i) \wr K_m$, $i = 1, 2$, are isomorphic Cayley digraphs of G. Additionally, by Theorem 5.1.8 or its digraph analogue, we have $\mathrm{Aut}(\mathrm{Cay}(G, T_i)) \cong \mathrm{Aut}(\mathrm{Cay}(G/H, S_i)) \wr \mathcal{S}_m$, $i = 1, 2$. If $\mathrm{Cay}(G/H, S_1) \cong \Gamma_1 \wr K_\ell$ for some Γ_1 and $\ell \geq 2$, then $\mathrm{Cay}(G/H, S_2) \cong \Gamma_2 \wr K_\ell$ for some Γ_2. In this case, define $T_i = \{gh : gH \in S_i, h \in H\}$, $i = 1, 2$. Then $\mathrm{Cay}(G, T_i) \cong \mathrm{Cay}(G/H, S_i) \wr \bar{K}_m$, $i = 1, 2$, are isomorphic Cayley digraphs of G. As before, by Theorem 5.1.8 or its digraph analogue, we have $\mathrm{Aut}(G, T_i) \cong \mathrm{Aut}(\mathrm{Cay}(G/H, S_i)) \wr \mathcal{S}_m$, $i = 1, 2$. In either case, $\mathrm{Cay}(G, T_i)$, $i = 1, 2$, are isomorphic Cayley digraphs of G such that $\mathrm{Aut}(\mathrm{Cay}(G, T_i)) \cong \mathrm{Aut}(\mathrm{Cay}(G/H, S_i)) \wr \mathcal{S}_m$, $i = 1, 2$.

Let $\hat{H} = \{h_L : h \in H\}$, and $\mathcal{B}$ be the block system of G_L that is the set of orbits of $\hat{H}$. We then have $\hat{H} \leq \mathrm{fix}_{\mathrm{Aut}(\mathrm{Cay}(G,T_i))}(\mathcal{B})$, for $i = 1, 2$. By Theorem 4.3.7, we have that $\mathcal{B}$ is the unique block system of $\mathrm{Aut}(\mathrm{Cay}(G, T_i))$, $i = 1, 2$, with blocks of size m. As G is a CI-group with respect to digraphs, there exists $\alpha \in \mathrm{Aut}(G)$ such that $\alpha(\mathrm{Cay}(G, T_1)) = \mathrm{Cay}(G, T_2)$. Then,

$$\alpha \mathrm{fix}_{\mathrm{Aut}(\mathrm{Cay}(G,T_1))}(\mathcal{B})\alpha^{-1} \trianglelefteq \mathrm{Aut}(\mathrm{Cay}(G, T_2)) = \alpha \mathrm{Aut}(\mathrm{Cay}(G, T_1))\alpha^{-1},$$

and the orbits of $\alpha \mathrm{fix}_{\mathrm{Aut}(\mathrm{Cay}(G,T_1))}(\mathcal{B})\alpha^{-1}$ have size m. As, by Theorem 2.2.7, the set of orbits of $\alpha \mathrm{fix}_{\mathrm{Aut}(\mathrm{Cay}(G,T_1))}(\mathcal{B})\alpha^{-1}$ is a block system of $\mathrm{Aut}(\mathrm{Cay}(G, T_2))$ with blocks of size m , we see

$$\alpha \mathrm{fix}_{\mathrm{Aut}(\mathrm{Cay}(G,T_1))}(\mathcal{B})\alpha^{-1} = \mathrm{fix}_{\mathrm{Aut}(\mathrm{Cay}(G,T_2))}(\mathcal{B}).$$

As $\hat{H} = \mathrm{fix}_{\mathrm{Aut}(\mathrm{Cay}(G,T_1))}(\mathcal{B}) \cap G_L$, we have

$$\begin{aligned} \alpha \hat{H} \alpha^{-1} &= \alpha(\mathrm{fix}_{\mathrm{Aut}(\mathrm{Cay}(G,T_1))}(\mathcal{B}) \cap G_L)\alpha^{-1} \\ &= \alpha \mathrm{fix}_{\mathrm{Aut}(\mathrm{Cay}(G_L,T_1))}(\mathcal{B})\alpha^{-1} \cap \alpha G_L \alpha^{-1} \\ &= \mathrm{fix}_{\mathrm{Aut}(\mathrm{Cay}(G,T_2))}(\mathcal{B}) \cap G_L \\ &= \hat{H}. \end{aligned}$$

So $\alpha(H) = H$, α induces an automorphism $\bar{\alpha}$ of G/H, and $\bar{\alpha}(\mathrm{Cay}(G/H, S_1)) = \mathrm{Cay}(G/H, S_2)$. Thus G/H is a CI-group with respect to digraphs. □

7.4 Non-CI-groups With Respect to Graphs

It is known that "most" groups are not CI-groups with respect to digraphs and graphs. This should not be too surprising in light of Theorems 7.3.11 and 7.3.12, which state that subgroups and quotients of CI-groups with respect to digraphs are CI-groups with respect to digraphs. For our purposes, in this section the contrapositives of these results are what we need, or that if a group G has a subgroup or quotient group that is not a CI-group with respect to digraphs, then it is not a CI-group with respect to digraphs. Perhaps the first deliberate attempt to construct a list of CI-groups with respect to digraphs was begun by Babai and Frankl (1978, 1979) who set out to prove that a CI-group with respect to digraphs is solvable, a project finished by Li (1999). The following result contained in Dobson et al. (2020b) summarizes the current state of knowledge on which groups may be CI-groups with respect to graphs. Before stating it, we will need to define some groups.

Definition 7.4.1 Let M be an abelian group of order m such that every Sylow p-subgroup of M is elementary abelian, and let $\exp(M)$ be the largest order of any element of M. Let $n \in \{2, 3, 4, 8\}$ be relatively prime to $|M|$. Set $E(M, n) = M \rtimes_\phi \mathbb{Z}_n$, where if n is even then $\phi(g) = g^{-1}$, while if $n = 3$ then $\phi(g) = g^\ell$, where ℓ is an integer satisfying $\ell^3 \equiv 1 \pmod{\exp(M)}$ and $\gcd(\ell(\ell - 1), \exp(M)) = 1$.

If $n = 2$, then m is odd. Also, $E(\mathbb{Z}_m, 2)$ is simply the dihedral group of order $2m$. If $n = 3$, and M is cyclic of prime order p, then $E(\mathbb{Z}_p, 3)$ is the unique nonabelian group of order $3p$.

Theorem 7.4.2 *Let G, M, and K be CI-groups with respect to graphs such that M and K are abelian, all Sylow subgroups of M are elementary abelian, and all Sylow subgroups of K are elementary abelian of order 9 or cyclic of prime order.*

(i) *If G does not contain elements of order 8 or 9, then $G = H_1 \times H_2 \times H_3$, where the orders of H_1, H_2, and H_3 are pairwise relatively prime, and*

  (a) *H_1 is an abelian group, and each Sylow p-subgroup of H_1 is isomorphic to $\mathbb{Z}_p^k$ for $k < 2p + 3$ or $\mathbb{Z}_4$;*

  (b) *H_2 is isomorphic to one of the groups $E(K, 2)$, $E(M, 3)$, $E(K, 4)$, A_4, or 1;*

  (c) *H_3 is isomorphic to one of the groups D_{10}, Q_8, or 1.*

(ii) *If G contains elements of order 8, then $G \cong E(K, 8)$ or $\mathbb{Z}_8$.*

(iii) *If G contains elements of order 9, then G is one of the groups $\mathbb{Z}_9 \rtimes \mathbb{Z}_2$, $\mathbb{Z}_9 \rtimes \mathbb{Z}_4$, $\mathbb{Z}_2^2 \rtimes \mathbb{Z}_9$, or $\mathbb{Z}_2^n \times \mathbb{Z}_9$, with $n \le 5$.*

Many groups listed in the above result are known to be CI-groups with respect to graphs. Holt and Royle (2020) found all CI-groups with respect to graphs of order at most 47, using their census of transitive groups of degree at most 47. This was extended to all groups of order at most 59 in Dobson et al. (2020b). Let p and q be distinct primes. The following groups are known to be CI-groups with respect to graphs (as are all of their subgroups by Theorem 7.3.11): In (i)(a), they are $\mathbb{Z}_{2n}$ (Muzychuk, 1995), $\mathbb{Z}_{4n}$ (Muzychuk, 1997) with n an odd square-free integer, $\mathbb{Z}_p^5$ (Feng and Kovács, 2018), $\mathbb{Z}_q \times \mathbb{Z}_p^4$ (Kovács and Ryabov, 2019), $\mathbb{Z}_2^3 \times \mathbb{Z}_p$ (Dobson and Spiga, 2013), $\mathbb{Z}_4 \times \mathbb{Z}_p^2$ (Ryabov, 2020), $\mathbb{Z}_2^5 \times \mathbb{Z}_p$ (Ryabov, 2021), and for some groups of special orders (Dobson, 2002, 2018). For (i)(b), $E(\mathbb{Z}_p, 2)$ (Babai, 1979), $E(\mathbb{Z}_p, 3)$ (Dobson, 1998), and $E(\mathbb{Z}_p, 4)$ and $E(\mathbb{Z}_p, 8)$ (Li et al., 2007). For direct products of these groups, some H_1 and $H_2 = E(M, 2)$ and $E(M, 3)$ for some special M (Dobson, 2002) and $Q_8 \times \mathbb{Z}_p$ (Somlai, 2015).

With the result that all CI-groups with respect to graphs are solvable in hand, to prove the Theorem 7.4.2 requires many constructions to shorten the list of possible CI-groups with respect to graphs. We will content ourselves here by giving some of the constructions for cyclic groups that are not CI-groups with respect to graphs. Our first such construction was by Babai and Frankl (1978) (whose construction works for all $n \geq 5$, $n \neq 6$, and not just primes) as well as Alspach and Parsons (1979).

Example 7.4.3 Let $p \geq 5$ be prime. Then $\mathbb{Z}_{p^2}$ is not a CI-group with respect to graphs.

Solution Let C_p be a cycle of length p, and $\Gamma = C_p \wr C_p$. Let $A = \{\pm p\}$ and $B = \{\pm 1 + kp : k \in \mathbb{Z}_p\}$. By Proposition 5.3.2 (with H of that result $\langle p \rangle$), we see that $\Gamma \cong \mathrm{Cay}(\mathbb{Z}_{p^2}, A \cup B)$. As $p \geq 5$ is prime, there exists $c \in \mathbb{Z}_p^*$ such that $c \neq \pm 1$. Then $cA = \{\pm cp\} \neq A$, and the graph $\mathrm{Cay}\left(\mathbb{Z}_{p^2}, cA \cup B\right)$ is isomorphic to $\mathrm{Cay}\left(\mathbb{Z}_{p^2}, A \cup B\right)$ via the map $i + jp \mapsto i + cjp$, where we are writing each element of $\mathbb{Z}_{p^2}$ uniquely in the form $i + jp$, $0 \leq i, j \leq p - 1$. Now suppose $\mathrm{Cay}\left(\mathbb{Z}_{p^2}, A \cup B\right)$ and $\mathrm{Cay}\left(\mathbb{Z}_{p^2}, cA \cup B\right)$ are isomorphic by a multiplier $m \in \mathbb{Z}_{p^2}^*$, with $m \cdot (A \cup B) = cA \cup B$. As $m \cdot (A \cup B) = m \cdot A \cup m \cdot B = cA \cup B$, we must have $mB = B$ as m maps units to units in $\mathbb{Z}_{p^2}$ and all of the units in $cA \cup B$ are in B. Hence $m \equiv \pm 1 \pmod{p}$. However, in that case $m \cdot A = A \neq cA$, a contradiction. □

The above example combined with Theorem 7.3.11 shows that $\mathbb{Z}_n$ is not a CI-group with respect to graphs (and hence digraphs) if $p^2 \mid n$ for any $p \geq 5$. In the case of $\mathbb{Z}_9$, it turns out that $\mathbb{Z}_9$ is a CI-group with respect to graphs, but not with respect to digraphs. In the case of digraphs, the preceding example can be modified to yield an appropriate example – see Exercise 7.4.4. For graphs, the preceding example can also be modified to show that $\mathbb{Z}_{27}$ is not a CI-group with respect to graphs – see Exercise 7.4.5. For powers of the even prime, we have the following result, first observed by Babai (1977, Corollary 3.3). The proof is left as Exercise 7.4.10.

Lemma 7.4.4 *Any group of order* 4 *is a CI-group.*

The following fact was first obtained by Hell and Kirkpatrick (Alspach and Parsons, 1979, p. 108).

Example 7.4.5 If $8|n$ and $n > 8$, then $\mathbb{Z}_n$ is not a CI-group with respect to graphs.

Solution In view of Theorem 7.3.11, it suffices to show that $\mathbb{Z}_{8p}$, p a prime, is not a CI-group with respect to graphs. Let $S = \{1, 8p-1, 4p+1, 4p-1, 2, 8p-2\}$, and $\Gamma_1 = \mathrm{Cay}\left(\mathbb{Z}_{8p}, S\right)$. Let $\beta \in \mathbb{Z}_{4p}$ such that $\beta \equiv 3 \pmod 4$ and $\beta \equiv 1 \pmod p$. Let $T = \{1, 8p-1, 4p+1, 4p-1, 2\beta, 8p-2\beta\}$, and $\Gamma_2 = \mathrm{Cay}\left(\mathbb{Z}_{8p}, T\right)$ (here 2β is considered as an element of $\mathbb{Z}_{8p}$ in the natural way). We will show that Γ_1 and Γ_2 are isomorphic but not by a multiplier.

First, observe that as $1 \in S$ if $mS = T$ for any $m \in \mathbb{Z}_{8p}^*$, then $m = 1, 4p+1, 8p-1$, or $4p-1$. It is easy to check that for such m, $mS = S$, in which case the map $x \mapsto mx$ is contained in $\mathrm{Aut}(\Gamma_1)$ by Lemma 1.2.15. It thus suffices to show that $\Gamma_1 \cong \Gamma_2$.

Define $\phi\colon \mathbb{Z}_{8p} \to \mathbb{Z}_{8p}$ by $\phi(2k) = 2(\beta k)$ and $\phi(2k+1) = 2(\beta k) + 1$, where $0 \le k \le 4p-1$. Let $\alpha = \rho^{-1}\phi^{-1}\rho\phi$. Define $\rho\colon \mathbb{Z}_{8p} \to \mathbb{Z}_{8p}$ by $\rho(i) = i+1$. Observe that $\alpha(2k) = 2k$ while $\alpha(2k+1) = 2k+1+2(\beta^{-1}-1)$ (do these computations!). Note that $\beta^{-1} - 1 \equiv 0 \pmod p$, while $\beta^{-1} - 1 \equiv 2 \pmod 4$. We conclude that $a = 2\left(\beta^{-1} - 1\right) \equiv 4p \pmod{8p}$. Noting α is self-inverse, we see $\alpha^{-1}\rho\alpha = \rho^{1+a}$. As $\alpha(0) = 0$, we have by Corollary 4.4.11 that $\alpha \in \mathrm{Aut}\left(\mathbb{Z}_{8p}\right)$. Thus $\alpha(i) = (4p+1)i$, and we have already seen that $\alpha(S) = S$. By Lemma 1.2.15, $\alpha \in \mathrm{Aut}(\Gamma_1)$. Thus $\phi^{-1}\rho\phi = \rho\alpha \in \mathrm{Aut}(\Gamma_1)$, and $\rho = \phi\rho\alpha\phi^{-1}$. Then $\phi\rho\alpha\phi^{-1} = \rho \in \phi(\Gamma_1)$, and $\phi(\Gamma_1)$ is a circulant graph of order $8p$ by Theorem 1.2.20. Then the neighbors of 0 in $\phi(\Gamma_1)$ are $\phi(S) = T$ (again, check the computations, but first show $2\beta \equiv 4p + 2 \pmod 8$), and so $\phi(\Gamma_1) = \Gamma_2$. □

Some additional groups that are not CI-groups with respect to graphs should be mentioned. Babai and Frankl (1978) conjectured that $\mathbb{Z}_p^n$ is a CI-group with respect to graphs for all primes p and integers $m \ge 1$. Nowitz (1992) showed this conjecture was false by showing $\mathbb{Z}_2^6$ is not a CI-group with respect to graphs. Muzychuk (2003), Spiga (2007), and Somlai (2011) successively considered the case of $\mathbb{Z}_p^n$, $p \geq 3$, and Somlai showed that $\mathbb{Z}_p^{2p+3}$ is not a CI-group with respect to graphs. More examples can be found in Li (2002) and its references.

7.5 Groups of Order a Product of Two Primes

In this section, we will consider the isomorphism problem for many groups of order a product of two primes.The solution to the isomorphism problem in these cases is usually accessible, while the complexity of the problem seems to grow as the number of prime factors increases. Additionally, one can see the basic "structure" of how to, in general, approach an isomorphism problem even if in these smaller cases many details are simplified.

It turns out that not many group are CI-groups. In fact, Pálfy (1987) has determined them all by showing that a group G of order n is a CI-group if and only if $|G| = 4$ or n, where $\gcd(n, \varphi(n)) = 1$ (here φ is Euler's phi function – see Exercise 4.5.2 for the definition of this function). We shall see the sufficiency part of Pálfy's theorem in Section 7.6. There are many more CI-groups with respect to graphs, so there must be some property or properties of graphs that allow for the construction of additional automorphisms. Typically, when one begins an isomorphism problem the only part of the automorphism group that one has is $\langle G_L, \phi^{-1}G_L\phi\rangle$ for some $\phi \in S_G$. That is, we start with a Cayley graph and an isomorphism, and then using this information, hopefully construct additional automorphisms of the Cayley graph if needed. We have actually already seen one such tool, namely Lemma 5.3.11.

The next result will allow us in many cases to simplify isomorphisms between Cayley digraphs.

Lemma 7.5.1 *Let G be a group, $S \subset G$, and $\alpha \in$ Aut(G). Then* Cay(G, S) *is a CI-digraph of G if and only if* $\alpha(\mathrm{Cay}(G, S)) = \mathrm{Cay}(G, \alpha(S))$ *is a CI-digraph of G.*

Proof That $\alpha(\mathrm{Cay}(G, S)) = \mathrm{Cay}(G, \alpha(S))$ is Lemma 1.2.15. If Cay(G, S) is a CI-digraph of G, and $S' \subseteq G$ such that $\mathrm{Cay}(G, S') \cong \mathrm{Cay}(G, \alpha(S))$, then $\mathrm{Cay}(G, S) \cong \mathrm{Cay}(G, S')$. As Cay$(G, S)$ is a CI-digraph of G, there exists $\beta \in$ Aut(G) such that $\beta(\mathrm{Cay}(G, S)) = \mathrm{Cay}(G, S')$. Then,

$$\beta\alpha^{-1}(\mathrm{Cay}(G, \alpha(S))) = \beta(\mathrm{Cay}(G, S)) = \mathrm{Cay}(G, S').$$

Conversely, if Cay$(G, \alpha(S))$ is a CI-digraph and $S' \subseteq G$ such that $\mathrm{Cay}(G, S) \cong \mathrm{Cay}(G, S')$, then $\mathrm{Cay}(G, \alpha(S)) \cong \mathrm{Cay}(G, S')$. As Cay$(G, \alpha(S))$ is a CI-digraph of G, there exists $\beta \in \mathrm{Aut}(G)$ such that $\beta(\mathrm{Cay}(G, \alpha(S))) = \mathrm{Cay}(G, S')$. Then $\beta\alpha(\mathrm{Cay}(G, S)) = \mathrm{Cay}(G, S')$ and the result follows. □

Typically, when showing a group G is a CI-group with respect to digraphs, one will show that G_L and $\phi^{-1}G_L\phi$ are always conjugate in $\mathrm{Aut}(\mathrm{Cay}(G, S))$ for every $S \subseteq G$. The desired conclusion then follows by Lemma 7.6.3. As a first step towards this, it is often the case that one will show that there is a $\delta \in \langle G_L, \phi^{-1}G_L\phi\rangle$ such that $\langle G_L, \delta^{-1}\phi^{-1}G_L\phi\delta\rangle$ has appropriate block systems. There are a variety of methods for doing this depending upon the group G. For example, if G is a Burnside group, then $\langle G_L, \phi^{-1}G_L\phi\rangle$ has a block system or is 2-transitive, in which case the digraph is either the complete graph or its complement and the isomorphism problem is trivial.

Definition 7.5.2 An **almost simple group** is a group G that contains a normal nonabelian simple group N and is contained in the automorphism group of N.

It is not the case that every group is a Burnside group and some nonabelian groups of order qp, q and p distinct primes are not Burnside groups – see Exercise 7.5.13 for an example of an almost simple group that is primitive and contains a regular nonabelian subgroup of order qp. While not all Burnside groups are known, all Burnside groups of square-free degree have been determined by Li and Seress (2005). For Cayley digraphs of order qp, we have the following result.

Lemma 7.5.3 *Let G be a group of order qp, where $q < p$ are distinct primes. For $\phi \in S_G$, there exists $\delta \in S_G$ such that $\langle G_L, \delta^{-1}\phi^{-1}G_L\phi\delta\rangle$ has a normal block system with blocks of size p.*

Proof Let ρ be a semiregular element of S_{qp} of order p with $\Delta_1, \ldots, \Delta_q$ the orbits of $\langle\rho\rangle$. The key fact for this proof is that $\langle \rho|_{\Delta_i} : 1 \le i \le q\rangle$ is the *unique* Sylow p-subgroup of S_{qp} that contains ρ. Indeed, $\langle \rho|_{\Delta_i} : 1 \le i \le q\rangle$ has order p^q (and in fact is isomorphic to $\mathbb{Z}_p^q$), and the only integers between 1 and qp that are multiples of p are $p, 2p, \ldots, qp$. Thus the largest power of p that divides $(qp)!$ is p^q, and so $\langle \rho|_{\Delta_i} : 1 \le i \le q\rangle$ is a Sylow p-subgroup of S_{qp}. To establish uniqueness, suppose Π is a Sylow p-subgroup of S_{qp} that contains ρ, and so $\Pi = \langle \sigma|_{\Omega_i} : 1 \le i \le q\rangle$ for some semiregular element σ of S_{qp} of order p with $\Omega_1, \ldots, \Omega_q$ the orbits of $\langle\sigma\rangle$ (that Π can be written in this form must be true as Π is conjugate to $\langle \rho|_{\Delta_i} : 1 \le i \le q\rangle$). As $\rho \in \Pi$, $\rho = \Pi_{i=1}^q (\sigma|_{\Omega_i})^{a_i}$ for some positive integers $a_1, \ldots, a_q$. Note that $a_i \not\equiv 0 \pmod p$ as ρ is semiregular, $1 \le i \le q$. We deduce that $(\sigma|_{\Omega_i})^{a_i} = \rho|_{\Delta_{j_i}}$, $1 \le i \le q$ and some $1 \le j_i \le q$. This implies that $\langle \rho|_{\Delta_i}\rangle \le \Pi$, and as the orders of these groups are the same,

$\Pi = \langle \rho|_{\Delta_i} : 1 \le i \le q\rangle$. Thus $\langle \rho|_{\Delta_i} : 1 \le i \le q\rangle$ is indeed the unique Sylow p-subgroup of $\mathcal{S}_{qp}$ that contains ρ.

Recall that, by a Sylow theorem, the number of Sylow p-subgroups of G divides $|G|$ and is of the form $kp + 1$ for some positive integer k. We conclude that G has a normal Sylow p-subgroup. Then G_L has a normal Sylow p-subgroup $P = \langle \rho\rangle$ and so by Theorem 2.2.7, there is a block system of G_L with blocks of size p. As G_L is regular and $\rho \neq 1$, ρ is semiregular.

Let Π_1 be a Sylow p-subgroup of $\left\langle G_L, \phi^{-1}G_L\phi\right\rangle$ that contains ρ, and Π_2 a Sylow p-subgroup of $\left\langle G_L, \phi^{-1}G_L\phi\right\rangle$ that contains $\phi^{-1}\rho\phi$. Then there exists $\delta \in \left\langle G_L, \phi^{-1}G_L\phi\right\rangle$ such that $\delta^{-1}\phi^{-1}\Pi_2\phi\delta = \Pi_1$. Then Π_1 contains semiregular elements ρ and $\delta^{-1}\phi^{-1}\rho\phi\delta$ and so by arguments above

$$\Pi_1 \le \langle \rho_{\Delta_i} : 1 \le i \le q\rangle = \left\langle \left(\delta^{-1}\phi^{-1}\rho\phi\delta\right)|_{\Omega_i} : 1 \le i \le q\right\rangle,$$

where the orbits of $\langle\rho\rangle$ and $\left\langle \delta^{-1}\phi^{-1}\rho\phi\delta\right\rangle$ are $\Delta_1, \ldots, \Delta_q$ and $\Omega_1, \ldots, \Omega_q$, respectively. We conclude that $\mathcal{B} = \{\Delta_i : 1 \le i \le q\} = \{\Omega_i : 1 \le i \le q\}$ are block systems of both G_L and $\delta^{-1}\phi^{-1}G_L\phi\delta$. Then $\mathcal{B}$ is a block system of $\left\langle G_L, \delta^{-1}\phi^{-1}G_L\phi\delta\right\rangle$ by Exercise 2.3.8. □

With this result we may already say something about the isomorphism problem for Cayley digraphs of groups of order qp. The next result says that if we have conjugated $\phi^{-1}G_L\phi$ by δ in such a way that the subgroup generated by G_L and $\delta^{-1}\phi^{-1}G_L\phi\delta$ has a normal block system, then $\phi\delta$ has a restricted form. Intuitively, if ϕ is an isomorphism between two Cayley digraphs, this makes the isomorphism $\phi\delta$ "nicer" than ϕ.

Lemma 7.5.4 *Let G be a group of square-free order with $H \trianglelefteq G$ and $\mathcal{B}$ be the block system of G_L that is the set of orbits of $\hat{H}_L = \{h_L : h \in H\}$. Let $\Gamma = \mathrm{Cay}(G, S)$ for some $S \subseteq G$. For $\phi \in \mathcal{S}_G$, the following are equivalent:*

(i) *$\phi^{-1}G_L\phi \le \mathrm{Aut}(\Gamma)$ and there exists $\delta \in \mathrm{Aut}(\Gamma)$ such that $\left\langle G_L, \delta^{-1}\phi^{-1}G_L\phi\delta\right\rangle$ has $\mathcal{B}$ as a block system,*
(ii) *$\phi(\Gamma)$ is a Cayley digraph of G and $\omega(\Gamma) = \phi(\Gamma)$ for some $\omega \in \mathcal{S}_{G/H} \wr \mathcal{S}_H$.*

Proof (i) $\implies$ (ii): As $\phi(\Gamma)$ has automorphism group $\phi\mathrm{Aut}(\Gamma)\phi^{-1}$ and so $G_L \le \mathrm{Aut}(\phi(\Gamma))$, the condition $\phi^{-1}G_L\phi \le \mathrm{Aut}(\Gamma)$ implies $\phi(\Gamma)$ is a Cayley digraph of G by Theorem 1.2.20. Then $\mathcal{B}$ is the unique normal block system of $\left\langle G_L, \delta^{-1}\phi^{-1}G_L\phi\delta\right\rangle$ by Exercise 2.3.7. By Exercise 4.3.4, we have $\omega = \phi\delta \in \mathcal{S}_{G/H} \wr \mathcal{S}_H$, and as $\delta \in \mathrm{Aut}(\Gamma)$, $\omega(\Gamma) = \phi(\Gamma)$.

(ii) $\implies$ (i): Suppose $\phi(\Gamma)$ is a Cayley digraph of G and $\omega(\Gamma) = \phi(\Gamma)$ for some $\omega \in \mathcal{S}_{G/H} \wr \mathcal{S}_H$. Then $G_L \le \mathrm{Aut}(\phi(\Gamma))$, and as $\mathrm{Aut}(\Gamma) = \phi^{-1}\mathrm{Aut}(\phi(\Gamma))\phi$, $\phi^{-1}G_L\phi \le \mathrm{Aut}(\Gamma)$. Additionally, $\delta = \phi^{-1}\omega \in \mathrm{Aut}(\Gamma)$, and $\delta^{-1}\phi^{-1}G_L\phi\delta =$

$\omega^{-1}G_L\omega$. As $\mathcal{B}$ is a block system of G_L and $\omega(\mathcal{B}) = \mathcal{B}$, it follows that $\mathcal{B}$ is a block system of $\omega^{-1}G_L\omega$. The result follows by Exercise 2.3.8. □

Our next result gives an implication of the previous result to the CI-problem.

Lemma 7.5.5 *Let G be a group of square-free order with H characteristic in G and $\mathcal{B}$ be the block system of G_L that is the set of orbits of $\hat{H}_L = \{h_L : h \in H\}$. Let $\Gamma = \mathrm{Cay}(G, S)$ for some $S \subseteq G$. Suppose that whenever $\phi \in S_G$ such that $\phi^{-1}G_L\phi \le \mathrm{Aut}(\Gamma)$ there exists $\delta \in \mathrm{Aut}(\Gamma)$ such that $\langle G_L, \delta^{-1}\phi^{-1}G_L\phi\delta\rangle$ has $\mathcal{B}$ as a block system. Then Γ is a CI-digraph of G if and only if there exists $\gamma \in \mathrm{Aut}(\Gamma) \cap (S_{G/H} \wr S_H)$ such that $\gamma^{-1}\delta^{-1}\phi^{-1}G_L\phi\delta\gamma = G_L$.*

Proof Suppose Γ is a CI-digraph of G. Then there exists $\gamma \in \mathrm{Aut}(\Gamma)$ such that $\gamma^{-1}\delta^{-1}\phi^{-1}G_L\phi\delta\gamma = G_L$. Then $\gamma\delta\phi \in G_L \rtimes \mathrm{Aut}(G)$ by Corollary 4.4.11. Also, $\mathcal{B}$ is a block system of G_L, and as H is characteristic in G, any element of $\alpha \in \mathrm{Aut}(G)$ maps cosets of H to cosets of H, and so $\alpha(\mathcal{B}) = \mathcal{B}$. We conclude that $\gamma\delta\phi \in S_{G/H} \wr S_H$, and as $\delta\phi \in S_{G/H} \wr S_H$, $\gamma \in S_{G/H} \wr S_H$ as required. The converse is trivial. □

The next result is the last result we will have on groups of order qp with $q \le p$ primes, where we will not have some additional hypothesis either on the relationship between q and p or specifying the group G.

Lemma 7.5.6 *Let G be a group of order qp where $q \le p$ are prime. If $\phi \in S_G$ such that $\langle G_L, \phi^{-1}G_L\phi\rangle$ has a block system $\mathcal{B}$ with blocks of size p, then there exists $\delta \in \langle G_L, \phi^{-1}G_L\phi\rangle$ such that $\delta^{-1}\phi^{-1}G_L\phi\delta/\mathcal{B} = G_L/\mathcal{B}$.*

Proof As $\langle G_L, \phi^{-1}G_L\phi\rangle/\mathcal{B} \le S_q$ and $q \le p$, it must be the case that $\mathcal{B}$ is a normal block system of both G_L and $\phi^{-1}G_L\phi$ as qp does not divide $q!$. Then $G_L/\mathcal{B}$ and $\phi^{-1}G_L\phi/\mathcal{B}$ are Sylow q-subgroups of $\langle G_L, \phi^{-1}G_L\phi\rangle/\mathcal{B}$ and so by a Sylow theorem there exists $\delta \in \langle G_L, \phi^{-1}G_L\phi\rangle$ such that

$$G_L/\mathcal{B} = \left(\delta^{-1}/\mathcal{B}\right)(\phi^{-1}/\mathcal{B})(G_L/\mathcal{B})(\phi/\mathcal{B})(\delta/\mathcal{B}) = \delta^{-1}\phi^{-1}G_L\phi\delta/\mathcal{B}. \quad \square$$

The reader may have observed that thus far in this section, all of our results and discussion has been permutation group theoretical, not combinatorial. Typically, one conjugates $\phi^{-1}G_L\phi$ to make it "as close as possible" to G_L first using permutation group-theoretic arguments, and only when one cannot reach G_L does one consider combinatorial arguments. This approach has the advantage of making clear where/if combinatorial information is needed, as the Cayley isomorphism problem has been considered for "combinatorial objects" other than digraphs (see, for example, Brand [1989]; Dobson [2003b]; Koike et al. [2019]; Muzychuk [2015]). An obvious question one might ask is whether or not combinatorial information is needed in order for $\langle G_L, \delta^{-1}\phi^{-1}G_L\phi\delta\rangle$ to have a block of prime size, even for common abelian groups. The answer to this question is yes – see Dobson and Spiga (2013, Corollary 2.4). This will be discussed in more detail in the next section.

Once appropriate block systems have been found, one can often use group theory to further restrict the form of a nicest isomorphism between the isomorphic Cayley digraphs. Again, there are a variety of tools that can used, with a main tool being Burnside's theorem (Theorem 4.4.2).

Lemma 7.5.7 *Let Γ be a Cayley digraph of a group G of order qp, where $q < p$ are distinct primes, with its (q, p)-metacirculant labeling. If Γ' is another Cayley digraph of G (with the same metacirculant labeling), then $\Gamma \cong \Gamma'$ if and only if $\phi(\Gamma) = \Gamma'$, where $\phi \in \mathrm{AGL}(1, q) \wr \mathrm{AGL}(1, p)$. Additionally, if a Sylow p-subgroup of $\mathrm{Aut}(\Gamma)$ has order p, then $\phi(i, j) = (ri, \beta j + b_i)$ for some $r \in \mathbb{Z}_q^*$, $\beta \in \mathbb{Z}_p^*$, and each $b_i \in \mathbb{Z}_p$.*

Proof In the proof of this lemma, we will employ a common notation gimmick. The gimmick is used as we will be doing many conjugations, and writing down each successive conjugation becomes notationally unwieldy. So, we will start with ϕ as an isomorphism. This is not the ϕ in the conclusion. We will then conjugate $\phi^{-1}G_L\phi$ by some map, say δ, but instead of writing $\phi\delta$ as the isomorphism and $\delta^{-1}\phi^{-1}G_L\phi\delta$, we write the same symbol ϕ and same group $\phi^{-1}G_L\phi$.

Let $\phi\colon \Gamma \to \Gamma'$ be an isomorphism. By Lemma 7.5.3, after an appropriate conjugation of $\phi^{-1}G\phi$ by an element of $\langle G, \phi^{-1}G\phi\rangle$, if necessary, we may assume without loss of generality that $\langle G, \phi^{-1}G\phi\rangle$ has a block system $\mathcal{B}$ with blocks of size p that is the set of orbits of the characteristic subgroup of G of order p. By Lemma 7.5.4 we may additionally assume $\phi \in S_q \wr S_p$, so that $\phi(i, j) = (\sigma(i), \omega_i(j))$ for some $\sigma \in S_q$ and each $\omega_i \in S_p$. By Lemma 7.5.6 after another conjugation, if necessary, we may assume $\langle G, \phi^{-1}G\phi\rangle / \mathcal{B} = G/\mathcal{B}$. As $G/\mathcal{B} \cong \mathbb{Z}_q$ is transitive, we see $\phi/\mathcal{B}$ (which is defined as $\phi \in S_q \wr S_p$) normalizes $G/\mathcal{B}$. Thus $\phi/\mathcal{B} \in \mathrm{AGL}(1, q)$ by Corollary 4.4.11 and so $\sigma(i) = ri + a$ for some $r \in \mathbb{Z}_q^*$ and $a \in \mathbb{Z}_q$. As $\tau \in \mathrm{Aut}(\Gamma')$ (as usual with the metacirculant labeling, $\tau(i, j) = (i + 1, \alpha j)$, where $\alpha \in \mathbb{Z}_p^*$), by replacing ϕ with $\phi\tau^{-a}$ we may assume without loss of generality that $a = 0$. Then $\sigma(i) = ri$. Finally, as $\langle G_L, \phi^{-1}G_L\phi\rangle / \mathcal{B} \le S_q$ and $q < p$, we see that $\langle\rho\rangle$ and $\phi^{-1}\langle\rho\rangle\phi$ are contained in $\mathrm{fix}_{\langle G_L, \phi^{-1}G_L\phi\rangle}(\mathcal{B})$. By a Sylow theorem, there exists $\omega \in \langle G_L, \phi^{-1}G_L\phi\rangle$ such that $\omega^{-1}\phi^{-1}\langle\rho\rangle\phi\omega$ is contained the same Sylow p-subgroup P of $\mathrm{fix}_{\langle G_L, \phi^{-1}G_L\phi\rangle}(\mathcal{B})$ as $\langle\rho\rangle$. Replacing ϕ with $\phi\omega$, we assume without loss of generality that $\phi^{-1}\langle\rho\rangle\phi$ is contained in P. As $P \le \langle\rho|_B : B \in \mathcal{B}\rangle$, this implies that $\phi^{-1}\rho\phi \in \langle\rho|_B : B \in \mathcal{B}\rangle$ and, consequently, that ϕ normalizes $\langle\rho|_B : B \in \mathcal{B}\rangle$. The result now follows by Lemma 6.5.17. The converse is trivial. □

We still have not really used any graph theory! That is, we have only used permutation group-theoretic techniques, so the above results hold for other combinatorial objects, provided they exist! What we mean by this last statement is that certain kinds of Cayley combinatorial objects do not always exist for every group G or even for every group G of order qp. This is especially true for combinatorial designs.

Example 7.5.8 Let $q < p$ be distinct primes with $q|(p-1)$. If $q \geq 3$ then the nonabelian group of order qp is not a CI-group with respect to digraphs.

Solution Define $\rho, \tau \colon \mathbb{Z}_q \times \mathbb{Z}_p \to \mathbb{Z}_q \times \mathbb{Z}_p$ by $\rho(i,j) = (i, j+1)$ and $\tau(i,j) = (i+1, \alpha j)$, where $\alpha \in \mathbb{Z}_p^*$ of order q. Then $\langle \rho, \tau \rangle$ is isomorphic to the nonabelian group of order qp. We are using the metacirculant labeling here. Define $\bar{\alpha} \colon \mathbb{Z}_q \times \mathbb{Z}_p \to \mathbb{Z}_q \times \mathbb{Z}_p$ by $\bar{\alpha}(i,j) = (i, \alpha j)$, and let $\mathcal{O}$ be an orbit of $\bar{\alpha}$ of order q contained in $B_1 = \{(1,j) : j \in \mathbb{Z}_p\}$. To define a (q,p)-metacirculant digraph Γ, we set $S_i = \emptyset$ if $i \neq 1$ and $S_1 = \mathcal{O}$. We will show that Γ is not isomorphic to a CI-digraph of $G = \langle \rho, \tau \rangle$.

First, observe that $\bar{\alpha} \in \mathrm{Aut}(\Gamma)$. This is easiest to verify by thinking of Γ as a Cayley digraph of G by Theorem 1.5.13 (so G is regular), observing that $\bar{\alpha} \in \mathrm{Aut}(G)$ by Corollary 4.4.11 (it normalizes G and fixes a point), and applying Corollary 1.2.16. It is then straightforward to verify that $H = \langle \rho, \tau\bar{\alpha} \rangle$ is a regular subgroup of $\mathrm{Aut}(\Gamma)$ isomorphic to G. We will show that H and G are not conjugate in $\mathrm{Aut}(\Gamma)$. This will establish our claim by Lemma 7.3.1 and the fact that G and H are conjugate in S_{qp} by Theorem 4.1.2. As a Sylow p-subgroup of G is characteristic in G, by Lemma 7.5.3 and Lemma 7.5.5 it suffices to show that G and H are not conjugate by an element of $K = \mathrm{Aut}(\Gamma) \cap (S_q \wr S_p)$.

Clearly, the set of orbits $\mathcal{B}$ of the unique Sylow p-group of G is a block system of K. Additionally, Γ is clearly not a nontrivial wreath product of a circulant digraph of order q and a circulant digraph of order p as $(0,0)$ has exactly q outneighbors in $\{(1,j) : j \in \mathbb{Z}_p\}$. By Lemma 5.3.11 we see that a Sylow p-subgroup of $\mathrm{fix}_K(\mathcal{B})$ has order p. By Lemma 7.5.7 it follows that if $k \in K$ then $k(i,j) = (ri + b, \beta j + b_i)$, $r \in \mathbb{Z}_q^*$, $b \in \mathbb{Z}_q$, $\beta \in \mathbb{Z}_p^*$, and $b_i \in \mathbb{Z}_p$. As $\Gamma/\mathcal{B}$ is a directed cycle with automorphism group $\mathbb{Z}_q$ and $K/\mathcal{B} \leq \mathrm{Aut}(\Gamma/\mathcal{B})$, we see that $r = 1$. It is then straightforward to verify that $k^{-1}Hk \neq G$. □

The only sense in which graph theory was used in the previous result was to show that there existed a Cayley digraph of the nonabelian group G of order qp such that it is a CI-digraph of G if and only if any two regular subgroups of $\langle \rho, \tau, \bar{\alpha} \rangle$ isomorphic to G are conjugate in $\langle \rho, \tau, \bar{\alpha} \rangle$. Interestingly, the "digraph" part of the previous example is required for $q = 3$, as the nonabelian group of order $3p$ is a CI-group with respect to *graphs* (Dobson, 1998, Theorem 21) (see also Li et al. (2007)). The next result was first shown by Babai (1977, Theorem 4.4) (for ternary relational structures, which we define in Definition 7.6.1).

Theorem 7.5.9 *The dihedral group D_{2p} of order $2p$ is a CI-group with respect to digraphs.*

Proof Let Γ be a Cayley digraph of $G = D_{2p}$. We use the metacirculant labeling of Γ. Define $\rho, \tau\colon \mathbb{Z}_2 \times \mathbb{Z}_p \to \mathbb{Z}_2 \times \mathbb{Z}_p$ by $\rho(i, j) = (i, j+1)$ and $\tau(i, j) = (i+1, -j)$. Then $D_{2p} \cong \langle \rho, \tau \rangle \leq \mathrm{Aut}(\Gamma)$. Let $\Gamma' \cong \Gamma$ be a Cayley digraph of G, also with its metacirculant labeling. By Lemma 7.5.7, there exists $\phi \in \mathrm{AGL}(1,2) \wr \mathrm{AGL}(1,p)$ such that $\phi(\Gamma) = \Gamma'$. Let $\phi(i, j) = (i + a, \beta_i j + b_i)$, $a \in \mathbb{Z}_2$, $\beta_i \in \mathbb{Z}_p^*$, and $b_i \in \mathbb{Z}_p$. As $\tau \in \mathrm{Aut}(\Gamma)$, replacing ϕ with $\phi\tau^{-1}$, we assume without loss of generality that $a = 0$. It is straightforward to check (see Exercise 7.5.2) that the map $(i, j) \mapsto (i, \beta j)$ is an automorphism of G for every $\beta \in \mathbb{Z}_p^*$, so by Lemma 7.5.1 we may, by replacing Γ' with its image under the map $(i, j) \mapsto \left(i, \beta^{-1} j\right)$, assume without loss of generality that $\beta_0 = 1$. Additionally, replacing ϕ with $\phi\rho^{-b_0}$ we may assume $b_0 = 0$. By Exercise 7.5.3, the map $(i, j) \mapsto (i, j + i)$ is an automorphism of G, so again by Lemma 7.5.1 we may assume without loss of generality that $b_1 = 0$. If $\beta_1 = 1$, then $\phi = 1$ is an automorphism of G and the result follows. So we assume $\beta_0 \neq 1$.

For $i \in \mathbb{Z}_2$ set $V_i = \{(i, j) : j \in \mathbb{Z}_p\}$, let $H = \left\langle \rho, \tau, \phi^{-1}\rho\phi, \phi^{-1}\tau\phi \right\rangle$, and $z_i = \rho|_{V_i}$, $i \in \mathbb{Z}_2$. As $\beta_0 \neq 1$, $\phi^{-1}\rho\phi = z_0 z_1^{\beta_1^{-1}} \neq \rho$, and a Sylow p-subgroup of $\mathrm{fix}_H(\mathcal{B})$ has order p^2. Calculating $\tau^{-1}\phi^{-1}\tau\phi(i, j) = \left(i, \beta_1^{(-1)^{i+1}} j\right)$, and applying Lemma 5.3.11 to $\tau^{-1}\phi^{-1}\tau\phi \in \mathrm{fix}_H(\mathcal{B})$, we see that $\left(\tau^{-1}\phi^{-1}\tau\phi\right)|_{B_1} = \phi \in \mathrm{Aut}(\Gamma)$. Then $\Gamma = \Gamma'$ are isomorphic by a group automorphism of D_{2p}. □

We now turn to the isomorphism problem for Cayley digraphs of $\mathbb{Z}_{qp}$. We first consider when $q \neq p$, and begin with two preliminary results. The first is really group theoretic, and so holds for all Cayley objects, not just digraphs.

Lemma 7.5.10 *Let $q < p$ be prime and $S \subseteq \mathbb{Z}_{qp}$ such that* $\mathrm{Aut}(\mathrm{Cay}(\mathbb{Z}_{qp}, S))$ *has a Sylow p-subgroup of order p. Then* $\mathrm{Cay}(\mathbb{Z}_{qp}, S)$ *is a CI-digraph of $\mathbb{Z}_{qp}$.*

Proof Set $G = \mathbb{Z}_q \times \mathbb{Z}_p \cong \mathbb{Z}_{qp}$, and $\Gamma = \mathrm{Cay}(G, S)$. Define $\rho, \tau\colon \mathbb{Z}_q \times \mathbb{Z}_p \to \mathbb{Z}_q \times \mathbb{Z}_p$ by $\rho(i, j) = (i, j+1)$ and $\tau(i, j) = (i+1, j)$. Then $\langle \rho, \tau \rangle = G_L \leq \mathrm{Aut}(\Gamma)$. Let $\Gamma' \cong \Gamma$ be a Cayley digraph of G. By Lemma 7.5.7, there exists $\phi \in \mathrm{AGL}(1,q) \wr \mathrm{AGL}(1,p)$ such that $\phi(\Gamma) = \Gamma'$. Let $H = \left\langle G_L, \phi^{-1}G_L\phi \right\rangle$. Then the set of orbits of $\langle \rho \rangle$ is a block system $\mathcal{B}$ of H with blocks of size p. If the action of $\mathrm{fix}_H(\mathcal{B})$ on $B \in \mathcal{B}$ is not faithful for some $B \in \mathcal{B}$, then the kernel K of this action is nontrivial. As $\mathrm{fix}_H(\mathcal{B})^B$ is primitive as $|B| = p$, and a normal subgroup of a primitive group is transitive by Theorem 2.4.7, we see that p divides $|K|$. Hence p^2 divides $|H|$, a contradiction. Thus $\mathrm{fix}_H(\mathcal{B})$ is faithful on $B \in \mathcal{B}$ for every $B \in \mathcal{B}$, in which case by Lemma 7.5.7 we have $\phi(i, j) = (ri, \beta j + b_i)$ for some $r \in \mathbb{Z}_q^*$, $\beta \in \mathbb{Z}_p^*$, and $b_i \in \mathbb{Z}_p$. As $(i, j) \mapsto \left(r^{-1}i, \beta^{-1}j\right)$ is an automorphism of G, by Lemma 7.5.1 we may assume that $r = \beta = 1$. Then

$$\tau^{-1}\phi^{-1}\tau\phi(i, j) = (i, j + b_i - b_{i+1}).$$

Suppose $\tau^{-1}\phi^{-1}\tau\phi \neq 1$. Then $b_i - b_{i+1} = c$ for some $c \in \mathbb{Z}_p^*$ and every $i \in \mathbb{Z}_q$, and $\phi^{-1}\tau\phi(i, j) = (i+1, j+c)$ has order q. Then $qc \equiv 0 \pmod p$, a contradiction. Thus $\tau^{-1}\phi^{-1}\tau\phi = 1$, so $b_i - b_{i+1} \equiv 0 \pmod p$ or $b_i \equiv b_{i+1} \pmod p$ for every $i \in \mathbb{Z}_q$. Then $\phi \in \langle\rho\rangle \leq \mathrm{Aut}(\Gamma)$, and Γ is a CI-digraph of G. □

In our next result, we will start with a digraph Γ but not a digraph Γ' isomorphic to Γ. However, many of the tools we have developed do make use of Γ'. If Γ is a Cayley digraph of G and $\phi \in S_G$ such that $\phi^{-1}G_L\phi \leq \mathrm{Aut}(\Gamma)$, then $\phi(\Gamma) \cong \Gamma$ is a Cayley digraph of G (this was shown in the proof of Lemma 7.3.1). So by giving $\phi \in S_G$ such that $\phi^{-1}G_L\phi \leq \mathrm{Aut}(\Gamma)$, we are implicitly defining Γ' and so may use the tools developed so far for that implicit isomorphic Cayley digraph $\phi(\Gamma)$. We will though mention $\phi(\Gamma) \cong \Gamma$, and employ a new technique. We have seen that if $\phi\colon \Gamma \to \Gamma'$ is an isomorphism, then $\phi\delta\colon \Gamma \to \Gamma'$ is an isomorphism for any $\delta \in \mathrm{Aut}(\Gamma)$. By the same token, $\delta\phi\colon \Gamma \to \Gamma'$ is also an isomorphism for any $\delta \in \mathrm{Aut}(\Gamma')$, and sometimes it is computationally easier to deal with – that will be the case below.

Lemma 7.5.11 *Let $q < p$ be prime. If $S \subseteq \mathbb{Z}_{qp}$ such that $\Gamma = \mathrm{Cay}(\mathbb{Z}_{qp}, S) \cong \Gamma_1 \wr \Gamma_2$, where Γ_1 and Γ_2 are circulant digraphs of order q and p, respectively, then Γ is a CI-digraph of $\mathbb{Z}_{qp}$.*

Proof We use the (q, p)-metacirculant labeling here and assume $\rho, \tau\colon \mathbb{Z}_q \times \mathbb{Z}_p \to \mathbb{Z}_q \times \mathbb{Z}_p$ given by $\rho(i, j) = (i, j + 1)$ and $\tau(i, j) = (i + 1, j)$ are contained in $\mathrm{Aut}(\Gamma)$. Let $\phi \in S_{qp}$ such that $\phi^{-1}(\mathbb{Z}_{qp})_L\phi \leq \mathrm{Aut}(\Gamma)$. By Lemma 7.5.7 we may assume $\phi \in \mathrm{AGL}(1, q) \wr \mathrm{AGL}(1, p)$, with $\phi(i, j) = (rj, \beta_i j + b_i)$, where $r \in \mathbb{Z}_q^*$, $\beta_i \in \mathbb{Z}_p^*$, and $b_i \in \mathbb{Z}_p$. As is usual, we may assume $r = 1$ as the map $(i, j) \mapsto \left(r^{-1}i, j\right)$ is in $\mathrm{Aut}(\mathbb{Z}_{qp})$. For $i \in \mathbb{Z}_q$, let $V_i = \{(i, j) : j \in \mathbb{Z}_p\}$, and $\mathcal{B} = \{V_i : i \in \mathbb{Z}_q\}$. Then $\mathcal{B}$ is the lexi-partition corresponding to $V(\Gamma_2)$, and so $z_i = \rho|_{V_i} \in \mathrm{Aut}(\Gamma)$ for every $V_i \in \mathcal{B}$. Then $z_i(i, j) = (i, j+1)$ and $z_i(i', j) = (i', j)$ if $i' \neq i$. The map $\delta\colon \mathbb{Z}_q \times \mathbb{Z}_p \to \mathbb{Z}_q \times \mathbb{Z}_p$ given by $\delta(i, j) \mapsto (i, j - b_i)$ is then contained in $\mathrm{Aut}(\Gamma)$ and $\mathrm{Aut}(\phi(\Gamma))$ by Lemma 6.5.17. Replacing ϕ with $\delta\phi$, we may assume that $b_i = 0$ for all $i \in \mathbb{Z}_p$. Hence $\phi(i, j) = (i, \beta_i j)$. Then $\tau^{-1}\phi^{-1}\tau\phi(i, j) = \left(i, \beta_i\beta_{i+1}^{-1}j\right) \in \mathrm{Aut}(\Gamma)$. For brevity, set $\omega = \tau^{-1}\phi^{-1}\tau\phi$ and $\gamma_i = \beta_i\beta_{i+1}^{-1}$, so $(i, j) \mapsto (i, \gamma_i j) \in \mathrm{Aut}(\Gamma)$.

We now employ what can be thought of as a "weight shifting" technique to $\phi^{-1}\tau\phi$. That is, we will conjugate $\phi^{-1}\tau\phi$ by an appropriate element δ_0 of $\mathrm{Aut}(\Gamma)$ so that $\delta_0^{-1}\phi^{-1}\tau\phi\delta_0(0, j) = (0, j)$, $\delta_0^{-1}\phi^{-1}\tau\phi\delta_0(p-1, j) = (0, \gamma_0\gamma_{p-1}j)$, and $\delta_0^{-1}\phi^{-1}\tau\phi\delta_0(i, j) = (i, \gamma_i j)$ if $i \neq 0, 1$. We think of this as shifting the "weight" γ_0 on V_0 to V_{p-1} (the reason it is V_{p-1} and not V_1 will be clear from the details). We will then repeat the procedure $q-2$ times, and shift all of the weight to V_1. As $\phi^{-1}\tau\phi(i, j) = (i, \gamma_i j)$ and has order q as it is a conjugate of τ, $\left(\phi^{-1}\tau\phi\right)^q(i, j) = \left(i, \Pi_{i=0}^{q-1} j\right) = (i, j)$. Hence $\Pi_{i=0}^{q-1}\gamma_i \equiv 1 \pmod{p}$. So once we have shifted all of the weight to V_1, we will have conjugated $\phi^{-1}\tau\phi$ to obtain τ. Here are the details.

As $\mathrm{Aut}(\Gamma)$ is a wreath product, we know that $\omega|_{V_0} \in \mathrm{Aut}(\Gamma)$. So $\omega|_{V_0}(0, j) = (0, \gamma_0 j)$ while $\omega|_{V_0}(i, j) = (i, j)$ if $i \neq 0$. Computations will show that

$$(\omega|_{V_0})\phi^{-1}\tau\phi(\omega|_{V_0})^{-1}(0, j) = (0, j),$$
$$(\omega|_{V_0})\phi^{-1}\tau\phi(\omega|_{V_0})^{-1}(p-1, j) = (p-1, \gamma_0\gamma_{p-1}j), \text{ and}$$
$$(\omega|_{V_0})\phi^{-1}\tau\phi(\omega|_{V_0})^{-1}(i, j) = (i, \gamma_i j),$$

if $i \neq 0, p-1$. Notationally, it is easiest to now just assume $\gamma_0 = 1$. We now conjugate $\phi^{-1}\tau\phi$ by $(\omega_{V_{p-1}})^{-1}$, which shifts the weight at V_{p-1} to V_{p-2}. Shifting the weights a total of $q-1$ times, we may assume without loss of generality that $\phi^{-1}\tau\phi = \tau$, or $\phi = 1$, and the result follows. □

By Theorem 5.1.8 and its digraph analogue, automorphism groups of vertex-transitive digraphs that are wreath products are well understood. However, Example 7.4.3 shows that digraphs that are Cayley digraphs of a group G and are wreath products need not be CI-graphs of G. In general, the behavior of wreath products with respect to the CI-problem is not well understood. In fact, it is not necessarily true (see, for example, Dobson and Morris [2009, Example 6.4]) that if $\mathrm{Cay}(G, S)$ and $\mathrm{Cay}(H, T)$ are CI-digraphs with respect to G and H, respectively, then $\mathrm{Cay}(G, S) \wr \mathrm{Cay}(H, T)$ is a CI-digraph of $G \times H$! See Dobson and Morris (2009) for some partial results along this line.

We now prove that $\mathbb{Z}_{qp}$ is a CI-group with respect to digraphs, a result independently proven by Alspach and Parsons (1979) using group-theoretic techniques, and by Klin and Pöschel (1981) using the method of Schur.

Theorem 7.5.12 *Let $q < p$ be prime. Then $\mathbb{Z}_{qp}$ is a CI-group with respect to digraphs.*

Proof Let $\Gamma = \mathrm{Cay}(\mathbb{Z}_{qp}, S)$. View $\mathbb{Z}_{qp}$ as $\mathbb{Z}_q \times \mathbb{Z}_p$. Define $\rho, \tau \colon \mathbb{Z}_q \times \mathbb{Z}_p \to \mathbb{Z}_q \times \mathbb{Z}_p$ by $\rho(i, j) = (i, j+1)$ and $\tau(i, j) = (i+1, j)$ so that $(\mathbb{Z}_q \times \mathbb{Z}_p)_L = \langle\rho, \tau\rangle$. Let $\Gamma' \cong \Gamma$ be a circulant digraph of order qp (with vertex set $\mathbb{Z}_q \times \mathbb{Z}_p$). By

Lemma 7.5.7, there exists $\phi \in \mathrm{AGL}(1,q) \wr \mathrm{AGL}(1,p)$ such that $\phi(\Gamma) = \Gamma'$. Let $H = \left\langle (\mathbb{Z}_q \times \mathbb{Z}_p)_L, \phi^{-1}(\mathbb{Z}_q \times \mathbb{Z}_p)_L\phi \right\rangle$. Then the set of orbits of $\langle \rho \rangle$ is a block system of H with blocks of size p. If the induced action of $\mathrm{fix}_H(\mathcal{B})$ on $B \in \mathcal{B}$ is not faithful, then, by Corollary 5.3.15 Γ is a nontrivial generalized wreath product, and so a nontrivial wreath product. The result follows by Lemma 7.5.11. If the action of $\mathrm{fix}_H(\mathcal{B})$ on $B \in \mathcal{B}$ is faithful, then a Sylow p-subgroup of $\mathrm{Aut}(\Gamma)$ has order p. The result follows by Lemma 7.5.10. □

The following result is due to Godsil (1983).

Theorem 7.5.13 *For p a prime, $\mathbb{Z}_p \times \mathbb{Z}_p$ is a CI-group with respect to digraphs.*

Proof Let Γ be a Cayley digraph of $G = \mathbb{Z}_p \times \mathbb{Z}_p$. Define $\rho, \tau \colon \mathbb{Z}_p \times \mathbb{Z}_p \to \mathbb{Z}_p \times \mathbb{Z}_p$ by $\rho(i,j) = (i, j+1)$ and $\tau(i,j) = (i+1, j)$, so that $(\mathbb{Z}_p \times \mathbb{Z}_p)_L = \langle \rho, \tau \rangle$. We wish to apply Lemma 7.3.1, and so let $\phi \in S_G$ such that $\phi^{-1}G_L\phi \le \mathrm{Aut}(\Gamma)$. Then G_L and $\phi^{-1}G_L\phi$ are contained in Sylow p-subgroups P_1 and P_2 of $\mathrm{Aut}(\Gamma)$, and so there exists $\delta_1 \in \mathrm{Aut}(\Gamma)$ such that $\delta_1^{-1}\phi^{-1}G_L\phi\delta_1 \le P_1$. Replacing ϕ with $\phi\delta$, we assume without loss of generality that G_L and $\phi^{-1}G_L\phi$ are contained in the same Sylow p-subgroup P of $\mathrm{Aut}(\Gamma)$. As P has a nontrivial center, P contains a central element β of order p, and as a transitive abelian group is self-centralizing by Corollary 2.2.18, $\beta \in G_L$. Then the set of orbits of $\langle \beta \rangle$ is a nontrivial block system $\mathcal{B}$ of P.

If the action of $\mathrm{fix}_P(\mathcal{B})$ on $B \in \mathcal{B}$ is faithful, then $|\mathrm{fix}_P(\mathcal{B})| = p$ so $|P| = p^2$ and $\phi^{-1}G_L\phi = G_L$. The result then follows by Lemma 7.3.1.

If the action of $\mathrm{fix}_P(\mathcal{B})$ on $B \in \mathcal{B}$ is not faithful, then Γ is a nontrivial generalized wreath product by Corollary 5.3.15 and so a nontrivial wreath product. Then $P \cong (\mathbb{Z}_p)_L \wr (\mathbb{Z}_p)_L$ by Theorem 5.1.8 (which also holds for digraphs). Now let $R \ne G$ be a regular subgroup of P isomorphic to $\mathbb{Z}_p^2$. As a transitive abelian group is self-centralizing by Corollary 2.2.18, we see $\mathrm{Z}(P) \le R$ and $\mathrm{Z}(P)$ has order p. We now do some things that will simplify the main computation. Let $\delta \in \mathrm{GL}(2,p)$ such that $\delta\beta\delta^{-1} = \rho$. Replacing Γ with $\delta(\Gamma)$ by Lemma 7.5.1 we may assume without loss of generality that $\beta = \rho$. Let $\tau' \in R$ such that $\tau'/\mathcal{B} \ne 1$. Raising τ' to an appropriate power, if necessary, we may assume without loss of generality that $\tau'/\mathcal{B} = \tau/\mathcal{B}$. Then $\tau'(i,j) = (i+1, j+b_i)$, where $b_i \in \mathcal{B}$. Turning to the main computation, as $\tau' \in R$, τ' has order p, and so $(\tau')^p(i,j) = \left(i, j + \sum_{i=0}^{p-1} b_i\right)$. Hence $\sum_{i=0}^{p-1} b_i \equiv 0 \pmod p$. By Exercise 7.5.10, there is $\delta \in P$ such that $\delta^{-1}\tau'\delta = \tau$. As $\rho \in \mathrm{Z}(P)$, and as a regular abelian group is self-centralizing by Corollary 2.2.18, $\rho \in \delta^{-1}R\delta$. So $\delta^{-1}R\delta = G$. The result follows by Lemma 7.3.1. □

7.6 Pálfy's Theorem

In this section we prove one direction of Pálfy's remarkable result that $\mathbb{Z}_n$ is a CI-group if and only if $\gcd(n, \varphi(n)) = 1$ or $n = 4$, where φ is Euler's phi function. Namely, we will show that if $\gcd(n, \varphi(n)) = 1$, then $\mathbb{Z}_n$ is a CI-group. The converse consists of constructing appropriate quaternary relational structures whose automorphism group contains nonconjugate regular subgroups isomorphic to G, for any other group G. Consequently, if G is not a CI-group, then it is not a CI-group for quaternary relational structures. The interested reader is referred to Pálfy (1987) for these constructions. It may seem that this result is inappropriate for a text studying symmetry in graphs, as in reality the result we will prove is almost entirely group theoretic. As we have seen, results concerning the isomorphism problem usually alternate between applying a group-theoretic argument and then a graph-theoretic argument, etc. There is an obvious question. When is use of a graph-theoretic argument strictly necessary? Pálfy's theorem answers that question by showing that if one wishes to solve the isomorphism problem for Cayley graphs of G, and G does not satisfy the hypothesis of Pálfy's theorem, then one *must* use facts that are derived from the graph structure.

The condition that $\gcd(n, \varphi(n)) = 1$ or $n = 4$ is of course a number-theoretic condition. But that is not the point *behind* the condition. Of course, $n = 4$ is just a small, exceptional value, and so we ignore it for this discussion. The condition that $\gcd(n, \varphi(n)) = 1$ is actually a condition on $\mathrm{Aut}(\mathbb{Z}_n)$. Remember that $\varphi(n)$ is the number of positive integers less than n and relatively prime to n. Thus it counts the number of generators of $\mathbb{Z}_n$, and the maps $x \mapsto mx$, with $m \in \mathbb{Z}_n^*$, are the automorphisms of $\mathbb{Z}_n$. Thus the condition that $\gcd(n, \varphi(n)) = 1$ also means that $\gcd(n, |\mathrm{Aut}(\mathbb{Z}_n)|) = 1$! This is the group-theoretic condition that will be exploited, which is more concisely expressed as a number-theoretic condition. Indeed, as Pálfy pointed out in his paper, the condition that $\gcd(n, \varphi(n)) = 1$ occurs if and only if every group of order n is cyclic (Scott, 1987, 9.2.7).

Before turning to our last "basic tools," we will need to prove the most general form of Lemma 7.3.1, which is also a "basic tool." These last basic tools are intuitively results that can be obtained only using permutation group theory. In order to derive the most general form of Lemma 7.3.1, we must show that every transitive group is the automorphism group of some "combinatorial structure." The "combinatorial structures" that we will need are k**-ary relational structures** introduced by Wielandt (1969).

Definition 7.6.1 Let Ω be a set. A k**-ary relational structure** on Ω is an ordered pair (Ω, U), where $U \subseteq \Omega^k = \Pi_{i=1}^k \Omega$. An **automorphism** of a k-ary relational structure $X = (\Omega, U)$ is a bijection $g\colon \Omega \to \Omega$ such that $g(U) = U$. The set of all automorphisms of X is a group, called the **automorphism group** of X, and denoted $\mathrm{Aut}(X)$.

A group $G \leq \mathcal{S}_\Omega$ is called k-**closed** if G is the intersection of the automorphism groups of some set of k-ary relational structures. The k-**closure of** G, denoted $G^{(k)}$, is the intersection of all k-closed subgroups of $\mathcal{S}_\Omega$ that contain G. Clearly $G^{(k)}$ is k-closed. A **color** k-**ary relational structure** is a k-ary relational structure in which each element of U has been assigned a color. An **automorphism** of a color k-ary relational structure $W = (\Omega, U)$ is a bijection $g\colon \Omega \to \Omega$ such that an element $u \in U$ is colored with color i if and only if $g(u) \in U$ is colored with color i. The **automorphism group** of W, denoted $\mathrm{Aut}(W)$, is the group of all automorphisms of W.

So our definition of 2-closure earlier in Section 5.2 agrees with our definition of k-closure when $k = 2$. Also, k-ary relational structures can be thought of as natural generalizations of digraphs where the "edges" have more than two vertices. Hence they are also generalizations of k-uniform hypergraphs (which we will not define here). The next result gives a sufficient condition for a group G to be k-closed.

Theorem 7.6.2 *Let G be a group such that the point-wise stabilizer F of the $k-1$ points in $S = \{s_1, \ldots, s_{k-1}\}$ is trivial. Then G is the automorphism group of a color k-ary relational structure.*

Proof Let $U_1, \ldots, U_t$ be the orbits of G acting canonically on Ω^k. Define a color k-ary relational structure X by letting $U = \cup_{i=1}^t U_i$ and $u \in U$ is colored with color i if and only if $u \in U_i$, $1 \leq i \leq t$. Clearly $G \leq \mathrm{Aut}(X)$. Let $\ell \in \mathrm{Aut}(X)$, $s \in X$, and $1 \leq i \leq t$ such that $(s_1, \ldots, s_{k-1}, s) \in U_i$. As $\ell \in \mathrm{Aut}(X)$, $\ell(s_1, \ldots, s_{k-1}, s) \in U$ and is of color i. By the definition of U_i, there exists $g_s \in G$ such that $g_s(s_1, \ldots, s_{k-1}, s) = \ell(s_1, \ldots, s_{k-1}, s)$ so that $\ell^{-1} g_s(s_1, \ldots, s_{k-1}, s) = (s_1, \ldots, s_{k-1}, s)$, and $\ell^{-1} g_s(s) = s$. As $F = 1$, $\ell^{-1} g_s = 1$, and $g_s = \ell \in G$. Hence $\mathrm{Aut}(X) = G$. □

Lemma 7.6.3 *For a group G, the following are equivalent:*

(i) *G is a CI-group,*
(ii) *whenever $\phi \in \mathcal{S}_G$, then G and $\phi^{-1} G \phi$ are conjugate in $\langle G, \phi^{-1} G \phi \rangle$.*

Proof (i) ⇒ (ii) Suppose G is a CI-group. Let $\phi \in \mathcal{S}_G$. Then there exists an integer $k \leq |G|$ and $s_1, \ldots, s_k \in G$ such that the point-wise stabilizer of the set $S = \{s_1, \ldots, s_k\}$ in $H = \langle G_L, \phi^{-1} G_L \phi \rangle$ is trivial. By Theorem 7.6.2 we have that H is the automorphism group of a color k-ary relational structure X. As G is a CI-group, G is a CI-group with respect to color k-ary relational structures.

By Lemma 7.3.1, G_L and $\phi^{-1}G_L\phi$ are conjugate in $\mathrm{Aut}(X) = \left\langle G_L, \phi^{-1}G_L\phi \right\rangle$. (ii) $\Rightarrow$ (i) is straightforward and left as Exercise 7.6.5. □

We now turn to the proof of Pálfy's Theorem. The overall structure of the proof we present is similar to that of Pálfy. Namely, we will reduce the primitive case to the imprimitive case. Much of the imprimitive case is handled very similarly to how Pálfy did, but parts are also taken from Dobson (2006b). We begin with the primitive case, and first consider when $\mathrm{soc}(G) = \mathrm{PSL}(d, q)$, which Zsigmondy's theorem (due to Zsigmondy (1892)) will settle for most $d \geq 2$ and q a prime power. We remark that we could also use more powerful results such as Theorem 5.4.14 instead, but we wish to bring Zsigmondy's theorem to the readers attention.

Theorem 7.6.4 (Zsigmondy's theorem) *If $a > b > 0$ are relatively prime integers, then for any integer $k > 1$, there is a prime p such that p divides $a^k - b^k$ but does not divide $a^\ell - b^\ell$ for any positive integer $\ell < k$, with the following exceptions:*

(i) *$a = 2, b = 1$, and $k = 6$, or*
(ii) *$a + b$ is a power of two, and $k = 2$.*

Lemma 7.6.5 *Let n be square-free and $\mathrm{PSL}(d, q)$ be 2-transitive of degree $n = \left(q^d - 1\right)/(q-1)$. Then there exists a prime divisor p of n such that a Sylow p-subgroup of $\mathrm{P\Gamma L}(d, q)$ has order p.*

Proof Let $q = q_0^f$. Then (see pages 160 and 195)

$$|\mathrm{P\Gamma L}(d, q)| \text{ divides } f\Pi_{i=0}^{d-1}\left(q^d - q^i\right) = fq^{d(d-1)/2}\Pi_{i=0}^{d-1}\left(q^{d-i} - 1\right).$$

Applying Zsigmondy's theorem with $k = df$, $a = q_0$, and $b = 1$, there exists a prime divisor p of n and $q^d - 1 = q_0^{df} - 1$ that does not divide $q_0^\ell - 1$ for any $\ell < df$, or the exceptional cases when $a = 2, b = 1$, and $k = 6$, or $a + b$ is a power of two, and $k = 2$ occur.

If $q^d - 1 \equiv 0 \pmod p$ but $q_0^\ell - 1 \not\equiv 0 \pmod p$ for any $\ell < df$, then $q^d \equiv 1 \pmod p$ but $q_0^\ell \not\equiv 1 \pmod p$ for any $\ell < df$. Then p is relatively prime to $q^\ell - 1$ for every $1 \leq \ell \leq d - 1$, and as $q_0^{df} = q^d \equiv 1 \pmod p$, we see $p \neq q_0$. Finally, if p divides f, then $f = rp$, $r \geq 1$, and $q_0^f = q_0^{pr} = (q_0^p)^r \equiv q_0^r \pmod p$ by Fermat's little theorem. This implies that $q_0^f \equiv q_0^r \pmod p$ and so $q_0^r\left(q_0^{f-r} - 1\right) \equiv 0 \pmod p$. As $f - r < fd$, we see $q_0^{f-r} - 1 \not\equiv 0 \pmod p$, and so $q_0^r \equiv 0 \pmod p$ and $q_0 = p$, a contradiction. Then p is relatively prime to $fq^{d(d-1)/2}\Pi_{i=1}^{d-1}\left(q^{d-i} - 1\right)$, and so the highest power or p that divides $|\mathrm{P\Gamma L}(d, q)|$ is the highest power of p that divides $q^d - 1$. As n is square-free, $\left(q^d - 1\right)/(q-1)$

is square-free. As $\gcd(q-1,p)=1$, the highest power of p that divides q^d-1 is the highest power of p that divides $\left(q^d-1\right)/(q-1)$, and that power is the first power. The result follows in this case.

In the exceptional cases, if $a=2$, $b=1$, and $k=6$, then the only possible groups are $P\Gamma L(6,2)$, $P\Gamma L(3,4)$, and $P\Gamma L(2,8)$, which have degrees 63, 21, and 9, respectively. As n is square-free, the only possibility is $P\Gamma L(3,4)$, which has order $2^7\cdot 3^3\cdot 5\cdot 7$, the result follows with $p=7$. If q_0+1 is a power of 2 and $k=2$, then $n=q_0+1$, which is not square-free, a contradiction. □

With the above result in hand, we now prove the lemma, which will allow us reduce the primitive case to the imprimitive case when the primitive group has socle $PSL(d,q)$.

Lemma 7.6.6 *Let q be a prime power, $d\geq 2$, and $n=\left(q^d-1\right)/(q-1)$ be square-free. Let $G\leq S_n$ be a group with $\mathrm{soc}(G)=PSL(d,q)$, and $\rho,\gamma\in G$ be n-cycles. Then there exists $\delta\in G$ such that $\left\langle\rho,\delta^{-1}\gamma\delta\right\rangle$ has a block system with blocks of prime size.*

Proof As stated after the proof of Lemma 5.4.13, we have $G\leq P\Gamma L(d,q)$. By Lemma 7.6.5 there is a prime divisor p of n such that a Sylow p-subgroup of $P\Gamma L(d,q)$, and so of G, has order p. Then $\langle\rho^{n/p}\rangle$ and $\langle\gamma^{n/p}\rangle$ are Sylow p-subgroups of G, and so there exists $\delta\in G$ such that $\delta^{-1}\langle\gamma^{n/p}\rangle\delta=\langle\rho^{n/p}\rangle$. As $\langle\rho^{n/p}\rangle\trianglelefteq\langle\rho\rangle$ and $\delta^{-1}\langle\gamma^{n/p}\rangle\delta\trianglelefteq\delta^{-1}\langle\gamma\rangle\delta$, we see the set of orbits of $\langle\rho^{n/p}\rangle$ are a block system $\mathcal{B}$ of $\langle\rho\rangle$ and $\delta^{-1}\langle\gamma\rangle\delta$. By Exercise 2.3.8, we have $\mathcal{B}$ is a block system of $\left\langle\langle\rho\rangle,\delta^{-1}\langle\gamma\rangle\delta\right\rangle=\langle\rho,\delta^{-1}\gamma\delta\rangle$. As the orbits of $\langle\rho^{n/p}\rangle$ have prime order p, the result follows. □

The above results are essentially exactly how Pálfy proceeded. The next result, while maintaining the structure of Pálfy's approach, is different. Pálfy himself proved (1987, Lemma 3) that if n is odd, then there are either one or two conjugacy classes of regular cyclic subgroups of A_n, and there is one if and only if n is square-free. The following lemma will be sufficient for our purposes, but can also be used for groups other than cyclic groups.

Lemma 7.6.7 *Let $p|n$ be prime, $G_1,G_2\leq A_n$ be transitive and have normal Sylow p-subgroups P_1 and P_2 with orbits of size p. Then there exists $\delta\in A_n$ such that $\left\langle G_1,\delta^{-1}G_2\delta\right\rangle$ has a block system that is the set of orbits of P_1.*

Proof As $P_i\trianglelefteq G_i$, there is a block system $\mathcal{B}_i$ of G_i with blocks of size p that is the set of orbits of P_i, $i=1,2$. Let $\mathcal{B}_i=\{B_{i,j}:j\in\mathbb{Z}_{n/p}\}$. As A_n is $(n-2)$-transitive by Exercise 2.4.5, there exists $\delta^{-1}\in A_n$ such that $\delta^{-1}(B_{2,j})=B_{1,j}$ for every $0\leq j\leq n/p-2$. But that implies that $\delta^{-1}(B_{2,n/p-1})=B_{1,n/p-2}$ and so $\delta^{-1}(B_{2,j})=B_{1,j}$ for every $j\in\mathbb{Z}_{n/p}$. Then the set of orbits of $\delta^{-1}P_2\delta$ is $\mathcal{B}_1$, and so $\mathcal{B}_1$ is a block system of $\delta^{-1}G_2\delta$. Then $\mathcal{B}_1$ is a block system of $\left\langle G_1, \delta^{-1}G_2\delta\right\rangle$ by Exercise 2.3.8. □

The following result reduces the primitive case in Pálfy's theorem to the imprimitive case, and is the only part of the proof that depends upon CFSG.

Lemma 7.6.8 *Let n be square-free, and $\rho, \gamma \in S_n$ be n-cycles such that $\langle \rho, \gamma\rangle$ is primitive. Then there exists $\delta \in \langle \rho, \gamma\rangle$ such that $\left\langle \rho, \delta^{-1}\gamma\delta\right\rangle$ has a block system with blocks of prime size $p|n$.*

Proof If $G = \langle \rho, \gamma\rangle = S_n$, then clearly $\langle\gamma\rangle$ is conjugate to $\langle\rho\rangle$ in G, and the result is trivial. If $G = A_n$, then the result follows by Lemma 7.6.7, while by Lemma 7.6.6 the result follows if $\mathrm{PSL}(d, q) \le G \le \mathrm{P\Gamma L}(d, q)$. Clearly the result is trivial if n is itself prime. These groups are all of the primitive groups of square-free degree that contain a regular cyclic subgroup. This follows as a consequence of Theorem 5.4.14 (Li, 2003, Corollary 1.2), or independently (Jones, 2002, Theorem 1.2). □

Inductively applying the preceding result one can show that there exists $\delta \in \langle \rho, \gamma\rangle$ such that $\langle \rho, \delta^{-1}\gamma\delta\rangle$ is **normally m-step imprimitive** (Exercise 7.6.3).

Definition 7.6.9 Let $n = p_1^{a_1}p_2^{a_2}\cdots p_r^{a_r}$ be the prime power decomposition of n, and define $\Omega\colon \mathbb{N} \to \mathbb{N}$ by $\Omega(n) = \sum_{i=1}^{r} a_i$ (so $m = \Omega(n)$ is the number of prime divisors of n with repetition allowed). A transitive group $G \le S_n$ is **m-step imprimitive** if there exists a sequence of block systems $\mathcal{B}_0 < \mathcal{B}_1 < \cdots < \mathcal{B}_m$ of G, where $\mathcal{B}_0$ is the set of all singleton sets and $\mathcal{B}_m$ is $\{\mathbb{Z}_n\}$, and if $B_{i+1} \in \mathcal{B}_{i+1}$ and $B_i \in \mathcal{B}_i$, then $|B_{i+1}|/|B_i|$ is prime, $0 \le i \le m-1$. (Technically this last condition is not strictly necessary and $\mathcal{B}_{i+1}$ is not, by definition, equal to $\mathcal{B}_i$, but we list it anyway to emphasize the property.) If, in addition, each $\mathcal{B}_i$ is normal, then we say that G is **normally m-step imprimitive**.

Muzychuk (1999, Theorem 4.9) has shown much more. With n and m as in Definition 7.6.9, he has shown that there exists $\delta \in \langle\rho, \gamma\rangle$ such that $\left\langle \rho, \delta^{-1}\gamma\delta\right\rangle$ is normally m-step imprimtiive and that for each $0 \le i \le m-2$, $|B_{i+2}|/|B_{i+1}| \le |B_{i+1}|/|B_i|$ where $B_i \in \mathcal{B}_i$, $B_{i+1} \in \mathcal{B}_{i+1}$, and $B_{i+2} \in \mathcal{B}_{i+2}$. One may wonder if this can always be done for abelian groups (we first ask about abelian groups as by Theorem 2.2.19 all block systems are normal). That is, is it true that if G is an abelian group and $\phi \in S_G$, then there exists $\delta \in \left\langle G_L, \phi^{-1}G_L\phi\right\rangle$ such that $\left\langle G_L, \delta^{-1}\phi^{-1}G_L\phi\delta\right\rangle$ is normally m-step imprimitive? The answer is actually "No." Dobson (2010, Theorem 6.14) has shown that $\mathbb{Z}_2^3 \times \mathbb{Z}_p$ is a CI-group with respect to 3-ary relational structures provided that whenever $\phi \in S_{\mathbb{Z}_2^3\times\mathbb{Z}_p}$ there exists $\delta \in \left\langle \left(\mathbb{Z}_2^3 \times \mathbb{Z}_p\right)_L, \phi^{-1}\left(\mathbb{Z}_2^3 \times \mathbb{Z}_p\right)_L\phi\right\rangle$ such that $\left\langle \left(\mathbb{Z}_2^3 \times \mathbb{Z}_p\right)_L, \delta^{-1}\phi^{-1}\left(\mathbb{Z}_2^3 \times \mathbb{Z}_p\right)_L\phi\delta\right\rangle$ is normally 4-step imprimitive. We remark that the condition given in Dobson (2010, Theorem 6.14) is that $p \geq 11$, but it was also shown that if $p \geq 11$, then such a δ exists (Dobson, 2010, Lemma 6.1), and that is all that is needed for the proof of Dobson (2010, Theorem 6.14). Also, Dobson and Spiga (2013) have shown that for the groups $G = \mathbb{Z}_2^3 \times \mathbb{Z}_3$ and $\mathbb{Z}_2^3 \times \mathbb{Z}_7$, there are Cayley 3-ary relational structures X of G and $\phi \in S_G$ such that $\phi^{-1}G_L\phi \leq \mathrm{Aut}(X)$ and no such δ exists in $\mathrm{Aut}(X)$.

The next result is shown in more generality than we will need (as no extra work is involved). Intuitively, it says that if $\left\langle G_L, \phi^{-1}G_L\phi\right\rangle$ has blocks that are the left cosets of a CI-group, then we may conjugate $\phi^{-1}G_L\phi$ so that the new fixer of $\mathcal{B}$ in the subgroup generated by G_L and its new conjugate is "nice." This is *always* the case when the blocks are of prime size.

Lemma 7.6.10 *Let G be a group and suppose $\mathcal{B}$ is a normal block system of G_L and $\mathrm{fix}_{G_L}(\mathcal{B})$ is a CI-group. Suppose $\phi \in S_G$ such that $\mathcal{B}$ is also a normal block system of $\phi^{-1}G_L\phi$, and $\mathrm{fix}_{\phi^{-1}G_L\phi}(\mathcal{B}) \cong \mathrm{fix}_{G_L}(\mathcal{B})$. Then there exists $\delta \in \mathrm{fix}_{\langle G_L,\phi^{-1}G_L\phi\rangle}(\mathcal{B})$ such that $\mathrm{fix}_{\delta^{-1}\phi^{-1}G_L\phi\delta}(\mathcal{B})^B = \mathrm{fix}_{G_L}(\mathcal{B})^B$ for every $B \in \mathcal{B}$. Consequently,*

$$\langle \mathrm{fix}_{G_L}(\mathcal{B})|_B : B \in \mathcal{B}\rangle \cap \left\langle G_L, \delta^{-1}\phi^{-1}G_L\phi\delta\right\rangle \trianglelefteq \left\langle G_L, \delta^{-1}\phi^{-1}G_L\phi\delta\right\rangle.$$

Proof By Exercise 2.3.8, $\mathcal{B}$ is a block system of $\left\langle G_L, \phi^{-1}G_L\phi\right\rangle$. Fix $B \in \mathcal{B}$. As $\mathcal{B}$ is a normal block system of G_L and $\phi^{-1}G_L\phi$, both $\mathrm{fix}_{\phi^{-1}G_L\phi}(\mathcal{B})^B$ and $\mathrm{fix}_{G_L}(\mathcal{B})^B$ are transitive. As G_L is regular and $\mathcal{B}$ is normal, we see $\mathrm{fix}_{G_L}(\mathcal{B})^B$ is regular. Similarly, $\mathrm{fix}_{\phi^{-1}G_L\phi}(\mathcal{B})^B$ is also regular. As $\mathrm{fix}_{\phi^{-1}G_L\phi}(\mathcal{B}) \cong \mathrm{fix}_{G_L}(\mathcal{B})$, the permutation representations induced by the actions of $\mathrm{fix}_{G_L}(\mathcal{B})$ and $\mathrm{fix}_{\phi^{-1}G_L\phi}(\mathcal{B})$ on B are permutation equivalent representations by Theorem 4.1.8. By Proposition 4.1.6, $\mathrm{fix}_{G_L}(\mathcal{B})^B$ and $\mathrm{fix}_{\phi^{-1}G_L\phi}(\mathcal{B})^B$ are permutation isomorphic, and so by Theorem 4.1.2, $\mathrm{fix}_{G_L}(\mathcal{B})^B$ and $\mathrm{fix}_{\phi^{-1}G_L\phi}(\mathcal{B})^B$ are conjugate in S_B.

Let $|\mathcal{B}| = m$ and $\mathcal{B} = \{B_i : i \in \mathbb{Z}_m\}$. As $\mathrm{fix}_{G_L}(\mathcal{B})$ is a CI-group, by Lemma 7.6.3 there exists $\delta_0 \in \left\langle \mathrm{fix}_{G_L}(\mathcal{B}), \mathrm{fix}_{\phi^{-1}G_L\phi}(\mathcal{B})\right\rangle$ such that $\left(\delta_0^{-1}\mathrm{fix}_{\phi^{-1}G_L\phi}(\mathcal{B})\delta_0\right)^{B_0} = \mathrm{fix}_{G_L}(\mathcal{B})^{B_0}$. Note $\delta_0^{-1}\mathrm{fix}_{\phi^{-1}G_L\phi}(\mathcal{B})\delta_0 = \mathrm{fix}_{\delta_0^{-1}\phi^{-1}G_L\phi\delta_0}(\mathcal{B})$. Again by Lemma 7.6.3, there exists $\delta_1 \in \left\langle \mathrm{fix}_{G_L}(\mathcal{B}), \mathrm{fix}_{\delta_0^{-1}\phi^{-1}G_L\phi\delta_0}(\mathcal{B})\right\rangle$ such that $\mathrm{fix}_{\delta_1^{-1}\delta_0^{-1}\phi^{-1}G_L\phi\delta_0\delta_1}(\mathcal{B})^{B_1} = \mathrm{fix}_{G_L}(\mathcal{B})^{B_1}$. Additionally, as

$$\delta_1^{B_0} \in \left\langle \mathrm{fix}_{G_L}(\mathcal{B}), \mathrm{fix}_{\delta_0^{-1}\phi^{-1}G_L\phi\delta_0}(\mathcal{B})\right\rangle^{B_0} = \mathrm{fix}_{G_L}(\mathcal{B})^{B_0},$$

we have $\left(\delta_1^{-1}\mathrm{fix}_{\delta_0^{-1}\phi^{-1}G_L\phi\delta_0}(\mathcal{B})\delta_1\right)^{B_0} = \mathrm{fix}_{G_L}(\mathcal{B})^{B_0}$. Continuing this process inductively, we find $\delta \in \left\langle \mathrm{fix}_{G_L}(\mathcal{B}), \mathrm{fix}_{\phi^{-1}G_L\phi}(\mathcal{B})\right\rangle$ such that

$$\mathrm{fix}_{\delta^{-1}\phi^{-1}G_L\phi\delta}(\mathcal{B})^B = \mathrm{fix}_{G_L}(\mathcal{B})^B$$

for all $B \in \mathcal{B}$. Then G_L and $\delta^{-1}\phi^{-1}G_L\phi\delta$ both normalize $\langle \mathrm{fix}_{G_L}(\mathcal{B})|_B : B \in \mathcal{B}\rangle$, and so $\langle \mathrm{fix}_{G_L}(\mathcal{B})|_B : B \in \mathcal{B}\rangle \cap \langle G_L, \delta^{-1}\phi^{-1}G_L\phi\delta\rangle \trianglelefteq \langle G_L, \delta^{-1}\phi^{-1}G_L\phi\delta\rangle$. □

We now give some terminology that will be used throughout the rest of this section. Let $\rho\colon \mathbb{Z}_n \to \mathbb{Z}_n$ by $\rho(i) = i+1 \pmod{n}$, so that $\langle\rho\rangle = (\mathbb{Z}_n)_L$. Let $\gamma \in \mathcal{S}_n$ be any n-cycle such that $G = \langle\rho,\gamma\rangle$ has a block system $\mathcal{B}$ with m blocks of size k (so $n = mk$). Let $\rho^m = z_0z_1\cdots z_{m-1}$ where each z_i is a k-cycle that permutes i. That is, $z_i = (i, i+m, i+2m, \ldots, i+(k-1)m)$. Finally, let $K = \langle z_i : i \in \mathbb{Z}_m\rangle$. Our next result deviates from Pálfy's proof and is taken from Dobson (2006b).

Lemma 7.6.11 *If G has $\mathcal{B}$ as above, $\gamma^m \in K$, $\mathbb{Z}_m$ is a CI-group, and* $\gcd(m, k\cdot\varphi(k)) = 1$, *then $\langle\gamma\rangle$ is conjugate to $\langle\rho\rangle$ in G.*

Proof As $\langle\rho\rangle$ and $\langle\gamma\rangle$ are abelian, by Theorem 2.2.19 $\mathcal{B}$ is the set of orbits of the normal subgroups of $\langle\rho\rangle$ and $\langle\gamma\rangle$ of order k. Hence $\langle\rho^m\rangle = \mathrm{fix}_{\langle\rho\rangle}(\mathcal{B})$, $\langle\gamma^m\rangle = \mathrm{fix}_{\langle\gamma\rangle}(\mathcal{B})$, and both $\langle\rho\rangle/\mathcal{B}$ and $\langle\gamma\rangle/\mathcal{B}$ are cyclic of order m. As $\mathbb{Z}_m$ is a CI-group, by Lemma 7.6.3, there exists $\delta \in \langle\rho,\gamma\rangle/\mathcal{B}$ such that $\delta^{-1}\langle\gamma\rangle\delta/\mathcal{B} = \langle\rho\rangle/\mathcal{B}$. We thus assume without loss of generality that $\langle\gamma\rangle/\mathcal{B} = \langle\rho\rangle/\mathcal{B}$.

For $i \in \mathbb{Z}_m$, we have $\rho^{-1}z_i\rho = z_{i-1}$. As $\gamma^m \in K$, $\gamma^{-1}z_i\gamma = z^{a_i}_{\sigma(i)}$, for some $\sigma \in \mathcal{S}_m$ and $a_i \in \mathbb{Z}_m$. Hence both ρ and γ normalize K. By Lemma 6.5.17, we see that $\langle\rho,\gamma\rangle$ is permutation isomorphic to a subgroup of $\mathcal{S}_m \wr N$, where $N = \left\{x \mapsto ax+b : a \in \mathbb{Z}_k^*, b \in \mathbb{Z}_k\right\} \le \mathcal{S}_k$ (note we have permutation isomorphic here as the set being permuted here and that in Lemma 6.5.17 are different). As $\langle\gamma\rangle/\mathcal{B} = \langle\rho\rangle/\mathcal{B}$, we see $\langle\rho,\gamma\rangle$ is permutation isomorphic to a subgroup of $\mathbb{Z}_m \wr N$. Hence $\langle\rho,\gamma\rangle$ is solvable.

By Exercise 4.2.3, $\mathbb{Z}_m \wr N$ has order $m\cdot[k\cdot\varphi(k)]^m$. As $\gcd(m, k\cdot\varphi(k)) = 1$, we have $\gcd(m, [k\cdot\varphi(k)]^m) = 1$. Let π be the set of primes dividing m. Then $\langle\rho^k\rangle$ and $\langle\gamma^k\rangle$ are Hall π-subgroups of $\langle\rho,\gamma\rangle$, and are conjugate in $\langle\rho,\gamma\rangle$ by Hall's theorem (Dummit and Foote, 2004, Theorem 19.2.8). So we may assume that $\langle\rho^k\rangle = \langle\gamma^k\rangle$. As K is abelian, γ^m and ρ^m commute. As $\langle\gamma^k\rangle = \langle\rho^k\rangle$ and γ^m commutes with γ^k, γ^m and ρ^k commute. As $\langle\rho^m,\rho^k\rangle = \langle\rho\rangle$ is a transitive abelian group that is self-centralizing, $\gamma^m \in \langle\rho\rangle$. As $\langle\gamma^k\rangle \le \langle\rho\rangle$, we have $\langle\gamma\rangle \le \langle\rho\rangle$ so $\langle\gamma\rangle = \langle\rho\rangle$. □

Theorem 7.6.12 *If n is a positive integer and* $\gcd(n,\varphi(n)) = 1$, *then $\mathbb{Z}_n$ is a CI-group.*

Proof We proceed by induction on $r = \Omega(n)$. The case $r = 1$ is Theorem 7.3.3. Assume the result holds for all n with $\gcd(n,\varphi(n)) = 1$ such that $\Omega(n) = r-1$. Let n be an integer with $\gcd(n,\varphi(n)) = 1$ and $\Omega(n) = r$, and $\gamma \in \mathcal{S}_n$ be any n-cycle – so $\langle\gamma\rangle$ is conjugate to $\langle\rho\rangle$ in $\mathcal{S}_n$.

If $\langle\rho,\gamma\rangle$ is primitive, then by Lemma 7.6.8, there exists $\delta_1 \in \langle\rho,\gamma\rangle$ such that $\left\langle\rho, \delta_1^{-1}\gamma\delta_1\right\rangle$ has a block system with blocks of prime size $p|n$. It thus suffices to show the result holds if $\langle\rho,\gamma\rangle$ is imprimitive.

If $\langle\rho,\gamma\rangle$ is imprimitive, with a block system $\mathcal{B}$ of m blocks of size k, then by the induction hypothesis $\mathbb{Z}_k$ and $\mathbb{Z}_m$ are CI-groups. By Theorem 2.2.19, $\mathcal{B}$ is a normal block system of $\langle\rho\rangle$ and $\langle\gamma\rangle$. By Lemma 7.6.10, there exists $\delta_1 \in \langle\rho,\gamma\rangle$ such that $\delta_1^{-1}\gamma^m\delta_1 \in K$. By Lemma 7.6.11, there exists $\delta_2 \in \left\langle\rho,\delta_1^{-1}\gamma\delta_1\right\rangle$ such that $\delta_2^{-1}\delta_1^{-1}\langle\gamma\rangle\delta_1\delta_2 = \langle\rho\rangle$. By Lemma 7.6.3 we have $\mathbb{Z}_n$ is a CI-group and the result follows by induction. □

It may seem that Pálfy's theorem finishes off a particular line of research, but this is not at all the case. In fact, Muzychuk (1999) showed that if $m = p_1 \dots p_r$ satisfies $\gcd(m,\varphi(m)) = 1$ and $n = p_1^{a_1}\cdots p_r^{a_r}$, $a_i \geq 1$, then the isomorphism problem for Cayley combinatorial objects of $\mathbb{Z}_n$ reduces to the isomorphism problem for Cayley combinatorial objects of the Sylow p_i-subgroups of $\mathbb{Z}_n$. Later, Dobson (2003a) gave a sufficient condition to show that the same results holds for abelian groups of order n, and for nilpotent groups of order n (Dobson, 2014).

7.7 Isomorphisms of Circulant Digraphs

In this section, we discuss a result of Muzychuk (2004) that completely solves the isomorphism problem for circulant digraphs of any order. The proof of this result uses the method of Schur, and so is beyond the scope of this text. One can think of Muzychuk's theorem as having two parts. The first part "reduces" (we will explain later precisely what this means) the isomorphism problem for circulant digraphs of any order to the prime power case, while the second part solves the isomorphism problem for circulant digraphs of prime power order. As the isomorphism problem is reduced to the prime power case, we focus on solving that problem first.

The solution in the prime power case will require quite a bit of terminology and notation, so it may be useful to summarize how one would use Muzychuk's theorem to determine whether two circulant digraphs Γ_1 and Γ_2 of order p^k, p a prime and $k \geq 1$, are isomorphic, and then how we will develop the terminology and notation. For all but the last step when we check isomorphism, we will actually only use one of the digraphs. So suppose Γ_1 has connection set S_1. We will first determine a partition, called a **key partition**, of $\mathbb{Z}_{p^k}$ based upon its connection set S_1. While we will not show this fact, in the odd prime power case the key partition is in fact the orbits of the stabilizer of 0 of a Sylow p-subgroup P of $\mathrm{Aut}(\Gamma_1)$ that contains $(\mathbb{Z}_{p^k})_L$ (for general n as opposed to prime powers, the interpretation of the key partition is more complicated, which is one reason we focus on prime powers). It turns out (and again we will not show this fact) that for odd primes p, this partition is uniquely determined by P. So a key partition gives a Sylow p-subgroup of the automorphism group of a circulant digraph of order p^k that contains $(\mathbb{Z}_{p^k})_L$ and vice versa when p is odd. We will use the key partition simply because it can be represented in a more compact way (as a **primary key**) than constructing the Sylow p-subgroup of $\mathrm{Aut}(\Gamma_1)$ that contains $(\mathbb{Z}_{p^k})_L$. The possible isomorphisms from Γ_1 to any other circulant digraph of order p^k are then calculated using the key partition – these are called **genuine generalized multipliers for** S_1. The adjective "genuine" is used here as while we will define a **generalized multiplier**, not all generalized multipliers need to be checked for a specific circulant digraph. Muzychuk's theorem gives that Γ_2 is isomorphic to Γ_1 if and only if $\gamma(S_1) = S_2$ (the connection set of Γ_2) for some genuine generalized multiplier of S_1, which is then easy to check as there are not many genuine generalized multipliers (no more than $\varphi(p^k) = p^k - p^{k-1}$). So in reality, what Muzychuk's theorem does is give all bijections for which the image of Γ_1 is a circulant digraph, we calculate the connections sets of those digraphs, and see if S_2 is one of them. If so Γ_1 and Γ_2 are isomorphic and if not they are not isomorphic.

We will proceed by first defining primary keys, and then show how to construct a primary key partition from a primary key (again, this is more or less equivalent to constructing a Sylow p-subgroup of the automorphism group of a circulant digraph that contains $(\mathbb{Z}_{p^k})_L$ when p is odd). We then define generalized multipliers and genuine generalized multipliers, with genuine generalized multipliers depending upon the primary key. Finally, let us remark that while we will only consider digraphs, Muzychuk's theorem actually holds for colored circulant digraphs. We should also point out that our notation and terminology are slightly different from that used in Muzychuk (2004)

7.7.1 Primary Keys and Primary Key Partitions

Definition 7.7.1 Let p be prime and $n \geq 1$ an integer. Define a **primary key space** $\mathbf{K}_{p^n}$ to be the set of all integer vectors $(k_1, \ldots, k_n)$ satisfying the following two properties:

(i) $k_i < i$ for each $1 \leq i \leq n$, and
(ii) $k_{i-1} \leq k_i$ for each $2 \leq i \leq n$.

A vector in $\mathbf{K}_{p^n}$ is called a **primary key**.

Note that the primary key space $\mathbf{K}_{p^n}$ is independent of the choice of p, and that as $k_i < i$, we have $k_1 = 0$.

Example 7.7.2 $\mathbf{K}_p = \{(0)\}$, $\mathbf{K}_{p^2} = \{(0,0),(0,1)\}$,

$$\mathbf{K}_{p^3} = \{(0,0,0),(0,0,1),(0,0,2),(0,1,1),(0,1,2)\},$$

while $\mathbf{K}_{p^4}$ is the set of the following vectors:

$$\begin{array}{ccc} (0,0,0,0) & (0,0,1,1) & (0,1,1,1) \\ (0,0,0,1) & (0,0,1,2) & (0,1,1,2) \\ (0,0,0,2) & (0,0,1,3) & (0,1,1,3) \\ (0,0,0,3) & (0,0,2,2) & (0,1,2,2) \\ & (0,0,2,3) & (0,1,2,3). \end{array}$$

Note that, for example, $(0,0,3,3) \notin \mathbf{K}_{p^4}$ as $k_3 = 3$.

Definition 7.7.3 Let $\mathbf{k}$ be a primary key. For $g \in \mathbb{Z}_{p^n}\backslash\{0\}$, define $b(g) = k$, where $|g| = p^k$, and let $C_g = g + \langle p^{n-k_{b(g)}}\rangle$.

Clearly each C_g is a coset of some subgroup of $\mathbb{Z}_{p^n}$.

Example 7.7.4 Let $p^n = 9$, $\mathbf{k} = (0,1)$, and $g = 6$. Then $C_6 = \{6\}$.

Solution Note $|g| = 9/\gcd(6,9) = 3 = 3^1$, and so $b(6) = 1$ and $k_{b(6)} = k_1 = 0$. As $C_g = g + \left\langle p^{n-k_{b(g)}}\right\rangle$, $C_6 = 6 + \left\langle 3^{2-0}\right\rangle = \{6\}$. □

Example 7.7.5 Let $p^n = 27$, $\mathbf{k} = (0,1,1)$, and $g = 15$. Then $C_{15} = 15 + \langle 9\rangle$.

Solution Note $|g| = 27/\gcd(15,27) = 9 = 3^2$, and so $b(15) = 2$ and $k_{b(15)} = k_2 = 1$ (recall that $\mathbf{k} = (k_1,k_2,k_3)$ in this case). As $C_g = g + \left\langle p^{n-k_{b(g)}}\right\rangle$, $C_{15} = 15 + \left\langle 3^{3-1}\right\rangle = 15 + \langle 9\rangle$. □

Example 7.7.6 Let $p^n = 16$, $\mathbf{k} = (0,1,2,2)$, and $g = 8$. Then $C_8 = \{8\}$.

Solution Note $|8| = 16/\gcd(8,16) = 2 = 2^1$, and so $b(8) = 1$ and $k_{b(8)} = k_1 = 0$. As $C_g = g + \left\langle p^{n-k_{b(g)}}\right\rangle$, $C_8 = \{8\} + \left\langle 2^{4-0}\right\rangle = \{8\}$. □

Lemma 7.7.7 *Any element of C_g has the same order as g. Furthermore, if $\mathbf{k}$ is a primary key then $\Sigma(\mathbf{k}) = \{\{0\}, C_g : g \in \mathbb{Z}_{p^n}\backslash\{0\}\}$ is a partition of $\mathbb{Z}_{p^n}$.*

Proof As $|g| = p^n/\gcd(g,p^n)$, we see that $|g| = p^b$ if and only if p^{n-b} divides g but p^{n-b+1} does not. If $h \in C_g = g + \left\langle p^{n-k_b}\right\rangle$, then $h = g + rp^{n-k_b}$ for some r. As $k_b < b$, it follows that $n - k_b > n - b$ and so p^{n-k_b} is a multiple of p^{n-b}. Then $h = p^{n-b}\left(g/p^{n-b} + rp^{b-k_b}\right)$ while $\gcd\left(g/p^{n-b}, p\right) = 1$ and $\gcd\left(p, p^{b-k_b}\right) = p$. Hence p^{n-b} divides h but p^{n-b+1} does not, and $|h| = |g|$.

Clearly $g \in C_g$ if $g \neq 0$, and so $\cup\Sigma(\mathbf{k}) = \mathbb{Z}_{p^n}$. It is also clear $0 \notin C_g$ for any $g \in \mathbb{Z}_{p^n}\backslash\{0\}$, and so in order to show that $\Sigma(\mathbf{k})$ is a partition of $\mathbb{Z}_{p^n}$, it suffices to show that if $g \in C_h$ and $g \in C_\ell$ for $h, \ell \in \mathbb{Z}_{p^n}$, then $C_h = C_\ell$. Now C_h is a coset of some subgroup $H \leq \mathbb{Z}_{p^n}$ while C_ℓ is a coset of some subgroup $L \leq \mathbb{Z}_{p^n}$. As g is in both C_h and C_ℓ, we see by the first part of this result that $|g| = |h| = |\ell|$, in which case $H = L$ as $\mathbb{Z}_{p^n}$ has a unique subgroup of order $|g|$. Then C_h and C_ℓ are cells in the partition of $\mathbb{Z}_{p^n}$ that is the set of cosets of H (or L), and so are either disjoint or equal. As they are not disjoint, they are equal and the result follows. □

Definition 7.7.8 The partition $\Sigma(\mathbf{k})$ of the previous result will be called the **primary key partition** corresponding to $\mathbf{k}$.

Example 7.7.9 Find the primary key partition of $\mathbb{Z}_9$ corresponding to the primary key $\mathbf{k} = (0, 1)$.

Solution As $1, 2, 4, 5, 7, 8 \in \mathbb{Z}_9$ all have order 9, for $g \in \{1, 2, 4, 5, 7, 8\}$ we have $b(g) = 2$, so $k_{b(g)} = 1$, and $n - k_{b(g)} = 2 - 1 = 1$. Then $C_g = g + \left\langle p^{n-k_{b(g)}} \right\rangle = g + \left\langle 3^1 \right\rangle = g + \langle 3 \rangle$. So $C_1 = \{1, 4, 7\}$ while $C_2 = \{2, 5, 8\}$. Also, $C_1 = C_4 = C_7$ and $C_2 = C_5 = C_8$. Each of $3, 6 \in \mathbb{Z}_9$ has order 3, in which case $b(g) = 1$, $k_{b(g)} = 0$, and $n - k_{b(g)} = 2 - 0$ for $g \in \{3, 6\}$. Then $C_g = g + \left\langle p^{n-k_{b(g)}} \right\rangle = g + \left\langle 3^2 \right\rangle = g + \langle 0 \rangle = \{g\}$. Hence $\Sigma(0, 1) = \{\{0\}, \{3\}, \{6\}, \{1, 4, 7\}, \{2, 5, 8\}\}$. □

Example 7.7.10 Find the primary key partition of $\mathbb{Z}_8$ corresponding to the primary key $\mathbf{k} = (0, 1, 2)$.

Solution As $1, 3, 5, 7 \in \mathbb{Z}_8$ all have order 8, we have $b(g) = 3$, $k_{b(g)} = 2$, and $n - k_{b(g)} = 3 - 2 = 1$ for $g \in \{1, 3, 5, 7\}$. Then $C_1 = 1 + \left\langle p^{n-k_{b(1)}}] \right\rangle = 1 + \left\{2^1\right\} = 1 + \{0, 2, 4, 6\} = \{1, 3, 5, 7\}$. Hence $C_1 = C_3 = C_5 = C_7$. The elements $2, 6 \in \mathbb{Z}_8$ have order 4, so $b(g) = 2$, $k_{b(g)} = 1$, and $n - k_{b(g)} = 3 - 1 = 2$ for $g \in \{2, 6\}$. Then $C_g = g + \left\langle p^{n-k_{b(g)}} \right\rangle = 2 + \left\langle 2^2 \right\rangle = 2 + \langle 4 \rangle = \{2, 6\}$. Finally, 4 has order 2, in which case $b(g) = 1$, $k_{b(g)} = 0$, and $n - k_{b(g)} = 3 - 0$. Then $C_4 = 4 + \left\langle 2^3 \right\rangle = 4 + \{0\} = \{4\}$. Hence $\Sigma(0, 1, 2) = \{\{0\}, \{4\}, \{2, 6\}, \{1, 3, 5, 7\}\}$. □

Example 7.7.11 Find the primary key partition of $\mathbb{Z}_8$ corresponding to the key $\mathbf{k} = (0, 1, 1)$.

Solution As $1, 3, 5, 7 \in \mathbb{Z}_8$ all have order 8, we have $b(g) = 3$, $k_{b(g)} = 1$, and $n - k_{b(g)} = 3 - 1 = 2$ for $g \in \{1, 3, 5, 7\}$. Then $C_1 = 1 + \left\langle 2^{n-k_{b(g)}} \right\rangle = 1 + \langle 4 \rangle = 1 + \{0, 4\} = \{1, 5\} = C_5$ and $C_3 = 3 + \langle 4 \rangle = 3 + \{0, 4\} = \{3, 7\} = C_7$. Notice that for other values of $g \in \mathbb{Z}_8$, the cell C_g is the same as in the previous example. Hence $\Sigma(0, 1, 1) = \{\{0\}, \{4\}, \{2, 6\}, \{1, 5\}, \{3, 7\}\}$. □

Table 7.1 gives all primary key partitions of $\mathbb{Z}_8$. We leave to the reader as an exercise (Exercise 7.7.4) the verification of the remaining entries in Table 7.1. While we will not have need of this fact, it is also left to the exercises (Exercise 7.7.5 and 7.7.6) to verify that unless a primary key is $(0, 0, 0)$, a circulant digraph of order 8 whose connection set is a subset of a primary key partition is always a generalized wreath product, and a similar statement is true in general.

Primary key	Primary key partition
$(0,0,0)$	$\{\{0\},\{1\},\{2\},\{3\},\{4\},\{5\},\{6\},\{7\}\}$
$(0,0,1)$	$\{\{0\},\{2\},\{4\},\{6\},\{1,5\},\{3,7\}\}$
$(0,0,2)$	$\{\{0\},\{2\},\{4\},\{6\},\{1,3,5,7\}\}$
$(0,1,1)$	$\{\{0\},\{4\},\{2,6\},\{1,5\},\{3,7\}\}$
$(0,1,2)$	$\{\{0\},\{4\},\{2,6\},\{1,3,5,7\}\}$

Table 7.1 *The keys and key partitions of* $\mathbb{Z}_8$.

7.7.2 Generalized Multipliers

Notice that in the previous examples the partition $\Sigma(0,1,1)$ is a refinement of $\Sigma(0,1,2)$, so $\Sigma(0,1,1) \leq \Sigma(0,1,2)$. Similarly, we may compare two primary keys by setting $\mathbf{k} = (k_1,\ldots,k_n) \leq (\ell_1,\ldots,\ell_n) = \boldsymbol{\ell}$ if and only if $k_i \leq \ell_i$, $1 \leq i \leq n$. It is not hard to show that if $\mathbf{k} \leq \boldsymbol{\ell}$, then $\Sigma(\mathbf{k}) \leq \Sigma(\boldsymbol{\ell})$ (Exercise 7.7.14). There is a largest primary key $(0,1,\ldots,n-1)$, and a smallest primary key $(0,\ldots,0)$. The primary key space lattice of $\mathbb{Z}_{p^3}$ is given in Figure 7.2.

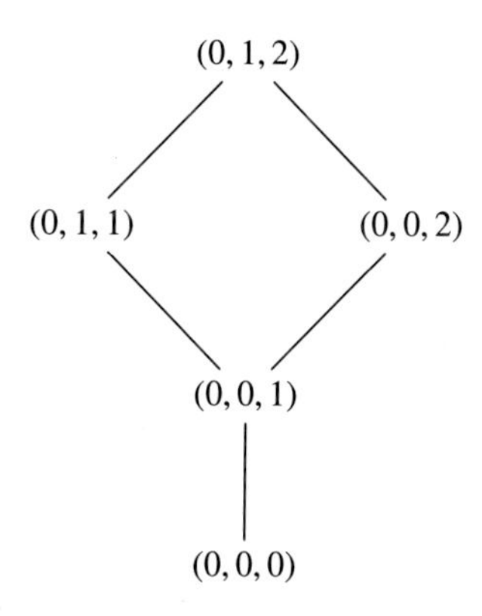

Figure 7.2 The primary key space lattice of $\mathbf{K}_{p^3}$.

We now turn to the bijections that will be isomorphisms between circulant digraphs of prime power order, which are called **generalized multipliers**. The idea is that while a multiplier simply multiplies elements of $\mathbb{Z}_{p^n}$ by a unit in $\mathbb{Z}_{p^n}$, we will want a finer way of multiplying by units. This can be accomplished in the following way. First, if $x \in \mathbb{Z}_{p^n}$, we write x uniquely as

$$x = x_0 + x_1 p + x_2 p^2 + \cdots + x_{n-1} p^{n-1},$$

where $0 \le x_i \le p - 1$ for each $0 \le i \le n - 1$, its p**-adic expansion**.

Example 7.7.12 We may write the element $52 \in \mathbb{Z}_{81}$ as $52 = 1 + 2 \cdot 3 + 2 \cdot 9 + 1 \cdot 27$. Note that the p-adic expansion of x modulo p^n is easy to calculate. Just use the division algorithm for $\mathbb{Z}$ inductively to first write $52 = 25 + 1 \cdot 27$, use the division algorithm to write the remainder $25 = 7 + 2 \cdot 9$, then write the remainder $7 = 1 + 2 \cdot 3$. In a similar fashion, we write 109 in $\mathbb{Z}_{128}$ as $109 = 1 + 0 \cdot 2 + 1 \cdot 2^2 + 1 \cdot 2^3 + 0 \cdot 2^4 + 1 \cdot 2^5 + 1 \cdot 2^6$.

We then multiply each x_i by some m_{n-i} where m_{n-i} is a positive integer relatively prime to p. Continuing our example of $52 \in \mathbb{Z}_{81}$, if $m_1 = 2$, $m_2 = 4$, $m_3 = 11$, and $m_4 = 28$, we obtain

$$\begin{aligned} & m_4 \cdot 1 + (m_3 \cdot 2) \cdot 3 + (m_2 \cdot 2) \cdot 9 + (m_1 \cdot 1) \cdot 27 \\ &= 28 \cdot 1 + (11 \cdot 2) \cdot 3 + (4 \cdot 2) \cdot 9 + (2 \cdot 1) \cdot 27 \\ &= 28 + 22 \cdot 3 + 8 \cdot 9 + 2 \cdot 27 = 220 \equiv 58 \pmod{81}. \end{aligned}$$

Note that arithmetic in the above computation is performed modulo 81, and in general, modulo p^n. We now state this formally as a definition.

Definition 7.7.13 Let p be prime and n a positive integer. Define the set $\mathbb{Z}_{p^n}^{**}$ of **primary generalized multipliers** to be the set of all vectors $\{(m_1, \ldots, m_n) : m_i \in \mathbb{Z}^+ \text{ such that } \gcd(m_i, p) = 1\}$. For each $\vec{m} \in \mathbb{Z}_{p^n}^{**}$, define a function $f_{\vec{m}} \colon \mathbb{Z}_{p^n} \to \mathbb{Z}_{p^n}$ by

$$f_{\vec{m}}(x) = f_{\vec{m}}\left(\sum_{i=0}^{n-1} x_i p^i\right) = \sum_{i=0}^{n-1} m_{n\ i} x_i p^i.$$

The computation immediately preceding the above definition is computing $f_{\vec{m}}(52)$ for the generalized multiplier $(2, 4, 11, 28)$ in $\mathbb{Z}_{81}$ using the language of the previous definition.

Example 7.7.14 Let $p^n = 2^7 = 128$ and $\vec{m} = (13, 15, 31, 9, 3, 3, 1)$. Compute $f_{\vec{m}}(109)$.

Solution We saw in Example 7.7.12 that the 2-adic expansion of 109 modulo 128 is $109 = 1 + 0 \cdot 2 + 1 \cdot 2^2 + 1 \cdot 2^3 + 0 \cdot 2^4 + 1 \cdot 2^5 + 1 \cdot 2^6$. Then

$$\begin{aligned} f_{\vec{m}}(x) &= f_{\vec{m}}\left(\sum_{i=0}^{n-1} x_i p^i\right) = \sum_{i=0}^{n-1} m_{n-i} x_i p^i \\ &= m_7 x_0 + m_6 x_1 \cdot 2 + m_5 x_2 \cdot 4 + m_4 x_3 \cdot 8 + m_3 x_4 \cdot 16 \\ &\quad + m_2 x_5 \cdot 32 + m_1 x_6 \cdot 64 \\ &= 1 \cdot 1 + 3 \cdot 0 \cdot 2 + 3 \cdot 1 \cdot 2^2 + 9 \cdot 1 \cdot 2^3 + 31 \cdot 0 \cdot 2^4 \\ &\quad + 15 \cdot 1 \cdot 2^5 + 13 \cdot 1 \cdot 2^6 \\ &= 1 + 0 + 12 + 72 + 0 + 480 + 832 = 1397 \equiv 117 \pmod{128}. \end{aligned}$$ □

Lemma 7.7.15 *For $\vec{m} \in \mathbb{Z}_{p^n}^{**}$, the function $f_{\vec{m}}\colon \mathbb{Z}_{p^n} \to \mathbb{Z}_{p^n}$ is a well-defined bijection.*

Proof It is clear that $f_{\vec{m}}$ is well defined. Let $\vec{m} = (m_1, \ldots, m_n)$, and suppose $f_{\vec{m}}(x) = f_{\vec{m}}(y)$, with $x = \sum_{i=0}^{n-1} x_i p^i$ and $y = \sum_{i=0}^{n-1} y_i p^i$ the p-adic expansions of x and y in $\mathbb{Z}_{p^n}$, respectively. We proceed by induction on n. If $n = 1$, then $f_{\vec{m}}$ is a multiplier, and so a bijection. Assume $f_{\vec{m}}$ is a bijection for every $\vec{m} \in \mathbb{Z}_{p^{n-1}}^{**}$, and let $\vec{m} \in \mathbb{Z}_{p^n}^{**}$. By our assumption that $f_{\vec{m}}(x) = f_{\vec{m}}(y)$, we have.

$$\sum_{i=0}^{n-1} m_{n-i} x_i p^i = \sum_{i=0}^{n-1} m_{n-i} y_i p^i.$$

As $m_n^{-1} f_{\vec{m}}(x) = m_n^{-1} f_{\vec{m}}(y)$, it is also true that $x_0 \equiv m_n^{-1} f_{\vec{m}}(x) \equiv m_n^{-1} f_{\vec{m}}(y) \equiv y_0 \pmod{p}$. As $0 \le x_0, y_0 \le p - 1$, we see $x_0 = y_0$. Similarly, $f_{\vec{m}}(x) - m_n x_0 = f_{\vec{m}}(y) - m_n y_0$, and $f_{\vec{m}}(x) - m_n x_0 \equiv f_{\vec{m}}(y) - m_n y_0 \equiv 0 \pmod{p}$. Also, $f_{\vec{m}}(x) - m_n x_0 = f_{\vec{m}}(y) - m_n y_0$ if and only if $(f_{\vec{m}}(x) - m_n x_0)/p \equiv (f_{\vec{m}}(y) - m_n y_0)/p \pmod{p^{n-1}}$. But then $f_{(m_1,\ldots,m_{n-1})}(x_1, \ldots, x_{n-1}) = (f_{\vec{m}}(x) - m_n x_0)/p$ and $f_{(m_1,\ldots,m_{n-1})}(y_1, \ldots, y_{n-1}) = (f_{\vec{m}}(y) - m_n x_0)/p$, and by the induction hypothesis $(x_1, \ldots, x_{n-1}) = (y_1, \ldots, y_{n-1})$. Thus $x = y$ and the result follows by induction. □

Example 7.7.16 Let $p = 2$ and $n = 3$. Find the cycle decomposition of the bijection $f_{(1,3,5)}$.

Solution First, $f_{(1,3,5)}(x) = f_{(1,3,5)}\left(\sum_{i=0}^{2} x_i 2^i\right) = 5x_0 + 3x_1 \cdot 2 + 1x_2 \cdot 4$ where $x \in \mathbb{Z}_8$ has 2-adic expansion $\sum_{i=0}^{2} x_i 2^i$. As a primary generalized multiplier always fixes 0, $f_{(1,3,5)}$ also fixes 0. Then 1 has 2-adic expansion 1, and so $f_{(1,3,5)}(1) = 5$. As 5 has 2-adic expansion $1 + 1 \cdot 4$, we see $f_{(1,3,5)}(5) = 5 \cdot 1 + 1 \cdot 1 \cdot 4 \equiv 1 \pmod{8}$. Similar computations give results as given in Table 7.2.

Hence $f_{(1,3,5)}$ has cycle decomposition $(1, 5)(2, 6)$. □

i	2-adic representation	$f_{(1,3,5)}(i)$
1	1	$5 \cdot 1 \equiv 5 \pmod 8$
2	$1 \cdot 2$	$3 \cdot 2 \equiv 6 \pmod 8$
3	$1 + 1 \cdot 2$	$5 \cdot 1 + 3 \cdot 1 \cdot 2 \equiv 3 \pmod 8$
4	$1 \cdot 4$	$1 \cdot 4 \equiv 4 \pmod 8$
5	$1 + 1 \cdot 4$	$5 \cdot 1 + 1 \cdot 1 \cdot 4 \equiv 1 \pmod 8$
6	$1 \cdot 2 + 1 \cdot 4$	$3 \cdot 1 \cdot 2 + 1 \cdot 1 \cdot 4 \equiv 2 \pmod 8$
7	$1 + 1 \cdot 2 + 1 \cdot 4$	$5 \cdot 1 + 3 \cdot 1 \cdot 2 + 1 \cdot 1 \cdot 4 \equiv 7 \pmod 8$

Table 7.2 *The values of the bijection $f_{(1,3,5)}$.*

While all of the isomorphisms that need to be tested to determine isomorphism between circulant digraphs will be generalized multipliers, not all generalized multiplies need to be checked. We now restrict our consideration to this smaller set of generalized multipliers.

Definition 7.7.17 Let $\mathbf{k} = (k_1, \ldots, k_n) \in \mathbf{K}_{p^n}$ be a primary key. Define the set of all **primary genuine generalized multipliers corresponding to the key k**, denoted $\mathbb{Z}^{**}_{p^n}(\mathbf{k})$, to be the set of all primary generalized multipliers $\vec{m}$ in $\mathbb{Z}^{**}_{p^n}$ that satisfy the following two conditions:

(i) $m_\delta \equiv m_{\delta-1} \left(\bmod p^{\delta-k_\delta-1}\right)$, $2 \leq \delta \leq n$, and
(ii) $m_\delta \in \left\{1, \ldots, p^{\delta-k_\delta} - 1\right\}$, $1 \leq \delta \leq n$.

Example 7.7.18 Let $\mathbf{k} = (0, 1) \in \mathbf{K}_{p^2}$. Calculate $\mathbb{Z}^{**}_{p^2}(\mathbf{k})$.

Solution Definition 7.7.17 (i) implies that $m_2 \equiv m_1 \left(\bmod p^{2-1-1}\right)$ or $m_2 \equiv m_1 \pmod 1$, which is always true. In this case Condition (i) of Definition 7.7.17 places no restrictions on m_1 and m_2. Condition (ii) of Definition 7.7.17 gives that $1 \leq m_1 \leq p-1$ and $1 \leq m_2 \leq p-1$. Then $\mathbb{Z}^{**}_{p^2}(\mathbf{k}) = \{(m_1, m_2) : 1 \leq m_1, m_2 \leq p-1\}$. □

Example 7.7.19 Let $\mathbf{k} = (0, 0, 1) \in \mathbf{K}_{p^3}$. Calculate $\mathbb{Z}^{**}_{p^3}(\mathbf{k})$.

Solution First, Definition 7.7.17 (i) implies that

- $m_2 \equiv m_1 \left(\bmod p^{2-0-1}\right)$ or $m_2 \equiv m_1 \pmod p$ (with $\delta = 2$), and
- $m_3 \equiv m_2 \left(\bmod p^{3-1-1}\right)$ or $m_3 \equiv m_2 \pmod p$ (with $\delta = 3$).

So $m_1 \equiv m_2 \equiv m_3 \pmod p$. Next, condition (ii) of Definition 7.7.17 implies that

- $1 \leq m_1 \leq p-1$ with $\delta = 1$ and $k_1 = 0$,

- $1 \leq m_2 \leq p^2 - 1$ with $\delta = 2$ and $k_2 = 0$,
- $1 \leq m_3 \leq p^2 - 1$ with $\delta = 3$ and $k_3 = 1$.

If $1 \leq m_1 \leq p-1$, then $m_2 = m_1 + ap$ and $m_3 = m_1 + bp$, where $0 \leq a, b \leq p-1$. The bounds for a and b are obtained in the obvious way. For the bound on a, for example, we use $1 \leq m_1 + ap \leq p^2 - 1$ or $1 \leq p - 1 + ap \leq p^2 - 1$. Solving, we obtain $2/p - 1 \leq a \leq p - 1$ from which we obtain $0 \leq a \leq p - 1$. Note that for m_2 and m_3 we are giving them in their p-adic expansion. Then $\mathbb{Z}^{**}_{p^3}(0, 0, 1) = \{(m_1, m_1 + ap, m_1 + bp) : 1 \leq m_1 \leq p-1 \text{ and } 0 \leq a, b \leq p-1\}$. □

Example 7.7.20 Let $\mathbf{k} = (0, 0, 1, 1) \in \mathbf{K}_{p^4}$. Calculate $\mathbb{Z}^{**}_{p^4}(\mathbf{k})$.

Solution First, condition (i) of Definition 7.7.17 implies that

- $m_2 \equiv m_1 \left(\text{mod } p^{2-0-1}\right)$ or $m_2 \equiv m_1 \pmod{p}$ (with $\delta = 2$),
- $m_3 \equiv m_2 \left(\text{mod } p^{3-1-1}\right)$ or $m_3 \equiv m_2 \pmod{p}$ (with $\delta = 3$), and
- $m_4 \equiv m_3 \left(\text{mod } p^{4-1-1}\right)$ or $m_4 \equiv m_3 \pmod{p^2}$ (with $\delta = 4$).

So $m_1 \equiv m_2 \equiv m_3 \pmod{p}$ and $m_4 \equiv m_3 \left(\text{mod } p^2\right)$. Next, condition (ii) of Definition 7.7.17 implies that

- $1 \leq m_1 \leq p - 1$ with $\delta = 1$ and $k_1 = 0$,
- $1 \leq m_2 \leq p^2 - 1$ with $\delta = 2$ and $k_2 = 0$,
- $1 \leq m_3 \leq p^2 - 1$ with $\delta = 3$ and $k_3 = 1$, and
- $1 \leq m_4 \leq p^3 - 1$ with $\delta = 4$ and $k_4 = 1$.

If $1 \leq m_1 \leq p-1$, then $m_2 = m_1 + ap$ and $m_3 = m_1 + bp$, where $0 \leq a, b \leq p-1$, and $m_4 = m_1 + bp + cp^2$, where $0 \leq c \leq p-1$. Then $\mathbb{Z}^{**}_{p^3}(0, 0, 1, 1) = \left\{\left(m_1, m_1 + ap, m_1 + bp, m_1 + bp + cp^2\right) : 1 \leq m_1 \leq p - 1 \text{ and } 0 \leq a, b, c \leq p - 1\right\}$. □

7.7.3 The Main Theorem in the Prime Power Case

Before stating the solution to the isomorphism problem for circulant digraphs of prime power order, we need two last terms that tie all of the other terms together. First, it is clear that one will start with two circulant digraphs, and so we will need to know the relationship between the terms we have already seen and circulant digraphs.

Definition 7.7.21 Let $\mathbf{u} = (u_1, \ldots, u_n)$ and $\mathbf{v} = (v_1, \ldots, v_n)$ be vectors in $\mathbf{K}_{p^n}$. Define the **meet** of $\mathbf{u}$ and $\mathbf{v}$, denoted $\mathbf{u} \wedge \mathbf{v}$, to be the vector

$$(\min\{u_1, v_1\}, \min\{u_2, v_2\}, \ldots, \min\{u_n, v_n\}).$$

Let P and Q be two partitions of a set X. We define the **join** of P and Q, denoted $P \vee Q$, to be the smallest partition that is refined by both P and Q.

Ordinarily, the term "meet" is used for partitions, not vectors. However, this terminology makes sense here, as we are using vectors to represent partitions.

Lemma 7.7.22 *Let* $\mathbf{k}$ *and* $\boldsymbol{\ell}$ *be primary keys in* $\mathbf{K}_{p^n}$*. Then* $\Sigma(\mathbf{k}) \vee \Sigma(\boldsymbol{\ell}) = \Sigma(\mathbf{k} \wedge \boldsymbol{\ell})$.

Proof Let $\mathbf{k} = (k_1, \ldots k_n)$, $\boldsymbol{\ell} = (\ell_1, \ldots, \ell_n)$, and $C_g \in \Sigma(\mathbf{k})$, $C_{g'} \in \Sigma(\boldsymbol{\ell})$ with $g \in C_g \cap C_{g'}$. Then both C_g and $C_{g'}$ are cosets of some subgroups H and L of $\mathbb{Z}_{p^n}$, respectively. As $\mathbb{Z}_{p^n}$ is cyclic of prime power order, either $H \leq L$ or $L \leq H$. Without loss of generality we assume $H \leq L$. Then $C_g \cap C_{g'} = C_g$, a coset of H in $\mathbb{Z}_{p^n}$ of order $p^{n-k_{b(g)}}$. As $H \leq L$, it follows that $p^{n-\ell_{b(g)}} \geq p^{n-k_{b(g)}}$, and so $\min\left\{k_{b(g)}, \ell_{b(g)}\right\} = \ell_{b(g)}$. As $\Sigma(\mathbf{k}) \vee \Sigma(\boldsymbol{\ell})$ is the smallest partition refined by both $\Sigma(\mathbf{k})$ and $\Sigma(\ell)$, the cell that contains g is also a coset of L. The result then follows. □

There are two points to the preceding lemma. First, the join of two primary key partitions is again a primary key partition. Second, if Δ is any partition of $\mathbb{Z}_{p^n}$, then there is a unique largest (in the sense of refinement) primary key partition that is a refinement of Δ. This follows as the trivial partition consisting of singleton sets is certainly a primary key partition, so Δ has a refinement that is a primary key partition. Suppose there are two primary key partitions $\Sigma(\mathbf{k})$ and $\Sigma(\boldsymbol{\ell})$ that are the largest primary key partitions refining Δ. Let $g \in \mathbb{Z}_{p^n}$, $g \neq 0$, and C_g and C'_g be the cells of $\Sigma(\mathbf{k})$ and $\Sigma(\boldsymbol{\ell})$, respectively, that contain g. Then C_g and C'_g are cosets of subgroups $H, L \leq \mathbb{Z}_{p^n}$ as in the proof above, and so either $H \leq L$ or $L \leq H$. As both are the largest primary key partitions refining δ, $H = L$, and $\Sigma(\mathbf{k}) = \Sigma(\boldsymbol{\ell})$.

Definition 7.7.23 Let Δ be a partition of $\mathbb{Z}_{p^n}$. The primary key for which $\Sigma(\mathbf{k})$ is the unique largest (in the sense of refinement) primary key partition that is a refinement of Δ is the **key of** Δ and denoted $\mathbf{k}(\Delta)$. Let $\mathrm{Cay}(\mathbb{Z}_{p^n}, S)$ be a circulant digraph. The **key of** $\mathrm{Cay}(\mathbb{Z}_{p^n}, S)$ is defined to be the key of the partition $\{S, \mathbb{Z}_{p^n} \backslash S\}$.

Example 7.7.24 Find the primary key $\mathbf{k}$ of $\mathrm{Cay}(\mathbb{Z}_8, \{1, 2, 5\})$.

Solution By definition, we need to find the largest primary key partition that is a refinement of $P = \{\{1, 2, 5\}, \{0, 3, 4, 6, 7\}\}$. Examining Table 7.1 and observing that $\{\{1, 5\}, \{3, 7\}, \{2\}, \{0, 4, 6\}\}$ is a refinement of P (so we are at this point only considering the odd elements of $\mathbb{Z}_8$), it is clear that $\mathbf{k} = (0, 0, 1)$ or $(0, 1, 1)$. Considering the even elements of $\mathbb{Z}_8$, it is now apparent that the primary key of $\mathrm{Cay}(\mathbb{Z}_8, \{1, 2, 5\})$ is $\mathbf{k} = (0, 0, 1)$. □

There is an easy inductive algorithm to calculate the primary key of a partition Π of $\mathbb{Z}_{p^n}$. To fix notation, let Π have primary key $\mathbf{k} = (k_1, \ldots, k_n)$.

To begin, notice that $\langle p \rangle$ contains every element that is not relatively prime to p. This means that every element not in $\langle p \rangle$ has the same order p^n. Also, for an element g of order p^n, C_g is the coset of $\left\langle p^{n-k_n} \right\rangle$ that contains g. As the primary key partition of the primary key of Π must refine Π, if $g \in D_g$ is the cell of Π that contains g, then $C_g \subseteq D_g$. Let $g \in \mathbb{Z}_{p^n} \backslash \langle p \rangle$ such that $|D_g|$ is as small as possible. Then $C_g = D_g$, and, as $C_g = g + \left\langle p^{n-k_n} \right\rangle$, k_n must be the largest nonnegative integer for which $D_g = g + \left\langle p^{n-k_n} \right\rangle$, $g \in \mathbb{Z}_{p^n} \backslash \langle p \rangle$. So $k_n \leq n - 1$ (as in the definition of a primary key).

Calculating k_{n-1} (and inductively all other values of k_r, $r \leq n-1$), is quite similar. Note that $\left\langle p^2 \right\rangle \leq \langle p \rangle$, and $\left\langle p^2 \right\rangle$ contains every element of $\langle p \rangle$ that is not relatively prime to p^2. So every element of $\langle p \rangle$ not in $\left\langle p^2 \right\rangle$ has the same order p^{n-1}. For an element g of order p^{n-1}, C_g is the coset of $\left\langle p^{n-k_{n-1}} \right\rangle$ that contains g. As the primary key partition of the primary key of Π must refine Π, k_{n-1} must be the largest nonnegative integer ℓ_{n-1} for which the partition Π' of $\langle p \rangle \backslash \left\langle p^2 \right\rangle$, that is the elements of Π contained in $\langle p \rangle \backslash \left\langle p^2 \right\rangle$ is a union of cosets of $\left\langle p^{k_{n-1}} \right\rangle$ *subject to the condition that* $k_{n-1} \leq k_n$. Thus $k_{n-1} = \min\{\ell_{n-1}, k_n\}$. Notice that instead of computing in Π and $\mathbb{Z}_{p^n}$ as in this paragraph, we could just as easily replace the partition Π'_1 of $\langle p \rangle$ induced by Π with the partition Π_1 of $\mathbb{Z}_{p^{n-1}}$ obtained from Π'_1 by replacing each element x of each cell with x/p. That is, setting $S'_1 = S \cap \langle p \rangle$ and $S_1 = \{x/p : x \in S'_1\}$, we have $\Pi_1 = \left\{S_1, \mathbb{Z}_{p^{n-1}} \backslash S_1\right\}$. Then calculating ℓ_{n-1} as in the previous paragraph, and setting $k_{n-1} = \min\{k_{n-1}, k_n\}$. So $\mathbf{k}$ can be calculated by repeating this process until some $k_i = 0$, which will always occur after at most $n-1$ iterations of this process, i.e., when $\left\langle p^{n-1} \right\rangle$ of prime order is considered.

Definition 7.7.25 Let X be a set and Π a partition of X. Let $Y \subseteq X$. The partition of Y **induced** by Π is the partition $\{Y \cap Z : Z \in \Pi\}$.

Example 7.7.26 Calculate the primary key of $S = \{1, 2, 3, 4, 5, 7\} \subset \mathbb{Z}_8$.

Solution As we are working in $\mathbb{Z}_8$ and $8 = 2^3$, our primary key has the form $(0, k_1, k_2)$. The partition Π we wish to refine with a primary key partition is $\Pi = \{S, \mathbb{Z}_8 \backslash S\} = \{\{1, 2, 3, 4, 5, 7\}, \{0, 6\}\}$. To begin, we consider the partition of the elements *not* in $\langle 2 \rangle = \{0, 2, 4, 6\}$ induced by Π. This is the partition $\{\{1, 3, 5, 7\}, \emptyset\} = \{\{1, 3, 5, 7\}\}$ of $\mathbb{Z}_8 \backslash \langle 2 \rangle$. We see that this is in fact the set containing $1 + \langle 2 \rangle$, so $2^{n-k_2} = 2^{3-k_2} = 2$, and so $k_2 = 2$. To calculate k_1, we only consider the partition of $\langle 2 \rangle$ induced by Π, which is $\{\{2, 4\}, \{0, 6\}\}$. Neither of the sets $\{2, 4\}$ or $\{0, 6\}$ contain cosets of any subgroup of $\langle 2 \rangle$ other than cosets of $\{0\}$. Hence $2^{n-k_1} = 2^{3-k_1} \equiv 0 \pmod 8$, and $k_1 = 0$. So $\mathbf{k} = (0, 0, 2)$. □

Example 7.7.27 Calculate the primary key of $S = \{1, 5\} \subset \mathbb{Z}_8$.

Solution As we are working in $\mathbb{Z}_8$ and $8 = 2^3$, our primary key has the form $(0, k_1, k_2)$. The partition we wish to refine with a primary key partition is $\{S, \mathbb{Z}_8 \backslash S\} = \{\{1,5\}, \{0,2,3,4,6,7\}\}$. To begin, we consider the partition of the elements *not* in $\langle 2 \rangle = \{0,2,4,6\}$ induced by Π. This is the partition $\{\{1,5\},\{3,7\}\}$ of $\{1,3,5,7\}$. This is not the set $\{1,3,5,7\}$ (or a coset of $\langle 2 \rangle$), but is two cosets of $\left\langle 2^2 \right\rangle = \langle 4 \rangle$, namely $1 + \langle 4 \rangle$ and $3 + \langle 4 \rangle$. Hence $2^{n-k_2} = 2^{3-k_2} = 4$, and so $k_2 = 1$. To calculate k_1, we only examine the partition of $\langle 2 \rangle$ induced by Π, which is $\{\emptyset, \{0,2,4,6\}\} = \{\{0,2,4,6\}\}$. This set contains the coset $\langle 2 \rangle$ and so $2^{n-\ell_1} = 2$ or $\ell_1 = 2$. Then $k_1 = \min\{\ell_1, n_2\} = 1$. Hence $\mathbf{k} = (0,1,1)$. □

Example 7.7.28 Calculate the primary key of $S = \{1,2,3\} \subset \mathbb{Z}_8$.

Solution As we are working in $\mathbb{Z}_8$ and $8 = 2^3$, our primary key has the form $(0, k_1, k_2)$. The partition we wish to refine with a primary key partition is $\{S, \mathbb{Z}_8 \backslash S\} = \{\{1,2,3\}, \{0,4,5,6,7\}\}$. To begin, we consider the partition of the elements *not* in $\langle 2 \rangle = \{0,2,4,6\}$ induced by Π. This is the partition $\{\{1,3\},\{5,7\}\}$ of $\{1,3,5,7\}$. We see that no set in this partition contains any coset of any nontrivial subgroup of $\mathbb{Z}_8$, and so only contains cosets of the trivial subgroup $\{0\}$. Then $2^{3-k_2} = 2^3 \equiv 0 \pmod 8$, and so $k_2 = 0$. As keys are non-negative and monotonically increasing, $k_1 = 0$ as well. Hence $\mathbf{k} = (0,0,0)$. □

We are now ready to state our final definition, and then give the statement of the main theorem in the prime power case.

Definition 7.7.29 Let $\mathrm{Cay}\left(\mathbb{Z}_{p^n}, S\right)$ have primary key $\mathbf{k}$. The **solving set of** $\mathrm{Cay}\left(\mathbb{Z}_{p^n}, S\right)$ is defined to be the set $P(\mathbf{k}) = \left\{f_{\vec{m}} : \vec{m} \in \mathbb{Z}_{p^n}^{**}(\mathbf{k})\right\}$. That is, the solving set of $\mathrm{Cay}\left(\mathbb{Z}_{p^n}, S\right)$ is the set of all genuine generalized multipliers corresponding to the primary key of $\mathrm{Cay}\left(\mathbb{Z}_{p^n}, S\right)$.

Theorem 7.7.30 *Let p be prime, n a positive integer, and* $\mathrm{Cay}\left(\mathbb{Z}_{p^n}, S\right)$ *and* $\mathrm{Cay}\left(\mathbb{Z}_{p^n}, S'\right)$ *be circulant digraphs with primary keys* $\mathbf{k}$ *and* $\mathbf{k}'$*, respectively. Then*

(i) *if* $\mathbf{k} \neq \mathbf{k}'$*, then* $\mathrm{Cay}\left(\mathbb{Z}_{p^n}, S\right)$ *is not isomorphic to* $\mathrm{Cay}\left(\mathbb{Z}_{p^n}, S'\right)$*,*
(ii) *if* $\mathbf{k} = \mathbf{k}'$*, then the following are equivalent:*
  (a) $\mathrm{Cay}\left(\mathbb{Z}_{p^n}, S\right)$ *and* $\mathrm{Cay}\left(\mathbb{Z}_{p^n}, S'\right)$ *are isomorphic,*
  (b) $f_{\vec{m}}\left(\mathrm{Cay}\left(\mathbb{Z}_{p^n}, S\right)\right) = \mathrm{Cay}\left(\mathbb{Z}_{p^n}, S'\right)$ *for some* $f_{\vec{m}} \in P(\mathbf{k})$*, and*
  (c) $f_{\vec{m}}(S) = S'$ *for some* $f_{\vec{m}} \in P(\mathbf{k})$*.*

Example 7.7.31 The digraph $\mathrm{Cay}(\mathbb{Z}_8, \{1,2,5\})$ is isomorphic to the digraphs $\mathrm{Cay}(\mathbb{Z}_8, \{1,5,6\})$ and $\mathrm{Cay}(\mathbb{Z}_8, \{2,3,7\})$, but not by multipliers.

Solution We have seen in Example 7.7.24 that $\mathrm{Cay}(\mathbb{Z}_8, \{1, 2, 5\})$ has primary key $(0, 0, 1)$, and in Example 7.7.19 that $\mathbb{Z}^{**}_{p^3}(0, 0, 1) = \{(m_1, m_1 + ap, m_1 + bp) : m_1 \in \mathbb{Z}^*_p, a, b \in \mathbb{Z}_p\}$, where p is prime. In the case $p = 2$, it follows that $m_1 = 1$ and so $P(0, 0, 1) = \{f_{1,1,1}, f_{1,3,1}, f_{1,1,3}, f_{1,3,3}\}$. It is left as Exercise 7.7.13 to verify that $f_{1,1,1}(\{1, 2, 5\}) = \{1, 2, 5\}$, $f_{1,3,1}(\{1, 2, 5\}) = \{1, 5, 6\}$, $f_{1,1,3}(\{1, 2, 5\}) = \{2, 3, 7\}$, and $f_{1,3,3}(\{1, 2, 5\}) = \{3, 6, 7\}$. Then $\mathrm{Cay}(\mathbb{Z}_8, \{1, 2, 5\})$ is isomorphic to exactly those circulant digraphs whose connection sets are either $\{1, 2, 5\}$, $\{1, 5, 6\}$, $\{2, 3, 7\}$, and $\{3, 6, 7\}$. As $m \cdot \{1, 2, 5\} = \{1, 2, 5\}$ or $\{3, 6, 7\}$ for $m \in \mathbb{Z}^*_8$, we have that $\mathrm{Cay}(\mathbb{Z}_8, \{1, 5, 6\})$ and $\mathrm{Cay}(\mathbb{Z}_8, \{2, 3, 7\})$ are not isomorphic to $\mathrm{Cay}(\mathbb{Z}_8, \{1, 2, 5\})$ by a multiplier in $\mathbb{Z}_8$. We remark that this is Elspas and Turner's original example showing that $\mathbb{Z}_8$ is not a CI-group with respect to digraphs, as mentioned in Section 7.2. □

7.7.4 The Main Theorem

With the solution to the prime power case in hand, we now turn to the solution of the isomorphism problem of circulant digraphs of any order n. In this case, just as $\mathbb{Z}_n$ may be written as a direct product $\Pi^r_{i=1}\mathbb{Z}_{p_i^{k_i}}$, where n has prime power decomposition $n = p_1^{k_1} \cdots p_r^{k_r}$, there are **keys, key partitions**, and **generalized multipliers**, each of which is a direct product of primary keys, primary key partitions, and primary generalized multipliers corresponding to the prime power decomposition of n. All such definitions are stated next.

Definition 7.7.32 Let n be a positive integer with prime power decomposition $n = p_1^{k_1} \cdot p_2^{k_2} \cdots p_r^{k_r}$. Define the **key space** $\mathbf{K}_n$ to be $\Pi^r_{i=1}\mathbf{K}_{p_i^{k_i}}$. That is, $\mathbf{K}_n$ is the direct product of primary keys. We will also call $\mathbf{K}_n$ a **composite key**. Similarly, if $\Sigma(\mathbf{k}_i)$ is a primary key partition of $\mathbb{Z}_{p_i^{k_i}}$, then a **key partition of** $\mathbb{Z}_n$ is $\Pi^r_{i=1}\Sigma(\mathbf{k}_i)$. Define a set of **generalized multipliers**, denoted $\mathbb{Z}^{**}_n$, as $\Pi^r_{i=1}\mathbb{Z}^{**}_{p_i^{k_i}}$. For each $\vec{m} = (\vec{m}_1, \vec{m}_2, \ldots, \vec{m}_r) \in \mathbb{Z}^{**}_n$, where each $\vec{m}_i \in \mathbb{Z}^{**}_{p_i^{k_i}}$ defines a function $f_{\vec{m}} : \mathbb{Z}_n \to \mathbb{Z}_n$ by

$$f_{\vec{m}}(x_1, \ldots, x_r) = (f_{\vec{m}_1}(x_1), f_{\vec{m}_2}(x_2), \ldots, f_{\vec{m}_r}(x_r)).$$

For a key $\mathbf{k} = (\mathbf{k}_1, \mathbf{k}_2, \ldots, \mathbf{k}_r)$ where $\mathbf{k}_i \in \mathbf{K}_{p_i^{k_i}}$ is a primary key, define the set of all **genuine generalized multipliers corresponding to the key k**, denoted $\mathbb{Z}^{**}_n(\mathbf{k})$, to be $\Pi^r_{i=1}\mathbb{Z}^{**}_{p_i^{k_i}}(\mathbf{k}_i)$. Now let Δ be a partition of $\mathbb{Z}_n$. The partition for which $\Sigma(\mathbf{k})$ is the unique largest (in the sense of refinement) primary key partition that is a refinement of Δ is the **key of** Δ and denoted $\mathbf{k}(\Delta)$. Let $\mathrm{Cay}(\mathbb{Z}_n, S)$ be a circulant digraph. The **key of** $\mathrm{Cay}(\mathbb{Z}_n, S)$ is defined to be the key of the partition $\{S, \mathbb{Z}_n \backslash S\}$. Finally, let $\mathrm{Cay}(\mathbb{Z}_n, S)$ have key $\mathbf{k}$. The **solving set of** $\mathrm{Cay}(\mathbb{Z}_n, S)$ is defined to be the set $P(\mathbf{k}) = \{f_{\vec{m}} : \vec{m} \in \mathbb{Z}_n^{**}(\mathbf{k})\}$.

We now give the solution to the isomorphism problem for circulant digraphs.

Theorem 7.7.33 *Let n be a positive integer,* $\mathrm{Cay}(\mathbb{Z}_n, S)$ *and* $\mathrm{Cay}(\mathbb{Z}_n, S')$ *circulant digraphs with keys* $\mathbf{k}$ *and* $\mathbf{k}'$, *respectively. Then.*

(i) *if* $\mathbf{k} \neq \mathbf{k}'$, *then* $\mathrm{Cay}(\mathbb{Z}_n, S)$ *is not isomorphic to* $\mathrm{Cay}(\mathbb{Z}_n, S')$,

(ii) *if* $\mathbf{k} = \mathbf{k}'$, *then the following are equivalent:*

(a) $\mathrm{Cay}(\mathbb{Z}_n, S)$ *and* $\mathrm{Cay}(\mathbb{Z}_n, S')$ *are isomorphic,*

(b) $f_{\vec{m}}(\mathrm{Cay}(\mathbb{Z}_n, S)) = \mathrm{Cay}(\mathbb{Z}_n, S')$ *for some* $f_{\vec{m}} \in P(\mathbf{k})$, *and*

(c) $f_{\vec{m}}(S) = S'$ *for some* $f_{\vec{m}} \in P(\mathbf{k})$.

Example 7.7.34 Let $S = \{3, 4, 9, 10, 12, 15, 16\}$. Calculate a solving set for the circulant digraph $\mathrm{Cay}(\mathbb{Z}_2 \times \mathbb{Z}_9, S)$. Is this digraph a CI-digraph of $\mathbb{Z}_{18}$?

Solution As the key in the solution to the isomorphism problem for circulant digraphs given in Theorem 7.7.33 is the product of primary keys, it is more convenient to consider $\mathbb{Z}_{18}$ as $\mathbb{Z}_2 \times \mathbb{Z}_9$, in which case the connection set also needs to be written as a subset of $\mathbb{Z}_2 \times \mathbb{Z}_9$, and this can be done using the Chinese remainder theorem (Dummit and Foote, 2004, Corollary 7.6.18). In our case we have,

$$S = \{(0, 1), (0, 3), (0, 4), (0, 7), (1, 0), (1, 3), (1, 6)\}.$$

As the primary key spaces for $p^n = 2$ and 9 are $\mathbf{K}_2 = \{(0)\}$ and $\mathbf{K}_9 = \{(0, 0), (0, 1)\}$, the primary key partitions of $\mathbb{Z}_2$ and $\mathbb{Z}_9$ are $\{\{0\}, \{1\}\}$, and $\{\{i\} : i \in \mathbb{Z}_9\}$ and $\{\{0\}, \{3\}, \{6\}, \{1, 4, 7\}, \{2, 5, 8\}\}$ respectively. Thus,

$$\mathbf{K}_{18} = \{((0), (0, 0)), ((0), (0, 1))\},$$

with corresponding key partitions $\{\{(i, j)\} : i \in \mathbb{Z}_2, j \in \mathbb{Z}_9\}$, and

$$T = \{\{(i, j)\}, \{(i, k) : k \in 1 + \langle 3\rangle\}, \{(i, \ell) : \ell \in 2 + \langle 3\rangle\} : i \in \mathbb{Z}_2, j \in \langle 3\rangle\}.$$

As $T < S$, we see that $\mathrm{Cay}(\mathbb{Z}_2 \times \mathbb{Z}_9, S)$ has key $\mathbf{k} = ((0), (0, 1))$. The function $f_{(0)} \colon \mathbb{Z}_2 \to \mathbb{Z}_2$ is the identity, while $f_{(0,1)} \colon \mathbb{Z}_9 \to \mathbb{Z}_9$ is given by

$$f_{(0,1)}(x_1 + 3x_2) = m_1 x_1 + 3m_2 x_2,$$

where $m_2 \equiv m_1 \left(\mathrm{mod}\ p^{2-1-1}\right)$, $m_1 \in \mathbb{Z}_3^*$, and $m_2 \in \mathbb{Z}_3^*$. Thus the genuine primary generalized multipliers corresponding to the key $(0, 1)$ are

$$x_1 + 3x_2 \mapsto x_1 + 3x_2, \text{ which is the identity,}$$
$$x_1 + 6x_2, \text{ which has cycle decomposition } (3,6)(4,7)(5,8),$$
$$2x_1 + 3x_2, \text{ which has cycle decomposition } (1,2,4,5,7,8),$$
$$2x_1 + 6x_2, \text{ which has cycle decomposition } (1,2,4,8,7,5)(3,6).$$

Hence a solving set for $\mathrm{Cay}(\mathbb{Z}_2 \times \mathbb{Z}_9, S)$ is the set of the following genuine generalized multipliers:

$$(x, y_1 + 3y_2) \mapsto (x, y_1 + 3y_2), \text{ applied to } S \text{ is } S,$$
$$\mapsto (x, y_1 + 6y_2), \text{ applied to } S \text{ is}$$
$$\{(0,1),(0,4),(0,6),(0,7),(1,0),(1,3),(1,6)\},$$
$$\mapsto (x, 2y_1 + 3y_2), \text{ applied to } S \text{ is}$$
$$\{(0,2),(0,3),(0,5),(0,8),(1,0),(1,3),(1,6)\},$$
$$\mapsto (x, 2y_1 + 6y_2), \text{ applied to } S \text{ is}$$
$$\{(0,2),(0,5),(0,6),(0,8),(1,0),(1,3),(1,6)\}.$$

It is easy to check that only two of these four isomorphic circulant digraphs of order 18 are isomorphic by a multiplier, and so we conclude that $\mathrm{Cay}(\mathbb{Z}_2 \times \mathbb{Z}_9, S)$ is not a CI-digraph of $\mathbb{Z}_{18}$. □

7.8 Exercises

7.2.1 Let $S = \{1, 2, 7, 9, 14, 15\}$. Find $K, L \leq \mathbb{Z}_{16}$ such that $\mathrm{Cay}(\mathbb{Z}_{16}, S)$ is a (K, L)-generalized wreath product. Note $\mathrm{Cay}(\mathbb{Z}_{16}, S)$ is not a CI-graph of $\mathbb{Z}_{16}$ by Example 7.2.1.

7.3.1 Let X and Y be combinatorial objects of order p^k, p a prime and $k \geq 1$, in some class $\mathcal{K}$ of combinatorial objects. If a Sylow p-subgroup P of X is also a Sylow p-subgroup of Y, show that X and Y are isomorphic if and only if they are isomorphic by $\alpha \in N_{S_{p^k}}(P)$.

7.4.1 Is $\mathrm{Cay}(\mathbb{Z}_{17}, \{1, 3, 5, 7, 11\}) \cong \mathrm{Cay}(\mathbb{Z}_{17}, \{3, 4, 10, 11, 12\})$? Explain.

7.4.2 Is $\mathrm{Cay}(\mathbb{Z}_{13}, \{2, 5, 7, 9\}) \cong \mathrm{Cay}(\mathbb{Z}_{13}, \{1, 3, 5, 12\})$? Explain.

7.4.3 Show that $\mathrm{Cay}(G, S)$ is a CI-graph of G if and only if its complement $\overline{\mathrm{Cay}(G, S)} = \mathrm{Cay}(G, G \backslash (S \cup \{1_G\}))$ is a CI-graph of G.

7.4.4 Modify Example 7.4.3 to show that $\mathbb{Z}_9$ is not a CI-group with respect to digraphs.

7.4.5 Modify Example 7.4.3 to show that $\mathbb{Z}_{27}$ is not a CI-group with respect to graphs.

7.4.6 Using the notation given in the paragraph after Definition 7.3.6, show that $\alpha(\Gamma) = \bar{\Gamma} = \mathrm{Cay}(G, S_1)$.

7.4.7 Using the notation in the paragraph after Definition 7.3.6, show that $\alpha(\Gamma) \cong \Gamma$.

7.4.8 Let p be prime and $r \in \mathbb{Z}_p^*$ such that $|r| = p - 1$. Let $S = \left\{r^{2k} : 1 \le k \le (p-1)/2\right\}$, and $\Gamma = \mathrm{Cay}(\mathbb{Z}_p, S)$. Let $T = \mathbb{Z}_p^* \backslash S$. Show that $\bar{\Gamma} = \mathrm{Cay}(\mathbb{Z}_p, T)$ and find an isomorphism between these two (di)graphs. Determine arithmetic conditions on p that determine whether Γ is a graph or a digraph.

7.4.9 Show that there do not exist three edge disjoint subgraphs of K_{10} isomorphic to the Petersen graph.

7.4.10 Show that both $\mathbb{Z}_4$ and $\mathbb{Z}_2 \times \mathbb{Z}_2$ are CI-groups. (Hint: Observe that a Sylow 2-subgroup T of S_4 has order 8. Does T contain a conjugate of every regular subgroup of S_4? By counting elements of T, show that it contains unique regular subgroups isomorphic to $\mathbb{Z}_2^2$ and $\mathbb{Z}_4$.)

7.5.1 Let $m < p$ be integers with p a prime, and G a group of order mp. Show that if $\phi \in S_G$ then there exists $\delta \in \left\langle G_L, \phi^{-1} G_L \phi \right\rangle$ such that $\left\langle G_L, \delta^{-1}\phi^{-1} G_L \phi\delta \right\rangle$ is imprimitive.

7.5.2 Show that the bijection of $\mathbb{Z}_q \times \mathbb{Z}_p$ given by $(i, j) \mapsto (i, \beta j), \beta \in \mathbb{Z}_p^*$ is an automorphism of D_{2p} with its metacirculant labeling by showing that β normalizes D_{2p}, fixes $(0, 0)$, and applying Corollary 4.4.11.

7.5.3 Let $q < p$ be prime with $q|(p - 1)$ and $\alpha \in \mathbb{Z}_p^*$ of order q. Define $\rho, \tau \colon \mathbb{Z}_q \times \mathbb{Z}_p \to \mathbb{Z}_q \times \mathbb{Z}_p$ by $\rho(i, j) = (i, j + 1)$, $\tau(i, j) = (i + 1, \alpha j)$, and $\hat{\rho}(i, j) = \left(i, j + \alpha^i\right)$. Show that $\hat{\rho}$ is an automorphism of $\langle \rho, \tau \rangle$.

7.5.4 Let $q < p$ be prime with $q|(p - 1)$ and $\alpha \in \mathbb{Z}_p^*$ of order q. Define $\rho, \tau, \hat{\rho}, \hat{\tau} \colon \mathbb{Z}_q \times \mathbb{Z}_p \to \mathbb{Z}_q \times \mathbb{Z}_p$ by $\rho(i, j) = (i, j + 1)$, $\tau(i, j) = (i + 1, \alpha j)$, $\hat{\rho}(i, j) = \left(i, j + \alpha^i\right)$, and $\hat{\tau} = (i + 1, j)$. Show that $\langle \hat{\rho}, \hat{\tau} \rangle$ centralizes $\langle \rho, \tau \rangle$, and that $\langle \hat{\rho}, \hat{\tau} \rangle \cong \langle \rho, \tau \rangle$.

7.5.5 Generalize Corollary 2.2.18 by showing that the centralizer in S_G of G_L is $G_R = \{x \mapsto xg : g \in G\}$, the right regular representation of G.

7.5.6 Let p be prime. Show that if a circulant digraph of order p^2 is not a CI-digraph of $\mathbb{Z}_{p^2}$, then it is a wreath product of two circulant digraphs of prime order. Deduce that such a digraph is also isomorphic to a Cayley digraph of $\mathbb{Z}_p \times \mathbb{Z}_p$.

7.5.7 Use Exercise 7.5.6 to show that $\mathbb{Z}_9$ is a CI-group with respect to *graphs* by considering every possible wreath product of two circulant graphs of order 3.

7.5.8 Let p be prime. Show that a circulant digraph of order p^2 with automorphism group $A \wr S_p$ is a CI-digraph of $\mathbb{Z}_{p^2}$.

7.5.9 Prove the following generalization of Lemma 7.5.6: Let G be a group with $\mathcal{B}$ a normal block system of G_L. Suppose $\phi \in S_G$, and $\mathcal{B}$ is also a block system of $\langle G_L, \phi^{-1}G_L\phi\rangle$. If $\phi^{-1}G_L\phi/\mathcal{B}$ is permutation isomorphic to $G_L/\mathcal{B}$ and $G_L/\mathcal{B}$ is a CI-group, then there exists $\delta \in \langle G_L, \phi^{-1}G_L\phi\rangle$ such that $\delta^{-1}\phi^{-1}G_L\phi\delta/\mathcal{B} = G_L/\mathcal{B}$.

7.5.10 Let p be prime, and define $\tau, \tau' : \mathbb{Z}_p^2 \to \mathbb{Z}_p^2$ by $\tau(i,j) = (i+1,j)$, and $\tau'(i,j) = (i+1, j+b_i)$, where $b_i \in \mathbb{Z}_p$ are such that $\sum_{i=0}^{p-1} b_i \equiv 0 \pmod p$. Use a weight shifting argument, as in Lemma 7.5.11, to show that there exists $\delta \in (\mathbb{Z}_p)_L \wr (\mathbb{Z}_p)_L$ such that $\delta^{-1}\tau'\delta = \tau$. Here, addition will play the role multiplication did in Lemma 7.5.11.

The purpose of the remaining exercises on Section 7.5 is to give an example of a nonabelian group of order qp, q and p distinct primes, that is not a Burnside group.

7.5.11 Let $n \geq 3$ and let S_n act on the set of all 2-subsets of $\{1, 2, \ldots, n\}$ (recall from Section 1.10 that for $n = 5$ this action can be used to give the automorphism group of the Petersen graph). Show that S_n in this action is simply primitive.

7.5.12 Use the orbit-stabilizer theorem to prove the following, which is from Marušič and Scapellato (1994b, Lemma 2.2): Let G be transitive of degree n and order ns. If $H \leq G$ such that n divides $|H|$ and $\gcd([G : H], |H|) = 1$, then H is transitive.

7.5.13 Using the previous exercise, show that AGL$(1,7)$ contains a regular subgroup of order 21 in its natural action on the 2-subsets of $\{1,2,3,4,5,6,7\}$. Deduce that the action of S_7 on the 2-subsets of $\{1,2,3,4,5,6,7\}$ is simply primitive and contains a regular subgroup, and thus that the nonabelian group of order 21 is not a Burnside group.

7.6.1 Show that a transitive nilpotent group of degree n is normally m-step imprimitive, where $m = \Omega(n)$.

7.6.2 Let G be a transitive group of degree n with $\Omega(n) = m$ and $\mathcal{B}_0 \prec \cdots \prec \mathcal{B}_m$ be block systems of G (so G is m-step imprimitive). For $1 \leq i \leq m-1$, show that $G/\mathcal{B}_i$ is $(m-i)$-step imprimitive. Additionally, if each $\mathcal{B}_i$, $0 \leq i \leq m$ is normal, show that $G/\mathcal{B}_i$ is normally $(m-i)$-step imprimitive.

7.6.3 If $H_1, H_2 \leq S_n$ are regular cyclic subgroups and n is square-free, show that there exists $\delta \in \langle H_1, H_2\rangle$ such that $\left\langle H_1, \delta^{-1}H_2\delta\right\rangle$ is normally m-step imprimitive, where $m = \Omega(n)$.

7.6.4 Let $q < p$ be prime, and $\phi \in S_{pq}$ such that $\left\langle (\mathbb{Z}_{pq})_L, \phi^{-1}(\mathbb{Z}_{pq})_L\phi\right\rangle$ has a block system with p blocks of size q. Show that $(\mathbb{Z}_{pq})_L$ and $\phi^{-1}(\mathbb{Z}_{pq})_L\phi$ are conjugate in $\left\langle (\mathbb{Z}_{pq})_L, \phi^{-1}(\mathbb{Z}_{pq})_L\phi\right\rangle$.

7.6.5 Show that (2) ⇒ (1) in Lemma 7.6.3.

7.6.6 Use Lemma 7.6.3 to simplify the proof of Theorem 7.3.3.

7.6.7 Show that $\left(q^d - 1\right)/(q-1) = \sum_{i=0}^{d-1} q^i$ factors if d is composite. Conclude that if $\left(q^d - 1\right)/(q-1)$ is prime, then d is prime.

7.6.8 Let d and q be integers with $\gcd(d, q-1) = r > 1$. Write $q = \ell r + 1$ for some ℓ and use the binomial theorem to show that $\left(q^d - 1\right)/(q-1) = \sum_{i=1}^{d-1} q^i$ is divisible by r.

7.7.1 Determine $\mathbf{K}_{p^4}$. (Hint: There are 42 primary keys in $\mathbf{K}_{p^4}$.)

7.7.2 Let $p^n = 125$, $\mathbf{k} = (0,1,2)$, and $g = 50$. Find C_{50}.

7.7.3 Let $p^n = 32$, $\mathbf{k} = (0,1,2,3,3)$, and $g = 20$. Find C_{20}.

7.7.4 Verify that the primary key partitions of $\mathbb{Z}_8$ corresponding to the primary keys $(0,0,0), (0,0,1)$, and $(0,0,2)$ given in Table 7.1 are correct, and then draw the lattice of primary key partitions of $\mathbb{Z}_8$.

7.7.5 Verify that a circulant digraph of order 8 whose primary key is not $(0,0,0)$ is a generalized wreath product.

7.7.6 Let p be a prime, $k \geq 2$, and $\Gamma = \mathrm{Cay}\left(\mathbb{Z}_{p^k}, S\right)$ a Cayley digraph of $\mathbb{Z}_{p^k}$. Show that unless the primary key of Γ is a $(0,0,\ldots,0)$, then Γ is a generalized wreath product. (Hint: First show it suffices to only consider the primary key $(0,\ldots,0,1)$.)

7.7.7 Find the primary key partitions of $(0,1,2,3)$ and $(0,1,1,2)$ in $\mathbb{Z}_{16}$.

7.7.8 Let $p = 2$ and $n = 3$. Find the cycle decomposition of $f_{(1,3,5)}$.

7.7.9 Let $p = 3$ and $n = 3$. Find the cycle decomposition of $f_{(2,5,11)}$.

7.7.10 Let $\mathbf{k} = (0,1,1) \in \mathbf{K}\left(p^3\right)$. Calculate $\mathbb{Z}_{p^3}^{**}(\mathbf{k})$.

7.7.11 Let $\mathbf{k} = (0,1,2) \in \mathbf{K}\left(p^3\right)$. Calculate $\mathbb{Z}_{p^3}^{**}(\mathbf{k})$.

7.7.12 Find the key of $\mathrm{Cay}(\mathbb{Z}_8, \{1,2,5,6\})$.

7.7.13 Using the notation of Example 7.7.31, verify that $f_{1,1,1}(\{1,2,5\}) = \{1,2,5\}$, $f_{1,3,1}(\{1,2,5\}) = \{1,5,6\}$, $f_{1,1,3}(\{1,2,5\}) = \{2,3,7\}$, and that $f_{1,3,3}(\{1,2,5\}) = \{3,6,7\}$.

7.7.14 Show that if $\mathbf{k} \leq \boldsymbol{\ell}$, then $\Sigma(\mathbf{k}) \leq \Sigma(\boldsymbol{\ell})$.

7.7.15 Find a solving set for $\mathrm{Cay}(\mathbb{Z}_9, \{3,4,7,8\})$. Is it a CI-digraph of $\mathbb{Z}_9$?

7.7.16 Show that $\mathbb{Z}_{18}$ is a CI-group with respect to *graphs*.

7.7.17 Let p be prime and $\mathbf{k}_2 = (0,0)$ and $\mathbf{k}_3 = (0,0,0)$. Show that for $k = 2$ and 3, a generalized multiplier in $\mathbb{Z}^{**}_{p^k}(\mathbf{k}_k)$ is a multiplier.

7.7.18 Let p be prime and $\mathbf{k} = (0,0,0,\ldots,0)$ (with k 0s). Show that a generalized multiplier in $\mathbb{Z}^{**}_{p^k}(\mathbf{k})$ is a multiplier.

7.7.19 Let p be prime, and $k \geq 2$. Use Exercises 7.7.6 and 7.7.18 to show that if $\mathrm{Cay}\left(\mathbb{Z}_{p^k}, S\right)$ is not a CI-digraph of $\mathbb{Z}_{p^k}$, then $\mathrm{Cay}\left(\mathbb{Z}_{p^k}, S\right)$ is a generalized wreath product.

7.7.20 Let p be prime and $k \geq 2$. Show that $\Gamma = \mathrm{Cay}\left(\mathbb{Z}_{p^k}, S\right)$ is isomorphic to $\Gamma_1 = \mathrm{Cay}\left(\mathbb{Z}_{p^{k-i}}, S_1\right) \wr \mathrm{Cay}\left(\mathbb{Z}_{p^i}, S_2\right)$ if and only if Γ has primary key $\mathbf{k} = (k_1, \ldots, k_n)$ and $k_{i+1} = i$ for some $1 \leq i \leq n$.

7.7.21 Let $S_1 \subset \mathbb{Z}_{p^i}$ and $S_2 \subset \mathbb{Z}_{p^j}$. Let $\Gamma_1 = \mathrm{Cay}\left(\mathbb{Z}_{p^i}, S_1\right)$, and $\Gamma_2 = \mathrm{Cay}\left(\mathbb{Z}_{p^j}, S_2\right)$. Let $\mathbf{k}_1 = (k_1, \ldots, k_i)$ be the primary key of Γ_1 and $\mathbf{k}_2 = (\ell_1, \ldots, \ell_j)$ be the primary key of Γ_2. Show that $\Gamma_1 \wr \Gamma_2$ has key $(\ell_1, \ldots, \ell_j, k_1 + j, \ldots, k_i + j)$.

7.7.22 Use Theorem 7.7.33 to show that if n is square-free, then $\mathbb{Z}_n$ is a CI-group with respect to digraphs.

7.7.23 Use Theorem 7.7.33 to show that if $n = 4m$ and m is odd and square-free, then $\mathbb{Z}_n$ is a CI-group with respect to digraphs.

7.7.24 Let n be a positive integer. Show that n is a CI-group with respect to digraphs if and only if $n = m, 2m, 4m$ where m is odd and square-free.

8

Automorphism Groups of Vertex-Transitive Graphs

In this chapter, we discuss automorphism groups of vertex-transitive digraphs. We begin with normal Cayley digraphs, which are known to be "almost all" Cayley digraphs. We finish Section 8.1 with a characterization of normal abelian Cayley graphs of valency 3. We then, in Section 8.2, discuss the transfer, which will be needed in Section 8.3, where we determine the automorphism groups of circulant digraphs of order p^2, p a prime, as well as develop tools to recognize when a vertex-transitive digraph has automorphism group isomorphic to a direct product of two permutation groups. We continue this development in Section 8.4 and then determine the automorphism groups of Cayley digraphs of $\mathbb{Z}_p^2$. Finally, in Section 8.5, we provide a mainly expository section where we discuss automorphism groups of circulant digraphs.

8.1 Normal Cayley Digraphs

An obvious question one might ask is what is a "typical" automorphism group of a Cayley graph of G? If one takes the point of view that symmetry is rare, then the natural guess would be "as small as possible." This has been conjectured by Babai and Godsil (1982) as well as by Imrich and Lovász.

Conjecture 8.1.1 *Let G be a finite group of order n that is neither abelian nor generalized dicyclic. Then almost all Cayley graphs of G have automorphism group G_L. That is, the proportion of self-inverse subsets S of G such that* $\mathrm{Cay}(G, S)$ *is a GRR goes to* 1 *as* $n \to \infty$.

Recall from Section 5.2 that a Cayley graph of G whose automorphism group is G_L is a GRR, while a Cayley digraph of G whose automorphism group is G_L

is a DRR. Also, a Cayley graph of an abelian group that is not an elementary abelian 2-group is not a GRR (Corollary 1.2.18), nor is a Cayley graph of a generalized dicyclic group (Exercise 5.2.8). In the case where G is an abelian group or a generalized dicyclic group, it is known that almost all of these graphs have automorphism group as small as possible – see Dobson et al. (2016) and Morris et al. (2015), respectively. For digraphs, the situation is somewhat simpler, as except for the five groups $\mathbb{Z}_2^2$, $\mathbb{Z}_2^3$, Q_8, $\mathbb{Z}_3^2$, and $\mathbb{Z}_2^4$ (Babai, 1980), every group has a DRR (Babai has also shown that every infinite group has a DRR (Babai 1978)). The corresponding conjecture for digraphs was made by Babai and Godsil (1982).

Conjecture 8.1.2 *Let G be a finite group. Then almost all Cayley digraphs of G have automorphism group G_L. That is, the proportion of subsets S of G such that* $\mathrm{Cay}(G,S)$ *is a DRR goes to* 1 *as $n \to \infty$.*

Astonishingly, Morris and Spiga (2018a) have verified this conjecture!

If one wishes to find a Cayley digraph $\mathrm{Cay}(G,S)$ of a group G whose automorphism group is not G_L, this is not too difficult. Simply choose S that is a union of orbits of an automorphism α of G. Then by Lemma 1.2.15 we have $\alpha \in \mathrm{Aut}(\mathrm{Cay}(G,S))$. Of course, it need not be the case that $\mathrm{Aut}(\mathrm{Cay}(G,S)) = \langle G_L, \alpha\rangle$ (Exercise 8.1.2). If $\mathrm{Aut}(\mathrm{Cay}(G,S)) = \langle G_L, \alpha\rangle$, then by Corollary 4.4.11 $G_L \trianglelefteq \mathrm{Aut}(\mathrm{Cay}(G,S))$.

Definition 8.1.3 A Cayley digraph $\mathrm{Cay}(G,S)$ of G is a **normal** Cayley digraph of G if $G_L \trianglelefteq \mathrm{Aut}(\mathrm{Cay}(G,S))$.

Clearly any GRR or DRR of G is a normal Cayley graph or digraph.

Example 8.1.4 A complete graph K_n is a normal Cayley graph of a group G if and only if $n = 2$, 3, or 4 and $G = \mathbb{Z}_2$, $\mathbb{Z}_3$, or $\mathbb{Z}_2^2$, respectively.

Proof Of course, $\mathrm{Aut}(K_n) = \mathcal{S}_n$. If $n \geq 5$, then A_n is simple (Dummit and Foote, 2004, Theorem 4.6.24), and so the only proper nontrivial normal subgroup of $\mathcal{S}_n$ is A_n. As A_n is not regular, K_n is not a normal Cayley graph of any group. If $n = 4$, then $|\mathrm{AGL}(2,2)| = 24 = |\mathcal{S}_4|$ and $\mathrm{AGL}(2,2) \cong \mathcal{S}_4$. Then K_4 is a normal Cayley graph of $\mathbb{Z}_2^2$ but not $\mathbb{Z}_4$ as $\mathrm{Aut}(\mathbb{Z}_4)$ has order 2. Finally, if $n = 3$, then $\mathcal{S}_3$ has a normal subgroup of order 3 and K_3 is normal, while $\mathrm{Aut}(K_2) = \mathbb{Z}_2$ is regular. Thus K_2 and K_3 are normal Cayley graphs of $\mathbb{Z}_2$ and $\mathbb{Z}_3$, respectively. □

Example 8.1.5 Every circulant digraph of prime order $p \geq 5$ with the exception of K_p and its complement is a normal Cayley digraph of $\mathbb{Z}_p$.

Solution By Theorem 4.4.5 a circulant digraph Γ of prime order $p \geq 5$ that is not K_p or its complement has automorphism group contained in $\mathrm{AGL}(1, p)$. Thus $(\mathbb{Z}_p)_L \trianglelefteq \mathrm{Aut}(\Gamma)$ by the comment following Corollary 4.4.11. □

Example 8.1.6 The Heawood graph is not a normal Cayley digraph of the dihedral group of order 14.

Solution By Exercise 4.1.2, the Heawood graph is a Cayley graph of D_{14}, and by Corollary 4.5.8 has automorphism group $G = \mathrm{PGL}(3, 2) \rtimes \mathbb{Z}_2$. As $\mathrm{PGL}(3, 2) \trianglelefteq G$ is of index 2, every Sylow 7-subgroup of G is contained in $\mathrm{PGL}(3, 2)$. As $\mathrm{PGL}(3, 2)$ is a 2-transitive nonabelian almost simple group (Alperin and Bell, 1995, Theorem 2.6.8) of degree 7, it does not have a normal Sylow 7-subgroup. Thus G does not have a normal Sylow 7-subgroup. However, a Sylow 7-subgroup P of D_{14} is characteristic, and so $P \trianglelefteq (D_{14})_L \rtimes \mathrm{Aut}(D_{14}) = N_{S_{14}}((D_{14})_L)$ by Corollary 4.4.11. □

The following result characterizes normal Cayley digraphs.

Lemma 8.1.7 *Let G be a group, $S \subseteq G$, and $\Gamma = \mathrm{Cay}(G, S)$. The following are equivalent:*

(i) Γ *is a normal Cayley digraph of* G,
(ii) $\mathrm{Stab}_{\mathrm{Aut}(\Gamma)}(1_G) \leq \mathrm{Aut}(G)$,
(iii) $\mathrm{Stab}_{\mathrm{Aut}(\Gamma)}(1_G) = \mathrm{Aut}(G, S)$, *and*
(iv) $\mathrm{Aut}(\Gamma) = G_L \rtimes \mathrm{Aut}(G, S)$.

Proof Note that $G_L \trianglelefteq \mathrm{Aut}(\Gamma)$ if and only if $\mathrm{Aut}(\Gamma) \leq N_{S_G}(G_L) = G_L \rtimes \mathrm{Aut}(G)$ by Corollary 4.4.11. As G_L is regular, this occurs if and only if $\mathrm{Stab}_{\mathrm{Aut}(\Gamma)}(1_G) \leq \mathrm{Aut}(G)$. This last statement occurs if and only if $\mathrm{Stab}_{\mathrm{Aut}(\Gamma)}(1_G) = \mathrm{Aut}(G, S)$, which is true if and only if $\mathrm{Aut}(\Gamma) = G_L \rtimes \mathrm{Aut}(G, S)$. □

So if $\mathrm{Cay}(G, S)$ is a normal Cayley digraph of G, then $\mathrm{Aut}(\mathrm{Cay}(G, S)) = \{g_L\alpha : \alpha \in \mathrm{Aut}(G, S), g \in G\}$.

Example 8.1.8 Let n be odd and composite, A an abelian group of order n, and $1 < H \leq K < A$. Any (H, K)-generalized wreath product is not a normal Cayley digraph of A.

Solution Let $\mathcal{B}$ and C be the block systems of A_L that are the sets of orbits of $\langle h_L : h \in H\rangle$ and $\langle k_L : k \in K\rangle$, respectively. If Γ is an (H, K)-generalized wreath product, then $\mathrm{fix}_{A_L}(\mathcal{B}) = \langle h_L : h \in H\rangle$ and $h_L|_C \in \mathrm{Aut}(\Gamma)$ for every $h \in H$ and $C \in C$ by Theorem 5.3.6. Let $m = |C|$ where $C \in C$. Then $h_L|_C$ has $n - m > n/2$ fixed points. It is straightforward to show that the set of fixed points of an automorphism of a group G is a subgroup of G (Exercise 8.1.3), and so an automorphism of any group G of order n has at most $n/2$ fixed points. Hence $h_L|_C \notin \mathrm{Aut}(G)$ and if $1 \in C$, stabilizes the identity. Thus Γ is not a normal Cayley digraph of A by Lemma 8.1.7. □

The groups G that do not have normal Cayley graphs or digraphs of G were determined by Wang, Wang, and Xu (Wang et al., 1998).

Theorem 8.1.9 *Let G be a finite group. Then G has a normal Cayley graph unless $G \cong \mathbb{Z}_2 \times \mathbb{Z}_4$ or $G \cong Q_8 \times \mathbb{Z}_2^r$, $r \geq 0$. Additionally, every finite group G has a normal Cayley digraph.*

Similar to Babai and Godsil (Conjectures 8.1.1 and 8.1.2), Mingyao Xu has made the apparently weaker conjecture (1998, Conjecture 1) that almost all Cayley graphs and digraphs are normal Cayley graphs and digraphs when he introduced the notions of normal Cayley graphs and digraphs.

Conjecture 8.1.10 *Denote by $\mathcal{G}_n$ the set of all groups of order n. Let*

$$f(n) = \min_{G \in \mathcal{G}_n} \frac{\text{the number of normal Cayley digraphs of } G}{\text{the number of Cayley digraphs of } G}$$

and

$$\tilde{f}(n) = \min_{G \in \mathcal{G}_n, G \neq Q_8 \times \mathbb{Z}_2^r} \frac{\text{the number of normal Cayley graphs of } G}{\text{the number of Cayley graphs of } G}.$$

Then

$$\lim_{n \to \infty} f(n) = \lim_{n \to \infty} \tilde{f}(n) = 1.$$

Of course, the result of Morris and Spiga mentioned above that shows that almost all digraphs are DRRs implies that Xu's conjecture is true for digraphs. Spiga (2021) has shown that the two remaining conjectures are equivalent! That is, that almost all Cayley graphs are GRRs if and only if $\lim_{n\to\infty} \tilde{f}(n) = 1$.

The remainder of this section is a discussion of a characterization of normal Cayley graphs of abelian groups of valency at most 4, by Baik, Feng, Sim, and Xu (Baik et al., 1998). We will prove their main tool, explicitly prove the case for valency 3, and discuss the more complicated proof for valency 4. Our proof mostly follows theirs, and we begin with a useful sufficient condition for a Cayley digraph of an abelian group A to be a normal Cayley digraph.

Lemma 8.1.11 *Let $\Gamma = \mathrm{Cay}(A, S)$ be a connected Cayley graph of an abelian group A. Suppose that if $x, y, z, u \in S$ with $1 \neq xy = zu$, then $\{x, y\} = \{z, u\}$. Then $\mathrm{Cay}(A, S)$ is a normal Cayley graph of A.*

Proof Let $\phi \in \text{Stab}_{\text{Aut}(\Gamma)}(1_A)$. As Γ is connected, $\langle S \rangle = A$ by Lemma 1.2.14. It suffices to show that $\phi(s_1 \cdots s_r) = \phi(s_1)\phi(s_2)\cdots\phi(s_r)$ for any finite product of elements $s_1, \ldots, s_r$ in S. We show this by induction on r. If $r = 1$, this is trivial, so we assume the result holds for $r \geq 1$, and let $s_1, \ldots, s_{r+1} \in S$, so that $r + 1 \geq 2$. Set $w = s_1 \cdots s_{r-1}$, and notice that $\phi(S) = S$.

Suppose $s_r s_{r+1} \neq 1$ and $s_r \neq s_{r+1}$. Then $C = w, ws_r, ws_r s_{r+1}, ws_{r+1}, w$ is a 4-cycle in Γ, as A is abelian. Suppose $w, y, ws_r s_{r+1}, z, w$ is a 4-cycle in Γ. Then $w^{-1}y, y^{-1}ws_r s_{r+1} \in S$ and $w^{-1}y \cdot y^{-1}ws_r s_{r+1} = s_r s_{r+1} \neq 1$, and so by hypothesis $\left\{w^{-1}y, y^{-1}ws_r s_{r+1}\right\} = \{s_r, s_{r+1}\}$. We conclude that there is a unique 4-cycle in Γ that contains w and $ws_r s_{r+1}$. Also, if w, ws_r, x, ws_{r+1}, w is a 4-cycle in Γ, then there exists $t_1, t_2 \in S$ with $x = ws_r t_1 = ws_{r+1}t_2$. Hence $w^{-1}x = s_r t_1 = s_{r+1}t_2$. Note $w^{-1}x \neq 1$ as otherwise $w = x$ and w, ws_r, x, ws_{r+1}, w is not a 4-cycle. So by hypothesis, $\{s_r, t_1\} = \{s_{r+1}, t_2\}$. As $s_r \neq s_{r+1}$, $t_1 = s_{r+1}$ and $x = ws_r s_{r+1}$. Hence Γ also contains a unique 4-cycle that contains w, ws_r, and ws_{r+1}.

We now set $w = s_1 \cdots s_{r-1}$ if $r + 1 \geq 3$, and $w = 1$ if $r + 1 = 2$, and consider

$$\phi(C) = \phi(w), \phi(ws_r), \phi(ws_r s_{r+1}), \phi(ws_{r+1}), \phi(w).$$

By the induction hypothesis and choice of w, we have $\phi(ws_r) = \phi(w)\phi(s_r)$ and $\phi(ws_{r+1}) = \phi(w)\phi(s_{r+1})$. Then,

$$\phi(C) = \phi(w), \phi(w)\phi(s_r), \phi(ws_r s_{r+1}), \phi(w)\phi(s_{r+1}), \phi(w),$$

and so $\phi(s_r), \phi(s_{r+1}) \in S$. As the unique 4-cycle in Γ that contains $\phi(w)$, $\phi(w)\phi(s_r)$, and $\phi(w)\phi(s_{r+1})$ is

$$\phi(w), \phi(w)\phi(s_r), \phi(w)\phi(s_r)\phi(s_{r+1}), \phi(w)\phi(s_{r+1}), \phi(w),$$

we see that $\phi(ws_r s_{r+1}) = \phi(w)\phi(s_r)\phi(s_{r+1})$. By induction, $\phi(w) = \phi(s_1)\cdots\phi(s_{r-1})$ and so $\phi(s_1 \cdots s_{r+1}) = \phi(s_1)\cdots\phi(s_{r+1})$, as required.

Now assume $s_r s_{r+1} \neq 1$ but $s_r = s_{r+1}$. As $s_r s_{r+1} \neq 1$, the walk w, ws_r, ws_r^2 is a path P in Γ. Suppose P is contained in a 4-cycle C in Γ. Then there exists $x \in A$ such that $C = P, x, w$ and so there exists $t_1, t_2 \in S$ with $x = wt_1 = ws_r^2 t_2$. Equivalently, $s_r^{-1}t_1 = s_r t_2$. If $s_r^{-1}t_1 = s_r t_2 = 1$, then $t_1 = s_r$ and $t_2 = s_r^{-1}$. But then $x = ws_r$ and C is not a 4-cycle. So $s_r^{-1}t_1 = s_r t_2 \neq 1$. By hypothesis, $\left\{s_r^{-1}, t_1\right\} = \{s_r, t_2\}$ and as $s_r \neq s_r^{-1}$ (as $s_r s_{r+1} \neq 1$), it follows that $t_2 = s_r^{-1}$ as before, a contradiction. As we have already seen above that if $s_r \neq s_{r+1}$ and $s_r s_{r+1} \neq 1$, then $ws_r s_{r+1}$ is contained in a 4-cycle, we see a path P of length 2 is not contained in a 4-cycle if and only if $P = w, ws_r, ws_r^2$ for some $s_r \in S$ with $s_r^2 \neq 1$. We now set $w = s_1 \cdots s_{r-1}$. Then $\phi(P) = \phi(w), \phi(ws_r), \phi\left(ws_r^2\right) = \phi(w), \phi(w)\phi(s_r), \phi\left(ws_r^2\right)$ is a path of length 2 not contained in a 4-cycle. We conclude that $\phi\left(ws_r^2\right) = \phi(w)\phi(s_r)\phi(s_r)$ and so $\phi(s_1 \cdots s_{r+1}) = \phi(s_1)\cdots\phi(s_{r+1})$, as required.

Finally, suppose $s_r s_{r+1} = 1$. Then $1, s_r, s_{r+1}$ is a walk in Γ of length 2 from 1 to 1, and $\phi(1), \phi(s_r), \phi(s_{r+1})$ is also a walk of length 2 from 1 to 1 as $\phi(1) = 1$. So $\phi(s_r)\phi(s_{r+1}) = 1$. Hence $\phi(s_r s_{r+1}) = \phi(1) = 1 = \phi(s_r)\phi(s_{r+1})$. Then

$$\begin{aligned}\phi(s_1 \cdots s_{r+1}) &= \phi(s_1 \cdots s_{r-1}) = \phi(s_1) \cdots \phi(s_{r-1}) \\ &= \phi(s_1) \cdots \phi(s_{r-1})\phi(s_r)\phi(s_{r+1}),\end{aligned}$$

and the result follows by induction. □

Example 8.1.12 The hypercube Q_n is a normal Cayley graph of $\mathbb{Z}_2^n$.

Solution Q_n isomorphic to $\mathrm{Cay}\left(\mathbb{Z}_2^n, S\right)$, where S is the set of the n vectors that are 0 in every coordinate except one. Let $x, y, z, u \in S$ with $(0, \ldots, 0) \neq x + y = z + u$. Then $x + y$ is not zero in exactly two coordinates, say the ith and jth. Also x and y, and z and u, are 0 in every coordinate except one each and these are either the ith and jth or jth and ith. Hence $\{x, y\} = \{u, z\}$ and Q_n is a normal Cayley graph of $\mathbb{Z}_2^n$ by Lemma 8.1.11. □

Theorem 8.1.13 *Let Γ be a connected Cayley graph of an abelian group A of valency at most 3. Then Γ is a normal Cayley graph of A unless:*

(i) $\Gamma = K_4$ *and* $A = \mathbb{Z}_4$, *or*
(ii) $\Gamma = K_{3,3}$ *and* $A = \mathbb{Z}_6$, *or*
(iii) $\Gamma = Q_3$ *and* $A = \mathbb{Z}_2 \times \mathbb{Z}_4$.

Proof As Γ is connected, $\langle S \rangle = A$ by Lemma 1.2.14. If Γ has valency at most 1 then $\Gamma = K_1$ or K_2, which are GRRs. If Γ has valency 2, then Γ is a cycle whose automorphism group is a dihedral group. Then Γ is a normal Cayley graph of the cyclic group, and the dihedral group contains no other transitive abelian subgroups. We thus assume Γ has valency 3.

Suppose first that all elements of S are involutions. Then $A = \mathbb{Z}_2^n$ for some $n \geq 2$. If S is a minimal generating set of $\mathbb{Z}_2^n$, then, viewing each element of S as a vector in $\mathbb{Z}_2^n$, we see that S is linearly independent and a basis for $\mathbb{Z}_2^n$. As Γ is cubic, $n = 3$. Then there exists $A \in \mathrm{Aut}\left(\mathbb{Z}_2^3\right) = \mathrm{GL}(3, 2)$ such that $A(S)$ is the canonical basis for $\mathbb{Z}_2^3$, and so $A(\Gamma)$ is the cube. By Example 8.1.12, Γ is a normal Cayley graph of $\mathbb{Z}_2^3$. If S is not a minimal generating set of $\mathbb{Z}_2^n$, then S is linearly dependent and so $n = 2$ as $|S| = 3$. So $\Gamma = K_4$, which by Example 8.1.4 is a normal Cayley graph of $\mathbb{Z}_2^2$.

The only other possibility is that $S = \left\{a, a^{-1}, s\right\}$ where s is an involution but a is not. By examining every possible choice of two elements of $S^2\backslash\{1\} = \left\{a^2, as, a^{-2}, a^{-1}s\right\}$, two elements of $S^2\backslash\{1\}$ are equal only if $s = a^{\pm 3}$ or $a^2 = a^{-2}$. If no two elements of $S^2\backslash\{1\}$ are equal, then, by Lemma 8.1.11, we have that

Γ is a normal Cayley graph of A. If $s = a^{\pm 3}$, then $|a| = 6$ and $A = \mathbb{Z}_6$ as s is an involution. Then $\Gamma \cong \text{Cay}(\mathbb{Z}_6, \{1,3,5\}) \cong K_{3,3} \cong K_2 \wr \bar{K}_3$. As $N_{S_6}((\mathbb{Z}_6)_L)$ has order 12 and $\text{Aut}(\Gamma) \cong \mathbb{Z}_2 \wr S_3$ has order 72, we see Γ is not a normal Cayley graph of $\mathbb{Z}_6$ and (ii) follows. If $a^2 = a^{-2}$ then $|a| = 4$. If $s \in \langle a \rangle$ then $A = \mathbb{Z}_4$ and $\Gamma = K_4$. As $N_{S_4}((\mathbb{Z}_4)_L)$ has order 8 and $|\text{Aut}(\Gamma)| = 24$, Γ is not a normal Cayley graph of $\mathbb{Z}_4$ and (i) follows. If $s \notin \langle a \rangle$, then $A \cong \mathbb{Z}_2 \times \mathbb{Z}_4$, $S = \{(0,1),(0,3),(1,0)\}$, and $\Gamma \cong Q_3$. Then $S_3 \wr \mathbb{Z}_2^3 \cong S_2 \times S_4 \leq \text{Aut}(\Gamma)$ and $S_2 \times S_4$ does not normalize $(\mathbb{Z}_2 \times \mathbb{Z}_4)_L$. So Γ is not a normal Cayley graph of $\mathbb{Z}_2 \times \mathbb{Z}_4$ and (iii) follows. □

The proof for valency 4 is similar but with more cases. Most cases are handled using Lemma 8.1.11, except there are several families of Cayley graphs of abelian groups that need to be shown to be normal directly. Some are more easily handled using techniques from Section 8.5, and so are left as exercises in that section (Exercises 8.1.6, 8.1.7, and 8.1.8). For the rest of the details, see Baik et al. (1998). Similarly, Baik, Feng, and Sim (Baik et al., 2000) determined all nonnormal Cayley graphs of abelian groups of valency 5.

In Example 8.5.3 we give an example of a Cayley graph Γ of a group G that is a normal Cayley graph of G, but for which there is a regular group of automorphisms H, not isomorphic to G, which is not normal in G (note Q_3 also has this property). Indeed, there are *isomorphic* regular groups G and H for which there is a graph Γ whose automorphism group contains both G and H with G normal in $\text{Aut}(\Gamma)$ but H is *not* normal in $\text{Aut}(\Gamma)$. Put another way, Γ isomorphic to Cayley graphs Γ_1 and Γ_2 of G and H, and Γ_1 is a normal Cayley graph of G while Γ_2 is not a normal Cayley graph of H. The first example of this phenomenon was found by Giudici and Smith (2010) for the group $\mathbb{Z}_6^2$, while Royle found another for the group $\mathbb{Z}_2^6$ (2008). The first infinite family was found by Bamberg and Giudici (2011). See Xu (2017) for additional constructions.

8.2 The Transfer

Throughout this section, let G be a group and $H \leq G$ **an abelian subgroup**. The transfer is a map from G to H, and arises in a natural way, which we now discuss. First consider G_L, the left regular representation of G. The left coset action of G on G/H is transitive, and multiplying a left coset of aH on the left by $g \in G$, or applying g_L to aH, permutes G/H. We would like to write down a "formula" for this permutation of G/H. We are immediately presented with many choices, as each left coset can be represented by any element it contains. We wish to choose and fix one element from each left coset.

The transfer is usually used to show that groups are not simple. We will use it in the next section to prove Lemma 8.3.6. Another well-known application is to prove Burnside's transfer theorem. As we will have need of this result in Chapter 11, and it is not too much extra work, we will go ahead and give a proof at the end of this section.

Definition 8.2.1 Let G be a group and $H \leq G$ be abelian of index n. A **transversal** of G/H is a set of left coset representatives $\{x_0, \ldots, x_{n-1}\}$ of G/H. That is, $G/H = \{x_i H : i \in \mathbb{Z}_n\}$.

Definition 8.2.2 Let G be a group with $H \leq G$ abelian. Let $\{x_i : i \in \mathbb{Z}_n\}$ be a transversal of G/H, and $g \in G$. Then G permutes the left cosets of H in G via left multiplication. So for $g \in G$ and $i \in \mathbb{Z}_n$,

$$gx_i H = x_{j(i)} H,$$

for some $j \in S_n$. Then

$$gx_i = x_{j(i)} h_i,$$

for some $h_i \in H$. Define $v\colon G \to H$ by $v(g) = \prod_{i=0}^{n-1} h_i$. We call v the **transfer** of G into H.

For each coset $x_i H$, choose $k_i \in H$ and consider the map k given by $k(x_i) \mapsto x_i k_i H$. Notice that all this has done is change the left coset representatives from x_i to $x_i k_i$, and that any choice of left coset representatives can be obtained from such a map. That is, k maps a transversal to a transversal, and any transversal can be obtained from $\{x_i : i \in \mathbb{Z}_n\}$ as the image of an appropriate map k, i.e., appropriate choices of k_is.

Lemma 8.2.3 *Let G be a group with $H \leq G$ abelian. The transfer $v\colon G \to H$ is independent of the choice of transversal. That is, v is the same map regardless of the choice of $\{x_i : i \in \mathbb{Z}_n\}$.*

Proof Let $k(\{x_i : i \in \mathbb{Z}_n\})$ be another transversal of G/H, with $y_i = k(x_i) = x_i k_i$, where $k_i \in H$. Then $x_i = y_i k_i^{-1}$, and

$$gy_i = gx_i k_i = x_{j(i)} h_i k_i = y_{j(i)} k_{j(i)}^{-1} h_i k_i.$$

As H is abelian,

$$\prod_{i=0}^{n-1} k_{j(i)}^{-1} h_i k_i = \left(\prod_{i=0}^{n-1} k_{j(i)}^{-1}\right)\left(\prod_{i=0}^{n-1} h_i\right)\left(\prod_{i=0}^{n-1} k_i\right) = \prod_{i=0}^{n-1} h_i.$$ □

Notice that we definitely used the assumption that H is abelian in the previous result, and it should be clear that the previous result may fail for nonabelian H. The next result also makes use of the hypothesis that H is abelian.

Lemma 8.2.4 *Let G be a group and $H \leq G$ be abelian. Then the transfer $v\colon G \to H$ is a homomorphism.*

Proof Let $\{x_i : i \in \mathbb{Z}_n\}$ be a transversal of G/H, and $g_1, g_2 \in G$ with $g_1 x_i = x_{j(i)} h_i$ and $g_2 x_i = x_{k(i)} h'_i$, where $h_i, h'_i \in H$, and $j, k \in S_n$. As $\prod_{i=0}^{n-1} h_{k(i)} = \prod_{i=0}^{n-1} h_i$, we see that

$$v(g_1 g_2) = \prod_{i=0}^{n-1} (h_{k(i)} h'_i) = \left(\prod_{i=0}^{n-1} h_{k(i)}\right)\left(\prod_{i=0}^{n-1} h'_i\right) = \left(\prod_{i=0}^{n-1} h_i\right)\left(\prod_{i=0}^{n-1} h'_i\right) = v(g_1)v(g_2),$$

with the second equality following from the fact that H is abelian. □

The transfer can be considered in a more general context where H is nonabelian. In this case it is necessary that the transfer v is defined from G to H/H', where H' is the commutator subgroup of H, as H/H' is abelian. See Exercises 8.2.3 and 8.2.4.

Example 8.2.5 If $H = 1$ then the transfer is the trivial map while if $H = G$ the transfer is the identity map.

Solution If $H = 1$, then there is a unique transversal of the left cosets of H in G, namely $\{\{x_i\} : x_i \in G\}$. If $g \in G$, with $gx_i H = x_{j(i)} H$ for $j \in S_n$, then $gx_i = x_{j(i)}$, and so $h_i = 1$. Then $v(g) = 1$.

If $H = G$, then there is a unique left coset of H in G, namely G itself. For $g \in G$, $gG = G$ and choosing the left coset representative of G to be 1, which we may do by Lemma 8.2.3, we have that $g1 = 1g$ and so $v(g) = g$. □

The preceding example makes it clear that the possibly interesting choices for H are proper, nontrivial abelian subgroups of H. For such a subgroup, by the first isomorphism theorem it will always be the case that $\mathrm{Ker}(v) \neq 1$. Of course, if $\mathrm{Ker}(v) = G$, then the transfer is the trivial map, and so to be useful, we need to find elements $g \in G$ for which $v(g) \neq 1$. The next result leads to a sufficient condition for this to be the case.

Lemma 8.2.6 *Let G be a group and $H \leq G$ be abelian. Let $g \in G$ and $n_1, \ldots, n_r$ be the sizes of the orbits of g in the left coset action of G on G/H, and $v\colon G \to H$ the transfer. Then $\sum_{i=1}^{r} n_i = [G : H]$, and there exists $g_1, \ldots, g_r \in G$ such that $g_i g^{n_i} g_i^{-1} \in H$ for $1 \leq i \leq r$, and*

$$v(g) = \prod_{i=1}^{r} g_i g^{n_i} g_i^{-1}.$$

Proof That $\sum_{i=1}^{r} n_i = [G : H]$ follows from the fact that left multiplication by $g \in G$ induces a permutation of the set of left cosets of H in G. Now, let g have cycle decomposition

$$(x_{1,1}H, \ldots, x_{1,n_1}H)(x_{2,1}H, \ldots, x_{2,n_2}H)\cdots(x_{r,1}H, \ldots, x_{r,n_r}H).$$

For $1 \leq i \leq r$ and $0 \leq j \leq n_i - 1$, set $y_{i,1} = x_{i,1}$, and let $y_{i,j} = g^j x_{i,1}$. Then $\{y_{i,j} : 1 \leq i \leq r, 0 \leq j \leq n_i - 1\}$ is a transversal of G/H. We use this transversal to calculate $v(g)$, which we may do by Lemma 8.2.3. For $1 \leq i \leq r$ and $0 \leq j \leq n_i - 2$, we have $gy_{i,j} = g\left(g^j x_{i,1}\right) = g^{j+1}x_{i,1} = y_{i,j+1}$. We conclude that $h_{i,j} = 1$ for $1 \leq i \leq r$ and $0 \leq j \leq n_i - 2$. Notice that

$$g^{n_i}x_{i,1} = gg^{n_i-1}x_{i,1} = gy_{i,n_i} = x_{i,1}h_{i,n_i}.$$

For $1 \leq i \leq r$, let $g_i = x_{i,1}^{-1}$. Then

$$g_i g^{n_i} g_i^{-1} = x_{i,1}^{-1} g^{n_i} x_{i,1} = x_{i,1}^{-1} x_{i,1} h_{i,n_i} = h_{i,n_i} \in H,$$

and

$$\prod_{i=1}^{r} g_i g^{n_i} g_i^{-1} = \prod_{i=1}^{r} h_{i,n_i} = \prod_{i=1}^{r}\prod_{j=0}^{n_i-1} h_{i,j} = v(g). \qquad \square$$

Corollary 8.2.7 *Let G be a group, $H \leq G$ be abelian such that* $\gcd(|H|, [G : H]) = 1$, *and $g \in H$ such that $g \in Z(G)$ and $g \neq 1$. Then $v(g) \neq 1$.*

Proof Let r be the number of orbits of g acting on the left cosets of H in G by left multiplication. By Lemma 8.2.6, there exists $g_1, \ldots, g_r$ such that $v(g) = \prod_{i=1}^{r} g_i g^{n_i} g_i^{-1}$. As $g \in Z(G)$, we have $g_i g^{n_i} g_i^{-1} = g^{n_i}$, so that $v(g) = \prod_{i=1}^{r} g^{n_i} = g^{\sum_{i=1}^{r} n_i} = g^{[G:H]}$ by Lemma 8.2.6. As $\gcd(|H|, [G : H]) = 1$ and $g \neq 1$ is in H, we see that $v(g) = g^{[G:H]} \neq 1$. $\square$

Lemma 8.2.8 *Let G be a group with abelian Sylow p-subgroup P. Any two elements of P that are conjugate in G are conjugate in $N_G(P)$.*

Proof Let $x, y \in P$ with $y = gxg^{-1}$ for some $g \in G$. Note that P centralizes x as P is abelian and so gPg^{-1} centralizes $gxg^{-1} = y$. Of course, P also centralizes y, so P and gPg^{-1} are Sylow p-subgroups of $C_G(y)$, the centralizer of y in G. Then there exists $h \in C_G(y)$ such that $hgPg^{-1}h^{-1} = P$, and $hgxg^{-1}h^{-1} = hyh^{-1} = y$. Of course, $hg \in N_G(P)$ and the result follows. $\square$

One of the main applications of the transfer is the following result proven by Burnside and known as Burnside's transfer theorem. We will need one more term before stating it.

Definition 8.2.9 Let G be a group. A **normal p-complement** of G is a normal subgroup $H \leq G$ such that for some Sylow p-subgroup of G, we have $G = PH$ and $P \cap H = 1$.

Theorem 8.2.10 (Burnside's transfer theorem) *Let G be a group with Sylow p-subgroup P. If P is contained in the center of its normalizer in G, then G has a normal p-complement.*

Proof Let $v\colon G \to P$ be the transfer. Let $g \in G$. Consider the action of g on the left cosets of P in G as in Lemma 8.2.6, with g having r orbits of size $n_1, \ldots, n_r$. For $1 \leq i \leq r$ there exists $g_i \in G$ such that $g_i g^{n_i} g_i^{-1} \in P$ and $v(g) = \prod_{i=1}^r g_i g^{n_i} g_i^{-1}$. Now assume $g \in P$. By Lemma 8.2.8, there exists $h_i \in N_G(P)$ such that $h_i g^{n_i} h_i^{-1} = g_i g^{n_i} g_i^{-1}$. As P is contained in the center of its normalizer, $h_i g^{n_i} h_i^{-1} = g^{n_i}$. Then

$$v(g) = \prod_{i=1}^{r} g_i g^{n_i} g_i^{-1} = \prod_{i=1}^{r} g^{n_i} = g^{\sum_{i=1}^r n_i} = g^{[G:P]}.$$

We now show that $v(G) = P$. Indeed, as $g \in P$ and $\gcd([G : P], |P|) = 1$, $[G : P]$ is a unit in $\mathbb{Z}_{|g|}$. Hence there exists $u \in \mathbb{Z}_{|g|}$ such that $u[G : P] = 1$. Then $v(g^u) = g$ and $v(G) = P$. Let $H = \mathrm{Ker}(v)$. Then $H \trianglelefteq G$, and as v is onto P and P is a Sylow p-subgroup of G, H has order relatively prime to p. Hence $P \cap H = 1$, and as $|G| = |P| \cdot |H|$, we see $PH = G$. So H is a normal p-complement of G. □

Corollary 8.2.11 *Let G be a transitive group of prime degree p with $P \leq G$ a regular cyclic subgroup. Then either $G = P$ or $N_G(P) \neq P$.*

Proof A Sylow p-subgroup of G has order p as a Sylow p-subgroup of S_p has order p. Thus P is a Sylow p-subgroup of G. If $G \neq P$ and $N_G(P) = P$, then P is contained in the center of its normalizer in G. Then P has a normal p-complement K by Burnside's transfer theorem. As $G \neq P$, $K \neq 1$. Then the set of orbits of K is a block system $\mathcal{B}$ whose blocks are nontrivial and are of order relatively prime to p. However, as p is prime, there are no nontrivial blocks of G of order relatively prime to p, a contradiction establishing the result. □

We have only developed those properties of the transfer that we will need. The reader interested in more information about the transfer is referred to Passman (1968, Section 12) and Isaacs (2008, Chapters 5 and 10).

8.3 More Tools and Automorphism Groups of Circulants of Order p^2

We begin our consideration of the automorphism groups of vertex-transitive digraphs of order p^2, p an odd prime, with the following result, which more or less lists the various possibilities for the automorphism group of such a digraph. We observe that the result could easily be condensed, but we prefer the form given as it lists the various cases for which different techniques are required.

Lemma 8.3.1 *Let Γ be a Cayley digraph of an abelian group of order p^2, where p is an odd prime. Then one of the following is true:*

(i) $\mathrm{Aut}(\Gamma)$ *is primitive,*
(ii) $\mathrm{Aut}(\Gamma)$ *has a normal block system $\mathcal{B}$ and $\mathrm{fix}_{\mathrm{Aut}(\Gamma)}(\mathcal{B})$ is not faithful in its induced action on $B \in \mathcal{B}$. Then there exists vertex-transitive digraphs Γ_1 and Γ_2 of order p such that* $\mathrm{Aut}(\Gamma) \cong \mathrm{Aut}(\Gamma_1) \wr \mathrm{Aut}(\Gamma_2)$,
(iii) $\mathrm{Aut}(\Gamma)$ *has a normal block system $\mathcal{B}$ and $\mathrm{fix}_{\mathrm{Aut}(\Gamma)}(\mathcal{B})$ is faithful in its induced action on $B \in \mathcal{B}$. Then one of the following is true:*
 (a) $\mathrm{fix}_{\mathrm{Aut}(\Gamma)}(\mathcal{B})$ *is semiregular of order p,*
 (b) $\mathrm{fix}_{\mathrm{Aut}(\Gamma)}(\mathcal{B})$ *contains a normal Sylow p-subgroup but is not semiregular, or*
 (c) $\mathrm{fix}_{\mathrm{Aut}(\Gamma)}(\mathcal{B})^B$ *is a 2-transitive nonabelian almost simple group.*

Proof We may assume without loss of generality that $\mathrm{Aut}(\Gamma)$ is imprimitive. By Theorem 2.2.19, $\mathrm{Aut}(\Gamma)$ has a normal block system $\mathcal{B}$. Then $\mathrm{fix}_{\mathrm{Aut}(\Gamma)}(\mathcal{B})$ is either faithful in its induced action on $B \in \mathcal{B}$ or it is not. If not, then, by Corollary 5.3.15 Γ is isomorphic to a nontrivial generalized wreath product, and as Γ has order p^2, is necessarily isomorphic to a wreath product $\Gamma_1 \wr \Gamma_2$, where Γ_1 and Γ_2 are circulant digraphs of order p. As $\mathrm{Aut}(\Gamma)$ is imprimitive, Γ_1 and Γ_2 cannot both be complete graphs or their complements by Theorem 5.1.8 or its digraph analogue, and (ii) follows by Theorem 5.1.8. If $\mathrm{fix}_{\mathrm{Aut}(\Gamma)}(\mathcal{B})$ is faithful on $B \in \mathcal{B}$, then $\mathrm{fix}_{\mathrm{Aut}(\Gamma)}(\mathcal{B})^B$ is a transitive group of prime degree. The result follows by Theorem 4.4.18. □

The previous result in many ways outlines the approach that we will take in this section and the next. Let Γ be a vertex-transitive graph of order p^2 with automorphism group G. First, if G is primitive, it is usually found using the O'Nan–Scott theorem and CFSG. If G is imprimitive with block system $\mathcal{B}$, then we know the blocks are of order p. We handle the three cases given in Lemma 8.3.1 using different techniques. The easiest is (ii), as we have already studied wreath products of digraphs and their automorphism groups. The case when $\mathrm{fix}_G(\mathcal{B})$ is semiregular is often the most difficult. In most of the other subcases given in (iii), the automorphism groups turns out to involve a natural direct product. We first develop the permutation group theory that we will need for these automorphism groups. This part of the section does not depend on the order of Γ, and so we consider it in more generality.

It is often the case that a vertex-transitive digraph Γ has automorphism group permutation isomorphic to a subgroup of $\mathcal{S}_k \times \mathcal{S}_m$, where $k, m \geq 2$ are integers and the action is the canonical one. That is, after an appropriate relabeling of $V(\Gamma)$ with elements of $\mathbb{Z}_k \times \mathbb{Z}_m$, $\mathrm{Aut}(\Gamma) \leq \mathcal{S}_k \times \mathcal{S}_m$ with action $(g, h)(i, j) = (g(i), h(j))$, where $g \in \mathcal{S}_k$, $h \in \mathcal{S}_m$, $i \in \mathbb{Z}_k$, and $j \in \mathbb{Z}_m$. We now develop tools that allow us to recognize this in certain situations.

Definition 8.3.2 Let $\mathcal{B}$ and $\mathcal{C}$ be block systems of the transitive group G such that $|B \cap C| = 1$ for every $B \in \mathcal{B}$ and $C \in \mathcal{C}$. We say that $\mathcal{B}$ and $\mathcal{C}$ are **orthogonal block systems of** G.

Observe that if $\mathcal{B}$ and $\mathcal{C}$ are orthogonal and $\mathcal{B}$ has m blocks of size k, then $\mathcal{C}$ has k blocks of size m.

Lemma 8.3.3 *Let $n = mk$ and $G \leq S_n$ such that G is transitive. Then G has orthogonal block systems $\mathcal{B}$ and $\mathcal{C}$ of m blocks of size k and k blocks of size m, respectively, if and only if G is permutation equivalent to a faithful permutation representation of G whose image is a subgroup of $\mathcal{S}_k \times \mathcal{S}_m$ in its natural action on $\mathbb{Z}_k \times \mathbb{Z}_m$.*

Proof Suppose G has orthogonal block systems $\mathcal{B}$ and $\mathcal{C}$ of m blocks of size k and k blocks of size m, respectively. Define $\pi \colon G \to \mathcal{S}_{\mathcal{B}\times\mathcal{C}}$ by $\pi(g)(B, C) = (g(B), g(C))$. If $g \in \mathrm{Ker}(\pi)$, then g fixes every block of both $\mathcal{B}$ and $\mathcal{C}$. As $|B \cap C| = 1$ for every $B \in \mathcal{B}$ and $C \in \mathcal{C}$, and there are exactly $mk = n$ such intersections, $\mathrm{Ker}(\pi) = 1$ and π is a faithful permutation representation of G. Let $B \in \mathcal{B}$ and $C \in \mathcal{C}$. If $g \in G$ stabilizes (B, C), then $g(B) = B$ and $g(C) = C$. Let $B \cap C = \{x\}$. Then $g(x) = x$. Conversely, if $g(x) = x$, then there exists $B \in \mathcal{B}$ and $C \in \mathcal{C}$ such that $x \in B \cap C$. Then $g(B, C) = (B, C)$ so $\mathrm{Stab}_G(x) = \mathrm{Stab}_{\pi(G)}((B, C))$. It then follows by Theorem 4.1.8 that π is permutation equivalent to the identity permutation representation of G. As $|\mathcal{B}| = m$ and $|\mathcal{C}| = k$, the result follows.

Conversely, suppose G is permutation equivalent to a faithful representation of G whose image is a subgroup of $\mathcal{S}_k \times \mathcal{S}_m$ in its natural action on $\mathbb{Z}_k \times \mathbb{Z}_m$. We identify G with its image, and so assume that $G \leq \mathcal{S}_k \times \mathcal{S}_m$. It suffices to show that for $i \in \mathbb{Z}_k$, the set $\{(i, j) : j \in \mathbb{Z}_m\}$ is a block of G (as an analogous argument will give that for $j \in \mathbb{Z}_m$ the set $\{(i, j) : i \in \mathbb{Z}_k\}$ is a block of G, and clearly intersections of blocks of those types are singletons). This is immediate: if $(\alpha,\beta) \in G$ with $\alpha \in \mathcal{S}_k$ and $\beta \in \mathcal{S}_m$, then

$$(\alpha,\beta)(\{(i,j) : j \in \mathbb{Z}_m\}) = \{(\alpha(i),\beta(j)) : j \in \mathbb{Z}_m\} = \{(\alpha(i),j) : j \in \mathbb{Z}_m\}.$$

Then $\{(\alpha(i),j) : j \in \mathbb{Z}_m\} \cap \{(i,j) : j \in \mathbb{Z}_m\} \neq \emptyset$ if and only if $\alpha(i) = i$ if and only if the two sets are equal, and $\{(i,j) : j \in \mathbb{Z}_m\}$ is a block of G. □

Lemma 8.3.4 *Let $n \geq 2$ and $G \leq \mathcal{S}_n$ be transitive with normal block system $\mathcal{B}$. Define an equivalence relation $\equiv$ on $\mathbb{Z}_n$ by $v_1 \equiv v_2$ if and only if $\mathrm{Stab}_{\mathrm{fix}_G(\mathcal{B})}(v_1) = \mathrm{Stab}_{\mathrm{fix}_G(\mathcal{B})}(v_2)$. Then the set of equivalence classes $\mathcal{C}$ of $\equiv$ is a block system of G. Additionally,*

(i) *$\mathcal{C}$ is one block of size n if and only if $\mathrm{fix}_G(\mathcal{B})$ is semiregular, and*
(ii) *assume $\mathrm{fix}_G(\mathcal{B})^B$ is primitive but not abelian for $B \in \mathcal{B}$. Then $|B \cap C| \leq 1$ for every $B \in \mathcal{B}$ and $C \in \mathcal{C}$. Let $\mathcal{D}$ be the $\mathcal{B}$-fixer block system of G as defined in Exercise 5.3.5, fix $x \in B \in \mathcal{B}$ and let $D \in \mathcal{D}$ such that $B \subseteq D$. Then $\mathcal{C} < \mathcal{D}$ and $B' \in \mathcal{B}$ contains an element $y \equiv x$ if and only if the permutation representation induced by the action of $\mathrm{fix}_G(\mathcal{B})$ on B is permutation equivalent to the permutation representation induced by the action of $\mathrm{fix}_G(\mathcal{B})$ on B'.*

Proof As conjugation by an element of G maps the stabilizer of a point in $\mathrm{fix}_G(\mathcal{B})$ to the stabilizer of a point in $\mathrm{fix}_G(\mathcal{B})$, $\equiv$ is a G-congruence, and so by Theorem 2.3.2 the equivalence classes of $\equiv$ are blocks of G.

(i) $\mathcal{C}$ has one block of size n if and only if $\mathrm{Stab}_{\mathrm{fix}_G(\mathcal{B})}(x) = \mathrm{Stab}_{\mathrm{fix}(\mathcal{B})}(y)$ for every $x,y \in \mathbb{Z}_n$ that, as G is a permutation group, occurs if and only if $\mathrm{Stab}_{\mathrm{fix}_G(\mathcal{B})}(x) = 1$ for all $x \in \mathbb{Z}_n$.

(ii) For the first part, let $B \in \mathcal{B}$ and suppose $x,y \in B$ with $x \equiv y$. First suppose $\mathrm{Stab}_{\mathrm{fix}_G(\mathcal{B})^B}(x) = 1$. Then $\mathrm{fix}_G(\mathcal{B})^B$ is regular, and if it has a proper nontrivial subgroup, it is not primitive by Theorem 2.3.7. Hence $\mathrm{fix}_G(\mathcal{B})^B$ is cyclic of prime order and so abelian, a contradiction. Thus $\mathrm{Stab}_{\mathrm{fix}_G(\mathcal{B})^B}(x) \neq 1$. Then there exists $z \in B$ such that $z \not\equiv x$. By the first part of this lemma applied to $\mathrm{fix}_G(\mathcal{B})^B$, $B \cap C$ is a nontrivial block of $\mathrm{fix}_G(\mathcal{B})^B$, contradicting the hypothesis that $\mathrm{fix}_G(\mathcal{B})^B$ is primitive.

For the second part, suppose $B' \subseteq D' \in \mathcal{D}$ with $D' \neq D$. Then there exists $H \trianglelefteq \mathrm{fix}_G(\mathcal{B})$ such that H^B is transitive, but $H^{B'}$ is not. As $\mathrm{fix}_G(\mathcal{B})$ is primitive on B' by Theorem 2.4.7, we have $H^{B'} = 1$. Thus $H \leq \mathrm{Stab}_{\mathrm{fix}_G(\mathcal{B})}(y)$ for every $y \in B'$. Hence $x \not\equiv y$. So if $x \equiv y$ then $y \in D$ and $\mathcal{C} \leq \mathcal{D}$. Also, $\mathcal{C} < \mathcal{D}$ as $\mathcal{D}$ is a union of blocks of $\mathcal{B}$. Finally, by Theorem 4.1.8 we have that the permutation representation induced by the action of $\mathrm{fix}_G(\mathcal{B})$ on $B \in \mathcal{B}$ is permutation equivalent to the permutation representation induced by the action of $\mathrm{fix}_G(\mathcal{B})$ on $B' \in \mathcal{B}$ if and only if the stabilizer of a point $x \in B$ in $\mathrm{fix}_G(\mathcal{B})$ is the stabilizer of a point $y \in B'$, or equivalently, if and only if $x \equiv y$. □

The previous result can be quite useful. Notice that Lemma 8.3.4 (i) seemingly places no restrictions on the quotient $G/\mathcal{B}$ as if $\mathcal{B}$ has m blocks, $H \leq S_m$ is transitive, and $K \cong \mathrm{fix}_G(\mathcal{B})$, then $H \times K_L$ satisfies Lemma 8.3.4 (i). We use the word "seemingly" here as while it is true that $\mathrm{fix}_G(\mathcal{B})$ being semiregular imposes no constraint on $G/\mathcal{B}$, it may be that there are other conditions that, when combined with $\mathrm{fix}_G(\mathcal{B})$ being semiregular, *do* place severe restrictions on $G/\mathcal{B}$. We will see such an example in Theorem 8.3.10.

Lemma 8.3.4 (ii) gives the restriction that the quotient is imprimitive unless $\mathcal{B} = \mathcal{D}$. In that case though, if Γ is a vertex-transitive digraph and $G \leq \mathrm{Aut}(\Gamma)$, then it can be shown that the hypothesis of Theorem 4.2.15 holds and so Γ can be written as a nontrivial wreath product.

Combining the two previous lemmas we have sufficient conditions for a transitive permutation group G to have a pair of orthogonal blocks, which by Lemma 8.3.3 means it has an equivalent representation contained in a direct product of two symmetric groups.

Lemma 8.3.5 *Let $G \leq S_n$ be transitive with block system $\mathcal{B}$ whose blocks are of size k. Suppose*

- $\mathrm{fix}_G(\mathcal{B})$ *is not semiregular,*
- *the induced action of* $\mathrm{fix}_G(\mathcal{B})$ *on $B \in \mathcal{B}$ is faithful for every $B \in \mathcal{B}$, and*
- $\mathrm{fix}_G(\mathcal{B})^B$ *is primitive.*

Then G is permutation equivalent to a subgroup of $\mathcal{S}_{n/k} \times \mathcal{S}_k$ if and only if $\mathrm{fix}_G(\mathcal{B})^B$ *and* $\mathrm{fix}_G(\mathcal{B})^{B'}$ *are permutation equivalent representations of* $\mathrm{fix}_G(\mathcal{B})$ *for every $B, B' \in \mathcal{B}$.*

Proof As $\mathrm{fix}_G(\mathcal{B})$ is not semiregular, Lemma 8.3.4 (ii) holds. As $\mathrm{fix}_G(\mathcal{B})^B$ is primitive but $\mathrm{fix}_G(\mathcal{B})$ is not semiregular and acts faithfully on $B \in \mathcal{B}$, $\mathrm{fix}_G(\mathcal{B})^B$ is not abelian. Let C be as in Lemma 8.3.4. Then $|B \cap C| \leq 1$ for every $B \in \mathcal{B}$ and $C \in C$. As the induced action of $\mathrm{fix}_G(\mathcal{B})$ on $B \in \mathcal{B}$ is faithful, $\mathcal{D}$ has one block. By Lemma 8.3.3, G is permutation equivalent to a subgroup of $\mathcal{S}_{n/k} \times \mathcal{S}_k$ if and only if $\mathcal{B}$ and C are orthogonal block systems, which by Lemma 8.3.4 (ii) is true if and only if $\mathrm{fix}_G(\mathcal{B})^B$ and $\mathrm{fix}_G(\mathcal{B})^{B'}$ are permutation equivalent representations of $\mathrm{fix}_G(\mathcal{B})$ for every $B, B' \in \mathcal{B}$. □

We have now finished the general results for direct products (at least in this section), and next consider the implications of the above result for transitive groups of degree p^2, with p an odd prime.

Lemma 8.3.6 *Let p be an odd prime and $G \le S_{p^2}$ have a normal block system $\mathcal{B}$ with blocks of size p. If $\mathrm{fix}_G(\mathcal{B})$ is faithful on $B \in \mathcal{B}$ but $\mathrm{fix}_G(\mathcal{B})$ is not semiregular, then G is permutation equivalent to a subgroup of $S_p \times S_p$.*

Proof As p is prime, $\mathrm{fix}_G(\mathcal{B})^B$ is primitive by Example 2.4.6. If the permutation representation induced by the action of $\mathrm{fix}_G(\mathcal{B})$ on B is permutation equivalent to the permutation representation induced by the action of $\mathrm{fix}_G(\mathcal{B})$ on B' for every $B, B' \in \mathcal{B}$, then the result follows by Lemma 8.3.5. Assume the action of $\mathrm{fix}_G(\mathcal{B})$ on $B \in \mathcal{B}$ is not equivalent to the action of $\mathrm{fix}_G(\mathcal{B})$ on $B' \in \mathcal{B}$. By Exercise 4.5.6, it follows that $\mathrm{fix}_G(\mathcal{B})^B$ is not permutation isomorphic to a subgroup of $\mathrm{AGL}(1, p)$, $B \in \mathcal{B}$. It then follows by Theorem 4.4.18 that $\mathrm{fix}_G(\mathcal{B})^B$ is a 2-transitive nonabelian almost simple group $B \in \mathcal{B}$.

Define the equivalence relation $\equiv$ on $\mathbb{Z}_{p^2}$ as in Lemma 8.3.4. By Lemma 8.3.4, the equivalence classes of $\equiv$ form a block system C of G, and each block of C contains at most one element of $B \in \mathcal{B}$. Hence the blocks of C have order at most p. As the permutation representation induced by $\mathrm{fix}_G(\mathcal{B})$ in its action on B is not equivalent to the permutation representation induced by the action of $\mathrm{fix}_G(\mathcal{B})$ on B', it follows that the blocks of C have order less than p, and consequently must be singletons. This implies that $\mathrm{fix}_G(\mathcal{B})$ has at least p inequivalent permutation representations of degree p. We have already mentioned several times that one consequence of CFSG is that all 2-transitive groups are known. We have also mentioned that the number of inequivalent permutation representation of each group has been found, and it turns out that there are at most two inequivalent permutation representations for each 2-transitive group with nonabelian socle. This information can be found in Cameron (1981, Theorem 5.3). This is a contradiction as p is odd, establishing the result. □

We now combine the above results with the transfer. This will allow us to handle Lemma 8.3.1 (iii)(a), where the fixer is semiregular. We will need an additional definition first.

Definition 8.3.7 Let G be a group and $H \le G$. Define the **normal closure of H in G**, denoted H^G, to be $\langle ghg^{-1} : h \in H, g \in G\rangle$. The group H^G is the largest normal subgroup of G that contains H (Exercise 8.3.1).

Lemma 8.3.8 *Let p be an odd prime, and $G \le S_{p^2}$ have a block system $\mathcal{B}$ such that $\mathrm{fix}_G(\mathcal{B})$ is semiregular of order p, $G/\mathcal{B}$ is nonsolvable, and G contains a regular abelian subgroup A. Then $H = A^G \trianglelefteq G$ is transitive and H has a block system C orthogonal to $\mathcal{B}$.*

Proof As $\mathcal{B}$ is a block system of G, by the embedding theorem we may assume $G \leq \mathcal{S}_p \wr \mathcal{S}_p$. As $\mathrm{fix}_G(\mathcal{B})$ is semiregular of order p, we may also assume the map $\rho\colon \mathbb{Z}_p \times \mathbb{Z}_p \to \mathbb{Z}_p \times \mathbb{Z}_p$ given by $\rho(i,j) = (i, j+1)$ generates $\mathrm{fix}_G(\mathcal{B})$. Then $G \leq \{(i,j) \mapsto (\sigma(i), \alpha j + b_i) : \sigma \in \mathcal{S}_p, \alpha \in \mathbb{Z}_p^*, b_i \in \mathbb{Z}_p\}$ by Lemma 6.5.17. As $\mathrm{fix}_G(\mathcal{B}) = \langle\rho\rangle$, $\rho \in A$. Then there exists $\tau \in G$ such that $\tau(i,j) = (\delta(i), j + b_i)$ where $\delta \in \mathcal{S}_p$ has order p, and $A = \langle\tau,\rho\rangle$ (note we do not assume $\rho \notin \langle\tau\rangle$). A straightforward computation will show that $A^G = H \leq \left\{(i,j) \mapsto (\sigma(i), j + b_i) : \sigma \in \mathcal{S}_p, b_i \in \mathbb{Z}_p\right\}$ (Exercise 8.3.2). It is then easy to see that ρ commutes with every element of H. Note that this statement is not true in G if some $\alpha \neq 1$, so we consider the normal closure as it has nontrivial center. Observe that as $G/\mathcal{B}$ is nonsolvable, by Theorem 4.4.18 $G/\mathcal{B}$ is a nonabelian almost simple group, in which case $H/\mathcal{B}$ is also a nonabelian almost simple group.

Let $\nu\colon H \to A$ be the transfer. Note that $\mathrm{fix}_H(\mathcal{B})$ is an abelian p-group contained in the center of H. Additionally, as $\mathrm{fix}_H(\mathcal{B})$ is semiregular of order p and $H/\mathcal{B} \leq \mathcal{S}_p$, we see that A is a Sylow p-subgroup of H. Then $\nu(\rho) \neq 1$ by Corollary 8.2.7. Also, as $H/\mathcal{B}$ is a nonabelian group, $K = \mathrm{Ker}(\nu) \neq 1$. The set of orbits of K is a block system C of H. As $\nu(H)$ is nontrivial, a Sylow p-subgroup of K has order p. Then C has blocks of size p that are distinct from the blocks of $\mathcal{B}$, and are necessarily orthogonal to $\mathcal{B}$. □

We now have the tools necessary to determine the automorphism groups of circulant digraphs of order p^2, p a prime. We choose to prove one preliminary result first, as it generalizes Burnside's theorem (Theorem 4.4.2). Note that Burnside's theorem has the following alternative form:

Theorem 8.3.9 *Let $G \leq \mathcal{S}_p$ be transitive. Then a Sylow p-subgroup P of G is regular and cyclic, and either $P \trianglelefteq G$ or G is 2-transitive.*

This form directly generalizes to degree p^2. We also remind the reader that generalizations of Burnside's theorem are discussed at the end of Section 4.4.

Theorem 8.3.10 *Let $G \leq \mathcal{S}_{p^2}$ contain a regular cyclic Sylow p-subgroup P. Then either $P \trianglelefteq G$ or G is 2-transitive.*

Proof Let $G \leq \mathcal{S}_{p^2}$ have a regular cyclic Sylow p-subgroup P. If $p = 2$, then the result follows by inspection of the subgroups of $\mathcal{S}_4$. So we assume p is odd. As, by Theorem 5.3.17 $\mathbb{Z}_{p^2}$ is a Burnside group, G is either 2-transitive or imprimitive. So we assume G is imprimitive. As P is regular, $\mathrm{fix}_G(\mathcal{B})$ acts faithfully on $B \in \mathcal{B}$. (We have seen the justification for this before – if not, then as $\mathrm{fix}_G(\mathcal{B})^B$ is primitive, the kernel K of the action will be transitive on some block of $\mathcal{B}$. This gives a larger Sylow p-subgroup as then p divides $|K|$ if

$K \neq 1$.) Also, $\mathbb{Z}_{p^2}$ has a unique subgroup of order p, and so there is no block system with blocks of size p orthogonal to $\mathcal{B}$. Then $\text{fix}_G(\mathcal{B})$ is semiregular by Lemma 8.3.6. Also, by Lemma 8.3.8 we have $G/\mathcal{B}$ is solvable. Then $P \trianglelefteq G$. □

As mentioned earlier, the case where the fixer is semiregular is often a hard case. The above result can be generalized (Dobson, 2005), but not to cover every situation. So this is one situation where we will use the hypothesis that the digraph has order p^2.

Theorem 8.3.11 *Let Γ be a circulant digraph of order p^2, p a prime. Then one of the following is true:*

(i) $\Gamma = K_{p^2}$ *or its complement and* $\text{Aut}(\Gamma) = S_{p^2}$,
(ii) $\Gamma = \Gamma_1 \wr \Gamma_2$ *and* $\text{Aut}(\Gamma) = \text{Aut}(\Gamma_1) \wr \text{Aut}(\Gamma_2)$, *where,* Γ_1 *and* Γ_2 *are circulant digraphs of order* p, *or*
(iii) Γ *is a normal Cayley digraph of* $\mathbb{Z}_{p^2}$ *and* $\left(\mathbb{Z}_{p^2}\right)_L \trianglelefteq \text{Aut}(\Gamma)$.

Proof If $\text{Aut}(\Gamma)$ is 2-transitive, then Γ is either K_{p^2} or its complement and (i) follows. Otherwise, let $\rho\colon \mathbb{Z}_{p^2} \to \mathbb{Z}_{p^2}$ by $\rho(i) = i + 1 \pmod{p^2}$, so that $\langle\rho\rangle = \left(\mathbb{Z}_{p^2}\right)_L$ is a regular subgroup of $\text{Aut}(\Gamma)$ of order p^2. By Theorem 5.3.17 $\text{Aut}(\Gamma)$ has a block system $\mathcal{B}$ with p blocks of size p that are the set of orbits of the unique subgroup $\langle\rho^p\rangle$ of order p of $\langle\rho\rangle$. If $\text{fix}_G(\mathcal{B})$ is not faithful on $B \in \mathcal{B}$, then by Lemma 8.3.1 there exist circulant digraphs Γ_1 and Γ_2 of order p such that $\text{Aut}(\Gamma) = \text{Aut}(\Gamma_1) \wr \text{Aut}(\Gamma_2)$ and (ii) follows by Theorem 5.1.8. If $\text{fix}_G(\mathcal{B})$ is faithful on $B \in \mathcal{B}$ then a Sylow p-subgroup of $\text{fix}_G(\mathcal{B})$ has order p, in which case $\langle\rho\rangle$ is a Sylow p-subgroup of G. Then (iii) follows by Theorem 8.3.10. □

We now give an algorithm for computing $\text{Aut}\left(\text{Cay}\left(\mathbb{Z}_{p^2}, S\right)\right)$, where p is a prime and $S \subset \mathbb{Z}_{p^2}$. Note that in this algorithm, other than for recognizing a complete graph or its complement, we really only need to know $\text{Aut}\left(\mathbb{Z}_{p^2}, S\right)$. We begin with a preliminary result that may itself be interesting.

Lemma 8.3.12 *Let p be a prime and $S \subset \mathbb{Z}_{p^2}$. Then $\Gamma = \text{Cay}(G, S)$ is isomorphic to a nontrivial wreath product if and only if $(1 + p)S = S$.*

Proof If $(1 + p)S = S$, then the map $x \mapsto (1 + p)x$ is in $\text{Aut}(\Gamma)$ by Corollary 1.2.16. Then a Sylow p-subgroup of $\text{Aut}(\Gamma)$ has order at least p^3 as $1 + p$ generates the unique subgroup of $\mathbb{Z}^*_{p^2}$ of order p (see Exercise 8.3.3). Applying Lemma 5.3.11 it follows that Γ is a nontrivial wreath product.

Conversely, if Γ is a nontrivial wreath product, it can be written $\Gamma \cong \Gamma_1 \wr \Gamma_2$, where Γ_1 and Γ_2 are circulant digraphs of order p. Then $\mathbb{Z}_p \wr \mathbb{Z}_p \leq \text{Aut}(\Gamma)$ by Lemma 4.2.14, and, by Example 4.2.8, $\text{Aut}(\Gamma)$ contains a Sylow p-subgroup P of S_{p^2}. We may assume $\left(\mathbb{Z}_{p^2}\right)_L \leq P$. By a Sylow theorem $N_P\left(\left(\mathbb{Z}_{p^2}\right)_L\right) \neq \left(\mathbb{Z}_{p^2}\right)_L$, and so P contains a subgroup N of order at least p^3 that normalizes $\left(\mathbb{Z}_{p^2}\right)_L$. As a Sylow p-subgroup of $N_{S_{p^2}}\left(\left(\mathbb{Z}_{p^2}\right)_L\right)$ has order p^3 and is normal in $N_{S_{p^2}}\left(\left(\mathbb{Z}_{p^2}\right)_L\right)$ the map $x \mapsto (1+p)x$ is contained in $\mathrm{Aut}(\Gamma)$, and so $(1+p)S = S$ by Corollary 1.2.16. □

Algorithm 8.3.13 Let $\Gamma = \mathrm{Cay}\left(\mathbb{Z}_{p^2}, S\right)$ for some $S \subset \mathbb{Z}_p$. The following algorithm will compute $\mathrm{Aut}(\Gamma)$:

- If $S = \mathbb{Z}_{p^2}\setminus\{0\}$ or $\emptyset$, then $\mathrm{Aut}(\Gamma) = \mathcal{S}_{p^2}$. Otherwise,
- if $(1+p)S = S$, then let $S_1 = \{a : 1 \leq a \leq p-1 \text{ and } a + \langle p \rangle \subseteq S\}$ and $S_2 = \{j : jp \in S\}$. Then $\Gamma \cong \mathrm{Cay}\left(\mathbb{Z}_p, S_1\right) \wr \mathrm{Cay}\left(\mathbb{Z}_p, S_2\right)$ and viewing each element of $\mathbb{Z}_{p^2}$ uniquely as $i + jp$ where $i, j \in \mathbb{Z}_p$, $\mathrm{Aut}(\Gamma) = \left\{i + jp \mapsto \sigma(i) + \delta_i(j)p : \sigma \in \mathrm{Aut}\left(\mathrm{Cay}\left(\mathbb{Z}_p, S_1\right)\right), \delta_i \in \mathrm{Aut}\left(\mathrm{Cay}\left(\mathbb{Z}_p, S_2\right)\right)\right\}$,
- if $(1+p)S \neq S$, then determine $\mathrm{Aut}\left(\mathbb{Z}_{p^2}, S\right)$, which is a p'-subgroup of $\mathbb{Z}_{p^2}^*$. Then $\mathrm{Aut}(\Gamma) = \left\{x \mapsto ax + b : a \in \mathrm{Aut}\left(\mathbb{Z}_{p^2}, S\right), b \in \mathbb{Z}_{p^2}\right\}$.

Proof Of course if $S = \mathbb{Z}_{p^2}\setminus\{0\}$ or $\emptyset$ then $\Gamma = K_{p^2}$ or its complement, respectively, and $\mathrm{Aut}(\Gamma) = \mathcal{S}_{p^2}$.

If $(1+p)S = S$, then by Lemma 8.3.12 we have Γ can be written as a nontrivial wreath product, so $\mathrm{Aut}(\Gamma)$ is imprimitive. Let $\mathcal{B}$ be the block system of $\left(\mathbb{Z}_{p^2}\right)_L$ that is the set of cosets of $B = \langle p \rangle$ in $\mathbb{Z}_{p^2}$. As it is the only nontrivial block system of $\left(\mathbb{Z}_{p^2}\right)_L$, it is a block system of $\mathrm{Aut}(\Gamma)$. By Theorem 5.3.6 (ii) and (iii), we see that $\mathrm{Cay}\left(\mathbb{Z}_{p^2}, S\setminus B\right) \cong \mathrm{Cay}\left(\mathbb{Z}_{p^2}/B, \{a + B \subseteq S\}\right) \wr \mathrm{Cay}(B, \emptyset)$ and $\mathrm{Cay}\left(\mathbb{Z}_{p^2}, S \cap B\right) \cong \mathrm{Cay}\left(\mathbb{Z}_{p^2}/B, \emptyset\right) \wr \mathrm{Cay}(B, S \cap B)$. Clearly $\mathrm{Cay}\left(\mathbb{Z}_{p^2}/B, \{a + B \subseteq S\}\right) \cong \mathrm{Cay}\left(\mathbb{Z}_p, S_1\right)$ and $\mathrm{Cay}(B, S \cap B) \cong \mathrm{Cay}\left(\mathbb{Z}_p, S_2\right)$. As $G = \mathrm{Aut}\left(\mathrm{Cay}\left(\mathbb{Z}_p, S_1\right) \wr \mathrm{Cay}\left(\mathbb{Z}_p, S_2\right)\right) = \mathrm{Aut}\left(\mathrm{Cay}\left(\mathbb{Z}_p, S_1\right)\right) \wr \mathrm{Aut}\left(\mathrm{Cay}\left(\mathbb{Z}_p, S_2\right)\right)$ and the lexi-partition of G corresponding to $V\left(\mathrm{Cay}\left(\mathbb{Z}_p, S_2\right)\right)$ is $\left\{\{i\} \times \mathbb{Z}_p : i \in \mathbb{Z}_p\right\}$, the map from $\mathbb{Z}_p^2$ to $\mathbb{Z}_{p^2}$ given by $(i, j) \mapsto i + jp$, is an isomorphism between $\mathrm{Cay}\left(\mathbb{Z}_p, S_1\right) \wr \mathrm{Cay}\left(\mathbb{Z}_p, S_2\right)$ and Γ. Hence $\mathrm{Aut}(\Gamma) = \left\{i + jp \mapsto \sigma(i) + \delta_i(j)p : \sigma \in \mathrm{Aut}\left(\mathrm{Cay}\left(\mathbb{Z}_p, S_1\right)\right), \delta_i \in \mathrm{Aut}\left(\mathrm{Cay}\left(\mathbb{Z}_p, S_2\right)\right)\right\}$.

If $(1+p)S \neq S$, then, by Lemma 8.3.12, Γ is not isomorphic to a nontrivial wreath product. By Lemma 8.1.7, we have $\mathrm{Aut}(\Gamma) = \left\{x \mapsto ax + b : a \in \mathrm{Aut}\left(\mathbb{Z}_{p^2}, S\right), b \in \mathbb{Z}_{p^2}\right\}$. As $x \mapsto (1+p)x$ generates the unique Sylow p-subgroup of $\left\{x \mapsto ax : a \in \mathbb{Z}_{p^2}^*\right\}$ and is not in $\mathrm{Aut}\left(\mathbb{Z}_{p^2}, S\right)$, $\mathrm{Aut}\left(\mathbb{Z}_{p^2}, S\right)$ is a p'-group. □

Recall that an algorithm for calculating the automorphism group of a circulant digraph of prime order is given in Algorithm 4.4.6.

Example 8.3.14 Let $\Gamma = \mathrm{Cay}(\mathbb{Z}_9, \{1, 3, 4, 6, 7\})$. Then $\mathrm{Aut}(\Gamma) \cong \mathbb{Z}_3 \wr \mathcal{S}_3$.

Solution Clearly Γ is neither complete nor the complement of a complete graph. Here, $1 + p = 4$ and $4 \cdot \{1, 3, 4, 6, 7\} = \{4, 3, 7, 6, 1\}$, so Γ is a wreath product. Also, $1 + \langle 3 \rangle$ is the only coset of $\langle 3 \rangle$ contained in S, so we have $\Gamma \cong \mathrm{Cay}(\mathbb{Z}_3, \{1\}) \wr \mathrm{Cay}(\mathbb{Z}_3, \{1, 2\})$. As $\mathrm{Cay}(\mathbb{Z}_3, \{1\})$ is a directed cycle with automorphism group $(\mathbb{Z}_3)_L$ and $\mathrm{Cay}(\mathbb{Z}_3, \{1, 2\}) = K_3$ has automorphism group S_3, we see that $\mathrm{Aut}(\Gamma) = \{i + 3j \mapsto i + a + 3\sigma_i(j) : a \in \mathbb{Z}_3, \sigma_i \in S_3\}$. Of course, it is also the case that $\mathrm{Aut}(\Gamma) \cong \mathbb{Z}_3 \wr S_3$. □

Example 8.3.15 Let $S = \{1, 5, 7, 10, 15, 18, 20, 24\}$, and $\Gamma = \mathrm{Cay}(\mathbb{Z}_{25}, S)$. Then $\mathrm{Aut}(\Gamma) = \{x \mapsto ax + b : a = 1, 7, 18, 24, b \in \mathbb{Z}_{25}\}$.

Solution Clearly Γ is neither complete nor the complement of a complete graph. Here, $1 + p = 6$, and $6 \cdot S \neq S$ as $1 \in S$ but $6 \notin S$. Thus Γ is not a nontrivial wreath product, and we only need check which $a \in \mathbb{Z}_{25}^*$ of order dividing 4 map S to S. As $1 \in S$ and $1, 7, 18, 24$ are the only units in S, we only need to check which of $a = 1, 7, 18, 24$ fix S. As $\{1, 7, 18, 24\} = \langle 7 \rangle$ (multiplicatively), checking $7 \cdot S = S$ gives $\mathrm{Aut}(\Gamma) = \{x \mapsto ax + b : a = 1, 7, 18, 24, b \in \mathbb{Z}_{25}\}$. Note that $\mathrm{Aut}(\Gamma)/\mathcal{B} = \mathrm{AGL}(1, 5)$ is 2-transitive, and that $\Gamma[H] = K_5$, where $H = \langle 5 \rangle$ and $\mathcal{B}$ is the set of cosets of H in $(\mathbb{Z}_{25})_L$. □

8.4 Automorphism Groups of Cayley Digraphs of $\mathbb{Z}_p^2$

We will first need some additional results before turning to the automorphism groups of Cayley digraphs of $\mathbb{Z}_p^2$, first determined in Dobson and Witte (2002). In particular, in Lemma 8.4.2 we give a sufficient condition to recognize that a permutation group is naturally a direct product with a factor that is a symmetric group. We then determine the groups that are the automorphism groups of Cayley digraphs of $\mathbb{Z}_p^2$ in Theorem 8.4.3, and then find an algorithm to compute the automorphism group of a Cayley digraph of $\mathbb{Z}_p^2$ in Algorithm 8.4.4. So some of the work in this section is a continuation of the results in the previous section regarding permutation groups that can be written as a natural direct product.

Lemma 8.4.1 *Let $m, k \geq 2$ be integers, $H \leq S_m$ be transitive, $K \leq S_k$ be 2-transitive, and $G = H \times K$ with its canonical action on $\mathbb{Z}_m \times \mathbb{Z}_k$. Let $\mathcal{B}$ be the block system of G that is the set of orbits of $L = 1_{S_m} \times K \trianglelefteq G$. The suborbits of* $\mathrm{Stab}_L(0, 0)$ *contained in $\{i\} \times \mathbb{Z}_k$ are $\{(i, 0)\}$ and $\{(i, j) : j \in \mathbb{Z}_k, j \neq 0\}$ for every $i \in \mathbb{Z}_m$.*

Proof Let $i \in \mathbb{Z}_m$. As $G = H \times K$ with its canonical action on $\mathbb{Z}_m \times \mathbb{Z}_k$, we have $\mathrm{Stab}_L(0, 0) = \mathrm{Stab}_L(i, 0)$, and, by Lemma 4.4.3, $\mathrm{Stab}_L(0, 0)^{B_i}$ is transitive on $B_i \backslash \{(i,0)\}$ as K is 2-transitive, where $B_i \in \mathcal{B}$ with $(i,0) \in B_i$. Thus $\{(i,0)\}$ is a suborbit of L in $\{i\} \times \mathbb{Z}_k$, and $\{(i,j') : j' \in \mathbb{Z}_k, j' \neq 0\}$ is also suborbit of $\{i\} \times \mathbb{Z}_k$. The result follows as every element in $\mathbb{Z}_m \times \mathbb{Z}_k$ is in a suborbit we have found. □

Lemma 8.4.2 *Let $k, m \geq 2$ be integers, $G \leq S_m \times S_k$ with the canonical action be transitive, and $\pi\colon G \to S_m$ be the projection in the first coordinate. Set $H = \pi(G)$, and let $\mathcal{B}$ be the block system of G with $\mathcal{B} = \{\{i\} \times \mathbb{Z}_k : i \in \mathbb{Z}_m\}$. If* $\mathrm{Ker}(\pi)^B$ *is 2-transitive for some (equivalently, all) $B \in \mathcal{B}$, then $G^{(2)} = H^{(2)} \times S_k$.*

Proof Let K be the projection of G in the second coordinate. As $G \leq H \times K$, we have $G^{(2)} \leq H^{(2)} \times K^{(2)}$ by Theorem 5.2.11 (ii). If $1_{S_m} \times S_k \leq G^{(2)}$, then the result will follow as then $H \times S_k \leq G^{(2)}$ and so $G^{(2)} = H^{(2)} \times S_k$. Also, $1_{S_m} \times S_k \leq G^{(2)}$ if and only if $\mathrm{Stab}_{G^{(2)}}(0,0) \geq 1_{S_m} \times \mathrm{Stab}_{S_k}(0)$. This is what we will show.

Let Γ be an orbital digraph of G, with $(i,x) \in N_\Gamma(0,0)$, the neighbors of $(0,0)$ in Γ. Let $\ell \in 1_{S_m} \times \mathrm{Stab}_{S_k}(0)$. We wish to show that $\ell(i,x) \in N_\Gamma(0,0)$. Let $B = \{i\} \times \mathbb{Z}_k \in \mathcal{B}$. By Lemma 8.4.1, $\mathrm{Stab}_K(0,0)^B$ has two suborbits $\{(i,0)\}$ and $\{(i,j) : j \in \mathbb{Z}_k, j \neq 0\}$. Also, $N_i = N_\Gamma(0,0) \cap B$ is a union of some set of suborbits of $\mathrm{Stab}_K(0,0)^B$. If $N_i = \{(i,0)\}$ then $x = 0$ and $\ell(i,0) = (i,0)$. Hence $\ell(i,x) \in N_\Gamma(0,0)$. If $N_i = \{(i,j) : j \in \mathbb{Z}_k, j \neq 0\}$, then $x \neq 0$ and as $\ell(i,0) = (i,0)$, $\ell(i,x) \neq (i,0)$. Hence $\ell(i,x) \in N_\Gamma(0,0)$. The only remaining case (as $N_i \neq \emptyset$) is that $N_i = B$, in which case $\ell(i,x) \in N_i$ trivially. □

We now determine the automorphism groups of Cayley digraphs of $\mathbb{Z}_p^2$.

Theorem 8.4.3 *Let Γ be a Cayley digraph of $\mathbb{Z}_p^2$, where p is prime. Then there exists $\alpha \in \mathrm{Aut}\left(\mathbb{Z}_p^2\right)$ such that one of the following is true:*

(i) *Γ is the complete graph or its complement and $\mathrm{Aut}(\Gamma) = S_{p^2}$,*
(ii) *Γ is a normal Cayley digraph of $\mathbb{Z}_p^2$ and $\mathrm{Aut}(\Gamma) < \mathrm{AGL}(2,p)$,*
(iii) *Γ is isomorphic to a nontrivial wreath product of two circulant digraphs of order p, and $\alpha\mathrm{Aut}(\Gamma)\alpha^{-1} = \mathrm{Aut}(\Gamma_1) \wr \mathrm{Aut}(\Gamma_2)$, for some circulant digraphs Γ_1 and Γ_2 of order p*
(iv) *Γ is isomorphic to the Hamming graph $H(2,p)$ or its complement, and $\alpha\mathrm{Aut}(\Gamma)\alpha^{-1} = S_2 \wr S_p$ is primitive, or*
(v) *$\alpha\mathrm{Aut}(\Gamma)\alpha^{-1} = A \times S_p$, where $A < \mathrm{AGL}(1,p)$ or $A = S_p$.*

Proof Set $G = \mathrm{Aut}(\Gamma)$. If G is 2-transitive then, of course, Γ is a complete graph or its complement, and $G = S_{p^2}$. This is (i). If G is simply primitive, then by the O'Nan–Scott theorem and order arguments (see Exercise 8.4.1) $G \leq \mathrm{AGL}(2,p)$, $G \leq S_2 \wr T$, where T is a 2-transitive nonabelian simple group of degree p, or G is a nonabelian almost simple group.

If $G \leq \mathrm{AGL}(2,p)$, then $G < \mathrm{AGL}(2,p)$ as $\mathrm{AGL}(2,p)$ is itself 2-transitive by Exercise 2.4.9 and (ii) holds. If G is simply primitive, then by Theorem 4.1.13

there is a nonabelian simple group T that contains a subgroup of index p^2. Using CFSG, Guralnick (1983) has determined all nonabelian simple groups with a subgroup of prime power index. None of them have index p^2. Otherwise, $T \times T \leq \mathcal{S}_2 \wr T$, and $T \times T$ is permutation isomorphic to a subgroup of $\mathcal{S}_p \times \mathcal{S}_p$. By Lemma 8.4.2, G contains a subgroup permutation isomorphic to $\mathcal{S}_p \times \mathcal{S}_p$, or equivalently, that $T = A_p$.

Now let $H \leq G$ be permutation isomorphic to $\mathcal{S}_p \times \mathcal{S}_p$. The reason we cannot simply conclude that $H = \mathcal{S}_p \times \mathcal{S}_p$ is that the blocks of H may not be the same as the blocks of $\mathcal{S}_p \times \mathcal{S}_p$. By Theorem 2.3.7, $\left(\mathbb{Z}_p^2\right)_L$ has exactly $p+1$ block systems with blocks of size p, one for each subgroup of order p. One can also think of the subgroups of order p as being lines through the origin in $\mathbb{F}_p^2$, and the block systems are these lines together with $p-1$ translates of the line through the origin. The lines through the origin in $\mathbb{F}_p^2$ can also be thought of as the subspaces of dimension 1 in the vector space $\mathbb{F}_p^2$. The block systems with blocks of size p of H can be any two different block systems of $\left(\mathbb{Z}_p^2\right)_L$ chosen from the $p+1$ such block systems, and any two different block systems are orthogonal. Let $\mathcal{B}$ and C be orthogonal block systems of H with blocks of size p. Then there exists $\alpha \in \mathrm{Aut}\left(\mathbb{Z}_p^2\right) = \mathrm{GL}(2,p)$ such that $\alpha(\mathcal{B}) = \left\{(0,i) : i \in \mathbb{Z}_p\right\}$ and $\alpha(C) = \left\{(i,0) : i \in \mathbb{Z}_p\right\}$. Then $\mathcal{S}_p \times \mathcal{S}_p = \alpha H \alpha^{-1}$.

It is not hard to see, using Lemma 8.4.1, that $\mathcal{S}_p \times \mathcal{S}_p$ has four suborbits, namely $\{(0,0)\}$, $\left\{(0,j) : j \in \mathbb{Z}_p^*\right\}$, $\left\{(j,0) : j \in \mathbb{Z}_p^*\right\}$, and $O_2 = \left\{(i,j) : i,j \in \mathbb{Z}_p^*\right\}$. Define $\tau\colon \mathbb{Z}_p \times \mathbb{Z}_p \to \mathbb{Z}_p \times \mathbb{Z}_p$ by $\tau(i,j) = (j,i)$. Then τ maps $\left\{(0,j) : j \in \mathbb{Z}_p^*\right\}$ to $\left\{(j,0) : j \in \mathbb{Z}_p^*\right\}$ and fixes the other two suborbits of $\mathcal{S}_p \times \mathcal{S}_p$. Thus $\mathcal{S}_2 \wr \mathcal{S}_p = \left\langle \tau, \mathcal{S}_p \times \mathcal{S}_p \right\rangle$ has three suborbits, namely, $\{(0,0)\}$, $O_1 = \{(i,j) : i = 0 \text{ or } j = 0 \text{ but } (i,j) \neq (0,0)\}$, and O_2. As $\mathrm{Cay}\left(\mathbb{Z}_p^2, O_1\right) = H(2,p)$ and $\mathrm{Cay}\left(\mathbb{Z}_p^2, O_2\right)$ is its complement, (iv) follows.

Finally, we will only sketch the imprimitive case as, with one exception, it is similar to arguments we have already seen. The remaining details are left as Exercise 8.4.2. First, if $\mathrm{fix}_G(\mathcal{B})$ does not act faithfully on some $B \in \mathcal{B}$ then Lemma 8.3.1 and an argument similar to one above gives (iii). Second, if $\mathrm{fix}_G(\mathcal{B})$ is faithful but is not semiregular, then Lemma 8.3.6, Lemma 8.4.2, and arguments as above give (v). Finally, if $\mathrm{fix}_G(\mathcal{B})$ is faithful and semiregular, then $\left(\mathbb{Z}_p^2\right)_L \trianglelefteq G$ if $G/\mathcal{B}$ is solvable, and otherwise Lemma 8.3.8 reduces this case to one above. □

Recognition of wreath products by checking an appropriate automorphism of $\mathbb{Z}_p^2$ (as in Lemma 8.3.12 for $\mathbb{Z}_{p^2}$) is more difficult. This is because $\left(\mathbb{Z}_{p^2}\right)_L$ only has one block system with blocks of size p, while $\left(\mathbb{Z}_p^2\right)_L$ has $p+1$ (this is the same as the number of projective points in $\mathbb{F}_p^2$). For the block system

$\mathcal{B} = \left\{(i, j) : j \in \mathbb{Z}_p\} : i \in \mathbb{Z}_p\right\}$, the natural map to check is $(i, j) \mapsto (i, j + i)$ – it is natural as it fixes $\mathcal{B}$, has order p, and normalizes $\left(\mathbb{Z}_p^2\right)_L$ (Exercise 8.4.6). There are then p conjugates of this map that would also have to be checked. We choose not to write this down explicitly. Instead, we will recognize them from the order of $\mathrm{Aut}\left(\mathbb{Z}_p^2, S\right)$, which has to be found anyway.

We now give an algorithm to calculate the full automorphism group of Cayley digraphs of $\mathbb{Z}_p^2$.

Algorithm 8.4.4 Let p be prime, $S \subset \mathbb{Z}_p^2$, and $\Gamma = \mathrm{Cay}\left(\mathbb{Z}_p^2, S\right)$. The following algorithm will compute $\mathrm{Aut}(\Gamma)$:

(i) If $S = \mathbb{Z}_p^2\backslash\{(0,0)\}$ or $\emptyset$, then $\Gamma \cong K_{p^2}$ or its complement, and $\mathrm{Aut}(\Gamma) = \mathcal{S}_{\mathbb{Z}_p^2}$. Otherwise,

(ii) calculate $\mathrm{Aut}\left(\mathbb{Z}_p^2, S\right) = \{A \in \mathrm{GL}(2, p) : A(S) = S\}$. Then find all subgroups $K_1, \ldots, K_r$ of $\mathbb{Z}_p^2$ of order p invariant under $\mathrm{Aut}\left(\mathbb{Z}_p^2, S\right)$.

  (a) If K_1 does not exist (so no K_i exists), then $\mathrm{Aut}(\Gamma)$ is primitive. Check if $\alpha(\Gamma)$ is the Hamming graph $H(2, p)$ or its complement for some $\alpha \in \mathrm{GL}(2, p)$. If so, then $\mathrm{Aut}(\Gamma) = \alpha^{-1}\mathcal{S}_2 \wr \mathcal{S}_p\alpha$. Otherwise, $\mathrm{Aut}(\Gamma) = \left(\mathbb{Z}_p^2\right)_L \rtimes \mathrm{Aut}\left(\mathbb{Z}_p^2, S\right)$.

  (b) If K_1 exists, then the set of orbits of K_1 is a block system $\mathcal{B}$ of $\mathrm{Aut}(\Gamma)$, and there is $\alpha \in \mathrm{GL}(2, p)$ such that $\alpha(\mathcal{B}) = \left\{\left\{(i, j) : j \in \mathbb{Z}_p\right\} : i \in \mathbb{Z}_p\right\}$.

    (1) If there exists $\beta \in \mathrm{Aut}\left(\mathbb{Z}_p^2, S\right)$ of order p, then Γ is a nontrivial wreath product. A Sylow p-subgroup P of $\left(\mathbb{Z}_p^2\right)_L \rtimes \mathrm{Aut}\left(\mathbb{Z}_p^2, S\right)$ that contains β has order p^3, and P has a unique block system with blocks of size p. Let $S_1 = \left\{i : (i, j) \in \alpha(S) \text{ for every } j \in \mathbb{Z}_p\right\}$ and $S_2 = \{j : (0, j) \in \alpha(S)\}$. Then $\alpha(\Gamma) = \mathrm{Cay}\left(\mathbb{Z}_p, S_1\right) \wr \mathrm{Cay}\left(\mathbb{Z}_p, S_2\right)$ and $\mathrm{Aut}(\Gamma) = \alpha^{-1}\mathrm{Cay}\left(\mathbb{Z}_p, S_1\right) \wr \mathrm{Cay}\left(\mathbb{Z}_p, S_2\right)\alpha$.

    (2) If no $\beta \in \mathrm{Aut}\left(\mathbb{Z}_p^2, S\right)$ of order p exists, then calculate $F(K_i) = \left\{\delta \in \mathrm{Aut}\left(\mathbb{Z}_p^2, S\right) : \delta(O) = O \text{ for every orbit } O \text{ of } K_i\right\}$ for every $1 \leq i \leq r$.

      (A) If $|F(K_i)| \neq p - 1$ for any $1 \leq i \leq r$, then $\mathrm{Aut}(\Gamma) = \left(\mathbb{Z}_p^2\right)_L \rtimes \mathrm{Aut}\left(\mathbb{Z}_p^2, S\right)$.

      (B) If $|F(K_i)| = p - 1$ for some $1 \leq i \leq r$, then $r = 2$. Let C be the set of orbits of K_2. Choose α so that in addition to $\alpha(\mathcal{B}) = \left\{\left\{(i, j) : j \in \mathbb{Z}_p\right\} : i \in \mathbb{Z}_p\right\}$, we also have $\alpha(C) = \left\{\left\{(i, j) : i \in \mathbb{Z}_p\right\} : j \in \mathbb{Z}_p\right\}$. If $|F(K_2)| \neq p - 1$, then $\mathrm{Aut}(\Gamma) = \alpha^{-1}A \times \mathcal{S}_p\alpha$, where $A = F(K_2) < \mathrm{AGL}(1, p)$. Otherwise, $\mathrm{Aut}(\Gamma) = \alpha^{-1}\mathcal{S}_p \times \mathcal{S}_p\alpha$.

Proof Let $G = \mathrm{Aut}(\Gamma)$, $H = \mathrm{Aut}\left(\mathbb{Z}_p^2, S\right)$, and $N = \left(\mathbb{Z}_p^2\right)_L \rtimes \mathrm{Aut}\left(\mathbb{Z}_p^2, S\right)$. If $S = \mathbb{Z}_p^2 \setminus \{(0,0)\}$ or $\emptyset$, then Γ is a complete graph or its complement, in which case $G = \mathcal{S}_{\mathbb{Z}_p^2}$. This is (i).

If no K_i exists, then $N \leq \mathrm{Aut}(\Gamma)$ has no block systems with blocks of size p by Theorem 2.2.19. Then G is primitive, and so, by Theorem 8.4.3, either $G \leq \mathrm{AGL}(2,p)$ or $G \cong \alpha^{-1}\mathcal{S}_2 \wr \mathcal{S}_p\alpha$ for some $\alpha \in \mathrm{Aut}\left(\mathbb{Z}_p^2\right) = \mathrm{GL}(2,p)$. There are only two digraphs with automorphism group $\mathcal{S}_2 \wr \mathcal{S}_p$ (this was shown in Theorem 8.4.3). One of these is the Hamming graph $H(2,p)$, by Theorem 5.4.4, while the other is its complement, as $H(2,p)$ is not self-complementary by Exercise 5.4.5. If Γ is isomorphic to either of these graphs, then $\mathrm{Aut}(\Gamma) = \alpha^{-1}\mathcal{S}_2 \wr \mathcal{S}_p\alpha$. Otherwise, $G \leq \mathrm{AGL}(2,p)$ and $G = N$. This is (ii)(a).

Now suppose K_1 exists. The set of orbits of K_1 is a block system $\mathcal{B}$ of G. So G is not primitive. Let $\alpha \in \mathrm{GL}(2,p)$ such that $\alpha(\mathcal{B}) = \left\{\left\{(i,j) : j \in \mathbb{Z}_p\right\} : i \in \mathbb{Z}_p\right\}$.

If there exists $\beta \in H$ of order p such that $\beta(K_i) = K_i$ for some $1 \leq i \leq r$, then a Sylow p-subgroup of N that contains β has order p^3 as p^4 does not divide the order of $\mathrm{AGL}(2,p) \geq N$, and β has a fixed point. Then N has a block system $\mathcal{B}$ with blocks of size p that is the set of orbits of K_i, and $\mathrm{fix}_N(\mathcal{B})$ has order divisible by p^2. By Corollary 5.3.15 Γ is isomorphic to a nontrivial wreath product. Let $S_1 = \left\{i : (i,j) \in \alpha(S) \text{ for every } j \in \mathbb{Z}_p\right\}$ and $S_2 = \{j : (0,j) \in \alpha(S)\}$. Arguments as in the proof of Algorithm 8.3.13 (see Exercise 8.4.3) give that $\alpha(\Gamma) = \mathrm{Cay}\left(\mathbb{Z}_p, S_1\right) \wr \mathrm{Cay}\left(\mathbb{Z}_p, S_2\right)$ and $\mathrm{Aut}(\Gamma) = \alpha^{-1}\mathrm{Cay}\left(\mathbb{Z}_p, S_1\right) \wr \mathrm{Cay}\left(\mathbb{Z}_p, S_2\right)\alpha$. The uniqueness of $\mathcal{B}$ follows from Theorem 4.3.7 (this gives $r = 1$ as well). This is (ii)(b)(1).

If no such β exists and Γ can be written as a nontrivial wreath product, then a Sylow p-subgroup P of G has order p^{p+1} by Lemma 4.2.14 and Example 4.2.8. Also, $\mathcal{B}$ is unique by Theorem 4.3.7. Then $N_P\left(\left(\mathbb{Z}_p^2\right)_L\right) \neq \left(\mathbb{Z}_p^2\right)_L$, and so $\mathrm{fix}_H(\mathcal{B})$ has a Sylow p-subgroup of order p^2. Then there is $\beta \in N$ of order p, a contradiction. By Theorem 8.4.3, the only remaining possibilities for G are $G < \mathrm{AGL}(2,p)$ or G is a direct product with a factor isomorphic to $\mathcal{S}_p$. Note that if $p = 3$, then $G \leq \mathrm{AGL}(2,p)$ as $\mathcal{S}_3 = \mathrm{AGL}(1,3)$. So we may additionally assume $p \geq 5$, and so A_p is simple. Under these hypotheses, we next show that G is a direct product with a factor isomorphic to $\mathcal{S}_p$ if and only if $|F(K_i)| = p-1$ for some $1 \leq i \leq r$.

If G is a direct product with a factor isomorphic to $\mathcal{S}_p$, then G contains a transitive subgroup isomorphic to $\left(\mathbb{Z}_p\right)_L \times \mathrm{AGL}(1,p)$, in which case $|F(K_i)| = p-1$ for some i. Conversely, if some $F(K_i)$ has order $p-1$, then let $\mathcal{D}$ be the set of orbits of K_i. Set $M = \left\langle\left(\mathbb{Z}_p^2\right)_L, F(K_i)\right\rangle$. Then $\mathcal{D}$ is a block system of M. Also $\mathrm{fix}_M(\mathcal{D})$ in its action on $D \in \mathcal{D}$ is permutation isomorphic to $\mathrm{AGL}(1,p)$, which is 2-transitive. As a Sylow p-subgroup of M has order p^2, $\mathrm{fix}_M(\mathcal{D})$ is faithful in its action on $D \in \mathcal{D}$. By Exercise 4.5.6, the permutation representation induced by the action of $\mathrm{fix}_M(\mathcal{D})$ on $D \in \mathcal{D}$ is permutation equivalent to the permutation representation induced by the action of $\mathrm{fix}_M(\mathcal{D})$ on $D' \in \mathcal{D}$. Then $\mathrm{fix}_{M^{(2)}}(\mathcal{D}) \cong \mathcal{S}_p$ by Lemma 8.4.2. Thus G is a direct product with a factor isomorphic to $\mathcal{S}_p$ as G is a direct product and $M^{(2)} \leq G$.

If $|F(K_i)| < p - 1$ for every $1 \leq i \leq r$, we see that $G < \mathrm{AGL}(2, p)$. This is (ii)(b)(2).

If $|F(K_i)| = p - 1$ for some $1 \leq i \leq r$, then G is a direct product with some factor isomorphic to $\mathcal{S}_p$. Suppose G has three distinct nontrivial block systems. Then Theorem 8.4.3 (v) holds. As every block system of G is normal by Theorem 2.2.19, we see that $\mathrm{soc}(G)$ contains at least three different factors. But this is not possible as $\mathrm{soc}(G) \cong \mathbb{Z}_p \times A_p$ or $A_p \times A_p$. Thus $r = 2$. Without loss of generality, assume $|F(K_1)| = p - 1$, let $\mathcal{B}$ be the set of orbits of K_1, and C be the set of orbits of K_2. Choose $\alpha \in \mathrm{Aut}\left(\mathbb{Z}_p^2\right)$ such that $\alpha(\mathcal{B}) = \left\{\left\{(i, j) : j \in \mathbb{Z}_p\right\} : i \in \mathbb{Z}_p\right\}$ and $\alpha(C) = \left\{\left\{(i, j) : i \in \mathbb{Z}_p\right\} : j \in \mathbb{Z}_p\right\}$. Then $\alpha G \alpha^{-1} = \mathcal{S}_p \times \mathcal{S}_p$ if and only if $|F(K_2)| = p - 1$, and the result follows with $A = F(K_2)$. □

Unfortunately, there are no good examples of this algorithm that are practical to do by hand. The smallest case when $p = 2$ is trivial. For $p = 3$, $\mathrm{GL}(2, 3)$ has order $\left(3^2 - 1\right)\left(3^2 - 3\right) = 48$, and we would need to check the condition $A(S) = S$ for all 48 elements $A \in \mathrm{GL}(2, 3)$. Other values of p are worse!

In addition to circulant digraphs, discussed in the next section, there are a few other groups G for which the automorphism groups of Cayley digraphs of G are known. Let p be prime. See Dobson and Kovács (2009) and Dobson (2012) for $G = \mathbb{Z}_p^3$ and $\mathbb{Z}_p \times \mathbb{Z}_{p^2}$, respectively. Let q be prime and distinct from p. The automorphism groups of Fermat digraphs of order qp can be found in Dobson et al. (2020a), where automorphism groups of digraphs whose automorphism group is an almost simple group but not primitive of order qp can also be found. See Praeger and Xu (1993) for graphs of order qp with primitive automorphism group (and also Dobson et al. (2020a)). Finally, automorphism groups of metacirculant digraphs of order qp are in Dobson (2006a). We will see in Section 9.2 that these results together imply the automorphism group of every vertex-transitive digraph of order qp are known, and so for all G of order qp.

8.5 Automorphism Groups of Circulant Digraphs

The material in this section is largely expository, and gives the strongest known result concerning automorphism groups of circulant digraphs. The main result is a translation into group-theoretic language of results proven using the method of Schur by Leung and Man (1996, 1998), and Evdokimov and Ponomarenko

(2002). The statement of Theorem 8.5.1 is contained in Li (2005, Theorem 2.3) in which Li determined all arc-transitive circulant digraphs. The exact same characterization of arc-transitive circulants was obtained independently at approximately the same time using the same method by Kovács (2004), and is proven in Theorem 8.5.8.

We have seen four classes of automorphism groups of circulant digraphs thus far. First and most trivially, there is the complete graph and its complement, whose automorphism group is the symmetric group. Second, there are normal circulant digraphs. Third, there are generalized wreath products – recall that in Example 5.3.18 we saw a generalized wreath product that is not isomorphic to a wreath product. Finally, there are digraphs whose automorphism group contains a symmetric group as a factor. Our main result says that these are the only possibilities.

Theorem 8.5.1 *Let* $\Gamma = \mathrm{Cay}(\mathbb{Z}_n, S)$ *be a circulant digraph of order n. Then one of the following is true:*

(i) Γ *is a complete graph or its complement and* $\mathrm{Aut}(\Gamma) = \mathcal{S}_n$,
(ii) Γ *is a normal Cayley digraph of* $\mathbb{Z}_n$, *and* $(\mathbb{Z}_n)_L \trianglelefteq \mathrm{Aut}(\Gamma)$,
(iii) *there are* $K \leq L < \mathbb{Z}_n$ *such that* Γ *is a nontrivial* (K, L)*-generalized wreath product, the cosets of* K *are a block system of* $\mathrm{Aut}(\Gamma)$,

$$\mathrm{Aut}(\Gamma) \geq \mathrm{Aut}(\mathrm{Cay}(\mathbb{Z}_n, S \setminus L)) \cap \mathrm{Aut}(\mathrm{Cay}(\mathbb{Z}_n, S \cap L)),$$

and every connected orbital digraph of Γ *is isomorphic to a wreath product* $\Gamma_1 \wr \bar{K}_k$ *with* $k = |K|$ *and with lexi-partition corresponding to* $V(\bar{K}_k)$, *the cosets of* K *in* $\mathbb{Z}_n$, *or*
(iv) *there are pairwise relatively prime divisors* $n_1, \ldots, n_r$ *of* n *such that* $n = n_1 \cdots n_r$, *and*

$$\mathrm{Aut}(\Gamma) = T_1 \times T_2 \times \cdots \times T_{r-1} \times T_r,$$

where $T_i \cong \mathcal{S}_{n_i}$, $1 \leq i \leq r-1$, *and* T_r *is isomorphic to a 2-closed subgroup of* $N_{S_{n_r}}\left((\mathbb{Z}_{n_r})_L\right)$ *that contains* $(\mathbb{Z}_{n_r})_L$.

Some comments are in order:

- One could shorten the statement as technically both (i) and (ii) are contained in (iv).
- We do not have $\mathrm{Aut}(\Gamma) = \mathcal{S}_{n_1} \times \cdots \times \mathcal{S}_{n_{r-1}} \times T_r$ simply because the vertex set of Γ is $\mathbb{Z}_n$ not $\mathbb{Z}_{n_1} \times \cdots \times \mathbb{Z}_{n_r}$.
- One may assume that $n_i \geq 4$ for every $1 \leq i \leq r-1$ as $\mathcal{S}_2 = (\mathbb{Z}_2)_L$, $\mathcal{S}_3 = \mathrm{AGL}(1,3)$, and $(\mathbb{Z}_3)_L \trianglelefteq \mathrm{AGL}(1,3)$.
- Ponomarenko (2005) used this result to find a polynomial time algorithm to calculate the automorphism group of a circulant digraph.

- Condition (iii) is not the formulation given in Li (2005, Theorem 2.3), but is equivalent by Bhoumik et al. (2014, Corollary 2.10) and Theorem 5.3.6.
- In the special case of circulant digraphs of square-free order n, a stronger version of this result with equality in

$$\mathrm{Aut}(\Gamma) \geq \mathrm{Aut}(\mathrm{Cay}(\mathbb{Z}_n, S\setminus L)) \cap \mathrm{Aut}(\mathrm{Cay}(\mathbb{Z}_n, S \cap L)),$$

was independently proven using group-theoretic techniques (Dobson and Morris, 2005, Theorem 3.5).

Our next result will allow us to recognize circulant digraphs that have automorphism group a direct product with some factor a symmetric group from its connection set.

Lemma 8.5.2 *Let $m, k \geq 2$ be positive integers, $\Gamma = \mathrm{Cay}(\mathbb{Z}_m \times \mathbb{Z}_k, S)$ and $G = (\mathbb{Z}_m)_L \times \mathcal{S}_k$ with the canonical action. Let $K = \langle(0,1)\rangle \leq \mathbb{Z}_m \times \mathbb{Z}_k$. Then $G \leq \mathrm{Aut}(\Gamma)$ if and only if $S \cap K = K\setminus\{(0,0)\}$ or $\emptyset$, and $((a,0)+K) \cap S = (a,0)+K$, $((a,0)+K)\setminus\{(a,0)\}$, $\{(a,0)\}$, or $\emptyset$ for every $a \in \mathbb{Z}_m\setminus\{0\}$.*

Proof Let $\mathcal{B} = \{(i,j) : i \in \mathbb{Z}_m\} : j \in \mathbb{Z}_k\}$, so $\mathcal{B}$ is the block system of G that is the set of orbits of $\mathcal{S}_k$, where throughout this proof $\mathcal{S}_k$ is always viewed as an internal subgroup of G.

Suppose $G \leq \mathrm{Aut}(\Gamma)$. By Lemma 8.4.1, we have $\mathrm{Stab}_G(0,0)$ has suborbits $\{(a,0)\}$ and $\{(a,j) : j \in \mathbb{Z}_m, j \neq 0\}$ in $(a,0)+K$, $a \in \mathbb{Z}_m$. As $S \cap ((a,0)+K)$ is a union of suborbits of $\mathrm{Stab}_G(0,0)$ that are contained in $(a,0)+K$, this implies $S \cap ((a,0)+K)$ is $K\setminus\{(0,0)\}$ or $\emptyset$ if $a = 0$ while $(a,0)+K$, $((a,0)+K)\setminus\{(a,0)\}$, $\{(a,0)\}$, or $\emptyset$ if $a \neq 0$.

Conversely, suppose $S \cap K = K\setminus\{(0,0)\}$ or $\emptyset$, and $((a,0)+K)\cap S = (a,0)+K$, $((a,0)+K)\setminus\{(a,0)\}$, $\{(a,0)\}$, or $\emptyset$ for every $a \in \mathbb{Z}_m\setminus\{0\}$. Let $\sigma \in \mathcal{S}_k$ and consider the map $g\colon \mathbb{Z}_m \times \mathbb{Z}_k \to \mathbb{Z}_m \times \mathbb{Z}_k$ given by $g(i,j) = (i, \sigma(j))$. It suffices to show that $g \in \mathrm{Aut}(\Gamma)$. Let $((x_1,y_1),(x_2,y_2)) \in A(\Gamma)$. As $(\mathbb{Z}_n)_L \leq \mathrm{Aut}(\Gamma)$ it thus suffices to show that $g((0,0),(x,y)) \in A(\Gamma)$ for every $(x,y) \in S$. Note $g((0,0),(x,y)) = ((0,\sigma(0))),(x,\sigma(y))$. If $x = 0$, then $y \neq 0$, and so $S \cap K = K\setminus\{(0,0)\}$. Then $g((0,0),(0,y)) \in A(\Gamma)$ as $\sigma(0) \neq \sigma(y)$. If $x \neq 0$, then $((x,0)+K)\cap S \neq \emptyset$. Also, trivially $g((0,0),(x,y)) \in A(\Gamma)$ unless $((x,0)+K) \cap S = ((x,0)+K)\setminus\{(x,0)\}$ or $\{(x,0)\}$. If $((x,0)+K)\cap S = ((x,0)+K)\setminus\{(x,0)\}$ then $y \neq 0$ and so $\sigma(0) \neq \sigma(y)$ and $((0,\sigma(0)),(x,\sigma(y))) \in A(\Gamma)$. If $((x,0)+K) \cap S = \{(x,0)\}$ then $y = 0$ and $g((0,0),(x,y)) = (0,\sigma(0)),(x,\sigma(y))$ has difference $(x,0) \in S$, and so $g((0,0),(x,y)) \in A(\Gamma)$. □

Suppose in the previous result $\gcd(m,k) = 1$ and so $\mathbb{Z}_m \times \mathbb{Z}_k \cong \mathbb{Z}_n$. If one writes Γ as a Cayley digraph of $\mathbb{Z}_n$, then $(a,0)$ and K in the statement should be replaced by $\hat{a}$ and H, respectively, where $\hat{a} \in \mathbb{Z}_n$ is the unique element of

$\mathbb{Z}_n$ such that $\hat{a} \equiv a \pmod{m}$ and $\hat{a} \equiv 0 \pmod{k}$, and H is the unique subgroup of $\mathbb{Z}_n$ of order k (so $\langle m \rangle$). Such an $\hat{a}$ exists by the Chinese remainder theorem (Dummit and Foote, 2004, Corollary 7.6.18) as $\gcd(m,k) = 1$.

Example 8.5.3 Let $S = \{\pm 1, \pm 16, \pm 26, \pm 31, \pm 36\}$. The circulant graph $\Gamma = \mathrm{Cay}(\mathbb{Z}_{55}, S)$ has automorphism group isomorphic to $(\mathbb{Z}_{55})_L \rtimes \mathbb{Z}_{10}$, is a normal Cayley graph of $\mathbb{Z}_{55}$, and has a regular subgroup isomorphic to the nonabelian group G of order 55, which is not normal in $\mathrm{Aut}(\Gamma)$.

Solution First observe that $|(1 + \langle 5 \rangle) \cap S| = 5$ while $(3 + \langle 11 \rangle) \cap S = \{36\}$ but $36 \not\equiv 0 \pmod{5}$ and $36 \not\equiv 0 \pmod{11}$. By Lemma 8.5.2, we see that $\mathrm{Aut}(\Gamma)$ contains no subgroup isomorphic to either $\mathcal{S}_5 \times K$ or $H \times \mathcal{S}_{11}$ for any transitive $K \leq \mathcal{S}_{11}$ or $H \leq \mathcal{S}_5$. Similarly, by Proposition 5.3.2, Γ is also not isomorphic to a nontrivial wreath product. By Theorem 8.5.1 we have that Γ is a normal Cayley graph of $\mathbb{Z}_{55}$. Checking that $\mathrm{Aut}(\mathbb{Z}_{55}, S) = \langle -26 \rangle$, by Lemma 8.1.7 $\mathrm{Aut}(\Gamma) = \{x \mapsto ax + b : a \in \langle -26 \rangle, b \in \mathbb{Z}_{55}\} \cong (\mathbb{Z}_{55})_L \rtimes \mathbb{Z}_{10}$ and has order $55 \cdot 10$. To see that $\mathrm{Aut}(\Gamma)$ contains a regular subgroup isomorphic to G, let $\alpha, \beta, \rho \in \mathrm{Aut}(\Gamma)$ be such that $\alpha \in (\mathbb{Z}_{55})_L$ is of order 5, $\rho \in (\mathbb{Z}_{55})_L$ is of order 11, and $\beta\colon \mathbb{Z}_{55} \to \mathbb{Z}_{55}$ is given by $\beta(x) = 26x$. Set $\tau = \alpha\beta$. Then $\langle \rho, \tau \rangle \cong G \leq \mathrm{Aut}(\Gamma)$ is transitive of order 55, but the map $x \mapsto -x$ is in $\mathrm{Aut}(\Gamma)$ and does not normalize $\langle \rho, \tau \rangle$. □

For reasons that will be clear once we have seen Theorem 8.5.8, attention in the literature has been given to graphs that satisfy the hypothesis of Lemma 8.5.2 when the conclusions that $S \cap K = \emptyset$ and $S \cap (a + K) = (a + K) \backslash \{\hat{a}\}$ or $\emptyset$ hold. We now want a digraph product that will produce digraphs that satisfy this property. While we will be focused on circulants as just described, the product we want can be defined for any two digraphs, and our definition reflects this.

Definition 8.5.4 Let Γ_1 and Γ_2 be digraphs. The **deleted wreath product** of Γ_1 and Γ_2 has vertex set $V(\Gamma_1) \times V(\Gamma_2)$ and arc set $\{((x_1, y_1), (x_2, y_2)) : (x_1, x_2) \in A(\Gamma_1)$ and $y_1 \neq y_2$, or $x_1 = x_2$ and $(y_1, y_2) \in A(\Gamma_2)\}$, and is denoted $\Gamma_1 \wr_d \Gamma_2$.

These digraphs are called deleted wreath products as the idea behind their definition is to first form the wreath product of Γ_1 and Γ_2, and then delete the 1-factor $\{(x_1, y_1)(x_2, y_1) : y_1 \in V(\Gamma_2)\}$ between $\{x_1\} \times V(\Gamma_2)$ and $\{x_2\} \times V(\Gamma_2)$ whenever $(x_1, x_2) \in A(\Gamma_1)$.

Example 8.5.5 Construct the deleted wreath product $K_2 \wr_d \bar{K}_5$.

Solution This graph has vertex set $\mathbb{Z}_2 \times \mathbb{Z}_5$, and edge set $\{(0, i)(1, j) : i, j \in \mathbb{Z}_5, i \neq j\}$. We see then that it is $\mathrm{Cay}(\mathbb{Z}_2 \times \mathbb{Z}_5, \{(1, j) : j = 1, 2, 3, 4\})$. This graph

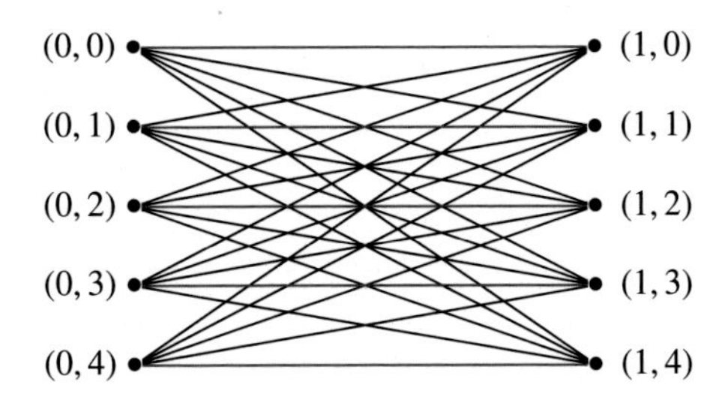

Figure 8.1 The deleted wreath product $K_2 \wr_d \bar{K}_5$.

is shown in Figure 8.1, where the edges are in black. The edges in red are the edges of $K_2 \wr \bar{K}_5$ that are deleted to obtain $K_2 \wr_d \bar{K}_5$. □

Example 8.5.6 Construct the deleted wreath product $C_{10} \wr_d \bar{K}_3$.

Solution The vertex set is $\mathbb{Z}_{10} \times \mathbb{Z}_3$, and edge set $\{(i, j)(i \pm 1, \ell) : j, \ell \in \mathbb{Z}_3, j \neq \ell\}$. It is $\mathrm{Cay}(\mathbb{Z}_{10} \times \mathbb{Z}_3, \{(\pm 1, j) : j = 1, 2\})$. It is shown in Figure 8.2, where the edges of $C_{10} \wr_d \bar{K}_3$ are in black, while the edges of $C_{10} \wr \bar{K}_3$ that are deleted to obtain $C_{10} \wr_d \bar{K}_3$ are in red. □

Example 8.5.7 Let $S = \{1, 3, 6, 8, 11, 13, 14, 16, 18, 23, 26, 31, 33\} \subset \mathbb{Z}_{35}$, $\Gamma = \mathrm{Cay}(\mathbb{Z}_{35}, S)$, and $G = \mathrm{Aut}(\Gamma)$. Show that G has no transitive subgroup isomorphic to $A \times S_5$ but does have a transitive subgroup isomorphic to $S_7 \times B$, where $A \leq S_7$ and $B \leq S_5$.

Solution Using the notation of Lemma 8.5.2 we see the only choices for K are the subgroup of $\mathbb{Z}_{35}$ of order 5, which is $\{0, 7, 14, 21, 28\}$, or the subgroup of order 7, which is $\{0, 5, 10, 15, 20, 25, 30\}$. We first need to check the sizes of the intersection of S with cosets of K. For $K = \langle 7 \rangle$, the coset of $\langle 7 \rangle$ that contains 4 is $\{4, 11, 18, 25, 32\}$ whose intersection with S has order 2. So $K \neq \langle 7 \rangle$ as this intersection must have order $0, 1, 4$, or 5 by Lemma 8.5.2. For $K = \langle 5 \rangle$, the coset of $\langle 5 \rangle$ that contains 1 is $\{1, 6, 11, 16, 21, 26, 31\}$ whose intersection with S has order 6, the only element missing being $21 \equiv 0 \pmod 7$. The coset of $\langle 5 \rangle$ that contains 3 is $\{3, 8, 13, 18, 23, 28, 33\}$ whose intersection with S has order 6, the only missing element being $28 \equiv 0 \pmod 7$. The coset of $\langle 5 \rangle$ that contains 4 is $\{4, 9, 14, 19, 24, 29, 34\}$ whose intersection with S has order 1, with the only element present being $14 \equiv 0 \pmod 7$. This accounts for every coset of $\langle 5 \rangle$ that intersects S, so $S_7 \times \mathbb{Z}_5 \leq \mathrm{Aut}(\Gamma)$ by Lemma 8.5.2. □

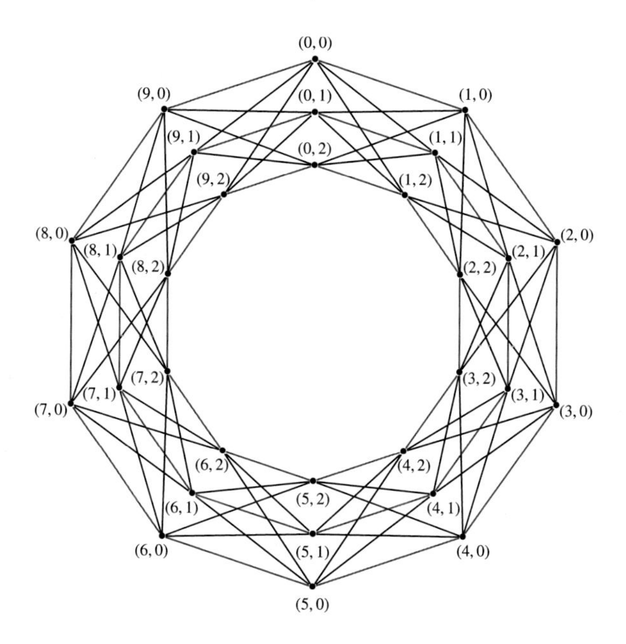

Figure 8.2 The deleted wreath product $C_{10} \wr_d \bar{K}_3$.

A generalized wreath product graph is given in Example 5.3.7 that has automorphism group $\mathcal{S}_2 \times \mathcal{S}_4$. This graph is also a deleted wreath product – see Figure 5.3 to make this clear. In fact, any deleted wreath product whose automorphism group is $A \times \mathcal{S}_m$ with m composite is also a generalized wreath product (Exercise 8.5.6). So the classes of generalized wreath products and deleted wreath products overlap.

A word about notation in the literature for deleted wreath products. Let Γ be a vertex-transitive digraph and $m \geq 2$ a positive integer. Recall that by $m\Gamma$ we mean a digraph that consists of m vertex-disjoint copies of the digraph Γ. It is common to see $\Gamma \wr_d \bar{K}_m$ written and defined as $\Gamma \wr \bar{K}_m - m\Gamma$, presumably to emphasize that intuitively we form the wreath product and then remove appropriate edges. This definition, however, should be avoided, as the digraph $m\Gamma$, while unique up to isomorphism, is not a unique *labelled* digraph, and so $\Gamma \wr \bar{K}_m - m\Gamma$ is not a unique or even necessarily vertex-transitive digraph!

We now turn to the classification of arc-transitive circulants given by Kovács (2004) and Li (2005). We remark that in the case of deleted wreath products, Kovács gives more information concerning these digraphs, and also enumerates arc-transitive circulants (Kovács, 2004, Theorem 4).

Theorem 8.5.8 *Let* $\Gamma = \mathrm{Cay}(\mathbb{Z}_n, S)$ *be a connected arc-transitive circulant. Then one of the following is true:*

(i) Γ *is a complete graph,*
(ii) Γ *is a normal circulant digraph,*
(iii) *there exists a connected arc-transitive circulant* Γ' *of order* m *that divides* n *and* $\Gamma \cong \Gamma' \wr \bar{K}_{n/m}$*, or*
(iv) *there exists a connected arc-transitive circulant* Γ' *of order* m *that divides* n *with* m *and* n/m *relatively prime, and* $\Gamma \cong \Gamma' \wr_d \bar{K}_{n/m}$*.*

Proof We assume Γ is neither complete nor a normal circulant. Also, it cannot be the case that there exists a nontrivial block B of $\mathrm{Aut}(\Gamma)$ such that some arc of Γ has both endpoints in B by Proposition 3.6.7 (ii).

Suppose Γ is a (K, L)-generalized wreath product for $K \leq L < (\mathbb{Z}_n)_L$ and the cosets of K are a block system of $\mathrm{Aut}(\Gamma)$. By Theorem 3.2.8 (ii) we have that Γ is an orbital digraph of $\mathrm{Aut}(\Gamma)$. By Theorem 8.5.1 (iii) we have that $\Gamma = \Gamma' \wr \bar{K}_{n/m}$, where $\mathbb{Z}_n/K$ has order m. Finally, Γ' is arc-transitive by Proposition 3.6.7 (i).

By Theorem 8.5.1, the only remaining possibility is that there are pairwise relatively prime divisors $n_1, \ldots, n_r$ of n such that $n = n_1 \cdots n_r$, and

$$\mathrm{Aut}(\Gamma) = T_1 \times T_2 \times \cdots \times T_{r-1} \times T_r,$$

where $T_i \cong \mathcal{S}_{n_i}$, $1 \leq i \leq r-1$, and T_r is isomorphic to a 2-closed subgroup A of $N_{\mathcal{S}_{n_r}}\left((\mathbb{Z}_{n_r})_L\right)$ that contains $(\mathbb{Z}_{n_r})_L$. As $\mathbb{Z}_n \cong \mathbb{Z}_{n_1} \times \cdots \times \mathbb{Z}_{n_r} = G$, we may assume without loss of generality that $\Gamma = \mathrm{Cay}(G, S_1)$ and $\mathrm{Aut}(\Gamma) = \mathcal{S}_{n_1} \times \mathcal{S}_{n_2} \times \cdots \times \mathcal{S}_{n_{r-1}} \times A$. Additionally, $r \geq 2$ as Γ is neither a normal circulant nor a complete graph. Let $\mathcal{B}$ be the block system of $\mathrm{Aut}(\Gamma)$ that is the set of orbits of $\mathcal{S}_{n_1}$ (as an internal subgroup of $\mathrm{Aut}(\Gamma)$), in which case $\mathrm{fix}_{\mathrm{Aut}(\Gamma)}(\mathcal{B}) = \mathcal{S}_{n_1}$. There is no arc of Γ with both endpoints in any block of $\mathcal{B}$, and $\mathrm{fix}_{\mathrm{Aut}(\Gamma)}(\mathcal{B})$ acts faithfully on each block of $\mathcal{B}$. Additionally, the permutation representation induced by the action of $\mathrm{fix}_{\mathrm{Aut}(\Gamma)}(\mathcal{B})$ on $B \in \mathcal{B}$ is permutation equivalent to the permutation representation induced by the action of $\mathrm{fix}_{\mathrm{Aut}(\Gamma)}(\mathcal{B})$ on $B' \in \mathcal{B}$. Define an equivalence relation $\equiv$ on G by $i \equiv j$ if and only if $\mathrm{Stab}_{\mathcal{S}_{n_1}}(i) = \mathrm{Stab}_{\mathcal{S}_{n_1}}(j)$, $i, j \in G$. By Lemma 8.3.4, the set of equivalence classes of $\equiv$ is a block system $\mathcal{C}$ of $\mathrm{Aut}(\Gamma)$, and each $C \in \mathcal{C}$ contains at most one element of a fixed $B \in \mathcal{B}$. Thus $\mathcal{C}$ has n_1 blocks of size n/n_1, and so these blocks are the cosets of the subgroup $\langle n_1 \rangle \leq \mathbb{Z}_n$. As the action of $\mathcal{S}_{n_1}$ on $B \in \mathcal{B}$ is 2-transitive, $\mathrm{Stab}_{\mathcal{S}_{n_1}}(i)$ has two orbits on each $B \in \mathcal{B}$, with one consisting of $\{j\}$ where $j \in B$ and $j \equiv i$, and the other $B \backslash \{j\}$.

Now, as Γ is arc-transitive, $\Gamma' = \Gamma/\mathcal{B} \cong \mathrm{Cay}(\mathbb{Z}_{n/n_1}, T)$ for some $T \subset \mathbb{Z}_{n/n_1}$, is also arc-transitive by Proposition 3.6.7 (i). This implies that the number of arcs from a block $B \in \mathcal{B}$ to any different block $B' \in \mathcal{B}$ is independent of the choice of B and B', provided there is at least one such arc. Fix $B, B' \in \mathcal{B}$ such that some vertex of B is outadjacent to some vertex of B', and let $i \in B$ with $j \in B'$ such that $j \equiv i$. As $\mathrm{Stab}_{S_{n_1}}(i)$ has $\{j\}$ and $B' \backslash \{j\}$ as orbits, the number of arcs from B to B' is either n_1 if the only vertex i is outadjacent to in B' is j, $n_1(n_1 - 1)$ if $i \in B$ is outadjacent to all vertices in $B' \backslash \{j\}$, or n_1^2 if i is outadjacent to every vertex of B'. If the only vertex i is outadjacent to in B' is j, then every arc of Γ lies within a block of C, and Γ is not connected, a contradiction. If $i \in B$ is outadjacent to every vertex in B', then Γ is a nontrivial wreath product by Theorem 4.2.15, and so does not have automorphism group of the form $\mathcal{S}_{n_1} \times \mathcal{S}_{n_2} \times \cdots \times \mathcal{S}_{n_{r-1}} \times A$. Thus i is outadjacent to all vertices in $B' \backslash \{j\}$, and, setting $i = 0$, we see that S is a union of cosets of $\langle n/n_1 \rangle$ (these are the blocks of $\mathcal{B}$) with their unique element equivalent to 0 modulo n_1 removed. Then $\Gamma \cong \Gamma' \wr_d \bar{K}_{n_1}$ and the result follows with $k = n_1$. □

Alspach, Conder, Marušič, and Xu classified s-arc-transitive circulants for $s \geq 2$ in Alspach et al. (1996). Their proof is elementary, and does not depend on the techniques in this section. Also, there has been some work on s-arc-transitive Cayley graphs of dihedral groups. See Du et al. (2008); Marušič (2003) for results on 2-arc-transitive and Kovács et al. (2011, 2013) for arc-transitive Cayley graphs of dihedral groups.

8.6 Exercises

8.1.1 Show that the Hamming graph $H(2, n)$ is a normal Cayley graph of $\mathbb{Z}_n^2$ if and only if $n \leq 4$.

8.1.2 Let $p \geq 5$ be prime. Find $1 \neq \alpha \in \mathrm{Aut}(\mathbb{Z}_p)$ and $S \subset \mathbb{Z}_p$, $S \neq \mathbb{Z}_p^*$, such that S is a union of orbits of α but $\mathrm{Aut}(\mathrm{Cay}(\mathbb{Z}_p, S)) \neq \langle (\mathbb{Z}_p)_L, \alpha \rangle$.

8.1.3 Let G be a group. Show that the set of fixed points of $\alpha \in \mathrm{Aut}(G)$ is a subgroup of G.

8.1.4 Show that $Q_8 \times \mathbb{Z}_2^r$ is a generalized dicyclic group for all $r \geq 0$.

8.1.5 Let $m = 2$ or 3. Show that $\Gamma = \mathrm{Cay}(\mathbb{Z}_{4m}, \{\pm 1, \pm m\})$ is a normal Cayley graph of $\mathbb{Z}_{4m}$.

8.1.6 Let $m \geq 3$ be odd. Show that $\Gamma = \mathrm{Cay}(\mathbb{Z}_{4m}, \{\pm 2, \pm m\})$ is a normal Cayley graph of $\mathbb{Z}_{4m}$.

8.1.7 Let $m \geq 2$. Show that $\Gamma = \mathrm{Cay}(\mathbb{Z}_2 \times \mathbb{Z}_{4m}, \{(1, \pm m), (0, \pm 1)\})$ is a normal Cayley graph of $\mathbb{Z}_2 \times \mathbb{Z}_{4m}$.

8.1.8 Let $n \geq 11$. Show that $\Gamma = \mathrm{Cay}(\mathbb{Z}_n, \{\pm 1, \pm 3\})$ is a normal Cayley graph of $\mathbb{Z}_n$.

8.1.9 Let G_1 and G_2 be groups with $S_1 \subset G_1$ and $S_2 \subset G_2$ be self-inverse. Show that

$$\mathrm{Cay}\left(G_1 \times G_2, \left(S_1 \times \left\{1_{G_2}\right\}\right) \cup \left(\left\{1_{G_1}\right\} \times S_2\right)\right) \cong \mathrm{Cay}(G_1, S_1) \Box \mathrm{Cay}(G_2, S_2).$$

8.1.10 Let G_1 and G_2 be groups with $S_1 \subset G_1$ and $S_2 \subset G_2$. Show that if $\mathrm{Cay}\left(G_1 \times G_2, \left(S_1 \times \left\{1_{G_2}\right\}\right) \cup \left(\left\{1_{G_1}\right\} \times S_2\right)\right)$ is normal, then $\mathrm{Cay}(G_1, S_1)$ and $\mathrm{Cay}(G_2, S_2)$ are normal.

8.1.11 Let G be a group and $\mathrm{Cay}(G, S)$ a normal Cayley graph of G. Show that $\alpha(G)$ is a normal Cayley graph of G for every $\alpha \in \mathrm{Aut}(G)$.

8.1.12 Show that $K_2 \Box K_4$ is isomorphic to $\bar{Q}_3$.

8.2.1 Show that the following are equivalent:

(i) G has a normal p-complement,
(ii) G has a normal p'-subgroup of index a power of p,
(iii) The product of any two p'-elements of G is again a p'-element of G.

8.2.2 Let p be prime, $G \leq S_p$ with Sylow p-subgroup P. Use Burnside's transfer theorem to show that if $N_G(P) = P$, then $G = P$.

The next definition will be used in the two exercises following it.

Definition 8.2.1 Let G be a group and $H \leq G$. Let $\{x_i : i \in \mathbb{Z}_n\}$ be a transversal of G/H, and $g \in G$. Then G permutes the left cosets of H in G via left multiplication. So for $g \in G$ and $i \in \mathbb{Z}_n$,

$$gx_iH = x_{j(i)}H,$$

for some $j \in S_n$. Then

$$gx_i = x_{j(i)}h_i,$$

for some $h_i \in H$. Define $v\colon G \to H/H'$ by $v(g) = \prod_{i=1}^{n} h_iH'$, where H' is the commutator subgroup of H. We call v the **transfer** of G into H.

8.2.3 Let G be a group with $H \leq G$. The transfer $v\colon G \to H/H'$ is independent of choice of transversal. That is, v is the same map regardless of choice of $\{x_i : i \in \mathbb{Z}_n\}$.

8.2.4 Let G be a group and $H \leq G$. Then the transfer $v\colon G \to H/H'$ is a homomorphism.

8.3.1 Let G be a group and $H \leq G$. Show that the normal closure of H in G is a normal subgroup of G and is the largest normal subgroup of G that contains H.

8.3.2 Let $g\colon \mathbb{Z}_m \times \mathbb{Z}_n \to \mathbb{Z}_m \times \mathbb{Z}_n$ by $g(i,j) = (\omega(i), \beta j + c_i)$, where $\omega \in \mathcal{S}_m$, $\beta \in \mathbb{Z}_n^*$, and $c_i \in \mathbb{Z}_n$. Find g^{-1}. Then show that if $r\colon \mathbb{Z}_m \times \mathbb{Z}_n \to \mathbb{Z}_m \times \mathbb{Z}_n$ is given by $r(i,j) = (\delta(i), j + d_i)$, $\delta \in \mathcal{S}_m$, $d_i \in \mathbb{Z}_n$, then $grg^{-1}(i,j) = (\sigma(i), j + b_i)$ for some $\sigma \in \mathcal{S}_m$, $b_i \in \mathbb{Z}_n$.

8.3.3 In $\mathbb{Z}_{p^2}^*$ show that $(1+p)^a = 1 + ap$, and so if $a \neq 0$, then $1 + ap$ generates the unique subgroup of $\mathbb{Z}_{p^2}^*$ of order p.

8.3.4 Prove the following analogue of Lemma 8.3.8: Let p be an odd prime, and q an odd prime less than p. Let $G \leq \mathcal{S}_{qp}$ have a block system $\mathcal{B}$ such that $\mathrm{fix}_G(\mathcal{B})$ is semiregular of order p, $G/\mathcal{B}$ is nonsolvable, and G contains a regular abelian subgroup A. Then $H = A^G \trianglelefteq G$ is transitive and H has a block system C orthogonal to $\mathcal{B}$. (We remark that this result is false if $q > p$ – see Dobson (2006a) and Marušič and Scapellato (1993).)

8.3.5 Calculate $\mathrm{Aut}(\mathbb{Z}_{49}, \{1, 7, 14, 18, 28, 30\})$.

8.3.6 Calculate $\mathrm{Aut}(\mathbb{Z}_{25}, \{1, 5, 6, 11, 16, 20, 21\})$.

8.3.7 Let $\Gamma = \mathrm{Cay}(\mathbb{Z}_{25}, S)$, where $S = \{1, 2, 3, 4, 21, 22, 23, 24\}$. Show that $\mathrm{Aut}(\Gamma) \cong D_{25}$. Additionally, let $\mathcal{B}$ be the block system of D_{25} with blocks of size 5 that is the set of cosets of $\langle 5\rangle$ in $(\mathbb{Z}_{25})_L$. Show that $\Gamma/\mathcal{B} = K_5$.

8.3.8 Let $G \leq \mathcal{S}_n$ be a transitive group with normal block system $\mathcal{B}$ such that $\mathrm{fix}_G(\mathcal{B})$ is nonabelian and $\mathrm{fix}_G(\mathcal{B})$ in its action on $B \in \mathcal{B}$ is primitive. Give a sufficient condition for the equivalence classes of $\equiv$ as defined in Lemma 8.3.4 to consist of n blocks of size 1.

8.4.1 Use the O'Nan–Scott theorem to show that a primitive group G of degree p^k, where p is prime, satisfies one of the following:

(i) $G \leq \mathrm{AGL}(k,p)$,

(ii) $G \leq \mathcal{S}_r \wr T$ for some divisor r of k and nonabelian simple group T of degree $p^{k/r}$, or

(iii) G is an almost simple group.

8.4.2 Use Lemmas 8.3.6, 8.3.8, and 8.4.2 to show that if Γ is a Cayley digraph of $\mathbb{Z}_p^2$ whose automorphism group is imprimitive, then there exists $\alpha \in \mathrm{Aut}\left(\mathbb{Z}_p^2\right)$ such that one of the following is true:

(i) $\mathrm{Aut}(\Gamma) < \mathrm{AGL}(2,p)$,

(ii) $\alpha\mathrm{Aut}(\Gamma)\alpha^{-1} = \mathrm{Aut}(\Gamma_1) \wr \mathrm{Aut}(\Gamma_2)$ for some circulant digraphs of order p,

(iii) $\alpha\mathrm{Aut}(\Gamma)\alpha^{-1} = \mathcal{S}_p \times K$, where $K \leq \mathcal{S}_p$ is 2-closed.

8.4.3 Let $\Gamma = \mathrm{Cay}\left(\mathbb{Z}_p^2, S\right)$ be neither complete nor the complement of a complete graph such that there is a subgroup $G \leq \mathrm{Aut}(\Gamma)$ that has a normal block system $\mathcal{B}$ consisting of the sets $\left\{B_i : i \in \mathbb{Z}_p\right\}$ where $B_i = \left\{(i, j) : j \in \mathbb{Z}_p\right\}$, and a Sylow p-subgroup of $\mathrm{fix}_G(\mathcal{B})$ has order at least p^2. Let $S_1 = \left\{i : (i, j) \in S \text{ for every } j \in \mathbb{Z}_p\right\}$ and $S_2 = \{j : (0, j) \in S\}$. Show that $\Gamma = \mathrm{Cay}\left(\mathbb{Z}_p, S_1\right) \wr \mathrm{Cay}\left(\mathbb{Z}_p, S_2\right)$ and $\mathrm{Aut}(\Gamma) = \mathrm{Cay}\left(\mathbb{Z}_p, S_1\right) \wr \mathrm{Cay}\left(\mathbb{Z}_p, S_2\right)$.

8.4.4 Let $G \leq \mathcal{S}_n \wr (\mathbb{Z}_k)_L$ so that if $g \in G$, then $g(i, j) = (\sigma(i), j + b_i)$, where $\sigma \in \mathcal{S}_n$ and $b_i \in \mathbb{Z}_k$. Define $m\colon G \to \mathbb{Z}_k$ be given by $m(g) = \sum_{i=0}^{n-1} b_i$. Verify that m is a homomorphism.

8.4.5 Show that every 2-closed permutation group of prime degree is permutation isomorphic to the automorphism group of a circulant digraph.

8.4.6 Let n be a positive integer. Define $\gamma\colon \mathbb{Z}_n^2 \to \mathbb{Z}_n^2$ by $\gamma(i, j) = (i, j + i)$. Show that γ has order n, fixes $\{\{(i, j) : j \in \mathbb{Z}_n\} : i \in \mathbb{Z}_n\}$, and normalizes $\left(\mathbb{Z}_n^2\right)_L$.

8.5.1 Does $\mathrm{Cay}(\mathbb{Z}_{15}, \{1, 3, 4, 6, 11, 13\})$ have a subgroup of automorphisms isomorphic to $\mathcal{S}_3 \times A$ or $B \times \mathcal{S}_5$, where $A \leq \mathcal{S}_5$ and $B \leq \mathcal{S}_3$?

8.5.2 Does $\mathrm{Cay}(\mathbb{Z}_{21}, \{1, 2, 4, 5, 7, 8, 11, 15, 18\})$ have a subgroup of automorphisms isomorphic to $\mathcal{S}_3 \times A$ or $B \times \mathcal{S}_7$, where $A \leq \mathcal{S}_7$ and $B \leq \mathcal{S}_3$?

8.5.3 Draw the deleted wreath product $K_3 \wr_d \bar{K}_5$.

8.5.4 Draw the deleted wreath product $C_5 \wr_d \bar{K}_4$.

8.5.5 Show that there exists a circulant digraph Γ and a positive integer m and a digraph D that is isomorphic to m copies of Γ such that $D \subseteq \Gamma$ and $\Gamma - D$ is not circulant.

8.5.6 Show that any deleted wreath product Γ whose automorphism group is of the form $A \times \mathcal{S}_m$, for $A \leq \mathcal{S}_n$ that contains a transitive abelian group and m composite, is also a generalized wreath product. (Hint: If $m = \ell k$ with $\ell, k > 1$, then $\mathcal{S}_\ell \wr \mathcal{S}_k \leq \mathcal{S}_m$ so $A \times (\mathcal{S}_\ell \wr \mathcal{S}_k) \leq \mathrm{Aut}(\Gamma)$. Apply Corollary 5.3.15.)

9

Classifying Vertex-Transitive Graphs

In this chapter we discuss classifications of vertex-transitive graphs. We will focus on classifying vertex-transitive graphs of particular, usually small, orders. Typically, by a classification, we mean a (shortest) list of minimal transitive subgroups such that every vertex-transitive graph of order n has a group on the list in its automorphism group. In some cases one will have a list of specific graphs, especially if the list is finite (this would usually be for more specialized classifications). Of course, often the minimal transitive subgroup is not explicitly given, but more often the name of the class constructed with the group is (for example instead of Cayley graphs of particular groups, perhaps metacirculant graphs, etc.). We remark that there are many results in the literature providing classifications of vertex-transitive graphs satisfying certain properties, so our meagre efforts here should not be thought of as the only approach one can take to classifying vertex-transitive graphs. See, for example, Alspach and Xu (1994); Chao (1971); Du and Xu (2000); Feng and Kwak (2007); Feng et al. (2007); Ivanov and Iofinova (1985), and papers that cite these references.

9.1 Digraphs of Prime Power Order

The easiest case to deal with is vertex-transitive digraphs of prime order, which, as we showed in Corollary 4.4.1, are all isomorphic to circulant digraphs. The proof of Corollary 4.4.1 is mainly group theoretic, as is the case for classifying vertex-transitive digraphs of prime-squared order.

Theorem 9.1.1 *Let p be prime. Every transitive group of degree p^2 contains a regular abelian subgroup. Consequently, every vertex-transitive digraph of order p^2 is isomorphic to a circulant digraph or a Cayley digraph of $\mathbb{Z}_p \times \mathbb{Z}_p$.*

Proof Let G be a transitive group of degree p^2. As G is transitive, it has a transitive Sylow p-subgroup P by Corollary 1.1.10. As P has nontrivial center $Z(P)$, and so contains an element $\rho \in Z(P)$ of order p, the set of orbits of $\langle\rho\rangle$ is a normal block system $\mathcal{B}$ of P with p blocks of size p. As P is transitive, there exists $\tau \in P$ such that $\tau/\mathcal{B}$ has order p. As $\tau \in P$, $|\tau| = p^2$ or p. If $|\tau| = p^2$, then $\langle\tau\rangle$ is transitive and cyclic. If $|\tau| = p$, then $\langle\rho, \tau\rangle$ is a transitive elementary abelian subgroup of order p^2. The result follows by Theorem 1.2.20. □

While Theorem 9.1.1 allows us to construct every vertex-transitive digraph of order p^2, it is somewhat deficient as we may wish to construct every vertex-transitive digraph of order p^2 *up to isomorphism*. Using Theorem 9.1.1 to construct all vertex-transitive digraphs of order p^2 will leave us many individual digraphs constructed multiple times. Of course, the results on the Cayley isomorphism problem from Chapter 7 will allow us to eliminate duplications within the lists of circulant digraphs (Theorem 7.7.30 – a solution only for digraphs of order p^2 was also found by Alspach and Parsons (1979)) as well as Cayley digraphs of $\mathbb{Z}_p^2$ (Theorem 7.5.13). The following result characterizes those vertex-transitive digraphs that are Cayley digraphs of both groups of order p^2, and will allow us to find digraphs that appear on both lists.

Proposition 9.1.2 *A transitive group G of degree p^2, p a prime, contains regular subgroups isomorphic to both $\mathbb{Z}_{p^2}$ and $\mathbb{Z}_p \times \mathbb{Z}_p$ if and only if a Sylow p-subgroup of G is isomorphic to $\mathbb{Z}_p \wr \mathbb{Z}_p$, a Sylow p-subgroup of $\mathcal{S}_{p^2}$. Consequently, a vertex-transitive digraph Γ of order p^2 is isomorphic to Cayley digraphs of both groups of order p^2 if and only if Γ is isomorphic to a wreath product $\Gamma_1 \wr \Gamma_2$ where Γ_1 and Γ_2 are circulant digraphs of order p.*

Proof By Example 4.2.8, we have that $\mathbb{Z}_p \wr \mathbb{Z}_p$ is a Sylow p-subgroup of $\mathcal{S}_{p^2}$. We view $\mathbb{Z}_p \wr \mathbb{Z}_p$ in its natural imprimitive action acting on $\mathbb{Z}_p \times \mathbb{Z}_p$, and so if $g \in \mathbb{Z}_p \wr \mathbb{Z}_p$, then $g(i, j) = (i + a, j + b_i)$, $a, b_i \in \mathbb{Z}_p$.

If a Sylow p-subgroup of G is $\mathbb{Z}_p \wr \mathbb{Z}_p$ then by Lemma 4.2.13, G has a regular subgroup isomorphic to $\mathbb{Z}_p \times \mathbb{Z}_p$. Also, the map $\rho\colon \mathbb{Z}_p^2 \to \mathbb{Z}_p^2$ given by $\rho(i, j) = (i + 1, j + b_i)$, where $b_0 = \cdots = b_{p-2} = 0$ and $b_{p-1} = 1$ is contained in G. This map has order p^2 as $\rho^p(i, j) = \left(i, j + \sum_{i=0}^{p-1} b_i\right) = (i, j + 1)$, and so is a regular cyclic subgroup. Hence P, and consequently G, contains a regular subgroup isomorphic to $\mathbb{Z}_{p^2}$.

Conversely, suppose G contains regular subgroups R_1 and R_2 isomorphic to $\mathbb{Z}_{p^2}$ and $\mathbb{Z}_p^2$, respectively. Then R_1 and R_2 are contained in Sylow p-subgroups P_1 and P_2 of G, respectively. Of course there exists $\delta \in G$ such that $\delta^{-1}P_2\delta = P_1$, so we may assume without loss of generality that R_1 and R_2 are contained in P_1. Define $\tau, \rho' : \mathbb{Z}_p^2 \to \mathbb{Z}_p^2$ by $\tau(i, j) = (i + 1, j)$ and $\rho'(i, j) = (i, j + 1)$, so $\langle \tau', \rho \rangle \cong \mathbb{Z}_p^2$. We assume without loss of generality that $R_2 = \langle \tau, \rho' \rangle$.

Now suppose $P_1 \not\cong \mathbb{Z}_p \wr \mathbb{Z}_p$. By a Sylow theorem, P_1 is not a Sylow p-subgroup of $N_{S_{p^2}}(P_1)$. This implies that there exists a maximal p-subgroup P properly contained in $\mathbb{Z}_p \wr \mathbb{Z}_p$ that contains P_1. Repeating the immediately preceding argument, the maximality of P implies that P is a normal subgroup of index p in $\mathbb{Z}_p \wr \mathbb{Z}_p$.

We now claim that if $a_0, \ldots, a_{p-1} \in \mathbb{Z}_p$ with $\sum_{i=0}^{p-1} a_i \equiv 0 \pmod p$, then the map $(i, j) \mapsto (i, j + a_i)$ is contained in P. Indeed, set $b_0 = 0$, $b_1 = -a_0$, and in general $b_i = -\sum_{j=0}^{i-1} a_j$. Define $\gamma : \mathbb{Z}_p^2 \to \mathbb{Z}_p^2$ by $\gamma(i, j) = (i, j + b_i)$. Then

$$\tau^{-1}\gamma^{-1}\tau\gamma(i, j) = (i, j + b_i - b_{i+1}) = (i, j + a_i),$$

which establishes our claim as $\tau, \gamma^{-1}\tau\gamma \in P$.

Clearly the set of maps $(i, j) \mapsto (i, j + a_i)$ with $\sum_{i=0}^{p-1} a_i \equiv 0 \pmod p$ is a subgroup P_0 of P, and $|P_0| = p^{p-1}$ as any of the $p - 1$ choices of $a_1, \ldots, a_{p-1}$ uniquely determines a_0. As $\tau \in P$ but $\tau \notin P_0$, $P = \langle \tau, P_0 \rangle$. Then P cannot contain any map of the form $(i, j) \mapsto (i, j + a_i)$ where $\sum_{i=0}^{p-1} a_i \not\equiv 0 \pmod p$. Let $R_1 = \langle \omega \rangle$. As $\omega \in P$, $\omega(i, j) = (i + b, j + a_i)$, where $b \in \mathbb{Z}_p^*$ and $a_i \in \mathbb{Z}_p$. Raising ω to an appropriate power, if necessary, we assume $b = 1$. As $|\omega| = p^2$, we see that $\sum_{i=0}^{p-1} a_i \not\equiv 0 \pmod p$. But then $\tau^{-1}\omega \notin P$, a contradiction, establishing the first part of this result. The second part now follows from Theorem 1.2.20. □

The previous result was first proven by Joseph (1995). Additionally, Joy Morris (1999) has shown amongst many other things that a circulant digraph of odd prime power order is a Cayley digraph of some other abelian group if and only if the digraph is a wreath product. Her result was extended to the prime 2 by Kovács and Servatius (2012). Vertex-transitive digraphs of order p^3 are also all isomorphic to Cayley digraphs (Marušič, 1985).

We have already mentioned that Alspach enumerated circulant digraphs and graphs of prime order (Alspach, 1973). Circulant digraphs and graphs of order $2n$ and $4n$, where n is odd and square-free, have also been enumerated (Alspach and Mishna, 2002). In the same paper, all unit circulant graphs and digraphs were enumerated, as well as all Cayley digraphs and graphs of $\mathbb{Z}_p^2$, p a prime.

Theorem 9.1.3 *Every vertex-transitive digraph of order p^3, p a prime, is isomorphic to a Cayley digraph of some group.*

Proof Let Γ be a vertex-transitive digraph of order p^3. By Theorem 1.2.20, to show that Γ is isomorphic to a Cayley digraph of G, it suffices to show that $\mathrm{Aut}(\Gamma)$ conains a regular subgroup isomorphic to G. By Corollary 1.1.10, a Sylow p-subgroup P of $\mathrm{Aut}(\Gamma)$ is transitive. Additionally, P has nontrivial center $Z(P)$, and so there exists $\alpha \in Z(P)$ such that $|\alpha| = p$. Then $\langle\alpha\rangle \trianglelefteq P$, and so the set of orbits of $\langle\alpha\rangle$ is a block system $\mathcal{B}$ of P with blocks of size p by Theorem 2.2.7. We then have $P/\mathcal{B}$ is a transitive group of degree p^2, and so by Theorem 9.1.1 we have that there exists a largest subgroup $P' \leq P$ such that $P'/\mathcal{B}$ is a regular group. By choice of P', $\alpha \in P'$ and so P' is transitive.

Define an equivalence relation $\equiv$ on $\mathcal{B}$ by $B \equiv B'$ if and only if whenever $\gamma \in \mathrm{fix}_{P'}(\mathcal{B})$ then γ^B is a p-cycle if and only if $\gamma^{B'}$ is also a p-cycle. By Lemma 5.3.11, the union of the equivalence classes of $\equiv$ is a block system $\mathcal{E}$ of P', and if $\gamma \in \mathrm{fix}_{P'}(\mathcal{B})$, then $\gamma|_E \in \mathrm{Aut}(\Gamma)$ for every $E \in \mathcal{E}$. By the maximality of P', we see $\gamma|_E \in P'$ for every $\gamma \in \mathrm{fix}_{P'}(\mathcal{B})$ and $E \in \mathcal{E}$. If there is one equivalence class of $\equiv$ (equivalently $\mathcal{E}$ has one block), then $\mathrm{fix}_{P'}(\mathcal{B}) = \langle\alpha\rangle$, and as $|P'/\mathcal{B}| = p^2$, we see that $|P'| = p^3$ and P' is regular. Thus Γ is isomorphic to a Cayley digraph of P'. If there are p^2 equivalence classes of $\equiv$ (equivalently, $\mathcal{E}$ has p^2 blocks and in fact $\mathcal{E} = \mathcal{B}$), it is easy to check that Γ is isomorphic to a wreath product $\Gamma_1 \wr \Gamma_2$, where Γ_1 is a vertex-transitive digraph of order p^2 and Γ_2 is a vertex-transitive digraph of order p. Applying Exercise 5.1.4 to Corollary 4.4.1 and Theorem 9.1.1, we see that Γ is isomorphic to a Cayley digraph of an abelian group.

If there are p equivalence classes of $\equiv$, then $\mathcal{E}$ has p blocks of size p^2. Then $\mathrm{fix}_{P'}(\mathcal{B}) = \langle\alpha|_E : E \in \mathcal{E}\rangle$, which has order p^p. As $P'/\mathcal{B}$ is regular, we see that $|P'| = p^{p+2}$. As $P'/\mathcal{E}$ is of degree p, there exists $\tau_1 \in P'$ such that $\tau_1/\mathcal{E}$ has order p. If $\tau_1/\mathcal{B}$ has order p^2, then $P'/\mathcal{B} = \langle\tau_1/\mathcal{B}\rangle$. As $\alpha \in Z(P')$, we see that τ_1 commutes with α, and so $\langle\tau_1, \alpha\rangle$ is a regular abelian subgroup. Then Γ is isomorphic to a Cayley digraph of either $\mathbb{Z}_{p^3}$ (if τ_1 has order p^3) or $\mathbb{Z}_{p^2} \times \mathbb{Z}_p$ (if τ_1 has order p^2), and the result follows. We thus assume $\tau_1/\mathcal{B}$ has order p.

We now claim that $\mathrm{Stab}_{P'}(E)^E$ is regular for every $E \in \mathcal{E}$. As $P'/\mathcal{B}$ is regular, any element of $\mathrm{Stab}_{P'}(E)$ that fixes a point must fix every block of $\mathcal{B}$, and so every element of $\mathrm{Stab}_{P'}(E)$ that fixes a point is contained in $\mathrm{fix}_{P'}(\mathcal{B}) = \langle\alpha|_E : E \in \mathcal{E}\rangle$. This latter group in its action on $E \in \mathcal{E}$ is semiregular, and so the only element of $\mathrm{Stab}_{P'}(E)^E$ that fixes a point is the identity, and our claim follows.

Now, as $\mathrm{Stab}_{P'}(\mathcal{E})^E$ is regular for each $E \in \mathcal{E}$, by the embedding theorem we see that P' is permutation isomorphic to a subgroup of $\mathbb{Z}_p \wr \mathbb{Z}_p^2$ or $\mathbb{Z}_p \wr \mathbb{Z}_{p^2}$. Let $\tau_2 \in P'$ such that $\langle\tau_2\rangle/\mathcal{B}$ has order p and is not contained in $\langle\tau_1\rangle/\mathcal{B}$.

If P' is permutation isomorphic to a subgroup of $\mathbb{Z}_p \wr \mathbb{Z}_p^2$, then we label the set that P' permutes with elements of $\mathbb{Z}_p^3$ in such a way that $\tau_1(i, j, k) = (i + 1, j, k + a_i)$, $\tau_2(i, j, k) = (i, j + 1, k + b_i)$, and $\alpha(i, j, k) = (i, j, k + 1)$, where $a_i, b_i \in \mathbb{Z}_p$. As $\mathrm{fix}_{P'}(\mathcal{B}) = \langle\alpha|_E : E \in \mathcal{E}\rangle$ and $\mathcal{E}$ is $\{\{(i, j, k) : j, k \in \mathbb{Z}_p\} : i \in \mathbb{Z}_p\}$, we see the maps $(i, j, k) \mapsto (i, j, k+a_i)$ and $(i, j, k) \mapsto (i, j, k+b_i)$ are contained in $\langle\alpha|_E : E \in \mathcal{E}\rangle$. We may thus assume without loss of generality that $a_i = b_i = 0$ for every $i \in \mathbb{Z}_p$, in which case $\langle\tau_1, \tau_2, \alpha\rangle \cong \mathbb{Z}_p^3$ and is transitive. Then Γ is isomorphic to a Cayley digraph of $\mathbb{Z}_p^3$. (We remark that in this case Γ is also isomorphic to a Cayley digraph of $\mathbb{Z}_p^2 \rtimes \mathbb{Z}_p$ – see Exercise 9.1.1.)

If P' is permutation isomorphic to a subgroup of $\mathbb{Z}_p \wr \mathbb{Z}_{p^2}$, then we label the set P' permutes with elements of $\mathbb{Z}_p \times \mathbb{Z}_{p^2}$ in such a way that $\tau_1(i, j) = (i + 1, j + b_i)$, and $\tau_2(i, j) = (i, j + c_i)$, where $b_i, c_i \in \mathbb{Z}_{p^2}$. As $\langle \tau_1, \tau_2 \rangle / \mathcal{B} \leq P'/\mathcal{B} \cong \mathbb{Z}_p \times \mathbb{Z}_p$, we may additionally assume that $b_i \equiv 0 \pmod p$, and that $c_i \equiv 1 \pmod p$ for every $i \in \mathbb{Z}_p$. Then $\tau_1(i, j) = (i + 1, j + pd_i)$ and $\tau_2(i, j) = (i, j + 1 + pe_i)$, where $0 \leq d_i, e_i \leq p - 1$. Straightforward computations will show that $\tau_2^p(i, j) = (i, j + p)$, so that $\tau_2^p \neq 1$ and $\tau_2^p \in \mathrm{fix}_{P'}(\mathcal{B})$. Then $\tau_2^p|_E \in P'$ for every $E \in \mathcal{E}$, and so the maps $(i, j) \mapsto (i, j + pd_i)$ and $(i, j) \mapsto (i, j + pe_i)$ are contained in P'. This implies that the maps $\tau_1', \tau_2' \colon \mathbb{Z}_p \times \mathbb{Z}_{p^2} \to \mathbb{Z}_p \times \mathbb{Z}_{p^2}$ given by $\tau_1'(i, j) = (i + 1, j)$ and $\tau_2'(i, j) = (i, j + 1)$ are contained in P'. We conclude that $\langle \tau_1', \tau_2' \rangle$ is regular and isomorphic to $\mathbb{Z}_p \times \mathbb{Z}_{p^2}$. So Γ is isomorphic to a Cayley digraph of $\mathbb{Z}_p \times \mathbb{Z}_{p^2}$. □

There are a couple of questions that now arise. First, in the statement of Theorem 9.1.3, is it necessary that the group under consideration be the automorphism group of a digraph? That is, is it true that every transitive permutation group of degree p^3 contains a regular subgroup? The answer is that there are indeed transitive groups of degree p^3 that do not contain a regular subgroup – see Exercise 9.1.2. Second, is it true that every vertex-transitive digraph of degree p^k is isomorphic to a Cayley digraph, $k \geq 1$? This question turns out to have a negative answer, and in fact there are $\left(p, p^3\right)$-metacirculant digraphs that are not isomorphic to Cayley digraphs. This was first observed by Marušič (1985). The proof that we present here is due to McKay and Praeger (1994, Theorem 6). Note that the existence of a non-Cayley vertex-transitive digraph of order p^4 implies the existence of a non-Cayley vertex-transitive digraph of order p^k, $k \geq 4$ (Exercise 9.1.3). Our first task will be to show that a particular graph is isomorphic to a GRR for the group $\mathbb{Z}_{p^2} \rtimes \mathbb{Z}_p$ (note that such a graph is isomorphic to a $(p, p^2, 1 + p)$-metacirculant graph whose automorphism group is $\langle \rho, \tau \rangle$, where $\rho, \tau \colon \mathbb{Z}_p \times \mathbb{Z}_{p^2} \to \mathbb{Z}_p \times \mathbb{Z}_{p^2}$ are given by $\rho(i, j) = (i, j + 1)$ and $\tau(i, j) = (i + 1, (1 + p)j)$). This graph will be used in the construction of a non-Cayley vertex-transitive graph of order p^4.

Lemma 9.1.4 *Let p be an odd prime, $\alpha = 1 + p$, $S_0 = \mathbb{Z}_{p^2} \backslash \{0\}$, $S_1 = \{0, 1\}$, and $S_{p-1} = -\alpha^{-1} S_1$. The $\left(p, p^2, \alpha, S_0, S_1, \emptyset, \ldots, \emptyset, S_{p-1}\right)$-metacirculant graph Γ has automorphism group $\langle \rho, \tau \rangle$, where $\rho, \tau \colon \mathbb{Z}_p \times \mathbb{Z}_{p^2} \to \mathbb{Z}_p \times \mathbb{Z}_{p^2}$ are given by $\rho(i, j) = (i, j + 1)$ and $\tau(i, j) = (i + 1, (1 + p)j)$. That is, Γ is isomorphic to a GRR of $\mathbb{Z}_{p^2} \rtimes \mathbb{Z}_p$.*

Proof First, a straightforward application of the binomial theorem gives $\alpha^p \equiv 1 \left(\bmod p^2\right)$, so $\alpha^p S_1 = S_1$ and $\alpha^p S_{p-1} = S_{p-1}$. By definition, Γ is a $\left(p, p^2, \alpha\right)$-metacirculant digraph, and $\langle\rho, \tau\rangle \leq \mathrm{Aut}(\Gamma)$ by Theorem 1.5.13. As $-S_0 = S_0$ and $S_{p-1} = -\alpha^{-1}S_1$ by Theorem 1.5.14, we have that Γ is a graph. As in Section 1.5, for $i \in \mathbb{Z}_p$ we set $V_i = \left\{(i, j) : j \in \mathbb{Z}_{p^2}\right\}$. As $S_0 = \mathbb{Z}_{p^2}\setminus\{0\}$, $\Gamma[V_i] \cong K_{p^2}$ for every $i \in \mathbb{Z}_p$.

Let $B \subset \mathbb{Z}_p \times \mathbb{Z}_{p^2}$ such that $\Gamma[B]$ is a clique of size p^2, and assume $B \neq V_i$ for some $i \in \mathbb{Z}_p$. Without loss of generality assume B contains a vertex of V_0. As a vertex $v \in V_0$ is adjacent to exactly four vertices not in V_0, $|B \cap V_0| \geq p^2 - 4 \geq 5$. As B is a clique, five vertices in B_0 have a common neighbor in B_1 or B_{p-1}, but this is not possible as each vertex in $B_{p-1} \cup B_1$ has two neighbors in B_0. So $\Gamma[V_i]$, $i \in \mathbb{Z}_p$, are the only cliques of size p^2 in Γ. By Exercise 2.2.3 (with $\Delta = K_{p^2}$) $\mathcal{B} = \left\{V_i : i \in \mathbb{Z}_p\right\}$ is a block system of $\mathrm{Aut}(\Gamma)$.

Now suppose C is a p-cycle in Γ that has a vertex contained in each V_i, $i \in \mathbb{Z}_p$. Then the edge in C from V_i to V_{i+1} is of the form $(i, j)(i+1, j)$ or $(i, j)\left(i+1, j+\alpha^i\right)$. Assume without loss of generality that $(0, 0) \in V(C)$. Then

$$C = (0,0)(1, c_1)(2, c_1+c_2)\cdots(p-1, c_1+c_2+\cdots+c_{p-1})(0, c_1+c_2+\cdots+c_{p-1}+c_0),$$

where for each i we have $c_i = 0$ or α^{i-1}. Additionally, as C is a p-cycle, $\sum_{i=0}^{p-1} c_i \equiv 0 \left(\bmod p^2\right)$. Hence either $c_i = 0$ for every $i \in \mathbb{Z}_p$, or for some nonempty subset T of $\left\{1, \alpha, \alpha^2, \ldots, \alpha^{p-1}\right\}$ we have $\sum_{t\in T} t \equiv 0 \left(\bmod p^2\right)$. In the latter case, as $\alpha^i \equiv 1 + ip \left(\bmod p^2\right)$, $\sum_{t\in T} t \equiv |T| \pmod p$. Thus $\sum_{t\in T} t \equiv 0 \left(\bmod p^2\right)$ only if $T = \left\{1, \alpha, \alpha^2, \ldots, \alpha^{p-1}\right\}$. But

$$\sum_{i=0}^{p-1} \alpha^i \equiv \sum_{i=0}^{p-1}(1 + ip) \equiv p + p\sum_{i=0}^{p-1} i \equiv p \left(\bmod p^2\right),$$

a contradiction. Thus $c_i = 0$ for every $i \in \mathbb{Z}_p$. We conclude that any element of $\mathrm{Aut}(\Gamma)$ maps such a p-cycle to a similar p-cycle, and so maps the set $\left\{(i, j) : i \in \mathbb{Z}_p\right\}$ to $\left\{(i, j') : i \in \mathbb{Z}_p\right\}$, where $j, j' \in \mathbb{Z}_{p^2}$. So $\left\{\left\{(i, j) : i \in \mathbb{Z}_p\right\} : j \in \mathbb{Z}_{p^2}\right\}$ is a block system C of $\mathrm{Aut}(\Gamma)$.

Let $E_1 = \left\{(i, j)(i+1, j) : i \in \mathbb{Z}_p, j \in \mathbb{Z}_{p^2}\right\}$ and $E_\alpha = \left\{(i, j)\left(i+1, j+\alpha^i\right) : i \in \mathbb{Z}_p, j \in \mathbb{Z}_{p^2}\right\}$. As C is a block system of $\mathrm{Aut}(\Gamma)$, E_1 is an orbit of $\mathrm{Aut}(\Gamma)$ acting on edges. As the only neighbors of a vertex in Γ not in E_1 or a clique of size p^2 (and as $\mathcal{B}$ is a block system of $\mathrm{Aut}(\Gamma)$ the set of edges contained in cliques in Γ of size p^2 are permuted amongst themselves by $\mathrm{Aut}(\Gamma)$), we see that E_α is also an edge orbit of $\mathrm{Aut}(\Gamma)$. Let W be the closed walk $(0,0), (1,1), (0,1), (1,2), \ldots, (1,0), (0,0)$. Then $V(W) = V_0 \cup V_1$. As the edges of W alternate between E_1 and E_α, the edges of $\gamma(W)$ also alternate between E_1 and E_α for every $\gamma \in \mathrm{Aut}(\Gamma)$.

If $\gamma \in \mathrm{Aut}(\Gamma)$ fixes $(0,0)$, it must map $(1,0)$ to $(1,0)$ or $(p-1,0)$ as E_1 is an edge orbit. In the first case, γ fixes every vertex of W, so γ fixes every vertex of $V_0 \cup V_1$. Successively applying τ to W, we see that $\gamma = 1$. In the second case, if $\gamma(1,0) = (p-1,0)$, then $\gamma(p-1,0) = (1,0)$. By the immediately preceding argument, we see that $\gamma^2 = 1$. Considering $\gamma(W)$ as before, we have $\gamma(0,j) = \left(0, -j\alpha^{p-1}\right)$. Then $\gamma(0,1) = \left(0, -\alpha^{p-1}\right)$ and $\gamma\left(0, -\alpha^{p-1}\right) = \left(0, \alpha^{2p-2}\right)$. As $\gamma^2 = 1$, we have $\alpha^{2p-2} \equiv 1 \pmod{p^2}$, but $\alpha^{2p-2} \equiv 1 - 2p \pmod{p^2}$, a contradiction. □

There is onc group-theoretic fact that we will need for our main construction.

Lemma 9.1.5 *Let p be an odd prime, and $\rho, \tau\colon \mathbb{Z}_p \times \mathbb{Z}_{p^2}$ by $\rho(i,j) = (i, j+1)$ and $\tau(i,j) = (i+1, (1+p)j)$. Then the only elements of $\langle \rho, \tau\rangle$ of order p are of the form $\tau^b\rho^{cp}$, where $b \in \mathbb{Z}_p, c \in \mathbb{Z}_p$, and one of b and c is nonzero.*

Proof Clearly $\rho^{p^2} = 1$ and a straightforward computation will show that $\tau^p = 1$. Also, $\tau\rho\tau^{-1}(i,j) = (i, j+1+p)$, and so $\tau\rho\tau^{-1} = \rho^{1+p}$. Thus $\tau\rho = \rho^{1+p}\tau$. As $\langle\rho\rangle \trianglelefteq \langle\rho,\tau\rangle$, we see that $\langle\rho,\tau\rangle = \langle\tau\rangle \cdot \langle\rho\rangle$, and so every element of $\langle\rho,\tau\rangle$ can be written as $\tau^b\rho^d$ for some $b \in \mathbb{Z}_p$ and $d \in \mathbb{Z}_{p^2}$. As $\tau\rho = \rho^{1+p}\tau$, an induction argument will show that $\tau^b\rho^d = \rho^{d(1+p)^b}\tau^b$ (Exercise 9.1.5). Now,

$$\begin{aligned}
\left(\tau^b\rho^d\right)^p &= \left(\tau^b\rho^d\right)\left(\tau^b\rho^d\right)\cdots\left(\tau^b\rho^d\right)\\
&= \rho^{d(1+p)^b}\tau^{2b}\rho^d\left(\tau^b\rho^d\right)\cdots\left(\tau^b\rho^d\right)\\
&= \rho^{d(1+p)^b}\rho^{d(1+p)^{2b}}\tau^{3b}\rho^i\left(\tau^b\rho^d\right)\cdots\left(\tau^b\rho^d\right)\\
&= \rho^{d(1+p)^b}\rho^{d(1+p)^{2b}}\cdots\rho^{d(1+p)^{pb}}\tau^{bp}\\
&= \rho^{\sum_{j=1}^p d(1+p)^{jb}}.
\end{aligned}$$

We conclude that $\tau^b\rho^d$ has order p if and only if $\sum_{j=1}^p d(1+p)^{jp} \equiv 0 \pmod{p^2}$. Also, as we saw in the last result, $(1+p)^p \equiv 1 \pmod{p^2}$ so $\sum_{j=1}^p d(1+p)^{jp} \equiv dp \pmod{p^2}$. Hence $\tau^b\rho^d$ has order dividing p if and only if $dp \equiv 0 \pmod{p^2}$, so if and only if $d \equiv 0 \pmod{p}$. That is, $\tau^b\rho^d$ has order dividing p if and only if $d = cp$, where $0 \le c \le p-1$. Of course, $\tau^b\rho^d$ has order 1 if and only if $b = 0 = d$, so at least onc of b and d is nonzero, and the result follows. □

Note this result implies that in $\langle\rho,\tau\rangle \cong \mathbb{Z}_{p^2} \rtimes \mathbb{Z}_p$, the product of an element of order p and an element of order p^2 is again an element of order p^2. We now give an example of a vertex-transitive graph of order p^4, p an odd prime, that is not isomorphic to a Cayley graph of any group of order p^4.

Theorem 9.1.6 *Let p be an odd prime, $\alpha = 1 + p \in \mathbb{Z}_{p^3}^*$, $S_0 = \left\{pk : k \in \mathbb{Z}_{p^2}, k \neq 0\right\}$, $S_1 = \left\{0, p, \alpha^{rp} : r \in \mathbb{Z}_p\right\}$, and $S_{p-1} = -\alpha^{-1}S_1$. Then the*

$\left(p, p^3, \alpha, S_0, S_1, \emptyset, \ldots, \emptyset, S_{p-1}\right)$-metacirculant graph Γ is not isomorphic to a Cayley graph of any group of order p^4.

Proof Note that α has multiplicative order p^2, and $\alpha^{rp} \equiv 1 + rp^2 \pmod{p^3}$. As $\alpha^p S_1 = \left(1 + p^2\right) S_1 = S_1$ and $\alpha^p S_{p-1} = S_{p-1}$, by definition Γ is a $\left(p, p^3, \alpha\right)$-metacirculant digraph. As $-S_0 = S_0$, and $S_{p-1} = -\alpha^{-1} S_1$, by Theorem 1.5.14 Γ is a $\left(p, p^3, \alpha\right)$-metacirculant graph. As usual for a metacirculant graph, we set $V_i = \left\{(i, j) : j \in \mathbb{Z}_{p^3}\right\}$. We first show that the p^2 sets $B_{i,j} = \left\{(i, j + pk) : k \in \mathbb{Z}_{p^2}\right\}$, $i, j \in \mathbb{Z}_p$, are the only cliques of Γ of order p^2.

Let C be a clique of Γ of size p^2. As $\Gamma[V_i]$ is a disjoint union of cliques on the sets $B_{i,j}$, $j \in \mathbb{Z}_p$, it suffices to show that C is contained in some V_i. Suppose otherwise, and let $i \in \mathbb{Z}_p$ with $V(C) \cap V_i \neq \emptyset$. As a vertex of V_i is only adjacent to vertices in V_{i-1}, V_i, and V_{i+1}, it must be the case that some V_i contains at least $p^2/3$ vertices in C. As every vertex in V_i is adjacent to at most $p+2$ vertices in each of V_{i+1} and V_{i-1} and C is a clique, we see that $p^2/3 \leq p+2$. Hence $p = 3$.

Assume without loss of generality that $|V_0 \cap C| \geq 3$ and $(0,0)$ is contained in C. Suppose C contains a vertex of V_1. If C contains $(0, kp)$ for some $k \in \mathbb{Z}_{p^2}$ such that $k \neq 0$, then the elements of V_1 that are neighbors of $(0, kp)$, are $(1, kp)$, $(1, (k+1)p)$, and $\left(1, kp + 1 + rp^2\right)$, $r \in \mathbb{Z}_p$. Perusing the lists of neighbors of $(0,0)$ and $(0, kp)$, $k \neq 0$, there are two common neighbors, namely $(1, 0)$ with $k = p^2 - p$, and $(1, p)$ with $k = 1$. As $(0, p)$ is not adjacent to $(1, 0)$, we see that V_0 contains at most two vertices of C, a contradiction. Thus C contains no vertices of V_1. A similar argument shows that C has no vertices in V_2, a contradiction. Thus $B_{i,j}$ are the only cliques of Γ of order p^2. This implies that $\mathcal{B} = \left\{B_{i,j} : i, j \in \mathbb{Z}_p\right\}$ is a block system of $\mathrm{Aut}(\Gamma)$.

Now, there are $2p^2$ edges between $B_{i,j}$ and $B_{i',j'}$ if $i - i' = \pm 1$ and $j = j'$ (corresponding to $0, p \in S_1$), p^3 edges if $i - i' = \pm 1$ and $j - j' = \pm 1$ (corresponding to $\alpha^{rp} \in S_1$, where $r \in \mathbb{Z}_p$), and no edges otherwise. This implies that $C_i = \cup_{j=0}^{p-1} B_{i,j}$ is a block of $\mathrm{Aut}(\Gamma)$. We set $C = \left\{C_i : i \in \mathbb{Z}_p\right\}$, so that C is a block system of $\mathrm{Aut}(\Gamma)$. Each of the graphs $\Gamma[C_i]$, $i \in \mathbb{Z}_p$, is isomorphic to the $\left(p, p^2, 1 + p, \mathbb{Z}_{p^2} - \{0\}, \{0, 1\}, \emptyset, \ldots, \emptyset, \left\{0, -\alpha^{-1}\right\}\right)$-metacirculant graph. By Lemma 9.1.4, we see that $\mathrm{fix}_{\mathrm{Aut}(\Gamma)}(C)^{C_i} = \mathrm{fix}_{\langle\rho,\tau\rangle}(C)^{C_i}$ and is isomorphic to $\mathbb{Z}_{p^2} \rtimes \mathbb{Z}_p$.

Suppose $R \leq \mathrm{Aut}(\Gamma)$ is regular. Then there exists $\gamma \in R$ such that $\gamma(0,0) = (1,0)$. Also, γ maps C_0 to C_0, and as $R/C \leq S_p$ is a transitive p-group, $R/C \cong \mathbb{Z}_p$ and so is regular. This implies that $\gamma \in \mathrm{fix}_{\mathrm{Aut}(\Gamma)}(C)$, and so $\gamma^{C_i} \in \mathrm{fix}_{\langle\rho,\tau\rangle}(C)^{C_i}$ for every $i \in \mathbb{Z}_p$. As $\mathrm{fix}_{\langle\rho,\tau\rangle}(C) = \langle\tau, \rho^p\rangle$, $\mathrm{fix}_{\langle\rho,\tau\rangle}(C)^{C_0} \cong \mathbb{Z}_{p^2} \rtimes \mathbb{Z}_p$, and $\tau(0,0) = (1,0)$, we see that $\gamma^{C_0} = \tau^{C_0}$ has order p. As γ is semiregular as R is regular, γ has order p, and so γ^{C_1} has order p. Note that $\tau \in \mathrm{fix}_{\mathrm{Aut}(\Gamma)}(C)$, and τ^{C_i} has order p^2 for every $i \neq 0$. Then $\gamma^{-1}\tau^{C_0} = 1$ while Lemma 9.1.5 implies $\left(\gamma^{-1}\tau\right)^{C_i}$

has order p^2 for every $i \neq 0$. But then $(0,0)$ is adjacent to at least p^2 elements of C_1, a contradiction. □

9.2 Graphs of Order a Product of Two Distinct Primes

When analyzing the automorphism groups of vertex-transitive graphs, at first glance it seems easier to deal with those that have an imprimitive group of automorphisms, as the embedding theorem shows that an imprimitive group is permutation isomorphic to a subgroup of a wreath product of groups of smaller degree. Presumably the reduction to groups of smaller degree makes the overall analysis easier. We will see in this section that this is sometimes the case, but in the next section we will see this is not always the case. Basically, this reduction to groups of smaller degree works fairly well when the number of factors of the degree is "small" and not so well when the number of factors of the degree is not "small." We saw this in the previous section as the classification problem for vertex-transitive graphs of order p^k became progressively harder as k grew, and is still open for $k = 4$! This approach is successful for vertex-transitive digraphs of order qp with $q < p$ distinct primes, and we will use the following strategy in their classification. As it turns out, the graphs whose automorphism group has a transitive subgroup with a nontrivial normal block system are easiest to deal with. So we split vertex-transitive graphs of order pq into three classes:

(i) Graphs with a group of automorphisms (not necessarily the full automorphism group) with a normal nontrivial block system.
(ii) Graphs with every transitive group of automorphisms (not necessarily the full automorphism group) with only nonnormal block systems with blocks of size q.
(iii) Graphs with no imprimitive group of automorphisms.

Graphs with a group of automorphisms with a normal nontrivial block system are the easiest to classify – they are all isomorphic to metacirculants (Theorem 9.2.2). This proof does not require CFSG. By Exercise 2.2.8, this implies that either Γ has no imprimitive group of automorphisms, or the only blocks are of size q with $q < p$. For graphs with imprimitive groups of automorphisms but only with blocks of size q, the case when $q = 2$ can be solved without the use of CFSG (Proposition 9.2.6). For odd primes q, CFSG is essential for the only known proofs. Finally, CFSG is also an integral part of the classification of graphs with only primitive groups of automorphisms. For the parts that involve CFSG, we only state and summarize the results. Our first goal is to see that every digraph with an imprimitive group with a normal block system is isomorphic to a metacirculant digraph (there is no extra work in showing this result for digraphs as opposed to only graphs). We begin with a technical fact that is useful to show that for a square-free integer n (not just a product of two primes), weak metacirculants of order n are isomorphic to metacirculants of order n (Exercise 9.2.1).

Lemma 9.2.1 *Let m and n be positive integers. Define $\rho\colon \mathbb{Z}_m \times \mathbb{Z}_n \to \mathbb{Z}_m \times \mathbb{Z}_n$ by $\rho(i,j) = (i, j+1)$. Then $Z_{S_{mn}}(\langle\rho\rangle) = \{(i,j) \mapsto (\sigma(i), j+b_i) : \sigma \in S_m, b_i \in \mathbb{Z}_n\} = S_m \wr (\mathbb{Z}_n)_L$.*

Proof Straightforward computations will show that every element of $\{(i,j) \mapsto (\sigma(i), j+b_i) : \sigma \in S_m, b_i \in \mathbb{Z}_n\}$ centralizes $\langle\rho\rangle$. Then $Z_{S_{mn}}(\langle\rho\rangle)$ is transitive, and $\langle\rho\rangle \trianglelefteq Z_{S_{mn}}(\langle\rho\rangle)$. So the set of orbits $\mathcal{B}$ of $\langle\rho\rangle$ is a block system of $Z_{S_{mn}}(\langle\rho\rangle)$. Let $B \in \mathcal{B}$, and $g \in \mathrm{Stab}_{Z_{S_{mn}}(\langle\rho\rangle)}(B)$. Then g^B commutes with $\langle\rho\rangle^B$, and as $\langle\rho\rangle^B$ is a regular cyclic group, it is self-centralizing by Corollary 2.2.18. Then $\mathrm{Stab}_{Z_{S_{mn}}(\langle\rho\rangle)}(B)^B \leq \langle\rho\rangle^B$, and so by the embedding theorem, $Z_{S_{mn}}(\langle\rho\rangle) \leq S_m \wr (\mathbb{Z}_n)_L$. As $S_m \wr (\mathbb{Z}_n)_L \leq Z_{S_{mn}}(\langle\rho\rangle)$, the result follows. □

Theorem 9.2.2 *A vertex-transitive digraph Γ of order qp, q and p distinct primes, is isomorphic to a (q,p)-metacirculant digraph if and only if there is $G \leq \mathrm{Aut}(\Gamma)$ that is transitive with a normal block system $\mathcal{B}$ with blocks of size p.*

Proof If Γ is isomorphic to a (q,p)-metacirculant digraph, then after an appropriate relabeling of $V(\Gamma)$, the maps $\rho, \tau\colon \mathbb{Z}_q \times \mathbb{Z}_p \to \mathbb{Z}_q \times \mathbb{Z}_p$ given by $\rho(i,j) = (i, j+1)$ and $\tau(i,j) = (i+1, \alpha j)$ for some $\alpha \in \mathbb{Z}_p^*$, are contained in $\mathrm{Aut}(\Gamma)$ by Theorem 1.5.11. Then the set $\mathcal{B}$ of orbits of $\langle\rho\rangle \trianglelefteq \langle\rho,\tau\rangle = G$ is a normal block system of G with blocks of size p.

Conversely, suppose $G \leq \mathrm{Aut}(\Gamma)$ is transitive and has $\mathcal{B}$ as a normal block system with blocks of size p. Then $G/\mathcal{B}$ is transitive, and there is $\tau \in G$ such that $\langle\tau\rangle/\mathcal{B}$ is cyclic of order q (and so regular). By raising τ to an appropriate power relatively prime to q, we may additionally assume that $|\tau|$ is a power of q. By Corollary 5.3.13, there exists $\rho \in \mathrm{fix}_{G^{(2)}}(\mathcal{B}) \leq \mathrm{Aut}(\Gamma)$ such that $\langle\rho\rangle$ is semiregular of order p. By Lemma 5.3.11, the $\mathcal{B}$-fixer block system $\mathcal{E}$ has blocks of size 1 or q, and as $\rho|_E \leq \mathrm{fix}_{G^{(2)}}(\mathcal{B})$ for every $E \in \mathcal{E}$, a Sylow p-subgroup P of $\mathrm{fix}_{G^{(2)}}(\mathcal{B})$ has order p or p^q.

If $|P| = p^q$, and there is an arc in Γ from some vertex of B to some vertex of $B' \neq B$, $B, B' \in \mathcal{B}$, then there is an arc from every vertex of B to every vertex of B'. By Theorems 4.4.1 and 4.2.15, Γ is isomorphic to the wreath product of a circulant digraph of order q and a circulant digraph of order p. By Exercise 5.1.4, Γ is isomorphic to a Cayley digraph of $\mathbb{Z}_q \times \mathbb{Z}_p$, and every such digraph is isomorphic to a (q,p)-metacirculant digraph by Exercise 1.5.5.

If $|P| = p$, then $P_1 = \langle\rho\rangle$ and $P_2 = \tau^{-1}\langle\rho\rangle\tau$ are Sylow p-subgroups of $\mathrm{fix}_{G^{(2)}}(\mathcal{B})$, and so there exists $\delta \in \mathrm{fix}_{G^{(2)}}(\mathcal{B})$ such that $\delta^{-1}P_2\delta = P_1$. Replacing τ with $\tau\delta$, if necessary, we assume without loss of generality that $\tau^{-1}P_1\tau = P_1$. By the embedding theorem we may label the vertex set of Γ with elements of $\mathbb{Z}_q \times \mathbb{Z}_p$ in such a way that $\rho(i,j) = (i, j+1)$, and $\tau(i,j) = (i+1, \omega_i(j))$, where $\omega_i \in \mathcal{S}_p$. Set $\tau^{-1}\rho\tau = \rho^a$, where $a \in \mathbb{Z}_p^*$. Define $\bar{a}\colon \mathbb{Z}_q \times \mathbb{Z}_p \to \mathbb{Z}_q \times \mathbb{Z}_p$ by $\bar{a}(i,j) = (i, aj)$. Then $\bar{a}^{-1}\rho^a\bar{a} = \rho$, and so $\tau\bar{a}$ centralizes $\langle\rho\rangle$. By Lemma 9.2.1, we see that $\tau\bar{a}(i,j) = (i+1, j+b_i)$, $b_i \in \mathbb{Z}_p$. Hence $\tau(i,j) = \left(i+1, a^{-1}j + c_i\right)$, $c_i \in \mathbb{Z}_p$.

Let $H = \left\langle \tau, z_k : k \in \mathbb{Z}_q\right\rangle$, where $z_k(i,j) = (i, j+\delta_{ik})$, where δ_{ik} is Kronecker's delta function. That is $\delta_{ik} = 1$ if $i = k$ and 0 otherwise. Note that $\left\langle z_k : k \in \mathbb{Z}_q\right\rangle \trianglelefteq H$ as $\tau z_i = z_{i+1}^{a^{-1}}\tau$, and $H/\left\langle z_k : k \in \mathbb{Z}_q\right\rangle \cong \langle\tau\rangle$. We conclude that $\langle\tau\rangle$ is a Sylow q-subgroup of H. Now let, $\tau'\colon \mathbb{Z}_q \times \mathbb{Z}_p \to \mathbb{Z}_q \times \mathbb{Z}_p$ by $\tau'(i,j) = \left(i+1, a^{-1}j\right)$. Then $\tau' \in H$. As $|\tau| = q^t$, $t \geq 1$, we have $|a| = q^t$ or $a = 1$ and so $|\tau'| = q^t = |\tau|$. Hence $\langle\tau'\rangle$ is a Sylow q-subgroup of H as well. Thus there exists $\gamma \in H$ such that $\gamma^{-1}\langle\tau\rangle\gamma = \langle\tau'\rangle$. Also, $\langle\rho\rangle \trianglelefteq H$, and so $\gamma^{-1}\langle\rho\rangle\gamma = \langle\rho\rangle$. Thus $\gamma^{-1}\langle\rho,\tau\rangle\gamma = \langle\rho,\tau'\rangle$. By Exercise 1.1.11, we see that $\langle\rho,\tau'\rangle \leq \mathrm{Aut}\left(\gamma^{-1}(\Gamma)\right)$. It then follows by Theorem 1.5.11 that $\gamma^{-1}(\Gamma)$ is a $\left(q, p, a^{-1}\right)$-metacirculant. □

The following result, in which we use the notation of Section 6.6, shows that some Fermat graphs are not isomorphic to graphs in any other class of graphs we have seen.

Theorem 9.2.3 *An orbital digraph Γ of $\mathrm{SL}\left(2, 2^k\right)$ acting on $\mathcal{D}_\ell$ of valency 2^k and order qp, where q and p are distinct primes, is a graph that is not isomorphic to a metacirculant graph of order qp.*

Proof Recall from Lemma 6.6.6 that Γ is a graph. Suppose Γ is isomorphic to a metacirculant graph. Let $p > q$. Then $p = 2^k + 1$, q divides $2^k - 1$, and $\mathrm{SL}\left(2, 2^k\right)$ has a block system with blocks of size q. Note that a (p,q)-metacirculant graph is isomorphic to a circulant graph by Exercise 1.5.6, and a (q,p)-metacirculant graph is also isomorphic to a circulant graph by Theorem 1.5.13 as $\gcd(p-1, q) = \gcd\left(2^k, 2^k - 1\right) = 1$ since $q \mid \left(2^k - 1\right)$. Thus Γ is isomorphic to a circulant graph. As $\mathbb{Z}_{qp}$ is a Burnside group by Theorem 5.3.17 and Γ is not complete, $\mathrm{Aut}(\Gamma)$ has a block system $\mathcal{B}$, and as $\mathrm{SL}\left(2, 2^k\right) \leq \mathrm{Aut}(\Gamma)$, $\mathcal{B}$ is a block system of $\mathrm{SL}\left(2, 2^k\right)$. If $\mathrm{SL}\left(2, 2^k\right)$ has a block system C with blocks of size p, then, by Exercise 2.2.8, C is a normal block system, and so $\mathrm{SL}\left(2, 2^k\right)$ has a normal subgroup of order at least $2^k + 1$, a contradiction. So the only block systems of $\mathrm{SL}\left(2, 2^k\right)$ have blocks of order q. We conclude that $\mathcal{B}$ has blocks of size q.

Examining the automorphism groups of circulant graphs given in Theorem 8.5.1, if Γ is a normal Cayley graph of $\mathbb{Z}_{qp}$ then clearly $\mathrm{Aut}(\Gamma)$ has blocks of size p, and similarly, by Lemma 8.3.3, if $\mathrm{Aut}(\Gamma) = \mathcal{S}_q \times A$ or $B \times \mathcal{S}_p$, $A \leq \mathcal{S}_p$ and $B \leq \mathcal{S}_q$. Hence $\mathrm{Aut}(\Gamma) \cong \mathrm{Aut}(\Gamma_1) \wr \mathrm{Aut}(\Gamma_2)$ for some circulant graphs Γ_1 and Γ_2 of order p and q respectively. By the embedding theorem we see that $\mathrm{SL}\left(2, 2^k\right) \leq \left(\mathrm{SL}\left(2, 2^k\right)/\mathcal{B}\right) \wr \mathbb{Z}_q$ and by Theorem 5.2.11, $\mathrm{SL}\left(2, 2^k\right)^{(2)} \leq \mathcal{S}_p \wr \mathbb{Z}_q \leq \mathrm{Aut}(\Gamma)$. But then Γ has valency $(p-1)q$, a contradiction. □

Our next goal is to prove that every vertex-transitive graph of order $2p$ has a group of automorphisms with blocks of size p. We will require an additional idea, the basis of which is contained in the following result whose straightforward proof is Exercise 9.2.2.

Lemma 9.2.4 *Let n be a positive integer, $\rho \in \mathcal{S}_{2n}$ be a semiregular element of order n with orbits X and Y, and $x_0 \in X$ and $y_0 \in Y$. Label the vertices of X with $\{x_i : i \in \mathbb{Z}_n\}$ and Y with $\{y_i : i \in \mathbb{Z}_n\}$ in such a way that*

$$\rho = (x_0, x_1, \ldots, x_{n-1})(y_0, y_1, \ldots, y_{n-1}).$$

Let Γ be a graph with $\rho \in \mathrm{Aut}(\Gamma)$. Then there exists $R, S \subseteq \mathbb{Z}_n \backslash \{0\}$ and $T \subseteq \mathbb{Z}_n$ such that

$$E(\Gamma) = \{x_i x_{i+r} : i \in \mathbb{Z}_n, r \in R\} \cup \{y_i y_{i+s} : i \in \mathbb{Z}_n, s \in S\} \cup \{x_i y_{i+t} : i \in \mathbb{Z}_n, t \in T\},$$

where addition in the subscripts is modulo n. Also, $R = -R$ and $S = -S$.

Note that in the above result we do *not* assume Γ is vertex-transitive. The graphs are analogous to circulant graphs, but instead of choosing the symmetry "built into" the automorphism group being a regular cyclic subgroup, the "built in" symmetry is a semiregular element with two orbits.

Definition 9.2.5 Let n be a positive integer, and Γ a graph with a semiregular element ρ of order n with two orbits. We call such a Γ a **bicirculant graph**. Let $\rho = (x_0, x_1, \ldots, x_{n-1})(y_0, y_1, \ldots, y_{n-1})$, $R, S \subseteq \mathbb{Z}_n \backslash \{0\}$, and $T \subseteq \mathbb{Z}_n$ be as in Lemma 9.2.4. We call $[R, S, T]$ a **symbol** of Γ with respect to ρ, x, and y.

The generalized Petersen graphs we studied in Section 6.5 are all examples of bicirculant graphs. The symbol of the bicirculant corresponds to the connection set of a circulant, in that the symbol has just enough information about the edges to construct the graph from the symbol and the "built in" symmetry. A definition analogous to Definition 9.2.5 can be given for any group G, not just cyclic groups; see Exercises 9.2.5 and 9.2.6. There is considerable literature regarding bicirculant graphs. See, for example, Antončič et al. (2015); Marušič (1988b); Pisanski (2007). We next prove (Marušič, 1981a, Theorem 6.2) that every vertex-transitive graph of order $2p$ with an imprimitive group of automorphisms has an imprimitive group of automorphisms with blocks of size p. The proof is CFSG-free.

Proposition 9.2.6 *Every vertex-transitive graph of order $2p$, p an odd prime, with an imprimitive group of automorphisms, has an imprimitive group of automorphisms with blocks of size p, and consequently is isomorphic to a $(2, p)$-metacirculant graph.*

Proof Let Γ be a vertex-transitive graph of order $2p$ with $G \leq \mathrm{Aut}(\Gamma)$ transitive with block system $\mathcal{B}$ with blocks of size 2. Assume G is maximal with $\mathcal{B}$ as a block system. It suffices to show that there is $H \leq \mathrm{Aut}(\Gamma)$ with a block system C with blocks of size p, as such a block system is normal as $H/C = S_2$. The result will then follow by Theorem 9.2.2.

Let $K = \mathrm{fix}_G(\mathcal{B})$. If $K \neq 1$, then Γ is isomorphic to a $(p, 2)$-metacirculant graph by Theorem 9.2.2. By Exercise 1.5.6, Γ is isomorphic to a circulant graph, and so contains a regular cyclic subgroup that has blocks of size p. So we assume $K = 1$.

Suppose G is solvable. Then $G/\mathcal{B} \cong G$ is a solvable group of degree p. Let $K \leq S_p$ be transitive and solvable, and $1 \neq N \trianglelefteq K$. The set of orbits $\mathcal{B}$ of N are a block system of K by Theorem 2.2.7, and the blocks of $\mathcal{B}$ are not singleton sets as $N \neq 1$. As the size of a block in $\mathcal{B}$ divides p, $\mathcal{B}$ is one block of size p. This implies that p divides the order of every nontrivial subgroup of K in every normal series of K. As K is solvable, this implies K, and so $G/\mathcal{B}$ as well, has a normal Sylow p-subgroup. Thus G has a block system with blocks of size p by Theorem 2.2.7. So we assume G is nonsolvable.

We now set some notation for the rest of the proof. Let $\rho \in \mathrm{Aut}(\Gamma)$ be of order p, and X and Y be the orbits of $\langle\rho\rangle$ on $V(\Gamma)$. Then there exist $x_0 \in X$ and $y_0 \in Y$ such that $\{x_0, y_0\} \in \mathcal{B}$. For each $i \in \mathbb{Z}_p$ let $x_i = \rho^i(x_0)$, $y_i = \rho^i(y_0)$, and $B_i = \{x_i, y_i\}$. Then $\mathcal{B} = \left\{B_i : i \in \mathbb{Z}_p\right\}$. Let τ be the permutation of $V(\Gamma)$ interchanging x_i with y_i for each $i \in \mathbb{Z}_p$ (τ need not be in G).

Now, by Theorem 4.4.2 we have that $G/\mathcal{B}$ is 2-transitive, and so $\Gamma/\mathcal{B} \cong K_p$. Also, all bipartite graphs $B(i, j) = \Gamma[B_i, B_j]$, $i, j \in \mathbb{Z}_p$, $i \neq j$, are isomorphic, and there exists an element of $G/\mathcal{B}$ that interchanges B_i and B_j. Hence $B(i, j) \ncong K_{1,2} + K_1$. Furthermore, $B(i, j)$ also cannot be isomorphic to $K_{2,2}$, $2K_2$, or $\bar{K}_4$. as in all these cases τ would be a nonidentity element of $K = \mathrm{fix}_G(\mathcal{B})$. Thus $B(i, j)$ is isomorphic either to P_3 or to $K_2 + 2K_1$.

Suppose that $B(i, j) = P_3$, and let $r, s \in \mathbb{Z}_p$ be distinct. As $G/\mathcal{B}$ is 2-transitive, each leaf of the path $B(i, j)$ is mapped by an element of G to a leaf of the path $B(r, s)$, and cannot be mapped by an element of G to the middle edge of the path $B(r, s)$. Thus there exists an edge orbit O of G such that $\Gamma[O] \cap B(i, j) = 2K_2$ for any two distinct $i, j \in \mathbb{Z}_p$. Hence there exists $S \subseteq \mathbb{Z}_p^*$ such that $\Gamma[O] = \left[S, S, \mathbb{Z}_p^* \backslash S\right]$. By Exercise 9.2.3, we have that $\{X, Y\}$

is a block system of $\mathrm{Aut}(\Gamma[O])$. As O is an edge orbit of G, it follows that G is contained in $\mathrm{Aut}(\Gamma[O])$, and so G has a block system with blocks of size p on $V(\Gamma)$.

If $B(i, j) = K_2 + 2K_1$ then a similar argument applied to $\bar{\Gamma}$ shows that G has a block system with blocks of size p on $V(\Gamma)$. □

We can now classify vertex-transitive graphs of order $2p$.

Theorem 9.2.7 *Every vertex-transitive graph of order $2p$, p a prime, is isomorphic to a $(2, p)$-metacirculant.*

Proof By Proposition 9.2.6, we need only show that there are no vertex-transitive graphs with the property that every transitive subgroup is primitive. By CFSG, it can be shown that A_5 and S_5 acting on the set of 2-subsets of a set of order 5 are the only simply primitive groups of degree $2p$ (see, for example, Liebeck and Saxl (1985a); Scott (1972) as well as Neumann (2009)). The corresponding graphs are the Petersen graph and its complement whose automorphism group contains a transitive and imprimitive subgroup of order 20 by Example 1.3.6. □

It would be interesting to see a proof of Theorem 9.2.7 that does not use CFSG. As a corollary to Theorem 9.2.7 we obtain the following algebraic description (Marušič, 1981a, Theorem 6.4) of vertex-transitive graphs of order $2p$ in terms of bicirculant graphs. The proof is left to the reader in Exercise 9.2.6.

Corollary 9.2.8 *A bicirculant graph Γ of order $2p$, p a prime, is vertex-transitive if and only if there exist $S = -S \subseteq \mathbb{Z}_p^*$, $T \subseteq \mathbb{Z}_p$, $a \in \mathbb{Z}_p^*$ and $b \in \mathbb{Z}_p$, such that $[S, aS, T]$ is a symbol of Γ, and either $T \in \{\emptyset, \mathbb{Z}_p\}$ or $a^2 S = S$ and $aT + b = -T$.*

We now consider vertex-transitive graphs such that every imprimitive group G of automorphisms has only blocks of size q, with q odd. We only sketch the strategy as it involves the use of CFSG. We can more or less follow the line of proof of Theorem 9.2.7 in all cases except when the quotient group G/K is nonsolvable and the kernel K is trivial. In this case knowing that a nonsolvable group of prime degree is necessarily 2-transitive does not suffice to carry through the analysis of all possible bipartite graphs induced by two adjacent blocks using solely elementary graph-theoretic and group-theoretic tools. We have to rely on the list of all 2-transitive groups of prime degree that can be extracted from CFSG. With a careful analysis of all such groups one can show that a graph with an imprimitive group of automorphisms with blocks of size q is either a (q, p)-metacirculant or is a Fermat graph (Marušič and Scapellato, 1992a).

Proposition 9.2.9 *Let $p > q$ be primes, and Γ a vertex-transitive graph of order pq such that no imprimitive subgroup of* $\mathrm{Aut}(\Gamma)$ *has blocks of size p. Then p is a Fermat prime, q is a divisor of $p - 2$ and Γ is isomorphic to a Fermat graph.*

We are left with vertex-transitive graphs for which every transitive subgroup of automorphisms is primitive. Liebeck and Saxl (1985a, Table 3) have determined all primitive groups of degree mp, $m < p$, from which we get a reduced list of simply primitive groups for m a prime by first excluding all those groups that are 2-transitive, and then excluding all those remaining groups that have an imprimitive subgroup. In the end we are left with generalized orbital graphs of the groups listed in Table 9.1. We remark that there is an additional family of primitive graphs of order $91 = 7 \cdot 13$ that was not covered in either Marušič and Scapellato (1994b) or in Praeger and Xu (1993). This is due to a missing case in Liebeck and Saxl (1985a, Table 3) (this is actually caused by a mysterious '+' in the table that should be a $\pm$). This missing case is the primitive groups of degree $91 = 7 \cdot 13$ with socle $\mathrm{PSL}(2, 13)$ acting on left cosets of A_4 and is included in Row 7 of Table 9.1. Finally, Theorem 9.2.10 below (first proved in Marušič and Scapellato (1994b, Theorem 2.1)) gives a complete classification of connected vertex-transitive graphs of order pq (see also Marušič and Scapellato (1992a,b, 1993, 1994a)).

Theorem 9.2.10 *A connected vertex-transitive graph of order pq, where p and q are odd primes, must be one of the following:*

(i) *a metacirculant,*
(ii) *a Fermat graph,*
(iii) *a generalized orbital graph of one of the groups in Table 9.1.*

We would also like to remark that vertex-transitive graphs of order pq with a primitive automorphism group or an imprimitive automorphism group with only blocks of size q were also characterized by Praeger and Xu (1993); Praeger et al. (1993); Wang and Xu (1993). They also include the full automorphism groups of graphs with primitive automorphism groups, as well as detailed information about the orbital graphs. They also determine all symmetric graphs of order qp in Praeger et al. (1993). We should point out that most proofs in the papers cited in this paragraph apply equally well to digraphs, but there are a few exceptions as well as some, usually small, errors (Dobson et al., 2020a).

Row	soc(G)	(p,q)	Action	Comment
1	$P\Omega^{\epsilon}(2d,2)$	$\left(2^d-\epsilon,2^{d-1}+\epsilon\right)$	singular 1-spaces	$\epsilon=+1: d$ Fermat prime $\epsilon=-1: d-1$ Mersenne prime
2	M_{22}	$(11,7)$	see Atlas	
3	A_7	$(7,5)$	triples	
4	PSL(2, 61)	$(61,31)$	cosets of A_5	
5	$\mathrm{PSL}\left(2,q^2\right)$	$\left(\frac{q^2+1}{2},q\right)$	cosets of PGL(2, q)	$q\geq 5$
6	PSL(2, p)	$\left(p,\frac{p+1}{2}\right)$	cosets of D_{p-1}	$p\equiv 1\ (mod\ 4)$ $p\geq 13$
7	PSL(2, 13)	$(13,7)$	cosets of A_4	missing in Liebeck and Saxl (1985a)

Table 9.1 *Primitive groups of degree pq without imprimitive subgroups and with nonisomorphic generalized orbital graphs.*

9.3 Circulant Digraphs of Order n with $\gcd(n,\varphi(n))=1$

In this section we will prove the most accessible result on classifying vertex-transitive digraphs with a transitive solvable group of automorphisms. We will show that every such digraph of order n with $\gcd(n,\varphi(n))=1$ is isomorphic to a circulant digraph (as usual, φ is Euler's phi function). Also, all such digraphs of square-free order n with $\gcd(n,\varphi(n))=q$ a prime have been classified (Dobson, 2008). This last result is much more difficult, and, in general, a classification of vertex-transitive graphs of square-free order with a transitive solvable group of automorphisms seems intractable at the current time. In contrast, Li and Seress have determined all primitive groups of square-free degree (2003) and then in Li and Seress (2005) they determined which of those primitive permutation groups contain a regular subgroup.

There is one additional result to mention. Certain values of order a product of three distinct primes have been considered by Hassani, Iranmanesh, and Praeger (1998), mainly to determine the so-called non-Cayley numbers of order a product of three distinct primes.

Definition 9.3.1 A **non-Cayley number** is a positive integer n for which there exists a vertex-transitive graph of order n that is not isomorphic to a

Cayley graph. Similarly, a **Cayley number** is a positive integer n for which all vertex-transitive graphs of order n are Cayley graphs.

We have seen that every vertex-transitive graph of order p (Theorem 4.4.1), p^2 (Theorem 9.1.1), and p^3 (Theorem 9.1.3) is a Cayley graph, so p, p^2, and p^3 are Cayley numbers for every prime p. As the Petersen graph is not isomorphic to a Cayley graph by Theorem 1.2.22, 10 is a non-Cayley number. Marušič (1983) posed the problem of determining all Cayley numbers:

Problem 9.3.2 Determine all positive integers n for which every vertex-transitive graph of order n is isomorphic to a Cayley graph.

It is easy to see that a multiple of a non-Cayley number is a non-Cayley number (Exercise 9.3.2). Much progress has been made on determining non-Cayley numbers (see, for example, Li and Seress (2005); McKay and Praeger (1994, 1996); Seress (1998)), and the only integers for which the problem has not been completely settled are square-free integers. There was some hope that the problem would be settled by using the multiplicative property of non-Cayley numbers to show that all products of a fixed finite number of primes are non-Cayley, but Dobson and Spiga (2017) showed this is not the case. That is, for any positive integer k, there is a product of k distinct primes that is a Cayley number. Theorem 9.3.5 is crucial to the proof of that result. To the proof! We begin with a standard group-theoretic result that will be useful here and elsewhere. See, for example, Gorenstein (1968, Theorem 1.5).

Theorem 9.3.3 *If H is a minimal normal subgroup of a group G, then either H is an elementary abelian p-group for some prime p or H is the direct product of isomorphic nonabelian simple groups.*

We will also need the following result from Dobson (2008, Theorem 1.9).

Theorem 9.3.4 *A normally m-step imprimitive group G of square-free degree n contains a transitive solvable subgroup. Conversely, if G is transitive and solvable of square-free degree, then G is normally m-step imprimitive.*

The next result is the main result of this section. The proof is the same as in Dobson (2000), although there the statement has the hypothesis that $\mathrm{Aut}(\Gamma)$ contains a transitive solvable subgroup. For square-free integers n, this is equivalent to $\mathrm{Aut}(\Gamma)$ having a normally m-step imprimitive subgroup by Theorem 9.3.4. The reason the proof is included here is to see the use of Pálfy's theorem. This illustrates the importance of the Cayley isomorphism problem, especially in its most general form of determining conjugacy classes of regular groups in permutation groups, to the problem of classifying vertex-transitive graphs.

Theorem 9.3.5 *Let n be a positive integer with $\gcd(n, \varphi(n)) = 1$ and $\Omega(n) = m$. A vertex-transitive digraph Γ of order n is isomorphic to a circulant digraph of order n if and only if* $\mathrm{Aut}(\Gamma)$ *contains a normally m-step imprimitive subgroup. Furthermore, if $G \le S_n$ is a transitive and normally m-step imprimitive group, then $G^{(2)}$ contains a regular cyclic subgroup.*

Proof If Γ is isomorphic to a circulant digraph of order n, then $(\mathbb{Z}_n)_L \le \mathrm{Aut}(\Gamma)$ and $(\mathbb{Z}_n)_L$ is normally m-step imprimitive as $\mathbb{Z}_n$ contains a unique subgroup of each order dividing n. Conversely, we will show that if G is a transitive and normally m-step imprimitive group of degree n, then $G^{(2)}$ contains a regular cyclic subgroup. The result will then follow by Theorem 1.2.20. By Theorem 9.3.4, we may assume without loss of generality that G is solvable. Let $H \le G^{(2)}$ be transitive and normally m-step imprimitive such that H contains a normal abelian subgroup K whose set of orbits $\mathcal{B}$ are of maximum cardinality k. Subject to the previous conditions, we assume H is maximal. As G is solvable, by Theorem 9.3.3 $H \ne 1$, and $\mathcal{B}$ is a block system of H by Theorem 2.2.7.

Towards a contradiction suppose K is not transitive. We first show there is $x \in H^{(2)}$ such that x is semiregular of order k, $\langle x, K\rangle$ is abelian, and the orbits of $\langle x\rangle$ are the same as the orbits of K. As K is abelian, K^B is abelian for every $B \in \mathcal{B}$. As B is an orbit of K, K^B is transitive on $B \in \mathcal{B}$, and, as a transitive abelian group is regular by Theorem 2.2.17, K^B is a regular abelian group of square-free degree and so cyclic. Now let $q|k$ be prime, and P_q a Sylow q-subgroup of K. As K is abelian, P_q is unique. As $K \trianglelefteq H$, $P_q \trianglelefteq H$, and so there is a normal block system $\mathcal{B}_q$ of H with blocks of size q. As $P_q \le K$, we see that $\mathcal{B}_q \le \mathcal{B}$.

Define the equivalence relation $\equiv$ as in Definition 5.3.10. As K^B is regular and cyclic, $P_q^{B_q}$ is semiregular and cyclic. We conclude that if $B_q, B_q' \in \mathcal{B}_q$ and $B_q, B_q' \subseteq B \in \mathcal{B}$, then γ^{B_q} is a q-cycle if and only if $\gamma^{B_q'}$ is a q-cycle for every nontrivial $\gamma \in P_q$. Then $\mathcal{E}_q$, the union of the equivalence classes of $\equiv$ as in Lemma 5.3.11, satisfies $\mathcal{B} \le \mathcal{E}_q$, and by Lemma 5.3.11 there is a semiregular element $x_q \in H^{(2)}$ and $(x_q)^B \in P_q^B$ for every $B \in \mathcal{B}$. Thus $(x_q)^B \in K^B$ and as K^B is abelian, $\langle x_q, K\rangle$ is abelian. As $\mathcal{B}_q \le \mathcal{B}$, the orbits of x_q are contained in the orbits of K.

Set $x = \Pi_{q|k \text{ is prime}} x_q$. Note that if q and p are primes dividing k, then $x_q^B, x_p^B \in K^B$ for every $B \in \mathcal{B}$, and so x has order k and is semiregular. As $\mathcal{B}_q \le \mathcal{B}$ for every prime $q|k$, we see that the set of orbits of $\langle x\rangle$ is $\mathcal{B}$.

As $H/\mathcal{B}$ is solvable it contains a nontrivial normal abelian p-subgroup $P_1/\mathcal{B}$ for some prime p by Theorem 9.3.3. Note that as n is square-free, $H/\mathcal{B}$ has a block system $\mathcal{C}/\mathcal{B}$ of $n/(kp)$ blocks of size p that is the set of orbits of $P_1/\mathcal{B}$. Hence H has $\mathcal{B} \prec \mathcal{C}$ as a block system with blocks of size kp. We will show that there exists abelian $M \le H^{(2)}$ such that M has $n/(kp)$ orbits of size kp, and $N_G(M)$ is transitive, contradicting our choice of H.

Let P be a Sylow p-subgroup of $\mathrm{fix}_H(C)$. As $C/\mathcal{B}$ is a block system of $H/\mathcal{B}$, by Theorem 5.2.10 $C/\mathcal{B}$ is a block system of $(H/\mathcal{B})^{(2)}$. Applying the argument in the third paragraph of this proof to $(H/\mathcal{B})^{(2)}$, there is a semiregular element $\bar{\rho} \in \mathrm{fix}_{(H/\mathcal{B})^{(2)}}(C/\mathcal{B})$ contained in the same Sylow p-subgroup of $\mathrm{fix}_{(H/\mathcal{B})^{(2)}}(C/\mathcal{B})$ as $P/\mathcal{B}$. Let $\rho \in \mathcal{S}_n$ such that $\rho/\mathcal{B} = \bar{\rho}$ and $\rho^{-1}x\rho = x$. We next show that $\rho^{-1}K\rho = K$.

We will show that $\rho^{-1}P_q\rho = P_q$, which will establish that $\rho^{-1}K\rho = K$, as K is abelian. By Lemma 5.3.11, $P_q \le \left\langle x^{k/q}|_E : E \in \mathcal{E}_q\right\rangle$. As the blocks of $G/\mathcal{B}$ are preserved by $\bar{\rho}$ and each $E \in \mathcal{E}_q$ is a union of blocks of $\mathcal{B}$, for each $E \in \mathcal{E}_q$ we have $\rho^{-1}x^{k/q}|_E\rho = x^{k/q}|_{E'}$, for some $E' \in \mathcal{E}_q$. Thus $\rho^{-1}P_q\rho = P_q$ so that $\rho^{-1}K\rho = K$.

Let $W = N_{\mathcal{S}_n}(K) \cap \left(\mathcal{S}_{n/(kp)} \wr \left(\mathcal{S}_p \wr \mathcal{S}_k\right)\right)$. Then $G \le W$, C is a block system of W, and as ρ normalizes K, $\rho \in W$. Let P_2 be a Sylow p-subgroup of $\mathrm{fix}_W(C)$ that contains ρ. By Lemma 6.5.17 and Corollary 4.4.11, for $B \in \mathcal{B}$ we have $\mathrm{Stab}_{N_{\mathcal{S}_n}(K)}(B)^B$ is permutation isomorphic to a subgroup of $(\mathbb{Z}_k)_L \rtimes \mathrm{Aut}(\mathbb{Z}_k)$ (as $W \le N_{\mathcal{S}_n}(K)$), and consequently has order dividing $k \cdot \varphi(k)$, which is relatively prime to p as $\gcd(n, \varphi(n)) = 1$. Suppose $\phi \in P_2$ and $\phi \notin \langle \rho|_C : C \in C\rangle$. As $P_2/\mathcal{B}$ is isomorphic to a subgroup of $1_{\mathcal{S}_{n/(mp)}} \wr (\mathbb{Z}_p)_L$, if $\phi^C \notin \left\langle \rho^C \right\rangle$ for some $C \in C$, then $\phi\rho^a$ fixes each block of $\mathcal{B}$ contained in C for some $a \in \mathbb{Z}_p$. Then $(\phi\rho^a)^C$ has order a power of p that divides $k \cdot \varphi(k)$, which is relatively prime to p. But then $\phi^C = (\rho^{-a})^C$, a contradiction. We conclude that $P_2 \le \langle \rho|_C : C \in C\rangle$. Let P_3 be a Sylow p-subgroup of $\mathrm{fix}_W(C)$ that contains P. Then there exists $\delta \in W$ such that $\delta^{-1}P_2\delta = P_3$. By replacing G with $\delta^{-1}G\delta$, we may assume $P_3 = P_2$. Hence $P \le \langle \rho|_C : C \in C\rangle$. Finally, $x \in \delta^{-1}K\delta$ as $\delta^{-1}K\delta = K$.

As ρ commutes with x and both ρ and x are contained in G (as G was replaced with $\delta^{-1}G\delta$), $\langle \rho, x\rangle$ is a semiregular cyclic subgroup with orbits of size pk. Let $M = \langle x|_C, \rho|_C : C \in C\rangle \cap G$. We will show that $N_G(M)$ is transitive. It suffices to show that $N_G(M)/C = G/C$. We will have then established the result as M is a normal abelian subgroup of $N_G(M)$ whose orbits are of order kp, contradicting our choice of H. Let $g \in G$. As $\mathbb{Z}_k$ is a CI-group, by Lemma 7.6.10 (here is where Pálfy's Theorem is used as Lemma 7.6.10 depends upon it) there exists $h_g \in \mathrm{fix}_G(C)$ such that $h_g^{-1}g^{-1}\langle \rho, x\rangle gh_g \le M$. Then $gh_g \in N_G(M)$ for every $g \in G$, $gh_g/C = g/C$ as $h_g \in \mathrm{fix}_G(C)$, and $N_G(M)$ is transitive as required. □

We leave the proof of the following characterization of the positive integers n for which every vertex-transitive graph of order n is a circulant graph as Exercise 9.3.3.

Corollary 9.3.6 *Every vertex-transitive graph Γ of order n is isomorphic to a circulant graph of order n if and only if* $\mathrm{Aut}(\Gamma)$ *contains a normally $\Omega(n)$-step imprimitive subgroup and $n = 4$, 6, or* $\gcd(n, \varphi(n)) = 1$.

9.4 Exercises

9.1.1 Using the notation of the proof of Theorem 9.1.3, show that if there are p equivalence classes of $\equiv$ and $P' \leq \mathbb{Z}_p \wr \mathbb{Z}_p^2$, then Γ is isomorphic to a Cayley digraph of $\mathbb{Z}_p^2 \rtimes \mathbb{Z}_p$.

9.1.2 Show that there exist transitive groups of degree p^3, p an odd prime, that do not contain a regular subgroup. (Hint: Let R be a regular group of degree p^2, and P a Sylow p-subgroup of $N_{S_{p^2}}(R)$. Show that P has order p^3, and that the group $P \rtimes_\phi \mathbb{Z}_p$ does not contain a regular subgroup, where $\phi \in \mathrm{Aut}(P)\backslash P$.)

9.1.3 Show that if there is a vertex-transitive digraph Γ of order n that is not isomorphic to a Cayley digraph, then there is a vertex-transitive digraph Γ' of order mn that is not isomorphic to a Cayley digraph, $m \geq 1$.

9.1.4 For the graph Γ as defined in Theorem 9.1.6, show that a clique of order p^2 in Γ that contains 0 cannot contain a vertex of $V_{p-1} = \left\{(p-1, j) : j \in \mathbb{Z}_{p^3}\right\}$.

9.1.5 Using the notation of Lemma 9.1.5, show by induction that $\tau^b\rho^d = \rho^{d(1+p)^b}\tau^b$.

9.2.1 Let n be a square-free integer. Show that every weak metacirculant digraph is isomorphic to a metacirculant digraph.

9.2.2 Prove Lemma 9.2.4.

9.2.3 For an odd integer n let ρ be a $(2, n)$-semiregular automorphism of a vertex-transitive graph Γ, let X and Y be the two orbits of ρ, let $x \in X$, $y \in Y$ and let the triple $[S, S, \mathbb{Z}_n^*\backslash S]$ be the symbol of Γ with respect to ρ, x and y. Show that $\{X, Y\}$ is a block system of $\mathrm{Aut}(\Gamma)$. (Hint: Show that the set of common neighbors of two elements of X is of even order, while the set of common neighbors of an element of X and an element of Y is of odd order. Then proceed by contradiction.)

9.2.4 Prove Corollary 9.2.8.

9.2.5 Let G be a group and $S \subseteq G$. Define the **Haar graph** of G with connection set S, denoted $\mathrm{Haar}(G, S)$, to be the graph with vertex set $\mathbb{Z}_2 \times G$ and edge set $\{\{(0, g), (1, gs)\} : g \in G, s \in S\}$. Show that $\mathrm{Aut}(\mathrm{Haar}(G, S))$ has a semiregular subgroup with two orbits isomorphic to G, and that $\mathrm{Haar}(G, S)$ is bipartite (this can be interpreted to mean that Haar graphs are bipartite analogues of Cayley graphs). For more information about Haar graphs, see Conder et al. (2018); Estélyi and Pisanski (2016); Feng et al. (2020); Hladnik et al. (2002); Koike and Kovács (2014).

9.2.6 Let G be a group, $L, R \leq G$, and S be a union of double cosets of R and L in G, namely, $S = \bigcup_i Rs_iL$. Define a bipartite graph $\Gamma = B(G, L, R, S)$ with bipartition $V(\Gamma) = G/L \cup G/R$ and edge set $E(\Gamma) = \{\{gL, gsR\} : g \in G, s \in S\}$. This graph is called the **bi-coset graph** with respect to L, R, and S. We call S the **connection set** of Γ. Show that $B(G, L, R, S)$ is a bipartite graph and that $\mathrm{Aut}(\mathcal{B}(G, L, R, S))$ contains a subgroup isomorphic to G with two orbits (this can be interpreted to mean that bi-coset graphs are bipartite analogues of double coset graphs). For more information about bi-coset graphs, see Du and Xu (2000).

9.3.1 Show that $S_m \wr G$ contains a regular subgroup if and only if G contains a regular subgroup. Compare this with Theorem 4.3.5.

9.3.2 Use the previous exercise to show that any multiple of a non-Cayley number is a non-Cayley number.

9.3.3 Prove Corollary 9.3.6.

10

Symmetric Graphs

Of course, when it comes to symmetric graphs the first valency of interest is valency 3. Cubic symmetric graphs (and cubic vertex-transitive graphs in general) have been a focal point of research for many decades. For example, Foster started collecting examples of small cubic symmetric graphs prior to 1934, maintaining a list of all such graphs. No claim of completeness was given for his list. In 1988 the then current version of the census, the so-called Foster census, was published by Bouwer, Chernoff, Monson and Star (Foster, 1988). The census contained data for graphs on up to 512 vertices and was remarkably complete, in that it only missed the graphs F408B, F432E, F448C, F480C, F480D, F486D, F512D, F512E, F512F, and F512G. Interestingly, the construction of F432E appears in the paper by Frucht (1952) as the first example of a cubic 1-arc-regular graph (see Exercise 10.5.1 for a construction of an infinite family of cubic 1-arc-regular graphs). Conder and Morton (1995) classified the cubic symmetric graphs on at most 240 vertices, and Conder and Dobcsányi (2002) extended the classifiction to 768 vertices. Conder extended the census to up to 2048 vertices (2006), and subsequently to up to $10,000$ vertices (2011).

There are many well-studied graphs among small order cubic symmetric graphs. To mention a few that we have seen so far: the Petersen graph and the Coxeter graph are the only two known cubic symmetric graphs that are not Hamiltonian (Theorems 3.4.1 and 6.1.5), while F8, F16, F20A, F20B, F24, and F48 (using the Foster census notation) are the remaining six symmetric generalized Petersen graphs (Theorem 6.5.14). There are also well-known cubic symmetric graphs we have not seen: for example Tutte's 8-cage F30 is the smallest 5-arc-regular cubic symmetric graph. The next sections are mainly devoted to an overview of recent developments in cubic symmetric graphs, but symmetric graphs of larger valencies will also be addressed. We begin with a classification of cubic symmetric graphs into 17 families.

10.1 The 17 Families of Cubic Symmetric Graphs

In this section we give an expository overview of work that has characterized cubic symmetric graphs into 17 families. The work was begun by Djoković and Miller (1980) who proved that a vertex stabilizer in an s-arc-regular subgroup of automorphisms of a cubic symmetric graph is isomorphic to $\mathbb{Z}_3, \mathcal{S}_3, \mathcal{S}_3 \times \mathbb{Z}_2, \mathcal{S}_4$, or $\mathcal{S}_4 \times \mathbb{Z}_2$ for $s = 1, 2, 3, 4$, or 5, respectively. They also proved that there is just one possibility for edge stabilizers if $s \in \{1, 3, 5\}$ and there are two such possibilities if $s \in \{2, 4\}$; see Table 10.1. In particular, for $s = 2$ the edge stabilizer is isomorphic to either $\mathbb{Z}_2 \times \mathbb{Z}_2$ or $\mathbb{Z}_4$, and for $s = 4$ the edge stabilizer is either isomorphic to the dihedral group D_{16} or to the quasi-dihedral group QD_{16} of order 16. The **quasi-dihedral group** QD_{2^n} has presentation $\left\langle r, s : r^{2^{n-1}} = s^2 = 1, srs = r^{2^{n-2}-1} \right\rangle$.

s	$\mathrm{Stab}_{\mathrm{Aut}(\Gamma)}(v)$	$\mathrm{Stab}_{\mathrm{Aut}(\Gamma)}(e)$
1	$\mathbb{Z}_3$	$\mathbb{Z}_2$
2	$\mathcal{S}_3$	$\mathbb{Z}_2^2$ or $\mathbb{Z}_4$
3	$\mathcal{S}_3 \times \mathbb{Z}_2$	D_8
4	$\mathcal{S}_4$	D_{16} or QD_{16}
5	$\mathcal{S}_4 \times \mathbb{Z}_2$	$(D_8 \times \mathbb{Z}_2) \rtimes \mathbb{Z}_2$

Table 10.1 *Vertex and edge stabilizers in cubic s-arc-regular graphs.*

They also showed that the automorphism group of any finite symmetric cubic graph is an epimorphic image of one of the following seven infinite groups:

$$G_1 = \left\langle h, a : h^3 = a^2 = 1 \right\rangle,$$
$$G_2^1 = \left\langle h, a, p : h^3 = a^2 = p^2 = 1, apa = p, php = h^{-1} \right\rangle,$$
$$G_2^2 = \left\langle h, a, p : h^3 = p^2 = 1, a^2 = p, php = h^{-1} \right\rangle,$$
$$G_3 = \left\langle h, a, p, q : h^3 = a^2 = p^2 = q^2 = 1, apa = q, qp = pq, ph = hp, \right.$$
$$\left. qhq = h^{-1} \right\rangle,$$
$$G_4^1 = \left\langle h, a, p, q, r : h^3 = a^2 = p^2 = q^2 = r^2 = 1, apa = p, aqa = r, \right.$$
$$\left. h^{-1}ph = q, h^{-1}qh = pq, rhr = h^{-1}, pq = qp, pr = rp, rq = pqr \right\rangle,$$
$$G_4^2 = \left\langle h, a, p, q, r : h^3 = p^2 = q^2 = r^2 = 1, a^2 = p, a^{-1}qa = r, h^{-1}ph = q, \right.$$
$$\left. h^{-1}qh = pq, rhr = h^{-1}, pq = qp, pr = rp, rq = pqr \right\rangle,$$
$$G_5 = \left\langle h, a, p, q, r, s : h^3 = a^2 = p^2 = q^2 = r^2 = s^2 = 1, apa = q, ara = s, \right.$$
$$h^{-1}ph = p, h^{-1}qh = r, h^{-1}rh = pqr, shs = h^{-1}, pq = qp, pr = rp,$$
$$\left. ps = sp, qr = rq, qs = sq, sr = pqrs \right\rangle.$$

This means that an arc-transitive subgroup of a cubic symmetric graph is a quotient of one of these seven groups by some normal torsion-free subgroup. Recall that a subgroup is **torsion-free** if the only element of finite order is the identity. In particular, an s-arc-regular subgroup of automorphisms of a cubic symmetric graph is a quotient group of a group G_s if $s \in \{1, 3, 5\}$, and of a group G_s^i, $i \in \{1, 2\}$, if $s \in \{2, 4\}$.

Conder and Nedela (2009) refined this classification and gave a complete characterization of types of cubic symmetric graphs according to the structure of arc-transitive subgroups of their automorphism groups. They additionally showed that each class contains infinitely many cubic graphs. For example, a cubic symmetric graph Γ is said to be of type $\left\{1, 2^1, 2^2, 3\right\}$ if its automorphism group is 3-arc-regular, contains two 2-arc-regular subgroups, of which one is a quotient of the group G_2^1 and one of G_2^2, and also contains a 1-arc-regular subgroup. All possible types are listed in the first column of Table 10.2.

Type	Odd automorphisms in Γ of order $2n$ exist if and only if	Bipartite?
$\{1\}$	n odd	Sometimes
$\left\{1, 2^1\right\}$	n odd, or $n = 2^{k-1}(2t+1)$ and Γ is a (2t+1)-Cayley graph on a cyclic group of order 2^k, where $k \geq 2$	Sometimes
$\left\{2^1\right\}$	n odd and Γ bipartite	Sometimes
$\left\{2^2\right\}$	never	Sometimes
$\left\{1, 2^1, 2^2, 3\right\}$	n odd	Always
$\left\{2^1, 2^2, 3\right\}$	n odd	Always
$\left\{2^1, 3\right\}$	n odd	Never
$\left\{2^2, 3\right\}$	n odd	Never
$\{3\}$	n odd and Γ bipartite	Sometimes
$\left\{1, 4^1\right\}$	n odd	Always
$\left\{4^1\right\}$	n odd and Γ bipartite	Sometimes
$\left\{4^2\right\}$	n odd	Sometimes
$\left\{1, 4^1, 4^2, 5\right\}$	n odd	Always
$\left\{4^1, 4^2, 5\right\}$	n odd	Always
$\left\{4^1, 5\right\}$	never	Never
$\left\{4^2, 5\right\}$	never	Never
$\{5\}$	n odd and Γ bipartite	Sometimes

Table 10.2 *All possible types of cubic symmetric graphs, the existence of odd automorphisms, and whether they are bipartite.*

This refined classification can be used to solve otherwise intractable problems, which we now illustrate. The problem we will discuss is a relatively new problem, posed in Hujdurović et al. (2016). It is a common result in a first group theory course that a transitive permutation group is either contained in the alternating group or exactly half of its elements are even permutations and half are odd permutations. Motivated by finding the full automorphism group of a graph, Hujdurović, Kutnar, and Marušič posed the following problem (Hujdurović et al., 2016).

Problem 10.1.1 Which vertex-transitive graphs have automorphism group with an odd permutation?

With this problem in mind, there is an obvious definition.

Definition 10.1.2 Let Γ be a graph and $\gamma \in \mathrm{Aut}(\Gamma)$. We say γ is an **odd automorphism** of Γ is γ is an odd permutation and an **even automorphism** otherwise. Similarly, a permutation group G is **even** if all of its elements are even permutations, and **odd** otherwise.

Table 10.2 also summarizes information from Kutnar and Marušič (2019), obtained using the refined characterization of automorphism groups of cubic symmetric graphs discussed above, on the existence of odd automorphisms in such graphs.

The next proposition gives necessary and sufficient conditions for the existence of odd automorphisms in 1-arc-regular subgroups of automorphisms of cubic symmetric graphs (implying row 1 of Table 10.2). For other types of cubic symmetric graphs see Kutnar and Marušič (2019).

Proposition 10.1.3 *Let Γ be a cubic symmetric graph of order $2n$ with a 1-arc-regular subgroup $G \leq \mathrm{Aut}(\Gamma)$. Then there exists an odd automorphism of Γ in G if and only if n is odd.*

Proof G is the quotient of G_1 with a normal torsion-free subgroup N of G_1 and G is generated by an element h of order 3 and an involution a. As N is torsion-free, aN and hN are elements of order 2 and 3, respectively, in G, and $G = \langle aN, hN\rangle$. As $|hN| = 3$, it is an even automorphism. So G is odd if and only if aN is odd. Let $v \in V(\Gamma)$. By Table 10.1, $\mathrm{Stab}_G(v) \cong \mathbb{Z}_3$, $v \in V(\Gamma)$, and so aN is semiregular. As $|V(\Gamma)| = 2n$, aN is a product of n transpositions. Hence aN is odd if and only if n is odd. □

10.2 Cubic Symmetric Graphs of Small Girth

In the study of cubic symmetric graphs, it was recognized immediately that symmetric graphs usually do not have small girth. For example, Tutte showed in his original paper (Tutte, 1947) that a cubic s-arc-transitive graph (recall from Theorem 3.1.5 that an arc-transitive cubic graph is s-arc-regular for some

$s \geq 1$) has girth at least $2s-2$ (which we stated as Theorem 3.1.9). The basis of this relationship between s-arc-transitivity and girth is easy to see: If a graph is s-arc-transitive it cannot have s-arcs that are contained in a cycle and not contained in a cycle, so intuitively such a graph either has small diameter or large girth.

As is usual in discrete mathematics, there are some small exceptions to the notion that a cubic arc-transitive graph has small diameter or large girth, and in a result that first appeared in Glover and Marušič (2007), we first determine all cubic arc-transitive graphs of girth at most 5 in Proposition 10.2.2. The rest of the section is mainly expository, and we survey what is known about cubic arc-transitive graphs of girth 6 or more, often in conjunction with the 17 families of cubic symmetric graphs discussed in the previous section. We begin with some terms that have proven useful in studying such graphs.

Definition 10.2.1 Let Γ be a graph and $G \leq \mathrm{Aut}(\Gamma)$. A walk $\vec{C} = u_0, \ldots, u_r$ in Γ is called G**-consistent** (or just **consistent** if $G = \mathrm{Aut}(\Gamma)$) if there exists $g \in G$ such that $g(u_i) = u_{i+1}$ for $i \in \{0, 1, \ldots, r-1\}$. The automorphism g is called a **shunt automorphism** for $\vec{C}$. If $\vec{C}$ is a simple closed walk then we say that $\vec{C}$ is a G**-consistent oriented cycle**. The underlying nonoriented cycle of $\vec{C}$ is called a G**-consistent cycle** and is denoted by C.

In our next result, we will deviate from our usual practice and allow multigraphs. The reason to do this is that in the next chapter we will consider quotient graphs of cubic symmetric graphs that are cubic symmetric graphs, and it turns out that exactly one is a multigraph. It is called the theta graph and denoted Θ_2. It has two vertices, joined by three edges (and so looks like the Greek symbol Θ).

Proposition 10.2.2 *Let Γ be a cubic symmetric graph of girth at most 5. Then Γ is isomorphic to Θ_2, K_4, $K_{3,3}$, the cube Q_3, the Petersen graph, or the dodecahedron.*

Proof Let $G = \mathrm{Aut}(\Gamma)$, and let $g(\Gamma)$ be the girth of Γ.

If $g(\Gamma) = 2$, then Γ is a multigraph, and it is clear that Θ_2 is the only cubic symmetric multigraph of girth 2.

Let $v \in V(\Gamma)$ have neighbors u_0, u_1, and u_2. As Γ is arc-transitive, there exists $\alpha \in G$ fixing v and cyclically permuting its neighbors, that is, $\alpha(u_i) = u_{i+1}$, $i \in \mathbb{Z}_3$.

If $g(\Gamma) = 3$, then each u_i is adjacent to the other two neighbors of v, and so $\Gamma \cong K_4$.

Suppose $g(\Gamma) = 4$. Then $\{u_0, u_1, u_2\}$ is an independent set in Γ, but there must exist a vertex, say, $x_{01} \neq v$, which is adjacent to both u_0 and u_1.

Suppose x_{01} is adjacent to u_2. As Γ is arc-transitive, there is an automorphism $\gamma \in \mathrm{Aut}(\Gamma)$ that interchanges v and u_0. Then $\gamma(x_{01})$ is a neighbor x of v. As the neighbors of v and x_{01} are the same, the neighbors of $\gamma(v) = u_0$ and $\gamma(x) = x_{01}$ are the same. As $\gamma(x) = u_1$ or u_2, u_0 has the same neighbors as one of u_1 and u_2, say u_1. Then $\alpha(u_1) = u_2$ so α maps the neighbors of u_1 to themselves. As $\alpha(u_2) = u_3$, the neighbors of u_1, u_2, and u_3 are the same. Hence $\Gamma \cong K_{3,3}$.

Suppose x_{01} is not adjacent to u_2. Set $x_{12} = \alpha(x_{01})$ and $x_{02} = \alpha^2(x_{01})$. Then x_{12} is adjacent to u_1 and u_2 while x_{02} is adjacent to u_0 and u_2. As x_{01} is not adjacent to u_2, we have that x_{01}, x_{02}, and x_{12} are pair-wise disjoint. Now observe that every 2-arc that begins with v is contained in a 4-cycle. As Γ is vertex-transitive, every 2-arc beginning with any vertex is contained in a 4-cycle. Thus x_{01}, u_0, x_{02} is contained in a 4-cycle, say $x_{01}, u_0, x_{02}, w, x_{01}$. As w is a neighbor of x_{01} and x_{02}, it must also be a neighbor of x_{12}. Hence Γ is the cube.

Suppose $g(\Gamma) = 5$. We first show that the order of Γ is divisible by 5. As Γ is vertex-transitive, v is contained in a 5-cycle in Γ. As $\alpha \in \mathrm{Aut}(\Gamma)$, each path $u_i v u_j$, $i \neq j$, is contained in a 5-cycle in Γ. This gives that each edge vu_i, $i \in \mathbb{Z}_3$, is contained in at least two 5-cycles. Also, as $g(\Gamma) = 5$, no path of length 3 can be contained in two different 5-cycles. Let $xy \in E(\Gamma)$ and consider the subgraph Δ of Γ consisting of xy and its incident edges. Every cycle in Γ that contains xy must contain a path of length 3 contained in Δ. As there are exactly four paths of length 3 in Δ, xy is contained in at most four 5-cycles. As Γ is arc-transitive, each edge of Γ is contained in the same number $2 \leq k \leq 4$ of 5-cycles in Γ. So the number of 5-cycles in Γ is $\left(\frac{3n}{2} \cdot k\right)/5 = 3nk/(10)$, where n is the order of Γ. As $\gcd(3k, 5) = 1$, $5|n$. As G is transitive, 5 also divides $|G|$.

We next show that Γ has a consistent 5-cycle. We first show that Γ has an automorphism of order 5 fixing a 5-cycle. By Corollary 3.1.6 we have $|G| = 3 \cdot 2^r \cdot n$, where r is maximum such that Γ is $(r+1)$-arc-transitive. Then $r \leq 2$ by Theorem 3.1.9. Let P be a Sylow 5-subgroup of G of order 5^ℓ. Hence 5^ℓ divides n and as there are $3nk/(10)$ 5-cycles in Γ, the number of 5-cycles is of the form $5^{\ell-1} \cdot m$, where $\gcd(5, m) = 1$. Let P act on the set of all 5-cycles in Γ in the canonical way. As P is transitive on each of its orbits and the sum of the lengths of the orbits is $5^{\ell-1} \cdot m$, there is an orbit $\mathcal{O}$ such that 5^ℓ does not divide its length. So $\mathrm{Stab}_P(\mathcal{O})$ contains an element ρ of order 5 that fixes some 5-cycle, say $C = v_0 v_1 v_2 v_3 v_4 v_0$, in Γ. As Γ is cubic, ρ is semiregular by Lemma 7.3.9. Then $\rho^{V(C)} \in \mathrm{Aut}(C)$ or, equivalently, $\rho^{V(C)} = (v_0, v_i, v_{2i}, v_{3i}, v_{4i})$ for some i. Raising ρ to an appropriate power we assume $\rho^{V(C)} = (v_0, v_1, v_2, v_3, v_4)$ and C is a consistent cycle.

Let u_i be the neighbor of v_i not in C, $i \in \mathbb{Z}_5$, so that $\rho(u_i) = u_{i+1}$ for all $i \in \mathbb{Z}_5$. If $U = \{u_i : i \in \mathbb{Z}_5\}$ induces a cycle then Γ has 10 vertices, and by the

Foster census the Petersen graph is the only cubic symmetric graph of order 10. We may thus assume U is an independent set of vertices. By comments at the beginning of this case, there is a 5-cycle D in Γ containing u_0v_0 and v_0v_1. It must then be the case that $v_1u_1 \in E(D)$, and so $D = u_0v_0v_1u_1w_0u_0$ for some vertex w_0. The orbit of ρ that contains w_0 then gives common neighbors w_i of u_i and u_{i+1}, $0 \leq i \leq 4$. The only vertices in $\Gamma[\{v_i, u_i, w_i : i \in \mathbb{Z}_5\}]$ of valency 2 are $\{w_i : i \in \mathbb{Z}_5\}$. Hence there are vertices $\{x_i : i \in \mathbb{Z}_5\}$ such that $w_ix_i \in E(\Gamma)$ but $x_i \notin \{v_i, u_i, w_i : i \in \mathbb{Z}_5\}$. This implies $\rho(x_i) = x_{i+1}$ for every $i \in \mathbb{Z}_5$. As above, w_0x_0 and w_0u_0 are contained in a 5-cycle that must be $w_0u_0w_4x_4x_0w_0$. This implies $x_ix_{i+1} \in E(\Gamma)$ for every $i \in \mathbb{Z}_5$. Then $\Gamma[\{v_i, u_i, w_i, x_i : i \in \mathbb{Z}_5\}]$ is regular of valency 3, and so is Γ. Finally, Γ is isomorphic to the dodecahedron. □

In contrast to Proposition 10.2.2 there are infinitely many cubic symmetric graphs of girth 6 (see Exercise 10.5.1). These graphs have received a lot of attention over the years. For example, Miller (1971, Theorem 2.2) characterized their automorphism groups as two generator groups satisfying certain relations. Further, Negami (1985) and Feng and Nedela (2006) described these graphs via certain embeddings on appropriate surfaces. In particular, it follows from the result of Feng and Nedela (2006) that every girth cycle in a cubic symmetric graph of girth $g \leq 7$, with the exception of the Möbius–Kantor graph, is a consistent cycle. The idea of consistent cycles together with a beautiful result of Conway (1971), stated next without proof, has proved to be quite useful when dealing with symmetric graphs.

Proposition 10.2.3 *Let Γ be an arc-transitive graph of valency $d \geq 2$, with $G \leq \mathrm{Aut}(\Gamma)$ arc-transitive. There are exactly $d - 1$ orbits in the action of G on the set of G-consistent oriented cycles in Γ.*

The first published proof of Proposition 10.2.3 was given by Biggs (1981), and a more general version was published in Miklavič et al. (2007a). For cubic symmetric graphs, Proposition 10.2.3 implies that there are two orbits of consistent cycles. For example, the Möbius–Kantor graph is of girth 6 and has consistent oriented cycles of length 8 and 12 (see Figure 10.1).

Coming back to cubic symmetric graphs of small girth, Conder and Nedela (2007) proved that such graphs of girth at most 9 are either 1-arc-regular, 2-arc-regular, or belong to a small family of exceptional graphs. In particular, the following useful results can be extracted from their work.

Proposition 10.2.4 *The girth of a cubic symmetric graph of type $\{2^2\}$, $\{3\}$, $\{4^2\}$, $\{1, 4^1, 4^2\}$, $\{4^1, 5\}$, $\{4^2, 5\}$, or $\{5\}$ is at least 10. Moreover, all but finitely many cubic symmetric graphs of girth less than 10 are quotients of the groups G_1 and G_2^1.*

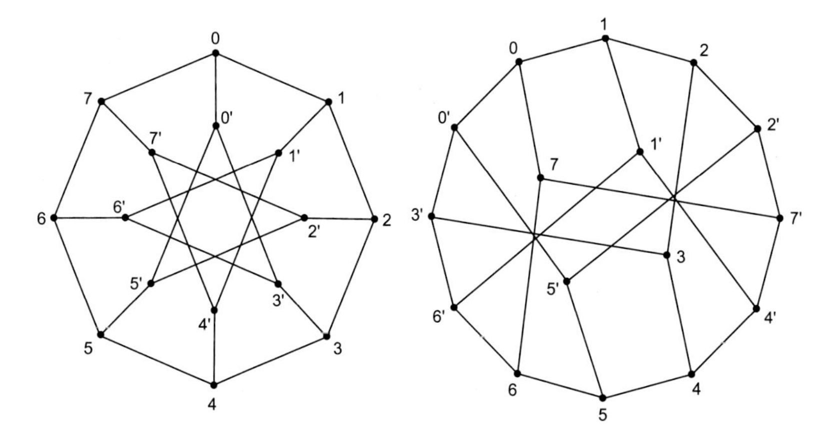

Figure 10.1 The Möbius–Kantor graph F16 drawn in two different ways indicating its consistent cycles. Representatives of the two orbits of consistent oriented cycles in F16 are $0, 1, 2, 3, 4, 5, 6, 7, 0$ and $0, 1, 2, 2', 7', 4', 4, 5, 6, 6', 3', 0', 0$.

Proposition 10.2.5 *With the exception of the Heawood graph, the Pappus graph, and the Desargues graph, which are of type* $\{1, 4^1\}$, $\{1, 2^1, 2^2, 3\}$, *and* $\{2^1, 2^2, 3\}$, *respectively, a cubic symmetric graph of girth* 6 *is a quotient of one of the groups* G_1 *and* G_2^1.

A full classification of cubic symmetric graphs of girth 6 was completed in Kutnar and Marušič (2009a). These results require several additional definitions of families of graphs, and so the interested reader is referred to the paper.

10.3 Rigid Cells

Let Γ be a vertex-transitive graph, and $\alpha \in \mathrm{Aut}(\Gamma)$ that fixes a point. Let F be the set of fixed points of α. In general, one would not expect that the subgraph Γ' induced by F would have any special properties. With cubic s-arc-transitive graphs though, it turns out that for some natural involutions these subgraphs have a particular form. After a preliminary definition, we will give all of these subgraphs, and then discuss how this happens.

Definition 10.3.1 Let Γ be a graph, and $\alpha \in \mathrm{Aut}(\Gamma)$ such that $V(\Gamma) \neq \mathrm{Fix}(\alpha) \neq \emptyset$. A component of $\Gamma[\mathrm{Fix}(\alpha)]$ is referred to as an α**-rigid cell**.

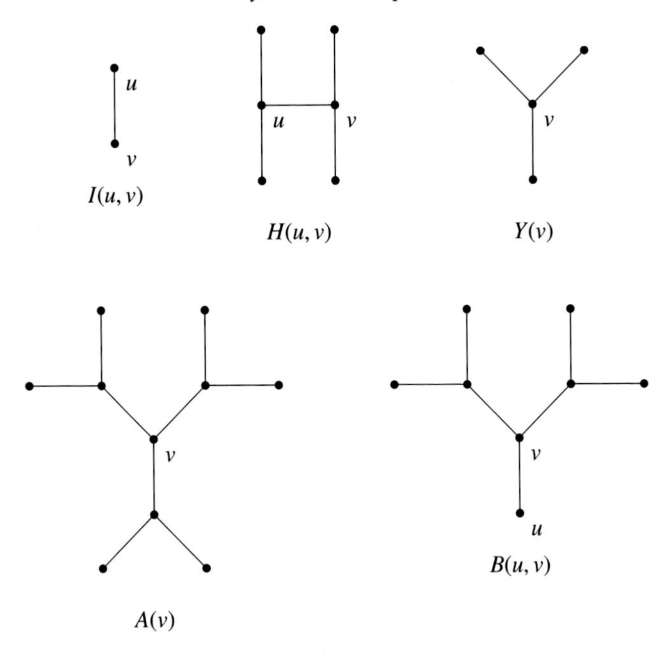

Figure 10.2 An I-tree, H-tree, Y-tree, A-tree, and B-tree.

Definition 10.3.2 Define an I-**tree**, H-**tree**, Y-**tree**, A-**tree**, and B-**tree** to be the trees as labeled in Figure 10.2, which we denote by $I(u,v)$, $H(u,v)$, $Y(v)$, $A(v)$, and $B(u,v)$, respectively.

Let Γ be a cubic s-arc-transitive graph and $\alpha \in \mathrm{Aut}(\Gamma)$ an involution. Our next result gives all of the theoretically possible α-rigid cells of Γ when $s = 3$, and shows that one of the two theoretically possible rigid cells is a rigid cell. It is left as Exercise 10.3.1 to verify that an α-rigid cell is an H-tree if $s = 4$, and as Exercise 10.3.2 that it is an A-tree or B-tree if $s = 5$. The cases $s = 1$ and 2 are given in Proposition 10.3.8, whose proof is Exercise 10.3.5.

Lemma 10.3.3 *Let Γ be a cubic 3-arc-transitive graph, with $G \leq \mathrm{Aut}(\Gamma)$ 3-arc-regular. Then there exists $\alpha \in G$ with a Y-tree an α-rigid cell. Conversely, if $\alpha \in G$ is an involution, then any α-rigid cell is a Y-tree or an I-tree.*

Proof We prove the “conversely” part first, so let $\alpha \in G$ be an involution, and $v \in \mathrm{Fix}(\alpha)$. Let the three neighbors of v be $v_1, v_2, v_3 \in V(\Gamma)$. As α is an involution, v must fix some neighbor of v, say v_1. If α also fixes v_2 and v_3, then it cannot fix any of the neighbors of v_1, v_2, and v_3 different from v, as otherwise α fixes a 3-arc. Then the connected component of $\mathrm{Fix}(\alpha)$ that contains v is a Y-tree. So we may assume α interchanges v_2 and v_3. Then α is either a transposition on the two neighbors $u_1, u_2 \in V(\Gamma)$ of v_1 different from v or it fixes them both. If α is a transposition on u_1 and u_2, then the connected component of $\Gamma[\mathrm{Fix}(\alpha)]$ that contains v is vv_1, an I-tree. Otherwise, each u_i has two neighbors different from v_1, and these must be interchanged by α as otherwise α fixes two different 3-arcs, contradicting G being 3-arc-regular. Hence $\Gamma[\{v, v_1, u_1, u_2\}]$ is a Y-tree.

To see that there exists an involution $\alpha \in G$ with Y-tree an α-rigid cell, note that in the beginning of the first part, we showed that if α fixes v and its three neighbors, then the Y-tree $Y(v)$ is an α-rigid cell. Let $A_1 = v_0, v_1, v_2, v_3$ be a 3-arc in Γ, and $v_4 \in V(\Gamma)$ such that $v_2v_4 \in E(\Gamma)$ and $v_1 \neq v_4 \neq v_3$. Set $A_2 = v_0, v_1, v_2, v_4$. As G is 3-arc transitive, there exists $\alpha \in G$ such that $\alpha(A_1) = A_2$. Since α fixes v_2 and v_1, α interchanges v_3 and v_4. As G is 3-arc-regular and α^2 fixes a 3-arc, α is an involution that fixes two, and so all, of the neighbors of v_1. Thus $Y(v_1)$ is an α-rigid cell. □

It was proven in Djoković and Miller (1980) (see Kutnar and Marušič (2019) for another proof) that a cubic symmetric graph does not have rigid cells isomorphic to the B-tree. Rigid cells of involutions that fix exactly $s-1$ vertices of an s-arc (as in the first part of Lemma 10.3.3) are given special names – see Exercise 10.3.3.

Definition 10.3.4 Let Γ be a cubic s-arc-regular graph, $s \geq 2$. We call an involution $\sigma \in \mathrm{Aut}(\Gamma)$ **canonical with respect to the pair of adjacent vertices** u **and** v of Γ if, for $s = 2$ it fixes the tree $I(u, v)$, or for $s = 4$ it fixes the tree $H(u, v)$. Similarly, we call it **canonical with respect to the vertex** v of Γ if for $s = 3$ it fixes the tree $Y(v)$, and for $s = 5$ it fixes the tree $A(v)$.

It follows from the structure of vertex stabilizers in cubic symmetric graphs in Table 10.1 that only automorphisms of order 2, 3, 4, and 6 can fix a vertex. The proposition below summarizes the structure of rigid cells of automorphisms of orders 3, 4, and 6. Its proof is left to the reader as Exercise 10.3.4. Until further notice, the results in this section are from Kutnar and Marušič (2019).

Proposition 10.3.5 *Let* Γ *be a cubic symmetric graph and let* $\alpha \in \mathrm{Aut}(\Gamma)$ *be an automorphism of* Γ *fixing a vertex.*

(i) *If* α *is of order* 3 *or* 6, *then all* α*-rigid cells are isolated vertices.*
(ii) *If* α *is of order* 4, *then all* α*-rigid cells are* I*-trees.*

Rigid cells of involutions are more complex. Although semiregular involutions usually exist in bipartite symmetric graphs (such as involutory "edge-flippers"), there are also nonbipartite symmetric cubic graphs with semiregular involutions, for example the dodecahedron.

After our final two definitions, we will give exact information on the rigid cells of involutions. From these results, it follows that with the exception of action types {3} and {5}, every involution in a vertex stabilizer has only one type of rigid cells. The proof of the first proposition is left to the reader in Exercise 10.3.5. The proof of the last is omitted, see Kutnar and Marušič (2019, Proposition 3.5) for this.

Definition 10.3.6 Let Γ be a cubic symmetric graph. Let $\mathcal{I}(\Gamma)$ be the set of all involutions in $\mathrm{Aut}(\Gamma)$ that fix at least one vertex of Γ, and $\mathcal{S}(\Gamma)$ be the set of all involutions in $\mathrm{Aut}(\Gamma)$ that are semiregular.

Definition 10.3.7 We say an involution in $\mathrm{Aut}(\Gamma)$ is

(i) an I-, Y-, H- or A-**involution** if all of its rigid cells are isomorphic to the I-, Y-, H-, or A-tree, respectively,
(ii) an M- **involution** if it has nonisomorphic rigid cells (so that its rigid cells are of **mixed** structure), and/or
(iii) an S-**involution** if it is semiregular.

Proposition 10.3.8 *Let Γ be an s-arc-regular cubic graph, where $s \in \{1,2\}$, and let $\alpha \in \mathcal{I}(\Gamma)$. Then $s = 2$ and α is an I-involution.*

Proposition 10.3.9 *Let Γ be a 3-arc-regular cubic graph, and let $\alpha \in \mathcal{I}(\Gamma)$. Then every α-rigid cell is an I-tree or a Y-tree, and if α is an M-involution then Γ has action type {3}.*

Proof First recall that every α-rigid cell is an I- or Y-tree by Lemma 10.3.3. Assume that Γ is not of type {3}. By Table 10.2 there is a 2-arc-regular subgroup $K \leq \mathrm{Aut}(\Gamma)$. Also, by Lemma 10.3.3, as Γ is 3-arc-regular, given any vertex $v \in V(\Gamma)$ there exists a canonical involution σ_v with $Y(v)$ a σ_v-rigid cell. We now show that the only rigid cells of canonical involutions are Y-trees. This is sufficient as if α has a Y-tree as an α-rigid cell, then α is a canonical involution for some vertex.

Towards a contradiction, suppose that there exists a canonical involution $\alpha \in \mathrm{Aut}(\Gamma)\backslash K$ with both the I-tree $I(u,w)$ and a Y-tree as α-rigid cells, where $u, w \in V(\Gamma)$. Note that as K is 2-arc-regular, there is no involution fixing a Y-tree point-wise in K. Let $u_1 \neq w \neq u_2$ be distinct neighbors of u and $w_1 \neq u \neq w_2$ distinct neighbors of w. As $I(u,w)$ is an α-rigid cell, $\alpha^{\{u_1,u_2\}} = (u_1, u_2)$ and $\alpha^{\{w_1,w_2\}} = (w_1, w_2)$. Let $\alpha_{uw} \in K$ be a canonical involution with respect to u and w. Then α_{uw} fixes both u and w, and interchanges w_1 and w_2, and also interchanges u_1 and u_2.Then $(\alpha_{uw})^{V(H(u,w))} = \alpha^{V(H(u,w))}$, and $(\alpha\alpha_{uw})^{V(H(u,w))} = 1$, so $\alpha\alpha_{uw} = 1$ as it fixes a 3-arc. Hence $\alpha = \alpha_{uw} \in K$, a contradiction. □

Proposition 10.3.10 *Let Γ be a cubic 4-arc-regular graph and let $\alpha \in \mathcal{I}(\Gamma)$. Then α is an H-involution.*

Proof Observe that an α-rigid cell R cannot contain a 4-arc. Also, on any s-arc in R, every neighbor of any vertex of valency 2 on the s-arc must also be fixed by α. Let s be the length of a longest s-arc in R. If $s = 3$, then R is the H-tree, if $s = 2$, then R is a Y-tree, and if $s = 1$, R is the I-tree. We now show that the last two possibilities cannot occur. Let u and v be adjacent vertices of Γ. As Γ is 4-arc-regular, there exists an involution $\alpha' \in \text{Aut}(\Gamma)$ with $H(u, v)$ as a rigid cell (for example, a canonical involution with respect to the pair of adjacent vertices u and v). Let v_1 and v_2 be the neighbors of v in Γ different from u, v_{i1}, v_{i2} the neighbors of v_i different from v, $i = 1, 2$.

Suppose some involution $\beta \in \text{Aut}(\Gamma)$ has a Y-tree as a rigid cell. As vertex stabilizers are conjugate, there is a conjugate β' of β with $Y(v)$ as a rigid cell. Clearly, $\beta' \neq \alpha'$ as they have different rigid cells containing v. As α' and β' both interchange v_{11} and v_{12} and v_{21} and v_{22}, it follows that $\alpha'\beta'$ fixes pointwise the B-tree $B(u, v)$, and thus a path of length 4. As $s = 4$ it follows that $\alpha' = \beta'$, a contradiction. Hence no involution can have a Y-tree as a rigid cell.

Suppose there exists an involution $\gamma \in \text{Aut}(\Gamma)$ with an I-tree as a rigid cell. As Γ is 4-arc-transitive, it is arc-transitive, and so edge-transitive. Thus edge stabilizers in $\text{Aut}(\Gamma)$ are conjugate in $\text{Aut}(\Gamma)$. Arguing as above, a conjugate γ' of γ has $I(v, v_1)$ as a rigid cell. Clearly, $\gamma' \neq \alpha'$ as they have different rigid cells containing vv_1. Note that as $H(u, v)$ is a rigid cell of α', and $I(v, v_1)$ is a rigid cell of γ', $(\alpha')^{\{v_{11},v_{12}\}} = (\gamma')^{\{v_{11},v_{12}\}}$, so $\alpha'\gamma'$ fixes v, v_1, v_{11}, and v_{12}. Observe that α' and γ' both interchange the set of neighbors of v_{11} that are not v_1 and the set of neighbors of v_{12} that are not v_1. Let N be the set of neighbors of v_{11} and v_{12} different from v_1. The actions of α' and γ' on N are different or the same. If the actions are the same, then $\alpha'\gamma'$ fixes all of the neighbors of v_{11} and v_{12}. This gives that $\alpha'\gamma'$ fixes a path of length 4, a contradiction. If the actions of α' and γ' on N are different, then $Y(v_1)$ is a rigid cell of $\alpha'\gamma'$. Also note that $\alpha'\gamma'$ interchanges u and v_2. We have already established that no involution in $\text{Aut}(\Gamma)$ has a Y-tree as a rigid cell, so $\alpha'\gamma'$ is not an involution. As we have already observed that an element that stabilizes a point in $\text{Aut}(\Gamma)$ has order 2, 3, 4, or 6, by Proposition 10.3.5 we see that $|\alpha'\gamma'| = 2$, a contradiction. □

Proposition 10.3.11 *Let Γ be a cubic 5-arc-regular graph and let $\alpha \in \mathcal{I}(\Gamma)$. The only α-rigid cells are H-trees and A-trees, and α is an M-involution only when Γ is of action type $\{5\}$.*

The action types {3} and {5} are all that remain to complete a characterization of rigid cells of involutions in cubic symmetric graphs. A first step was taken in Conder et al. (2020b), where the problem was completely solved for graphs of order 2 (mod 4). (In the theorem below $\Gamma + Y$ denotes the disjoint union of graphs Γ and Y.)

Theorem 10.3.12 *Let Γ be a connected symmetric cubic graph of order $2n$, where n is odd, and have action type {3} or {5}. Then all involutions in $\mathcal{I}(\Gamma)$ are conjugate in* Aut(Γ). *Additionally, every involution $\alpha \in \mathcal{I}(\Gamma)$ is an M-involution, and if $s = 3$ then $\Gamma[\mathrm{Fix}(\alpha)] \cong 3mI + 2mY$, while if $s = 5$ then $\Gamma[\mathrm{Fix}(\alpha)] \cong 3mH + 2mA$, for some positive integer m.*

For orders divisible by 4, the situation is different. In Example 10.3.13, we give examples of graphs with M-involutions, while in Examples 10.3.14 and 10.3.15, we give examples where no involutions in $\mathcal{I}(\Gamma)$ are M-involutions and contain more than one conjugacy class of involutions in $\mathcal{I}(\Gamma)$. These examples are the two smallest known graphs of types {3} and {5} without M-involutions, and partly because of their extremely large numbers of vertices, it is hard to say what the structural difference is between those graphs that have and those that do not have mixed involutions.

Example 10.3.13 The cubic graph C1012.2 of order 1,012 listed in Conder (2011) has action type {3}, with automorphism group isomorphic to $\mathrm{PSL}(2, 23) \times \mathbb{Z}_2$. This group contains three conjugacy classes of involutions, with two consisting of S-involutions, and one consisting of M-involutions whose rigid cells are six I-trees and four Y-trees. Similarly, there exists a cubic symmetric graph of order $|A_{48}|/\left(3 \cdot 2^4\right)$ with action type {5} and automorphism group isomorphic to A_{48}, with nonsemiregular involutions being M-involutions with rigid cells isomorphic to the H-tree and to the A-tree.

Example 10.3.14 By a computation in Magma similar to the computations that produced the graphs listed in Conder and Dobcsányi (2002); Conder and Nedela (2009), there exists a cubic symmetric graph with action type {3} and order 39,916,800, with automorphism group isomorphic to S_{12}. This group contains two conjugacy classes of involutions with fixed points. In one of these two classes, each involution fixes 15,360 vertices, which partition into 3,840 rigid cells, all isomorphic to the Y-tree, while in the other class, each involution fixes 2,304 vertices, which partition into 1,152 rigid cells, all isomorphic to the I-tree.

Example 10.3.15 Another computation in Magma produces a cubic symmetric graph with action type {5} and order 50,685,458,503,680,000, with automorphism group isomorphic to S_{20}. For this graph, there are two conjugacy classes in $\mathcal{I}[\Gamma]$, one consisting of A-involutions, and the other consisting of H-involutions.

10.4 Valency Preserving Quotienting

For certain problems, such as, for example, constructions of Hamilton paths or cycles, the standard quotients using reductions via normal subgroups are not always possible. Reductions via semiregular subgroups are then a possible and viable alternative. Asking the corresponding quotient graphs to retain as many properties of the original graphs as possible is a logical additional requirement if one is to successfully handle the ensuing analysis and complete the reduction. Staying within the class of simple graphs of the same valency seems like the least one would ask for. This motivates the following definition.

Definition 10.4.1 Let Γ be a vertex-transitive graph with semiregular automorphism ρ. We say ρ is a **simplicial** automorphism of Γ if all of its orbits are independent sets of vertices, and for every pair of different orbits B_1 and B_2 the bipartite graph $\Gamma[B_1, B_2]$ induced by B_1 and B_2 is either a 1-factor or has no edges.

The name "simplicial" automorphism is borrowed from topology, where a simplicial automorphism of a simplicial complex is an automorphism of the underlying topological space that preserves the combinatorial information of the complex. We have the following problem.

Problem 10.4.2 Determine the vertex-transitive graphs that have a simplicial automorphism.

While being a refinement of the semiregularity problem, this problem is somewhat restrictive as it only applies to a subclass of vertex-transitive graphs as the next result, whose proof is straightforward, shows.

Proposition 10.4.3 *Let ρ be an (m,n)-semiregular automorphism of a connected vertex-transitive graph of valency $d \geq m$. Then ρ is not simplicial.*

So vertex-transitive graphs of prime order do not have simplicial automorphisms!

Proposition 10.4.4 *A disconnected vertex-transitive graph has a simplicial automorphism.*

Proof A semiregular automorphism that cyclically permutes its connected components is simplicial. □

Example 10.4.5 The cube has simplicial and nonsimplicial automorphisms.

Solution The involution interchanging pairs of antipodal vertices in the cube (that is, vertices at distance 3) is simplicial and the quotient is isomorphic to the complete graph K_4. The cube also has semiregular involutions interchanging pairs of adjacent vertices as well as $(2, 4)$-semiregular automorphisms. The quotients with respect to these automorphisms all contain loops and thus none of these automorphisms is simplicial. □

Example 10.4.6 The Petersen graph only has semiregular automorphisms of order 5, and so has no simplicial automorphisms by Proposition 10.4.3.

Example 10.4.7 The dodecahedron has semiregular involutions interchanging antipodal vertices (that is, vertices at distance 4) that are simplicial with the quotient isomorphic to the Petersen graph. It also has $(4, 5)$-semiregular and $(2, 10)$-semiregular automorphisms but these are not simplicial.

Example 10.4.8 The line graph of the Petersen graph is regular of valency four and has semiregular automorphisms of orders 3 and 5. While the latter are not simplicial, semiregular automorphisms of order 3 are simplicial, with corresponding quotients isomorphic to the complete graph K_5. On the other hand, the wreath product $C_5 \wr K_2$ is regular of valency 5 and has semiregular automorphisms of orders 2 and 5, none of which are simplicial.

The definition of a simplicial automorphism translates into certain conditions, which we will see in Proposition 10.4.12, on consistent cycles and their generalization, the so-called $1/k$-consistent cyclets, which we now define.

Definition 10.4.9 Let Γ be a graph. A cycle C in Γ that is given an orientation $\vec{C} = (u_0, \ldots, u_{r-1}, u_0)$ with an identified initial vertex u_0, called a **root**, is called a **cyclet**. A cyclet is $(G, 1/k)$-**consistent** (or simply $1/k$-**consistent** if G is clear) if there is $\tau \in G \leq \mathrm{Aut}(\Gamma)$ such that $\tau(u_j) = u_{j+k}$ for every $j \in \mathbb{Z}_r$ (with arithmetic in the subscripts done in $\mathbb{Z}_r$).

The underlying cycle of a 1-consistent cyclet is a consistent cycle. For $k > 1$ each $1/k$-consistent cyclet $\vec{C}$ gives additional cyclets, namely $\tau^i\left(\vec{C}\right)$, $i \in \mathbb{Z}$ (we will not care about how many additional cyclets). Reversing the orientation of these cyclets gives more. The following result can be deduced from Miklavič et al. (2007b, Theorem 4.3).

Theorem 10.4.10 *Let $k \geq 1$ be an integer, Γ an arc-transitive graph of valency $d \geq 2$ and girth at least $2k+1$, and $G = \mathrm{Aut}(\Gamma)$. Let Ω be the set of all $(G, 1/k)$-consistent cyclets in Γ. Then G has exactly $(d-1)^k$ orbits in its action on Ω.*

We now give a sufficient condition for the existence of a simplicial automorphism in a cubic symmetric graph. We will first need a technical lemma about the lengths of $1/2$-consistent cyclets.

Lemma 10.4.11 *Let Γ be a cubic symmetric graph not isomorphic to K_4, $K_{3,3}$, or Q_3. Then there exist integers c_1, c_2, and c_3 such that every consistent cycle has lengths c_1 or c_2, and any $1/2$-consistent cyclet has length c_1, c_2, or c_3.*

Proof By Proposition 10.2.3 there are exactly two orbits of 1-consistent cyclets in the action of $\mathrm{Aut}(\Gamma)$ on the set of its 1-consistent cyclets. Clearly all 1-consistent cyclets in a single orbit have the same length. Thus c_1 and c_2 exist. As $\Gamma \ncong K_4$, $K_{3,3}$, or Q_3, Γ has girth at least 5 by Proposition 10.2.2. By Theorem 10.4.10, we see that $\mathrm{Aut}(\Gamma)$ has four distinct orbits on its $1/2$-consistent cyclets.

Each orbit of 1-consistent cyclets gives an orbit on $1/2$-consistent cyclets. Indeed, let $\vec{C} = (v_0, v_1, \ldots, v_{r-1}, v_0)$ be a 1-consistent cyclet and $\rho \in \mathrm{Aut}(\Gamma)$ with $\rho(v_i) = v_{i+1}$. Then $\rho^2(v_i) = v_{i+2}$ for all $0 \leq i \leq r-1$, so $\vec{C}$ is a $1/2$-consistent cyclet of length r. If there is $\iota \in \mathrm{Aut}(\Gamma)$ with $\iota(v_0, v_1 \ldots, v_{r-1} v_0) = (v_0, v_{r-1}, \ldots, v_1, v_0)$, then all 1-consistent cyclets whose underlying cycles are the same are in the same orbit. In this case it is possible that $c_1 \neq c_2$. If not, then $\vec{C}$ and the cyclet obtained from $\vec{C}$ by reversing its orientation are in different orbits, and $c_1 = c_2$.

As there are four orbits of $1/2$-consistent cyclets, there is a $1/2$-consistent cyclet $\vec{D} = (u_0, u_1, \ldots, u_{t-1}, u_0)$ that is not a 1-consistent cycle. As $\vec{D}$ is not a 1-consistent cyclet, the cyclet $\vec{D}' = (u_1, u_2, \ldots, u_0, u_1)$ is not in the same orbit as $\vec{D}$. As $\vec{D}$ is a $1/2$-consistent cyclet, so is $\vec{D}'$. Then the underlying cycle of every such $1/2$-consistent cycle has length $c_3 = t$, and the result follows. □

Proposition 10.4.12 *Let Γ be a cubic symmetric graph not isomorphic to K_4, $K_{3,3}$, or Q_3. Let c_1, c_2 be the lengths of the consistent cycles of Γ, and c_3 be the length of the $1/2$-consistent cyclets that are not 1-consistent cyclets. If ρ is a semiregular automorphism of Γ of odd prime order $p \notin \{c_1, c_2\}$, and $2p \notin \{c_1, c_2, c_3\}$, then ρ is simplicial.*

Proof As p is an odd prime and Γ is cubic, each orbit of $\langle\rho\rangle$ is either an independent set of p vertices or induces a cycle of length p. If the latter holds, the induced cycle is a consistent cycle with shunt automorphism ρ^i, for some

$i \in \mathbb{Z}_p$, and so $p \in \{c_1, c_2\}$, a contradiction. Hence all orbits of $\langle\rho\rangle$ induce independent sets of vertices.

Let B_1 and B_2 be distinct orbits of $\langle\rho\rangle$ with an edge x_1x_2 of Γ such that $x_i \in B_i$, $i = 1, 2$. If the bipartite graph $\Gamma[B_1, B_2]$ is not a 1-factor then this bipartite graph contains a cycle C of length $2p$ by Exercise 10.4.1. But then a rooted orientation of this $2p$-cycle is a $1/2$-consistent cyclet, contradicting the condition that $2p \notin \{c_1, c_2, c_3\}$. □

We now wish to discuss simplicial elements in cubic vertex-transitive graphs whose automorphism group is primitive. Recall that the notation $(G, G/H)$ is the image of the permutation representation obtained from the left multiplication action of G on the left cosets of H in G.

Theorem 10.4.13 *Let Γ be a cubic vertex-transitive graph with* $\mathrm{Aut}(\Gamma)$ *primitive. Then Γ is isomorphic to*

(i) *K_4, which is an orbital graph of $(A_4, A_4/\mathbb{Z}_3)$ and $(S_4, S_4/S_3)$, of type $\{2^2\}$,*
(ii) *the Petersen graph, which is an orbital graph of both $(A_5, A_5/S_3)$ and $(S_5, S_5/D_{12})$, of type $\{2^1, 3\}$,*
(iii) *the Coxeter graph, which is an orbital graph of* $(\mathrm{PGL}(2,7), \mathrm{PGL}(2,7)/D_{12})$, *of type $\{2^2, 3\}$,*
(iv) *the graph* F234B, *which is an orbital graph of both* $(\mathrm{SL}(3,3), \mathrm{SL}(3,3)/S_4)$ *and* $(\mathrm{Aut}(\mathrm{SL}(3,3)), \mathrm{Aut}(\mathrm{SL}(3,3))/S_4 \times \mathbb{Z}_2)$, *of type $\{4^2, 5\}$, or*
(v) *a graph in an infinite family of orbital graphs of* $(\mathrm{PSL}(2,p), \mathrm{PSL}(2,p)/S_4)$, *with $p \equiv \pm 1 \pmod{16}$ a prime, of type $\{4^1\}$.*

Proof Such a graph is necessarily symmetric and so is an orbital graph of a primitive group with a suborbit of length 3. The result then follows using Wong's list of primitive groups with a suborbit of length 3 (Wong, 1967) and the fact that only a cubic vertex-transitive graph is an orbital graph of a self-paired suborbit of length 3 (Wong, 1967). □

Example 10.4.14 The first graph in the infinite family in Theorem 10.4.13 (v) is the well-known H-graph F102 on 102 vertices with automorphism group isomorphic to $\mathrm{PSL}(2,17)$. It is one of the three remarkable graphs described and studied in Biggs (1973). Its consistent cycles are of lengths 9 and 17. Also, all semiregular automorphisms of this graph are of order 17. The quotient graph with respect to such an automorphism has six vertices with edges forming an H shape, explaining its name. So F102 has no simplicial automorphisms.

To finish the characterization of simplicial automorphisms in cubic symmetric graphs with primitive automorphism group, one is left with the graphs in

Theorem 10.4.13 (v) for $p \equiv \pm 1 \pmod{16}$ a prime $p > 17$. As $p > 3$, every automorphism π of order p (and thus a generator of a Sylow p-subgroup) is semiregular. We believe that every such automorphism is also simplicial. As $\langle \pi \rangle$ has more than two orbits, it follows that the valency induced by an orbit of $\langle \pi \rangle$ is either 0 or 2 and the valency of the bipartite graph induced by two orbits of $\langle \pi \rangle$ is either 0, 1, or 2. So π is simplicial when no orbit induces a cycle of length p and no pair of orbits with an edge between vertices in the orbits induces a bipartite graph isomorphic to a cycle of length $2p$. In the language of consistent cycles we conjecture that:

Conjecture 10.4.15 *Every graph in the infinite family given in Theorem 10.4.13* (v), *other than the H-graph, is simplicial. Specifically, they have no consistent cycle of length p and no $1/2$-consistent cycles of length $2p$.*

10.5 Symmetric Graphs of Larger Valencies

As we have seen, most of the research on symmetric graphs has been centered around cubic symmetric graphs. However, there are also many interesting results for symmetric graphs of larger valencies, especially for valencies 4 and 5. The essential property of cubic symmetric graphs that allows one to investigate these graphs in detail is the fact that the orders of their vertex stabilizers are bounded by $3 \cdot 2^4$ by Corollary 3.1.6 and Theorem 3.1.10. This result was generalized by Trofimov (1990, 1991) and Weiss (1979) to symmetric graphs of prime valencies.

Theorem 10.5.1 *Let Γ be a connected G-arc-transitive graph of prime valency p, and $v \in V(\Gamma)$. There exists a constant c_p depending only on p such that* $|\mathrm{Stab}_G(v)| \leq c_p$.

This result cannot be generalized to symmetric graphs of even valence.

Example 10.5.2 The graph $\Gamma = C_n \wr \bar{K}_2$, $n \geq 5$, is a connected quartic symmetric graph and $\mathrm{Aut}(\Gamma) = D_{2n} \wr \mathbb{Z}_2 \cong \mathbb{Z}_2^n \rtimes D_{2n}$ of order $2n \cdot 2^n$.

Before continuing, we will need another term (which you may have seen in Exercise 3.2.4).

Definition 10.5.3 Let Γ be a graph, $G \leq \mathrm{Aut}(\Gamma)$, and $v \in V(\Gamma)$. The permutation group induced by the action of $\mathrm{Stab}_G(v)$ on the neighbors of v is called the **local action of** G. It is usually denoted $\mathrm{Stab}_G(v)^{N_\Gamma(v)}$. If $\mathrm{Stab}_G(v)^{N_\Gamma(v)}$ is primitive, we say G is **locally primitive**. Finally, if $G = \mathrm{Aut}(\Gamma)$ then the local action of G is also called the **local action of** Γ.

Notice that the local action of G is independent, up to isomorphism, of choice of v, and if Γ is arc-transitive, then $\mathrm{Stab}_G(v)^{N_\Gamma(v)}$ is transitive by Theorem 3.2.8.

In Example 10.5.2 the local action of Γ is isomorphic to the dihedral group D_8. In a sequence of papers in the mid 1970s Gardiner (1973, 1974, 1976) considered s-arc-transitive graphs of order $p+1$, where p is a prime. Amongst other things, he showed that there is a constant c such that if Γ is a quartic arc-transitive graph whose local action is not D_8, then $|\mathrm{Stab}_{\mathrm{Aut}(\Gamma)}(v)| \leq c$ for every $v \in V(\Gamma)$ (here, think of D_8 as a Sylow 2-subgroup of $\mathcal{S}_4$, isomorphic to $\mathbb{Z}_2 \wr \mathbb{Z}_2$). There are many interesting families of such graphs, such as the so-called rose window graphs that we define at the end of this section. All five transitive subgroups of $\mathcal{S}_4$ are local actions in quartic symmetric graphs: $\mathbb{Z}_4$, $\mathbb{Z}_2^2$, D_8, A_4, and $\mathcal{S}_4$. The next proposition covers the first two cases. The proof is left to the reader in Exercise 10.5.6. As for the fourth and fifth cases, Gardiner proved that the order of the vertex stabilizer is bounded by $2^4 \cdot 3^6$.

Proposition 10.5.4 *Let Γ be a quartic symmetric graph such that the local action of Γ is isomorphic to $\mathbb{Z}_4$ or $\mathbb{Z}_2^2$, and $v \in V(\Gamma)$. Then Γ is 1-arc-regular and $\mathrm{Stab}_{\mathrm{Aut}(\Gamma)}(v) \cong \mathbb{Z}_4$ or $\mathbb{Z}_2^2$.*

Tutte showed (see Theorem 3.1.10) that a cubic symmetric graph is at most 5-arc-transitive. Weiss (1981b) generalized this result to symmetric graphs of larger valencies.

Theorem 10.5.5 (Weiss's Theorem) *Let Γ be a connected s-arc-transitive graph of valency $k \geq 3$. Then $s \leq 7$, and if $s = 7$ then $k = 3^t + 1$ for some t. Hence there are no finite 8-arc-transitive graphs of valency $k > 2$.*

The proof of Theorem 10.5.5 relies on CFSG. Namely, if Γ is s-arc-transitive for some $s \geq 2$, then Γ is 2-arc-transitive, and so the local action of Γ is 2-transitive. The proof then proceeds using Cameron's classification of finite 2-transitive groups (Cameron, 1981, Theorem 5.3), which uses CFSG.

We now discuss in more detail the various possibilities for the local action of quartic symmetric graphs given by Gardiner. We will consider when the local action is $\mathbb{Z}_4$ first, and will need some additional terminology.

Definition 10.5.6 Let G be a group with an involution a and an element x of order s such that G has presentation

$$\left\langle a, x : a^2 = x^s = (ax)^t = 1, \ldots \right\rangle,$$

for some positive integer t. Here, the "..." should be interpreted to mean that other relations are allowed. Such a group will be called a $(2, s, t)$**-generated group**, and its presentation will be called a $(2, s, t)$**-presentation**.

At first glance this appears to be a rather narrow class of groups, but contains, for example, “most” nonabelian simple groups (see the comment after Glover and Marušič [2007, Theorem 1.1] for the complete list). Also, if G is such a group, then $\mathrm{Cay}\left(G, \left\{a, x, x^{-1}\right\}\right)$ is a connected cubic graph. The extra information the relation $(ax)^t$ gives is that the graph contains cycles of length $2t$, namely for $y \in G$, $y, ya, y(ax), \ldots, y(ax)^{t-1}a, y$.

Our discussion of quartic symmetric graphs with point stabilizer $\mathbb{Z}_4$ requires information about Cayley maps, which is a separate area in its own right that we will not cover. While we will at least define Cayley maps, the appropriate background on Cayley maps is assumed for this discussion. The interested reader without the appropriate background should see Bonnington et al. (2002) and Jones and Singerman (1978).

Definition 10.5.7 Let G be a group, $S \subset G$ such that $S = S^{-1}$ and $1 \notin S$, and ρ a cyclic permutation of S. The **Cayley map** of G with respect to S and ρ, denoted $\mathrm{CM}(G, S, \rho)$, is a 2-cell embedding of the Cayley graph $\mathrm{Cay}(G, S)$ on an orientable surface for which the orientation-induced local ordering of the arcs incident with any vertex $g \in G$ is always the same as the order of generators in S induced by ρ.

In other words the neighbors of any vertex g are always spread counterclockwise around g in the order $\left(gs, g\rho(s), g\rho^2(s), \ldots, g\rho^{k-1}(s)\right)$. One thing should be emphasized about Cayley maps. They are, by definition, Cayley graphs with extra topological and algebraic properties. As such, most graph-theoretic terminology can be used when discussing Cayley maps without ambiguity. We will in fact talk about Hamilton cycles and paths in Cayley maps in the next chapter, and of course a Hamilton cycle or path in $\mathrm{CM}(G, S, \rho)$ is just a Hamilton cycle or path in $\mathrm{Cay}(G, S)$.

It was shown in Hujdurović et al. (2013, Proposition 7) that if Γ is a quartic symmetric graph with vertex stabilizer isomorphic to $\mathbb{Z}_4$, then $G = \mathrm{Aut}(\Gamma) = \left\langle a, x : a^2 = x^s = (ax)^4 = 1, \ldots\right\rangle$ is a $(2, s, 4)$-generated group where a flips (interchanges the endpoints of) an edge uv of Γ and $\mathrm{Stab}_G(v) = \langle ax\rangle$. We will use the next result to show $s = 4s_0$, where s_0 is odd, if Γ has odd order. We will see $(2, s, 4)$-generated groups again in Chapter 11; see Problem 11.5.13.

Lemma 10.5.8 *Let* $G = \left\langle a, x : a^2 = 1, x^s = 1, (ax)^t = 1, \ldots\right\rangle$ *be a group with a* $(2, s, t)$*-presentation where* $s = 2^i s_0$, s_0 *odd,* $t = 2^j t_0$, t_0 *odd, and let* $k = \max\{i, j\}$. *Then one of the following occurs:*

(i) $i = 0$, $j = 1$ *or* $i = 1$, $j = 0$; *or*
(ii) $k = i = j \geq 2$; *or*
(iii) $|G|$ *is divisible by* 2^{k+1}.

Proof Observe that 2^k divides $|G|$. Assume first that the group G is of order $|G| = 2^k n$ where n is odd. Consider the Cayley graph $\Gamma = \mathrm{Cay}\left(G, \left\{a, x, x^{-1}\right\}\right)$ embedded into a closed orientable surface of genus g whose set of faces $F(\Gamma)$ consists of s-gons and $2t$-gons. Their respective numbers are $|G|/s$ and $|G|/t$ as each vertex is in one s-gon and two $2t$-gons. By computation, the Euler characteristic of the surface in question is

$$\begin{aligned}\chi = 2 - 2g &= |V(\Gamma)| - |E(\Gamma)| + |F(\Gamma)| \\ &= 2^k n - 3 \cdot 2^{k-1} n + \frac{2^k n}{2^i s_0} + \frac{2^k n}{2^j t_0} \\ &= -2^{k-1} n + \frac{2^k n}{2^i s_0} + \frac{2^k n}{2^j t_0}.\end{aligned}$$

But n is odd. Therefore, as χ is even, assuming that $k = 1$ we have that the first term above is odd and hence $(i, j) \in \{(1, 0), (0, 1)\}$. If $k \geq 2$ then the first term is even and so the remaining two terms are of the same parity. But $k = \max\{i, j\}$ and so both terms are odd. Therefore $i = j = k$, completing the proof. □

Corollary 10.5.9 *Let Γ be a quartic symmetric graph of odd order with vertex stabilizer $\mathbb{Z}_4$. Then* $\mathrm{Aut}(\Gamma)$ *is a* $(2, 4s_0, 4)$*-generated group where s_0 is odd.*

Proof Lemma 10.5.8 (i) cannot occur as $t = 4$. As Γ has odd order and vertex-stabilizer $\mathbb{Z}_4$, $|\mathrm{Aut}(\Gamma)|$ is not divisible by 8, and so Lemma 10.5.8 (iii) cannot occur. This leaves only Lemma 10.5.8 (ii), and the result follows. □

As for quartic symmetric graphs with local action isomorphic to D_8, there are many known constructions in the literature. For example, most of the symmetric rose window graphs, defined below, are of this kind.

Definition 10.5.10 Given natural numbers $n \geq 3$ and $1 \leq a, r \leq n-1$, the **rose window graph** $R_n(a, r)$, has vertex set $\{x_i : i \in \mathbb{Z}_n\} \cup \{y_i : i \in \mathbb{Z}_n\}$ and edge set $\{\{x_i, x_{i+1}\} : i \in \mathbb{Z}_n\} \cup \{\{y_i, y_{i+r}\} : i \in \mathbb{Z}_n\} \cup \{\{x_i, y_i\} : i \in \mathbb{Z}_n\} \cup \{\{x_{i+a}, y_i\} : i \in \mathbb{Z}_n\}$.

Observe that the rose window graph $R_n(a, r)$ contains the generalized Petersen graph $\mathrm{GP}(n, r)$ as a spanning subgraph. These graphs were introduced by Wilson (2008), where he identified the following four families of edge-transitive rose window graphs $R_n(a, r)$:

(i) $R_n(2, 1)$;
(ii) $R_{2m}(m - 2, m - 1)$;
(iii) $R_{12m}(3m + 2, 3m - 1)$ and $R_{12m}(3m - 2, 3m + 1)$;
(iv) $R_{2m}(2b, r)$, where $b^2 = \pm 1 \pmod{m}$, $2 \leq 2b \leq m$, and $r \in \{1, m - 1\}$ is odd.

He conjectured (Wilson, 2008, Conjecture 11) that these graphs are all of the symmetric rose window graphs. This conjecture was verified in Kovács et al. (2010). For example, every graph in family (i) for $n > 4$ has local action isomorphic to D_8 (Kovács et al., 2010, Lemma 3.2).

As mentioned above, quartic symmetric graphs with local action isomorphic to either A_4 or $\mathcal{S}_4$ have vertex stabilizers whose orders are bounded by $2^4 \cdot 3^6$. Among such graphs, those with a primitive automorphism group have received particular attention, mainly because of the general research interest in primitive permutation groups with small suborbits that goes back to the 1960s and 1970s (Li et al. (2004); Neumann (1977); Quirin (1971); Sims (1968); Wong (1967)). Of several infinite families of such graphs we would like to single out the family of graphs that are orbital graphs of $(\mathrm{PSL}(2, p), \mathrm{PSL}(2, p)/\mathcal{S}_4)$, with $p \equiv \pm 1 \pmod 8$ a prime. Computation by Magma shows that the smallest graph occurs for $p = 23$ with automorphism group isomorphic to $\mathrm{PSL}(2, 23)$. The smallest quartic symmetric graph with primitive automorphism group is an orbital graph of the action of A_7 on the 35 subsets of size 3 from a set of seven elements. The graph is the odd graph O_4 (see Exercise 10.5.8). The local action of the vertex stabilizer is not faithful and is isomorphic to $\mathcal{S}_4$ with vertex stabilizer isomorphic to $(\mathbb{Z}_3 \times A_4) \rtimes \mathbb{Z}_2$. As for vertex stabilizers isomorphic to A_4, one is an orbital graph of $(\mathrm{PSL}(2, 13), \mathrm{PSL}(2, 13)/A_4)$ with 91 vertices with automorphism group isomorphic to $\mathrm{PSL}(2, 13)$.

Weiss conjectured in 1978 that symmetric locally primitive graphs have vertex stabilizers of bounded order Weiss (1981a). We refer the reader to Conder et al. (2000); Potočnik et al. (2012, 2014); Praeger et al. (2012) and Verret (2009) for the current status of this conjecture.

Conjecture 10.5.11 (The Weiss conjecture) *For a symmetric locally primitive graph, the order of the vertex stabilizer is bounded above by some function of the valency of the graph.*

10.6 Exercises

10.3.1 Let Γ be a connected cubic graph with 4-arc regular subgroup G of automorphisms. Show that if $\alpha \in G$ is a nonsemiregular involution, then an α-rigid cell of α is an I-tree, Y-tree, or H-tree.

10.3.2 Let Γ be a connected cubic 5-arc-regular graph. Show that if $\alpha \in G$ is a nonsemiregular involution, then an α-rigid cell is one of the graphs in Figure 10.2.

10.3.3 Let Γ be a cubic s-arc-regular graph, $s \geq 2$, and $\alpha \in \mathrm{Aut}(\Gamma)$ an involution that maps the s-arc $v_0, v_1, \dots, v_s$ in Γ to $v_0, v_1, \dots, v_{s-1}, v_{s+1}$, where v_{s+1} is the neighbor in Γ of v_{s-1} different from v_{s-2} and v_s. Let R be the rigid cell of α that contains v_0. Show that if $s = 2$, then R is an I-tree, if $s = 4$, then R is an H-tree, and if $s = 5$, then R is an A-tree or a B-tree.

10.3.4 Prove Proposition 10.3.5. (Hint: For (ii), Suppose α is an element of order 4 that fixes v. Show α^2 has $Y(v)$ as an α^2-rigid cell. Consider the vertices in Γ at distance 2 in Γ from v.)

10.3.5 Prove Proposition 10.3.8.

10.3.6 Prove Proposition 10.3.11.

10.4.1 Let p be a prime, and $\rho \in \mathcal{S}_{2p}$ a semiregular element of order p. Let Γ be a bicirculant graph with symbol $[\emptyset, \emptyset, T]$ with respect to ρ and vertices x, y chosen from different orbits of $\langle\rho\rangle$. Show that if $|T| = 2$, then Γ is a Hamilton cycle.

10.5.1 Prove that every bicirculant of order $2p$ with the symbol $[\emptyset, \emptyset, \{0, 1, r+1\}]$, where $p \geq 13$ is a prime and $r^3 \equiv \pm 1 \pmod p$, is a cubic symmetric graph of girth 6. (Hint: Analyze the distribution of 6-cycles.)

10.5.2 Find all rigid cells in connected cubic symmetric graphs up to order 30.

10.5.3 Find all semiregular automorphisms in all connected cubic symmetric graphs of order up to 30. Which of them are simplicial?

10.5.4 Find all consistent cycles and 1/2-consistent cycles in all connected cubic symmetric graphs up to order 30.

10.5.5 Prove that the graphs given in Example 10.5.2 are symmetric with automorphism groups isomorphic to $\mathbb{Z}_2^n \rtimes D_{2n}$.

10.5.6 Prove Proposition 10.5.4.

10.5.7 Give quartic symmetric graphs in Frucht's notation that are orbital graphs of following primitive groups:

(i) $(\mathbb{Z}_{13} \rtimes \mathbb{Z}_4, \mathbb{Z}_{13} \rtimes \mathbb{Z}_4/\mathbb{Z}_4)$,
(ii) $\left(\mathbb{Z}_5^2 \rtimes D_8, \mathbb{Z}_5^2 \rtimes D_8/D_8\right)$,
(iii) $(\mathrm{PSL}(2, 13), \mathrm{PSL}(2, 13)/A_4)$, and
(iv) $(\mathrm{PSL}(2, 23), \mathrm{PSL}(2, 23)/\mathcal{S}_4)$.

10.5.8 Prove that the action of A_7 on subsets of size 3 from a set of size 7 has six different generalized orbital graphs, one of which is quartic with local action isomorphic to A_4.

11

Hamiltonicity

11.1 Historic Perspectives

The main theme of this chapter is the following question asked by Lovász in 1969, tying together two seemingly unrelated graph-theoretic concepts: traversability and symmetry (Guy, 1970, Problem 11).

Problem 11.1.1 Construct a finite connected vertex-transitive graph that has no Hamilton path.

See Problem 3.3.1 for the original formulation of this problem.

Pak and Radoičić (2009) noted that Lovász himself originally conceived Problem 11.1.1 as a special case of what was then an open problem of Gallai (1968), which asked whether all longest paths in simple connected graphs must have a common vertex. In the special case of vertex-transitive graphs this would imply that all such longest paths must have every vertex in common, implying that vertex-transitive graphs contain a Hamilton path. Gallai's problem was later shown to have a negative answer (Walther, 1969). A counter-example with 12 vertices was constructed by Walther and Voss (1974); see Figure 11.1.

Arguably, however, the general problem of finding Hamilton paths (and cycles) in vertex-transitive graphs may be much older, as it can be traced back to bell ringing, Gray codes (which are Hamilton cycles in the hypercube $\mathbb{Z}_2^n$) and the knight's tour of a chessboard (Ahrens, 1901, p. 80) and Breckman (1956); Conway et al. (1989); Euler (1766); Gray (1953).

The importance of Problem 11.1.1, which we call the **Lovász problem**, goes beyond its mere graph-theoretic content for, over the last few decades, it has served as a catalyst for much of the increased interest in the study of vertex-transitive graphs, and, more generally, transitive group actions. The

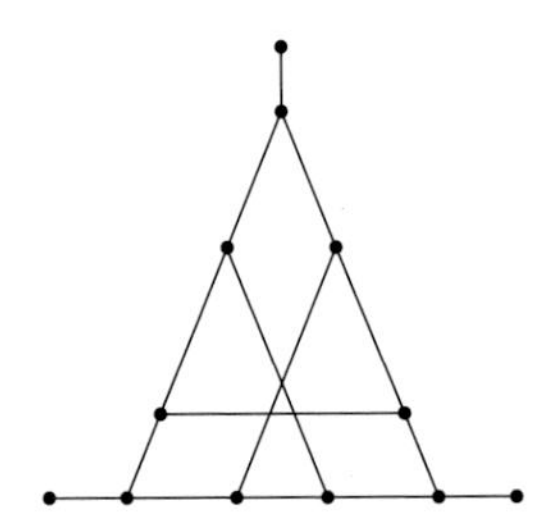

Figure 11.1 The smallest graph as a negative answer to Gallai's problem.

inherent symmetry makes vertex-transitive graphs well behaved, and so it is not surprising that many of them have Hamilton paths and cycles.

As we have seen, only five connected vertex-transitive graphs that do not have Hamilton cycles are known. These are K_2, the Petersen graph (Theorem 3.4.1), the Coxeter graph (Theorem 6.1.5), and the truncations of the Petersen graph and Coxeter graph (Corollary 3.4.6). All of these graphs have a Hamilton path. In addition, none of the four exceptional graphs with at least three vertices is a Cayley graph, which has led to a folklore conjecture that every connected Cayley graph has a Hamilton cycle (Conjecture 3.3.2).

According to Babai (Graham et al., 1995, p. 1472), the question about Hamilton paths and cycles in connected Cayley graphs can be traced back to 1959 and Rapaport-Strasser (1959).

One reason the Lovász problem and its Cayley version are intriguing is that, other than our inability to find connected vertex-transitive graphs without a Hamilton path, there is no particular reason to believe that such graphs do not exist. This has sparked considerable interest in the mathematical community producing conjectures and counter conjectures. For example, Babai believes that there exist infinitely many connected vertex-transitive graphs, even Cayley graphs, without Hamilton cycles, or more precisely without "long cycles" – see Conjecture 3.3.4.

On the contrary, Bermond (1978a) and Thomassen (1991) believe that only finitely many connected vertex-transitive graphs without a Hamilton cycle exist. Interestingly, Babai has proven the best-known general result regarding long cycles in connected vertex-transitive graphs, which we give in the next result (Babai, 1979). We will need another term for the proof.

Definition 11.1.2 A **hypergraph** is a pair $\mathcal{H} = (V, \mathcal{F})$ where V is a non-empty set, called the **vertex set** of $\mathcal{H}$, and $\mathcal{F}$ is a family of subsets of V, called the **edge set** of $\mathcal{H}$.

Theorem 11.1.3 *Every connected vertex-transitive graph with $n \geq 4$ vertices contains a cycle of length greater than $\sqrt{3n}$.*

Proof Let Γ be a connected d-regular vertex-transitive graph of order $n \geq 4$, and r be the length of a longest cycle in Γ. First note that $r \geq 3$.

By Watkins (1970, Theorem 3) (see also Godsil and Royle (2001, Theorem 3.4.2)), Γ is at least $2(d+1)/3$ connected. Hence Γ is either a cycle or it is 3-connected. If Γ is a cycle, its longest cycle is of length $n > \sqrt{3n}$. So we assume Γ is 3-connected. Let $C = c_0c_1 \ldots c_{r-1}c_0$ and $D = d_0d_1 \ldots d_{r-1}d_0$ be two of the longest cycles in Γ. We will first prove that C and D have at least three vertices in common.

Suppose C and D have no vertex in common. As Γ is connected, there is a path from some vertex of C to some vertex of D. Choose such a path P_1 so that it contains exactly one vertex of C and one vertex of D. We may assume these vertices are c_0 and d_0 as C and D are cycles. As Γ is 3-connected, $\Gamma\backslash\{c_0, d_0\}$ is connected. Again, choose a path P_2 in $\Gamma\backslash\{c_0, d_0\}$ from a vertex c_i of C to a vertex d_j of D that contains exactly one vertex of C and one vertex of D. Of course $C\backslash\{c_0, c_i\}$ is either one or two paths $Q_1 = c_1 \ldots c_{i-1}$ and $Q_2 = c_{i+1} \ldots c_{r-1}$ and, similarly, $D\backslash\{d_0, d_j\}$ is either one or two paths $R_1 = d_1 \ldots d_{j-1}$ and $R_2 = d_{j+1} \ldots d_{r-1}$. Choosing the longer Q_ℓ of Q_1 and Q_2 and the longer R_m of R_1 and R_2 (or the single path Q_ℓ or R_m if there is only one path), we have that $P_1 \cup P_2 \cup Q_\ell \cup R_m$ is a cycle in Γ of length longer than r, a contradiction.

Suppose C and D have one common vertex. The proof is very similar to that in the previous paragraph. Without loss of generality we can assume that $c_0 = d_0$. As Γ is 3-connected (in fact for this argument it suffices that Γ is 2-connected) there exists a path P from c_i to d_j between $C\backslash\{c_0\}$ and $D\backslash\{c_0\}$ that does not contain c_0 nor any other vertex of C or D. As above, both $C\backslash\{c_0, c_i\}$ and $D\backslash\{d_0, d_j\}$ are either one or two paths. Choose Q_ℓ and R_m as in the previous paragraph. Then $P \cup Q_\ell \cup R_m$ is a cycle in Γ of length longer than r, a contradiction.

Suppose C and D have exactly two vertices in common, say $c_0 = d_0$ and $c_k = d_l$. Observe that this pair of common vertices must be antipodal in both cycles C and D, for otherwise a cycle of length more than r would exist in Γ. This implies that r is even and $r = 2k = 2l$. There are four edge disjoint k-paths $P_1, \ldots, P_4$ connecting c_0 and c_k. Set $\Delta = (C \cup D)\backslash\{c_0, c_k\}$. Then Δ is a vertex-disjoint union of four paths $Q_1, \ldots, Q_4$ with $Q_i \subset P_i$, $1 \leq i \leq 4$. As Γ is 3-connected, there is a path in $\Gamma\backslash\{c_0, c_k\}$ connecting some vertex of Q_i in Δ to some vertex of Q_j in Δ, and we can choose such a path P so that it is vertex-disjoint from the other two components (paths) in Δ. Then a longest path Q in

$P \cup P_i \cup P_j$ from c_0 to c_k is of length at least $k + 1$, and $Q \cup P_\ell$, $i \neq \ell \neq j$ is a cycle in Γ of length at least $2k + 1 \geq r$, a contradiction.

Let $\mathcal{F}$ be the set of all longest cycles in Γ. Define the hypergraph $\mathcal{H} = (V, \mathcal{F})$. Clearly, $|E| = r$ for every $E \in \mathcal{F}$. As Γ is vertex-transitive, $\mathcal{H}$ is regular, that is, every vertex of $\mathcal{H}$ is contained in the same number t of edges of $\mathcal{H}$ (longest cycles in Γ). As any two longest cycles in Γ have at least three vertices in common, it follows that for $E, F \in \mathcal{F}$ we have $|E \cap F| \geq 3$.

Fix $E \in \mathcal{F}$. We count the ordered pairs (v, F), where $v \in E$ and $F \in \mathcal{F}$ is distinct from E, in two ways. In the first way, we choose the vertex v first and then the edge F. There are r vertices in E, and there are $t - 1$ edges that contain a vertex of E distinct from E. So there are $r(t - 1)$ such ordered pairs. Counting the second way, we choose an edge first, and then a vertex. There are $|\mathcal{F}| - 1$ choices for the edge F, and as $|E \cap F| \geq 3$, after choosing E there are at least three choices for v. So there are at least $3(|\mathcal{F}| - 1)$ such ordered pairs. Hence $r(t - 1) \geq 3(|\mathcal{F}| - 1)$. The obvious analogue of the Handshaking lemma for t-regular hypergraphs gives $|\mathcal{F}| = tn/r$. Substituting, we have

$$r(t - 1) \geq 3\left(\frac{tn}{r} - 1\right),$$

or, equivalently, $t\left(r^2 - 3n\right) \geq r(r-3)$. As $r \geq 3$, $r(r-3) \geq 0$ and so $t\left(r^2 - 3n\right) \geq 0$, or $r^2 \geq 3n$. The result follows. □

Various results on the Lovász problem, together with its Cayley graph variant, will be discussed at length in this chapter. We first give the promised second proof that the Petersen graph is not Hamiltonian.

Theorem 11.1.4 *The Petersen graph is not Hamiltonian.*

Proof First, recall that the Petersen graph P has girth 5. Assume P is Hamiltonian, with Hamilton cycle $C = v_0v_1 \ldots v_9v_0$. There are then five additional edges connecting five pairs of vertices on this cycle. As P has girth 5 the additional neighbor of a given vertex v is a vertex v' antipodal to v or one of the two neighbors of v' on the Hamilton cycle. Also, not all edges not on C can be incident with antipodal vertices, as otherwise P contains a 4-cycle. We may thus assume, without loss of generality, that v_0 is adjacent to v_6. Now the third neighbor of v_1 could be one of the three vertices v_5, v_6, and v_7. But v_6 is already the neighbor of v_0, and both v_5 and v_7 would give a cycle of length 4, a contradiction. □

There are two interesting features of this proof. First, the proof did not use transitivity, or any other property of the Petersen graph other than it has valency 3 and girth 5! So the exact same proof shows that a cubic graph of girth 5 on

10 vertices is not Hamiltonian. This, however, is not stronger, as the Petersen graph is the only cubic graph of girth 5 on 10 vertices.

The following theorem on upper bounds on the eigenvalues of the Laplacian matrix of a graph is due to Mohar (1992, Corollary 3.5) (see also van den Heuvel (1995)). It can also be used to show that the Petersen graph and the Coxeter graph are not Hamiltonian.

Definition 11.1.5 The **Laplacian matrix** of a d-regular graph Γ is $dI - A(\Gamma)$, where $A(\Gamma)$ is the adjacency matrix of Γ and I is the identity matrix.

Theorem 11.1.6 *Let Γ be a cubic graph of order n and let $\lambda_1 \leq \lambda_2 \leq \cdots \leq \lambda_n$ be the eigenvalues of its Laplacian matrix. Fix $k \leq n$. Suppose that there is an index i, $1 \leq i \leq 2k$, such that*

(i) *$i \leq k$ and $\lambda_i > 4 - 2\cos(2\pi\lfloor i/2 \rfloor/k)$, or*
(ii) *$\frac{5}{2}n + 1 - 2k \leq i \leq n$ and $\lambda_i < 4 - 2\cos\left(2\pi\lfloor \frac{1}{4}(5n + 2 - 2i)\rfloor/k\right)$.*

Then Γ does not contain a cycle of length k.

Example 11.1.7 Following Mohar (1992, 3.6 Example), the Laplacian spectrum of the Petersen graph is $0, 2, 2, 2, 2, 2, 5, 5, 5, 5$, and for $k = 10$ and $\lambda_7 = 5$ we have $\lambda_7 > 4 - 2\cos(3\pi/5) = 4.618034$. By Theorem 11.1.6, the Petersen graph does not have a cycle of length 10 so is not Hamiltonian.

Example 11.1.8 We have already seen that the Coxeter graph is not Hamiltonian in Theorem 6.1.5. We now use Theorem 11.1.6 to give another, much shorter proof. As communicated by Mohar, this was first done by Royle. The characteristic polynomial of the Laplacian matrix of the Coxeter graph is

$$x(x-1)^8(x-4)^7\left(x^2 - 8x + 14\right)^6,$$

implying that the Laplacian spectrum of the Coxeter graph is

$$[0]^1, [1]^8, \left[4 - \sqrt{2}\right]^6, [4]^7, \left[4 + \sqrt{2}\right]^6$$

(here $[a]^t$ means a is an eigenvalue of the Laplacian matrix t times). For $k = 28$ and $\lambda_{23} = 4 + \sqrt{2}$, we have $\lambda_{23} > 4 - 2\cos(22\pi/28) \approx 4.781831$, and by Theorem 11.1.6, the Coxeter graph does not have a cycle of length 28.

The main obstacle to making substantial progress on the Lovász problem is a lack of structural results for vertex-transitive graphs. This obstacle has motivated many new problems seeking such structural results. An example of this is the semiregularity problem that we saw in Section 3.8 and will see again in Chapter 12, which was posed for exactly this reason.

In all known results on the existence of Hamilton cycles in vertex-transitive graphs, the methods are constructive. In practice, a variety of approaches are used to find Hamilton cycles in vertex-transitive graphs, more or less depending on what will work! These include **lifting cycles** (the spiral path argument we saw in Section 3.7 is an example of this), which we will study next, the order of the graph that we will see in Section 11.3, and even embedding graphs in surfaces! This last topic we will examine in Section 11.5.

11.2 The Lifting Cycle Technique

This approach is based on taking the quotient with respect to a block system of a group of automorphisms or, sometimes, with respect to the set of orbits of a semiregular automorphism, preferably one of prime order. Provided the quotient multigraph contains a Hamilton cycle it is sometimes possible to lift this cycle to construct a Hamilton cycle in the original graph. Typically, with these techniques the number of edges between two orbits is important. Thus we need a more general notion of a quotient graph.

Definition 11.2.1 Let Γ be a graph and $K \leq \mathrm{Aut}(\Gamma)$ be such that every orbit of K has the same length, and $\mathcal{P}$ the set of orbits of K. Let $A, B \in \mathcal{P}$. By $d(A, B)$ we denote the valency of $\Gamma[A, B]$. (Note that the graph $\Gamma[A, B]$ is regular.) The **quotient multigraph corresponding to** K is the multigraph Γ/K whose vertex set is $\mathcal{P}$ and in which $A, B \in \mathcal{P}$ are joined by $d(A, B)$ edges (so multiple edges are definitely allowed!). Note that the quotient graph $\Gamma/\mathcal{P}$ is precisely the underlying simple graph of Γ/K. If K is a cyclic group generated by an element $\rho \in K$, then Γ/K is denoted by Γ/ρ.

Note that if $K \trianglelefteq G$, where $G \leq \mathrm{Aut}(\Gamma)$ is transitive, then the set of orbits of K is a block system of G, and the quotient graph $\Gamma/\mathcal{P}$ corresponding to $\mathcal{P}$ is simply our usual block quotient graph. Clearly the valency of the quotient multigraph Γ/K is the same as the valency of Γ.

Example 11.2.2 The Pappus graph is Hamiltonian.

Solution Consider the Pappus graph in Figure 11.2. It is the Levi graph of the Pappus configuration shown in Figure 6.12, and the points in the configuration have the same coloring as vertices in the graph. By inspection, the following semiregular permutation is an automorphism of the Pappus graph:

$$\pi = (p_0, p_3, p_6)(p_1, p_4, p_7)(p_2, p_5, p_8)(l_0, l_3, l_6)(l_1, l_4, l_7)(l_2, l_5, l_8).$$

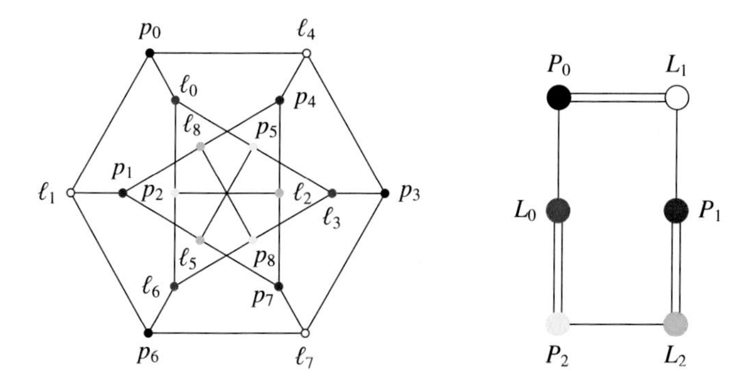

Figure 11.2 (Left) The Pappus graph and (Right) its quotient multigraph with respect to the $(6, 3)$-semiregular automorphism π.

Denote the orbits of π by $P_i = \{p_i, p_{i+3}, p_{i+6}\}$ and $L_i = \{l_i, l_{i+3}, l_{i+6}\}$, where $i \in \{0, 1, 2\}$. Its quotient multigraph with respect to π is a 6-cycle,

$$C = L_0, P_0, L_1, P_1, L_2, P_2, L_0,$$

with alternating double edges P_0L_1, P_1L_2, and P_2L_0. By Lemma 3.7.3, the Pappus graph has a Hamilton cycle. □

Note that the crucial step in the spiral path argument is that there are two edges between some vertices in the quotient multigraph. In general though, the quotient multigraph has only an edge between vertices, and hence we do not get a Hamilton cycle so easily. Instead more ingenious approaches are required.

The idea of lifting cycles was first introduced in Marušič (1983) to prove that Cayley graphs of $\mathbb{Z}_p \rtimes A$ have Hamilton cycles, where A is abelian of odd order and p is prime (see Theorem 11.2.11). This result was later used to show Hamilton cycles exist in connected vertex-transitive graphs with automorphism groups containing a transitive subgroup whose commutator subgroup is cyclic of prime power order (with the Petersen graph being the only counterexample); see Dobson et al. (1998); Durnberger (1983); Keating and Witte (1985); Witte (1982, 1986) (see also Ghaderpour and Morris (2014); Morris (2016; 2018) for recent results on the Lovász problem restricted to special commutator subgroups).

To give an example of the lifting cycle technique we will prove that every Cayley graph of $\mathbb{Z}_p \rtimes A$, where A is abelian of odd order, has a Hamilton cycle. The proof is taken from Marušič (1983), but reorganized to make the lifting cycle technique more apparent and earlier in the proof. It is in Lemma 11.2.9.

While we have constructed Hamilton cycles in Cayley graphs in Chapter 3, our next results require a more natural notation. Typically when constructing a cycle or path in a Cayley graph $\mathrm{Cay}(G, S)$, it is easiest when sequentially constructing such a cycle or path to not write the label of the next vertex, but to write the element of the connection set that is used to produce an edge between the current vertex and the next vertex. For example, suppose one starts at the identity, and then uses the elements $s_1, \ldots, s_r$ to produce a walk in $\mathrm{Cay}(G, S)$. Using our normal graph-theoretic notation for a walk, this would be $1, s_1, s_1 s_2, s_1 s_2 s_3, \ldots, s_1 \cdots s_r$. It is much simpler to write $[s_1, \ldots, s_r]$ to represent this walk, which is what we shall do. Formally,

Definition 11.2.3 Let G be a group and $S \subset G$. A **sequence** on S is a sequence all of whose terms are elements of S. Let $\mathcal{S} = [s_1, s_2, \ldots, s_r]$ be a sequence on S. We call the walk $1, s_1, s_1 s_2, \ldots, s_1 \cdots s_r$ the walk **corresponding** to the sequence $\mathcal{S}$. The ***i*th partial product** $\pi_i(\mathcal{S})$ of $\mathcal{S}$ is $s_1 s_2 \cdots s_i$ and $\pi(\mathcal{S}) = \pi_r(\mathcal{S})$. Let $\mathcal{T} = [t_1, t_2, \ldots, t_q]$ be a sequence on S. The **product** $\mathcal{S}\mathcal{T}$ is defined to be the sequence $[s_1, s_2, \ldots, s_r, t_1, t_2, \ldots, t_q]$.

Note that the product $\mathcal{S}\mathcal{T}$ corresponds to the usual concatenation of the two walks $1, s_1, s_1 s_2, \ldots, s_1 \cdots s_r$ and $s_1 \cdots s_r t_1, \ldots, s_1 \cdots s_r t_1 \cdots t_q$.

The proof of Theorem 11.2.11 is quite long, so a sketch of the structure of the proof will be useful. Let G be group whose commutator subgroup K is a cyclic group of prime order p. Let $\mathcal{B}$ be the block system of G_L that is the set of left cosets of K in G. Let $\Gamma = \mathrm{Cay}(G, S)$ be a Cayley graph of G, and let $\tilde{\Gamma} = \Gamma/\mathcal{B}$. Then $\tilde{\Gamma} = \mathrm{Cay}\left(G/K, \tilde{S}\right)$, $\tilde{S} \subseteq G/K$, say of order n. There are two cases depending on whether $S \cap K$ is empty or not. The lifting cycle technique applies in the first case. We use the fact that Cayley graphs of abelian groups are Hamiltonian with many Hamilton cycles. We may assume that there exists a sequence $\mathcal{S} = [s_1, s_2, \ldots, s_n]$ of elements in S such that $\tilde{\mathcal{S}}$ induces a Hamilton cycle in $\tilde{\Gamma}$, which corresponds to a walk in Γ starting and ending at a given orbit of K and meeting every other orbit of K exactly once. When $\pi(\mathcal{S}) = s_1 s_2 \cdots s_n = 1$, this walk is a cycle in Γ, and when $\pi(\mathcal{S}) \neq 1$ the walk is a spiral path in Γ, in which case repeating the sequence p times we end up with a Hamilton cycle $[s_1, s_2, \ldots, s_n]^p$ in Γ by Lemma 3.7.2. The crucial part in the proof is showing there are two sequences $\mathcal{S}$ and $\mathcal{T}$ of elements in S exist such that $\pi(\mathcal{S}) \neq \pi(\mathcal{T})$ and such that both $\tilde{\mathcal{S}}$ and $\tilde{\mathcal{T}}$ induce a Hamilton cycle in $\tilde{\Gamma}$. Note that this suffices to give a Hamilton cycle in Γ as the order of K is prime. (The same approach would also work for an arbitrary cyclic group K provided we can find a sequence $\mathcal{S}$ such that $\pi(\mathcal{S})$ is a generator of K.)

We begin with some group-theoretic facts.

Lemma 11.2.4 *Let G be a group with commutator subgroup K of prime order p and odd index. Then $p \neq 2$.*

Proof Suppose $p = 2$. As $K \trianglelefteq G$ and as K is cyclic of order 2, $K \leq Z(G)$. As G/K is odd, K is a Sylow 2-subgroup of G and is in the center of its normalizer in G. By Burnside's transfer theorem, there exists $N \trianglelefteq G$ such that $G = KN$. Then $G/K \cong N$ is abelian as K is the commutator subgroup of G, and so $G \cong K \times N$ is abelian. But then $K = 1$, a contradiction. So $p \geq 3$. □

Lemma 11.2.5 *Let G be a group with commutator subgroup K of odd prime order p and odd index. Then for each $g \in G$ there exists an integer $1 \leq d_g \leq p - 1$ such that $g^{-1}kg = k^{d_g}$ for all $k \in K$, and*

(i) $d_{gg'} \equiv d_g d_{g'} \pmod p$ *and if* $n = |g|$, *then* $d_g^n \equiv 1 \pmod p$, *and*
(ii) *if* $g \notin C_G(K)$ *then* $|gk| = |g|$ *for each* $k \in K$.

Proof Let $k \in K$ such that $\langle k \rangle = K$, and $g \in G$ with $n = |g|$. Then $g^{-1}kg = k^{d_g}$ for some $1 \leq d_g \leq p - 1$. Let i be an integer, and $k^i \in K$. Then $g^{-1}k^i g = \left(g^{-1}kg\right)^i = \left(k^{d_g}\right)^i = \left(k^i\right)^{d_g}$, and so d_g is independent of choice of generator of K.
(i) A simple computation shows that $d_{gg'} \equiv d_g d_{g'} \pmod p$ for every $g, g' \in G$. Then $d_g^n \equiv d_1 \equiv 1 \pmod p$.
(ii) Suppose $g \notin C_G(K)$. Then $gk \neq kg$, $k(gk) = (kg)k \neq (gk)k$, and so $gk \notin C_G(K)$. As $g \notin C_G(K)$, we have $d_g \not\equiv 1 \pmod p$. Let $e(g) = \sum_{i=0}^{n-1} d_g^i$. Then $(d_g - 1)e(g) = d_g^n - 1 \equiv 0 \pmod p$ as $d_g^n \equiv 1 \pmod p$. Thus $e(g) \equiv 0 \pmod p$. As $kg = gk^{d_g}$, a straightforward induction argument gives $(gk)^n = g^n k^{e(g)} = 1$. Thus $|gk| \leq |g|$. Analogously, as $gk \notin C_G(K)$ we have $\left|k^{-1}(kg)\right| \leq |kg|$. □

Definition 11.2.6 Let G be a group, $S \subset G$, and $\mathcal{S} = [s_1, s_2, \ldots, s_r]$ be a sequence on S. We say that $\mathcal{S}$ is **Hamiltonian** if $r = |G|$, $\pi(\mathcal{S}) = 1$, and the partial products $\pi_i(\mathcal{S})$, $i \in \{1, 2, \ldots, |G|-1\}$, are all distinct nonidentity elements of G. By $\hat{\mathcal{S}}$, we mean the sequence $[s_2, s_3, \ldots, s_r]$ and by $\bar{\mathcal{S}}$, the sequence $[s_1, s_2, \ldots, s_{r-1}]$. We denote the element s_r by $\ell_{\mathcal{S}}$ (so the ℓ stands for "last"). Finally, $(\mathcal{S}, \mathcal{T})$ denotes the sequence $[t_1]\hat{\mathcal{S}}^{-1}[t_2]\hat{\mathcal{S}} \cdots [t_{q-2}]\hat{\mathcal{S}}^{-1}[t_{q-1}]\hat{\mathcal{S}}$ if $q \geq 3$ is odd, if $q \geq 4$ is even the sequence $[t_1]\hat{\mathcal{S}}^{-1}[t_2]\hat{\mathcal{S}} \cdots [t_{q-3}]\hat{\mathcal{S}}^{-1}[t_{q-2}]\hat{\mathcal{S}}$, and the empty sequence if $q \in \{1, 2\}$.

Of course, a sequence is Hamiltonian if and only if the walk corresponding to $\mathcal{S}$ is a Hamilton cycle in $\mathrm{Cay}(G, S)$. We leave the straightforward proof of this fact to Exercise 11.2.1.

Lemma 11.2.7 *Let G be a group of order n and $S \subset G$. The sequence $[s_1, \ldots, s_n]$ is Hamiltonian if and only if the walk $1, s_1, s_1 s_2, \ldots, s_1 \cdots s_n$ is a Hamilton cycle in* $\mathrm{Cay}(G, S)$.

Let $k \in \mathbb{Z}_p$ be a generator. Then the sequence $[k]^p = [k, k, \ldots, k]$ (with p elements) is Hamiltonian. This argument can be translated inductively into the more general setting of all connected Cayley graphs of abelian groups. As mentioned in Section 3.5, this was first done in the early 1970s by Pelikán. The proof is given in Theorem 3.5.4. Another proof was given in Marušič (1983), as a consequence of the following more general result (Marušič, 1983, Lemma 3.1), the proof of which is left as Exercise 11.2.2.

Lemma 11.2.8 *Let G be an abelian group with a generating set S and let S' be a nonempty subset of S (that does not necessarily generate G). If $\mathcal{S}$ and $\mathcal{T}$ are Hamiltonian sequences in $\langle S' \rangle$ with $\ell_{\mathcal{S}} = \ell_{\mathcal{T}}$, then there exists a sequence Q on G such that $\overline{\mathcal{S}}Q$ and $\overline{\mathcal{T}}Q$ are Hamiltonian.*

We are now ready to lift!

Lemma 11.2.9 *Let $G = K \rtimes A$ with A abelian of odd order and $K \cong \mathbb{Z}_p$ for some prime p. If S is a minimal generating set of G such that $S \cap K = \emptyset$, then* $\mathrm{Cay}(G, S)$ *is Hamiltonian.*

Proof If G is abelian, then the result follows by Theorem 3.5.4. Otherwise, K is the commutator subgroup of G. As mentioned above, our general strategy is the following. We will first consider the quotient graph $\mathrm{Cay}(G/K, \{sK : s \in S\})$. As K is the commutator subgroup of G, G/K is an abelian group, and so as $\mathrm{Cay}(G, S)$ is connected, we know $\mathrm{Cay}(G/K, \{sK : s \in S\})$ has Hamilton cycles by Theorem 3.5.4. We then lift appropriate Hamilton cycles in $\mathrm{Cay}(G/K, \{sK : s \in S\})$ to a Hamilton cycle in $\mathrm{Cay}(G, S)$. So to begin with, we will be working in $\mathrm{Cay}(G/K, \{sK : s \in S\})$.

As G is nonabelian and Γ is connected, there are $x, y \in S$ such that $xy \neq yx$. Let n be the multiplicative order of the coset xK in G/K. Then n is odd as $|G/K|$ is odd, and $xK \neq K$ as $S \cap K = \emptyset$. So $n \geq 3$. Similarly, the multiplicative order of yK in G/K is also at least 3. For each sK, where $s \in S$, choose $s' \in S \cap sK$. In addition, if $xK \neq yK \neq x^{-1}K$, then we insist that $x' = x$, $\left(x^{-1}\right)' = x^{-1}$, $y' = y$, and $\left(y^{-1}\right)' = y^{-1}$. For a sequence $\mathcal{R} = [r_1, r_2, \ldots, r_t]$ of elements of $\{sK : s \in S\}$, we denote the sequence $[r_1', r_2', \ldots, r_t']$ by $\mathcal{R}'$.

Suppose $yK \in \langle xK \rangle \leq G/K$, and let i be the least positive integer such that $yK = (xK)^i$. Let $\mathcal{A} = \left[(xK)^{-1}\right]^n$ and let $\mathcal{B} = \mathcal{A}$ if $i = 1$ or $n - 1$ and

$$\mathcal{B} = [yK][xK]^{n-i-1}[yK]\left[(xK)^{-1}\right]^{i-1} = \left[(xK)^i\right][xK]^{n-i-1}\left[(xK)^i\right]\left[(xK)^{-1}\right]^{i-1}$$

otherwise. Clearly, the sequences $\mathcal{A}$ and $\mathcal{B}$ are Hamiltonian sequences in the graph $\mathrm{Cay}(\langle xK \rangle, \{xK, yK\})$ and $\ell_{\mathcal{A}} = \ell_{\mathcal{B}} = (xK)^{-1}$. By Lemma 11.2.8, there exists a sequence Q on G/K such that both $\mathcal{S} = \bar{\mathcal{A}}Q$ and $\mathcal{T} = \bar{\mathcal{B}}Q$ are Hamiltonian sequences in $\mathrm{Cay}(G/K, \{sK : s \in S\})$. As $yK = (xK)^i = x^iK$, we have $y \in x^iK$. As $xy \neq yx$, there exists $k \in K$, $k \neq 1$, such that $y = x^ik$. If $x \in C_G(K)$, then $yx = x^ikx = xk^i = xy$, a contradiction. Thus $x \notin C_G(K)$ and so by Lemma 11.2.5 (ii), $|x| = |xK| = n$. By Lemma 11.2.5 (i), $d_x^n \equiv 1 \pmod p$ and as n is odd we have

$$d_x \not\equiv -1 \pmod p.$$

Suppose $i \in \{1, n-1\}$, and consequently $\mathcal{A} = \mathcal{B}$. Let $\tilde{\mathcal{S}} = \left[x^{-1}\right]^{n-1} Q'$, and

$$\tilde{\mathcal{T}} = \begin{cases} \left[y^{-1}\right]\left[x^{-1}\right]^{n-2} Q' & \text{if } i = 1, \\ [y]\left[x^{-1}\right]^{n-2} Q' & \text{if } i = n-1. \end{cases}$$

Then $\tilde{\mathcal{S}}$ and $\tilde{\mathcal{T}}$ are walks in $\mathrm{Cay}(G, S)$ whose quotients are the Hamilton cycle in $\mathrm{Cay}(G/K, \{sK : s \in S\})$ induced by $\mathcal{S}$. As $xy \neq yx$, $x \neq y, y^{-1}$ and so $x^{-1}x^{-n+2}\pi(Q')$ is not equal to both $yx^{-n+2}\pi(Q')$ and $y^{-1}x^{-n+2}\pi(Q')$. Thus $\pi(\tilde{\mathcal{S}}) \neq \pi(\tilde{\mathcal{T}})$. Then one of the natural walks in $\mathrm{Cay}(G, S)$ corresponding to $\mathcal{S}$ and $\mathcal{T}$ is a spiral path in $\mathrm{Cay}(G, S)$, and the result follows by Lemma 3.7.2.

Suppose $i \notin \{1, n-1\}$. Then,

$$\begin{aligned} \pi(\mathcal{T}')\pi(\mathcal{S}')^{-1} &= yx^{n-i-1}yx^{-i+2}\pi(Q')\pi(Q')^{-1}x^{n-1} \\ &= yx^{-i-1}yx^{-i+1} = kx^{-1}kx = k^{1+d_x} \neq 1, \end{aligned}$$

as $d_x \not\equiv -1 \pmod p$. Thus $\pi(\mathcal{S}') \neq \pi(\mathcal{T}')$ and so, as above, one of the natural walks in $\mathrm{Cay}(G, S)$ corresponding to $\mathcal{S}'$ and $\mathcal{T}'$ is a spiral path in $\mathrm{Cay}(G, S)$, and the result follows by Lemma 3.7.2.

A similar argument shows that if $xK \in \langle yK \rangle \leq G/K$, then $\mathrm{Cay}(G, S)$ is Hamiltonian.

Suppose $xK \notin \langle yK \rangle$ and $yK \notin \langle xK \rangle$. As $G = K \rtimes A$, we may write $x = ak_1$ and $y = bk_2$, for some $k_1, k_2 \in K$ and $a, b \in A$. If $x, y \in C_G(K)$ then, as A and K are abelian, we have

$$xy = ak_1y = ayk_1 = abk_2k_1 = bak_1k_2 = bxk_2 = bk_2x = yx,$$

which contradicts $xy \neq yx$. So, without loss of generality, we assume $x \notin C_G(K)$.

Let m be the smallest positive integer such that $yK^m \in \langle xK \rangle$. Then m is odd and at least 3. Let

$$\mathcal{A} = [xK]^{n-1}([xK]^n, [yK]^m)[xK]\left[yK^{-1}\right]^{m-1},$$

and $\mathcal{B}$ be

$$[xK]^{n-1}\left([xK]^n, [yK]^{m-2}\right)[yK]\left[xK^{-1}\right]^{n-3}[yK][xK]^{n-1}\left[yK^{-1}\right].\left[xK^{-1}\right]\left[yK^{-1}\right]^{m-2}.$$

Then $\mathcal{A}$ and $\mathcal{B}$ are Hamilton cycles in $\mathrm{Cay}(\langle xK, yK \rangle, \{xK, yK\})$ and $l_{\mathcal{A}} = l_{\mathcal{B}}$. Hence, by Lemma 11.2.8, there exists a sequence Q on H such that $\mathcal{S} = \bar{\mathcal{A}}Q$

and $\mathcal{T} = \bar{\mathcal{B}}Q$ are Hamilton cycles in $\mathrm{Cay}(G/K, \{sK : s \in S\})$. As $xy \neq yx$, $xyx^{-1}y^{-1} \in K\backslash\{1\}$, and as $x \notin C_G(K)$, it follows that $\left(xyx^{-1}y^{-1}\right)x \neq x\left(xyx^{-1}y^{-1}\right)$. Thus, $x^{-2}\left(xyx^{-1}y^{-1}\right) \neq x^{-1}\left(xyx^{-1}y^{-1}\right)x^{-1}$, or

$$x^{-1}yx^{-1}y^{-1} \neq yx^{-1}y^{-1}x^{-1}.$$

As $x \notin C_G(K)$, by Lemma 11.2.5 (i) we have $|x| = n$. Thus,

$$x^{n-1}yx^{-1}y^{-1} \neq yx^{n-1}y^{-1}x^{-1},$$

and so

$$\pi\left(\bar{\mathcal{A}}'\right) = zx^{-1}yx^{n-1}y^{-1}w \neq zyx^{n-1}y^{-1}x^{-1}w = \pi\left(\bar{\mathcal{B}}'\right)$$

where $z = x^{n-1}\left(yx^{-n+2}yx^{n-2}\right)^{(m-3)/2} yx^{-n+3}$ and $w = y^{-m+3}$. Hence,

$$\pi(\mathcal{S}') = \pi\left(\bar{\mathcal{A}}'\right)\pi(Q') \neq \pi\left(\bar{\mathcal{B}}'\right)\pi(Q') = \pi(\mathcal{T}'),$$

and thus either $\mathcal{S}'$ or $\mathcal{T}'$ are spiral paths in $\mathrm{Cay}(G, S)$, and so $\mathrm{Cay}(G, S)$ is Hamiltonian by Lemma 3.7.2. □

Lemma 11.2.10 *Let G be a group with commutator subgroup K of prime order and odd index. If S is a minimal generating set of G such that $S \cap K \neq \emptyset$, then* $\mathrm{Cay}\left(G, S \cup S^{-1}\right)$ *is Hamiltonian.*

Proof As $\langle S \rangle = G$, $\Gamma = \mathrm{Cay}\left(G, S \cup S^{-1}\right)$ is connected by Lemma 1.2.14. Let $k \in S \cap K$ (so $\langle k \rangle = K$), $S' = \left(S \cup S^{-1}\right) \setminus \left\{k, k^{-1}\right\}$, and $F = \langle S' \rangle$. As S is a minimal generating set of G, we have $K \cap F = 1$. As $K \trianglelefteq G$, $KF = G$ is a group. By definition, $G = K \rtimes F$. As K is the commutator subgroup of G, $F \cong G/K$ is abelian. By Theorem 3.5.4, $\mathrm{Cay}(F, S')$ has a Hamilton cycle, so there is a Hamiltonian sequence $\mathcal{S} = [s_1, s_2, \ldots, s_r]$ for $\mathrm{Cay}(F, S')$.

Let $\mathcal{T} = \left[k^{-1}\right]^{p-1}[s_r]\left[k^{-1}\right]^{p-1}[s_{r-1}]\cdots\left[k^{-1}\right]^{p-1}[s_1]$. As $K \cap F = 1$, for $1 \leq i \leq pr - 1$ the partial products $\pi_i(\mathcal{T})$ are all distinct nontrivial elements of G. This is easiest to see by considering the walk corresponding to $\mathcal{T}$. First, observe that the walk corresponding to $\left[k^{-1}\right]^{p-1}$ is $1, k^{-1}, k^{-2}, \ldots, k^{-(p-1)}$, which, as $\langle k \rangle = K$, is a Hamilton path in K. Of course, G/K is a block system $\mathcal{B}$ of G_L, and $\pi_1\left(\mathcal{S}^{-1}\right), \pi_2\left(\mathcal{S}^{-1}\right), \ldots, \pi_r\left(\mathcal{S}^{-1}\right)$ is a transversal of G/K as $\mathcal{S}$, and consequently, $\mathcal{S}^{-1}$ is a Hamiltonian sequence in $\mathrm{Cay}(F, S') \cong \Gamma/\mathcal{B}$. Thus the walk corresponding to T first gives a Hamilton path in the block K, then moves to block $s_r K$ where it traverses a Hamilton path in the block $s_r K$, etc. Arguing inductively, the walk corresponding to

$$[k^{-1}]^{p-1}[s_{r-1}][k^{-1}]^{p-1}[s_{r-2}]\cdots[k^{-1}]^{p-1},$$

(note this is $\mathcal{T}$ with $[s_1]$ removed) is a Hamilton path in Γ, and for $1 \le i \le pr-1$ the partial products $\pi_i(\mathcal{T})$ are all distinct nonidentity elements of G. It only remains to show we can extend this path to a cycle by showing its last vertex is adjacent to 1 using $s_1 \in S \cup S^{-1}$.

Let $z = \sum_{f \in F} d_f$. As G is nonabelian, there exists $f' \in F \backslash C_G(K)$, and, as F is a group, $\{ff' : f \in F\} = F$. By Lemma 11.2.5 (i),

$$z = \sum_{f \in F} d_{ff'} \equiv \sum_{f \in F} d_f d_{f'} \equiv z d_{f'} \pmod p,$$

or $z(d_{f'} - 1) \equiv 0 \pmod p$. As $f' \notin C_G(K)$, $d_{f'} \neq 1$, and so $d_{f'} \not\equiv 1 \pmod p$. Hence $z \equiv 0 \pmod p$. As the walk corresponding to $\mathcal{S}$ is a Hamilton cycle in $\mathrm{Cay}(F, S')$, we have $\{\pi_i(\mathcal{S}) : 1 \le i \le r\} = F$. This gives that $\sum_{i=1}^{r} d_{\pi_i(\mathcal{S})} = z$, and so

$$\pi(\mathcal{T}) = ks_r ks_{r-1} \cdots ks_1 = s_r s_{r-1} \cdots s_1 k^z = \pi(\mathcal{S}) k^z = 1.$$

Hence the walk in Γ corresponding to $\mathcal{T}$ is a Hamilton cycle. Note that it is in the middle equality in the previous displayed equation that shows why, in $\mathcal{T}$, the sequence $\mathcal{S}$ is reversed. As $g^{-1}kg = k^{d_g}$, $kg = gk^{d_g}$, and we have

$$ks_r ks_{r-1} \cdots ks_1 = ks_r ks_{r-1} \cdots s_2 s_1 k^{d_{s_1}} = ks_r ks_{r-1} \cdots s_3 s_2 s_1 k^{d_{s_1 s_2}} k^{d_{s_1}}.$$

An induction argument then gives $ks_r ks_{r-1} \cdots ks_1 = s_r s_{r-1} \cdots s_1 k^z$. □

Combining Lemmas 11.2.9 and 11.2.10, we have the following result.

Theorem 11.2.11 *Let $G \cong K \rtimes A$, where A is abelian of odd order and K is of prime order. Then every connected Cayley graph of G is Hamiltonian.*

This simple idea of lifting Hamilton cycles in a suitable quotient has been widely used in many results on Hamilton paths and cycles in vertex-transitive graphs. Some details will be given in the next section, but also see Exercise 11.3.1 for a concrete application of Theorem 11.2.11.

In view of the comments in the paragraph immediately after Example 11.2.2, the essential question on lifting Hamilton cycles is the case where every two adjacent orbits in the quotient multigraph are joined by a single edge. In other words, when is a covering graph of a Hamiltonian graph also Hamiltonian?

Problem 11.2.12 Let Γ be a vertex-transitive graph with a Hamilton cycle and let $\mathrm{Cov}(\Gamma)$ be a connected vertex-transitive cover of Γ. Does $\mathrm{Cov}(\Gamma)$ have a Hamilton cycle that is a lift of a Hamilton cycle in Γ?

The importance of this general question lies in the fact that these sorts of problems are encountered in many special cases when one is searching for Hamilton cycles. As an illustration, Witte's theorem (Witte, 1986) showing that every Cayley (di)graph of a p-group is Hamiltonian could be generalized to arbitrary vertex-transitive graphs of prime power order provided one could prove that for a prime p, a connected regular $\mathbb{Z}_p$-cover of a Hamiltonian vertex-transitive graph of order a power of p, is Hamiltonian. Then an inductive argument as follows could be applied. Let Γ be a connected vertex-transitive graph of order p^k and assume that every connected vertex-transitive graph of order p^{k-1} is Hamiltonian. Let P be a Sylow p-subgroup of $\mathrm{Aut}(\Gamma)$. Then P is transitive on $V(\Gamma)$ by Corollary 1.1.10. Let $\rho \in Z(P)$ be of prime order. Then $\langle\rho\rangle \trianglelefteq P$, and the set of its orbits is a normal block system $\mathcal{B}$ of P. As Γ is connected, $\Gamma/\mathcal{B}$ is a vertex-transitive graph of prime power order p^{k-1}, and may be assumed to be Hamiltonian. There are two cases, depending on whether the orbits of ρ are independent sets or not. If not we can find a Hamilton cycle in Γ as a cycle in a Cartesian product of two cycles. If so, then we may assume that Γ is a cover of Γ_ρ. We are left with the following problem.

Problem 11.2.13 Let p be a prime. Is a connected regular $\mathbb{Z}_p$-cover of a Hamiltonian vertex-transitive graph of order p^k Hamiltonian?

It is known that connected vertex-transitive graphs of order p^k, p a prime and $k \leq 5$, are Hamiltonian. Connected vertex-transitive graphs of orders p, p^2, and p^3 are isomorphic to Cayley graphs by Theorems 4.4.1, 9.1.1, and 9.1.3 and so are Hamiltonian by Witte's result (Witte, 1986) (see also Marušič (1985)) and Exercise 11.3.1. Such graphs of orders p^4 and p^5 were shown to be Hamiltonian in Chen (1998) and Zhang (2015), respectively.

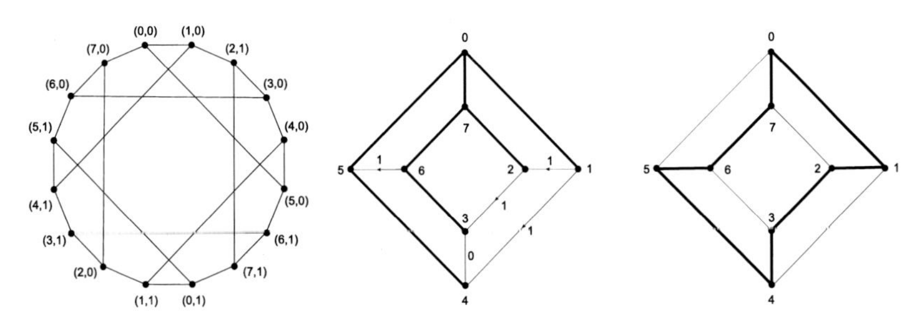

Figure 11.3 (Left) The Möbius–Kantor graph, (Middle) voltage assignment in the cube giving rise to the Möbius–Kantor graph and (Right) a Hamilton cycle in the cube.

Example 11.2.14 The Möbius–Kantor graph, shown in the left-hand side of Figure 11.3, is the unique cubic arc-transitive graph of order 16 by the Foster census. It is a regular $\mathbb{Z}_2$-cover of the cube, and it can be reconstructed by assigning zero voltages to the edges of the spanning tree shown in bold in the middle of Figure 11.3 (see, for example, Feng and Wang (2003)). The Hamilton cycle in the cube, shown in the right-hand graph in Figure 11.3, has voltage 1, and so lifts to a Hamilton cycle in the Möbius–Kantor graph by Lemma 3.7.2.

Motivated by Example 11.2.14 we propose the following question.

Problem 11.2.15 Is a connected regular $\mathbb{Z}_2$-cover of a Hamiltonian vertex-transitive graph also Hamiltonian?

11.3 Vertex-Transitive Graphs of Particular Orders

At the moment, the Lovász problem seems intractable. As is usual in mathematics for intractable problems, results have been proven showing vertex-transitive graphs have Hamilton paths or cycles that satisfy other conditions that restrict the problem to subclasses of vertex-transitive graphs. One such restriction is to consider vertex-transitive graphs of certain, usually "small," order. For example, connected vertex-transitive graphs of orders kp with $k \leq 6$, p^j with $j \leq 5$, $10p$ with $p \geq 11$ (except for generalized orbital graphs of $(\mathrm{PSL}(2,k), \mathrm{PSL}(2,k)/\mathbb{Z}_k \rtimes \mathbb{Z}_{(k-1)/10})$), $2p^2$, and pq where p and q are primes, have Hamilton paths – see Alspach (1979); Chen (1998); Du et al. (2018); Kutnar and Marušič (2008a); Kutnar et al. (2012b); Kutnar and Šparl (2009); Marušič (1985, 1987); Marušič and Parsons (1982, 1983); Zhang (2015). Furthermore, for all of these families, except for the graphs of orders $6p$ and $10p$ (and of course for the Petersen graph, the truncation of the Petersen graph, and the Coxeter graph), it is also known that they contain a Hamilton cycle – see Alspach (1979); Chen (1998); Du et al. (2018); Kutnar and Marušič (2008a); Kutnar and Šparl (2009); Marušič (1985, 1987); Marušič and Parsons (1982, 1983); Turner (1967). For some other partial results on the topic see Alspach et al. (1985); Kutnar and Šparl (2009); Marušič (1988a).

The most general and certainly the most complex result of this type is a very recent completion of the Lovász problem for vertex-transitive graphs of order a product of two prime numbers (Du et al., 2018). The classification of these graphs is discussed in Section 9.2. Here, we present the main ideas involved in the proof of this result, with parts of the proof given in full detail. We first deal with metacirculants and Fermat graphs, after handling the case when $qp = 2p$ separately, which was done in Alspach (1979).

Proposition 11.3.1 *Let p be a prime. Apart from the Petersen graph, a connected vertex-transitive graph of order $2p$ contains a Hamilton cycle.*

Proof Let Γ be a connected vertex-transitive graph of order $2p$. If $p = 2$, then the only connected vertex-transitive graphs of order 4 are the cycle of length 4 and K_4. As both are clearly Hamiltonian, we assume $p \geq 3$. By Theorem 3.8.6,

Γ has a semiregular automorphism π of order p. Let U and V be the orbits of $\langle\pi\rangle$, and $\Gamma' = \Gamma[U, V]$. Then $\pi \in \mathrm{Aut}(\Gamma')$, and as π cyclically permutes the elements of U and V, the valency of every vertex of U is the same, as is the valency of every vertex of V. This then implies that Γ' is regular of valency ν. As Γ' is regular, the graph $\Gamma'' = \Gamma\backslash E(\Gamma')$ is also regular of valency μ, and its components are two isomorphic circulant graphs of prime order by Theorem 4.4.1.

If $\nu \geq 2$, then Γ' contains a Hamilton cycle by Exercise 10.4.1, and so does Γ. If $\nu = 1$, then $\mu \geq 2$. If $\mu \geq 4$ then by Theorem 3.5.6, $\Sigma[U]$ and $\Sigma[V]$ are Hamilton connected. By Exercise 11.2.3, Γ is Hamiltonian. So $\mu = 2$. Then Γ is isomorphic to a $\mathrm{GP}(p, k)$ for some positive integer k by Exercise 6.5.7. The result then follows by Theorem 6.5.19 after recalling that $\mathrm{GP}(p, 2)$ is not vertex-transitive for $p \geq 7$ by Theorem 6.5.15. □

The graphs given in Theorem 9.2.10 (i) and (ii) were shown to be Hamiltonian in Alspach and Parsons (1982b); Marušič (1981b, 1992), respectively. For the next two results, we use the notation of metacirculant graphs from Section 1.5.

Lemma 11.3.2 *Let Γ be a connected (q, p)-metacirculant graph, where q and p are distinct primes. If $|S_0| > 2$ and $q, p \geq 3$, then Γ is Hamiltonian.*

Proof As Γ is connected, Γ/ρ is connected. As Γ/ρ is vertex-transitive, it is isomorphic to a circulant graph of order q by Theorem 4.4.1, and consequently is Hamiltonian by Theorem 3.5.4 as $q \geq 3$. Let $P = V_{i_0}V_{i_1}\ldots V_{i_{q-1}}V_{i_0}$ be a Hamilton cycle in Γ/ρ. Corresponding to each edge $V_{i_j}V_{i_{j+1}}$ in P, choose an edge $(i_j, k_j)(i_{j+1}, \ell_{j+1})$ in Γ. Note that if $(i_{j+1}, \ell_{j+1}) = (i_{j+1}, k_{j+1})$, then replacing the edge $(i_{j+1}, k_{j+1})(i_{j+2}, \ell_{j+2})$ with its image under ρ, we may assume without loss of generality that $(i_{j+1}, \ell_{j+1}) \neq (i_{j+1}, k_{j+1})$. Performing this replacement, if necessary, sequentially, we may assume without loss of generality that $(i_{j+1}, \ell_{j+1}) \neq (i_{j+1}, k_{j+1})$ for every $j \in \mathbb{Z}_q$. Now, $\Gamma[V_i]$ is a circulant graph of prime order p of valency at least 3 as $|S_0| \geq 3$. Also, as $p \geq 3$, $\Gamma[V_i]$ is a connected graph that is not bipartite. By Theorem 3.5.6 we see that $\Gamma[V_i]$ is Hamilton connected. Let P_{i_j} be a Hamilton path in $\Gamma[V_{i_j}]$ from (i_j, ℓ_j) to (i_j, k_j). Then $(i_0, k_0)(i_1, \ell_1)P_{i_1}(i_1, k_1)(i_2, \ell_2)P_{i_2}\ldots P_{i_{q-1}}(i_{q-1}, k_{q-1})(i_0, \ell_{i_0})P_{i_0}$ is a Hamilton cycle in Γ. □

Proposition 11.3.3 *Let $q < p$ be prime. A connected $(q, p, S_0, \ldots, S_{q-1})$-metacirculant graph Γ, other than the Petersen graph, is Hamiltonian.*

Proof We may assume by Proposition 11.3.1 that $q \neq 2$. The result follows by Theorem 3.5.4 if Γ is isomorphic to a circulant graph. Otherwise, if Γ is isomorphic to a Cayley graph, then the result follows by Theorem 11.2.11 as $\langle\rho\rangle$ is the commutator subgroup of $\langle\rho,\tau\rangle$ and is semiregular of order p. By Theorem 1.5.13 we may assume q^2 divides $|\alpha|$. Then $\tau^q(i,j) = (i,\alpha^q j)$ fixes $(i,0)$ for $i \in \mathbb{Z}_q$, and every other orbit of τ^q has at least q elements. If $|S_0| \neq \emptyset$, then $|S_0| \geq q$ and the result follows by Lemma 11.3.2. So we assume $S_0 = \emptyset$. If $S_i \not\subseteq \{0\}$ for some $1 \leq i \leq q-1$, then $|S_i| \geq q$. Let $\mathcal{B}$ be the set of orbits of $\langle\rho\rangle$. As every edge of $\Gamma/\mathcal{B}$ is contained in a Hamilton cycle by Corollary 3.5.7, Γ is Hamiltonian by Corollary 3.7.4. Otherwise, for $1 \leq i \leq q-1$ we have $S_i \subseteq \{0\}$ and so Γ is disconnected with components $V_j = \{(i,j) : i \in \mathbb{Z}_q\}$, $j \in \mathbb{Z}_p$, a contradiction. □

There are times when one would like to lift a Hamilton cycle in a quotient graph $\Gamma/\mathcal{B}$ to Γ, but perhaps $\Gamma/\mathcal{B}$ does not have a Hamilton cycle that lifts to a Hamilton cycle in Γ, or one can not find such a cycle. In this case, sometimes one can lift a smaller cycle in $\Gamma/\mathcal{B}$ to a cycle in Γ, and then somehow extend this cycle in Γ to a Hamilton cycle. For example, in Marušič and Scapellato (1992a) it was shown that a Fermat graph X of order qp is a $\mathbb{Z}_q$-cover of K_p, with $p = 2^{2^s} + 1$ a Fermat prime and $q|(p-2)$ prime. In Marušič (1992), a non-hamiltonian cycle C' was found in K_p that lifts to a cycle C in X. The cycle C has the property that for each vertex v of X not on C, there are a pair of vertices on C that are adjacent in C and are both neighbors of v. It was also shown that these pairs are all different, and so C can be extended to a Hamilton cycle in X.

Theorem 11.3.4 *Let q and p be distinct primes such that $p = 2^{2^s} + 1$ is a Fermat prime and $q|(p-2)$ is prime. Fermat graphs of order qp are Hamiltonian.*

We now consider the graphs in Theorem 9.2.10 (iii). This was finished in Du et al. (2020) and combines a variety of graph-theoretic, group-theoretic, and number-theoretic tools. Let Γ be a vertex-transitive graph of order qp, with $p > q$. For example, by Theorem 3.8.6 Aut(Γ) contains a semiregular automorphism ρ of prime order. This allows an application of the lifting cycle technique in Section 11.2, provided the quotient graph Γ/ρ has a Hamilton cycle. As Γ/ρ need not be vertex-transitive, classical graph theory results are often used. In particular, two results are noteworthy for the work discussed here. The first is Jackson's theorem (Jackson, 1980), improved by Zhu et al. (1985, 1986a,b), which is useful for graphs with large enough valency.

Theorem 11.3.5 *Let Γ be a 2-connected k-regular graph of order n. Then Γ is Hamiltonian if*

(i) $n \leq 3k + 1$ *and* Γ *is not isomorphic to the Petersen graph;*

(ii) $n \leq 3k + 3$ *and* $k \geq 6$.

The second is Chvátal's theorem (Chvátal, 1972).

Theorem 11.3.6 *Let* Γ *be a graph of order n. For a given positive integer i let* $S_i = \{x \in V(\Gamma)\colon \deg(x) \leq i\}$. *If for every* $i < n/2$ *either* $|S_i| \leq i-1$ *or* $|S_{n-i-1}| \leq n - i - 1$ *then* Γ *contains a Hamilton cycle.*

We need to show that all generalized orbital graphs of the primitive groups given in Table 9.1 have a Hamilton cycle. As these groups are primitive, by Theorem 2.4.12 their orbital graphs are connected, so it suffices to show the orbital graphs of these groups are Hamiltonian. The next result uses finite simple groups that we have not defined.

Proposition 11.3.7 *Orbital graphs of primitive groups in rows 1–3 of Table 9.1 are Hamiltonian.*

Proof First consider the rank 3 action of the orthogonal group $P\Omega^{\epsilon}(2d, 2)$ on singular 1-spaces, for $\epsilon \in \{+1, -1\}$, in row 1 of Table 9.1. The primes p and q are as follows: $p = 2^d - 1$ and $q = 2^{d-1} + 1$ for $P\Omega^{+}(2d, 2)$ and $p = 2^d + 1$ and $q = 2^{d-1} - 1$ for $P\Omega^{-}(2d, 2)$. In the first case, the valencies of the two orbital graphs are $\left(2^{d-1} - 1\right)\left(2^{d-1} + 2\right) = q^2 - q - 2$ and $2^{2d-2} = q^2 - 2q + 1$. Similarly, in the second case the valencies of the two graphs are $\left(2^{d-1} + 1\right)\left(2^{d-2} - 2\right) = q^2 + q - 2$ and $2^{2d-2} = q^2 + 2q + 1$. It is straightforward to see that for each of these graphs the valency exceeds one third of the corresponding order, and so, by Theorem 11.3.5, the result follows.

Now consider the action of M_{22} of rank 3 and degree 77 in row 2 of Table 9.1. The suborbits are of lengths 1, 16, and 60. Again, Theorem 11.3.5 applies to the graph of valency 60. For the graph Γ of valency 16, the quotient corresponding to a $(7, 11)$-semiregular automorphism ρ is isomorphic to K_7 and contains multiple edges. An appropriate Hamilton cycle is then chosen in Γ/ρ and by Corollary 3.7.4, Γ is Hamiltonian.

Finally, consider A_7 acting on triples in $\{1, 2, \ldots, 7\}$ in row 3 of Table 9.1. The image of its permutation representation has suborbits of lengths 1, 4, 12, and 18. By Theorem 11.3.5, there is a Hamilton cycle in each of these graphs except for the odd graph O_4 of valency 4. As mentioned in Section 6.2, all of the odd graphs other than the Petersen graph are Hamiltonian. Using the lifting cycle technique we sketch a proof that O_4 is Hamiltonian. The odd graph O_4 has a $(7, 5)$-semiregular automorphism whose corresponding quotient multigraph contains a 7-cycle with a double edge (see Figure 11.4), and thus

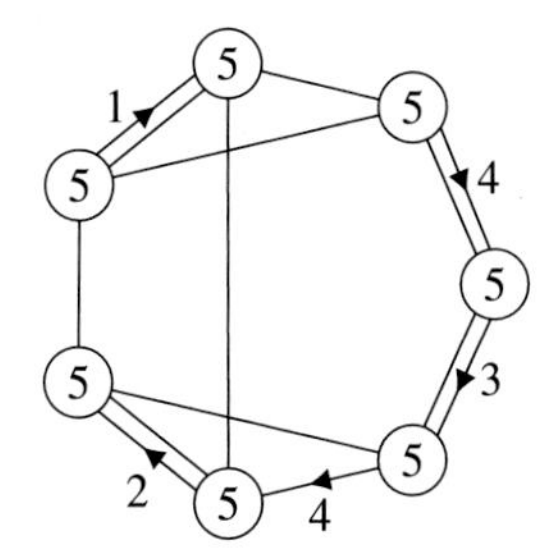

Figure 11.4 The odd graph O_4 given in Frucht's notation with respect to a $(7,5)$-semiregular automorphism ρ where undirected lines carry label 0.

O_4 is Hamiltonian by Corollary 3.7.4. We remark that Balaban (1972) first showed that O_4 is Hamiltonian. □

Proposition 11.3.8 *The orbital graphs of* $(\mathrm{PSL}(2,61), \mathrm{PSL}(2,61)/A_5)$ *of order* $1891 = 31 \cdot 61$ *in row 4 of Table 9.1 are Hamiltonian.*

Proof This case was settled using Magma. Here, we simply summarize the information that was obtained. This action has 41 suborbits, 33 of which are self-paired and 8 are not self-paired. More precisely, there is one suborbit of length 1, one of length 6, one of length 10, two of length 12, four of length 20, five of length 30, and twenty-seven of length 60. Of the latter, 8 are non-self-paired. That all 36 of these orbital graphs are Hamiltonian was checked. □

Proposition 11.3.9 *The orbital graphs of* $(\mathrm{PSL}(2,13), \mathrm{PSL}(2,13)/A_4)$ *of order* $91 = 7 \cdot 13$ *in row 7 of Table 9.1 are Hamiltonian.*

Proof Ultimately, Magma was also used here as in the previous result, but the information regarding suborbits are given in Marušič and Scapellato (1994c). There is one suborbit of length 1, three of length 4, one of length 6, and six of length 12. Of the latter, two are non-self-paired. Magma was used to check that up to isomorphism, there are only two graphs of valency 4 and only three graphs of valency 12. For each of these seven graphs we can find a semiregular automorphism whose quotient contains a Hamilton cycle that lifts to a Hamilton cycle in the original graph. Except for one graph of valency 4, where this automorphism is $(13,7)$-semiregular, the automorphism used for the lifting cycle technique is $(7,13)$-semiregular. □

We now discuss the orbital graphs in rows 5 and 6 of Table 9.1. The method used to show these orbital graphs are Hamiltonian is for the most part based on the lifting cycle technique. As a vertex-transitive graph of order

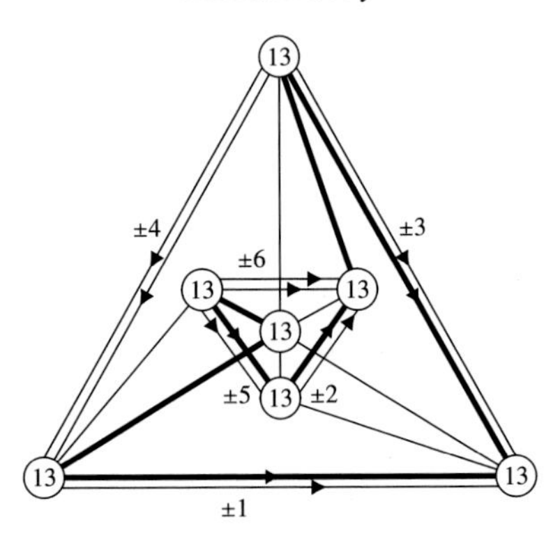

Figure 11.5 An orbital graph of $(\mathrm{PSL}(2,13), \mathrm{PSL}(2,13)/D_{12})$ in Frucht's notation with respect to a $(7,13)$-semiregular automorphism ρ, where undirected lines carry label 0. The bold edges form a Hamilton cycle.

pq, $q < p$ primes, always has a (q,p)-semiregular automorphism, the lifting cycle technique can be applied to these graphs provided appropriate Hamilton cycles can be found in the corresponding quotients. However, it turns out that in orbital graphs of the groups in row 5 of Table 9.1, a (p,q)-semiregular automorphism is used. As one would expect, finding Hamilton cycles in the quotient is the main obstacle to finding Hamilton cycles in these orbital graphs. To do this, many specific tools are used, most notably Chvátal's theorem, and certain results on polynomials in finite fields (see Du et al. (2018) for the details). In the next example we illustrate this method on one of the orbital graphs from the group in row 6 of Table 9.1 of smallest possible order $91 = 7 \cdot 13$ (Kutnar and Marušič, 2009b, Example 3.2).

Example 11.3.10 The orbital graph Γ of $(\mathrm{PSL}(2,13), \mathrm{PSL}(2,13)/D_{12})$ of a non-self-paired suborbit of size 3 contains a $(7,13)$-semiregular automorphism ρ, represented in Frucht's notation in Figure 11.5. As the quotient graph Γ/ρ has a Hamilton cycle containing a double edge and as 13 is a prime, this cycle lifts to a Hamilton cycle in Γ.

We are now ready to prove the main result of this section.

Theorem 11.3.11 *With the exception of the Petersen graph, a connected vertex-transitive graph of order pq, where p and q are primes, contains a Hamilton cycle.*

Proof Let Γ be a vertex-transitive graph of order qp. If $q = p$, then Γ is Hamiltonian by Theorems 9.1.1 and 3.5.4. If $q \neq p$, then by Propositions 11.3.1

and 11.3.3, and Theorems 11.3.4 and 9.2.10, the result follows unless Aut(Γ) contains no imprimitive subgroups. Combining Theorems 9.2.10, Propositions 11.3.7, 11.3.8, 11.3.9, and the discussion following Proposition 11.3.9, all of these graphs are Hamiltonian, with the exception of the orbital graphs of $\left(\mathrm{PSL}\left(2, q^2\right), \mathrm{PSL}\left(2, q^2\right)/\mathrm{PGL}(2, q)\right)$ in row 5 of Table 9.1. As mentioned earlier, with a proof too long to give here, Du, Kutnar, and Marušič showed in Du et al. (2020) that all such orbital graphs are Hamiltonian, and the result follows. □

11.4 Specific Approaches for Cayley Graphs

While we will see some proofs, the rest of this chapter is mainly expository as many of the results are beyond the scope of this book. Nonetheless, we can give the reader at least the flavor of the results, as well as discuss the tools that are used to prove them. For Cayley graphs, most of the results proved thus far depend on various restrictions imposed on the regular subgroups of the automorphism groups – see Alspach (1989); Alspach et al. (2010); Alspach and Zhang (1989); Curran and Gallian (1996); Curran et al. (2012); Dobson et al. (1998); Ghaderpour and Witte (2011, 2014); Glover et al. (2009, 2012); Glover and Marušič (2007); Keating and Witte (1985); Kutnar and Marušič (2009b); Kutnar et al. (2012b); Marušič (1983, 1985); Morris (2018); Witte (1982, 1986); Witte Morris (2015). For some groups it is known that connected Cayley graphs of these groups have Hamilton cycles – see Alspach et al. (1990); Alspach and Qin (2001); Kriloff and Lay (2014); Marušič (1983); Morris (2016). However, even for the class of dihedral groups the question remains open. The best result in this respect is that every connected Cayley graph on a generalized dihedral group of order divisible by 4 has a Hamilton cycle (see also Alspach et al. (2010)). Three other results of note are a theorem of Witte (1982, Theorem 3.1) showing that every group G with minimal generating set of size d contains a generating set of size less than $4d^2$ such that the corresponding Cayley graph has a Hamilton cycle, a theorem of Pak and Radoičić (2009) showing that every group G has a generating set of size at most $\log_2|G|$ for which the corresponding Cayley graph has a Hamilton cycle, and a theorem of Krivelevich and Sudakov (2003) showing that for every $c > 0$ and large enough n, a Cayley graph formed by choosing a set of $c \log^5 n$ generators randomly from a given group of order n almost surely has a Hamilton cycle.

With a combination of algebraic and topological methods a Hamilton path, and in some cases a Hamilton cycle, was proved to exist in cubic Cayley graphs of $(2, s, 3)$-generated groups (Glover and Marušič (2007); Glover et al. (2009, 2012)). The strategy in obtaining this result is presented in Section 11.5. There are several other results obtained based on restrictions on the generating sets (see survey articles Curran and Gallian (1996); Kutnar and Marušič (2009b); Witte and Gallian (1984); and Proposition 11.4.3). The following result of Pak and Radoičić (2009, Lemma 2), which is not given in the survey articles mentioned above, is also worth noting here.

Proposition 11.4.1 *Let G be a finite group, generated by an involution a and an element x. Let $b = a^x = x^{-1}ax$. Then the Cayley graph* $\mathrm{Cay}\left(G, \left\{a, b, x, x^{-1}\right\}\right)$ *contains a Hamilton cycle.*

For connected vertex-transitive graphs of particular orders, the Lovász problem is solved only for a few very specific forms (see the first paragraph in Section 11.3). The situation is quite different for Cayley graphs. Namely, it follows from a series of papers, each of which proves Cayley graphs of specific orders are Hamiltonian, such as p^k, kp, etc., that every connected Cayley graph of order less than 144 is Hamiltonian – see Chen and Quimpo (1983); Curran et al. (2012); Ghaderpour and Witte (2011); Ghaderpour and Morris (2012); Kutnar et al. (2012a); Witte (1986); Witte Morris and Wilk (2020). (Some of the results are obtained via a computer-assisted proof.) The following result is a combination of results in Curran et al. (2012); Ghaderpour and Witte (2011); Ghaderpour and Morris (2012); Kutnar et al. (2012a); Witte Morris and Wilk (2020).

Proposition 11.4.2 *Let G be a finite group, p a prime, and $1 \leq k < 48$. If $|G| = kp > 2$, then every connected Cayley graph of G is Hamiltonian.*

A most useful and promising approach in the search for Hamilton cycles in Cayley graphs is based on the idea of embedding a Cayley graph on an appropriate surface, and then finding a region made up of faces of the embedding (a "tree of faces") such that the boundary of the region is a Hamilton cycle. As an illustration, Figure 11.6 gives the corresponding regions of the five Platonic solids.

The embedding on surfaces approach was used in Sjerve and Cherkassoff (1994, Theorem 5.1) to show that a cubic Cayley graph of a group generated by three involutions, of which two commute, has a Hamilton cycle.

Proposition 11.4.3 *Let $G = \left\langle a, b, c : a^2 = b^2 = c^2 = 1, ab = ba, (ac)^s = 1, (bc)^t = 1, \ldots \right\rangle$. Then* $\mathrm{Cay}(G, \{a, b, c\})$ *has a Hamilton cycle.*

Proof We embed the graph $\mathrm{Cay}(G, \{a, b, c\})$ on an appropriate surface giving a map with 4-gons, $2s$-gons and $2t$-gons as the corresponding faces. The $2s$-gons are given by the relation $(ac)^s = 1$, the $2t$-gons are given by $(bc)^t$, while

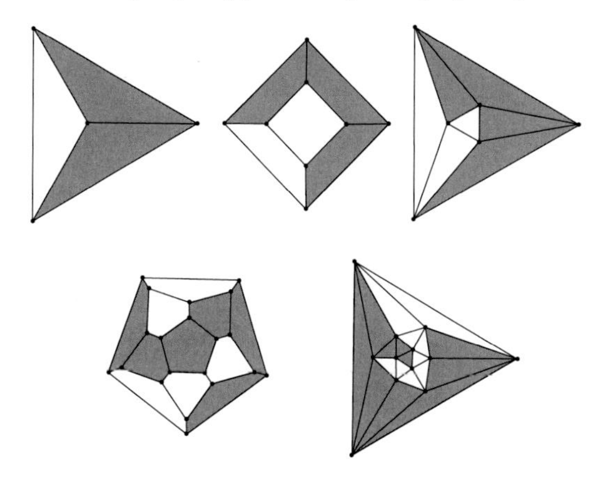

Figure 11.6 Hamilton cycles in the five Platonic solids.

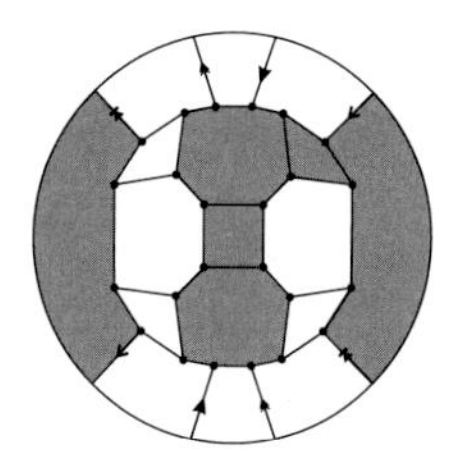

Figure 11.7 A Hamilton cycle in $\mathrm{Cay}(S_4, \{a,b,c\})$, where $a = (1,2)(3,4)$, $b = (1,2)$ and $c = (1,3)$.

the 4-gons come from $(ab)^2 = 2$. We find a Hamilton cycle as the boundary of a tree of faces of this map. As a first step, color all the $2t$-gons with the same color. The boundary of all these colored faces contains all vertices of the graph but in $|G|/2t$ cycles. As the next step pick any 4-gon and color it. Now the boundary of the union of all colored faces will have one less component, but will still involve all vertices. Continuing this way by repeatedly coloring an appropriate 4-gon at each step, we find a tree of faces consisting of all $2t$-gons and some 4-gons the boundary of which is a Hamilton cycle in the graph. □

We now give two examples illustrating Proposition 11.4.3.

Example 11.4.4 Let $a = (1,2)(3,4)$, $b = (1,2)$, and $c = (1,3)$ be involutions in the symmetric group S_4. Then $S_4 = \langle a,b,c\rangle$, $ab = ba$, $(ac)^4 = 1$, and $(bc)^3 = 1$. The Cayley graph $\mathrm{Cay}(S_4, \{a,b,c\})$ embeds on a projective plane, a nonorientable surface of genus 1, with six 4-gons, four 6-gons, and three 8-gons. With the approach laid out in the proof of Proposition 11.4.3 we can find a tree of faces consisting of all three 8-gons and two 4-gons whose boundary is a Hamilton cycle in $\mathrm{Cay}(S_4, \{a, b, c\})$ (see Figure 11.7).

Example 11.4.5 The Möbius–Kantor graph Γ is Hamiltonian.

Solution This was also shown in Example 11.2.14. Embed Γ into the double torus with octagonal faces as shown in Figure 11.8 (also see Marušič and Pisanski (2000)). The right-hand side in Figure 11.8 shows a tree of faces whose boundary is a Hamilton cycle in Γ, whereas the left-hand side of Figure 11.8 shows a graph of faces that is not a tree but also with a boundary giving a Hamilton cycle in Γ. □

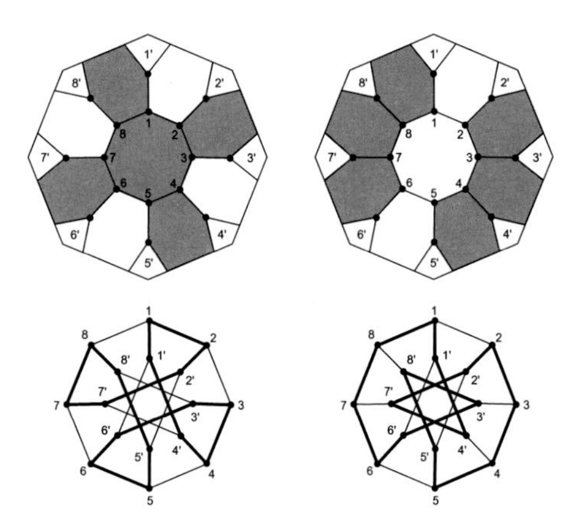

Figure 11.8 Two different Hamilton cycles in the Möbius–Kantor graph given as a boundary of colored faces of the Möbius–Kantor graph embedding into the double torus.

Among embeddings of graphs on surfaces, it is the Cayley maps that deserve special attention. Observe that the graph embedding from Example 11.4.4 is not a Cayley map. In fact, none of the embeddings in the proof of Proposition 11.4.3 is a Cayley map. Namely, two adjacent vertices on a 4-gon have a different local orientation of arcs. We now give an example of constructing a Hamilton cycle with a Cayley map.

Example 11.4.6 Let $G = \mathrm{PSL}(2, p) = \left\langle a, x : a^2 = x^p = (ax)^3 = 1 \right\rangle$, and $\Gamma = \mathrm{Cay}(G, S)$. Then Γ is Hamiltonian.

Solution Let $\rho = \left(a, x, x^{-1}\right)$. The faces of $\mathrm{CM}(G, S, \rho)$ are hexagons and p-gons. Glover and Yang (1996) constructed a Hamilton cycle in such a graph as a boundary of a colored tree of faces in the corresponding Cayley map consisting of $\left(p^2 - 5\right)/2$ p-gons and $((p + 1)/2)^2 - 2$ hexagons. □

The techniques used in Glover and Yang (1996) together with additional tools, both group-theoretic and graph-theoretic, were used in Glover and Marušič (2007) to extend the embedding on surfaces approach to Cayley graphs of arbitrary finite $(2, s, 3)$-generated groups, that is, to groups with a presentation $\langle a, x : a^2 = x^s = (ax)^3, \ldots \rangle$ (also see Glover and Marušič (2007); Glover et al. (2009; 2012)). Details of these constructions are given next.

11.5 Embeddings on Surfaces

A successful generalization of the embedding on surfaces method to Cayley graphs of $(2, s, 3)$-generated groups (Theorem 11.5.12) is based on two purely graph-theoretic results. The first, by Payan and Sakarovitch (1975), deals with maximum sizes of vertex subsets in cubic graphs inducing acyclic subgraphs (Proposition 11.5.6). The second, by Nedela and Škoviera (1995b), shows the edge cyclic connectivity of a cubic vertex-transitive graph is its girth (Proposition 11.5.10).

Definition 11.5.1 Given a graph (or more generally a loopless multigraph) Γ, a subset U of $V(\Gamma)$ is **cyclically stable** if the induced subgraph $\Gamma[U]$ is acyclic (a forest). The size $|U|$ of a maximum cyclically stable subset U of $V(\Gamma)$ is said to be the **cyclic stability number** of Γ.

The previous definition is from Payan and Sakarovitch (1975), a paper that is not readily available. Jaeger (1974) gave an upper bound on the cyclic stability number.

Proposition 11.5.2 *Let Γ be a cubic loopless multigraph of order n and let U be a maximum cyclically stable subset of $V(\Gamma)$. Then,*

$$|U| = (3n - 2c - 2e)/4, \tag{11.1}$$

where c is the number of connected components (trees) in $\Gamma[U]$ and e is the number of edges in $\Gamma[V(\Gamma)\backslash U]$. In particular, $|U| \leq (3n - 2)/4$.

Proof Let $V = V(\Gamma)$. First, for a cyclically stable subset T of V and a vertex v in $V\backslash T$ of valency at least two, $T \cup \{v\}$ is also cyclically stable as v has at most one neighbor in T. By the maximality of U, a vertex in $V\backslash U$ has at most one neighbor in $V\backslash U$, so each of the e edges in $\Gamma[V\backslash U]$ is an isolated edge. Let f be the number of edges in $\Gamma[U]$, and g be the number of edges with one end in U and the other in $V\backslash U$. Then we have that $f = |U| - c$, and as $3|U| = 2f + g$, we have $g = |U| + 2c$. Of course, $e + f + g = 3n/2$ and so $e + 2|U| + c = 3n/2$, giving us the desired expression for $|U|$. Now clearly, the maximum value for $|U|$ occurs when $e = 0$, that is, when $V\backslash U$ is an independent set of vertices, and when at the same time $c = 1$, that is when $\Gamma[U]$ is a tree. □

In order to explain the result of Payan and Sakarovitch, we need to introduce the concept of edge cyclic connectivity.

Definition 11.5.3 Let Γ be a connected graph. A subset $F \subseteq E(\Gamma)$ of edges of Γ is **cycle-separating** if $\Gamma\backslash F$ is disconnected and at least two of its components contain cycles. We say Γ is **cyclically k-edge-connected** if no set of fewer than k edges is cycle-separating in Γ.

For example, in the cube four edges are needed to separate two 4-cycles, and so the cube is cyclically k-edge-connected for $k \in \{1, 2, 3, 4\}$.

Definition 11.5.4 Let Γ be a graph. Define the **edge cyclic connectivity** of Γ to be the largest integer k, not exceeding the Betti number $|E(\Gamma)| - |V(\Gamma)| + 1$, for which Γ is cyclically k-edge-connected.

The condition that the edge cyclic connectivity of Γ does not exceed the Betti number $|E(\Gamma)| - |V(\Gamma)| + 1$ is required in order to deal with graphs without two edge disjoint cycles. This distinction is indeed necessary as, for example, the theta graph Θ_2, K_4, and $K_{3,3}$ have no cycle-separating sets of edges and without this condition would be cyclically k-edge-connected for all k. However, these graphs have edge cyclic connectivity of 2, 3, and 4, respectively.

Example 11.5.5 The edge cyclic connectivity of the Petersen graph is 5.

Solution The Petersen graph P has girth 5 and it contains exactly twelve 5-cycles by Exercise 2.5.2. Additionally, Aut(P) is transitive on the six pairs of disjoint 5-cycles by Exercise 2.5.3. The five edges joining such a pair of disjoint 5-cycles is a maximum cycle separating set. □

As proved by Payan and Sakarovitch (1975, Théorème 5), in a cubic cyclically 4-edge-connected graph the upper bound for its cyclic stability number given in Proposition 11.5.2 is always attained. More precisely, bearing in mind the expression for the cyclic stability number given in Equation (11.1), the following result may be deduced from Payan and Sakarovitch (1975, Théorème 5).

Proposition 11.5.6 *Let Γ be a cyclically 4-edge-connected cubic graph of order n, and let U be a maximum cyclically stable subset of $V(\Gamma)$. Then $|U| = \lfloor(3n - 2)/4\rfloor$. Additionally,*

(i) *if* $n \equiv 2\,(mod\,4)$ *then* $|U| = (3n-2)/4$, $\Gamma[U]$ *is a tree and* $V(\Gamma)\backslash U$ *is an independent set of vertices, and*

(ii) *if* $n \equiv 0\,(mod\,4)$ *then* $|U| = (3n-4)/4$, *and either* $\Gamma[U]$ *is a tree and* $V(\Gamma)\backslash U$ *induces a graph with a single edge, or* $\Gamma[U]$ *has two components and* $V(\Gamma)\backslash U$ *is an independent set of vertices.*

We now explain the connection between edge cyclic stability and Hamiltonicity. Let $G = \left\langle a, x : a^2 = x^s = (ax)^3 = 1, \ldots \right\rangle$ be a $(2, s, 3)$-generated group, $S = \left\{a, x, x^{-1}\right\}$, and $\Gamma = \mathrm{Cay}(G, S)$. The strategy is based on associating with Γ the Cayley map $\mathrm{CM}(G, S, \rho)$, where $\rho = \left(a, x, x^{-1}\right)$, with genus $g = 1 + (s-6)|G|/12s$. The faces of this map are $|G|/s$ disjoint s-gons (each one of which is a cycle of the form $g, gx, gx^2, \ldots, gx^{s-1}, g$, which has vertex set a left coset of $\langle x \rangle$ in G), and $|G|/3$ hexagons, with one s-gon and two hexagons meeting at each vertex. To see that the genus g is indeed $1 + (s-6)|G|/12s$ just plug $|V(\Gamma)| = |G|$, $|E(\Gamma)| = 3|G|/2$, and number of faces $|F(\Gamma)| = |G|(1/s + 1/3)$ into the Euler characteristic formula

$$\chi = 2 - 2g = |V(\Gamma)| - |E(\Gamma)| + |F(\Gamma)|.$$

The notation for S and ρ will be used throughout the rest of this section.

In the search for a Hamilton cycle (or path) in Γ, we look for a long tree of faces in $\mathrm{CM}(G, S, \rho)$ – a tree of faces whose boundary is either a Hamilton cycle in Γ or a cycle missing two adjacent vertices of Γ. This long tree of faces comes from a sufficiently large induced tree in an auxiliary graph of Γ.

Before turning to the definition of this graph, it is best to discuss a degenerate case not included in the general definition. This is when G is abelian, which only happens when $ax = xa$. Consequently, $s = 6$ and $G \cong \mathbb{Z}_6$. The graph is then $\mathrm{Cay}(\mathbb{Z}_6, \{1, 3, 5\})$, which has two hexagons intersecting in three edges. It is obviously Hamiltonian.

Definition 11.5.7 Let $G = \left\langle a, x : a^2 = x^s = (ax)^3 = 1, \ldots \right\rangle$ be nonabelian, and $\Gamma = \mathrm{Cay}(G, S)$. Define the **hexagon graph of** Γ, denoted $\mathrm{Hex}(\Gamma)$, to be the graph with vertex set the set of hexagons of $\mathrm{CM}(G, S, \rho)$ from the relation $(ax)^3 = 1$ (that is, of the form $g, ga, gax, gaxa, gaxax, gaxaxa, g$), and two such hexagons are adjacent if and only if they share an edge.

The hexagon H that contains the edge $\left\{1, x^{-1}\right\}$, as well as its neighbors in $\mathrm{Hex}(\Gamma)$, are shown in Figure 11.9. Consider the induced action of G_L on hexagons, and let $\pi \colon G_L \mapsto S_{V(\mathrm{Hex}(\Gamma))}$ be the permutation representation induced by this action. Then $\pi(G_L) \leq \mathrm{Aut}(\mathrm{Hex}(\Gamma))$. Also, $\mathrm{Stab}_H(\pi(G_L)) = \langle \pi((ax)_L) \rangle \cong \mathbb{Z}_3$ (note that the only possibilities to stabilize this hexagon in G_L are images of multiplication on the left by elements on the hexagon). It is

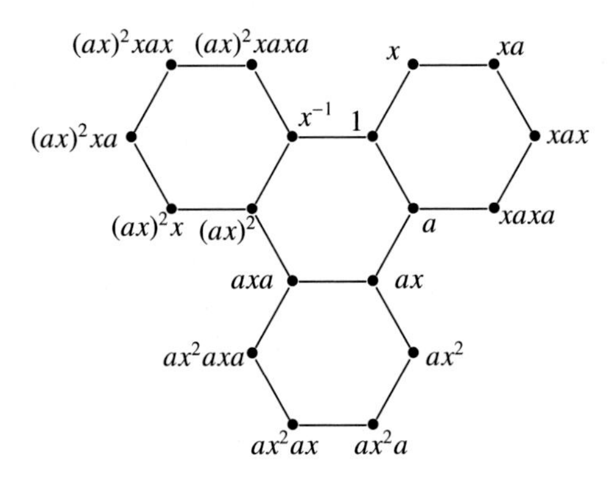

Figure 11.9 The local structure of $\mathrm{CM}(G, S, \rho)$ illustrating a vertex and its three neighbors in $\mathrm{Hex}(\Gamma)$.

then easy to see that $\pi(G_L)$ is edge-transitive, and the map a_L in Γ fixes the edge $\{1, a\}$ and interchanges H and the other hexagon that contains 1. This means that $\pi(a_L)$ interchanges the endpoints of the edge of $\mathrm{Hex}(\Gamma)$ between the two hexagons that contains 1, and so by Theorem 3.2.10 $\mathrm{Hex}(\Gamma)$ is $\pi(G_L)$-arc-transitive. As $|\mathrm{Stab}_H(\pi(G_L))| = 3$, we see that $|V(\mathrm{Hex}(\Gamma))| = |G|/3$, and so $\pi(G_L)$ is regular on the arcs of $\mathrm{Hex}(\Gamma)$.

Example 11.5.8 Let $G = \mathbb{Z}_{13} \rtimes \mathbb{Z}_6 = \left\langle y, z : y^6 = z^{13} = 1, z^y = z^4 \right\rangle$. Note that G has (2, 6, 3)-presentation $\left\langle a, x : a^2 = x^6 = (ax)^3 = 1, \ldots \right\rangle$, where $a = y^3$ and $x = y^3 z^{-1} y^2 z$. In the bottom of Figure 11.10 we show a tree of hexagons whose boundary is a Hamilton cycle in the toroidal Cayley map $\mathrm{CM}(G, S, \rho)$. The

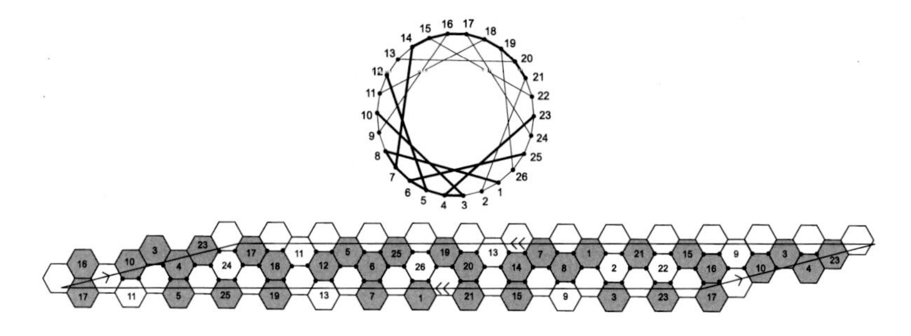

Figure 11.10 (Bottom) A Hamilton tree of faces in a toroidal Cayley map of $\mathbb{Z}_{13} \rtimes \mathbb{Z}_6$ with respect to a (2, 6, 3)-presentation whose boundary is a Hamilton cycle, and (Top) the associated hexagon graph.

upper graph shows the same tree in Hex(Cay(G, S)), which is isomorphic to F26. The correspondence between hexagons in the Cayley map and the vertices in the hexagon graph is indicated by the numbers from 1 to 26. The group G can be written as a subgroup of $\mathcal{S}_{13}$ with $a = (0,8)(1,7)(2,6)(3,5)(9,12)(10,11)$ and $x = (0,1,11,7,6,9)(2,8,3,5,12,4)$.

In Example 11.5.8, it is precisely the fact that the vertex set of Hex(Γ) can be partitioned into two subsets, with one inducing a tree, and its complement an independent set, that gives a Hamilton cycle in Γ for the $(2,6,3)$-presentation of $\mathbb{Z}_{13} \rtimes \mathbb{Z}_6$. The proof of the next result is left as Exercise 11.5.4.

Lemma 11.5.9 *Let G be a $(2,s,3)$-group with $ax \neq xa$. Let $\Gamma = \mathrm{Cay}(G,S)$, T be an induced tree in* Hex(Γ) *such that its complement is an independent set in* Hex(Γ)*, and H the set of all vertices of Γ contained in hexagons corresponding to the vertices of T. The boundary of the tree of faces in* CM(G, S, ρ) *corresponding to T is a Hamilton cycle in $\Gamma[H]$.*

So, if $|G|$, and hence the order of Hex(Γ), is congruent to 2 modulo 4, then Proposition 11.5.6 (i) suffices if Hex(Γ) is cyclically 4-edge-connected. On the other hand, if $|G|$, and hence the order of Hex(Γ), is divisible by 4 (and cyclically 4-edge-connected), then we can construct a Hamilton path in Γ by Proposition 11.5.6 (ii), as we shall see later. To deal with the edge cyclic connectivity of Hex(Γ), we next state an important result of Nedela and Škoviera (1995b, Theorem 17).

Proposition 11.5.10 *The cyclic edge connectivity of a cubic vertex-transitive graph Γ is its girth.*

As a hexagon graph is a cubic arc-transitive graph, it is vertex-transitive by Proposition 3.2.12, and so its cyclic edge connectivity is its girth. It will soon be clear that a lower bound on the girth is needed, and by Proposition 10.2.2 such a graph has girth at least 6, with a few exceptions. We now show that for cubic arc-transitive graphs of girth 6, a maximum cyclically stable set may always be chosen to induce a tree, giving a refinement of Proposition 11.5.6. The result is from Glover and Marušič (2007).

Proposition 11.5.11 *Let Γ be a cubic arc-transitive graph of order $n \equiv 0 \pmod 4$, not isomorphic to K_4, the cube Q_3, or the dodecahedron. Then there exists a cyclically stable subset U of $V(\Gamma)$ that induces a tree and $V(\Gamma)\backslash U$ induces a graph with a single edge.*

Proof By Proposition 10.2.2 Γ has girth at least 6, and hence, by Proposition 11.5.10, Γ is cyclically 6-edge-connected.

Let u and v be adjacent vertices of Γ, $u_1 \neq v \neq u_2$ be neighbors of u, and $v_1 \neq u \neq v_2$ be neighbors of v. Note that u_1, u_2, v_1, and v_2 must all be distinct as otherwise Γ has girth 3. Set $X = \Gamma \backslash \{u, v\}$. Then X has $n - 2$ vertices with u_1, u_2, v_1, and v_2 of valency 2, and all other vertices of valency 3. Modify X by replacing the four paths of length 2 through u_1, u_2, v_1, and v_2 by a single edge, to obtain the graph Y. Then Y is cubic of order $n - 6$.

We now show Y is cyclically 4-edge-connected. Take two vertex- and edge-disjoint cycles C_1 and C_2 in Y. If C_1 and C_2 are cycles in Γ, then there are at least six edges in Γ separating C_1 and C_2, and so at least five edges in Y separating C_1 and C_2. Otherwise, at least one of C_1 and C_2 contains an edge replacing a path of length 2 in X to obtain Y, so we assume C_1 contains such an edge. Then C_1 contains at least one of the four edges that replaced the four paths of length 2 through u_1, u_2, v_1, and v_2. So C_1 gives a cycle C_1' in X by replacing each edge of C_1 not in X by its corresponding path of length 2. Similarly, if C_2 is not in Γ, then there is a cycle C_2' in X corresponding to C_2'. If C_2 is contained in Γ, we set $C_2' = C_2$. In any case, C_1' and C_2' are vertex- and edge-disjoint cycles in Γ.

As Γ is cyclically 6-edge connected, the deletion of a set U of at least 6 edges is required to separate C_1' and C_2' in Γ. Any path P in Γ separating C_1' from C_2' not in Y must contain a vertex of Γ not in Y. Also, no neighbors of u or v are contained in Y, so if u or v is in P, then some other vertex distinct from u and v, but not in Y, must be contained in P. As no edge in U is on C_1' or C_2', we conclude at least one of uu_1, u_2u, v_1v, and v_2v, must be contained in U, and at most two such edges (one for each cycle $C_i' \neq C_i$, $i = 1, 2$), are contained in U. Thus Y is cyclically 4-edge connected.

As $|V(Y)| = n - 6$ and $n \equiv 0 \pmod 4$, we have $|V(Y)| \equiv 2 \pmod 4$. By Proposition 11.5.6 (i), there exists a maximum cyclically stable subset R of $V(Y)$ inducing a tree and such that its complement $V(Y)\backslash R$ is an independent set of vertices. A maximum cyclically stable subset U of $V(\Gamma)$ is now $U = R \cup \{u_1, u_2, v_1, v_2\}$ (Exercise 11.5.5). The edge uv is thus the single edge of the graph induced on the complement $V(\Gamma)\backslash U$. □

We now state the main result of this section showing that cubic Cayley graphs of $(2, s, 3)$-generated groups have Hamilton paths (Glover and Marušič, 2007, Theorem 1.1). Additionally, if the group is of order congruent to 2 modulo 4 such graphs are Hamiltonian, and if the order of the group is divisible by 4 such graphs have a cycle missing only two vertices.

Theorem 11.5.12 *Let $s \geq 3$ be an integer and $G = \left\langle a, x : a^2 = 1, x^s = 1, (ax)^3 = 1, \ldots \right\rangle$ be a $(2, s, 3)$-generated group. Then $\Gamma = \mathrm{Cay}(G, S)$ is*

Hamiltonian if $|G|$ (and thus also s) is congruent to 2 modulo 4, and has a cycle of length $|G| - 2$, and thus necessarily a Hamilton path, if $|G| \equiv 0 \pmod 4$.

Proof We have already discussed that if $ax = xa$ and G is abelian, then $\Gamma \cong \mathrm{Cay}(\mathbb{Z}_6, \{1, 3, 5\})$ and is Hamiltonian. We thus assume that $ax \neq xa$ and G is nonabelian.

If $\mathrm{Hex}(\Gamma)$ is isomorphic to K_4, Q_3, or the dodecahedron, then G acts regularly on $A(\mathrm{Hex}(\Gamma))$, and $|G| = 12, 24$, or 60, respectively. Then G is isomorphic to A_4, S_4, or A_5, respectively. We consider each case separately.

If $\mathrm{Hex}(\Gamma) \cong K_4$, then $G = A_4$ and we may assume without loss of generality that $a = (0, 1)(2, 3)$ and $x = (0, 1, 2)$. In Figure 11.11 we show a Hamilton tree of faces, including two triangles, whose boundary is a Hamilton cycle in Γ.

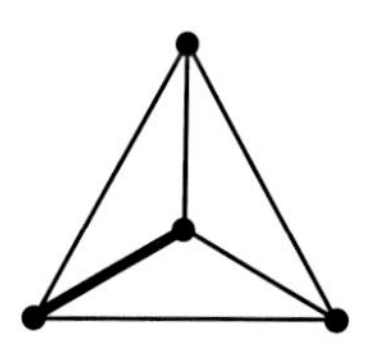
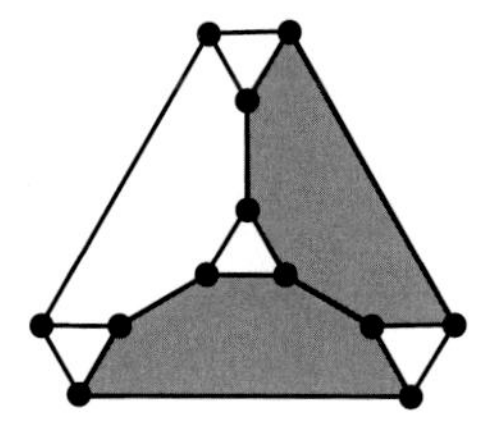
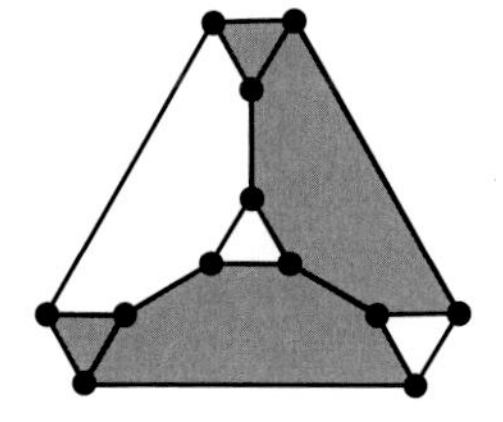

Figure 11.11 (Middle) A tree of faces in the spherical Cayley map $\mathrm{CM}(A_4, \{a, x, x^{-1}\}, \rho)$, where $a = (0, 1)(2, 3)$ and $x = (0, 1, 2)$, whose boundary is a cycle missing two vertices, (Left) the associated Hexagon graph, and (Right) a modified Hamilton tree of faces.

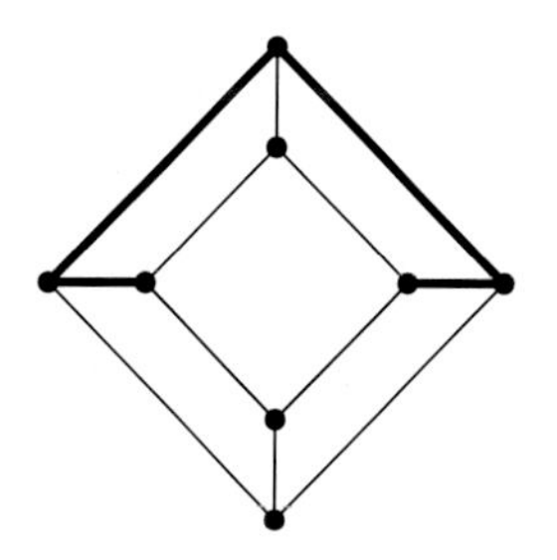
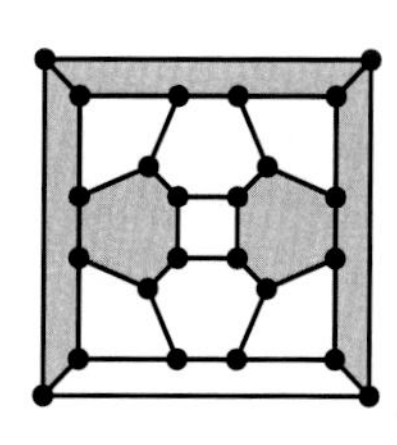
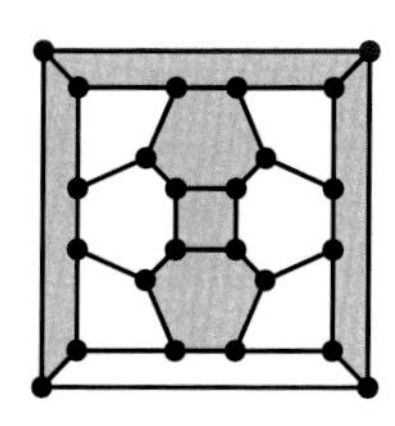

Figure 11.12 (Left) The boundary in a tree of faces in the spherical Cayley map $\mathrm{CM}(S_4, S, \rho)$ is a cycle missing two vertices, (Middle) the associated Hexagon graph, and (Right) a modified Hamilton tree of faces.

If $\mathrm{Hex}(\Gamma) \cong Q_3$, then $G = S_4$ and we may assume without loss of generality that $a = (0, 1)$ and $x = (0, 1, 2, 3)$. In the middle graph of Figure 11.12 we show a tree of hexagons, whose boundary is a cycle missing only two vertices in the spherical Cayley map $\mathrm{CM}(S_4, S, \rho)$. The left-hand graph shows this same tree in $\mathrm{Hex}(\Gamma) \cong Q_3$, the cube, and the right-hand shows a modified tree of faces, including a square, whose boundary is a Hamilton cycle in Γ.

If $\mathrm{Hex}(\Gamma)$ is isomorphic to the dodecahedron, then $G \cong A_5$ and we may assume without loss of generality that $a = (1, 2)(3, 4)$ and $x = (0, 1, 2, 3, 4)$. In the middle graph of Figure 11.13 we show a tree of hexagons, whose boundary is a cycle missing only two vertices in the spherical Cayley map $\mathrm{CM}(A_5, S, \rho)$. The left-hand graph shows this same tree in $\mathrm{Hex}(\Gamma)$, the dodecahedron, and the right-hand graph shows a Hamilton tree of faces, including two pentagons, whose boundary is a Hamilton cycle in this map.

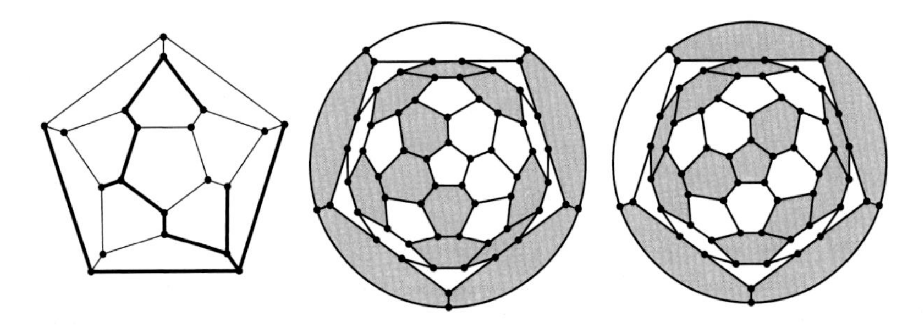

Figure 11.13 (Middle) The boundary in a tree of faces in the spherical Cayley map $\mathrm{CM}(A_5, S, \rho)$ is a cycle missing two vertices, (Left) the associated Hexagon graph, and (Right) a modified Hamilton tree of faces.

If $\mathrm{Hex}(\Gamma) \not\cong K_4$, Q_3, or the dodecahedron then by Proposition 10.2.2 we have the girth of Γ is at least 6. By Proposition 11.5.10 the cyclic edge connectivity of Γ is also at least 6.

Suppose $|G| \equiv 0 \pmod 4$. By Proposition 11.5.6 there is a induced tree in $\mathrm{Hex}(\Gamma)$ whose complement is an independent set of vertices. By Lemma 11.5.9 Γ has a Hamilton cycle.

Suppose $|G| \equiv 2 \pmod 4$. By Proposition 11.5.11 there is an induced tree in $\mathrm{Hex}(\Gamma)$ whose complement in $\mathrm{Hex}(\Gamma)$ induces a graph with a single edge. By Lemma 11.5.9, Γ contains a Hamilton cycle containing every vertex of Γ except for the intersection of two hexagons that meet at an edge – so of order $|G| - 2$. In particular, as Γ is connected it has a Hamilton path. □

The embedding on surfaces method described in this section gives Hamilton cycles in cubic Cayley graphs $\mathrm{Cay}(G, S)$ of $(2, s, 3)$-generated groups G if $|G| \equiv 2 \pmod 4$. If $|G| \equiv 0 \pmod 4$, further generalization of the method – via certain modifications of the Hexagon graphs giving a Hamiltonian tree of faces in $\mathrm{CM}(G, S, \rho)$ also containing s-gons – was successfully applied in Glover et al. (2009, 2012) to Hamilton cycles if s is odd or divisible by 4. In both cases a Hamilton cycle is a boundary of a region not consisting entirely of hexagons: two s-gons are needed in the first case and one s-gon in the second case.

In the remaining case of $s \equiv 2 \pmod 4$, new ideas are needed to show such graphs are Hamiltonian, as we now show. A necessary condition for a Hamilton tree of faces in these Cayley maps is given by the so-called HEAD equation (Glover and Yang, 1996, Proposition 1.2), which says that if there is a Hamilton cycle given as the boundary of a tree of k s-gons and l hexagons in the Cayley map of Γ, then

$$4l + (s-2)k = |G| - 2.$$

If $s \equiv 2 \pmod 4$ and $|G| \equiv 0 \pmod 4$ then this equation has no solution, and thus a Hamilton cycle cannot be produced using this method.

On the other hand there may very well be ways for possible generalizations of the embedding on surface method to arbitrary cubic Cayley graphs Γ of $(2, s, t)$-generated groups, with $t \geq 4$ and connection sets S. In this case, an arc-transitive graph of valency t with a cyclic vertex stabilizer whose vertex set is all the $2t$-gons in the corresponding Cayley map, an analogue of the Hexagon graph, is associated with Γ. This suggests that finding Hamilton paths/cycles in these graphs will depend on finding analogues of decomposition results, such as Proposition 11.5.11, for regular graphs (in particular arc-transitive graphs of valency $t \geq 4$ with cyclic vertex stabilizers) into induced forests whose complements are independent sets or have a small number of edges (see Example 11.5.15). This is a new area of research of independent interest from its potential applications to the Lovász problem. In summary, we propose the following problems.

Problem 11.5.13 Let $G = \left\langle a, x : a^2 = 1, x^s = 1, (ax)^t = 1, \dots \right\rangle$ be a $(2, s, t)$-generated group, and $S = \left\{a, x, x^{-1}\right\}$. Construct a longest possible cycle/path in $\mathrm{Cay}(G, S)$.

Problem 11.5.14 Let Γ be an arc-transitive graph of valency at least 4 with a 1-arc-regular subgroup of automorphisms with cyclic vertex stabilizer. Decompose $V(\Gamma)$ into a set inducing a tree, whose complement induces a graph with as few edges as possible. Proposition 11.5.6 shows that for cubic graphs the fewest edges possible is 1.

Example 11.5.15 The right-hand side of Figure 11.14 shows a tree of octagons (a path of length 2) whose boundary is a cycle missing only two pairs of adjacent vertices in the spherical Cayley map $\mathrm{CM}(S_4, S, \rho)$. Note S_4

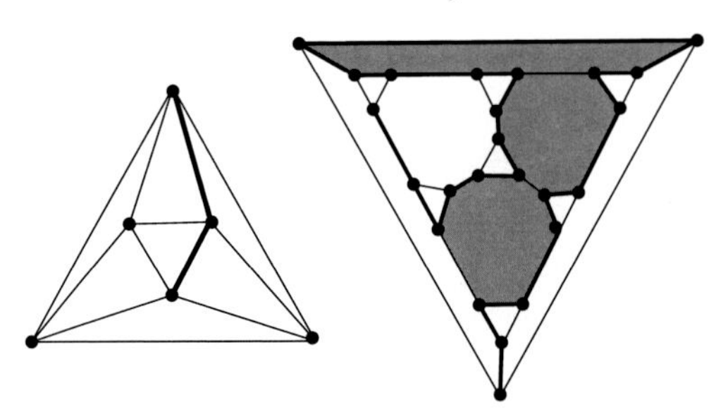

Figure 11.14 (Right) A tree of octagons in the spherical Cayley map CM($\mathcal{S}_4, S, \rho$) with respect to a (2, 3, 4)-presentation whose boundary is a cycle missing two pairs of adjacent vertices, and (Left) the associated octagon graph.

has a (2, 3, 4)-presentation $\langle a, x : a^2 = x^3 = (ax)^4 = 1, \ldots \rangle$, where $a = (1, 2)$ and $x = (0, 1, 3)$. The left-hand side shows this same tree in the corresponding "octagon" graph, an arc-transitive graph of valency 4. In order to produce a Hamilton path in Γ we start at one pair of adjacent vertices not on the boundary of this tree of octagons, then we move to the boundary itself and, after visiting all the vertices on the boundary, we move to the remaining pair of adjacent vertices outside the boundary (see Figure 11.14).

The following example shows that this and similar approaches may be successful for more general families of cubic Cayley graphs, and perhaps cubic vertex-transitive graphs in general.

Example 11.5.16 Let $D_{48} = \langle \rho, \tau : \tau^2 = \rho^{24} = (\rho\tau)^2 = 1 \rangle$, and $S = \{\tau, \tau\rho^3, \tau\rho^{11}\}$. The right-hand side of Figure 11.15 shows a tree of hexagons whose boundary is a cycle missing only two adjacent vertices in the toroidal Cayley map CM(D_{48}, S, ρ). Clearly the boundary of this cycle is a Hamilton

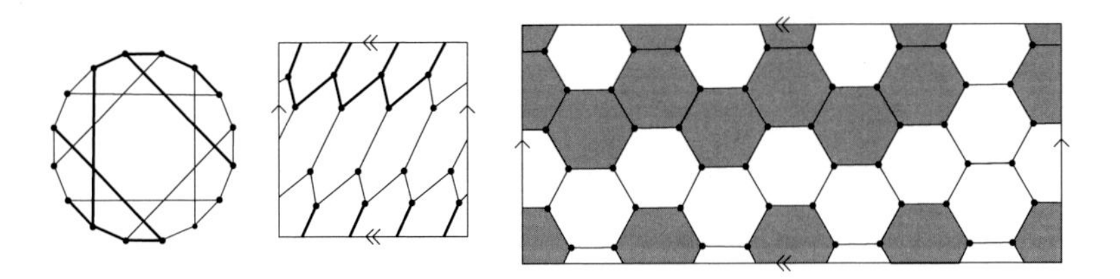

Figure 11.15 (Right) A tree of faces in the toroidal Cayley map of a Cayley graph of D_{48} with respect to a generating set consisting of three involutions missing two adjacent vertices, and the corresponding induced tree in the Möbius–Kantor graph shown (Left) in the plane and (Middle) on the torus.

path in Γ. In the other two graphs in Figure 11.15, this same tree is shown in its corresponding hexagon graph, the Möbius–Kantor graph.

11.6 A Final Observation

As observed in Section 11.5, the hexagon graph of a Cayley map of a $(2, s, 3)$-group is a cubic arc-transitive graph with a 1-arc-regular subgroup, and so it is of type $\{1\}$, $\left\{1, 2^1\right\}$, $\left\{1, 2^1, 2^2, 3\right\}$, $\left\{1, 4^1\right\}$ or $\left\{1, 4^1, 4^2, 5\right\}$ as in Table 10.2. The converse is also true. Every cubic arc-transitive graph with a 1-arc-regular subgroup is isomorphic to the hexagon graph of some Cayley graph of a $(2, s, 3)$-generated group. And just as the "Payan and Sakarovitch" decomposition property (as in Section 11.5) in cubic arc-transitive graphs can be used to show cubic Cayley graphs of $(2, s, 3)$-generated groups with a 1-arc-regular subgroup are Hamiltonian, one wonders whether these graphs can be shown to be Hamiltonian using certain properties of the corresponding Cayley graphs. Namely, by Theorem 3.4.5, an arc-transitive graph Γ with a 1-arc-regular $(2, s, 3)$-generated subgroup G being Hamiltonian is equivalent to its truncation $T(\Gamma)$ being Hamiltonian. Of course, $T(\Gamma) \cong \mathrm{Cay}(G, S')$, where $S' = \left\{a, ax, (ax)^{-1}\right\}$. The corresponding Cayley map has the same genus as the Cayley map associated with a $(2, s, 3)$-presentation of this group for which, by Theorem 11.5.12, we do have a Hamilton path/cycle. So, if one showed that a Cayley graph of a $(2, 3, s)$-generated group, and hence every cubic arc-transitive graph with a 1-arc-regular action, has a Hamilton path/cycle, this would provide a first step towards a possible solution of the Lovász problem for all cubic arc-transitive graphs. We therefore propose the following problem.

Problem 11.6.1 Construct a Hamilton path/cycle in a Cayley graph of a $(2, 3, s)$-generated group.

11.7 Exercises

11.2.1 Let G be a group and $S \subset G$. Show that $1, s_1, s_1 s_2, \ldots, s_1 \cdots s_r$ is a Hamilton cycle in $\mathrm{Cay}(G, S)$ if and only if $[s_1, \ldots, s_r]$ is a Hamiltonian sequence on S.

11.2.2 Prove Lemma 11.2.8.

11.2.3 Let Γ be a graph with edges e_1 and e_2 such that $\Gamma \backslash \{e_1, e_2\}$ has two Hamilton connected components. Show that Γ is Hamiltonian.

11.2.4 Let Γ be a graph with an (m, p)-semiregular automorphism ρ, where p is a prime, and $\mathcal{P}$ the set of orbits of ρ. Let C be a cycle of length k in the quotient graph $\Gamma/\mathcal{P}$, and W the set of all paths in Γ whose projection in $\Gamma/\mathcal{P}$ is C. Then either W contains a cycle of length kp or it is p disjoint k-cycles. In the latter case every edge of C induces a subgraph of Γ isomorphic to pK_2. (This is a reformulation of Marušič and Parsons (1982, Lemma 5).)

11.3.1 Prove that every connected vertex-transitive graph of order p^k, where p is a prime and $k \leq 3$, contains a Hamilton cycle. (Hint: Use Theorem 9.1.3 and Theorem 11.2.11.)

11.3.2 Prove Propositions 11.3.8 and 11.3.9 using Magma.

11.3.3 Prove that the H-graph has a Hamilton cycle.

11.5.1 Find the cyclic edge connectivity of all cubic arc-transitive graphs up to order 30.

11.5.2 Prove that every connected Cayley graph of S_4 or A_5 is Hamiltonian.

11.5.3 Show that every connected cubic arc-transitive graph of order up to 40 with the exceptions of the Petersen graph and the Coxeter graph has a Hamilton cycle. (Hint: Use the lifting cyclic technique where appropriate.)

11.5.4 Prove Lemma 11.5.9.

11.5.5 Using the notation of the proof of Proposition 11.5.11, show that $U = R \cup \{u_1, u_2, v_1, v_2\}$ is a maximum cyclically stable subset of $V(\Gamma)$.

12

Semiregularity

12.1 Historic Perspectives

Definition 12.1.1 Let G be a transitive permutation group on a finite set X of cardinality at least 2. An element of G is a **derangement** if it has no fixed points.

A semiregular element is the special case of a derangement where all cycles have the same length (this need not be the case for a derangement). By a classical result of Jordan (1872, Théorème I), a transitive permutation group of degree at least 2 always contains derangements.

Proposition 12.1.2 *A transitive permutation group G on a finite set X of cardinality at least* 2 *contains derangements.*

Proof By the orbit-counting lemma we have $|G| = \sum_{g\in G} |\mathrm{Fix}(g)|$. If G has no derangements then, as the identity fixes all points of X, we have

$$|G| = \sum_{g\in G} |\mathrm{Fix}(g)| \geq |X| + (|G| - 1) \cdot 1 = |G| + |X| - 1,$$

a contradiction as $|X| \geq 2$. □

By a theorem of Fein, Kantor and Schacher, every transitive permutation group contains a derangement of prime power order (Fein et al., 1981, Theorem 1). Their proof uses CFSG. As they say *"Our proof of Theorem 1 is unpleasant. We reduce quickly to the case where G is a finite simple group acting primitively on Ω. Invoking the classification of finite simple groups, we proceed to eliminate potential counterexamples to the theorem by a tedious case by case verification. ... It would be desirable to have a direct proof of this result."* (Fein et al., 1981, p. 41). Many results in algebraic graph theory, including some we have seen in this book, use CFSG. It would be desirable to find "a classification of finite simple groups free" (CFSG-free) proof of these results. We saw in Theorem 3.8.13 that if we were to replace the requirement that a derangement is of "prime power order" with that of being of "prime order" the result would no longer be true as it would then contain a semiregular element of prime order.

Of course, a derangement of prime order exists if and only if a semiregular derangement exists, and so Theorem 3.8.13 shows there are transitive permutation groups without semiregular derangements. Recall from Definition 3.8.2 that such a group is called **elusive**. As mentioned in Section 3.8, the name is intended to suggest that these groups are quite rare. The construction given in Theorem 3.8.13 is the first known construction of such groups. In Example 12.1.3 below, we give a construction from Dobson and Marušič (2011) of two elusive groups isomorphic to the group $\mathrm{AGL}(1,9)$ (from Theorem 3.8.13). The construction gives a taste of the relationship between codes (certain vector subspaces) and elusive groups. In Theorem 12.1.6 we show how to build new elusive groups from old elusive groups. Also, infinite families of elusive groups were given in Cameron et al. (2002); Giudici (2003, 2007).

Example 12.1.3 Define $z_i, \tau\colon \mathbb{Z}_4 \times \mathbb{Z}_3 \to \mathbb{Z}_4 \times \mathbb{Z}_3$ by $z_i(i,j) = (i, j+1)$, $z_i(k,j) = (k,j)$ if $i \neq k$, and $\tau(i,j) = (i+1, \alpha_i j)$, where $\alpha_0 = \alpha_1 = \alpha_2 = 1$ and $\alpha_3 = 2$. Let $\gamma_1 = z_1 z_2 z_3^2$ and $\gamma_2 = z_1 z_2^2 z_3^2$. The groups $\langle \tau, \gamma_1\rangle$ and $\langle \tau, \gamma_2\rangle$ are transitive solvable elusive groups of order 72 and degree 12.

Solution Let $H = \langle z_i\colon i \in \mathbb{Z}_4\rangle$. Define $v\colon H \to \mathbb{F}_3^4$ by $v\left(z_0^{a_0} z_1^{a_1} z_2^{a_2} z_3^{a_3}\right) = (a_0, a_1, a_2, a_3)$. Computations show that τ normalizes H, and conjugation of H by τ induces a linear transformation $t\colon \mathbb{F}_3^4 \to \mathbb{F}_3^4$ whose matrix with respect to the standard basis of $\mathbb{F}_3^4$ is

$$A = \begin{bmatrix} 0 & 0 & 0 & 2 \\ 1 & 0 & 0 & 0 \\ 0 & 1 & 0 & 0 \\ 0 & 0 & 1 & 0 \end{bmatrix}.$$

The characteristic and minimal polynomial of A (with entries in $\mathbb{F}_3$) is $x^4 + 1 = \left(x^2 + x + 2\right)\left(x^2 + 2x + 2\right)$. So $\left\langle t^i\left((0,1,1,2)\right) : i \in \mathbb{Z}\right\rangle$ and $\left\langle t^i\left((0,1,2,2)\right) : i \in \mathbb{Z}\right\rangle$ are invariant subspaces of t, each with dimension 2. By repeatedly applying t to $(0,1,1,2)$ and $(0,1,2,2)$, the elements of these subspaces are the elements in the following arrays:

$$\begin{matrix} (0,0,0,0) & (0,1,1,2) & (1,0,1,1) & (2,1,0,1) & (2,2,1,0) \\ & (0,2,2,1) & (2,0,2,2) & (1,2,0,2) & (1,1,2,0) \end{matrix},$$

and

$$\begin{matrix} (0,0,0,0) & (0,1,2,2) & (1,0,1,2) & (1,1,0,1) & (2,1,1,0) \\ & (0,2,1,1) & (2,0,2,1) & (2,2,0,2) & (1,2,2,0) \end{matrix}.$$

This gives that $K_i = \left\langle \tau^{-j}\gamma_i\tau^j : j \in \mathbb{Z} \right\rangle$ is a normal subgroup of $\langle \tau, \gamma_i \rangle$ of order 9, $i \in \{1,2\}$, and is a Sylow 3-subgroup of $\langle \tau, \gamma_i \rangle$ as $\langle \tau, \gamma_i \rangle / K_i \cong \langle \tau \rangle$ has order 8. Thus $|\langle \tau, \gamma_i \rangle| = 72$.

The set of orbits of K_i is a block system $\mathcal{B}$ of $\langle \tau, \gamma_i \rangle$ with four blocks of size 3 (note the orbits of K_1 and K_2 are the same). As $\langle \tau, \gamma_i \rangle / \mathcal{B} \cong \mathbb{Z}_4$, by the embedding theorem $\langle \tau, \gamma_i \rangle \leq \mathbb{Z}_4 \wr S_3$, which is solvable. So $\langle \tau, \gamma_i \rangle$ is solvable. Finally, if $g \in K_i$, then $v(g)$ is one of the vectors given above, and so g has a fixed point as every vector is 0 in some coordinate. Otherwise, $g \in P_2$, a Sylow 2-subgroup of $\langle \tau, \gamma_i \rangle$. Then P_2 is conjugate to $\langle \tau \rangle$ and so cyclic. But $\langle \tau \rangle$ contains a unique subgroup of order 2, which has three fixed points. Hence $\langle \tau, \gamma_i \rangle$ is elusive. □

The following result is Marušič and Scapellato (1998, Lemma 2.1), and is a general tool for determining whether, for a group G with a core-free subgroup $H \leq G$, the permutation group $(G, G/H)$ is elusive. It shows this can be determined in the abstract group G, without computing the group $(G, G/H)$. We need some additional notation before stating the result.

Definition 12.1.4 Let G be a group, and $g \in G$. Denote the **conjugacy class of** g **in** G by $\mathrm{Cl}_G(g)$. That is, $\mathrm{Cl}_G(g) = \left\{ h^{-1}gh : h \in G \right\}$.

Lemma 12.1.5 *Let G be a group, and $H \leq G$ be core-free in G. Then $(G, G/H)$ is elusive if and only if for each $g \in G$ of prime order we have $H \cap \mathrm{Cl}_G(g) \neq \emptyset$.*

Proof Let $g \in G$ of prime order. By Exercise 1.3.1, $H = \mathrm{Stab}_{(G,G/H)}(H)$. Suppose $H \cap \mathrm{Cl}_G(g) \neq \emptyset$. Then there exists $k \in G$ such that $k^{-1}gk \in H$ and so fixes H. By Theorem 1.1.7, g fixes kH and so is not semiregular. As g is arbitrary of prime order, G has no semiregular elements. Conversely, suppose $(G, G/H)$ is elusive. Then g stabilizes some point kH, $k \in G$, and so $gkH = kH$ or $k^{-1}gkH = H$. Thus $\mathrm{Cl}_G(g) \cap H \neq \emptyset$. □

With examples of elusive groups in hand, more may be constructed using products we have already seen.

Theorem 12.1.6 *If $G \leq S_X$ and $H \leq S_Y$ are elusive, then*

(i) *$G \times H \leq S_X \times S_Y$ is elusive,*
(ii) *$G \wr H \leq S_{X \times Y}$ is elusive, and*
(iii) *$S_k \wr G \leq S_{X^k}$ with the product action is elusive.*

Proof We will show (i) and (iii). (ii) is similar to (i) and is left as Exercise 12.1.3.

(i) Let $(g, h) \in G \times H$ be of prime order p. This implies that both $|g|$ and $|h|$ divide p. As both G and H are elusive, g and h have fixed points $x \in X$ and $y \in Y$, respectively. Then $(g, h)(x, y) = (g(x), g(y)) = (x, y)$. As every element of prime order has a fixed point $G \times H$ is elusive.

(iii) Let $\delta \in \mathcal{S}_k \wr G$ be of prime order p. Then δ induces a permutation of order p on the k coordinates of X^k consisting of some number a of p-cycles. Then there exists $b \geq 0$ (with 0 a definite possibility if k is a multiple of p) such that $ap + b = k$ (so b is the number of coordinates fixed by δ). If $b > 0$, i.e., δ fixes some coordinate, then δ induces an element of order dividing p in every coordinate that it fixes. As G is elusive, this element fixes a point in which case δ fixes a point.

If $b = 0$, then $k = mp$ for some $m \geq 1$. Let $a_1, \ldots, a_p$ be p coordinates such that δ induces a p-cycle on these coordinates. It suffices to show that the induced action of δ on these coordinates fixes a point. So it suffices to show that $\mathcal{S}_p \wr G$ is elusive if G is elusive.

Let $\delta \in \mathcal{S}_p \wr G$ be of prime order p with $\delta(i, j) = (h(i), g_i(j))$, $h \in \mathcal{S}_p$, and $g_i \in G$. Slightly differently than we did when we defined the product action, we view h as permuting the set $\mathbb{Z}_p$ (so we will count coordinates starting from 0, and perform arithmetic on the coordinates modulo p). Then $|h| = p$, so after relabeling, if necessary, we may assume without loss of generality that h has cycle decomposition $(p-1, p-2, \ldots, 1, 0)$. Then

$$\begin{aligned}\delta(y_0, \ldots, y_{p-1}) &= \left(g_{h^{-1}(0)}\left(y_{h^{-1}(0)}\right), \ldots, g_{h^{-1}(p-1)}\left(y_{h^{-1}(p-1)}\right)\right) \\ &= (g_1(y_1), g_2(y_2), \ldots, g_{p-1}(y_{p-1}), g_0(y_0)).\end{aligned}$$

Fix $y_0 \in X$. As $g_i \in G$ is a bijection, $1 \leq i \leq p-1$, there exists $y_i \in X$ such that $g_i(y_i) = y_{i-1}$. Then $\delta(y_0, \ldots, y_{p-1}) = (y_0, \ldots, y_{p-2}, g_0(y_0))$. Finally, as $\delta^p = 1$, $\Pi_{i=0}^{p-1} g_i = 1$, so $g_0 = g_{p-1}^{-1} g_{p-2}^{-1} \cdots g_1^{-1}$. As for $1 \leq i \leq p-1$, $g_i^{-1}(y_{i-1}) = y_i$, we have

$$g_0(y_0) = g_{p-1}^{-1} g_{p-2}^{-1} \cdots g_1^{-1}(y_0) = y_{p-1},$$

and δ fixes $(y_0, \ldots, y_{p-1})$. □

For all known elusive groups G, $G^{(2)}$ is not elusive (Cameron et al. (2002); Giudici (2003, 2007)). As an example of this, we give the following result.

Lemma 12.1.7 *Let p be a Mersenne prime. The 2-closure of* $\mathrm{AGL}\left(1, p^2\right)$ *in its faithful action on affine lines has a semiregular element of order p.*

Proof We saw in Theorem 3.8.13 that there are $(p+1)p$ affine lines. We have already seen that the subgroup T of $\mathrm{AGL}\left(1,p^2\right)$ consisting of translations is normal in $\mathrm{AGL}\left(1,p^2\right)$. Hence by Theorem 2.2.7 there is a normal block system $\mathcal{B}$ of $\mathrm{AGL}\left(1,p^2\right)$ that is the set of orbits of T. A block of $\mathcal{B}$ is a parallel class of lines of size p. That is, a block of $\mathcal{B}$ is an affine line through the origin and all of its translates. The result follows by Corollary 5.3.13. □

Our next result summarizes the relationships between the three classes of permutation groups we have seen when considering the semiregularity problem.

Proposition 12.1.8 *There exist elusive groups that are not 2-closed, as well as 2-closed transitive groups that are not automorphism groups of vertex-transitive digraphs.*

Proof By Theorem 3.8.13, for a Mersenne prime p, the group $\mathrm{AGL}\left(1,p^2\right)$ is elusive, but its 2-closure is not by Lemma 12.1.7. Hence $\mathrm{AGL}\left(1,p^2\right)^{(2)} \neq \mathrm{AGL}\left(1,p^2\right)$ establishing the first part of the result. For the second part, by Example 5.2.7 the group $\mathbb{Z}_2^2$ is 2-closed, but not the automorphism group of a vertex-transitive digraph. □

The now commonly accepted, and slightly more general, version of Problem 3.8.1 involves the whole class of 2-closed transitive groups, and is stated as a conjecture Cameron (1997). It is known as the polycirculant conjecture.

Conjecture 12.1.9 (The polycirculant conjecture) *All 2-closed groups contain a semiregular element.*

We would like to emphasize, however, that the evidence gathered so far suggests that the main difficulty in resolving the polycirculant conjecture is within the "vertex-transitive digraphs" part, and that the "2-closed groups" part is just an obvious natural generalization. In line with this comment it is the semiregularity problem that we concentrate on hereafter. Therefore, unless specifically stated otherwise, we will stay within the class of vertex-transitive digraphs for the rest of this chapter. Nevertheless, when appropriate, results involving semiregular elements in 2-closed groups, and more generally in transitive permutation groups, will be mentioned.

By Lemma 1.2.13, the regular subgroup G_L is contained in $\mathrm{Aut}(\mathrm{Cay}(G,S))$ for every $S \subset G$. So the automorphism group of every Cayley digraph contains a semiregular element. We now define a generalization of Cayley digraphs introduced in Marušič et al. (1992) for which every member of this family also has automorphism group with a semiregular element.

Definition 12.1.10 Let G be a group, $S \subset G$, and $\alpha \in \mathrm{Aut}(G)$ such that $\alpha^2 = 1$, and if $g \in G$ then $\alpha(g)g^{-1} \notin S$. The **generalized Cayley digraph** $\mathrm{GC}(G, S, \alpha)$ is the digraph with vertex set G, and arc set $\{(g, \alpha(g)s) : g \in G\}$.

Clearly Cayley digraphs are generalized Cayley digraphs for $\alpha = 1$. Not every generalized Cayley digraph is vertex-transitive (Marušič et al., 1992, Proposition 3.2), and even if a generalized Cayley digraph is vertex-transitive it need not be isomorphic to a Cayley digraph (Hujdurović et al., 2015). The smallest example of such a graph is the line graph of the Petersen graph. It is isomorphic to the generalized Cayley graph $\mathrm{GC}(\mathbb{Z}_{15}, S, \alpha)$ where $S = \{1, 2, 4, 8\}$ and $\alpha \in \mathrm{Aut}(\mathbb{Z}_{15})$ is defined by $\alpha(x) = 11x$. The following result is from Hujdurović et al. (2015, Theorem 3.4).

Proposition 12.1.11 *The automorphism group of every generalized Cayley digraph has a semiregular automorphism.*

Proof Let $G \neq 1$, $\alpha \in \mathrm{Aut}(G)$ such that $\alpha^2 = 1$, and $S \subset G$ such that and $S \cap \left\{\alpha(g)g^{-1} : g \in G\right\} = \emptyset$. Let $\Gamma = \mathrm{GC}(G, S, \alpha)$. Define $\omega_\alpha \colon G \to G$ by $\omega_\alpha(x) = \alpha(x)x^{-1}$, and let $\omega_\alpha(G) = \{\omega_\alpha(g) : g \in G\}$. The condition $S \cap \{\alpha(g)g^{-1} : g \in G\} = \emptyset$ is equivalent to $S \cap \omega_\alpha(G) = \emptyset$.

Suppose $\omega_\alpha(x) = \omega_\alpha(y)$ for some $x, y \in G$. That is, $\alpha(x)x^{-1} = \alpha(y)y^{-1}$. This is equivalent to $\alpha\left(y^{-1}x\right) = y^{-1}x$, or $y^{-1}x \in \mathrm{Fix}(\alpha)$. So $\omega_\alpha(x) = \omega_\alpha(y)$ is equivalent to $x \in y\mathrm{Fix}(\alpha)$. It is straightforward to show that as $\alpha \in \mathrm{Aut}(G)$, $\mathrm{Fix}(\alpha) \leq G$ (Exercise 12.1.5). Thus x and y have the same image under ω_α if and only if x and y belong to the same left coset of $\mathrm{Fix}(\alpha)$. Hence the size of the image of ω_α is the number of left cosets of $\mathrm{Fix}(\alpha)$ in G. That is, $|\omega_\alpha(G)| = |G|/|\mathrm{Fix}(\alpha)|$. Also, as $\omega_\alpha(G) \cap S = \emptyset$ we have $S \subseteq G\backslash\omega_\alpha(G)$ and so,

$$|S| \leq |G| - \frac{|G|}{|\mathrm{Fix}(\alpha)|}. \tag{12.1}$$

Now suppose $g \in \mathrm{Fix}(\alpha)$. Then $\alpha(g) = g$ and so,

$$\alpha(gx)^{-1}gy = \alpha(gx)^{-1}\alpha(g)y = \alpha\left((gx)^{-1}g\right)y = \alpha\left(x^{-1}\right)y,$$

for all $x, y \in G$. Hence $(gx, gy) \in A(\Gamma)$ if and only if $(x, y) \in A(\Gamma)$, implying that $g_L \in \mathrm{Aut}(\Gamma)$. Thus $\{g_L : g \in \mathrm{Fix}(\alpha)\} \leq \mathrm{Aut}(\Gamma)$. As G_L is regular, g_L is semiregular, and so the result follows or $\mathrm{Fix}(\alpha) = \{1\}$. If $\mathrm{Fix}(\alpha) = 1$, then by Equation (12.1) we have $S = \emptyset$. Thus Γ is totally disconnected, and $\mathrm{Aut}(\Gamma) = S_G$ has a semiregular automorphism. □

The next three sections are devoted, respectively, to the three lines of approach to the semiregularity problem that have been taken so far: *the order of a graph*, *the valency of a graph*, and *special types of group actions*.

12.2 Orders of (Di)graphs

We saw in Theorem 3.8.3 that a transitive group of prime power degree has a semiregular element. We also saw in Theorem 3.8.6 that any transitive group of order mp, $m < p$, contains a semiregular element. Here, we give Marušič's original proof of this result as promised in Section 3.4.

Theorem 12.2.1 *Let p be a prime and m be an integer such that $1 \leq m < p$. Then every transitive permutation group of degree mp contains a semiregular element of order p.*

Proof Let $G \leq S_{mp}$ be transitive, and let P be a Sylow p-subgroup of G. By Theorem 1.1.9, the orbits of P have size p, and so P has m orbits of size p. If $\alpha \in P$ and O is an orbit of P, then α^O is either the identity or it is $(1, p)$-semiregular.

Let $\beta \in P$ such that β has the maximum number n of orbits of size p. If $n = m$, then β is semiregular, so we assume $n < m$. Let $O_1, \ldots, O_n$ be all orbits of β of length p, and let O_{n+1} be an orbit of P for which $\beta^{O_{n+1}} = 1$. Let $\gamma \in P$ such that $\gamma^{O_{n+1}} \neq 1$. As $P^O \cong \mathbb{Z}_p$ for every orbit O of P, there is at most one $a_i \in \mathbb{Z}_p$ for which $(\beta\gamma^{a_i})^{O_i} = 1$, $1 \leq i \leq n$. As $n < m$, there is $a \in \mathbb{Z}_p$ distinct from all a_i, $1 \leq i \leq n$. Then $(\beta\gamma^a)^{P_i}$ is $(1, p)$-semiregular for all $1 \leq i \leq n + 1$, contradicting our choice of β. □

The next natural step would be to consider vertex-transitive digraphs of square-free order. This was done in the larger context of 2-closed groups in Dobson et al. (2007, Theorem 4.1). Parts of the proof use CFSF.

Theorem 12.2.2 *Every 2-closed group of square-free degree is nonelusive.*

As for non-square-free orders of vertex-transitive (di)graphs the situation seems to be much more complex. Recently, 2-closed groups of degree p^2q and p^2qr, where p, q, and r are distinct primes, were proved to contain semiregular elements (Arezoomand and Ghasemi, 2021). Again, CFSG was essential in parts of the proof.

Theorem 12.2.3 *Every 2-closed group of degree p^2q and p^2qr has a semiregular element.*

Before this result were two papers showing that vertex-transitive digraphs of orders $2p^2$ (Marušič and Scapellato, 1998, Theorem 4.1) and $3p^2$ (Marušič, 2018, Theorem 1.) have semiregular elements. The proofs are CFSG-free.

We end this section with a result about the existence of semiregular elements in solvable groups of specific degrees. The following result is a special case of Dobson and Marušič (2011, Theorem 4.8).

Proposition 12.2.4 *Let n be a positive integer such that* $\gcd(n, \phi(n)) = 1$. *Then every transitive solvable group of degree n contains a semiregular element.*

12.3 Valency of Graphs

The main result of this section is Corollary 12.3.6, which shows that every cubic vertex-transitive graph has a semiregular automorphism. We first consider the more general context of vertex-transitive graphs of prime valency. The next result is part of Marušič and Scapellato (1998, Lemma 3.2).

Lemma 12.3.1 *Let Γ be a connected vertex-transitive graph of prime valency p, $v \in V(\Gamma)$, and $G \leq \mathrm{Aut}(\Gamma)$ be transitive. Then $|\mathrm{Stab}_G(v)|$ is not divisible by p^2.*

Proof Suppose p divides $|\mathrm{Stab}_G(v)|$. Let P be a Sylow p-subgroup of $\mathrm{Stab}_G(v)$. We will first show that any element $\gamma \in P$ that fixes a point in $N_\Gamma(v)$ is the identity. Let $u \in V(\Gamma)$. We proceed by induction on $d = \mathrm{dist}_\Gamma(u, v)$, which is finite as Γ is connected. If $\mathrm{dist}_\Gamma(u, v) = 0$ then $u = v$ and u is fixed by γ. If $\mathrm{dist}_\Gamma(u, v) = 1$, then u is a neighbor of v in Γ. If $\gamma(u) \neq u$, then u is contained in an orbit of γ of size at least p, so that v has at least $p + 1$ neighbors, a contradiction. Assume that $d \geq 2$ and γ fixes every vertex u such that $\mathrm{dist}_\Gamma(u, v) \leq d - 1$, and $u \in V(\Gamma)$ be of distance d from v. Let z and w be the two vertices immediately preceding u on a path of length d from v to u (with $wu \in E(\Gamma)$). Of course, z and w are at distance at most $d - 1$ from v, and so by the induction hypothesis we have γ fixes both z and w. But then γ stabilizes w and fixes a vertex in $N_\Gamma(w)$, namely z, and by arguments above we have that γ fixes all of the neighbors of w in Γ. In particular, γ fixes u and by induction $\gamma = 1$.

Now consider the natural restriction map $f\colon P \to S_{N_\Gamma(v)}$. Then f is a homomorphism, and as any element of P that fixes v and one element of $N_\Gamma(v)$ is necessarily the identity, f is one-to-one. Thus P is isomorphic to a subgroup of S_p, and so necessarily has order p. □

Theorem 12.3.2 *Let p be a prime and let Γ be a connected vertex-transitive graph of valency p whose automorphism group contains a transitive solvable subgroup G of order divisible by p. Then Γ has a semiregular automorphism.*

Proof As p divides $|G|$, G contains an element of order p. If this element is semiregular, we are finished. Otherwise, it fixes a point. By Lemma 7.3.9, p is the largest prime dividing $|\mathrm{Stab}_G(v)|$. A minimal normal subgroup M of G

is elementary abelian by Theorem 9.3.3, so $M \cong \mathbb{Z}_q^k$ for some prime q. We may assume that M is intransitive as otherwise Γ is of order q^k, and Γ has a semiregular automorphism by Theorem 3.8.3. Let $\mathcal{B}$ be the block system of G that is the set of orbits of M. Note that $\mathcal{B}$ has blocks of order q^m for some positive integer m. Then M^B is a transitive abelian subgroup, and so by Theorem 2.2.17 is regular. Thus every element of M^B is semiregular.

Suppose the induced action of M on $B \in \mathcal{B}$ is faithful for every $B \in \mathcal{B}$. So the kernel of the action of M on $B \in \mathcal{B}$ is trivial for every $B \in \mathcal{B}$. As M^B is regular, this gives that if $\rho \in M$ and $\rho^B \neq 1$, then ρ is semiregular. We thus assume that the kernel K of the action of M on $B \in \mathcal{B}$ is not trivial. Let $B' \in \mathcal{B}$ such that $K^{B'} \neq 1$. As Γ is connected, we may assume B and B' are adjacent in $\Gamma/\mathcal{B}$. This also implies that for every $v \in V(\Gamma)$, $\mathrm{Stab}_G(v)$ contains elements of order q as well as elements of order p. Fix a vertex $v \in B$. We consider the cases where $q \neq p$ and $q = p$ separately.

Suppose $q \neq p$. Then there is $\rho \in M$ such that $\rho^B = 1$ but $\rho^{B'} \neq 1$. As $M^{B'}$ is semiregular and elementary abelian, every orbit of $\rho^{B'}$ has order q. Hence the valency of $\Gamma[B, B']$ is a multiple of q. As $q \neq p$, there is $\gamma \in \mathrm{Stab}_G(v)$ of order p. Let $w \in V(\Gamma)$ such that $\gamma(w) \neq w$. As Γ is connected, there is a vw-path P in Γ. As Γ fixes v but does not fix w, there is a first vertex x on P such that x is fixed by γ but its successor y on P is not. Let O be the orbit of γ that contains y. As $|\gamma| = p$, $|O| = p$, and every neighbor of x in Γ is contained in O. By Theorem 3.2.9, we have that Γ is edge-transitive, and so by the same result, there is $\delta \in \mathrm{Stab}_G(v)$ of order p cyclically permuting the p neighbors $w_0, w_1, \ldots, w_{p-1}$ of v. Either these vertices all belong to B' or they belong to pairwise distinct blocks. The latter, however, cannot occur as B' contains at least q neighbors of v. It follows that each w_i belongs to B'. As Γ is connected, $\mathcal{B} = \{B, B'\}$ and so Γ is a bipartite graph with bipartition $\{B, B'\}$. Since $m \geq 1$, there are elements $\sigma, \tau \in M$ of order q such that σ is semiregular on B and trivial on B' and τ is semiregular on B' and trivial on B. Then $\sigma\tau$ is semiregular.

Suppose $q = p$. Then $|M \cap \mathrm{Stab}_G(v)| \geq p$, so by Lemma 12.3.1 $|M \cap \mathrm{Stab}_G(v)| = p$. Thus if $u \in \mathrm{Fix}(\alpha) \cap \mathrm{Fix}(\beta)$ for $\alpha, \beta \in M \backslash \{1\}$, then $\beta \in \langle \alpha \rangle$, and so $\mathrm{Fix}(\beta) = \mathrm{Fix}(\alpha)$. It follows that $\{\mathrm{Fix}(\alpha) : \alpha \in M \setminus \{1\}\}$ is a partition $\mathcal{P}$ of $V(\Gamma)$. Now, as Γ is connected there exist adjacent vertices u and v belonging to different cells $\mathrm{Fix}(\alpha)$ and $\mathrm{Fix}(\beta)$, respectively, of $\mathcal{P}$. Then every vertex of $\{\alpha^i(v) \colon i \in \mathbb{Z}_p\}$ is adjacent to every vertex of $\{\beta^i(v) : i \in \mathbb{Z}_p\}$. Hence $\Gamma[\{\alpha^i(v), \beta^i(v) : i \in \mathbb{Z}_p\}] \cong K_{p,p}$ has valency p. As Γ is connected and regular of valency p, we have $\Gamma \cong K_{p,p}$, and so Γ has a semiregular automorphism. □

The next three corollaries, of which Corollary 12.3.6 completes the case of cubic graphs, are more or less direct consequences of Theorem 12.3.2.

Corollary 12.3.3 *A graph Γ of prime valency with a solvable edge- and vertex-transitive group of automorphisms G has a semiregular automorphism.*

Proof Let $v \in V(\Gamma)$. By Theorem 3.2.9 $\mathrm{Stab}_G(v)$, and so G, contains an element of order p. The result follows by Theorem 12.3.2. □

The following result of Burnside is frequently useful, and is called **Burnside's $p - q$ theorem** (see Dummit and Foote (2004, Theorem 19.2.1) for a proof of this result).

Theorem 12.3.4 *Let p and q be prime. Every $\{p, q\}$-group is solvable.*

Corollary 12.3.5 *Let p and q be primes, and let Γ be a connected vertex-transitive graph of valency p whose automorphism group contains a transitive $\{p, q\}$-group G. Then Γ has a semiregular automorphism.*

Proof As G is a $\{p, q\}$-group, $|G| = p^a \cdot q^b$ for some $a, b \geq 0$. If $ab = 0$, then Γ is of prime power order and the result follows by Theorem 3.8.3. Otherwise, by Burnside's $p - q$ theorem we have that G is solvable, and the result follows by Theorem 12.3.2. □

We are now ready to show that every cubic vertex-transitive graph has a semiregular element, a result first proven in Marušič and Scapellato (1998, Theorem 3.3).

Corollary 12.3.6 *Every cubic vertex-transitive graph Γ has a semiregular automorphism.*

Proof By Lemma 7.3.9, $\mathrm{Aut}(\Gamma)$ contains a semiregular element of order p, or $|\mathrm{Aut}(\Gamma)|$ is not divisible by a prime $p > 3$. In the later case, $\mathrm{Aut}(\Gamma)$ is a $\{2, 3\}$-group and the result follows by Corollary 12.3.5 with $q = 2$ and $p = 3$. □

The approach used in the proof of Theorem 12.3.2 was generalized in Dobson et al. (2007, Theorem 4.2) to graphs of valency $p + 1$ whose automorphism group contains a transitive $\{2, p\}$-group, and was applied in the special case $p = 3$ to quartic vertex-transitive graphs (Dobson et al., 2007, Theorem 1.1). We state this result without proof.

Theorem 12.3.7 *Let p be an odd prime and let Γ be a connected vertex-transitive graph of valency $p + 1$ with $G \leq \mathrm{Aut}(\Gamma)$ a transitive $\{2, p\}$-group. Then Γ has a semiregular automorphism.*

The proof of the next result is identical to that of Corollary 12.3.6.

Corollary 12.3.8 *Every quartic vertex-transitive graph Γ has a semiregular automorphism.*

An obvious next goal would be to prove that every **quintic** (of valency 5) vertex-transitive graph Γ has a semiregular automorphism. By Lemma 7.3.9, either this is true or 5 is the largest prime dividing $|\mathrm{Aut}(\Gamma)|$. That is, we may assume $\mathrm{Aut}(\Gamma)$ is a $\{2,3,5\}$-group. If $\mathrm{Aut}(\Gamma)$ is solvable, then Γ has a semiregular automorphism in all cases expect possibly when $\mathrm{Aut}(\Gamma)$ is a $\{2,3\}$-group by Theorem 12.3.2. When $\mathrm{Aut}(\Gamma)$ is nonsolvable then Theorem 12.3.4 implies that 5 divides $|\mathrm{Aut}(\Gamma)|$, and so $\mathrm{Aut}(\Gamma)$ contains an element of order 5. This element is either semiregular or fixes a vertex. In the latter case, Γ is arc-transitive and Γ has a semiregular element by a result of Giudici and Xu (2007, Theorem 1.1), stated below as Corollary 12.3.11. So to prove semiregular automorphisms exist for all quintic vertex-transitive graphs only graphs whose automorphism groups are $\{2,3\}$-groups remain. This is by no means an easy task, confirming a somewhat surprising (and definitely counter-intuitive) fact that it is the solvable and not the nonsolvable groups that present the main difficulty in obtaining a complete solution to the semiregularity problem. Giudici and Xu's result mentioned above follows from a stronger result that relies on CFSG, and so will not be proven here. Before stating this result, we need an additional term.

Definition 12.3.9 A graph Γ is **locally quasiprimitive** if its local action is quasiprimitive.

Proposition 12.3.10 *A graph Γ with a vertex-transitive locally quasiprimitive group G of automorphisms has a semiregular automorphism.*

Corollary 12.3.11 *Any arc-transitive graph of prime valency or any 2-arc-transitive graph has semiregular automorphism.*

Proof As a primitive group is quasiprimitive, a graph Γ that is locally primitive has a semiregular element. In particular, as any group of prime degree is primitive by Example 2.4.6, any arc-transitive group of prime valency has a semiregular automorphism. Also, as a 2-transitive group is primitive, any 2-arc-transitive graph has a semiregular element by Exercise 3.1.10. □

We end this section by stating, without proof, other known results concerning semiregular automorphisms of graphs with restricted valency. The case for arc-transitive graphs of valency twice a prime was recently settled in Giudici and Verret (2020, Theorem 1.2).

Proposition 12.3.12 *Every arc-transitive graph of valency $2p$, p a prime, has a semiregular automorphism.*

Further, Xu (2008, Theorem 1.1) has a partial result for arc-transitive graphs of valency qp.

Proposition 12.3.13 *Let Γ be a finite connected arc-transitive graph of valency pq, where p and q are primes. Suppose further that* $\mathrm{Aut}(\Gamma)$ *has a nonabelian minimal normal subgroup M that has at least* 3 *orbits on the vertex set. Then Γ has a nonidentity semiregular automorphism.*

Finally, Verret (2015, Theorem 1.2) has shown that arc-transitive graphs of valency 8 have semiregular automorphisms.

Proposition 12.3.14 *Every arc-transitive graph of valency* 8 *has a semiregular automorphism.*

12.4 Types of Group Actions

The third approach to the semiregularity problem taken thus far is based on examining special types of permutation groups. Two results in this respect have already been mentioned in Section 12.3 (Propositions 12.3.10 and 12.3.13). Perhaps the most important work on the subject is due to Giudici (2003, Theorem 1.2) who proved, using CFSG, that with the exception of a certain family of groups associated with M_{11}, all quasiprimitive permutation groups contain semiregular elements. He also showed (Proposition 12.4.1) that those that do not, have 2-closures that do. As with other problems whose solutions require CFSG, it would be of interest to find a CFSG-free argument.

Proposition 12.4.1 *Every* 2*-closed permutation group with at least one transitive minimal normal subgroup has a semiregular automorphism.*

Definition 12.4.2 A **biquasiprimitive** permutation group is a transitive permutation group for which every nontrivial normal subgroup has at most two orbits and there is some normal subgroup with precisely two orbits.

Giudici and Xu (2007) extended the above results by first classifying all biquasiprimitive groups without semiregular elements (which we will not state here), proving the following result (Giudici and Xu, 2007, Theorem 1.4).

Proposition 12.4.3 *Every* 2*-closed biquasiprimitive permutation group contains a semiregular element.*

To end this section we turn our attention to semiregular automorphisms in **distance-transitive** graphs.

Definition 12.4.4 A graph Γ is said to be **distance-transitive** if, for any two ordered pairs of vertices (u, v) and (u', v') with $\text{dist}(u, v) = \text{dist}(u', v')$, there is $\gamma \in \text{Aut}(\Gamma)$ with $\gamma(u) = u'$ and $\gamma(v) = v'$.

Distance-transitive graphs were introduced by Biggs and Smith (1971). They showed that there are exactly twelve cubic distance-transitive graphs – most of which we have seen. The examples we have seen are K_4, $K_{3,3}$, the cube, the Petersen graph, the Heawood graph, the Pappus graph, the dodecahedron, and the Desargues graph. Using Proposition 12.4.1 and strong constraints on the structure of distance-transitive graphs with imprimitive automorphism group found by Smith (1971), Kutnar and Šparl (2010, Theorem 1.1) showed that distance transitive graphs have semiregular automorphisms.

Theorem 12.4.5 *The automorphism group of a distance transitive has a semiregular automorphism.*

12.5 Further Directions

The results presented in previous sections provide a good start to resolving the semiregularity problem for nonsolvable groups. This suggests a shift of emphasis to solvable groups is needed if one is to hope for a solution of the semiregularity problem. For our discussion here, let Γ be a vertex-transitive graph, $G \leq \text{Aut}(\Gamma)$ be transitive, M be a minimal normal subgroup of G, and $\mathcal{B}$ the block system of G that is the set of orbits of M. By Propositions 12.4.1 and 12.4.3, if M has at most two orbits then $\text{Aut}(\Gamma)$ has a semiregular element. So we need only consider when M has at least three orbits. The corresponding normal quotient graph is smaller and vertex-transitive. This suggests that, at least in principle, an induction type of argument could be successful (see Exercise 12.2.1). It seems reasonable to begin by first considering solvable groups, where the normal subgroup reduction has obvious specific advantages.

If G is solvable, then by Theorem 9.3.3 $M = \mathbb{Z}_p^m$ for some prime p and $m \geq 1$. Initial work in this direction has been done in Dobson and Marušič (2011) where it was proved, amongst other things, that every transitive solvable group of degree n such that $\gcd(n, \phi(n)) = 1$ contains a semiregular element (see Proposition 12.2.4). This result was obtained as a consequence of a more general result relating (non)elusive groups with certain cyclic codes over fields of prime order. (Application of cyclic codes over fields of prime order in the wider context of transitive permutation groups first appeared in Dobson and Witte (2002).)

12.6 Exercises

12.1.1 Show that a generalized Cayley digraph $\mathrm{GC}(G, S, \alpha)$ is a graph if and only if $\alpha(S) = S^{-1}$.

12.1.2 Show that the line graph of the Petersen graph is isomorphic to the graph $\mathrm{GC}(\mathbb{Z}_{15}, S, \alpha)$ where $S = \{1, 2, 4, 8\}$ and $\alpha \in \mathrm{Aut}(\mathbb{Z}_{15})$ is defined by $\alpha(x) = 11x$.

12.1.3 Let $G \leq S_X$ and $H \leq S_Y$ be elusive. Show that $G \wr H \leq S_{X \times Y}$ is elusive.

12.1.4 Let $G \leq S_X$ and $H \leq S_Y$. Show that $G \times H \leq S_{X \times Y}$ is not elusive if and only if either G or H is not elusive. Deduce that $G \wr H \leq S_{X \times Y}$ is not elusive if G or H is not elusive.

12.1.5 Let G be a group and $\alpha \in \mathrm{Aut}(G)$. Show that $\mathrm{Fix}(\alpha) \leq G$.

12.2.1 Prove that a vertex-transitive digraph whose automorphism group has all Sylow subgroups cyclic, has a semiregular automorphism. (Hint: Use Theorem 3.8.3, and also see Kutnar and Marušič (2008b).)

12.2.2 Construct at least one semiregular automorphism for every cubic arc-transitive graph of order up to 40.

12.2.3 Prove that every vertex-transitive graph of order 12 has a semiregular automorphism.

12.2.4 A nontrivial automorphism α of a graph Γ is **quasi-semiregular** if all of its orbits are of the same length with the exception of one orbit that has length 1. Prove that K_4, the Petersen graph, and the Coxeter graphs are the only cubic arc-transitive quasi-semiregular graphs. (Hint: See Feng et al. (2019).)

13

Graphs with Other Types of Symmetry Half-arc-transitive Graphs and Semisymmetric Graphs

13.1 Historic Perspectives

Tutte (1966) proved that the automorphism group of a vertex- and edge-transitive graph of odd valency must act transitively on the arc set, and hence the graph is symmetric. He also asked if a vertex- and edge-transitive graph of even valency was also necessarily arc-transitive. In the terminology we have developed, Tutte asked if half-arc-transitive graphs exist. Of course we have already seen that the Holt graph is half-arc-transitive, but the first example was provided earlier by Bouwer (1970). He constructed a $2k$-valent half-arc-transitive graph for every $k \geq 2$. He defined a wider class of graphs as follows.

Definition 13.1.1 Let m be a positive integer, $k \geq 2$, and n an integer such that $2^m \equiv 1 \pmod{n}$. Define a graph $B(k,m,n)$ with vertex set $V(B(k,m,n)) = \mathbb{Z}_m \times \mathbb{Z}_n^{k-1}$ and two vertices are adjacent if they can be written in the form

$$(i, a_2, a_3, \ldots, a_k) \text{ and } (i+1, b_2, b_3, \ldots, b_k),$$

where $i \in \mathbb{Z}_m$, and $a_j, b_j \in \mathbb{Z}_n$ for each $2 \leq j \leq k$, and, additionally, either $b_j = a_j$ for each $2 \leq j \leq k$ or there exists exactly one $2 \leq \ell \leq k$ for which $b_\ell \neq a_\ell$, in which case $b_\ell = a_\ell + 2^i$. We refer to a graph $B(k,m,n)$ as a **Bouwer graph**.

The condition that $2^m \equiv 1 \pmod{n}$ means 2 is a unit modulo n, and consequently that n is odd.

Let $\mathbf{e}_j$ be the element of $\mathbb{Z}_n^{k-1}$ with 1 in the jth coordinate and 0s elsewhere. Note that two adjacent vertices in $B(k,m,n)$ can then be written as $(i,\mathbf{a})(i+1,\mathbf{a})$ or $(i,\mathbf{a})(i+1,\mathbf{a}+2^i\mathbf{e}_j)$, where $\mathbf{a} \in \mathbb{Z}_n^{k-1}$. That the graph $B(k,m,n)$ is regular of valency $2k$ and is vertex- and edge-transitive are left as Exercises 13.1.1

and 13.1.2. Bouwer (1970, Proposition 2) gave the first examples of half-arc-transitive graphs.

Proposition 13.1.2 *For $k \geq 2$, the graphs $B(k, 6, 9)$ are half-arc-transitive.*

The proof of Proposition 13.1.2 is rather technical. It uses a detailed analysis of different types of 6-cycles in the graphs, similar to the use of 8-cycles in the determination of the automorphism groups of the generalized Petersen graphs in Section 6.5. This analysis will show that $\mathcal{B} = \left\{\{i\} \times \mathbb{Z}_9^{k-1} : i \in \mathbb{Z}_6\right\}$, is a block system of $\text{Aut}(B(k, 6, 9))$. Also, $B(k, 6, 9)/\mathcal{B}$ is a cycle of length 6. The key reason that $B(k, 6, 9)$ is not arc-transitive is that Bouwer shows that $\text{Aut}(B(k, 6, 9))/\mathcal{B}$ is isomorphic to $\mathbb{Z}_6$, not to the dihedral group D_{12}. For the details, the reader is referred to Bouwer (1970).

It turns out that almost all of the Bouwer graphs $B(k, m, n)$ are half-arc-transitive. Conder and Žitnik (2016) have recently shown that the only ones that are arc-transitive satisfy $n = 3$, $(k, n) = (2, 5)$, or are one of three exceptions, with $(k, m, n) = (2, 3, 7)$ or $(2, 6, 7)$ or $(2, 6, 21)$. Additionally, Bouwer graphs have been generalized and used to obtain more families of half-arc-transitive graphs by Ramos Rivera and Šparl (2017).

Before leaving the early examples it is worth mentioning that an example of a graph that could be used to easily construct a half-arc-transitive graph was found by Frucht (1952). This is a greatly overlooked construction of the cubic 1-arc-regular graph F432E by Frucht (see also the end of Section 9.1). As observed in Marušič and Xu (1997), the line graph of a cubic 1-arc-regular graph is a half-arc-transitive graph of valency 4 (Exercise 13.1.4). Therefore the line graph of F432E is, in a sense, the first example of a half-arc-transitive graph.

At first glance, the problem of finding half-arc-transitive graphs seems to be quite similar to the problem of finding arc-transitive graphs. In reality, determining whether a graph is half-arc-transitive is much harder than determining whether a graph is arc-transitive. This is because in order to show a graph Γ is arc-transitive, one needs to only find a subgroup $G \leq \text{Aut}(\Gamma)$ that is transitive on the arcs of Γ – one does not need the full automorphism group. For half-arc-transitive graphs, one must show it is edge-transitive but then also show there is no automorphism interchanging the endpoints of an edge (by Theorem 3.2.11). Often, the way this is done is to compute the full automorphism group of the graph (although technically this is not strictly required) and then check by inspection that no such automorphism exists.

Recall from Example 3.2.14 and the paragraph immediately preceding that example that the Holt graph is the smallest half-arc-transitive graph. In Example 3.2.14 it was constructed as a Cayley graph of $\mathbb{Z}_9 \rtimes \mathbb{Z}_3$.

There is a concise description of the Holt graph, as well as Bouwer's smallest graph $B(2,6,9)$, as part of a much larger family of half-arc-transitive graphs of valency 4 (Alspach et al., 1994).

In fact, research on half-arc-transitive graphs gained new momentum in the 1990s, going in a few distinct directions. Various results obtained from a geometric approach based on certain substructures of these graphs, the so-called alternets, will be discussed in the next section. In Section 13.4, the connection between half-arc-transitive graphs and *semisymmetric graphs* (see, for example, Marušič and Potočnik (2002)), that is, graphs that are edge- but not vertex-transitive will be given.

13.2 Half-arc-transitive Graphs: A Geometric Approach

We now introduce an approach to studying G-half-arc-transitive graphs of a geometric flavor. As mentioned in the previous section, half-arc-transitive graphs can be difficult to study, and so it should not be surprising that the approach requires a fairly extensive setup before we can discuss an example. We begin with a general digraph term that will prove convenient.

Definition 13.2.1 For an arc (u,v) of a digraph Γ, the vertices u and v are called its **tail** and **head**, respectively.

Now let Γ be a G-half-arc-transitive graph. Choose an edge $uv \in E(\Gamma)$ and let G act on the directed arc (u,v), and separately on (v,u). As G is half-arc-transitive, this gives two oppositely oriented digraphs Γ_1 and Γ_2 in the sense that $(x,y) \in A(\Gamma_1)$ if and only if $(y,x) \in A(\Gamma_2)$. Note that Γ_1 and Γ_2 are paired orbital digraphs of G. We choose one of these orientations of Γ (it does not matter for what follows which one), and denote it by $\mathbf{D_G(\Gamma)}$.

Definition 13.2.2 Let Γ be a G-half-arc-transitive graph. Define a relation R on $A(D_G(\Gamma))$ by aRb if and only if the heads or tails of a and b are the same. We define the **reachability relation** on $D_G(\Gamma)$ to be the transitive closure of the relation R.

Clearly R is reflexive and symmetric, but not necessarily transitive. Recall the transitive closure of a relation T on a set X is the smallest relation on X that is transitive and contains T. As $X \times X$ contains T and is transitive, the transitive closure of T is the intersection of all transitive relations on X that contain T, and so always exists. Thus the reachability relation is an equivalence relation, and its equivalence classes are independent of the choice of $D_G(\Gamma)$ (which is why we do not care which G-induced orientation of Γ is chosen for $D_G(\Gamma)$).

Definition 13.2.3 The subdigraph of $D_G(\Gamma)$ induced by an equivalence class of the reachability relation is called a **G-alternating cycle** (or just an **alternating cycle** if $G = \mathrm{Aut}(\Gamma)$) if Γ is of valency 4, and in general is called a **G-alternet**.

Lemma 13.2.4 *Let Γ be a G-half-arc-transitive graph. The set of G-alternets of Γ are a block system of the transitive action of G on $E(\Gamma)$, and consequently G-alternets are isomorphic.*

Proof If $g \in G$, then aRb if and only if $g(a)Rg(b)$. As G is arc-transitive on $D_G(\Gamma)$, the reachability relation is a G-congruence in its transitive action on $A(D_G(\Gamma))$, and, by Theorem 2.3.2, the set of alternets of Γ are blocks of G in its action on the arcs of $D_G(\Gamma)$. As there is a one-to-one correspondence between the arcs of $D_G(\Gamma)$ and the edges of Γ, the alternets are blocks of G acting on the edges of Γ. This also gives that G-alternets are isomorphic as digraphs, and so contain the same number of arcs and vertices. □

Definition 13.2.5 Let Γ be G-half-arc-transitive. The **length of a G-alternet** is the number of vertices contained in it. If there are at least two G-alternets, then all of them have the same even length, half of which is called the **G-radius of Γ**. Let $\mathcal{A} = \{A_i : i \in \{1, 2, \ldots, m\}\}$ be the set of G-alternets in Γ. For each $1 \le i \le m$, define the **head set** H_i and **tail set** T_i to be the sets of all vertices that are heads or tails, respectively, of arcs of A_i.

Note that if $m \ge 2$ then $T_i \cap H_i = \emptyset$. The set of head sets as well as the set of tail sets partition $V(\Gamma)$, and, as G is transitive on $\mathcal{A}$, they are block systems of G. Set $A_{i,j} = H_i \cap T_j$ $1 \le i \ne j \le m$. If $A_{i,j} \ne \emptyset$, then as H_i and T_j are blocks of G, $A_{i,j}$ is also a block of G. So if $A_{i,j} \ne A_{i',j'}$ are both nonempty, then $|A_{i,j}| = |A_{i',j'}|$.

Definition 13.2.6 If $A_{i,j} \ne \emptyset$, then we call $A_{i,j}$ a **G-attachment set of Γ**. Also, $|A_{i,j}|$ is the **G-attachment number of Γ**. If $A_{i,j}$ is a singleton, we say Γ is **G-loosely attached**. If $A_{i,j} = H_i = T_j$, then Γ is **G-tightly attached**.

For all of the above definitions the symbol G is omitted when $G = \mathrm{Aut}(\Gamma)$. We can finally give an example!

Example 13.2.7 Consider the Holt graph, viewed as a Cayley graph of $\langle \rho, \tau \rangle$ as in Example 3.2.14. It has three alternating cycles, each of length 18, and these are the subgraphs induced by the sets obtained by choosing two of the

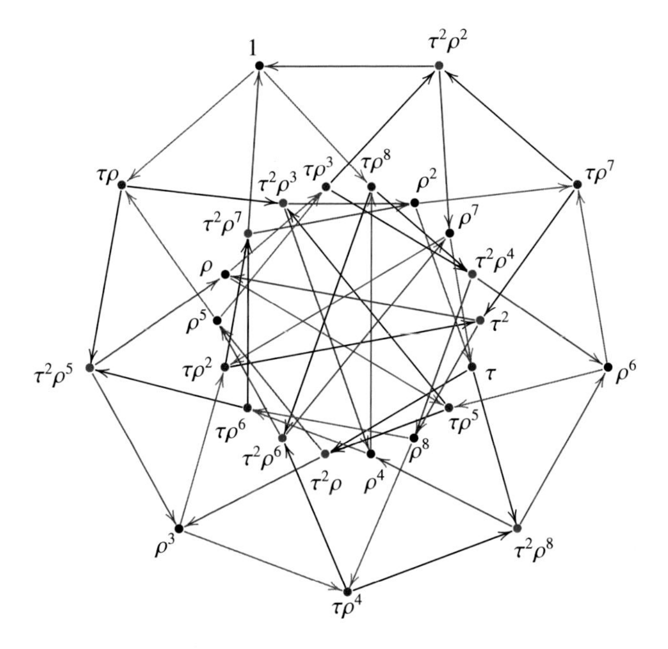

Figure 13.1 A natural orientation of the Holt graph with alternets.

three left cosets of $\langle\rho\rangle$ in $\langle\rho,\tau\rangle$. Additionally, the Holt graph is tightly attached with radius 9.

Solution Most of the verification for this example is contained in the digraph shown in Figure 13.1. This digraph (without anything colored) is $D_G(\Gamma)$ where the edge chosen to construct the digraph was $1(\tau\rho)$ and oriented $(1,\tau\rho)$. The arcs of a same color represent an alternating cycle. They are calculated by choosing two arcs with the same head or tail, and then inductively extending an already constructed directed walk W by adding an arc in the following way, if possible. If there is a vertex v in W whose in- or outvalence in W is one (if no such vertex exists then stop as an alternet has been found), then there is an arc $(w,v) \in D_G(\Gamma)$ not in W. We then extend the walk W by the arc (w,v). A similar process occurs if the outvalence of v in W is one. This process is repeated until every arc of $D_G(\Gamma)$ is contained in some alternet. By inspection (as the reader you should check that the arcs colored the same color do indeed form an 18-cycle), the vertices are colored so that all the vertices of each left coset of $\langle\rho\rangle$ in $\langle\rho,\tau\rangle$ are the same color. There is no relation between colored vertices and arcs of the same color. The rest of the example follows by inspection. □

The automorphism groups of the Holt graph and the graphs in Bouwer's family are imprimitive. Supposedly, this prompted both Holt (1981) and Holton (1982) to ask if a half-arc-transitive graph with a primitive automorphism group exists. The first such graph was found as a byproduct of the classification of vertex-transitive graphs of order a product of two distinct primes having primitive automorphism group (Praeger and Xu, 1993), which was discussed in Section 9.2. As a half-arc-transitive graph with primitive automorphism group necessarily has only one alternet, half-arc-transitive graphs with one alternet exist. In Li et al. (2004, Theorem 1.4) a construction of an infinite family of half-arc-transitive graphs with primitive automorphism group was given, with smallest valency 14. In the same result it was shown that no half-arc-transitive graphs with primitive automorphism group of valency less than or equal to 8 exist. Finally, in Fawcett et al. (2018), it was shown that no such graphs of valency 10 exist, but there are infinitely many such graphs of valency 12.

What can we say about graphs with a small number of alternates? The case of one alternet, which corresponds to the reachability relation being universal, has been given considerable attention, leaving many open questions. As mentioned above, half-arc-transitive graphs with a primitive automorphism group are necessarily of this kind. More recently, an infinite family of half-arc-transitive graphs of valency 12, with one alternet and an imprimitive group of automorphisms was constructed in Kutnar et al. (2010), and an infinite family of half-arc-transitive graphs of valency 6 with one alternet were given in Jajcay et al. (2019).

Three examples of half-arc-transitive graphs of orders 112, 144, and 162, valency 6, and two alternets were found (Hujdurović et al., 2014, Example 2.4) using Magma. Half-arc-transitive graphs with $n \geq 3$ alternets were given in Marušič (1998), where the classification of quartic tightly attached half-arc-transitive graphs of odd radius is given (see Šparl (2008) for the classification of such graphs of even radius).

Our next goal is to show that half-arc-transitive graphs with a "small" number of alternets are always tightly attached (Hujdurović et al., 2014).

Theorem 13.2.8 *Let Γ be a G-half-arc-transitive graph with two alternets. Then G contains a normal subgroup M of index 2 having two orbits on vertices as well as on edges of Γ. Also, Γ is G-tightly attached.*

Proof Let A_1 and A_2 be the two alternets of Γ, and $\mathcal{A} = \{A_1, A_2\}$. Considering the action of G on $E(\Gamma)$, we see that G is transitive, $\mathcal{A}$ is a block system of G, and $G/\mathcal{A} \leq S_2$. So $G/\mathcal{A} \cong S_2$. Set $M = \mathrm{fix}_G(\mathcal{A})$. Then $M \trianglelefteq G$, $[G : M] = 2$, and M has two orbits on $E(\Gamma)$.

Let H_1, H_2 and T_1, T_2 be the head set and the tail set of arcs in A_1 and A_2, respectively. As $H_1 \cap T_1 = \emptyset$, and $H_1 \subset T_1 \cup T_2$, we conclude that $H_1 = T_2$, and similarly that $H_2 = T_1$. So Γ is G-tightly attached. Also, M fixes both H_1 and $T_1 = H_2$ setwise. As G is transitive on $V(\Gamma)$ the orbits of M on $V(\Gamma)$ are H_1 and H_2, and the result follows. □

Theorem 13.2.9 *Let Γ be a G-half-arc-transitive graph with three alternets. Then Γ is G-tightly attached.*

Proof For $1 \le i \le 3$, let A_i be the G-alternets with tail sets T_i and head sets H_i. Recall that $H_i \cap T_i = \emptyset$ for every $1 \le i \le 3$, and both sets $\{H_1, H_2, H_3\}$ and $\{T_1, T_2, T_3\}$ are block systems of G.

Suppose that Γ is not G-tightly attached. Then $A_{1,2} = H_1 \cap T_2 \neq \emptyset$ and $A_{1,3} = H_1 \cap T_3 \neq \emptyset$. Let $v \in A_{1,2}$ and $N = N^+_{D_G(\Gamma)}(v)$. By Theorem 3.2.8, $\mathrm{Stab}_G(v)$ is transitive on N. As $v \in A_{1,2}$, it follows that $\mathrm{Stab}_G(v)$ fixes H_1 and T_2 setwise. Then $\mathrm{Stab}_G(v)$ fixes the alternets A_1 and A_2. So either $N \subseteq T_2$ or $N \subseteq T_3$. An analogous argument shows that the set of out-neighbors of every vertex in $A_{1,3}$ is a subset of T_2 or T_3, and so this holds for every vertex of H_1.

Now, let $u \in T_1$ be such that $N_1 = N^+_{D_G(\Gamma)}(u) \subseteq T_2$. Then the set N_2 of all in-neighbors of vertices in N_1 also have their out-neighbors in T_2. Namely, if this was not the case then there would exist a vertex in T_1 with some of its out-neighbors in T_2 and some in T_3, which is not possible. Repeating the same argument, one can see that all of the vertices in T_1 have their out-neighbors in T_2. So $H_1 = T_2$, a contradiction. □

The Holt graph has three alternets. In the example below, an infinite family of half-arc-transitive graphs with three alternets with arbitrarily large valencies is given (see Alspach and Xu (1994, Theorem 2.5) for the proof).

Example 13.2.10 Let $p \equiv 1 \pmod 3$ be prime, $d > 1$ be a divisor of $(p-1)/3$ and let $S = \langle s \rangle$ be the subgroup of $\mathbb{Z}_p^*$ of order d. Let $r \in \mathbb{Z}_p^* \backslash S$ be such that $r^3 \in S$. Denote the $(3, p, r, \emptyset, S)$-metacirculant graph by $M(d; 3, p)$. If $(d, p) \neq (2, 7)$ or $(3, 19)$ then $M(d; 3, p)$ is a half-arc-transitive graph of order $3p$ and valency $2d$ that has three alternets. This graph is independent of the choice of r. Also, $\mathrm{Aut}(M(d; 3, p)) \cong \mathbb{Z}_p \rtimes \mathbb{Z}_{3d}$ and is regular on the edge set of $M(d; 3, p)$.

In Hujdurović et al. (2014) examples of G-half-arc-transitive graphs are given with four or five alternets that are not tightly attached, for some group of automorphisms G. Additionally, there is a partition $\mathcal{P}$ of their vertex set for which the corresponding quotient is the rose window graph $R_6(5, 4)$, and the graph Γ_5 defined next, respectively. See Hujdurović et al. (2014, Example 3.1) for the proof.

Example 13.2.11 For a positive integer $n \geq 4$, let Γ_n be the graph with vertex set $V(\Gamma_n) = \{(i, j) : i \in \mathbb{Z}_n, j \in \mathbb{Z}_n \backslash \{i\}\}$ and edge set

$$E(\Gamma_n) = \{\{(i, j), (k, i)\} : (i, j) \in V(\Gamma_n), k \in \mathbb{Z}_n, k \neq i, k \neq j\}.$$

Then Γ_n is a connected G-half-arc-transitive graph of valency $2(n-2)$ that is G-loosely attached for the image of a transitive and faithful permutation representation G of $\mathcal{S}_n$.

13.3 Quartic Half-arc-transitive Graphs

In this section we give a brief summary of some structural results on automorphism groups of quartic half-arc-transitive graphs. Quartic half-arc-transitive graphs, and quartic graphs having half-arc-transitive group actions in general, have been the most active research topic on half-arc-transitivity. One of many interesting questions is: What is the structure of vertex stabilizers in connected quartic half-arc-transitive graphs? Clearly, a vertex stabilizer is a 2-group (Exercise 13.3.1), and, for quite some time, only constructions with vertex stabilizers isomorphic to $\mathbb{Z}_2$ were known. The first example of a connected quartic half-arc-transitive graph with a larger vertex stabilizer was given in Malnič and Marušič (2002) using the following idea. We will need an additional term.

Definition 13.3.1 A digraph Γ is **balanced** if the in- and outvalence of each vertex is the same.

Definition 13.3.2 Let Γ be a quartic graph with a fixed balanced orientation, denoted by $\vec{\Gamma}$. Define $\vec{\mathrm{Pl}}(\Gamma)$ to be the digraph with vertex set $A\left(\vec{\Gamma}\right)$ and arc set $A(\mathrm{Pl}(\Gamma)) = \{(x, y) : u, v, w \text{ is a } 2-\text{arc in } \Gamma \text{ with } x = (u, v) \text{ and } y = (v, w)\}$. We define the **partial line graph of** Γ, denoted $\mathrm{Pl}(\Gamma)$, to be the underlying simple graph of $\vec{\mathrm{Pl}}(\Gamma)$.

As the in- and outvalence of each vertex of $\vec{\Gamma}$ is 2, $\mathrm{Pl}(\Gamma)$ is quartic. As an arc $x \in A(\Gamma)$ is the first arc of two 2-arcs and the last arc of two 2-arcs, $\vec{\mathrm{Pl}}(\Gamma)$ is balanced. So $\mathrm{Pl}(\Gamma)$ is a quartic graph with a balanced orientation. Note that the arc set of $\vec{\mathrm{Pl}}(\Gamma)$ decomposes into alternating 4-cycles no two of which intersect in more than one vertex, as Γ is a simple graph - see Figures 13.2 and 13.3.

Thinking of Pl as an operation, to reverse $\mathrm{Pl}(\Gamma)$ (so to obtain $\vec{\Gamma}$ from $\vec{\mathrm{Pl}}(\Gamma)$ and consequently obtain Γ from $\mathrm{Pl}(\Gamma)$), we will need a larger view of $\mathrm{Pl}(\Gamma)$. In Figure 13.3, a neighborhood around the oriented edge vw in Γ is shown on the left-hand side of the figure. On the right-hand side are the two alternating

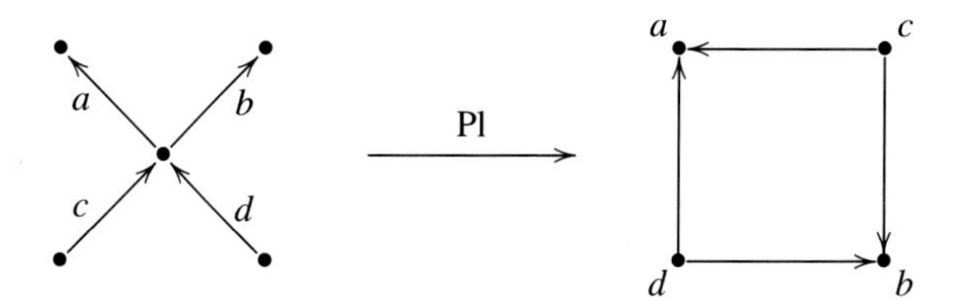

Figure 13.2 Pl at a vertex.

4-cycles C_v and C_w in $\mathrm{Pl}(\Gamma)$ that contain the arc $a = (v, w)$. We will identify the vertices of Γ with their corresponding alternating 4-cycles in the oriented graph $\mathrm{Pl}(\Gamma)$. We orient the edge vw in Γ from v to w if and only if the two arcs in $\mathrm{Pl}(\Gamma)$ whose tails are in $C_v \cap C_w$ have heads on C_w.

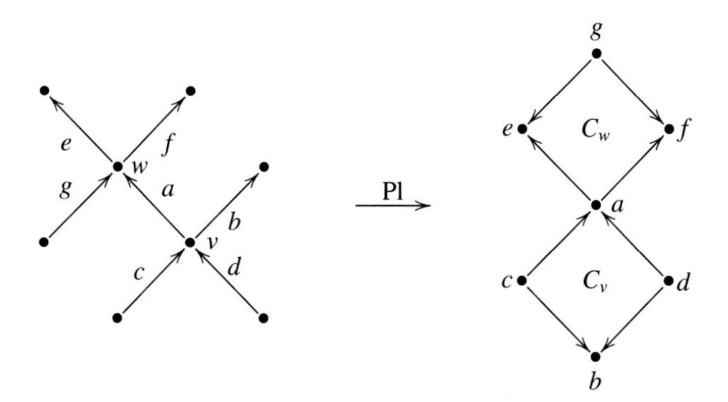

Figure 13.3 The operation Pl at a neighborhood of an arc.

Definition 13.3.3 Let Δ be a quartic graph with a balanced orientation $\vec{\Delta}$ that has the property that its alternating cycles are of length 4 and meet in at most one vertex and partition $A\left(\vec{\Delta}\right)$. Define a digraph $\vec{\mathrm{Al}}(\Delta)$ by $V\left(\vec{\mathrm{Al}}(\Delta)\right)$ is the set of alternating cycles of Δ, and (C_1, C_2) is an arc of $\vec{\mathrm{Al}}(\Delta)$ if and only if C_1 and C_2 are alternating 4-cycles in Δ that meet at a vertex, and the two arcs in $\vec{\Delta}$ whose tails are in $C_1 \cap C_2$ have heads on C_2. We define $\mathrm{Al}(\Delta)$ to be the underlying simple graph of $\vec{\mathrm{Al}}(\Delta)$.

Observe that $\mathrm{Al}\left(\mathrm{Pl}\left(\vec{\Gamma}\right)\right) = \vec{\Gamma}$ for every quartic graph with balanced orientation $\vec{\Gamma}$. Moreover, $\mathrm{Pl}\left(\mathrm{Al}\left(\vec{\Delta}\right)\right) = \vec{\Delta}$ as long as the digraph $\vec{\Delta}$ satisfies the hypothesis of Definition 13.3.3. The following result was proved in Marušič and Nedela (2001b, Proposition 2.1) (see also Marušič (2005, Proposition 2.2.)).

Proposition 13.3.4 *If Γ is a quartic graph with balanced orientation $\vec{\Gamma}$, then*

$$\mathrm{Aut}(\mathrm{Pl}(\Gamma)) = \mathrm{Aut}(\Gamma).$$

Conversely, let Δ be quartic graph with balanced orientation $\vec{\Delta}$ whose alternating cycles have length 4, no two intersect in more than one vertex, and they decompose the edge set. Then $\mathrm{Aut}(\mathrm{Al}(\Delta)) = \mathrm{Aut}(\Delta)$.

Proof Let $\alpha \in \mathrm{Aut}(\Gamma)$. As $\alpha(E(\Gamma)) = E(\Gamma)$, α also permutes the vertices of $\mathrm{Pl}(\Gamma)$. Also, α applied to a 2-arc in Γ is a 2-arc in Γ, so $\alpha \in \mathrm{Aut}(\mathrm{Pl}(\Gamma))$. Hence $\mathrm{Aut}(\Gamma) \leq \mathrm{Aut}(\mathrm{Pl}(\Gamma))$. To see that $\mathrm{Aut}(\mathrm{Pl}(\Gamma)) \leq \mathrm{Aut}(\Gamma)$, let $\alpha \in \mathrm{Aut}(\mathrm{Pl}(\Gamma))$. Then α maps an alternating 4-cycle in $\mathrm{Pl}(\Gamma)$ to an alternating 4-cycle in $\mathrm{Pl}(\Gamma)$, and maps any two intersecting at a vertex to two that intersect at a vertex. As the alternating 4-cycles of $\mathrm{Pl}(\Gamma)$ correspond to vertices of Γ, α maps vertices in Γ to vertices in Γ and preserves adjacency. Also, $\alpha(A(\mathrm{Pl}(\Gamma))) = A(\mathrm{Pl}(\Gamma))$, and so maps 2-arcs of Γ to 2-arcs of Γ. Hence α preserves the arcs of Γ, and so α is induced by the action of an automorphism of Γ on $A(\Gamma)$. Hence $\mathrm{Aut}(\mathrm{Pl}(\Gamma)) = \mathrm{Aut}(\Gamma)$. The second statement now follows as Al and Pl are inverses of each other. □

These two operations can also be applied to an (undirected) vertex-transitive graph Γ (of any valence) whenever an accompanying oriented graph is (perhaps implicitly) associated with Γ. We have seen the example of a quartic graph with a half-arc-transitive group of automorphisms and its two accompanying balanced oriented graphs. There is another example, namely a Cayley graph with connection set consisting of two noninvolutory generators, for each of which one of the two possible orientations is given. This was used in Malnič and Marušič (2002, Theorems 1.1 and 1.2) for the construction of an infinite family of quartic half-arc-transitive graphs with vertex stabilizer $\mathbb{Z}_2 \times \mathbb{Z}_2$.

Theorem 13.3.5 *Let $n = 2k + 1 \geq 17$, $a = (0, 1, \ldots, 2k) \in A_n$, $t = (0, 2)(4, 7) \in A_n$, and $b = tat^{-1}$. Then $\langle a, b\rangle = A_n$ and*

(i) $\Gamma_n = \mathrm{Cay}\left(A_n, \left\{a^{\pm 1}, b^{\pm 1}\right\}\right)$ *is a quartic half-arc-transitive graph with alternating cycles of length 4, vertex stabilizers isomorphic to $\mathbb{Z}_2$ and with* $\mathrm{Aut}(\Gamma_n) \cong A_n \times \mathbb{Z}_2$,

(ii) $\Delta_n = \mathrm{Al}(\Gamma_n)$ *is a quartic half-arc-transitive graph with vertex stabilizers isomorphic to $\mathbb{Z}_2 \times \mathbb{Z}_2$ and with* $\mathrm{Aut}(\Delta_n) \cong A_n \times \mathbb{Z}_2$.

This idea is also the basis of a construction of a family of quartic half-arc-transitive graphs with arbitrarily large vertex stabilizers, all elementary abelian 2-groups (Marušič, 2005). The first example of a connected quartic half-arc-transitive graph with nonabelian vertex stabilizers was found in Conder and

Marušič (2003). It has order 10,752 and was constructed from a transitive permutation group of degree 32 and order 86,016 with dihedral point stabilizer D_8 of order 8 having a non-self-paired suborbit. More recently, an example of a quartic half-arc-transitive graph with vertex stabilizer $D_8 \times \mathbb{Z}_2$, so neither abelian nor dihedral, was constructed in Conder et al. (2015). These constructions give partial answers to some open problems from Marušič and Nedela (2001a) on the question of which 2-groups can appear as vertex stabilizers of connected quartic half-arc-transitive graphs. It was shown in Marušič and Nedela (2001a) that such a vertex stabilizer is contained in a subclass of 2-groups of nilpotency class at most 2. See also Conder and Marušič (2003); Conder et al. (2015); Malnič and Marušič (2002); Marušič (2005); Marušič and Nedela (2001a,b).

A consequence of the next result (Marušič, 1998, Proposition 2.4.) is that the reachability relation is never universal in quartic half-arc-transitive graphs.

Proposition 13.3.6 *Let Γ be a quartic G-half-arc-transitive graph for some subgroup G of* $\mathrm{Aut}(\Gamma)$. *Then there exists an integer $r \geq 2$ such that all G-alternating cycles of Γ have length $2r$ and form a partition of $E(\Gamma)$. Additionally, one of the following is true:*

(i) *Γ has exactly two G-alternating cycles, both spanning $V(\Gamma)$, and Γ is arc-transitive. This occurs if and only if $\Gamma \cong \mathrm{Cay}(\mathbb{Z}_{2r}, \{\pm 1, \pm s\})$ for some odd $s \in \mathbb{Z}_{2r}^* \setminus \{1, -1\}$ such that $s^2 \pm 1 = 0$, or*

(ii) *Γ has at least three G-alternating cycles, all of which are induced cycles, and if C is a G-alternating cycle of Γ then the transitive constituent $(\mathrm{Stab}_G V(C))^{V(C)}$ has two orbits consisting of every other vertex of C, and is isomorphic to $\mathbb{Z}_2^2$ if $r = 2$, and to D_{2r}, the dihedral group of order $2r$, if $r \geq 3$.*

As mentioned earlier, the classification of all tightly attached quartic half-arc-transitive graphs has been completed (Marušič, 1998; Šparl, 2008). The importance as well as beauty of this result can be appreciated by knowing that following Marušič and Praeger (1999, Theorem 1.1) every quartic half-arc-transitive graph is either tightly attached or a cover of a loosely attached or antipodaly attached graph whose automorphism group contains subgroup that is half-arc-transitive. A quartic graph with a half-arc-transitive action of a group G is said to be G-**antipodally attached** if for any two adjacent G-alternating cycles their intersection is the set of two vertices antipodal on these cycles. Finally, Marušič and Praeger (1999, Theorem 1.1) has been generalized by Ramos Rivera and Šparl (2019) so that the base graph of the cover can be obtained from the cover as a quotient of a normal cyclic group.

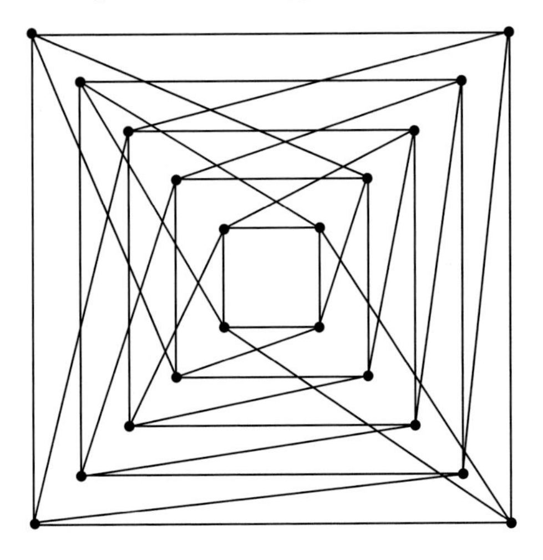

Figure 13.4 The Folkman graph.

13.4 Semisymmetric Graphs

This section is devoted to **semisymmetric graphs**, that is, regular graphs that are edge-transitive but are not vertex-transitive. The study of semisymmetric graphs was initiated in the 1960s by Folkman (1967) who gave a construction of several infinite families of such graphs. Among them is a graph on 20 vertices that was later proved to be the smallest semisymmetric graph and is now known as the Folkman graph (see Figure 13.4). Inspired by Folkman's work and a number of open problems posed in Folkman (1967), the study of semisymmetric graphs has received considerable attention over the years (see, for example, Cara et al. (2014); Conder et al. (2006); Du and Marušič (1999); Du and Wang (2015); Han and Lu (2013); Klin et al. (2012); Lipschutz and Xu (2002); Malnič et al. (2007b); Marušič and Potočnik (2002); Monson et al. (2007); Potočnik and Wilson (2016); Wang and Du (2014); Wang et al. (2014); Wang and Chen (2008); Wilson (2003); Zhou and Feng (2010)). Below, we give a generalization of his constructions from Marušič and Potočnik (2001).

Definition 13.4.1 Let X be a set, and $H \leq S_X$ be transitive. Let O be an orbital digraph of H, and $\tau \in N_{S_X}(H)$ such that there is $k \geq 2$ with $\tau^k \in H$. Let $\mathcal{B} = \{B_x : x \in X\}$ and for each $j \in \mathbb{Z}_k$, let $X_{0,j} = \{x_{0,j} : x \in X\}$. Hence $\mathcal{B}, X_{0,0}, \ldots, X_{0,k-1}$ are $k+1$ sets of size $|X|$. Define the graph $\Delta = \Delta(H, X, O, \tau, k)$ to have vertex set $\mathcal{B} \cup X_{0,0} \cup \cdots \cup X_{0,k-1}$ and edge set $\left\{x_{0,j}B_y : (x, y) \in \tau^j(O)\right\}$. The **generalized Folkman graph** $\mathcal{F}(H, X, O, \tau, k)$ is obtained from $\Delta(H, X, O,$

$\tau, k)$ by replacing each B_x with k vertices $x_{1,0}, x_{1,1}, \ldots, x_{1,k-1}$ each adjacent to the neighbors of B_x. For $j \in \mathbb{Z}_k$ we let $X_{1,j}$ be the set $\{x_{1,j} : x \in X\}$.

By Exercise 1.4.16, τ permutes the set of orbital digraphs of H. So in Δ, the neighbors of B_y are the in-neighbors of y in $\tau^j(O)$. Many of the basic properties of the generalized Folkman graphs are in the next result.

Lemma 13.4.2 *A generalized Folkman graph $\mathcal{F} = \mathcal{F}(H, X, O, \tau, k)$ is regular, bipartite, edge-transitive, and its automorphism group has at most two orbits, namely the sets in its bipartition or their union.*

Proof It is clear that $\mathcal{F}$ is bipartite. Each vertex $B_x \in \mathcal{B}$ has valency kr in Δ, where O is r-regular. So each vertex of $X_{1,j}$ has valency kr in $\mathcal{F}$ as well, $j \in \mathbb{Z}_k$. Each vertex $x_{0,j}$ is adjacent to r vertices in $\mathcal{B}$, and so has valency kr in $\mathcal{F}$. Each element of H induces an automorphism of $\mathcal{F}$ as if $h \in H$, then $h(x, y) = (h(x), h(y)) \in A\left(\tau^j(O)\right)$. In Δ, we have $h(x_{0,j}B_y) = h(x)_{0,j}B_{h(y)}$, and in $\mathcal{F}$ we have $h(x_{0,j}y_{1,\ell}) = h(x)_{0,j}h(y)_{1,\ell}$. In particular, H stabilizes the sets $X_{i,j}$, $i \in \mathbb{Z}_2$, $j \in \mathbb{Z}_k$. Also, τ stabilizes the sets $X_{1,j}$ and cyclically permutes the sets $X_{0,j}$, $j \in \mathbb{Z}_k$. For every $x \in X$ the vertices $x_{1,j}$, $j \in \mathbb{Z}_k$, have the same set of neighbors in $\mathcal{F}$. It follows that for each $x \in X$, there is $H_x \leq \mathrm{Aut}(\mathcal{F})$ such that $H_x^{X_{1,j}} \cong S_{|X|}$ and is the identity on the other vertices of $V(\mathcal{F})$. The group generated by H, τ, and all H_x, $x \in X$, is transitive on $E(\mathcal{F})$ and has the two orbits $\bigcup_{j \in \mathbb{Z}_k} X_{0,j}$ and $\bigcup_{j \in \mathbb{Z}_k} X_{1,j}$ on $V(\Gamma)$. □

Note that for each vertex in $\bigcup_{j \in \mathbb{Z}_k} X_{1,j}$ there are at least $k - 1$ other vertices in $\bigcup_{j \in \mathbb{Z}_k} X_{1,j}$ sharing the same set of neighbors in $\mathcal{F}$. The next result, whose proof is left as Exercise 13.3.5, gives a sufficient condition for a generalized Folkman graph to be semisymmetric.

Proposition 13.4.3 *If no k distinct vertices in $\bigcup_{j \in \mathbb{Z}_k} X_{0,j}$ have the same set of neighbors in the graph $\Delta(H, X, O, \tau, k)$, then the generalized Folkman graph $\mathcal{F}(H, X, O, \tau, k)$ is semisymmetric.*

The case $k = 2$ was analyzed in Marušič and Potočnik (2001) in detail, resulting in several new families of semisymmetric graphs of order a multiple of 4. One of these families is given in Proposition 13.4.4, namely a family of quartic semisymmetric graphs of order $4p$, $p \equiv 1 \pmod 4$ a prime. The smallest case $p = 5$ is the Folkman graph.

Proposition 13.4.4 *Let $p \equiv 1 \pmod 4$ be a prime, let $r \in \mathbb{Z}_p^*$ such that $r^2 \equiv 1 \pmod p$, and let $\Gamma = \mathcal{F}\left(\mathbb{Z}_p, \mathbb{Z}_p, \{(i, i \pm 1) : i \in \mathbb{Z}_p\}, \tau, 2\right)$ where τ maps according to the rule $\tau(i) = ri$. Then Γ is a quartic semisymmetric graph.*

The proof of Proposition 13.4.4 is left to the reader in Exercise 13.3.6. The next result gives a relationship between cubic 1-arc-regular graphs and quartic half-arc-transitive graphs with semisymmetric graphs.

Lemma 13.4.5 *Let Γ be a cubic symmetric graph with $G \leq H \leq \mathrm{Aut}(\Gamma)$ such that Γ is $(G, 1)$-arc-regular and $(H, 2)$-arc-regular. Then there exists a non-self-paired orbital $\mathcal{O}$ of G acting on $V(L(\Gamma))$ and $\tau \in H \backslash G$ such that $\mathcal{F}(G, V(L(\Gamma)), \mathcal{O}, \tau, 2)$ is a well-defined generalized Folkman graph.*

Proof We have already mentioned that $L(\Gamma)$ is a quartic G-half-arc-transitive graph of girth 3 (Marušič and Xu, 1997, Proposition 1.1). Similarly, H is 1-arc-regular on $L(\Gamma)$. Then G is contained in H as a subgroup of index 2. Let $e \in E(L(\Gamma))$, and let G act on the directed arc (u, v), and separately on (v, u) to obtain the digraphs Γ_1 and Γ_2 (we remark that one of these is $D_G(\Gamma)$ as in Section 13.2). So Γ_1 and Γ_2 are orbital digraphs of G acting on $V(L(\Gamma))$. Set $\mathcal{O} = A(\Gamma_1)$. Then $A(\Gamma_2) = \{(j, i) : (i, j) \in \mathcal{O}\}$, so $\mathcal{O}$ is non-self-paired. Let $\tau \in H \backslash G$. Then $\tau(\mathcal{O}) = A(\Gamma_2)$ and $\tau^2 \in G$. We can thus construct a generalized Folkman graph $\mathcal{F}(G, V(L(\Gamma)), \mathcal{O}, \tau, 2)$. □

Let Γ satisfy the hypothesis of Lemma 13.4.5. Using the notation in that lemma, in Marušič and Potočnik (2002, Theorem 2.1), sufficient conditions for $\mathcal{F}(H, V(L(\Gamma)), \mathcal{O}, \tau, 2)$ to be semisymmetric are given. In particular, if H is primitive on $V(L(\Gamma))$, or if $L(\Gamma)$ has neither alternating nor parallel 4-cycles, then $\mathcal{F}(H, V(L(\Gamma)), \mathcal{O}, \tau, 2)$ is semisymmetric (a parallel 4-cycle consists of two directed paths of length 2). As by Proposition 10.2.2 all cubic symmetric graphs of order greater than 8 have girth greater than 4, $L(\Gamma)$ does not have 4-cycles. Hence Marušič and Potočnik (2002, Theorem 2.1) implies that $\mathcal{F}(H, V(L(\Gamma)), \mathcal{O}, \tau, 2)$ is semisymmetric. Recall from Section 10.1 that having 1-arc-regular and 2-arc-regular subgroups of automorphisms means Γ must be of type $\{1, 2^1\}$ or of type $\{1, 2^1, 2^2, 3\}$.

One of the problems posed in Folkman (1967) asks for the construction of semisymmetric graphs of prime valency. Bouwer (1968) pointed out that the Gray graph, first discovered by Gray in 1932, is a cubic semisymmetric graph. More recently, an infinite family of cubic semisymmetric graphs was constructed in Marušič (2000) – see Theorem 13.4.6. The idea behind the construction is fairly simple. We start with a quartic Cayley graph of the alternating group A_n with connection set consisting of two elements of order 3 and their inverses. The triangles in this Cayley graph induced by these two permutations force the graph to be the line graph of a cubic graph – the **anti-line graph** of the original Cayley graph. With a careful choice of the two permutations of order 3 this anti-line graph turns out to be semisymmetric (see Marušič (2000, Theorem 1.1) for the details).

Theorem 13.4.6 *Let $n = 3k \geq 9$ and $a_n, b_n \in S_n$ be*

$$a_n = (0,1,2)(3,4,5)(6,7,8)\cdots(3k-3,3k-2,3k-1)$$

and

$$b_n = (0,1,5)(3,4,8)(6,7,11)\cdots(3k-6,3k-5,3k-1)(2)(3k-3)(3k-2).$$

Then $\langle a_n, b_n\rangle = A_n$ and the anti-line graph Γ_n of $\Delta_n = \mathrm{Cay}\left(A_n, \left\{a_n^{\pm 1}, b_n^{\pm 1}\right\}\right)$, is a cubic semisymmetric graph with edge stabilizer isomorphic to $\mathbb{Z}_2$. Moreover, $\mathrm{Aut}(\Gamma_n) = \mathrm{Aut}(\Delta_n)$ *is isomorphic to $A_n \times \mathbb{Z}_2$ for n even and to S_n for n odd.*

Finally, we mention that Conder, Malnič, Marušič, and Potočnik have obtained a census of the semisymmetric cubic graphs of order up to 768 (Conder et al., 2006).

13.5 Exercises

13.1.1 Show that $B(k,m,n)$ is $2k$-regular.

13.1.2 Show the following maps are automorphisms of $B(k,m,n)$. Do not forget to show that the maps are bijections. Then show $B(k,m,n)$ is vertex-transitive and edge-transitive.

(i) $\theta(i,a_2,\ldots,a_k) = (i,a_3,a_4,\ldots,a_k,a_2)$,

(ii) $\tau(i,a_2,\ldots,a_k) = (i+1,2a_2,\ldots,2a_k)$, and

(iii) $\psi(i,a_2,a_3,\ldots,a_k) = (i,2i-1-(a_2+a_3+\cdots+a_k),a_3,\ldots,a_k)$.

13.1.3 With the notation of Exercise 13.1.2, show that $\langle\theta,\tau\rangle$ is regular, and so $B(k,m,n)$ is isomorphic to a Cayley graph of $\mathbb{Z}_n^{k-1} \rtimes \mathbb{Z}_m$.

13.1.4 Show that the line graph of a 1-arc-regular cubic graph is a quartic half-arc-transitive graph.

13.1.5 Prove that the graph $B(2,6,9)$ in Bouwer's family of graphs is a half-arc-transitive graph. (Hint: Observe that $\mathrm{Aut}(B(2,6,9))$ distinguishes between paths of length 2 in the graph that have end vertices in different blocks of $\mathcal{B}$ and those that have end vertices in the same block. Also, there are two types of 6-cycles in the graph. The first type has one vertex in each of the six blocks, and the second type of 6-cycles are those having vertices in three consecutive blocks. Examine these two types of 6-cycles to show that $\mathrm{Aut}(B(2,6,9))/\mathcal{B}$ is cyclic, not dihedral.)

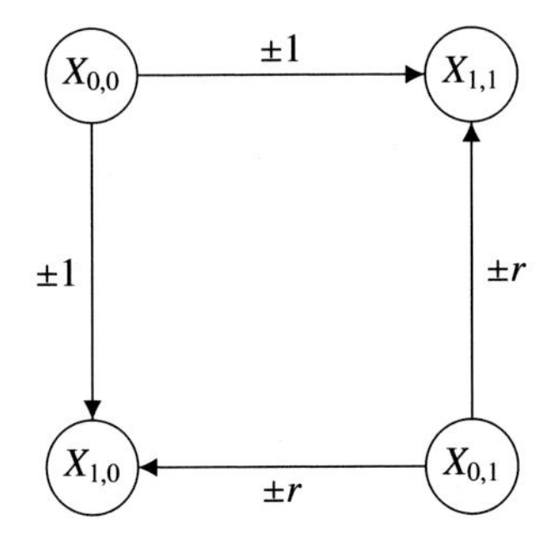

Figure 13.5 A generalized Folkman graph in Frucht notation.

13.3.1 Prove that a vertex stabilizer in a connected quartic half-arc-transitive graph is necessarily a 2-group.

13.3.2 Show that no bicirculant graph is half-arc-transitive.

13.3.3 Let Γ be a G-half-arc-transitive graph of valency 4 and let $v_0 v_1 \cdots v_{2k}$ be a G-alternating path of Γ of even length $2k$. Show that v_0 and v_{2k} are not adjacent in Γ.

13.3.4 The cube F8 has two 1-arc-regular subgroups, and so its line graph has two half-arc-transitive actions. Show that the corresponding attachment numbers are 2 and 3.

13.3.5 Prove Proposition 13.4.3.

13.3.6 Prove Proposition 13.4.4. (Hint: Figure 13.5 gives this graph in Frucht notation.)

13.3.7 Construct the two generalized Folkman graphs from the two orientations of the edge set of the line graph of the cube F8 as given in Exercise 13.3.4.

13.3.8 The Pappus graph F18 is a $\mathbb{Z}_3$-cover of F6, and both are of type $\left\{1, 2^1, 2^2, 3\right\}$. The graph F54 of type $\left\{1, 2^1\right\}$ is a $\mathbb{Z}_3$-cover of F18 where obviously some of the symmetry of F18 has been lost along the covering construction. The Gray graph is defined to be the Levi graph of the $3 \times 3 \times 3$ grid (viewed as a configuration) shown in Figure 13.6 (Bouwer, 1972). Prove that the Gray graph is also a $\mathbb{Z}_3$-cover of F18. It can be shown that the Gray graph is the smallest cubic semisymmetric graph. Interestingly, while vertex transitivity is lost with the covering construction, additional automorphisms are obtained so that the vertex stabilizers of the automorphism group of the Gray graph are of order $2^4 \cdot 3$, while the vertex stabilizers in the automorphism group of F18 are of order $2^2 \cdot 3$.

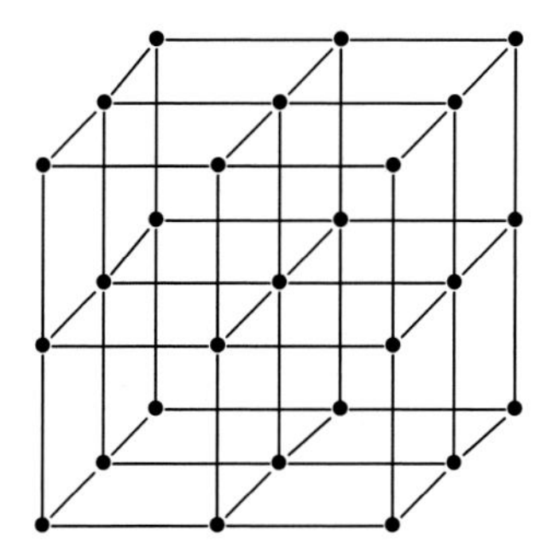

Figure 13.6 The 3x3x3 grid.

14

Fare You Well

Throughout this book a number of important research problems on symmetry in graphs, together with their mostly partial solutions, have been discussed. It seems that complete solutions to many problems are presently beyond us, and in all likelihood these problems will remain open for some time. In no way, however, should we be discouraged from continuing to address these problems. It is often the case that unsolved problems, providing they capture enough attention, stimulate the growth of a particular mathematical discipline. And turning toward the future of algebraic graph theory, we believe that two areas of research deserve special attention.

14.1 Algebraic Graph Theory: Towards CFSG-Free Arguments

The first area concerns the wide applications of CFSG in algebraic graph theory. We would like to argue that a somewhat more conservative outlook is needed. Undoubtedly, CFSG has had huge consequences in algebraic graph theory, as in all other areas where group actions are an essential research and investigating tool. Many problems on symmetry in graphs, previously impossible to deal with, have been solved thanks to CFSG. For example, the complete classification of vertex-transitive graphs of order a product of two primes, discussed in Section 9.2, would not have been possible without CFSG. Similarly, CFSG has been an essential tool in the proof that 2-closures of quasiprimitive groups contain semiregular elements (Giudici, 2003), thus essentially reducing the bulk of the semiregularity problem to solvable groups. In addressing solvable groups, however, CFSG is of no help at all. If one is ever to solve the semiregularity problem, then most likely a more combinatorial approach will have to be taken. In short, even CFSG can only take you so far.

Then there is the journey. If we happen to think that the journey to attain one's goal far exceeds the goal itself, then a CFSG solution to a particular problem might leave us with a nagging feeling of incompleteness, rather than an exalted feeling of joy that usually accompanies a successfully proven theorem. That is, CFSG is such a powerful tool that in certain situations, the results are, in a way, obtained almost effortlessly and for free, thus obscuring to a certain extent the structural content of the problem itself. On the other hand, there are situations where one needs to go through a long list of primitive groups. Due to the many technicalities involved in the process, this makes the arguments prone to minor inaccuracies or even major errors. One way or another, in the end one is left with a solution to the problem, but with not much understanding, not really knowing what goes/is going on behind the scenes.

It is therefore understandable that some people working in algebraic graph theory (and/or permutation groups) believe that CFSG should be used somewhat more cautiously and conservatively and, if possible, one should find a proof that is CFSG-free. For example, as mentioned in Chapter 12, every transitive permutation group contains a derangement of prime-power order by Fein et al. (1981, Theorem 1). Their proof requires CFSG. As they say, "... *It would be desirable to have a direct proof of this result.*" (Fein et al., 1981, p. 41).

Müller (2013) classified all primitive permutation groups that contain a permutation with two cycles in its cycle decomposition (not necessarily of the same length). All of these groups are of rank at most 3. This last statement generalizes both the classical result of Wielandt about primitive groups of degree $2p$, where p is a prime, being of rank at most 3 (Wielandt, 1964, Theorem 31.2), and a more recent generalization of Wielandt's theorem to primitive permutation groups containing a permutation with two cycles in its cycle decomposition being both of same prime-power length (Kovács et al., 2013) (also see Section 9.2). In contrast to Müller's classification, both of these results have been obtained without CFSG. It is then no surprise that Müller suggests "*...would be very interesting to have a direct proof, without appealing to classification of the finite simple groups*" (Müller, 2013, p. 375).

One of the consequences of CFSG is that A_5 and S_5 are the only simply primitive groups of degree $2p$, where p is prime. Thus, the Petersen graph and its complement are the only vertex-transitive graphs of order $2p$ with a simply primitive automorphism group. Hence, by Proposition 9.2.6, every vertex-transitive graph of order $2p$ is isomorphic to a $(2, p)$-metacirculant. We now discuss a possible approach to a CFSG-free proof of the above facts. If successful, this would also give us a CFSG-free refinement to Wielandt's theorem.

Problem 14.1.1 Find a CFSG-free proof of the nonexistence of simply primitive groups of degree $2p$, $p > 5$ a prime.

Let us start with a common sense observation. Such a CFSG-free proof presupposes two things at the very least. First, identifying one or possibly more combinatorial features of the corresponding vertex-transitive graphs so as to get sufficient structural information. Second, finding a clever way to exploit this information in order to come up with a nonexistence proof for primes greater than 5. Before proceeding, we will need an additional idea.

Definition 14.1.2 A **strongly regular graph** Γ with parameters (n, k, λ, μ) is a regular graph of order n and of valency k such that every two adjacent vertices have λ common neighbors and any two nonadjacent vertices have μ common neighbors. A strongly regular graph Γ is said to be **nontrivial** if both Γ and its complement $\bar{\Gamma}$ are connected.

Note that the complement $\bar{\Gamma}$ of a strongly regular graph Γ with parameters (n, k, λ, μ) is a strongly regular graph with parameters $(n, n-k-1, n-2-2k+\mu, n-2k+\lambda)$. The following equality for the parameters (n, k, λ, μ) of a strongly regular graph is well known (see, for example, Godsil and Royle (2001, Section 10.1))

$$k(k-\lambda-1) = \mu(n-1-k). \tag{14.1}$$

Lemma 14.1.3 *Let G be a transitive permutation group of rank 3. If Γ is an orbital digraph of G that is a graph, then Γ is strongly regular.*

Proof As Γ is an orbital digraph of G that is a graph, $\bar{\Gamma}$ is a graph and, as G has rank 3, is also an orbital digraph of G. Hence Γ and $\bar{\Gamma}$ are arc-transitive and so edge-transitive. Then every two adjacent vertices in Γ have the same number of common neighbors, and any two nonadjacent vertices (and so adjacent in $\bar{\Gamma}$, which is also edge-transitive) have the same number of common neighbors. As Γ and $\bar{\Gamma}$ are vertex-transitive, they are both regular and so both strongly regular. □

So, let Γ be an orbital digraph of a simply primitive group G of degree $2p$, p a prime. Wielandt (1964, Theorem 31.2 (B)) showed that $2p = (2m+1)^2+1$ for some $m \geq 1$. By the same result, G is rank 3 with nontrivial suborbits of length $m(2m+1)$ and $(m+1)(2m+1)$. By Exercise 1.4.10 the suborbits are self-paired, and so the orbital digraphs are graphs. By Lemma 14.1.3, Γ is strongly regular of valency $m(2m+1)$ or $(m+1)(2m+1)$.

The next result, proven in Marušič (1988b), characterizes strongly regular bicirculants of order $2p$. The proof given in Marušič (1988b) uses the fast

Fourier transform and convolutions. The result can be proven using the group algebra – see Kutnar et al. (2009).

Theorem 14.1.4 *Let Γ be a nontrivial strongly regular bicirculant graph of order $2p$, p a prime, with parameters $(2p, k, \lambda, \mu)$. Then there are subsets $S \subseteq \mathbb{Z}_p^*$ and $T \subseteq \mathbb{Z}_p$ such that a symbol of Γ is of the form $[S, \bar{S}, T]$, and, moreover, $k = |S| + |T|$, $|S| = (p-1)/2$, $|T| = \left(p \pm \sqrt{2p-1}\right)/2 = \mu$ and $\lambda = \mu - 1$.*

Unfortunately, knowing that the orbital graphs of simply primitive groups of degree $2p$ are strongly regular does not suffice to solve Problem 14.1.1. Namely, by Malnič et al. (2007a) there exist strongly regular vertex-transitive bicirculants of order 82 and 122 that are not edge-transitive, and so not orbital graphs of any group. Therefore additional tools need to be applied. At least three possible routes could be taken.

First we consider the action of a simply primitive group G of degree $2p = (2m+1)^2 + 1$, $p \geq 5$, on the edge set $E(\Gamma)$ of one of the two orbital graphs Γ of G. Then the action is either primitive, imprimitive with blocks of size relatively prime to p, or imprimitive with blocks of size divisible by p. While the first two possibilities are more demanding, the third is rather straightforward to deal with. The idea is to consider the action of the subgroup H of G generated by all Sylow p-subgroups of G. Even though H is not necessarily the whole of G, its action is still primitive on $V(\Gamma)$. An in-depth analysis gives us that the only such block system is a decomposition of the edge set of the complement of the Petersen graph into six 5-cycles.

Another approach to consider is consistent cycles, as in Definition 10.2.1, of the orbital graphs of G. As these two graphs are arc-transitive, Proposition 10.2.3 (also see Conway (1971)) implies that the number of orbits of such cycles are $2m^2 + m - 1$ and $2m^2 + 3m$, respectively. A more detailed analysis of different types of consistent cycles is worth pursuing. Of particular interest are consistent cycles of length p. Interestingly, the numbers of orbits of such cycles are the same for both graphs. This follows from the fact that all such cycles are induced on orbits of Sylow p-subgroups. But Sylow p-subgroups are conjugate, and therefore representatives of these orbits may be chosen from a single Sylow p-subgroup, for example, the one that gives that Γ is bicirculant with symbols $\left[S, \widehat{S}, T\right]$ or $\left[\widehat{S}, S, \bar{T}\right]$. It is then clear that the number of orbits of consistent p-cycles is the same for both graphs.

The action of G on the set of all Sylow p-subgroups by conjugation is of further interest. The stabilizer of this action is the normalizer $N_G(P)$ of a Sylow p-subgroup P (Dummit and Foote, 2004, Proposition 4.3.6). Suppose $P = N_G(P)$. By Burnside's transfer theorem, G has a normal p-complement, and so by Theorem 2.4.7, G is not primitive, a contradiction. Thus P is a proper subgroup of $N_G(P)$. We believe that a next step toward a CFSG-free proof of the nonexistence of simply primitive groups of degree $2p$, $p > 5$, should consist in proving that the index $[N_G(P)\colon P]$ is even. This would immediately imply that the set T in the bicirculant symbol $\left\lfloor S, \widehat{S}, T \right\rfloor$ can be chosen to be a symmetric subset of $\mathbb{Z}_p$, which would considerably simplify a further analysis of the structure of these graphs. As a final comment, we observe that if $G = \text{Aut}(\Gamma)$ contains an odd automorphism then $[N_G(P)\colon P]$ is necessarily even. This brings us to the content of the next section, which deals with the existence of odd automorphisms in vertex-transitive graphs.

14.2 Even/Odd Automorphisms Problem

Finding the full automorphism group is one of the fundamental objectives when dealing with symmetry properties of mathematical objects. Many of these objects naturally display certain inherently obvious symmetries. It is often the case, however, that certain additional symmetries, though hidden or difficult to grasp, are present. One would want to find a reason for their existence and a method for describing them. Along these lines the above question reads as follows.

Problem 14.2.1 Given a group H acting transitively on the set of vertices of a graph, determine whether H is its full automorphism group or not. When the answer is no, find a method to describe the additional automorphisms.

In other words, decide whether the group H embeds into a larger group G preserving the structure of the graph Γ in question. One possibility for addressing such group "extensions" is by studying whether $\text{Aut}(\Gamma)$ has an odd automorphism. If we were to study whether odd automorphisms exist in vertex-transitive graphs then it would be convenient to put the problem in the framework of orbital digraphs of transitive groups.

Problem 14.2.2 Given a transitive group H, possibly consisting of even permutations only, is there an orbital digraph of H with an odd automorphism?

As mentioned in Proposition 10.1.3, this problem has been solved for cubic symmetric graphs with complete information regarding the existence of odd automorphisms in the full automorphism group for each of the 17 types of cubic symmetric graphs.

Problem 14.2.2 can be viewed in a larger context of transitive permutation groups that only have even permutations. Let us call such groups **even**. Similarly, a transitive permutation group is called **odd** if it also contains odd permutations. Is it possible to determine when an even transitive permutation group has an orbital digraph whose automorphism group is odd? There are three possibilities, which we call layers, for the automorphism groups of the orbital digraphs of an even group H with regards to the automorphism groups of its orbital digraphs being even or odd:

Layer 1: The first layer consists of all transitive permutation groups H that are even but have an odd 2-closure $H^{(2)}$. This means that despite the fact that H is even, all orbital digraphs have a common odd automorphism. We refer to such groups as **closure-odd**. An example of this situation is the rank 3 action of A_5 on pairs of a 5-element set, with orbital graph the Petersen graph and its complement and S_5 as the full automorphism group.
Layer 2: The second layer consists of all transitive permutation groups H that are even and have even 2-closures $H^{(2)}$ but have at least one orbital digraph with odd automorphisms. We refer to such groups as **orbital-odd**. An example of such a group is the group $(\mathrm{SL}(2,16), \mathrm{SL}(2,16)/G)$, where $G \leq \mathrm{SL}(2,16)$ is of order 80 with orbital digraphs or graphs of two isomorphism classes: the first one is the union of 17 directed 3-cycles, and the second one is the Fermat graph $X(17, 3, \emptyset, \{0\})$, which is a 3-fold cover of K_{17}.
Layer 3: Finally, the third layer consists of all transitive permutation groups H that are even and have all orbital digraphs or graphs admitting only even automorphisms. We refer to such groups as **zero-odd**. For example, the group $(\mathrm{PSL}(2,17), \mathrm{PSL}(2,17)/S_4)$ acting transitively has even permutations only, and is isomorphic to the full automorphism group of every orbital digraph or graph associated with this action. Hence, $\mathrm{PSL}(2,17) \leq S_{102}$ is zero-odd.

In the above terminology we would like to propose the following general problem.

Problem 14.2.3 Given an even transitive group, determine to which of the Layers 1,2, or 3 it belongs.

References

Ádám, A. 1967. Research problem 2–10. *J. Combin. Theory*, **2**, 393.

Ahrens, W. 1901. *Mathematische Unterhaltungen und Spiele*. Teubner, Leipzig.

Alperin, J. L., and Bell, Rowen B. 1995. *Groups and Representations*. Graduate Texts in Mathematics, vol. 162. Springer-Verlag, New York.

Alspach, Brian. 1973. Point-symmetric graphs and digraphs of prime order and transitive permutation groups of prime degree. *J. Combin. Theory Ser. B*, **15**, 12–17.

Alspach, Brian. 1979. Hamiltonian cycles in vertex-transitive graphs of order $2p$. Pages 131–139 of: *Proceedings of the Tenth Southeastern Conference on Combinatorics, Graph Theory and Computing* (Florida Atlantic Univ., Boca Raton, FL, 1979). Congress. Numer., XXIII–XX. Utilitas Math., Winnipeg, MB.

Alspach, Brian. 1983. The classification of Hamiltonian generalized Petersen graphs. *J. Combin. Theory Ser. B*, **34**(3), 293–312.

Alspach, Brian. 1989. Lifting Hamilton cycles of quotient graphs. *Discrete Math.*, **78**(1–2), 25–36.

Alspach, Brian. 2013. Johnson graphs are Hamilton-connected. *Ars Math. Contemp.*, **6**(1), 21–23.

Alspach, Brian, and Dobson, Edward. 2015. On automorphism groups of graph truncations. *Ars Math. Contemp.*, **8**(1), 215–223.

Alspach, Brian, and Mishna, Marni. 2002. Enumeration of Cayley graphs and digraphs. *Discrete Math.*, **256**(3), 527–539.

Alspach, Brian, and Parsons, T. D. 1979. Isomorphism of circulant graphs and digraphs. *Discrete Math.*, **25**(2), 97–108.

Alspach, Brian, and Parsons, T. D. 1982a. A construction for vertex-transitive graphs. *Canad. J. Math.*, **34**(2), 307–318.

Alspach, Brian, and Parsons, T. D. 1982b. On Hamiltonian cycles in metacirculant graphs. Pages 1–7 of: *Algebraic and Geometric Combinatorics*. North-Holland Math. Stud., vol. 65. North-Holland, Amsterdam.

Alspach, Brian, and Qin, Yusheng. 2001. Hamilton-connected Cayley graphs on Hamiltonian groups. *European J. Combin.*, **22**(6), 777–787.

Alspach, Brian, and Xu, Ming Yao. 1994. 1/2-transitive graphs of order $3p$. *J. Algebraic Combin.*, **3**(4), 347–355.

Alspach, B., and Zhang, Cun Quan. 1989. Hamilton cycles in cubic Cayley graphs on dihedral groups. *Ars Combin.*, **28**, 101–108.

Alspach, Brian, Durnberger, Erich, and Parsons, T. D. 1985. Hamilton cycles in metacirculant graphs with prime cardinality blocks. Pages 27–34 of: *Cycles in Graphs (Burnaby, B.C., 1982)*. North-Holland Math. Stud., vol. 115. North-Holland, Amsterdam.

Alspach, Brian, Locke, Stephen C., and Witte, Dave. 1990. The Hamilton spaces of Cayley graphs on abelian groups. *Discrete Math.*, **82**(2), 113–126.

Alspach, Brian, Marušič, Dragan, and Nowitz, Lewis. 1994. Constructing graphs which are 1/2-transitive. *J. Austral. Math. Soc. Ser. A*, **56**(3), 391–402.

Alspach, Brian, Conder, Marston D. E., Marušič, Dragan, and Xu, Ming-Yao. 1996. A classification of 2-arc-transitive circulants. *J. Algebraic Combin.*, **5**(2), 83–86.

Alspach, Brian, Morris, Joy, and Vilfred, V. 1999. Self-complementary circulant graphs. *Ars Combin.*, **53**, 187–191.

Alspach, Brian, Chen, C. C., and Dean, Matthew. 2010. Hamilton paths in Cayley graphs on generalized dihedral groups. *Ars Math. Contemp.*, **3**(1), 29–47.

André, Jorge, Araújo, João, and Cameron, Peter J. 2016. The classification of partition homogeneous groups with applications to semigroup theory. *J. Algebra*, **452**, 288–310.

Antončič, Iva, Hujdurović, Ademir, and Kutnar, Klavdija. 2015. A classification of pentavalent arc-transitive bicirculants. *J. Algebraic Combin.*, **41**(3), 643–668.

Arezoomand, Majid, and Ghasemi, Mohsen. 2021. On 2-closed elusive permutation groups of degrees p^2q and p^2qr. *Communications in Algebra*, **49**(2), 614–620.

Artin, E. 1988. *Geometric Algebra*. Wiley Classics Library. John Wiley & Sons, Inc., New York. Reprint of the 1957 original, A Wiley-Interscience Publication.

Aschbacher, M. 1984. On the maximal subgroups of the finite classical groups. *Invent. Math.*, **76**(3), 469–514.

Babai, L. 1977. Isomorphism problem for a class of point-symmetric structures. *Acta Math. Acad. Sci. Hungar.*, **29**(3–4), 329–336.

Babai, L. 1978. Infinite digraphs with given regular automorphism groups. *J. Combin. Theory Ser. B*, **25**(1), 26–46.

Babai, László. 1979. Long cycles in vertex-transitive graphs. *J. Graph Theory*, **3**(3), 301–304.

Babai, L. 1980. Finite digraphs with given regular automorphism groups. *Period. Math. Hungar.*, **11**(4), 257–270.

Babai, László. 1995. Automorphism groups, isomorphism, reconstruction. Pages 1447–1540 of: *Handbook of Combinatorics,* vol. 1, 2. Elsevier, Amsterdam.

Babai, László. 2015. Graph Isomorphism in Quasipolynomial Time. *CoRR*, arXiv: abs/1512.03547.

Babai, László. 2016. Graph isomorphism in quasipolynomial time [extended abstract]. Pages 684–697 of: *STOC'16 – Proceedings of the 48th Annual ACM SIGACT Symposium on Theory of Computing*. ACM, New York.

Babai, L., and Frankl, P. 1978. Isomorphisms of Cayley graphs. I. Pages 35–52 of: *Combinatorics (Proc. Fifth Hungarian Colloq., Keszthely, 1976), Vol. I.* Colloq. Math. Soc. János Bolyai, vol. 18. North-Holland, Amsterdam.

Babai, L., and Frankl, P. 1979. Isomorphisms of Cayley graphs. II. *Acta Math. Acad. Sci. Hungar.*, **34**(1–2), 177–183.

Babai, László, and Godsil, Chris D. 1982. On the automorphism groups of almost all Cayley graphs. *European J. Combin.*, **3**(1), 9–15.

Baik, Young-Gheel, Feng, Yanquan, Sim, Hyo-Seob, and Xu, Mingyao. 1998. On the normality of Cayley graphs of abelian groups. *Algebra Colloq.*, **5**(3), 297–304.

Baik, Young-Gheel, Feng, Yanquan, and Sim, Hyo-Seob. 2000. The normality of Cayley graphs of finite abelian groups with valency 5. *Systems Sci. Math. Sci.*, **13**(4), 425–431.

Balaban, A. T. 1972. Chemical graphs, part XIII; Combinatorial patterns. *Rev. Roumaine Math. Pures Appl.*, **17**, 3–16.

Bamberg, John, and Giudici, Michael. 2011. Point regular groups of automorphisms of generalised quadrangles. *J. Combin. Theory Ser. A*, **118**(3), 1114–1128.

Barber, Rachel, and Dobson, Ted. 2022. *Recognizing vertex-transitive digraphs which are wreath products, double coset digraphs, and generalized wreath products.* Preprint.

Bays, S. 1930. Sur les systèmes cycliques de triples de Steiner différents pour N premier (ou puissance de nombre premier) de la forme $6n+1$. *Comment. Math. Helv.*, **2**(1), 294–306.

Bays, S. 1931. Sur les systèmes cycliques de triples de Steiner différents pour N premier de la forme $6n + 1$. *Comment. Math. Helv.*, **3**, 22–41, 122–147, 307–325.

Beaumont, R. A., and Peterson, R. P. 1955. Set-transitive permutation groups. *Canad. J. Math.*, **7**, 35–42.

Bermond, J.-C. 1978a. Hamiltonian decompositions of graphs, directed graphs and hypergraphs. *Ann. Discrete Math.*, **3**, 21–28. In *Advances in Graph Theory* (Cambridge Combinatorial Conf., Trinity College, Cambridge, 1977).

Bermond, J.-C. 1978b. Hamiltonian graphs. Pages 127–167 of: L.W. Beinke, R.J. Wilson (ed), *Selected Topics in Graph Theory*. Academic Press, London.

Bhoumik, Soumya, Dobson, Edward, and Morris, Joy. 2014. On the automorphism groups of almost all circulant graphs and digraphs. *Ars Math. Contemp.*, **7**(2), 487–506.

Biggs, Norman. 1973. Three remarkable graphs. *Canad. J. Math.*, **25**, 397–411.

Biggs, Norman. 1979. Some odd graph theory. Pages 71–81 of: *Second International Conference on Combinatorial Mathematics (New York, 1978)*. Ann. New York Acad. Sci., vol. 319. New York Acad. Sci., New York.

Biggs, N. L. 1981. Aspects of symmetry in graphs. Pages 27–35 of: *Algebraic Methods in Graph Theory,* Vol. I, II (Szeged, 1978). Colloq. Math. Soc. János Bolyai, vol. 25. North-Holland, Amsterdam-New York.

Biggs, N. L., and Smith, D. H. 1971. On trivalent graphs. *Bull. London Math. Soc.*, **3**, 155–158.

Boben, Marko, Pisanski, Tomaž, and Žitnik, Arjana. 2005. *I*-graphs and the corresponding configurations. *J. Combin. Des.*, **13**(6), 406–424.

Bollobás, Béla. 1998. *Modern Graph Theory*. Graduate Texts in Mathematics, vol. 184. Springer-Verlag, New York.

Bondy, J. A. 1972. Variations on the Hamiltonian theme. *Canad. Math. Bull.*, **15**, 57–62.

Bonnington, C. Paul, Conder, Marston, Morton, Margaret, and McKenna, Patricia. 2002. Embedding digraphs on orientable surfaces. *J. Combin. Theory Ser. B*, **85**(1), 1–20.

Boreham, T. G., Bouwer, I. Z., and Frucht, R. W. 1974. A useful family of bicubic graphs. Pages 213–225. Lecture Notes in Math., Vol. 406 of: *Graphs and Combinatorics* (Proc. Capital Conf., George Washington Univ., Washington, D.C., 1973). Berlin: Springer.

Bosma, Wieb, Cannon, John, and Playoust, Catherine. 1997. The Magma algebra system. I. The user language. *J. Symbolic Comput.*, **24**(3–4), 235–265. Computational algebra and number theory (London, 1993).

Bouwer, I. Z. 1968. An edge but not vertex transitive cubic graph. *Canad. Math. Bull.*, **11**, 533–535.

Bouwer, I. Z. 1970. Vertex and edge transitive, but not 1-transitive, graphs. *Canad. Math. Bull.*, **13**, 231–237.

Bouwer, I. Z. 1972. On edge but not vertex transitive regular graphs. *J. Combin. Theory Ser. B*, **12**, 32–40.

Brand, Neal. 1989. On the Bays–Lambossy theorem. *Discrete Math.*, **78**(3), 217–222.

Breckman, J. 1956. *Encoding Circuit*. US Patent 2733432. Issued Jan 31, 1956.

Brouwer, A. E., Cohen, A. M., and Neumaier, A. 1989. *Distance-Regular Graphs*. Ergebnisse der Mathematik und ihrer Grenzgebiete (3) [Results in Mathematics and Related Areas (3)], vol. 18. Springer-Verlag, Berlin.

Burnside, W. 1897. *Theory of Groups of Finite Order*. Cambridge University Press.

Burnside, W. 1901. On some properties of groups of odd order. *J. London Math. Soc.*, **33**, 162–185.

Burton, David M. 2010. *Elementary Number Theory*. Seventh ed. McGraw-Hill, New York.

Cameron, Peter J. 1981. Finite permutation groups and finite simple groups. *Bull. London Math. Soc.*, **13**(1), 1–22.

Cameron, P. J. (ed). 1997. Problems from the fifteenth British combinatorial conference. *Discrete Math.*, **167/168**, 605–615.

Cameron, Peter J., Giudici, Michael, Jones, Gareth A., et al. 2002. Transitive permutation groups without semiregular subgroups. *J. London Math. Soc. (2)*, **66**(2), 325–333.

Cara, Philippe, Rottey, Sara, and Van de Voorde, Geertrui. 2014. A construction for infinite families of semisymmetric graphs revealing their full automorphism group. *J. Algebraic Combin.*, **39**(4), 967–988.

Castagna, Frank, and Prins, Geert. 1972. Every generalized Petersen graph has a Tait coloring. *Pacific J. Math.*, **40**, 53–58.

Chao, Chong-yun. 1965. On groups and graphs. *Trans. Amer. Math. Soc.*, **118**, 488–497.

Chao, Chong-yun. 1971. On the classification of symmetric graphs with a prime number of vertices. *Trans. Amer. Math. Soc.*, **158**, 247–256.

Chen, C. C., and Quimpo, N. F. 1981. On strongly Hamiltonian abelian group graphs. Pages 23–34 of: *Combinatorial Mathematics, VIII* (Geelong, 1980). Lecture Notes in Math., vol. 884. Springer, Berlin.

Chen, C. C., and Quimpo, N. 1983. Hamiltonian Cayley graphs of order *pq*. Pages 1–5 of: *Combinatorial Mathematics, X* (Adelaide, 1982). Lecture Notes in Math., vol. 1036. Springer, Berlin.

Chen, Ya-Chen. 2003. Triangle-free Hamiltonian Kneser graphs. *J. Combin. Theory Ser. B*, **89**(1), 1–16.

Chen, Yu Qing. 1998. On Hamiltonicity of vertex-transitive graphs and digraphs of order p^4. *J. Combin. Theory Ser. B*, **72**(1), 110–121.

Chvátal, V. 1972. On Hamilton's ideals. *J. Combin. Theory Ser. B*, **12**, 163–168.

Conder, Marston. 2006. *Trivalent (Cubic) Symmetric Graphs on up to 2048 Vertices*. www.math.auckland.ac.nz/conder/symmcubic2048list.txt.

Conder, Marston. 2011. *Trivalent Symmetric Graphs on up to 10000 Vertices*. www.math.auckland.ac.nz/ conder/symmcubic10000list.txt.

Conder, Marston, and Dobcsányi, Peter. 2002. Trivalent symmetric graphs on up to 768 vertices. *J. Combin. Math. Combin. Comput.*, **40**, 41–63.

Conder, Marston D. E., and Marušič, Dragan. 2003. A tetravalent half-arc-transitive graph with non-abelian vertex stabilizer. *J. Combin. Theory Ser. B*, **88**(1), 67–76.

Conder, Marston, and Morton, Margaret. 1995. Classification of trivalent symmetric graphs of small order. *Australas. J. Combin.*, **11**, 139–149.

Conder, Marston, and Nedela, Roman. 2007. Symmetric cubic graphs of small girth. *J. Combin. Theory Ser. B*, **97**(5), 757–768.

Conder, Marston, and Nedela, Roman. 2009. A refined classification of symmetric cubic graphs. *J. Algebra*, **322**(3), 722–740.

Conder, Marston D. E., and Žitnik, Arjana. 2016. Half-arc-transitive graphs of arbitrary even valency greater than 2. *European J. Combin.*, **54**, 177–186.

Conder, Marston D., Li, Cai Heng, and Praeger, Cheryl E. 2000. On the Weiss conjecture for finite locally primitive graphs. *Proc. Edinburgh Math. Soc. (2)*, **43**(1), 129–138.

Conder, Marston, Malnič, Aleksander, Marušič, Dragan, and Potočnik, Primož. 2006. A census of semisymmetric cubic graphs on up to 768 vertices. *J. Algebraic Combin.*, **23**(3), 255–294.

Conder, Marston D. E., Potočnik, Primož, and Šparl, Primož. 2015. Some recent discoveries about half-arc-transitive graphs. *Ars Math. Contemp.*, **8**(1), 149–162.

Conder, Marston D. E., Estélyi, István, and Pisanski, Tomaž. 2018. Vertex-transitive Haar graphs that are not Cayley graphs. Pages 61–70 of: *Discrete Geometry and Symmetry*. Springer Proc. Math. Stat., vol. 234. Springer, Cham.

Conder, Marston, Zhou, Jin-Xin, Feng, Yan-Quan, and Zhang, Mi-Mi. 2020a. Edge-transitive bi-Cayley graphs. *J. Combin. Theory Ser. B*, **145**, 264–306.

Conder, Marston D. E., Hujdurović, A., Kutnar, K., and Marušič, D. 2020b. *Symmetric Cubic Graphs Via Rigid Cells*. https://doi.org/10.1007/s10801-020-00946-3.

Conway, J. H. 1971. *More About Symmetrical Graphs*. Talk given at the Second British Combinatorial Conference at Royal Holloway College.

Conway, J. H., Curtis, R. T., Norton, S. P., Parker, R. A., and Wilson, R. A. 1985. *Atlas of Finite Groups*. Oxford University Press, Eynsham. Maximal subgroups and ordinary characters for simple groups, With computational assistance from J. G. Thackray.

Conway, J. H., Sloane, N. J. A., and Wilks, Allan R. 1989. Gray codes for reflection groups. *Graphs Combin.*, **5**(4), 315–325.

Coxeter, H. S. M. 1950. Self-dual configurations and regular graphs. *Bull. Amer. Math. Soc.*, **56**, 413–455.

Coxeter, H. S. M. 1983. My graph. *Proc. London Math. Soc. (3)*, **46**(1), 117–136.

Coxeter, H. S. M. 1986. The generalized Petersen graph $G(24, 5)$. *Comput. Math. Appl. Part B*, **12**(3–4), 579–583. Symmetry: unifying human understanding, II.

Coxeter, H. S. M., and Moser, W. O. J. 1965. *Generators and Relations for Discrete Groups*. Second ed. Ergebnisse der Mathematik und ihrer Grenzgebiete, Neue Folge, Band 14. Springer-Verlag, Berlin.

Cuaresma, Maria Cristeta, Giudici, Michael, and Praeger, Cheryl E. 2008. Homogeneous factorisations of Johnson graphs. *Des. Codes Cryptogr.*, **46**(3), 303–327.

Curran, Stephen J., and Gallian, Joseph A. 1996. Hamiltonian cycles and paths in Cayley graphs and digraphs – a survey. *Discrete Math.*, **156**(1–3), 1–18.

Curran, Stephen J., Morris, Dave Witte, and Morris, Joy. 2012. Cayley graphs of order $16p$ are Hamiltonian. *Ars Math. Contemp.*, **5**(2), 185–211.

Dieudonné, Jean. 1980. *On the Automorphisms of the Classical Groups*. Memoirs of the American Mathematical Society, vol. 2. American Mathematical Society, Providence, RI. With a supplement by Loo Keng Hua [Luo Geng Hua], Reprint of the 1951 original.

Dixon, John D., and Mortimer, Brian. 1996. *Permutation Groups*. Graduate Texts in Mathematics, vol. 163. Springer-Verlag, New York.

Djoković, Dragomir Ž., and Miller, Gary L. 1980. Regular groups of automorphisms of cubic graphs. *J. Combin. Theory Ser. B*, **29**(2), 195–230.

Dobson, Edward. 1998. Isomorphism problem for metacirculant graphs of order a product of distinct primes. *Canad. J. Math.*, **50**(6), 1176–1188.

Dobson, Edward. 2000. Classification of vertex-transitive graphs of order a prime cubed. I. *Discrete Math.*, **224**(1–3), 99–106.

Dobson, Edward. 2002. On the Cayley isomorphism problem. *Discrete Math.*, **247**(1–3), 107–116.

Dobson, Edward. 2003a. On isomorphisms of abelian Cayley objects of certain orders. *Discrete Math.*, **266**(1–3), 203–215. The 18th British Combinatorial Conference (Brighton, 2001).

Dobson, Edward. 2003b. On the Cayley isomorphism problem for ternary relational structures. *J. Combin. Theory Ser. A*, **101**(2), 225–248.

Dobson, Edward. 2005. On groups of odd prime-power degree that contain a full cycle. *Discrete Math.*, **299**(1–3), 65–78.

Dobson, Edward. 2006a. Automorphism groups of metacirculant graphs of order a product of two distinct primes. *Combin. Probab. Comput.*, **15**(1–2), 105–130.

Dobson, Edward. 2006b. On the proof of a theorem of Pálfy. *Electron. J. Combin.*, **13**(1), Note 16, 4 pp. (electronic).

Dobson, Edward. 2008. On solvable groups and Cayley graphs. *J. Combin. Theory Ser. B*, **98**(6), 1193–1214.

Dobson, Edward. 2009. On overgroups of regular abelian p-groups. *Ars Math. Contemp.*, **2**(1), 59–76.

Dobson, Edward. 2010. The isomorphism problem for Cayley ternary relational structures for some abelian groups of order $8p$. *Discrete Math.*, **310**(21), 2895–2909.

Dobson, Edward. 2012. The full automorphism group of Cayley graphs of $\mathbb{Z}_p \times \mathbb{Z}_{p^2}$. *Electron. J. Combin.*, **19**(1), Paper 57, 17.

Dobson, Edward. 2014. On the Cayley isomorphism problem for Cayley objects of nilpotent groups of some orders. *Electron. J. Combin.*, **21**(3), Paper 3.8, 15.

Dobson, Ted. 2016. On isomorphisms of Marušič-Scapellato graphs. *Graphs Combin.*, **32**(3), 913–921.

Dobson, Ted. 2018. On the isomorphism problem for Cayley graphs of abelian groups whose Sylow subgroups are elementary abelian cyclic. *Electron. J. Combin.*, **25**(2), Paper No. 2.49, 22.

Dobson, Edward, and Kovács, István. 2009. Automorphism groups of Cayley digraphs of $\mathbb{Z}_p^3$. *Electron. J. Combin.*, **16**(1), Research Paper 149, 20.

Dobson, Edward, and Malnič, Aleksander. 2015. Groups that are transitive on all partitions of a given shape. *J. Algebraic Combin.*, **42**(2), 605–617.

Dobson, Edward, and Marušič, Dragan. 2011. On semiregular elements of solvable groups. *Comm. Algebra*, **39**(4), 1413–1426.

Dobson, Edward, and Morris, Joy. 2005. On automorphism groups of circulant digraphs of square-free order. *Discrete Math.*, **299**(1–3), 79–98.

Dobson, Edward, and Morris, Joy. 2009. Automorphism groups of wreath product digraphs. *Electron. J. Combin.*, **16**(1), Research Paper 17, 30.

Dobson, Edward, and Morris, Joy. 2015. Quotients of CI-groups are CI-groups. *Graphs Combin.*, **31**(3), 547–550.

Dobson, Edward, and Šajna, Mateja. 2004. Almost self-complementary circulant graphs. *Discrete Math.*, **278**(1–3), 23–44.

Dobson, Edward, and Spiga, Pablo. 2013. CI-groups with respect to ternary relational structures: new examples. *Ars Math. Contemp.*, **6**(2), 351–364.

Dobson, Ted, and Spiga, Pablo. 2017. Cayley numbers with arbitrarily many distinct prime factors. *J. Combin. Theory Ser. B*, **122**, 301–310.

Dobson, Edward, and Witte, Dave. 2002. Transitive permutation groups of prime-squared degree. *J. Algebraic Combin.*, **16**(1), 43–69.

Dobson, Edward, Gavlas, Heather, Morris, Joy, and Witte, Dave. 1998. Automorphism groups with cyclic commutator subgroup and Hamilton cycles. *Discrete Math.*, **189**(1–3), 69–78.

Dobson, Edward, Malnič, Aleksander, Marušič, Dragan, and Nowitz, Lewis A. 2007. Minimal normal subgroups of transitive permutation groups of square-free degree. *Discrete Math.*, **307**(3–5), 373–385.

Dobson, Edward, Li, Cai Heng, and Spiga, Pablo. 2012. Permutation groups containing a regular abelian Hall subgroup. *Comm. Algebra*, **40**(9), 3532–3539.

Dobson, Edward, Spiga, Pablo, and Verret, Gabriel. 2016. Cayley graphs on abelian groups. *Combinatorica*, **36**(4), 371–393.

Dobson, Ted, Hujdurović, Ademir, Kutnar, Klavdija, and Morris, Joy. 2020a. *Classification of Vertex-Transitive Digraphs Via Automorphism Group*. arXiv:2003.07894[math.CO].

Dobson, Ted, Muzychuk, Mikhail, and Spiga, Pablo. 2020b. *Generalised Dihedral CI-Groups*. arXiv:2008.00200[math.CO].

Doyle, John Kevin, Tucker, Thomas W., and Watkins, Mark E. 2018. Graphical Frobenius representations. *J. Algebraic Combin.*, **48**(3), 405–428.

Doyle, P. G. 1976. *On Transitive Graphs*. Senior Thesis, Harvard College.

Du, Shaofei, and Marušič, Dragan. 1999. An infinite family of biprimitive semisymmetric graphs. *J. Graph Theory*, **32**(3), 217–228.

Du, Shaofei, and Wang, Li. 2015. A classification of semisymmetric graphs of order $2p^3$: unfaithful case. *J. Algebraic Combin.*, **41**(2), 275–302.

Du, Shaofei, and Xu, Mingyao. 2000. A classification of semisymmetric graphs of order $2pq$. *Comm. Algebra*, **28**(6), 2685–2715.

Du, Shaofei, Malnič, Aleksander, and Marušič, Dragan. 2008. Classification of 2-arc-transitive dihedrants. *J. Combin. Theory Ser. B*, **98**(6), 1349–1372.

Du, Shaofei, Kutnar, Klavdija, and Marušič, Dragan. 2018. Hamilton cycles in vertex-transitive graphs of order a product of two primes, arXiv:1808.08553.

Du, Shaofei, Kutnar, Klavdija, and Marušič, Dragan. 2020. Hamilton cycles in primitive vertex-transitive graphs of order a product of two primes – the case $\mathrm{PSL}(2, q^2)$ acting on cosets of $\mathrm{PGL}(2, q)$. *Ars Math. Contemp.*, **19**(1), 1–15.

Dummit, D.S., and Foote, R.M. 1999. *Abstract Algebra*. Prentice Hall, Upper Saddle River, NJ.

Dummit, David S., and Foote, Richard M. 2004. *Abstract Algebra*. Third ed. John Wiley & Sons, Inc., Hoboken, NJ.

Durnberger, Erich. 1983. Connected Cayley graphs of semidirect products of cyclic groups of prime order by abelian groups are Hamiltonian. *Discrete Math.*, **46**(1), 55–68.

Eiben, Eduard, Jajcay, Robert, and Šparl, Primož. 2019. Symmetry properties of generalized graph truncations. *J. Combin. Theory Ser. B*, **137**, 291–315.

Elspas, Bernard, and Turner, James. 1970. Graphs with circulant adjacency matrices. *J. Combin. Theory*, **9**, 297–307.

Erdős, P., and Rényi, A. 1963. Asymmetric graphs. *Acta Math. Acad. Sci. Hungar.*, **14**, 295–315.

Erdős, P., Ko, Chao, and Rado, R. 1961. Intersection theorems for systems of finite sets. *Quart. J. Math. Oxford Ser. (2)*, **12**, 313–320.

Estélyi, István, and Pisanski, Tomaž. 2016. Which Haar graphs are Cayley graphs? *Electron. J. Combin.*, **23**(3), Paper 3.10, 13.

Euler, Leonard. 1766. *Solution d'une Question Curieuse Qui ne Paroit Soumise a Aucune Analyse*. Euler Archive – All Works. https://scholarlycommons.pacific.edu/euler-works/309.

Evdokimov, S. A., and Ponomarenko, I. N. 2002. Characterization of cyclotomic schemes and normal Schur rings over a cyclic group. *Algebra i Analiz*, **14**(2), 11–55.

Fawcett, Joanna B., Giudici, Michael, Li, Cai Heng, Praeger, Cheryl E., Royle, Gordon, and Verret, Gabriel. 2018. Primitive permutation groups with a suborbit of length 5 and vertex-primitive graphs of valency 5. *J. Combin. Theory Ser. A*, **157**, 247–266.

Fein, Burton, Kantor, William M., and Schacher, Murray. 1981. Relative Brauer groups. II. *J. Reine Angew. Math.*, **328**, 39–57.

Feng, Y.-Q., and Kovács, I. 2018. Elementary abelian groups of rank 5 are DCI-groups. *J. Combin. Theory Ser. A*, **157**(5), 162–204.

Feng, Yan-Quan, and Kwak, Jin Ho. 2007. Cubic symmetric graphs of order a small number times a prime or a prime square. *J. Combin. Theory Ser. B*, **97**(4), 627–646.

Feng, Yan-Quan, and Nedela, Roman. 2006. Symmetric cubic graphs of girth at most 7. *Acta Univ. M. Belii Ser. Math.*, 33–55.

Feng, Yan-Quan, and Wang, Kaishun. 2003. s-regular cyclic coverings of the three-dimensional hypercube Q_3. *European J. Combin.*, **24**(6), 719–731.

Feng, Yan-Quan, Wang, Kaishun, and Zhou, Chuixiang. 2007. Tetravalent half-transitive graphs of order $4p$. *European J. Combin.*, **28**(3), 726–733.

Feng, Yan-Quan, Hujdurović, Ademir, Kovács, István, Kutnar, Klavdija, and Marušič, Dragan. 2019. Quasi-semiregular automorphisms of cubic and tetravalent arc-transitive graphs. *Appl. Math. Comput.*, **353**, 329–337.

Feng, Yan-Quan, Kovács, István, Wang, Jie, and Yang, Da-Wei. 2020. Existence of non-Cayley Haar graphs. *European J. Combin.*, **89**, 103146, 12.

Folkman, Jon. 1967. Regular line-symmetric graphs. *J. Combin. Theory*, **3**, 215–232.

Foster, Ronald M. 1988. *The Foster Census*. Charles Babbage Research Centre, Winnipeg, MB. R. M. Foster's census of connected symmetric trivalent graphs, With a foreword by H. S. M. Coxeter, With a biographical preface by Seymour Schuster, With an introduction by I. Z. Bouwer, W. W. Chernoff, B. Monson and Z. Star, Edited and with a note by Bouwer.

Fronček, Dalibor, Rosa, Alexander, and Širáň, Jozef. 1996. The existence of selfcomplementary circulant graphs. *European J. Combin.*, **17**(7), 625–628.

Frucht, Roberto. 1936/37. Die gruppe des Petersen'schen Graphen und der Kantensysteme der regularen Polyeder. *Comment. Math. Helv.*, **9**, 217–223.

Frucht, Robert. 1952. A one-regular graph of degree three. *Canadian J. Math.*, **4**, 240–247.

Frucht, Roberto. 1970. How to describe a graph. *Ann. New York Acad. Sci.*, **175**, 159–167.

Frucht, Roberto, Graver, Jack E., and Watkins, Mark E. 1971. The groups of the generalized Petersen graphs. *Proc. Cambridge Philos. Soc.*, **70**, 211–218.

Gallai, T. 1968. On directed paths and circuits. Pages 115–118 of: *Theory of Graphs* (Proc. Colloq., Tihany, 1966). Academic Press, New York.

GAP. 2019. *GAP – Groups, Algorithms, and Programming, Version 4.10.2*. The GAP Group. Website. www.gap-system.org

Gardiner, A. 1973. Arc transitivity in graphs. *Quart. J. Math. Oxford Ser. (2)*, **24**, 399–407.

Gardiner, A. 1974. Arc transitivity in graphs. II. *Quart. J. Math. Oxford Ser. (2)*, **25**, 163–167.

Gardiner, A. 1976. Arc transitivity in graphs. III. *Quart. J. Math. Oxford Ser. (2)*, **27**(107), 313–323.

Ghaderpour, Ebrahim, and Morris, Dave Witte. 2011. Cayley graphs of order $27p$ are Hamiltonian. *Int. J. Comb.*, Art. ID 206930, 16.

Ghaderpour, Ebrahim, and Morris, Dave Witte. 2012. Cayley graphs of order $30p$ are Hamiltonian. *Discrete Math.*, **312**(24), 3614–3625.

Ghaderpour, Ebrahim, and Morris, Dave Witte. 2014. Cayley graphs on nilpotent groups with cyclic commutator subgroup are Hamiltonian. *Ars Math. Contemp.*, **7**(1), 55–72.

Giudici, Michael. 2003. Quasiprimitive groups with no fixed point free elements of prime order. *J. London Math. Soc. (2)*, **67**(1), 73–84.

Giudici, Michael. 2007. New constructions of groups without semiregular subgroups. *Comm. Algebra*, **35**(9), 2719–2730.

Giudici, Michael, and Smith, Murray R. 2010. A note on quotients of strongly regular graphs. *Ars Math. Contemp.*, **3**(2), 147–150.

Giudici, Michael, and Verret, Gabriel. 2020. Arc-transitive graphs of valency twice a prime admit a semiregular automorphism. *Ars Math. Contemp.*, **18**(1), 179–186.

Giudici, Michael, and Xu, Jing. 2007. All vertex-transitive locally-quasiprimitive graphs have a semiregular automorphism. *J. Algebraic Combin.*, **25**(2), 217–232.

Giudici, Michael, Li, Cai Heng, Potočnik, Primož, and Praeger, Cheryl E. 2007. Homogeneous factorisations of complete multipartite graphs. *Discrete Math.*, **307**(3–5), 415–431.

Giudici, Michael, Li, Cai Heng, Potočnik, Primož, and Praeger, Cheryl E. 2008. Homogeneous factorisations of graph products. *Discrete Math.*, **308**(16), 3652–3667.

Glover, Henry, and Marušič, Dragan. 2007. Hamiltonicity of cubic Cayley graphs. *J. Eur. Math. Soc. (JEMS)*, **9**(4), 775–787.

Glover, H. H., and Yang, T. Y. 1996. A Hamilton cycle in the Cayley graph of the $\langle 2, p, 3\rangle$ presentation of $\mathrm{PSL}_2(p)$. *Discrete Math.*, **160**(1-3), 149–163.

Glover, Henry H., Kutnar, Klavdija, and Marušič, Dragan. 2009. Hamiltonian cycles in cubic Cayley graphs: the $\langle 2, 4k, 3\rangle$ case. *J. Algebraic Combin.*, **30**(4), 447–475.

Glover, Henry H., Kutnar, Klavdija, Malnič, Aleksander, and Marušič, Dragan. 2012. Hamilton cycles in $(2, \mathrm{odd}, 3)$-Cayley graphs. *Proc. Lond. Math. Soc. (3)*, **104**(6), 1171–1197.

Godsil, C. D. 1980. More odd graph theory. *Discrete Math.*, **32**(2), 205–207.

Godsil, C. D. 1981a. GRRs for nonsolvable groups. Pages 221–239 of: *Algebraic Methods in Graph Theory, Vol. I, II (Szeged, 1978)*. Colloq. Math. Soc. János Bolyai, vol. 25. North-Holland, Amsterdam.

Godsil, C. D. 1981b. On the full automorphism group of a graph. *Combinatorica*, **1**(3), 243–256.

Godsil, C. D. 1983. On Cayley graph isomorphisms. *Ars Combin.*, **15**, 231–246.

Godsil, Chris, and Royle, Gordon. 2001. *Algebraic Graph Theory*. Graduate Texts in Mathematics, vol. 207. Springer-Verlag, New York.

Goldschmidt, David M. 1980. Automorphisms of trivalent graphs. *Ann. of Math. (2)*, **111**(2), 377–406.

Gorenstein, Daniel. 1968. *Finite Groups*. Harper & Row Publishers, New York.

Graham, R. L., Grötschel, M., and Lovász, L. (eds). 1995. *Handbook of Combinatorics. Vol. 1, 2*. Elsevier Science B.V., Amsterdam; MIT Press, Cambridge, MA.

Gray, F. 1953. *Pulse Code Communication*. US Patent 2632058. Issued March 17, 1953.

Gross, Fletcher. 1987. Conjugacy of odd order Hall subgroups. *Bull. London Math. Soc.*, **19**(4), 311–319.

Gross, Jonathan L., and Tucker, Thomas W. 1987. *Topological Graph Theory*. Wiley-Interscience Series in Discrete Mathematics and Optimization. John Wiley & Sons, Inc., New York. A Wiley-Interscience Publication.

Grünbaum, Branko. 2009. *Configurations of Points and Lines*. Graduate Studies in Mathematics, vol. 103. American Mathematical Society, Providence, RI.

Guralnick, Robert M. 1983. Subgroups of prime power index in a simple group. *J. Algebra*, **81**(2), 304–311.

Guy, Richard Kenneth (ed). 1970. Combinatorial structures and their applications. *Proceedings of the Calgary International Conference on Combinatorial Structures and Their Applications Held at the University of Calgary,* Calgary, Alberta, Canada, June, vol. 1969. Gordon and Breach Science Publishers, New York.

Han, Hua, and Lu, Zai Ping. 2013. Affine primitive groups and semisymmetric graphs. *Electron. J. Combin.*, **20**(2), P39.

Harary, Frank. 1959. On the group of the composition of two graphs. *Duke Math. J.*, **26**, 29–34.

Hassani, Akbar, Iranmanesh, Mohammad A., and Praeger, Cheryl E. 1998. On vertex-imprimitive graphs of order a product of three distinct odd primes. *J. Combin. Math. Combin. Comput.*, **28**, 187–213. Papers in honour of Anne Penfold Street.

Higman, D. G. 1967. Intersection matrices for finite permutation groups. *J. Algebra*, **6**, 22–42.

Hladnik, Milan, Marušič, Dragan, and Pisanski, Tomaž. 2002. Cyclic Haar graphs. *Discrete Math.*, **244**(1–3), 137–152.

Holt, D. F. 1981. A graph which is edge transitive but not arc transitive. *J. Graph Theory*, **5**(2), 201–204.

Holt, Derek, and Royle, Gordon. 2020. A census of small transitive groups and vertex-transitive graphs. *J. Symbolic Comput.*, **101**, 51–60.

Holton, D. 1982. Research problem 9. *Discrete Math.*, **38**(1), 125.

Holton, D. A., and Sheehan, J. 1993. *The Petersen Graph*. Australian Mathematical Society Lecture Series, vol. 7. Cambridge University Press, Cambridge.

Howard, Ben, Millson, John, Snowden, Andrew, and Vakil, Ravi. 2008. A description of the outer automorphism of S_6, and the invariants of six points in projective space. *J. Combin. Theory Ser. A*, **115**(7), 1296–1303.

Hua, Xiao-Hui, Feng, Yan-Quan, and Lee, Jaeun. 2011. Pentavalent symmetric graphs of order $2pq$. *Discrete Math.*, **311**(20), 2259–2267.

Hujdurović, Ademir, Kutnar, Klavdija, and Marušič, Dragan. 2013. On prime-valent symmetric bicirculants and Cayley snarks. Pages 196–203 of: *Geometric Science of Information*. Lecture Notes in Comput. Sci., vol. 8085. Springer, Heidelberg.

Hujdurović, Ademir, Kutnar, Klavdija, and Marušič, Dragan. 2014. Half-arc-transitive group actions with a small number of alternets. *J. Combin. Theory Ser. A*, **124**, 114–129.

Hujdurović, Ademir, Kutnar, Klavdija, and Marušič, Dragan. 2015. Vertex-transitive generalized Cayley graphs which are not Cayley graphs. *European J. Combin.*, **46**, 45–50.

Hujdurović, Ademir, Kutnar, Klavdija, and Marušič, Dragan. 2016. Odd automorphisms in vertex-transitive graphs. *Ars Math. Contemp.*, **10**(2), 427–437.

Imrich, Wilfried, Klavžar, Sandi, and Rall, Douglas F. 2008. *Topics in Graph Theory*. Wellesley, MA: A K Peters Ltd. Graphs and their Cartesian product.

Isaacs, I. Martin. 2008. *Finite Group Theory*. Graduate Studies in Mathematics, vol. 92. American Mathematical Society, Providence, RI.

Ivanov, A. A., and Iofinova, M. E. 1985. Biprimitive cubic graphs. Pages 123–134 of: *Investigations in the Algebraic Theory of Combinatorial Objects (Russian)*. Vsesoyuz. Nauchno-Issled. Inst. Sistem. Issled., Moscow.

Jackson, Bill. 1980. Hamilton cycles in regular 2-connected graphs. *J. Combin. Theory Ser. B*, **29**(1), 27–46.

Jaeger, F. 1974. On vertex-induced forests in cubic graphs. Pages 501–512. Congressus Numerantium, No. X of: *Proceedings of the Fifth Southeastern Conference on Combinatorics, Graph Theory and Computing* (Florida Atlantic Univ., Boca Raton, FL, 1974).

Jajcay, Robert, and Li, Cai Heng. 2001. Constructions of self-complementary circulants with no multiplicative isomorphisms. *European J. Combin.*, **22**(8), 1093–1100.

Jajcay, Robert, Miklavič, Štefko, Šparl, Primož, and Vasiljević, Gorazd. 2019. On certain edge-transitive bicirculants. *Electron. J. Combin.*, **26**(2), P2.6.

Janusz, Gerald, and Rotman, Joseph. 1982. Outer automorphisms of S_6. *Amer. Math. Monthly*, **89**(6), 407–410.

Jones, Gareth A. 1979. Abelian subgroups of simply primitive groups of degree p^3, where p is prime. *Quart. J. Math. Oxford Ser. (2)*, **30**(117), 53–76.

Jones, Gareth A. 2002. Cyclic regular subgroups of primitive permutation groups. *J. Group Theory*, **5**(4), 403–407.

Jones, Gareth A., and Jajcay, Robert. 2016. Cayley properties of merged Johnson graphs. *J. Algebraic Combin.*, **44**(4), 1047–1067.

Jones, Gareth A., and Singerman, David. 1978. Theory of maps on orientable surfaces. *Proc. London Math. Soc. (3)*, **37**(2), 273–307.

Jordan, C. 1872. Recherches sur les substitutions. *J. Math. Pures Appl.*, **17**, 351–367.

Joseph, Anne. 1995. The isomorphism problem for Cayley digraphs on groups of prime-squared order. *Discrete Math.*, **141**(1–3), 173–183.

Kaloujnine, Léo. 1948. La structure des p-groupes de Sylow des groupes symétriques finis. *Ann. Sci. École Norm. Sup. (3)*, **65**, 239–276.

Kalužnin, L. A., and Klin, M. H. 1976. Some numerical invariants of permutation groups. *Latviĭsk. Mat. Ežegodnik*, 81–99, 222.

Kantor, William M. 1985. Classification of 2-transitive symmetric designs. *Graphs Combin.*, **1**(2), 165–166.

Katona, G. O. H. 1972. A simple proof of the Erdős–Chao Ko–Rado theorem. *J. Combin. Theory Ser. B*, **13**, 183–184.

Keating, Kevin, and Witte, David. 1985. On Hamilton cycles in Cayley graphs in groups with cyclic commutator subgroup. Pages 89–102 of: *Cycles in Graphs* (Burnaby, B.C., 1982). North-Holland Math. Stud., vol. 115. North-Holland, Amsterdam.

Kleidman, Peter, and Liebeck, Martin. 1990. *The Subgroup Structure of the Finite Classical Groups*. London Mathematical Society Lecture Note Series, vol. 129. Cambridge University Press, Cambridge.

Klin, M. H., and Pöschel, R. 1981. The König problem, the isomorphism problem for cyclic graphs and the method of Schur rings. Pages 405–434 of: *Algebraic methods in graph theory, Vol. I, II* (Szeged, 1978). Colloq. Math. Soc. János Bolyai, vol. 25. North-Holland, Amsterdam.

Klin, Mikhail, Lauri, Josef, and Ziv-Av, Matan. 2012. Links between two semisymmetric graphs on 112 vertices via association schemes. *J. Symbolic Comput.*, **47**(10), 1175–1191.

Kneser, M. 1955. Aufgabe 300. *Jber. Deutsch. Math.-Verein*, **58**, 27.

Koike, Hiroki, and Kovács, István. 2014. Isomorphic tetravalent cyclic Haar graphs. *Ars Math. Contemp.*, **7**(1), 215–235.

Koike, Hiroki, Kovács, István, Marušič, Dragan, and Muzychuk, Mikhail. 2019. Cyclic groups are CI-groups for balanced configurations. *Des. Codes Cryptogr.*, **87**(6), 1227–1235.

Kotlov, Andrew, and Lovász, László. 1996. The rank and size of graphs. *J. Graph Theory*, **23**(2), 185–189.

Kovács, István. 2004. Classifying arc-transitive circulants. *J. Algebraic Combin.*, **20**(3), 353–358.

Kovács, István, and Ryabov, Grigory. 2022. The group $C_p^4 \times C_q$ is a DCI-group. *Discrete Math.*, **345**(3), Paper No. 112705, 15.

Kovács, István, and Servatius, Mary. 2012. On Cayley digraphs on nonisomorphic 2-groups. *J. Graph Theory*, **70**(4), 435–448.

Kovács, István, Kutnar, Klavdija, and Marušič, Dragan. 2010. Classification of edge-transitive rose window graphs. *J. Graph Theory*, **65**(3), 216–231.

Kovács, István, Marušič, Dragan, and Muzychuk, Mikhail E. 2011. On dihedrants admitting arc-regular group actions. *J. Algebraic Combin.*, **33**(3), 409–426.

Kovács, István, Marušič, Dragan, and Muzychuk, Mikhail. 2013. On G-arc-regular dihedrants and regular dihedral maps. *J. Algebraic Combin.*, **38**(2), 437–455.

Kowalewski, A. 1917. W.R. Hamilton's Dodeaederaufgabe als Buntordnungsproblem. *Sitzungsber. Akad. Wiss. Wien (Abt. IIa)*, **126**, 67–90, 963–1007.

Krasner, Marc, and Kaloujnine, Léo. 1951. Produit complet des groupes de permutations et problème de groupes. II. *Acta Sci. Math. (Szeged)*, **14**, 39–66.

Kriloff, Cathy, and Lay, Terry. 2014. Hamiltonian cycles in Cayley graphs of imprimitive complex reflection groups. *Discrete Math.*, **326**, 50–60.

Krivelevich, Michael, and Sudakov, Benny. 2003. Sparse pseudo-random graphs are Hamiltonian. *J. Graph Theory*, **42**(1), 17–33.

Kutnar, Klavdija, and Marušič, Dragan. 2008a. Hamiltonicity of vertex-transitive graphs of order $4p$. *European J. Combin.*, **29**(2), 423–438.

Kutnar, Klavdija, and Marušič, Dragan. 2008b. Recent trends and future directions in vertex-transitive graphs. *Ars Math. Contemp.*, **1**(2), 112–125.

Kutnar, Klavdija, and Marušič, Dragan. 2009a. A complete classification of cubic symmetric graphs of girth 6. *J. Combin. Theory Ser. B*, **99**(1), 162–184.

Kutnar, Klavdija, and Marušič, Dragan. 2009b. Hamilton cycles and paths in vertex-transitive graphs – current directions. *Discrete Math.*, **309**(17), 5491–5500.

Kutnar, Klavdija, and Marušič, Dragan. 2019. Odd extensions of transitive groups via symmetric graphs – the cubic case. *J. Combin. Theory Ser. B*, **136**, 170–192.

Kutnar, Klavdija, and Šparl, Primož. 2009. Hamilton paths and cycles in vertex-transitive graphs of order $6p$. *Discrete Math.*, **309**(17), 5444–5460.

Kutnar, Klavdija, and Šparl, Primož. 2010. Distance-transitive graphs admit semiregular automorphisms. *European J. Combin.*, **31**(1), 25–28.

Kutnar, Klavdija, Marušič, Dragan, Miklavič, Štefko, and Šparl, Primož. 2009. Strongly regular tri-Cayley graphs. *European J. Combin.*, **30**(4), 822–832.

Kutnar, Klavdija, Marušič, Dragan, and Šparl, Primož. 2010. An infinite family of half-arc-transitive graphs with universal reachability relation. *European J. Combin.*, **31**(7), 1725–1734.

Kutnar, K., Marušič, D., Morris, D. W., Morris, J., and Šparl, P. 2012a. Hamiltonian cycles in Cayley graphs whose order has few prime factors. *Ars Math. Contemp.*, **5**(1), 27–71.

Kutnar, Klavdija, Marušič, Dragan, and Zhang, Cui. 2012b. Hamilton paths in vertex-transitive graphs of order $10p$. *European J. Combin.*, **33**(6), 1043–1077.

Lambossy, P. 1931. Sur une manière de différencier les fonctions cycliques d'une forme donnée. *Comment. Math. Helv.*, **3**(1), 69–102.

Leung, Ka Hin, and Man, Shing Hing. 1996. On Schur rings over cyclic groups. II. *J. Algebra*, **183**(2), 273–285.

Leung, Ka Hin, and Man, Shing Hing. 1998. On Schur rings over cyclic groups. *Israel J. Math.*, **106**, 251–267.

Levi, F. W. 1942. *Finite Geometrical Systems*. University of Calcutta, Calcutta.

Li, Cai Heng. 1998. On isomorphisms of connected Cayley graphs. *Discrete Math.*, **178**(1–3), 109–122.

Li, Cai Heng. 1999. Finite CI-groups are soluble. *Bull. London Math. Soc.*, **31**(4), 419–423.

Li, Cai Heng. 2002. On isomorphisms of finite Cayley graphs – a survey. *Discrete Math.*, **256**(1–2), 301–334.

Li, Cai Heng. 2003. The finite primitive permutation groups containing an abelian regular subgroup. *Proc. London Math. Soc. (3)*, **87**(3), 725–747.

Li, Cai Heng. 2005. Permutation groups with a cyclic regular subgroup and arc transitive circulants. *J. Algebraic Combin.*, **21**(2), 131–136.

Li, Cai Heng, and Praeger, Cheryl E. 2002. Constructing homogeneous factorisations of complete graphs and digraphs. *Graphs Combin.*, **18**(4), 757–761.

Li, Cai Heng, and Seress, Ákos. 2003. The primitive permutation groups of squarefree degree. *Bull. London Math. Soc.*, **35**(5), 635–644.

Li, Cai Heng, and Seress, Ákos. 2005. On vertex-transitive non-Cayley graphs of square-free order. *Des. Codes Cryptogr.*, **34**(2–3), 265–281.

Li, Cai Heng, and Sim, Hyo-Seob. 2001. On half-transitive metacirculant graphs of prime-power order. *J. Combin. Theory Ser. B*, **81**(1), 45–57.

Li, Cai Heng, Lu, Zai Ping, and Marušič, Dragan. 2004. On primitive permutation groups with small suborbits and their orbital graphs. *J. Algebra*, **279**(2), 749–770.

Li, Cai Heng, Lu, Zai Ping, and Pálfy, P. P. 2007. Further restrictions on the structure of finite CI-groups. *J. Algebraic Combin.*, **26**(2), 161–181.

Li, Cai Heng, Lim, Tian Khoon, and Praeger, Cheryl E. 2009. Homogeneous factorisations of complete graphs with edge-transitive factors. *J. Algebraic Combin.*, **29**(1), 107–132.

Li, Cai Heng, Song, Shu Jiao, and Wang, Dian Jun. 2013. A characterization of metacirculants. *J. Combin. Theory Ser. A*, **120**(1), 39–48.

Li, Cai Heng, Sun, Shaohui, and Xu, Jing. 2014. Self-complementary circulants of prime-power order. *SIAM J. Discrete Math.*, **28**(1), 8–17.

Liebeck, Martin W., and Saxl, Jan. 1985a. Primitive permutation groups containing an element of large prime order. *J. London Math. Soc. (2)*, **31**(2), 237–249.

Liebeck, Martin W., and Saxl, Jan. 1985b. The primitive permutation groups of odd degree. *J. London Math. Soc. (2)*, **31**(2), 250–264.

Liebeck, Martin W., and Seitz, Gary M. 1990. Maximal subgroups of exceptional groups of Lie type, finite and algebraic. *Geom. Dedicata*, **35**(1–3), 353–387.

Liebeck, Martin W., and Seitz, Gary M. 1999. On finite subgroups of exceptional algebraic groups. *J. Reine Angew. Math.*, **515**, 25–72.

Liebeck, Martin W., and Seitz, Gary M. 2004. The maximal subgroups of positive dimension in exceptional algebraic groups. *Mem. Amer. Math. Soc.*, **169**(802), vi+227.

Liebeck, Martin W., and Seitz, Gary M. 2005. Maximal subgroups of large rank in exceptional groups of Lie type. *J. London Math. Soc. (2)*, **71**(2), 345–361.

Liebeck, Martin W., Praeger, Cheryl E., and Saxl, Jan. 1987. A classification of the maximal subgroups of the finite alternating and symmetric groups. *J. Algebra*, **111**(2), 365–383.

Liebeck, Martin W., Praeger, Cheryl E., and Saxl, Jan. 1988. On the O'Nan-Scott theorem for finite primitive permutation groups. *J. Austral. Math. Soc. Ser. A*, **44**(3), 389–396.

Liebeck, Martin W., Praeger, Cheryl E., and Saxl, Jan. 2010. Regular subgroups of primitive permutation groups. *Mem. Amer. Math. Soc.*, **203**(952), vi+74.

Lipschutz, Seymour, and Xu, Ming-Yao. 2002. Note on infinite families of trivalent semisymmetric graphs. *European J. Combin.*, **23**(6), 707–711.

Liskovets, Valery, and Pöschel, Reinhard. 2000. Non-Cayley-isomorphic self-complementary circulant graphs. *J. Graph Theory*, **34**(2), 128–141.

Livingstone, Donald, and Wagner, Ascher. 1965. Transitivity of finite permutation groups on unordered sets. *Math. Z.*, **90**, 393–403.

Lorimer, Peter. 1984. Vertex-transitive graphs: symmetric graphs of prime valency. *J. Graph Theory*, **8**(1), 55–68.

Lovász, L. 1978. Kneser's conjecture, chromatic number, and homotopy. *J. Combin. Theory Ser. A*, **25**(3), 319–324.

Lovász, L. 1979. *Combinatorial Problems and Exercises*. North-Holland Publishing Co., Amsterdam-New York.

Lovrečič Saražin, Marko. 1997. A note on the generalized Petersen graphs that are also Cayley graphs. *J. Combin. Theory Ser. B*, **69**(2), 226–229.

Luks, Eugene M. 1982. Isomorphism of graphs of bounded valence can be tested in polynomial time. *J. Comput. System Sci.*, **25**(1), 42–65.

Malnič, Aleksander, and Marušič, Dragan. 2002. Constructing $\frac{1}{2}$-arc-transitive graphs of valency 4 and vertex stabilizer $Z_2 \times Z_2$. *Discrete Math.*, **245**(1–3), 203–216.

Malnič, Aleksander, and Marušič, Dragan. 2002b. Constructing $\frac{1}{2}$-arc-transitive graphs of valency 4 and vertex stabilizer $Z_2 \times Z_2$. *Discrete Math.*, **245**(1-3), 203–216.

Malnič, Aleksander, Nedela, Roman, and Škoviera, Martin. 2000. Lifting graph automorphisms by voltage assignments. *European J. Combin.*, **21**(7), 927–947.

Malnič, Aleksander, Marušič, Dragan, and Šparl, Primož. 2007a. On strongly regular bicirculants. *European J. Combin.*, **28**(3), 891–900.

Malnič, Aleksander, Marušič, Dragan, Miklavič, Štefko, and Potočnik, Primož. 2007b. Semisymmetric elementary abelian covers of the Möbius–Kantor graph. *Discrete Math.*, **307**(17–18), 2156–2175.

Marušič, D. 1985. Vertex transitive graphs and digraphs of order p^k. Pages 115–128 of: *Cycles in Graphs* (Burnaby, BC, 1982). North-Holland Math. Stud., vol. 115. Amsterdam: North-Holland.

Marušič, Dragan. 1981a. On vertex symmetric digraphs. *Discrete Math.*, **36**(1), 69–81.

Marušič, Dragan. 1981b. *On Vertex Symmetric Digraphs*. PhD thesis, University of Reading.

Marušič, Dragan. 1983. Cayley properties of vertex symmetric graphs. *Ars Combin.*, **16**(B), 297–302.

Marušič, Dragan. 1987. Hamiltonian cycles in vertex symmetric graphs of order $2p^2$. *Discrete Math.*, **66**(1–2), 169–174.

Marušič, Dragan. 1988a. On vertex-transitive graphs of order *qp*. *J. Combin. Math. Combin. Comput.*, **4**, 97–114.

Marušič, Dragan. 1988b. Strongly regular bicirculants and tricirculants. *Ars Combin.*, **25**(C), 11–15. Eleventh British Combinatorial Conference (London, 1987).

Marušič, Dragan. 1992. Hamiltonicity of vertex-transitive pq-graphs. Pages 209–212 of: *Fourth Czechoslovakian Symposium on Combinatorics, Graphs and Complexity* (Prachatice, 1990). Ann. Discrete Math., vol. 51. North-Holland, Amsterdam.

Marušič, Dragan. 1998. Half-transitive group actions on finite graphs of valency 4. *J. Combin. Theory Ser. B*, **73**(1), 41–76.

Marušič, Dragan. 2000. Constructing cubic edge- but not vertex-transitive graphs. *J. Graph Theory*, **35**(2), 152–160.

Marušič, Dragan. 2003. On 2-arc-transitivity of Cayley graphs. *J. Combin. Theory Ser. B*, **87**(1), 162–196. Dedicated to Crispin St. J. A. Nash-Williams.

Marušič, Dragan. 2005. Quartic half-arc-transitive graphs with large vertex stabilizers. *Discrete Math.*, **299**(1–3), 180–193.

Marušič, Dragan. 2018. Semiregular automorphisms in vertex-transitive graphs of order $3p^2$. *Electron. J. Combin.*, **25**(2), P2.25.

Marušič, Dragan, and Nedela, Roman. 2001a. On the point stabilizers of transitive groups with non-self-paired suborbits of length 2. *J. Group Theory*, **4**(1), 19–43.

Marušič, Dragan, and Nedela, Roman. 2001b. Partial line graph operator and half-arc-transitive group actions. *Math. Slovaca*, **51**(3), 241–257.

Marušič, Dragan, and Parsons, T. D. 1982. Hamiltonian paths in vertex-symmetric graphs of order $5p$. *Discrete Math.*, **42**(2–3), 227–242.

Marušič, Dragan, and Parsons, T. D. 1983. Hamiltonian paths in vertex-symmetric graphs of order $4p$. *Discrete Math.*, **43**(1), 91–96.

Marušič, Dragan, and Pisanski, Tomaž. 2000. The remarkable generalized Petersen graph $G(8,3)$. *Math. Slovaca*, **50**(2), 117–121.

Marušič, Dragan, and Praeger, Cheryl E. 1999. Tetravalent graphs admitting half-transitive group actions: alternating cycles. *J. Combin. Theory Ser. B*, **75**(2), 188–205.

Marušič, Dragan, and Scapellato, Raffaele. 1992a. Characterizing vertex-transitive pq-graphs with an imprimitive automorphism subgroup. *J. Graph Theory*, **16**(4), 375–387.

Marušič, Dragan, and Potočnik, Primož. 2001. Semisymmetry of generalized Folkman graphs. *European J. Combin.*, **22**(3), 333–349.

Marušič, Dragan, and Potočnik, Primož. 2002. Bridging semisymmetric and half-arc-transitive actions on graphs. *European J. Combin.*, **23**(6), 719–732.

Marušič, Dragan, and Scapellato, Raffaele. 1992b. A class of non-Cayley vertex-transitive graphs associated with $\mathrm{PSL}(2,p)$. *Discrete Math.*, **109**(1–3), 161–170. Algebraic graph theory (Leibnitz, 1989).

Marušič, Dragan, and Scapellato, Raffaele. 1993. Imprimitive representations of $\mathrm{SL}(2,2^k)$. *J. Combin. Theory Ser. B*, **58**(1), 46–57.

Marušič, Dragan, and Scapellato, Raffaele. 1994a. A class of graphs arising from the action of $\mathrm{PSL}(2,q^2)$ on cosets of $\mathrm{PGL}(2,q)$. *Discrete Math.*, **134**(1–3), 99–110. Algebraic and topological methods in graph theory.

Marušič, D., and Scapellato, R. 1994b. Classifying vertex-transitive graphs whose order is a product of two primes. *Combinatorica*, **14**(2), 187–201.

Marušič, Dragan, and Scapellato, Raffaele. 1994c. Permutation groups with conjugacy complete stabilizers. *Discrete Math.*, **134**(1–3), 93–98.

Marušič, Dragan, and Scapellato, Raffaele. 1998. Permutation groups, vertex-transitive digraphs and semiregular automorphisms. *European J. Combin.*, **19**(6), 707–712.

Marušič, Dragan, and Šparl, Primož. 2008. On quartic half-arc-transitive metacirculants. *J. Algebraic Combin.*, **28**(3), 365–395.

Marušič, Dragan, and Xu, Ming-Yao. 1997. A $\frac{1}{2}$-transitive graph of valency 4 with a nonsolvable group of automorphisms. *J. Graph Theory*, **25**(2), 133–138.

Marušič, Dragan, Scapellato, Raffaele, and Zagaglia Salvi, Norma. 1992. Generalized Cayley graphs. *Discrete Math.*, **102**(3), 279–285.

McKay, Brendan D., and Praeger, Cheryl E. 1994. Vertex-transitive graphs which are not Cayley graphs. I. *J. Austral. Math. Soc. Ser. A*, **56**(1), 53–63.

McKay, Brendan D., and Praeger, Cheryl E. 1996. Vertex-transitive graphs that are not Cayley graphs. II. *J. Graph Theory*, **22**(4), 321–334.

McKay, Brendan D., and Royle, Gordon F. 1990. The transitive graphs with at most 26 vertices. *Ars Combin.*, **30**, 161–176.

Miklavič, Štefko, Potočnik, Primož, and Wilson, Steve. 2007a. Consistent cycles in graphs and digraphs. *Graphs Combin.*, **23**(2), 205–216.

Miklavič, Štefko, Potočnik, Primož, and Wilson, Steve. 2007b. Overlap in consistent cycles. *J. Graph Theory*, **55**(1), 55–71.

Miller, G. A. 1903. A fundamental theorem with respect to transitive substitution groups. *Bull. Amer. Math. Soc.*, **9**(10), 543–544.

Miller, G. A. 1915. Limits of the degree of transitivity of substitution groups. *Bull. Amer. Math. Soc.*, **22**(2), 68–71.

Miller, Robert C. 1971. The trivalent symmetric graphs of girth at most six. *J. Combin. Theory Ser. B*, **10**, 163–182.

Mohar, Bojan. 1992. A domain monotonicity theorem for graphs and Hamiltonicity. *Discrete Appl. Math.*, **36**(2), 169–177.

Monson, Barry, Pisanski, Tomaž, Schulte, Egon, and Weiss, Asia Ivić. 2007. Semisymmetric graphs from polytopes. *J. Combin. Theory Ser. A*, **114**(3), 421–435.

Morris, Dave Witte. 2016. Infinitely many nonsolvable groups whose Cayley graphs are hamiltonian. *J. Algebra Comb. Discrete Struct. Appl.*, **3**(1), 13–30.

Morris, Dave Witte. 2018. Cayley graphs on groups with commutator subgroup of order $2p$ are hamiltonian. *The Art of Discrete and Applied Math.*, **1**(1), #P04.

Morris, Joy. 1999. Isomorphic Cayley graphs on nonisomorphic groups. *J. Graph Theory*, **31**(4), 345–362.

Morris, Joy, and Spiga, Pablo. 2018a. Asymptotic enumeration of Cayley digraphs. *Israel J., Math.*, **242**(1), 401–459.

Morris, Joy, and Spiga, Pablo. 2018b. Classification of finite groups that admit an oriented regular representation. *Bull. Lond. Math. Soc.*, **50**(5), 811–831.

Morris, Joy, and Tymburski, Josh. 2018. Most rigid representations and Cayley index. *Art Discrete Appl. Math.*, **1**(1).

Morris, Joy, Spiga, Pablo, and Verret, Gabriel. 2015. Automorphisms of Cayley graphs on generalised dicyclic groups. *European J. Combin.*, **43**, 68–81.

Müller, Peter. 2013. Permutation groups with a cyclic two-orbits subgroup and monodromy groups of Laurent polynomials. *Ann. Sc. Norm. Super. Pisa Cl. Sci. (5)*, **12**(2), 369–438.

Mütze, Torsten, Nummenpalo, Jerri, and Walczak, Bartosz. 2018. Sparse Kneser graphs are Hamiltonian. Pages 912–919 of: *STOC'18 – Proceedings of the 50th Annual ACM SIGACT Symposium on Theory of Computing*. ACM, New York.

Muzychuk, Mikhail. 1995. Ádám's conjecture is true in the square-free case. *J. Combin. Theory Ser. A*, **72**(1), 118–134.

Muzychuk, Mikhail. 1997. On Ádám's conjecture for circulant graphs. *Discrete Math.*, **176**(1–3), 285–298.

Muzychuk, Mikhail. 1999. On the isomorphism problem for cyclic combinatorial objects. *Discrete Math.*, **197/198**, 589–606. 16th British Combinatorial Conference (London, 1997).

Muzychuk, M. 2003. An elementary abelian group of large rank is not a CI-group. *Discrete Math.*, **264**(1–3), 167–185. The 2000 Com^2MaC Conference on Association Schemes, Codes and Designs (Pohang).

Muzychuk, M. 2004. A solution of the isomorphism problem for circulant graphs. *Proc. London Math. Soc. (3)*, **88**(1), 1–41.

Muzychuk, Mikhail. 2015. A solution of an equivalence problem for semisimple cyclic codes. Pages 327–334 of: *Topics in Finite Fields*. Contemp. Math., vol. 632. Amer. Math. Soc., Providence, RI.

Muzychuk, Mikhail, and Ponomarenko, Ilia. 2009. Schur rings. *European J. Combin.*, **30**(6), 1526–1539.

Nedela, Roman, and Škoviera, Martin. 1995a. Which generalized Petersen graphs are Cayley graphs? *J. Graph Theory*, **19**(1), 1–11.

Nedela, Roman, and Škoviera, Martin. 1995b. Atoms of cyclic connectivity in cubic graphs. *Math. Slovaca*, **45**(5), 481–499.

Negami, S. 1985. *Uniqueness and Faithfulness of Embeddings of Graphs into Surfaces*. PhD thesis, Tokyo Inst. of Technology.

Neumann, Peter M. 1977. Finite permutation groups, edge-coloured graphs and matrices. Pages 82–118 of: *Topics in Group Theory and Computation (Proc. Summer School, University College, Galway, 1973)*.

Neumann, Peter M. 2009. Primitive permutation groups and their section-regular partitions. *Michigan Math. J.*, **58**(1), 309–322.

Nowitz, Lewis A. 1968. On the non-existence of graphs with transitive generalized dicyclic groups. *J. Combin. Theory*, **4**, 49–51.

Nowitz, Lewis A. 1992. A non-Cayley-invariant Cayley graph of the elementary abelian group of order 64. *Discrete Math.*, **110**(1–3), 223–228.

Ore, Oystein. 1962. *Theory of Graphs*. American Mathematical Society Colloquium Publications, Vol. XXXVIII. American Mathematical Society, Providence, RI.

Pak, Igor, and Radoičić, Radoš. 2009. Hamiltonian paths in Cayley graphs. *Discrete Math.*, **309**(17), 5501–5508.

Pálfy, P. P. 1987. Isomorphism problem for relational structures with a cyclic automorphism. *European J. Combin.*, **8**(1), 35–43.

Passman, Donald. 1968. *Permutation Groups*. W. A. Benjamin, Inc., New York-Amsterdam.

Payan, C., and Sakarovitch, M. 1975. Ensembles cycliquement stables et graphes cubiques. *Cahiers Centre Études Recherche Opér.*, **17**(2–4), 319–343.

Pisanski, Tomaž. 2007. A classification of cubic bicirculants. *Discrete Math.*, **307**(3–5), 567–578.

Pisanski, Tomaž, and Servatius, Brigitte. 2013. *Configurations From a Graphical Viewpoint*. Birkhäuser Advanced Texts: Basler Lehrbücher. [Birkhäuser Advanced Texts: Basel Textbooks]. Birkhäuser/Springer, New York.

Ponomarenko, I. N. 2005. Determination of the automorphism group of a circulant association scheme in polynomial time. *Zap. Nauchn. Sem. S.-Peterburg. Otdel. Mat. Inst. Steklov. (POMI)*, **321**(Vopr. Teor. Predst. Algebr. i Grupp. 12), 251–267, 301; English translation in: *J. Math. Sci.* (2006) **136**(3), 3972–3979.

Potočnik, Primož, and Šajna, Mateja. 2006. On almost self-complementary graphs. *Discrete Math.*, **306**(1), 107–123.

Potočnik, Primož, and Šajna, Mateja. 2007. Self-complementary two-graphs and almost self-complementary double covers. *European J. Combin.*, **28**(6), 1561–1574.

Potočnik, Primož, and Šajna, Mateja. 2009. Vertex-transitive self-complementary uniform hypergraphs. *European J. Combin.*, **30**(1), 327–337.

Potočnik, Primož, and Wilson, Steve. 2016. Linking rings structures and semisymmetric graphs: Cayley constructions. *European J. Combin.*, **51**, 84–98.

Potočnik, Primož, Spiga, Pablo, and Verret, Gabriel. 2012. On graph-restrictive permutation groups. *J. Combin. Theory Ser. B*, **102**(3), 820–831.

Potočnik, Primož, Spiga, Pablo, and Verret, Gabriel. 2013. Cubic vertex-transitive graphs on up to 1280 vertices. *J. Symbolic Comput.*, **50**, 465–477.

Potočnik, Primož, Spiga, Pablo, and Verret, Gabriel. 2014. On the order of arc-stabilisers in arc-transitive graphs with prescribed local group. *Trans. Amer. Math. Soc.*, **366**(7), 3729–3745.

Potočnik, Primož, Spiga, Pablo, and Verret, Gabriel. 2015. A census of 4-valent half-arc-transitive graphs and arc-transitive digraphs of valence two. *Ars Math. Contemp.*, **8**(1), 133–148.

Praeger, Cheryl E. 1985. Imprimitive symmetric graphs. *Ars Combin.*, **19**(A), 149–163.

Praeger, Cheryl E. 1993. An O'Nan–Scott theorem for finite quasiprimitive permutation groups and an application to 2-arc transitive graphs. *J. London Math. Soc. (2)*, **47**(2), 227–239.

Praeger, Cheryl E., and Xu, Ming Yao. 1993. Vertex-primitive graphs of order a product of two distinct primes. *J. Combin. Theory Ser. B*, **59**(2), 245–266.

Praeger, Cheryl, Li, Cai Heng, and Stringer, Linda. 2009. Common circulant homogeneous factorisations of the complete digraph. *Discrete Math.*, **309**(10), 3006–3012.

Praeger, Cheryl E., Spiga, Pablo, and Verret, Gabriel. 2012. Bounding the size of a vertex-stabiliser in a finite vertex-transitive graph. *J. Combin. Theory Ser. B*, **102**(3), 797–819.

Praeger, Cheryl E., Wang, Ru Ji, and Xu, Ming Yao. 1993. Symmetric graphs of order a product of two distinct primes. *J. Combin. Theory Ser. B*, **58**(2), 299–318.

Quirin, William L. 1971. Primitive permutation groups with small orbitals. *Math. Z.*, **122**, 267–274.

Ramos Rivera, Alejandra, and Šparl, Primož. 2017. The classification of half-arc-transitive generalizations of Bouwer graphs. *European J. Combin.*, **64**, 88–112.

Ramos Rivera, Alejandra, and Šparl, Primož. 2019. New structural results on tetravalent half-arc-transitive graphs. *J. Combin. Theory Ser. B*, **135**, 256–278.

Ramras, Mark, and Donovan, Elizabeth. 2011. The automorphism group of a Johnson graph. *Siam J. Discrete Math.*, **25**, 267–270.

Rapaport-Strasser, Elvira. 1959. Cayley color groups and Hamilton lines. *Scripta Math.*, **24**, 51–58.

Royle, Gordon F. 2008. A normal non-Cayley-invariant graph for the elementary abelian group of order 64. *J. Aust. Math. Soc.*, **85**(3), 347–351.

Ryabov, Grigory. 2020. The Cayley isomorphism property for the group $C_2^5 \times C_p$. *Ars Math. Contemp.*, **19**(2), 277–295.

Ryabov, Grigory. 2021. The Cayley isomorphism property for the group $C_4 \times C_p^2$. *Comm. Algebra*, **49**(4), 1788–1804.

Sabidussi, Gert. 1958. On a class of fixed-point-free graphs. *Proc. Amer. Math. Soc.*, **9**, 800–804.

Sabidussi, Gert. 1959. The composition of graphs. *Duke Math. J.*, **26**, 693–696.

Sabidussi, Gert. 1964. Vertex-transitive graphs. *Monatsh. Math.*, **68**, 426–438.

Sachs, Horst. 1962. Über selbstkomplementäre Graphen. *Publ. Math. Debrecen*, **9**, 270–288.

Schur, Issai. 1933. Zur Theorie der einfach transitiven Permutationsgruppen. *S. B. Preuss. Akad. Wiss., Phys.-Math. KI*, **70**, 598–623.

Scott, Leonard L. 1972. On permutation groups of degree 2 p. *Math. Z.*, **126**, 227–229.

Scott, W. R. 1987. *Group Theory*. Second ed. Dover Publications Inc., New York.

Seress, Ákos. 1998. On vertex-transitive, non-Cayley graphs of order pqr. *Discrete Math.*, **182**(1–3), 279–292.

Sims, Charles C. 1968. Graphs and finite permutation groups. II. *Math. Z.*, **103**, 276–281.

Sjerve, Denis, and Cherkassoff, Michael. 1994. On groups generated by three involutions, two of which commute. Pages 169–185 of: *The Hilton Symposium 1993 (Montreal, PQ)*. CRM Proc. Lecture Notes, vol. 6. Amer. Math. Soc., Providence, RI.

Smith, D. H. 1971. Primitive and imprimitive graphs. *Quart. J. Math. Oxford Ser. (2)*, **22**, 551–557.

Somlai, Gábor. 2011. Elementary abelian p-groups of rank $2p + 3$ are not CI-groups. *J. Algebraic Combin.*, **34**(3), 323–335.

Somlai, Gábor. 2015. The Cayley isomorphism property for groups of order $8p$. *Ars Math. Contemp.*, **8**(2), 433–444.

Song, Shu Jiao, Li, Cai Heng, and Zhang, Hua. 2014. Finite permutation groups with a regular dihedral subgroup, and edge-transitive dihedrants. *J. Algebra*, **399**, 948–959.

Šparl, Primož. 2008. A classification of tightly attached half-arc-transitive graphs of valency 4. *J. Combin. Theory Ser. B*, **98**(5), 1076–1108.

Spiga, Pablo. 2007. Elementary abelian p-groups of rank greater than or equal to $4p - 2$ are not CI-groups. *J. Algebraic Combin.*, **26**(3), 343–355.

Spiga, Pablo. 2018. On the existence of Frobenius digraphical representations. *Electron. J. Combin.*, **25**(2), P2.6.

Spiga, Pablo. 2021. On the equivalence between a conjecture of Babai–Godsil and a conjecture of Xu concerning the enumeration of Cayley graphs. *Art Discrete Appl. Math.*, **4**(1), P1.10.

Steimle, Alice, and Staton, William. 2009. The isomorphism classes of the generalized Petersen graphs. *Discrete Math.*, **309**(1), 231–237.

Suprunenko, D. A. 1985. Self-complementary graphs. *Kibernetika (Kiev)*, i, 1–6, 24, 133.

Thomassen, Carsten. 1991. Tilings of the torus and the Klein bottle and vertex-transitive graphs on a fixed surface. *Trans. Amer. Math. Soc.*, **323**(2), 605–635.

Trofimov, V. I. 1990. Vertex stabilizers of graphs with projective suborbits. *Dokl. Akad. Nauk SSSR*, **315**(3), 544–546.

Trofimov, V. I. 1991. Graphs with projective suborbits. *Izv. Akad. Nauk SSSR Ser. Mat.*, **55**(4), 890–916.

Turner, James. 1967. Point-symmetric graphs with a prime number of points. *J. Combin. Theory*, **3**, 136–145.

Tutte, W. T. 1947. A family of cubical graphs. *Proc. Cambridge Philos. Soc.*, **43**, 459–474.

Tutte, W. T. 1959. On the symmetry of cubic graphs. *Canad. J. Math.*, **11**, 621–624.

Tutte, W. T. 1960. A non-Hamiltonian graph. *Canad. Math. Bull.*, **3**, 1–5.

Tutte, W. T. 1966. *Connectivity in graphs.* Mathematical Expositions, No. 15. Toronto, Ont.: University of Toronto Press.

Tutte, W. T. 1998. *Graph Theory as I Have Known It.* Oxford Lecture Series in Mathematics and its Applications, vol. 11. The Clarendon Press, Oxford University Press, New York.

Tyshkevich, R. I., and Tan, Ngo Dak. 1987. A generalization of Babai's lemma on Cayley graphs. *Vestsī Akad. Navuk BSSR Ser. Fīz.-Mat. Navuk*, 29–32, 124.

van den Heuvel, J. 1995. Hamilton cycles and eigenvalues of graphs. *Linear Algebra Appl.*, **226/228**, 723–730.

Verret, Gabriel. 2009. On the order of arc-stabilizers in arc-transitive graphs. *Bull. Aust. Math. Soc.*, **80**(3), 498–505.

Verret, Gabriel. 2015. Arc-transitive graphs of valency 8 have a semiregular automorphism. *Ars Math. Contemp.*, **8**(1), 29–34.

Walther, Hansjoachim. 1969. Über die Nichtexistenz eines Knotenpunktes, durch den alle längsten Wege eines Graphen gehen. *J. Combin. Theory*, **6**, 1–6.

Walther, Hansjoachim, and Voss, Heinz-Jürgen. 1974. *Über Kreise in Graphen.* VEB Deutscher Verlag der Wissenschaften, Berlin.

Wang, Chang Qun, and Chen, Tie Sheng. 2008. Semisymmetric cubic graphs as regular covers of $K_{3,3}$. *Acta Math. Sin. (Engl. Ser.)*, **24**(3), 405–416.

Wang, Changqun, Wang, Dianjun, and Xu, Mingyao. 1998. Normal Cayley graphs of finite groups. *Sci. China Ser. A*, **41**(3), 242–251.

Wang, Li, and Du, Shaofei. 2014. Semisymmetric graphs of order $2p^3$. *European J. Combin.*, **36**, 393–405.

Wang, Li, Du, Shaofei, and Li, Xuewen. 2014. A class of semisymmetric graphs. *Ars Math. Contemp.*, **7**(1), 40–53.

Wang, Ru Ji, and Xu, Ming Yao. 1993. A classification of symmetric graphs of order $3p$. *J. Combin. Theory Ser. B*, **58**(2), 197–216.

Watkins, Mark E. 1969. A theorem on Tait colorings with an application to the generalized Petersen graphs. *J. Combin. Theory*, **6**, 152–164.

Watkins, Mark E. 1970. Connectivity of transitive graphs. *J. Combin. Theory*, **8**, 23–29.

Watkins, Mark E. 1990. Vertex-transitive graphs that are not Cayley graphs. Pages 243–256 of: *Cycles and Rays* (Montreal, PQ, 1987). NATO Adv. Sci. Inst. Ser. C Math. Phys. Sci., vol. 301. Kluwer Academic Publishers, Dordrecht.

Weiss, Richard. 1979. An application of p-factorization methods to symmetric graphs. *Math. Proc. Cambridge Philos. Soc.*, **85**(1), 43–48.

Weiss, R. 1981a. s-transitive graphs. Pages 827–847 of: *Algebraic Methods in Graph Theory, Vol. I, II (Szeged, 1978)*. Colloq. Math. Soc. János Bolyai, vol. 25. North-Holland, Amsterdam-New York.

Weiss, Richard. 1981b. The nonexistence of 8-transitive graphs. *Combinatorica*, **1**(3), 309–311.

Wielandt, Helmut. 1964. *Finite Permutation Groups*. Translated from the German by R. Bercov. Academic Press, New York.

Wielandt, H. 1969. *Permutation Groups Through Invariant Relations and Invariant Functions*. Lectures given at The Ohio State University, Columbus, Ohio.

Wielandt, Helmut. 1994. *Mathematische Werke/Mathematical Works, vol. 1*. Walter de Gruyter & Co., Berlin.

Wilson, Robert A. 2009. *The Finite Simple Groups*. Graduate Texts in Mathematics, vol. 251. Springer-Verlag London, Ltd., London.

Wilson, Robert, Walsh, Peter, Tripp, Jonathan, et al. (n.d.). ATLAS of Finite Group Representations - Version 3. http://brauer.maths.qmul.ac.uk/Atlas/v3/.

Wilson, Steve. 2003. A worthy family of semisymmetric graphs. *Discrete Math.*, **271**(1–3), 283–294.

Wilson, Steve. 2008. Rose window graphs. *Ars Math. Contemp.*, **1**(1), 7–19.

Wilson, Steve, and Potočnik, Primož. 2016. *Recipes for Edge-Transitive Tetravalent Graphs*.

Witte, Dave. 1986. Cayley digraphs of prime-power order are Hamiltonian. *J. Combin. Theory Ser. B*, **40**(1), 107–112.

Witte, David. 1982. On Hamiltonian circuits in Cayley diagrams. *Discrete Math.*, **38**(1), 99–108.

Witte, David, and Gallian, Joseph A. 1984. A survey: Hamiltonian cycles in Cayley graphs. *Discrete Math.*, **51**(3), 293–304.

Witte Morris, Dave. 2015. Odd-order Cayley graphs with commutator subgroup of order pq are Hamiltonian. *Ars Math. Contemp.*, **8**(1), 1–28.

Witte Morris, Dave, and Wilk, Kirsten. 2020. Cayley graphs of order kp are Hamiltonian for $k < 48$. *Art Discrete Appl. Math.*, **3**(2), P2.02.

Wong, Warren J. 1967. Determination of a class of primitive permutation groups. *Math. Z.*, **99**, 235–246.

Xu, Jing. 2008. Semiregular automorphisms of arc-transitive graphs with valency pq. *European J. Combin.*, **29**(3), 622–629.

Xu, Ming Yao. 1992. Half-transitive graphs of prime-cube order. *J. Algebraic Combin.*, **1**(3), 275–282.

Xu, Ming-Yao. 1998. Automorphism groups and isomorphisms of Cayley digraphs. *Discrete Math.*, **182**(1–3), 309–319.

Xu, Yian. 2017. On constructing normal and non-normal Cayley graphs. *Discrete Math.*, **340**(12), 2972–2977.

Zhang, Jun-Yang. 2015. Vertex-transitive digraphs of order p^5 are Hamiltonian. *Electron. J. Combin.*, **22**(1), P1.76.

Zhou, Jin-Xin. 2017. Edge-transitive almost self-complementary graphs. *J. Combin. Theory Ser. B*, **123**, 215–239.

Zhou, Jin-Xin, and Feng, Yan-Quan. 2010. Semisymmetric elementary abelian covers of the Heawood graph. *Discrete Math.*, **310**(24), 3658–3662.

Zhu, Yong Jin, Liu, Zhen Hong, and Yu, Zheng Guang. 1985. An improvement of Jackson's result on Hamilton cycles in 2-connected regular graphs. Pages 237–247 of: *Cycles in Graphs (Burnaby, B.C., 1982)*. North-Holland Math. Stud., vol. 115. North-Holland, Amsterdam.

Zhu, Yong Jin, Liu, Zhen Hong, and Yu, Zheng Guang. 1986a. 2-connected k-regular graphs [$k \geq 6$] on at most $3k + 3$ vertices are Hamiltonian. *J. Systems Sci. Math. Sci.*, **6**(1), 36–49.

Zhu, Yong Jin, Liu, Zhen Hong, and Yu, Zheng Guang. 1986b. 2-connected k-regular graphs [$k \geq 6$] on at most $3k+3$ vertices are Hamiltonian. II. *J. Systems Sci. Math. Sci.*, **6**(2), 136–145.

Zsigmondy, K. 1892. Zur Theorie der Potenzreste. *Monatsh. Math. Phys.*, **3**(1), 265–284.

Index of Graphs

Index of Symbols

Select Author Index

Index of Terms